ANNUAL REVIEW OF MICROBIOLOGY

EDITORIAL COMMITTEE (1997)

ANNUAL REVIEW OF MICROBIOLOGY

VOLUME 51, 1997

L. NICHOLAS ORNSTON, *Editor*
Yale University

ALBERT BALOWS, *Associate Editor*
Centers for Disease Control, Atlanta

E. PETER GREENBERG, *Associate Editor*
University of Iowa, Iowa City

http://annurev.org   science@annurev.org   415-493-4400

ANNUAL REVIEWS INC.   4139 EL CAMINO WAY   P.O. BOX 10139   PALO ALTO, CALIFORNIA 94303-0139

R̲ ANNUAL REVIEWS INC.
Palo Alto, California, USA

International Standard Serial Number: 0066-4227
International Standard Book Number: 0-8243-1151-5
Library of Congress Catalog Card Number: 49-432

∞ The paper used in this publication meets the minimum requirements of American National Standard for Information Sciences—Permanence of Paper for Printed Library Materials, ANSI Z39.48-1992.

TYPESET BY TECHBOOKS, FAIRFAX, VA
PRINTED AND BOUND IN THE UNITED STATES OF AMERICA

PREFACE 1997

One gift that gives pleasure in the giving is communication of the biological wealth to be found in the diversity of microorganisms. Lest the general picture be lost in an abundance of detail, it might be worth considering why microorganisms are so varied in their properties. The answer lies in evolutionary challenges presented by the physical dimensions of the smallest creatures. Limits in both size and information demand that each microorganism be a specialist. In the parlance of small business, the organism must define its core technology and bring it to the marketplace. And what a range of constantly shifting marketplaces is available! Unencumbered by the complexities presented by multicellularity, a microbial specialist can move swiftly into a newly defined niche. Collectively, these specialists have occupied habitats representing extremes of opportunities for life, and the game goes on. Rules of the marketplace dictate that small specialists are unlikely to survive alone, so it follows that microorganisms tend to live in consortia. A further key to their success is their ability to network, both physiologically and genetically. Thus, appreciation of the skills inherent in microbial survival primes enthusiastic interest in the interplay of forces that drive life in its elegant simplicity and in its full intricacy.

All of this makes for grand science, but it places burdens on the editorial committee that must decide which gifts from microbiology to share in the forthcoming volume. The deliberations of the committee benefited greatly from the wisdom of Mary Lidstrom and Graham Walker, and we note with regret that their terms on the committee have come to an end. With this volume, the torch of production editor has passed from Naomi Lubick to Renée Burgard to Bob Johnson. The exchanges were flawless, so deft that we were reminded how the illusion of biological constancy is achieved through constant change. Gratitude for making this volume possible must be expressed to all three production editors and to the authors who shared their insights into the intriguing lives of microorganisms.

NICK ORNSTON

Annual Review of Microbiology
Volume 51 (1997)

CONTENTS

OTHER REVIEWS OF INTEREST TO MICROBIOLOGISTS

From the *Annual Review of Biochemistry*, Volume 66, 1997:

Molybdenum-Cofactor-Containing Enzymes: Structure and Mechanism, C. Kisker, H. Schindelin, and D. C. Rees

Bacterial Cell Division and the Z Ring, J. Lutkenhaus and S. G. Addinall

Herpes Simplex Virus DNA Replication, P. E. Boehmer and I. R. Lehman

Models of Amyloid Seeding in Alzheimer's Disease and Scrapie: Mechanistic Truths and Physiological Consequences of the Time-Dependent Solubility of Amyloid Proteins, J. D. Harper and P. T. Lansbury, Jr.

Molecular Basis for Membrane Phospholipid Diversity: Why Are There So Many Lipids?, W. Dowhan

Polyadenylation of mRNA in Prokaryotes, N. Sarkar

Structure-Based Perspectives on B_{12}-Dependent Enzymes, M. L. Ludwig and R. G. Matthews

Target Site Selection in Transposition, N. Craig

From the *Annual Review of Entomology*, Volume 43, 1998:

Malaria Parasite Development in Mosquitoes, J. C. Beier

Parasites and Pathogens of Mites, G. Poinar, Jr., and R. Poinar

Ecological Considerations for the Environmental Impact Evaluation of Recombinant Baculovirus Insecticides, A. Richards, M. Matthews, and P. Christian

From the *Annual Review of Genetics*, Volume 31, 1997:

Antisense RNA Regulated Programmed Cell Death, K. Gerdes, A. P. Gultyaev, T. Franch, K. Pedersen, and N. D. Mikkelsen

Gene Amplification and Genome Plasticity in Prokaryotes, D. Romero and R. Palacios

Genetics of Prions, S. B. Prusiner and M. R. Scott

Viral Transactivating Proteins, J. Flint and T. Shenk

From the *Annual Review of Immunology*, Volume 15, 1997:

Lyme Disease: A Review of Aspects of Its Immunology and Immunopathogenesis, L. H. Sigal

From the *Annual Review of Plant Physiology and Plant Molecular Biology*, Volume 48, 1997:

Holger W. Jannasch

Annu. Rev. Microbiol. 1997. 51:1–45

SMALL IS POWERFUL: Recollections of a Microbiologist and Oceanographer

Holger W. Jannasch
Woods Hole Oceanographic Institution, Woods Hole, Massachusetts 02543

CONTENTS

PROLOGUE

Born into a certain time and social setting, each of us harbors personal biases despite our professional calling for scientific objectivity. This conflict between objectivity and subjectivity is perhaps what makes these invited self portraits interesting to colleagues, and even to younger scientists. In my own story, I have tried to balance accounts of my education, my work, and my coworkers with historical notes, anecdotes, and personal views. As space is limited, I hope that the necessary omissions are less noticeable to the reader than they are frustrating to the writer. Also I submit to the fact, convincingly documented by J Kotre (79), that personal remembrances have the tendency to turn into myths. I have tried hard not to let this happen, but I presume that some unconscious embellishment of certain memories, and selective amnesia for others, are part of everyone's true self.

1

0066-4227/97/1001-0001$08.00

Midway during my studies in general biology, I began to understand that all higher forms of life on our planet depend on microbes. Small is not only beautiful—considering the elegance that microbes permit in physiological and biochemical experiments—but also intrinsically powerful. Highly dispersible through their smallness, and enduring extreme conditions, microbes are ubiquitous in the biosphere, and they define its limits. With their enormous metabolic and genetic diversity, microbes controlled the chemical balance of the global environment more than two billion years before macro-organisms appeared on the scene. They made, and are still making, the planet livable for these more complex and fragile forms of life. In our anthropocentric way, we take the activities of microbes largely for granted, except when we experience them as diseases. I decided early that a medical profession was not for me, because I aspired to study microbes as allies, not as enemies. Beyond that, my early professional pathway was out of my conscious control and mostly directed by influential mentors. It took me from studies on microbial processes in soil and freshwaters to those in the largest part of the biosphere, the oceanic environment.

In retrospect, a strong driving factor must also have been an attraction toward interdisciplinary research, the collaboration with scientists of other backgrounds, the challenge of meeting them on their turf, and particularly the expansion of one's own inevitably narrow perspective toward a comprehensive ecological orientation. This might have been due to an early influence on me by A Thienemann, an eminent ecologist, and to CB van Niel, whom I met later during my career, and to whom I owe a sense of profound satisfaction for having become a microbiologist.

It must be an outgrowth of my family's history—the global spread of my Moravian ancestors—that living somewhere other than in the country of my birth never seemed difficult to me. To the contrary, it has always been a stimulant and a reason, possibly, for the exploratory flavor of my work, if not of my life style.

FAMILY, EDUCATION

The voyage of my life started from the peaceful port of the best *Elternhaus* I could have wished for. The word, poorly translated into "parental home," means much more than that. It included, in my case, a healthy atmosphere provided by my father's and mother's complementary and harmonizing personalities. During an eventful life, my father became the cofounder of a private school outside the town of Holzminden on the Weser River, in western Germany, where I was born in May of 1927. In 1930, my father—known by that time as a dedicated educator—was asked, during the brief remaining life of the Weimar Republic, to enter the academic world by joining a new and, even by modern standards, highly advanced enterprise for the education and training of elementary school teachers at Hamburg. This admirable endeavor was terminated when the Nazis came to power in 1933, and my father was demoted to the eastern province of Silesia, first as a school administrator and later as a teachers' teacher again. There I grew up, spending a blissful boyhood in the beautiful foothills of the northeastern Sudeten Mountains, until the misguided country started—and was crushed by—World War II.

In 1943, my school class, the equivalent of the 9th grade in the United States, was drafted to man anti-aircraft guns at Berlin. It was ironic that the battery happened to be located near the site where my father, having been born in Labrador as the son of a Moravian missionary, spent most of WWI as a "true-born British subject" in a German prison camp. How our family of four—my brother lives as a retired Director of Halifax's Maritime Museum in Nova Scotia—survived WWII and the turmoil of its end was simply a miracle, but not to be told here.

After a brief stint in the disintegrating German Navy, and a discharge from a short imprisonment by the British in mid-1945, I was able (in the quest to realize my childhood dream of becoming the German equivalent of a forest ranger) to get a job as a forestry apprentice, which required starting as a lumberjack. Only the lower career was open to me because there was no money to finish my secondary school education, three years of which were still missing for completion of the *Abitur*, the graduation exam. My parents were lost behind the Russian and Polish iron curtains when Silesia had become part of Poland. In fact, I had to attend a memorial service for them after a (fortunately false) report of their death in a refugee camp was received. They arrived in West Germany a year and a half after the war had ended. Prior to that, thanks to their friends in Holzminden, it was possible for me to interrupt my forestry apprenticeship to continue my secondary education at the same private boarding school where I was born. Beautifully located at the edge of the Solling hill's large beech forests, this new home felt like paradise to me.

As I was too old for a normal high school student, my self-motivation derived from exposure to disgruntled corporals rather than parents, or at least benevolent teachers, for the preceding two years. I soaked up everything that was offered: literature and composition classes, playing the viola in the school's orchestra and singing cantatas in the choir, participating in the staging of Sophocles' *Antigone* or von Weber's opera *Abu Hassan*. I remember a trip with our art teacher to study the cathedrals of Mainz, Worms, and Speyer along the Rhine river while sneaking illegally in and out of the French occupation zone, in order to sketch the Maria Laach monastery church and the Roman architecture of the Porta Nigra at Trier. I also spent parts of cold winter nights in a small observatory with a four-inch refractor that the school had been given by a benefactor. The only science taught was chemistry. Lacking all materials for experimental work, we studied the periodic system in great detail. I profited enormously from this background later, when inorganic chemistry became part of my studies.

Making up for lost childhood years may excuse my overly enthusiastic and romantic embrace of these pleasures, but their flavor never left me. To this day I often visit my old school, this erstwhile postwar haven, trying to connect to the present generation of pupils. Their attitudes and mental dispositions, quite naturally, are very different from those we held almost half a century earlier. Germany's economic recovery appears to have quenched much of the cultural thirst of the postwar years. The school choir and orchestra have lost their appeal. We also had fewer problems, it seems, and they were quite basic. For instance, being always hungry, we spent long hours in the forest collecting beech nuts for their oil, dreaming of the potato pancakes that were to follow.

Well before my graduation (passing the *Abitur* exam) in early 1948 at Holzminden, my parents had arrived at the nearby Göttingen as refugees, with the total allowance of 20 pounds of their belongings. My father was able to resume his work as a professor of education. I cherished the time I was to spend with them during subsequent years when I studied at Göttingen, and later during frequent visits from the United States. There is a profound blessing in a close lifetime parent-child relationship. Now my parents are both back at the Holzminden school that has, within its premises and overlooking the beautiful Weser valley, a tiny private cemetery for its founders and some of its teachers and students.

With the *Abitur* in my pocket, the higher forestry career was now open to me. By this time, however, the idea of spending my life as a *Forstmeister* in a village or small town had lost its appeal, and I applied as a biology student to the University of Göttingen. In 1948/1949 acceptance was still limited by a *numerus clausus*, and required evidence of some practical work experience and passing a qualification exam. Right after my graduation, therefore, I obtained a job as warden of a bird sanctuary on the small island of Scharhörn, which was nothing more than a dune—300×100 m^2 at high tide—at the outer edge of the intertidal flats on the German North Sea coast, 10 miles off the mainland. My solitary existence there lasted from April to November of 1948. I spent my time protecting eggs—especially of the rare tern, *Sterna sandwichensis*—from seafaring poachers in those hungry days, monitoring the numbers and offspring of nesting birds, making special observations on nesting behavior, and counting migrating species. Few visitors undertook the long and somewhat dangerous walk to this island at low tide. One of them was my future wife.

This was my first meeting with the ocean and, longing to be on one of those fishing steamers passing the island every day, I managed between semesters during subsequent years (1950–1954) to get temporary jobs aboard some of them. Without being able to do my full share of their exceedingly hard work, I replaced seamen who were grateful for the chance of taking leaves without losing their social benefits. My first trip was to the Dogger Bank in the western North Sea, where herring were still plentiful. I was seasick most of the time and swore never to do it again. Half a year later, I did it anyway, and subsequently joined five more 3- to 4-week voyages, fishing for cod and haddock near Iceland or for red perch near the Malangen coast off Tromsø, Norway. Why? I really cannot tell. It might have been a mix of romanticism and the personal challenge of being able to manage life among the seamen, whose company I enjoyed for their gentlemanly acceptance of a stranger, a different experience from the one I had had with the provincial mentality of my previous lumberjack coworkers. And after all, I was able to cover my tuition and other expenses from the relatively generous pay. Some additional income accrued from collecting and preserving specimens that came up from 500-m deep trawls, and then selling them (after due determination of their species) to a Göttingen company that mounted them for educational displays. Much later, I understood how educational these endeavors had been for me.

GÖTTINGEN, NAPLES, WADI NATRUN (1949–1956)

From the bird island I went straight to Göttingen in late 1948 and passed my entry exam at the University, not by answering anticipated questions on bird migration or nesting behavior, but by the successful linguistic analysis of the word asymptote. According to still-valid academic standards, a humanistic education including Latin and Greek was thought to provide a good basis for making a scientist.

Before starting with the 1949 spring semester, I took a job at the Zoological Institute. A wing of that building, housing a collection of marine invertebrates from the *Valdivia* Deep-Sea Expedition (1898–1899), had been hit by one of the few bombs dropped on Göttingen, meant for the nearby railroad tracks. Under the guidance of Professor D Beling, a distinguished zoologist, Russian refugee, and former Director of the Hydrobiological Institute at Kiev, I had to re-identify and label specimens from broken jars of deep-sea mollusks. My career as a zoologist seemed to be set.

In order to get ready for the intermediate exam after four or five semesters, I had to listen to all required lectures and pass all required courses, leaving no room for a choice of subjects. Yet I remember these first five semesters with special fondness: the camaraderie during the tough *practica* in basic physics and analytical chemistry; the many weekend excursions: long hikes with friends in the environs of Göttingen and the Harz mountains, most of my fellow students knowing much more of the local flora and fauna than I did. Being very impressed by a professor of geobotany, F Firbas, as my budding ideal of a scientist—and, I must admit, by his equally accomplished, highly admired and attractive assistant, Marghita von Rochow—I overcame my dislike of memorizing names involved in the systematics of higher plants and underwent a special exam to become one of the few accepted to join his coveted field trips. He was an accomplished pollen analyst and the author of a famous opus on the history of the European forests. In between, we listened to lectures of illustrious scientists who, during this leveling postwar atmosphere, mingled with lowly students before disappearing in the clouds again. Physicists were still dominating Göttingen, particularly those (Born, Heisenberg, Hahn, von Laue, Windaus, von Weizsäcker—Planck had just died) who were known through their much publicized "bugged" protective custody at Hall, England. Characteristic was von Weizsäcker's most formidably presented series of lectures on "The History of Nature" (127). Following a strong contemporary trend, he included a philosophical synthesis of the natural sciences and reflected on past and present failures of academia and the well-educated to stand up to their humanistic ideals in contemporary social and political issues—a continuing problem.

Feeling that I lacked training in biochemistry, I went for one memorable semester to Munich to take a course in Feodor Lynen's lab, listened to Karl von Frisch's artful lectures in zoology, and took a last required botany course with Renner, a plant geneticist. As at Göttingen, the postwar euphoria was still evident in Munich, although fading. We went to theaters, concerts, or cabarets almost every night. Students standing in line could fill the unsold seats for a pittance quickly, just before the curtain rose. I remember the most hilarious persiflages of classical literature and political parodies in Erich Kästner's cabaret, *Die Kleine Freiheit*, "The Little Freedom." Once, Orson Wells made a guest appearance with a string of single acts out of various famous plays, something unheard of in German theater. I adored Luise Ulrich, who used Bavarian slang in Shaw's *Pygmalion*. The conductor of the Göttingen University Choir had taken a freelance position at Munich a year before I got there, and he took me on again as a member (bass). We sang classical and some modern music in public concerts and for the radio, which required long evening rehearsals. Free weekends were used for skiing trips in the foothills of the Alps, easily reached by train. The semester ended with a fantastic excursion to Tunisia, organized by Professor F Gessner, for the study of salt-adapted plants.

When these first five semesters ended with the intermediate exam at Göttingen, possibly an equivalent to a master's degree without a written thesis, Professor H Autrum, the zoologist, asked me to see him for the discussion of a research project related to a doctoral dissertation. While he was suggesting a biochemical study of the retina pigments of *Ephestia* (flour moth), he casually asked whether I had anything else in mind. Indeed I had: the feeding of protozoa on bacteria, using field observations and experimental studies. He was not interested, and I found myself outside his door. I was told by my shocked colleagues that I had blown the chance of my life—my professional life anyway. Being selected as a graduate student by this man, who later indeed did become von Frisch's successor at Munich, was a great honor and meant a sure-fire career. (Although undoubtedly disappointed, Professor Autrum later helped me most kindly when I arrived at another critical crossroad.)

After some rewriting of my ideas for a thesis, Professor Rippel-Baldes, Director of the Institute for Microbiology and founder of the *Archives of Microbiology*, accepted me as a graduate student. Little did I know that the focus of my subsequent work would stay with me during later years and would make particular sense in marine studies. This work involved the uptake and growth efficiencies of microbes in highly dilute aqueous media, which make up about 80% of our planet's biosphere. Through Rippel-Baldes, I became part of what I informally call the Breslau school of microbiology. Breslau, now Polish Wroclaw, was the capital of Silesia and one of the most renowned German universities. Microbiology there was represented by Ferdinand Cohn, who discovered both bacterial spores and Robert Koch. Other early microbiologists at Breslau were CG Ehrenberg, who accompanied A von Humboldt to Russia and described *Spirillum volutans*, and later P Buchner, the father of research in microbial symbiosis.

During these early 1950s at Göttingen, I was part of a small group of enthusiastic students that founded a limnological research facility with the help of the above-mentioned Professor Beling. I became an apprentice of his wife, who had been the bacteriologist in his lab and, at the same time, at Kiev's municipal water supply installations. Shortly after her arrival at Göttingen, she found ready employment with the Sartorius Company, which, in addition to scientific instruments, produced membrane filters for use in chemistry. These filters were a byproduct of the work of R Zsygmondy of Göttingen University, who received the Nobel prize for research in macromolecular chemistry in 1925. As only a few people knew, the Russians acquired the patent rights and developed membrane filter techniques in hygienic and aquatic microbiology during the 1920s and the 1930s (4). After the war, Dr. Beling brought all of this knowledge to Göttingen, where the membrane filter section of the Sartorius Company began to flourish. In 1946, A Goetz, an American Army Officer and Caltech professor, acquired the rights from Sartorius and cofounded the Millipore Filter Company in the United States. It was often amusing to watch the perplexity of Millipore representatives when I told them that their filters were invented in Göttingen and that many of their advertised appliances were developed in Russia half a century ago. Today, of course, the filter materials have advanced far beyond those early collodium acetone mixtures.

In 1952, our student initiative to build a research establishment for freshwater biology had caught the eye of August Thienemann, Director of the Max Planck

Institute for Limnology at Plön, Holstein. I was deeply impressed by the rare combination of a generalist and specialist in this eminent scientist. On his recommendation and with the sponsorship of a Hessian aristocrat, a new extension of his Institute was opened for us at the little Hessian town of Schlitz. During the official opening, I remember being asked most kindly by Professor Otto Hahn—then President of the Max Planck Society (the first to split the atom, which got him a Nobel prize and virtually started the Manhattan Project)—what limnology was all about. As a true disciple of Thienemann, I answered to the effect that it represented comprehensive studies of inland water bodies involving physics, chemistry, and all facets of biology; this synoptic ecological concept being the most complex and difficult, but also the most important area of all natural sciences. A penny for his thoughts.

In 1953, Professor Rippel-Baldes patiently went along with my odd idea to obtain a fellowship for a three months' stay at the *Stazione Zoologica* at Naples. For the lack of travel funds I went on a borrowed decrepit motor bike, inviting all sorts of adventures that readily occurred. In retrospect, I appeared to have been guided by a forgiving providence that a seasoned traveler would not have deserved. The stay at Naples opened my eyes to the delightful atmosphere of the international scientific community. The director, Reinhard Dohrn, son of the well-known founder Anton Dohrn, received me, a mere graduate student, with the same kindness and collegial respect he accorded to the established professorial guests who, with seasoned projects, traditionally came from all over the world to work at the *Stazione*. Professor Califano, microbiologist at the University of Naples, supplied me with needed glassware and helped me to participate at the Third International Microbiology Congress at Rome (1953). During the concert at the opening ceremony, a gentleman sitting next to me glanced at my program signed with my name and said: "I have a friend in Germany of your not-so-common name." Instantly I knew that he must be (and he was) the Canadian soil microbiologist A G Lochead from Ottawa, cellmate of my father's at the above-mentioned Ruhleben WWI prison camp at Berlin. When I had told my father about my choice of microbiology, he turned out to be quite well informed through Lochead's many lectures to him during 1914–1918. At this congress I also met C B ZoBell, a well-known marine microbiologist, who kindly invited me, whenever ready, to work in his laboratory at La Jolla, California.

My stay at the *Stazione Zoologica* in Naples, then still the alleged permanent international congress in marine biology, widened my horizon in an exciting way and started a close association that lasted almost two decades. Life at the *Stazione* had a wonderful cultural flavor. Once, Reinhard Dohrn invited me to a dinner party while Wilhelm Kempff, then a famous pianist, was staying at his house. Afterward we went to Kempff's concert in the huge opera house, San Carlo, overfilled with music-loving Neapolitans. I never forgot this close view of the stressful life of a star musician. He appeared to have two entirely different personalities before and after his performance. During later years, teaching courses in marine microbiology at Naples, I incorporated a lecture on the history of the *Stazione*, based on Theodor Heuss' outstanding book (27), and visits to the major archaeological sites around Naples as well as its museums. I still cherish the friendships of Peter Dohrn, the son of Reinhard Dohrn, who lives in retirement near Rieti northeast of Rome, and the archivist of the *Stazione*, Christiane Groeben, who helped to establish an archival collection here at Woods Hole's Marine Biological Laboratory.

Another escape from Göttingen a year later (1954) took me to Egypt with hardly a penny, but with two good friends: Sepp Fittkau, a most gifted morphological systematist who later became the director of the Bavarian State's Zoological Collection at Munich, and his brother Hans, an amateur botanist who supplied us with an old Volkswagen Beetle. Fascinated by the desert, we went off the road between Alexandria and Cairo into the Wadi Natrun, the most western and ancient dried-up arm of the Nile delta, with a string of snow-white salt flats covered by blood-red puddles: halophilic and photosynthetic bacteria. Equipped with a microscope, pH paper, some chemicals for basic analyses, and lots of sample bottles, we examined all these new things while being closely watched by curious shepherds squatting around our tent, or leaning on their croziers with their flocks nearby, as in a biblical setting. Later, while asking me to add qualitative and quantitative ionic analyses, Professor Rippel-Baldes made me write a note on this saline microbial habitat as an exercise (36). This would, 34 years later, in wondrous ways, make me a member of a US–Chinese expedition to study microbial pigments in drying-up salt lakes of the high Trans-Himalayan region of Tibet (see below). After visiting Professor Abdel Malek, a microbiologist at the University of Cairo, we went south along the Nile valley at high water time. From today's viewpoint of worldwide tourism, our visits to the Egyptian archeological treasures of Karnak, Luxor, and Thebes may sound commonplace. To us, in 1954, they were not. Crossing the Arabian desert, we returned north along the Red Sea shore and visited Dr. Gora, a specialist on nudibranchs, at the Marine Station at Hurdhaga. There we experienced for the first time snorkeling in coral reefs. On our way home via Greece and Yugoslavia, there was not a cubic inch in our little car that went unused for water samples, pickled specimens, or sea shells. Our financial resources being exhausted before reaching friends at Munich, a kind Austrian numismatic garage attendant filled up our gas tank in exchange for a handful of fancy Egyptian coins.

Although full of more-or-less related and interesting observations and experimental results, my thesis work—finished in 1955 (35)—left me with an empty feeling: Is this all I would do for the rest of my life? This led (getting here somewhat ahead of my story) a few years later to an important event:

C B van Niel at Hopkins Marine Station, Pacific Grove, California, questioned me, point-blank, about my work. Still afflicted with unfulfilled expectations for a productive professional life, I could not prevent his penetrating queries from uncovering my dilemma. I said that any young carpenter had done more useful things at my age than I had accomplished so far. He appreciated that, but then suggested that we delay the discussion of this problem until after his summer course, to which I had been accepted (1958). It was only then, through van Niel's course, that I caught up with the carpenter and that my professional self-confidence was established.

Close to the time when I concluded my studies at Göttingen, Professor Thienemann invited me to his laboratory at Plön, Holstein, for a full day's discussion of microbiological research within limnology. I went away with the conviction that, under his leadership and with the nearness of the University of Kiel, I could not find a better place to work. Meeting Thienemann—and later the oceanographer Alfred Redfield and the ecologist Evelyn Hutchinson—corrected my youthful judgement of generalists. I learned to respect their ability to see beyond research details by means of experience and intuition. So-called hard data, which may be wrongly obtained or interpreted, can easily mislead the accuracy-ridden experimentalist. Both deductive research and the generalist's intuition can be seen as refined arts, but they are rarely combined in the same person. In Thienemann they were, and to me he seemed to personify the art of doing science.

To my good fortune, shortly after I attained my degree, he offered me a position as a microbiologist at his Max Planck Institute. Because there were no facilities yet for a microbiologist and a new building was just under construction, he told me to use the meantime well by the traditional "go in the world and learn your trade," or, as one would say nowadays, "seek a postdoctoral position." He gave me a personal contract with the assurance that I would have a job anytime I came home. So, when C E ZoBell renewed his earlier offer, now with a postdoctoral fellowship at the Scripps Institution of Oceanography at La Jolla, California, I gladly accepted.

LA JOLLA, PACIFIC GROVE, MADISON (1957–1959)

Our departure for the United States in January, 1957, was a major separation for my wife, our son of eight months, and me from our families. But it was meant to be just for one year. The 22-hour flight in a propeller-driven airplane (DC-6), with stops in Greenland ($-10°C$) and Winnipeg ($-35°C$) before landing at Los Angeles ($+28°C$), was most memorable, and the surroundings at La Jolla, everything being new to us, were nothing less than fantastic. A good friend from Naples, Michael Bernhard, helped us to find our way around. (A marine phytologist, he later became the Director of the Euratom Marine Laboratory at Lerici, Italy.) ZoBell's assistant, G E Jones, and his German-speaking wife, Edith, did their best to explain the many things foreign to us. Under palm trees and in bright summer-like sunshine, there were Christmas decorations of types we had never seen before. With colleagues like Bill Belser and his wife Nao, who showed us the flowering Anza Desert, and Jack Talling, a limnologist from the British Lake District, we soon felt at home. The friendliness and openness of everyone we met were overwhelming.

I was given a wonderful little lab for myself, but one thing was making a quick start difficult: my utter lack of spoken English. From high school, the humanistically oriented *gymnasium*, I had brought a knowledge of Latin and Greek to my schooling at Holzminden, whereas the students in my class had had six years of English. With all the other wonderful activities going on at that time, I had felt little inclination to pant behind the rest in a vain effort to catch up. And, when I heard about a ruling that, in my case of having lost years of schooling due to the war, I was allowed to skip one field in the final graduation exam, I readily chose to skip English. Now, in the United States, I was paying for this, and I will never forget the state of being totally exhausted every night after having tried all day, as if being afflicted by a mental shortcoming, to communicate. My wife's schooling in English had been much better, and our one-year-old son, standing up in his play pen on the lawn in front of our rented house and greeting everyone passing with "Hi," had no problems at all.

My work did not proceed too well. While building a mercury recycling apparatus for desalting seawater and using Moore's semiquantitative analysis of free amino acids by paper chromatography, I tried to determine threshold levels of organic nitrogen for growth of a variety of marine bacteria that I had isolated. ZoBell praised my efforts and my impressive looking apparatus. It remained to van Niel, whom I met shortly after, to kill the concept of my endeavor thoroughly within minutes of questioning my approach and going through literature with me that I had not known. Clearly, much more sophistication and analytical sensitivity were needed. G E Jones and I, however, published a paper critically comparing a variety of commonly used and new microscopic and cultural techniques for estimating the microbial biomass in seawater (59). Fluorescence microscopy and Strugger's acridine orange staining (113) did not work too well with membrane filters and transmission light microscopy. It was only 20 years later that, with the introduction of epifluorescence microscopy, a generally useful counting technique for microbial cells in natural waters emerged (28).

My first meeting with van Niel, already alluded to above, happened during late 1957. After I had delivered greetings from Professor Rippel-Baldes of Göttingen to his highly valued US representative on the editorial board of the *Archiv für Mikrobiologie*, and after van Niel's demolition of my efforts at

Scripps in an unforgettable three-hour discussion, I asked him quite naively whether it would be possible for me to attend his next year's summer course. A brief "no" was the answer, and then, "This course deals with basic microbiology and you, I understand, have a PhD in microbiology." By now convinced of it, I managed to say that our preceding discussion might have proven that I was still far from being a microbiologist. Looking me straight in the eye and with a wily, yet benevolent, smile he said, "You've got a point there," and accepted me in the course. Only later did I become aware of the privilege it was to become one of the 12 participants.

In June of 1958, after a half year's extension of my stay at La Jolla, I moved my family to Pacific Grove and found a house for rent near Hopkins Marine Station. Another course participant living there was Kjell Eimhjellen from Trondheim, Norway. A lifelong friendship developed between our families. The assistant was the zoology student A Farmanfarmaian, who, after a gallant attempt to reform teaching biological sciences in his native Iran during the late 1960s, became a well-known physiologist at Rutgers University. To describe this 1958 summer course and its group of students—especially van Niel's talent as a teacher, and his superb chronological presentations of various chapters of the microbial diversity leading up to his own artful laboratory demonstrations—would take a dissertation by itself. Much has been said in two obituaries (32, 33), a remembrance by N Pfennig (99), and a biographic memoir by his first students, HA Barker and RE Hungate (1). We, the course participants, became part of the Delft school of basic microbiology for life.

In 1966 I was proud to be invited by Roger Stanier and Mike Douduroff to contribute to a *Festschrift* Volume of the *Archiv für Mikrobiologie* for van Niel's 70th birthday (40). In contrast, I cannot forget an utter defeat that happened shortly after. In 1967, Harvard was to hire an applied microbiologist (a memorable affair I witnessed as a member of the ad hoc committee, headed by Harvey Brooks and attended by the President, N Pusey). The Civil Engineering Department at Berkeley was to follow suit, and as a possible candidate, I was invited to give a seminar. Having had just presented a paper on applications of continuous culture kinetics in Werner Stumm's group at Harvard's Department of Applied Physics, I confidently planned to give a similar talk, rather loftily worked over. Minutes before the start, van Niel appeared out of the blue, casually remarking that he saw my talk announced and thought it would be nice to combine this opportunity with a visit to his daughter's home at Berkeley. I wished fainting had been an option. Had I known that he would be there, I would have worked on my lecture day and night. Now I was hoping that the sky would fall in or for some other heavenly intervention. But no such reprieve occurred, and the ordeal started, with van Niel sitting right in the front row, the expression on his face telling and his pencil recording every slip in the use of clear English, every glitch in successive context, insufficient explanation of diagrams, and the like. My regular nervousness as a speaker usually leaves me after the first minutes. Not this time. I never forgot this wholly deserved disgrace. During lunch at the Faculty Club I realized that the lecture had been well received and that no one had noticed my distress—except van Niel. Then, over the hood of his car in the shade of an oak tree, we spent two hours going over his notes, point by point, a lesson that stuck with me to this day. At the end, possibly to cheer me up, he said that the elegance of continuous culture approaches, such as the detection of Gorini's derepression phenomenon (23) and the possibility of measuring the effect of light intensity and wave length on growth of photosynthetic bacteria, enticed him so much that, if he would get to work again in the lab, it would be with a chemostat or with its modification, the turbidostat.

He took questions from students very seriously, often incorporating them in his lectures. As a result, our questions became more thoughtful. He took pains to come up with satisfactory answers that often required him to go back to his large reprint collection. This applied to nonscientific matters as well. Working happily away in the lab, I used to whistle, not necessarily to the pleasure of others around me. One day I felt a tap on the shoulder: "What was that?" van Niel said. It was some catching tune from one of our Holzminden school orchestra pieces, but what exactly, I was not sure. Next morning I found a note on my desk, "Bach: *Double Concerto in D-minor*, second movement." He had gone home and, remembering the tune, went through a few records until he found it. If my mentioning of van Niel in this chapter appears out of proportion, it must be understood by the enormous impression he made on me during the most receptive phase of my growth as a microbiologist. His autobiographical sketch (124) is a key to his gifts as an educator. He taught me to understand the conduct of science as a privilege and an obligation at the same time, and to see the endless, sometimes frightening, opportunities it offered for accomplishment. I also owe to van Niel a sense of profound satisfaction and pleasure in understanding old and new concepts, successfully passing them on to students—and finally the joy of having, here and there, an original idea oneself. I felt greatly honored being asked to give the second van Niel Memorial Lecture two years after his death (1985) at Hopkins Marine Station.

Meeting AD Hasler from Madison, Wisconsin, at an AIBS meeting at Stanford University in late 1957 resulted in his invitation, before my return to Germany, for me to experience an American university that I, so far, had seen only during short visits. Thienemann, back in Plön, to whom I reported duly all of my exploits, approved gladly of this visit because Madison was also the cradle of American limnology and Hasler the living representative. So, after leaving Pacific Grove, we spent almost four weeks crossing the country by car, an unbelievable experience that contributed to our growing admiration and love of this country. Arriving at the beginning of a midwestern winter, I was given space in the large student Laboratory 107 in the Microbiology Building at the University of Wisconsin at Madison. One of my many lasting impressions was the unencumbered cooperation between university departments as I have not seen it before or since. I started to work on the effect of free oxygen on denitrification (37) and threshold levels for the inhibition of nitrate and nitrite reductases. Using all the help of the limnologists for field work, the project suddenly required parallel measurements of gases that could best be done by mass spectrometry. I was sent to the Chairman of the Biochemistry Department, Bob Burris. With no ado he permitted me, an utter newcomer, to use the Department's only mass spectrometer after just a brief instruction. Luckily nothing imploded. The youngest professor in the Microbiology Department was Harlyn Halvorson, whom I was later to meet on many occasions at Woods Hole. AD Hasler's kindness and confidence in me with my limnology-zoology background was unlimited and survived, in 1970, my turning down an offer of the EB Fred Endowed Professorship in microbiology at Madison. This position was assumed, in 1971, by Tom Brock.

Professor Fred had earned his doctoral degree from the University of Göttingen in 1911 under Alfred Koch, an early expert on microbial nitrogen fixation, the founder of the Institute for Agricultural Bacteriology, and the predecessor of Rippel-Baldes. During a later trip to the United States in 1961, as a representative of the University of Göttingen, I would present Professor Fred, who had been President of the University of Wisconsin, with his Golden Anniversary Diploma and a Latin laudatio. Through his teacher, HL Russell, who worked at Naples (24, 109) and Woods Hole (110) during his career, intriguing historical ties between these laboratories, as well as between those at Madison and Göttingen, became apparent.

During May of 1958, I attended the ASM meetings at Chicago where I met Hans Schlegel, then working with L Krampitz of Case Western Reserve University at Cleveland. He was greatly interested to hear about Göttingen, where he was being considered as the successor of Rippel-Baldes. It was the beginning of a lasting friendship and some joint writings (111, 112). At the 1959 ASM meeting at St. Louis I met Stanley Watson from the Woods Hole Oceanographic Institution, who suggested a visit to Woods Hole before my return to Germany. My wife and I took this trip after "parking" our son, then three years old, with my brother's family near Halifax, Nova Scotia. Woods Hole, as an oceanographic counterpart to La Jolla, impressed me with its smaller, more personal group of scientists and its beautiful New England surroundings: oaks and junipers instead of La Jolla's palms and eucalyptus, the jovial Columbus Iselin instead of the regal Roger Revelle.

BACK AT GÖTTINGEN (1960–1963)

About the time of my return to Germany, Professor Thienemann passed away. My visits with the new Director of the Max Planck Institute at Plön were first quite amiable, although I missed Thienemann's spirit sorely. Because the new building was not yet quite completed, my request for a temporary stay at Göttingen was granted. Hans Schlegel had taken over the Department of Microbiology at the University and kindly offered me some space in the old and quite cramped building at the Gosslerstrasse. I shared a tiny room, just holding two desks and a table for setting up a chemostat, with Norbert Pfennig, who was then Assistant Professor at the Institute. The following year was extraordinarily fruitful. Not only did I watch Norbert start his pioneering work on the photosynthetic bacteria, but our joint systematic study and everyday discussions on steady-state kinetics in chemostat cultures (101) meant a great leap forward for me. Now it became possible to determine microbial growth at low substrate levels and low population densities quantitatively.

In order to have me participate in some teaching activities, Hans Schlegel suggested a course in marine microbiology. As a suitable place for it, I thought of my beloved *Stazione Zoologica* at Naples, strongly believing in the *spiritus loci* that I had experienced there as a student. This first course in 1960 for 25 graduate students from Göttingen started a tradition that lasted for the next decade and a half. The Gulf of Naples provided a large diversity of microbial types, not to mention others from the *solfatare* of the *Campi Flegrei* and the hot springs on the island of Ischia. From the 70-m deep *Bocca Piccola* between Capri and Sorrento straight south, a depth of 1000 m could be reached in half an hour with small boats. Extended weekends were used for the archaeology of Naples' surroundings: Paestum, Cumae, Caserta, Pompei, and Herculaneum. Later when I had left Göttingen for the United States, the course continued and justified my staying on the faculty list of the University's Microbiology Department.

During a fortuitous visit with Professor Autrum at Munich in 1960, he supported my growing feelings that the limnological institute at Plön without Thienemann was not where my future lay, and he encouraged me to stay at Göttingen. Terminating my position with the Max Planck Society in 1960 was not easy, especially because there was no job available at Göttingen at the time. Subsequently a grant proposal was successful. It required, however, my preparations for the *habilitation*, a qualification procedure for professorship, common to all universities of German-speaking countries, which may involve lengthy research—ending with another thesis to be approved by the entire science faculty. A colloquium in front of that faculty is the critical part, usually followed by an extensive discussion in the nature of a general exam, often going way beyond one's field of expertise. Then, after a public special lecture at the University, the candidate may receive the *venia legendi*, the permission to read, that is, to give lectures. This procedure is meant to demonstrate the candidate's independence in research, his ability to advise graduate students, and to lecture well. Before the title *Privatdozent* is conferred on the candidate, however, he has to announce and present special lectures for two semesters attracting enough students. With Schlegel's backing, I was able to fulfill these demands by the end of the summer semester of 1963.

Earlier, in 1961, I was invited to participate in the first (and last) International Meeting in Marine Microbiology in Chicago, preceding the ASM meetings, and traveled on to the West Coast, paying a number of memorable visits. Syd Rittenberg at Los Angeles, whom I had met in van Niel's lab with June Lascelles and Hans Veldkamp in 1957, told me about a sulfate reducer he had isolated from a depth of 300 m below the sea floor from a core of the Mohole drilling project. After careful consideration, he threw the culture away, because he could not be certain where it came from. Obtaining uncontaminated samples from such cores has inherent problems. One wishes this type of scientific integrity was still prevalent in spite of today's increasing pressure to sell one's science. I agree with a quote, attributed to Carl Sagan, that "extraordinary claims require extraordinary evidence." Syd decided that he did not have enough of it. He was a top scientist.

While visiting Hopkins Marine Station, I reported to van Niel on my work at Göttingen, but what really excited him was looking at one of Pfennig's cultures of *Chromatium okenii* that I had brought with me. A year later, Norbert himself came and had a most productive year at Pacific Grove (100). On my way up the coast, at Davis, I visited the ever amiable and kind Bob Hungate, the earliest van Niel student, who almost single-handedly explored the microbiology of the ruminant's complex cellulose-fermenting digestive system and designed the necessary techniques to deal with extremely oxygen-sensitive bacteria. I also saw Jerry Marr there, whose work on growth kinetics had been very helpful to me. At Eugene, Oregon, I consulted with Aaron Novick, a former physicist who, together with Leo Szilard, had devised the chemostat (93) at the same time as—but independently from—Monod (86) in 1950. Referring to Herbert's help for biologists in understanding steady-state kinetics (26), he explained to me that, for a physicist, the mathematics were simply too elementary to merit inclusion in his publications.

This row of visits up the West Coast ended with a four-week stay in Erling Ordal's lab at Seattle. He was a true representative of Henrici's American school of aquatic microbiology, which I had encountered once before through EB Fred and E McCoy at Madison. In Ordal's lab, I built a microscopic flow-through chamber, a continuous culture for semiquantitatively studying growth of attached *Thiothrix* filaments as affected by different concentrations of hydrogen sulfide and acetate. My stay was too short and the culture chambers too few to obtain enough data for a publication, but the experience of daily discussions with E. Ordal, listening to lectures by Helen Whitely, and meeting the limnologists Yvette and Tom Edmondson as well as the oceanographers Richard Fleming and Francis Richards, was well worth the time I spent at Seattle.

Passing through Woods Hole once more on my way back to Germany, I was asked to give a seminar. To my surprise, I was taken out afterwards by two senior staff members, rather than, as is commonly done, by a group of students. While being treated with a martini or two, I enjoyed their company greatly at the Flying Bridge restaurant, never suspecting for a moment that I was being interviewed. Back at home a few weeks later, I received a letter from Woods Hole offering me a permanent position as a microbiologist. I regretfully declined because I had just accepted my *habilitation* stipend, and I explained that this open-ended commitment might take two years or even longer. Back came a letter from Paul Fye, the Director of the Institution and one of the two interviewers (the other one was Bostwick Ketchum), saying that they were willing to wait until my obligations were fulfilled. Attracted by the tone of this letter and the congenial colleagueship and working climate it promised, I accepted. A year later I got another letter from WHOI's comptroller telling me that the offered starting salary had been increased with everyone else's salary in consideration of the fact that I was surely working as hard as I would if I were at Woods Hole already. This happened once more; however, it was not the salary or the rank at which I was hired (and which I did not comprehend at that time) that mattered to me, instead it was the personal touch and liberal research atmosphere that I sensed. I was not mistaken and never regretted my decision. Shortly thereafter came an equally enticing offer from the University of Washington at Seattle, but, by that time, it was too late to consider it.

With the completion of my obligatory special lectures at Göttingen during the summer semester of 1963 (for a class of 15 students) came the qualification as *Privatdozent*. I met the required teaching for a second semester by giving another course in marine microbiology at Naples for graduate students from the Microbiology Department at Göttingen. Schlegel's generosity shielded me from the faculty's disappointment about my leaving for the United States after all their efforts on my behalf. As a *Privatdozent*, however, I was now on the list of possible candidates for a professorship and ready for an official call to return should a chair become vacant.

WOODS HOLE

Nevertheless, this second departure was more definite then our first one had been six years earlier, and, looking back, I realize how difficult it must have been for our parents. I had inherited Norbert Pfennig's technician, Jutta Koslowski, while Norbert was spending a year at van Niel's lab at Pacific Grove. Now Jutta was eager to come with us to the United States and, with Norbert's consent, I was glad to have her help. So, in October 1963, my wife and I, with our son, then seven years old, and Jutta boarded a shiny Norwegian freighter that also took our fully packed car into her hold. Leaving from Hamburg, we crossed the tail of a hurricane during a memorable voyage, arriving after ten days at New York and two days later at Woods Hole. By earlier correspondence, I had been able to make some suggestions for the layout of my laboratory in the brand-new Redfield Building. Just a short time after we arrived, a series of chemostats, equipped for running at extremely low flow rates of reliable constancy, was in operation.

I had well understood that WHOI was a private "soft money" research laboratory but had never spent a second thought on it. And indeed, having to take time only once every two or three years for the preparation of a proposal based on our progress, I never experienced a problem in being fully funded by NSF grants until 1988, when I was given hints to diversify my funding sources, something most colleagues had experienced much earlier. Not having been bothered by these problems for most of my tenure at WHOI, I felt lucky and productive and did not mind 10- to 12-hour days. During frequent travels, lecturing, meeting students and colleagues, and getting a feel for this country's large scientific community, I learned that Woods Hole represented a friendly and sociable portion of it, concentrated in this small village, a bridge to colleagues from all over the world, with a congenial research climate in which it is a pleasure to live and to work. Professionally, I was given the rare personal freedom to put my talents where I thought they fitted best and to free me from the kind of tasks that I knew I was not made for. Local flavors were provided by sailing an old cat boat and attending a daily breakfast club composed, for its best part, of nonscientists. The pleasure of wonderful and frequent chamber music I owed to colleagues such as Jelle Atema (flute), Bill Simmons (cello), and Gaspar Taroncher (harpsichord). I was very moved when Jelle, a former Rampal student, devoted a special evening to play my father's old-fashioned wooden traverse flute, by now 100 years old.

While introducing me at a seminar at the Scripps Institution of Oceanography in 1992, Angelo Carlucci, a friend for many years, surprised me by describing my microbial work at Woods Hole as divided into three general areas roughly coinciding with the past three decades: microbial growth kinetics in seawater, effects of low temperature and high pressure, and processes at hydrothermal vents, all dealing with the pelagic and deep ocean. In the following, I will make use of Angelo's categories, mixing in more-or-less unrelated activities and events as they occurred.

Microbial Growth Kinetics in Seawater

In one of van Niel's papers (125), I discovered a gratifying confirmation of my general approach: *Growth is the expression par excellence of the dynamic nature of living organisms. Among the general methods available for the scientific investigation of dynamic phenomena, the most useful ones are those that deal with kinetic aspects.* Perceiving seawater as a highly dilute medium, mainly carbon limited, the chemostat work was re-initiated in my new lab and proceeded well. When the concentration of growth-limiting substrates (a number of fatty and amino acids as carbon sources) in the reservoir (S_0) was lowered stepwise and steady states re-established, we observed that, at a given dilution rate and a still-considerable value of S_0, the culture was washed out of the chemostat (38, 41). This indicated that the mode of growth limitation had changed and appeared to depend on a minimum population density. It turned out that growth depended on a low redox potential that was controlled by population density. At a given aeration rate, lowering of the population density by lowering S_0 led to washout of the culture. Adding ascorbic acid or lowering the amount of oxygen in the aerating gas mixture allowed the population density to reach lower steady-state levels (41, 60). These results had interesting implications: seemingly optimal growth conditions in dense microbial populations can involve hidden cell-medium interactions that become apparent only at low population densities. Such cryptic feedback effects are not detectable in high density batch-cultures and are commonly not considered in continuous-culture kinetics. The data indicated that growth in typically well-oxygenated seawater may only occur in relative nutrient-rich micro-environments that permit high cell densities, such as aggregates of particles known as marine snow.

In steady-state cell suspensions we observed threshold concentrations of the growth-limiting substrate, or unused left-over concentrations, when growth ceased and washout of the populations occurred. These threshold levels depended on the imposed dilution rate as well as on sub-optimal growth conditions (46, 51). Because the former was known, the latter could be calculated in seawater and experimentally manipulated. The alleged recalcitrance of the organic carbon in deep and pelagic seawater may partly be explained by this threshold phenomenon, and changes of left-over concentration might occur as a result of changing environmental factors (temperature, oxygen concentration, etc) affecting the efficiency of microbial growth. While low in concentration, the total amount of organic carbon dissolved in seawater is on the same order of magnitude as CO_2 in the atmosphere (8). Therefore, any change in the efficiency of its microbial turnover through a change of temperature, redox potential, or the like, must have a tremendous and hitherto unappreciated effect on the carbon cycle, and thereby on the global climate. Our data also lend themselves to explain the persistence of man-made microbially degradable pollutants at high dilution in offshore seawater. In culturing microorganisms, our results helped to explain lag-phase phenomena and the need for starter populations (39, 49).

Trying to determine growth rates in natural seawater, where the concentrations and types of growth-limiting substrates are unknown, we ran chemostats at sea. Coarsely filtered (2–3 μm) seawater was inoculated with morphologically distinguishable isolates and the dilution rate lowered stepwise in an attempt to establish steady-state populations of the test organisms in the presence of the natural population. Washout occurred every time. When, however, the washout data were statistically treated and compared to the dilution rate, the calculated difference between washout and dilution rate equaled the actual growth rate of the test organism. This technique not only allowed us to measure microbial growth rates in natural seawater (50), but it also made it possible to assess the competition between test organisms and natural microbial populations under various conditions (49).

These experiments were run largely on research cruises "of opportunity." Several such cruises to the Peruvian coast and the Galapagos Islands in 1964 on Truman's former presidential yacht, the ex-*Williamsburg*, then named the *Anton Bruun*, and to the anoxic waters of the Venezualan Cariaco Trench in 1965 on the *Atlantis II*, taught me the necessary oceanography and technical hydrographic know-how. Sampling and working under tropical conditions with the typically psychrophilic marine bacteria, which cannot survive temperatures above 20°C, required special procedures, and bringing up box cores of sediment from depths of up to 8000 m in the Puerto Rico trench required expert help and collaboration with other users of such valuable sample material. Oceanographic cruises became a regular part of my Woods Hole life. Developing mutual respect and close working relationships with crew and officers, learning some basic navigation and the constellations of the southern sky from them, brings definite joy that is one of the best parts of oceanography. In 1960, prior to joining a ship in Trinidad for Cariaco Trench studies, I took the chance to visit my uncle's family at Paramaribo, Surinam. My father's brother did not become a Moravian missionary there, but instead used his excellent craftsman's training as a very successful architect and builder who never wanted to leave the tropics again. Some of my cousins are still living there.

Oligotrophy is found in many marine microbes from offshore and deep seawater. It is difficult to detect in batch-culture, however, because the lag and stationary phases overlap at low substrate concentrations (60). The chemostat permits enrichment, isolation, and growth studies of oligotrophic bacteria under suitable conditions (40). Characterized by extremely low growth constants (K_s and μ_m) on a relatively rich medium (10 mg/l lactate), such isolates produced only micro-colonies. However, chemostat cultures led to quick selection for big-colony mutants exhibiting increased growth constants (43). They were no longer inhibited by high substrate concentrations and had lost their growth efficiency at low substrate levels. The low versus high growth constants in oligotrophic and copiotrophic organisms (43, 47) was used for predicting the survival of *Escherichia coli* in seawater on the basis of substrate competition only (42). The various uses of growth kinetics in microbial ecology became the topic of a number of productive discussions (56, 60, 126), in particular, at a Dahlem Conference organized in 1978 by M Shilo, which included a section on microbial activity and survival at low nutrient levels (50).

After these first several years at Woods Hole, Jutta Koslowsky, my lab assistant, left my laboratory for the most acceptable of all reasons: She married a colleague of mine and moved with him to Chapel Hill, North Carolina. In 1968, she was replaced by Carl Wirsen, who has been with me ever since as a most valuable coworker. Although he is a member of the technical staff, he has grown into a scientist in his own right, attaining the rank of Research Specialist. In 1971, he was joined by Stephen Molyneaux as Laboratory Assistant, now a Senior Research Assistant. It has been a particular stroke of luck to find these two steady teammates. As time went on, they contributed many valuable ideas to our projects. They became experts in the preparation of scientific cruises and work at sea, abilities necessary in almost any branch of oceanography.

The idea of measuring a collective bacterial activity in natural populations has always dominated marine and freshwater microbiology. To parallel the commonly used $^{14}CO_2$-uptake measurements for estimating the photosynthetic productivity of phytoplankton, great efforts were made in finding a similar unifying parameter describing total bacterial activity, often translated into microbial biomass, by determining ATP, DNA/RNA ratios, or ultimately the incorporation of ^3H-labeled adenine or thymidine into DNA, as discussed in detail by Karl (78). I have followed these studies only from a distance, as I am not fully convinced that a useful common denominator exists for the metabolically highly diverse and flexible natural microbial communities. The recent discovery of a substantial group of archaea (16), made by the use of 16S rRNA probes, has shown that whole portions of natural microbial communities in seawater and their responses to a labeled precursor are unknown. Not yet being isolated, the particular metabolism of these new archaea in the oxygenated seawater column is still a conundrum.

In 1975 I had taken on a student, Russell Cuhel, who had the great idea to use the uptake of $^{35}SO_4$ as a measure of productivity in marine bacteria and algae. To give his plan a solid base, I asked him to take biochemistry with Gene Brown at MIT and to learn techniques for the study of algal and bacterial sulfur metabolism from Jerry Schiff at Brandeis. The knowledge he gained soon made it clear to him, better than I could, that sulfate uptake would not serve as a useful measurement of a general activity. Besides exhibiting unlimited bursts of energy, such as working in overnight stints in order not to miss a single point in a 48-hour growth curve, and being the life of the biology department throughout his stay, Russell produced excellent research; only two publications (13, 14) of his thesis are cited here. He is now a staff member at the Center for Great Lakes Studies at the University of Wisconsin at Milwaukee.

Psychro-, Baro-, and Oligocarbo-philic Bacteria

We observed early that the relative number of psychrophilic bacteria increased with depth in the ocean (129) and found that many of them did not survive temperatures between 20 and 25°C. Considering an analogy of microbial adaptation to increased pressure (barophily), we left work on barophilic bacteria for later, hoping that decompression-free sampling techniques would become available. Not being able to obtain truly barophilic bacteria from anyone had made me somewhat skeptical about their being real, until A Yayanos clearly demonstrated their existence in 1979 (136). Still, because all isolation procedures so far involved a decompression cycle, it appeared possible that the resulting cultures just represented those barophiles that survived decompression and that more pressure-sensitive organisms might still exist that are, accordingly, adapted to even higher pressures. Then, suddenly, an approach not hampered by this decompression problem was indicated to us by the accidental sinking and recovery of the Institution's research submersible ALVIN in 1968.

First, with two early friends from Pacific Grove, K Eimhjellen and A Farmanfarmaian, both at Woods Hole at the time of ALVIN's recovery in 1969, we documented the surprisingly limited decomposition of the submersible's lunch food after an 11-months' incubation period at 2000-m depth and 3–4°C (57). Far from being conclusive, this involuntary experiment nevertheless gave us the idea of depositing substrates on the deep-sea floor to be inoculated automatically in situ. Attached to the anchors of routinely deployed and recovered buoys on the Woods Hole–Bermuda transect at depths from 3600–5300 m, incubation periods varied from weeks to months. After retrieval growth and the state of substrate, transformation could be measured by various means (130). Although rates were indeed slower than those of controls kept at 2–3°C and normal pressure, the data of these field studies were limited to mere end-point measurements. Yet, they started a most productive collaboration with some of the Institution's engineers, especially Cliff Winget and Ken Doherty, ultimately leading to the development of decompression-free sampling devices that permitted time-course measurements in growth and uptake studies on natural populations (66, 73, 75). Then the logistical problem arose that, with only two such units, only two samples per cruise could be obtained, not enough to yield statistically significant data. This was solved by designing a different sampler that could be re-autoclaved at sea (63). From then on, the return of samples under pressure was limited only by the number of transfer units.

At this point, we were still not able to work with pure cultures of undecompressed deep-sea microbes. Then, after Craig Taylor conducted comparative growth experiments at hydrostatic and hyperbaric pressures (115), Ken Doherty designed a hyperbaric isolation chamber that could be precompressed to any desired in situ pressure with helium containing 0.5% oxygen. Undecompressed samples were introduced from transfer units and streaked on small square petri dishes under a window (74, 116). Subsequently a number of barophilic bacteria were isolated without ever submitting them to decompression. We observed that barophily, besides being characterized by membrane fluidity (133), varied greatly with the type of metabolism (66, 67). Key enzymes appeared to be barophilic or nonbarophilic, not the organisms as such. Some of the isolates tested survived decompression only for short periods of time. Still missing is an efficient technique for selecting microbes that do not survive decompression at all. Much of this work was funded by the Office of Naval Research, which, after the USSR's collapse and the end of the Cold War, switched its interest from offshore and deep-sea phenomena to those of shallow, coastal waters.

Barophilic growth appeared associated with psychrophily and oligotrophy, as typical characteristics of most deep-sea isolates (50, 60). As pointed out above, work on oligotrophic growth requires studies in steady-state cultures. Because pressurized chemostats did not exist, most work on microbial baro- and psychrophily was done with copiotrophs, organisms that require relatively high substrate levels and can be well studied in batch culture. When Ken Doherty found out that the design and construction of a pressurized continuous culture system could be greatly aided by equipment originally developed for high pressure liquid chromatography, the apparatus was built, tested, and used in work on growth kinetics of deep-sea oligotrophs (68). Because carbon sources are used as growth-limiting substrates, these organisms are better described as oligocarbophiles. Building a chemostat system for a maximum of 700 atm (71 MPa) had economic and ecological reasons. Less than 1% of the oceans' volume lies below a depth of 7000 m. Because all of our barophilic isolates are psychrophiles, the growth experiments were run at 3 and 8°C. The first data show that barophily and oligocarbophily are indeed closely linked (68).

During the years of work at Woods Hole, the access to engineering expertise and instrument development has been a unique and most pleasant experience, so valuable, in fact, that I turned down an offer of a "hard-money" professorship at Dalhousie University at Halifax in 1973. At that time we were in the middle of this pressure-related work, and engineering help would be available only through the Canadian Navy. However, as a non-Canadian, I had no access to its workshops and labs until I would qualify for citizenship some years later.

Oligotrophic bacteria occur, of course, in freshwater too. Being kindly invited, I introduced the kinetic approach to their study during one of the limnologists' international meetings in Jerusalem (44). It was 1968, and I also participated in the official opening of the Steinitz Laboratory of Marine Biology at Eilat where I met M Shilo, A Keynan, and Y Cohen, who later often visited and worked at Woods Hole.

Microbial Sulfur Oxidation and Deep-Sea Hydrothermal Vents

My interest in the microbial sulfur cycle started when observing the photosynthetic sulfide-oxidizing bacteria overlaying the black mud of the Wadi Natrun salt flats. As a consequence of the high sulfate content of seawater, biogeochemical transformations of sulfur are central in marine microbiology. The cycle is initiated wherever seawater or sediment turns anoxic and sulfate is reduced to sulfide during anaerobic respiration, to be oxidized again at anoxic/oxic interfaces. We were able to work in the Black Sea, the world's largest anoxic marine basin, three times: in 1969 on the research vessel *Atlantis* II, in 1975 on the *Chain*, and in 1988 on the *Knorr*. In 1975, the scientific crew included Hans Trüper of Bonn and Gijs Kuenen of Delft. Despite the additional, potentially explosive, inclusion of two Turkish guests, plus Yehuda Cohen from Eilat, Israel, and Fuad Hashwa from Amman, Jordan (along with Woods Hole colleagues of Greek and Armenian descent Ollie Zafiriou and Bob Gagosian), there could not have been a more congenial atmosphere. On a later cruise in 1988, when Bo Jørgensen and Henrik Fossing from Aarhus and Heribert Cypionka from Pfennig's lab at Konstanz came along, we observed that the O_2/H_2S interface had risen from a depth of 150 m, as measured in 1969 and 1975, to less than 100 m below the sea surface (69, 77, 87). First indicated by the detection of a bacteriochlorophyll (103), a layer of a green phototrophic bacterium was discovered just below the O_2/H_2S interface. A *Chlorobium* strain able to grow at very low light intensities was later isolated by J Overmann and H Cypionka in Pfennig's lab (95).

Very unexpectedly, in 1977, microbial sulfur oxidation was found to be a most important deep-sea process leading to dense animal communities in the absence of light. I have been incredibly lucky to become part of this discovery that nobody imagined beforehand—except GE Hutchinson, the American counterpart to A Thienemann. In a book on ecology and evolution (34), he had contemplated that sunlight might not be the only source of energy maintaining life on this planet but that its own internal heat "in theory would provide an alternative to incoming radiation though we have little precedent how an organism could use it." Now we do know, and the reason for this discovery was a close interdisciplinary collaboration between biologists, geochemists, and geophysicists—and a lot of luck.

While biology brewed the molecular revolution during the 1950s, a similarly consequential development was taking place in the geosciences: plate tectonics.

This theory became reality, and because most plate-spreading zones appeared to occur at mid-ocean ridges, oceanographers were the first in line to obtain proof of in situ volcanic activity. Dick van Herzen, a physical oceanographer from Woods Hole, and John Corliss, a geologist from Corvallis, Oregon, selected a tectonically active area north of the Galapagos Islands, dragged heat probes over the ocean floor, and used ALVIN to dive where heat anomalies were observed at depths around 2500 m. They not only found warm-water emissions but also dense populations of strange new animals clustered around these hydrothermal vents. The biomass was clearly too high to depend on a photosynthetic food supply at this depth, but the chemically reduced hydrothermal fluid contained hydrogen sulfide, indicating the possibility of a microbial chemolithoautotrophic production of an organic food source. Very excited, not surprisingly, I started writing a proposal as soon as the news reached Woods Hole. So did others, and two years later, in the spring of 1979, a group of biologists took ALVIN back to the same site.

Having seen the barren deep-sea floor during our in situ incubation studies on many ALVIN dives before, it was an overwhelming experience to fly, 2550 m deep, over dense beds of large mussels (*Bathymodiolus*) or even larger, up to 30-cm long, white clams (*Calyptogena*), or stands of hundreds of snow-white tube worms (*Riftia*) up to 2 m long and crowned with feather-like blood-red plumes (gills). We were struck by the thought, and its fundamental implications, that here solar energy, which is so prevalent in running life on our planet, appears to be largely replaced by terrestrial energy, liberated by the oxidation of reduced inorganic chemicals, with chemolithoautotrophic bacteria taking over the role of green plants. This was a powerful new concept (52, 53, 54, 64) and, in my mind, one of the major biological discoveries of the twentieth century. Communicating it to the geochemists (55, 61) was a delightful learning experience. Because the demand for oxygen in the desert-type deep-sea is low, there is a sufficient amount available for aerobic chemosynthesis—a term coined by Pfeffer in 1897 in analogy to photosynthesis (98).

Collecting water samples around the clams in order to assess the food source of these typical suspension feeders resulted in a puzzle. For a food source, the bacterial density in the surrounding water was much too low. Guessing that microbes might grow within their gills, conveniently flushed by oxygen- and hydrogen-sulfide–containing water, resulted in another blank. From there it was not far to look for an endosymbiotic association between these invertebrates and counterparts to the free-living chemosynthetic bacteria. While we started to work on the latter, H Felbeck and G Somero at Scripps, C Cavanaugh at Harvard, and J Childress at Santa Barbara began to study these symbiotic systems. Because none of the symbiotic microbes, identifiable by transmission electron microscopy, have yet been isolated, work has concentrated on their phylogenetic description based on 16S rRNA sequences (11, 17). Although work with gene probes will yield much information on the distribution of such symbionts and their free-living stages, clarification of their actual function will ultimately require isolates. Most likely a student, unencumbered by a history of failures, will someday do everything wrong, and as a result, break the log jam. Ceram says it well: "This is the sort of triumph reserved to the unbiased newcomer who leaps over the hurdles of conservatism at which the scholar shies" (12).

Our earlier work on the microbial oxidation of reduced sulfur compounds in the Black Sea and Cariaco Trench with Jon Tuttle (119, 120, 121) gave us a wonderful head start. After a morphological survey of the free-living and mat-forming bacteria (65), we isolated a large number of obligately and facultatively lithotrophic sulfur-oxidizing strains (72, 105, 107) and determined the stable carbon isotopic fractionation by one of the isolates (106). Interestingly enough, some isolates appeared on the phylogenetic tree right next to the cluster of chemosynthetic symbionts (11, 17). From the sheer visual impression, the symbiotic biomass production of tube worms and clams appears to be greater than that of free-living chemolithotrophs, but considering the huge areas of H_2S-exposed surfaces covered by microbial mats, visible only with the microscope, they may well produce as much or more biomass.

The first hyperthermophilic deep-sea organism, a methanogen, was isolated by John Leigh in Ralph Wolfe's lab (76) from one of our smoker samples. Later this isolate provided the first archaeal genome that was fully sequenced (9). K Stetter's brilliant and prolific discoveries of highly diverse hyperthermophiles were partly done from deep-sea smoker samples or hot sediments collected during our diving cruises (10, 20, 30, 31, 80, 102). Having him along or sending samples to him as well as to other colleagues was the best way to maximize the outcome of expensive ALVIN dives. We added a number of new heterotrophic, hyperthermophilic isolates ourselves, as we were primarily interested in their physiology and ecology (5, 58, 70, 71, 92). I agree with Ralph Wolfe's Second Law (135) that it is easy to make discoveries if you are the first one on the scene. Yet, it is also important who the first one is, and I greatly enjoy the still-productive friendship with Karl Stetter (7a).

Hydrothermal vent fields in the Guaymas Basin, Gulf of California, are covered by approximately 400 m of sediment and feature massive mats of *Beggiatoa* spp. with cells exhibiting huge "liquid vacuoles" (62, 91). Although mats grown in the lab at O_2/H_2S interfaces never reached a thickness of more than 0.5 mm (89), the natural Guaymas Basin mats can be up to a few centimeters thick (91). Using microsensors for measuring profiles of O_2, HS^-, pH, and temperature through these mats in situ, Gunderson & Jørgensen observed a fluctuating flow of the hydrothermal fluid and periodic movement of the O_2/H_2S interface that appeared to be the reason for this massive growth (25). Later, the Bremen group (see below) found that huge liquid vacuoles were related to storage and transport of nitrate in *Thioploca* (21). Shortly after, a similar process was found in *Beggiatoa*, turning this traditional aerobe into a facultative anaerobe (82). Currently G Muyzer and A Teske are studying with us the molecular biodiversity of the aerobic sulfide-oxidizing bacteria at deep-sea vents and other marine habitats (88).

During these exciting 1970s and 1980s, the Biology Department at WHOI included five senior microbiologists. Working on new marine nitrifiers was Stan Watson, student of Ken Raper; on in situ measurements of microbial processes was Craig Taylor, student of R Wolfe; on marine cyanobacteria was John Waterbury, student of R Stanier; on protozoa was David Caron, student of J Sieburth; and on the projects detailed herein, my lab. After Stan Watson's death in 1995, we were reinforced for several years by Ed Delong from Norman Pace's lab, concentrating on new marine archaea, and Paul Dunlap, who had studied with Pete Greenberg and was then specializing in luminescent bacterial symbionts in squid. It was, and still is, a lively and stimulating group to be associated with.

Postdoctoral Fellows and Sabbatical Guests

The almost exclusive research atmosphere at WHOI always seemed to me particularly well suited for postdoctoral training. Before young scientists are exposed to the hassles of their first job, postdoctoral programs should allow them a few years of independence for unrestricted research. They should be weaned from an advisor's or mentor's influence, follow their own ideas, develop their own research programs, and assume responsibility for their own work. This may be the most important learning period in a scientist's career. It seems an easy task for a lab leader to have such independent workers around, and yet, productivity does not always come easily and some subtle and sympathetic guidance is usually required.

My first postdoc was Hans Trüper (1966–68) who worked on marine sulfate reducers and phototrophic bacteria (117, 118). Taking my place during a Red Sea cruise, he visited Professor Abdel Malek at Cairo and the Wadi Natrun, where he later did his profound work on the halophilic photosynthetic genus *Ectothiorhodospira*. In 1966, I was offered the position of full professor of microbiology at the University of Bonn. It took me a week to consider and to decline, largely because of my ongoing research projects that I would have to give up, although it was quite clear that this declination might mean burning my professional bridges to Germany. My decision must have appeared quite ungrateful to my German colleagues, especially during a time when, during the 1970s and early 1980s, a number of new chairs in microbiology were created there, persuading most young German microbiologists staying in the United States at that time to return. That I was forgiven, however, I assume from my election to the Göttingen Academy of Sciences in 1984. The Bonn chair was later offered to Hans Trüper, who accepted it and has done extremely well in creating his own "school" on microbial sulfur metabolism. His student, Johannes Imhoff, is now the marine microbiologist at the Institut für Meereskunde at Kiel, Germany. In my lab, Hans Trüper was succeeded by Hans van Gemerden from the Netherlands, who stayed from 1967–1968 studying growth of a *Chromatium* isolate in the chemostat, calculating the saturation constants for sulfide and sulfur oxidation (123).

A brief cooperation with Ralph Mitchell in 1967 produced interesting observations on plaque-forming units on *E. coli* lawns on agar plates (84, 85). This study elicited a letter from Selman Waksman who had made similar observations in 1935 here at Woods Hole. The follow-up work, however, was stopped short when, just before his return to Rutgers University, an eager cleaning woman dumped his possible phage enrichments down the sink. During 1969–1971, P ("Happy") Pritchard worked with us on the effect of *Aufwuchs* on bacterial growth in steady state cultures adding sterile particle suspensions at constant density to the inflowing medium. He later amazed us by his illustrious career at the Environmental Protection Agency.

In 1973, we were joined by Craig Taylor from Ralph Wolfe's lab in Urbana. He was attracted by the many physico-chemical and technical problems involved in the work on retrieving and purifying deep-sea bacteria under hydrostatic or hyperbaric pressure, described above. In 1981 he was hired as an Assistant Scientist and moved to his own lab. Jon Tuttle stayed in my lab for a long and productive postdoctoral period (1974–1979), during which he studied the occurrence of sulfide-, sulfur-, and thiosulfate-oxidizing auto- and heterotrophs in the Black Sea (119, 120) and the Venezuelan Cariaco Trench (121), where oxic/anoxic interfaces provided reduced sulfur compounds in the marine environment. We found out later how well this work had prepared us, conceptually and technically, when chemoautolithotrophy had to be measured at hydrothermal deep-sea vents (122).

Ned Ruby stayed with us from 1980–1983, coming from Scripps, where he had studied microbial luminescence under Ken Nealson before accepting a postdoc position in Woody Hasting's lab at Harvard. To broaden his expertise in work with marine microbes, he isolated a variety of species of the genera *Thiobacillus* and *Thiomicrospira* (105, 107) from deep-sea vent samples. When ratios of stable sulfur isotopes became a means of identifying biocatalytically converted sulfur compounds, he measured the actual isotope fractionation by *Thiomicrospira* strain L-12 (106). Ned went on to broaden his knowledge of marine bacteria even further by accepting another postdoctoral stay with Syd Rittenberg at UCLA to work on *Bdellovibrio* spp. before he got a position at the nearby University of Southern California. He recently joined the University of Hawaii where he works jointly with Margaret McFall-Nai on luminescent microbial symbionts of squid.

Because of an increasing scarcity of organismically trained graduate students and postdocs in the wake of a dominating molecular microbiology, I was often asked for help in finding the right young faculty member or, for students, the right mentor. When, in 1978, I recommended Steve Giovannoni (then a student at Boston University and assistant in my MBL summer course) to Dick Castenholz at Eugene, Oregon, Dick recommended, in return, a postdoc to me. It was Doug Nelson, a most amiable and productive colleague who worked with us from 1981–1985. In preparation for an ALVIN diving cruise to the Guaymas Basin vent site in the Gulf of California, he achieved the first isolation of a marine *Beggiatoa* strain from a Woods Hole marsh (89), later also discovering its ability to fix nitrogen (90). He interrupted his stay twice: once to gain experience with the use of micro-sensors on *Beggiatoa* mats in Bo Jørgensen's laboratory at Aarhus, Denmark, and, at another time, to marry Audrey, thereby adding a wonderful member to our lab's extended family. In 1985 he joined the faculty of the Microbiology Department at Davis, California, where he continues his *Beggiatoa* work and has started new studies on chemolithotrophic symbiosis in deep-sea vent invertebrates.

From one of my last courses given at Naples in 1983, I returned with a sample taken from a coastal hot spring. Thanks to Shimshon Belkin from Israel, who had just joined us as a postdoc and stayed from 1983 through 1984, we won the competition with the Göttingen group that had taken home with them an aliquot of the same sample material. Shimshon was able to isolate two hyper-thermophilic strains from it, the archaeal *Thermococcus litoralis* (5), named later in Karl Stetter's lab (92), and the eubacterial *Thermotoga neapolitana* (7, 58). The former became particularly important as a commercial source of thermostable DNA-cleaving polymerases. The strange morphology of the genus *Thermotoga* had already been observed by Karl Stetter, who named it appropriately for its coat-like cell envelope. The strain *T. neapolitana* was studied later extensively by Kenneth Noll at Storrs, Connecticut. When Shimshon went on to Berkeley to take another postdoctoral position before his return to Israel, he worked jointly with Doug Nelson during a cruise on purified symbiont suspensions from vent mussels and tube worms (6).

Dennis Bazylinski was our next postdoc, staying from 1986 to 1989. Coming from Dick Blakemore's lab at the University of New Hampshire and a brief postdoc position with T Hollocher at Brandeis University, he combined the research areas of both of these labs: magnetotactic bacteria and denitrification. As a result, he isolated a magnetotactic bacterium, at that time only the second pure culture, that used nitrous oxide anaerobically as an electron acceptor (2). While accompanying us to the hydrothermal and oil-bearing sediments of the Guaymas Basin, he obtained data on the aerobic microbial utilization of naturally occurring hydrocarbons (3). From similar sample material, workers in Fritz Widdel's lab later isolated a moderately thermophilic sulfate-reducing bacterium that removed C_8–C_{11} fractions from crude oil with a stoichiometric production of sulfide (108).

Among many other temporary coworkers, of whom the limited space does not allow me to mention all, was Fred Goetz from Mankato State University, who tackled the breakdown of aromatic Guaymas Basin oils (22), and Carolyn Eberhard, who studied the bacterial oxidation of pyrite (18). This experimental study was started after our in situ observations and collections of iron sulfide and pyrite deposits at the Mid-Atlantic Ridge at a depth of 3600 m (132). The deposits are covered by bacterial mats that at certain locations appeared to serve as food for large populations of shrimps. A flavor of Russian microbiology was brought to us by Elena Odintsova from V M Gorlenko's lab for two years and, for a shorter period, Tatyana Sokolova from G Zavarzin's lab, both from the Russian Academy of Sciences. Elena was able to isolate a new *Thermothrix* species exhibiting an unusual pattern of obligately chemolithotrophic sulfide oxidation (94). Tatyana started enrichments of hydrogen-producing carboxydotrophic bacteria (114) from various marine samples that later led to successful isolations. A convert from general molecular biology to marine microbiology, Costantino Vetriani, from the University of Rome, worked this summer (1996) in my lab on archeal organisms in shallow and deep sediments.

The work atmosphere in our laboratory was greatly enriched by sabbatical guests such as Ercole Canale-Parola from Amherst, who showed us how to isolate *Sarcina ventriculi* at pH 2; Hideo Yonenaka from San Francisco, who tackled the enigmatic *Cristispira* in Woods Hole clams; Jane Gibson from Cornell, who set up continuous cultures of cyanobacteria; Fred Richards from Yale, who devised a new survey technique for barophilic bacteria; and, for a second time, Kjell Eimhjellen from Trontheim, who continued work on psychrophiles. When Maurits LaRiviére was a guest professor at Harvard, we attempted during many weekends to grow the typical O_2/H_2S interface organism, *Thiovulum*, in a continuous culture system. In contrast to that of thiobacilli, the growth rate of these large cells (approximately 20 μm in diameter) was too slow for any practical dilution rate in the presence of a competing chemical H_2S-oxidation— which is exactly the reason why *Thiovulum* stabilizes the O_2/H_2S interface by producing the characteristic slime veil (131). Although not necessarily manifested in publications, the daily exchanges of ideas and airing of large and small questions with congenial souls are the most productive means for progress, and are generally underestimated by funding authorities.

Probably my last postdoc is Andreas Teske, who comes from the new Max Planck Institute for Marine Microbiology at Bremen, Germany. One of his advisors, G Muyzer, had accompanied us on a deep-sea vent cruise for a study of the abundance and diversity of the aerobic sulfur oxidizing genus *Thiomicrospira* (88). The use of denaturing gradient gel electrophoresis of polymerase chain reaction (PCR)-amplified 16S rDNA fragments compares the diversity of phenotypically unknown phylotypes with that of pure-culture isolates. By extending a survey of the physiological and genetic diversity of marine sulfur-cycling microbes, Andreas introduces me now to a new chapter of microbial ecology that I had no part in developing, an experience that every scientist encounters sooner or later. Molecular phylogenetic approaches are rapidly evolving, and Andreas, knowledgeable and critical, is particularly well prepared to advance their application in microbial ecology.

Traveling

It must be clear by now that, during my professional pursuits, I have always greatly enjoyed opportunities to combine my job with a hobby by traveling. I find it exhilarating to be physically, with open eyes and all senses tuned in, at exotic and foreign places that, in terms of geography and geology, climate, as well as plant and animal life, and especially in terms of the region's human history, are very different from the part of the world where I happen to live. Out of many, let me mention just three of such travels I remember vividly.

In early 1970 I was able to join in a study of Lake Kivu in eastern Zaire. The lake is a freshwater counterpart to the Black Sea: Instead of sulfide, the water column below an oxic/anoxic interface 60 m under the surface to a depth of 450 m contains methane, supersaturated in the bottom waters. This African adventure started when Egon Degens, then a geologist at Woods Hole, gave a talk on the tectonic age of the African Rift Valley and the possibility of measuring it by studying sediment cores of the rift lakes. When he cited the large amounts of methane in Lake Kivu as evidence for tectonic venting, I suggested that the methane might also be biogenic. As a result of the ensuing discussion, he invited me to join two expeditions in the spring of 1971 and 1972. On the first part of the second trip, Ralph Wolfe, the most experienced microbiologist in methanogenic bacteria, came along and confirmed the abundance of these organisms in the lake. Later a new thermophilic methanogen, *Acetogenium kivui*, was isolated in his lab (81). Although we had to leave the question on the ratio of biogenic versus thermogenic methane to determinations of stable carbon isotope ratios, I did a rather simple study on the oxidation of methane in the oxic surface waters using a portable gas chromatograph and found that the methanogenesis appeared to balance its oxidation at the interface (48). This result rendered a sustained exploitation of this source of methane questionable. We continued to study also Lakes Edward and Albert to the north. I have most wonderful memories of this experience that had all the flavor of an old-fashioned expedition, such as loading heavy coring equipment at noontime right under the equator on an old rusty barge that was ready to sink, surrounded by curious pelicans and huge marabous, or hearing and seeing hippos stomping past our tents for their grazing grounds at night, when their predators do not hunt. On meeting by chance a Belgian anthropologist, we were able to watch a nomadic pigmy tribe in one of their temporary campgrounds deep within the Ruwenzori Mountain range.

In 1988, through the interest of the Chinese Academy of Sciences in the mineral contents of high-altitude Tibetan salt lakes, the kindness of Professor H Holland of Harvard, and my earlier experiences with the Egyptian salt flat microbiology, I became a member of a most interesting US–Chinese expedition to the central Tibetan Plateau. Starting at Lhasa, we traveled west by jeep, first using part of the road to Kathmandu, then, after crossing the Tsangpo River, we turned northwest (relying on space-shuttle photographs because of the lack of maps and roads) until we reached, at an elevation of 4400 m, a certain salt lake—Lake Zabuye—that had been chosen for geochemical, mineralogical, and biological studies. Collecting samples for later pigment analyses, isolating halophiles, and putting together an old-fashioned herbarium were my particular duties (29). I remember having been on a continuous and delightful high. An ardent reader as a boy of accounts of Sven Hedin's travels through these regions 100 years earlier, this opportunity appeared to me most fortunate, a gift from heaven. When we met nomads traveling with their packed yaks, it was exactly as in Sven Hedin's sketches and photographs. In order to justify the presence of an oceanographer in this part of the world, the geologist found me an ophiolite hill of serpentine boulders, formerly pillow lava extruded on the deep-sea floor. Metamorphosed by tremendous pressures at the suture between the Indian and Asian tectonic plates, it was lifted to these heights in just a few million years.

During an exploratory trip to the Canadian Arctic Archipelago in 1990 for an unsuccessful attempt to develop a research project on microbial aspects of global warming, I happened to meet Grant Duncan at the northernmost air base on Cornwallis Island. He was an arctic pilot and a master in handling the famous Twin Otter airplane. Unknown to me, he was also a well-known expert in the history of Arctic exploration and, as such, connected me in a most exciting way to a chapter in my family's history. On a brief trip to Beechey Island, I won his fancy when I was able to identify an ancient grave site. Next to three graves of members of the ill-fated Franklin Expedition who were buried there in 1846, there was a fourth one with an epitaph rendered unreadable by many years of Arctic weather. I could tell Duncan that it was the resting place of Mr. Morgan, carpenter of the ship Investigator, buried there in April of 1854 by my great grandfather. I knew this from a note in his handwritten diary, now translated and published (83). Mr. JA Miertsching was my grandmother's father and, as a Moravian missionary in Labrador, had been asked in 1849 by the Royal British Navy to join an expedition in search of Franklin as an officer and Eskimo interpreter. Leaving London in early 1850, passing through the Strait of Magellan into the Pacific Ocean and entering the Arctic through Bering Strait, his ship, the *Investigator*, had to be abandoned on the north shore of Banks Island at the western end of the Northwest Passage. Now, almost a century and a half later, through the pilot's kindness, I was able to visit this most desolate site where the crew of the ship survived for almost two years, 1851–1853, before being rescued at the last minute. We were able to land at a few other sites where the crew of the *Investigator* rested during their difficult journey by ship and by foot over the ice east to Beechey Island, and from there back to England in 1854. They were the first to circumnavigate the two American continents, and Captain McClure was later credited with the discovery of the Northwest Passage.

An invitation by the Australian Society of Microbiology to be the "Rubbo-Orator" in 1990 for an extensive lecture tour enabled me to see much of Tasmania and Australia. To Professor G Hempel I owe voyages to Spitsbergen and to the Antarctic. I am grateful that my profession afforded me such experiences, which enriched my life incredibly. I seem to harbor a longing for the last regions of our earth that humankind had no part in affecting. It was no less exciting to search the ocean floor 4000 m down with the submersible ALVIN than to be 4000 m up in the Tibetan Trans-Himalayas, not as a tourist but with a scientific project in mind and accompanied by colleagues with congenial minds. I cannot help feeling sorry for those who inherit a shrinking world that rapidly loses its diversity. There is now a Holiday Inn at Tschigatse.

Teaching at the Marine Biological Laboratory

Summer courses taught at the Marine Biological Laboratory (MBL) at Woods Hole have existed since 1888. Being asked, during the fall of 1970, to be the director of one of them turned out to be a most consequential event for me. On occasion I had been a guest lecturer in MBL courses, but directing one was different, a commitment to be carefully considered and not to be done alone. Both van Niel and Stanier, to whom I turned for advice, were encouraging and added to the number of colleagues whom I then called on for help. All I could promise to them was a cottage so that they could bring their families along, but no pay, and a summer of intense work while others took wholly deserved vacations. To my surprise, none of them turned me down, and I decided to accept the job. Looking back, it was one of the most felicitous decisions I have ever made. There was space for 20 students. They were selected from a large number of applicants, mostly highly motivated graduate students, a few postdocs and, from time to time, a professor on sabbatical. The group of four to five instructors changed from year to year because few could afford to come every summer. They were Dick Castenholz, Jane Gibson, Bob Hungate, Alex Keynan, Helge Larsen, Ed Leadbetter, Ken Nealson, Jeanne Poindexter, Syd Rittenberg, Moshe Shilo, and Ralph Wolfe. When the latter, the mainstay of the course, could not come, his former student, Craig Taylor, took over.

Students as well as lecturers had to learn how MBL courses are run: being together from the morning to often late at night, teaching with a minimum of formal lecturing but a maximum of student participation in discussions, often leading directly to laboratory experiments, some carefully planned, others spontaneous. For the instructors, this meant 12-hour days for at least six weeks, most weekends included, for collecting trips to the nearby salt marshes. Often students experienced for the first time that they could ask questions all day long and not let an answer go by without fully understanding it—10 contact hours a day. At the same time, they participated actively, passing on everything they knew to each other as well as to us. Most instructors found this unrestricted teaching a pleasure as well as a rare opportunity to design their own curriculum, incorporating brand-new concepts that an official exam-based curriculum might not allow. The students also felt this freedom from academic rules and bounds. Representative proponents and opponents of new concepts were invited for lectures and often stayed on for days. For an outsider, it was often difficult to keep students and instructors apart.

In 1975, I was glad to turn the course back to the ecologists from whom the course had been borrowed for a five-year term. Putting so much time into MBL teaching made it difficult to keep up with my research programs. By that time, however, the MBL had received pleas to keep the microbiology course going. I agreed under the condition that I would not have to do any fundraising because, during the mid 1970s, all NSF funding of summer courses had stopped. There was a memorable Woods Hole party of the National Academy (locals invited) where Professor Selman Waksman waved me aside and told me of his intent to fund the continuation of the course through the Foundation of Microbiology, originally established from his Nobel Prize award. He asked me to see him the next day at his house in Woods Hole for a discussion of details. This was very unexpected because, during brief visits in my lab almost every summer, he had been quite critical of the course's lack of emphasis on soil microbiology, actinomycetes, and the ecological role of antibiotics. My explanation had been that our limited curriculum was simply reflecting the instructors' areas of expertise. So I prepared myself to defend the course's present orientation, but when I called the next morning to ask for a convenient time to visit, I was given the sad news that Professor Waksman had passed away that very night. He had been 81 years old. Few microbiologists reach the degree of recognition he had during their lifetime.

Several weeks later, his son, Byron Waksman, immunologist at Yale University, who knew about his father's wishes, called me and asked to make a brief proposal to the trustees of the Foundation. The support was granted and the course continued in new, albeit smaller, quarters of the MBL's Lilly Building for a somewhat smaller group of students, 12 instead of 20. Ralph Wolfe stayed with me and so did Jeanne Poindexter, Moshe Shilo, Alex Keynan, Ed Leadbetter, and in 1978 again Bob Hungate and Helge Larsen. During that year, Jim Ebert was followed in the MBL directorship by Paul Gross, a cell biologist from Rochester. Out of the blue he asked me to report to an ad hoc committee and, as a result, in 1979 the microbiology course was given the status of a new regular MBL course. Because regular course directors are normally replaced every five years, I asked to be relieved after my eight years of service. Harlyn Halvorson from Brandeis University was willing to take over. By that time, however, it was spring 1979, and Harlyn asked me to stay on as a codirector. Moreover, because he had already arranged a sabbatical leave to Scotland for the following year, I had to direct the course in 1980, for the 10th and very last time, myself. This final summer turned out to be the most satisfying ever. We moved back into larger space of the Loeb Building, and Harlyn had provided already for generous new funding. The group of instructors, Y Cohen, J Gibson and J Poindexter, was reinforced by Bo Jørgensen, then from Aarhus, Denmark, teaching us the use of micro-electrodes, and by Gijs Kuenen from Delft, Holland, who brought along a former student, Jan Gottschal, both of them enriching and extending our teaching of chemostat techniques in microbial ecology. Craig Taylor dealt with the anaerobes, and Pete Greenberg with bacterial luminescence. Pete was a former student of the course, later to become an assistant and finally codirector with Ralph Wolfe for five subsequent years after Harlyn had finished his term in 1984.

Some of the lecturers, who came often or stayed with us for a longer period of time, were Ercole Canale-Parola, Arnie Demain, Jerry Ensign, Dick Hanson, (Sir) Hans Kornberg, Abdul Matin, Eugene Odum, Nick Ornston, Norbert Pfennig, Luigi Provasoli, Ned Ruby, Wolf Vishniac, and Meyer Wolin. It is a pity that there is no space here to talk more about their personal roles in this course or about some individual students who are now teaching courses of their own. I always tried to have seniors among the staff, such as Hungate, Rittenberg, Shilo, or Larsen, whose perspectives, in many cases, conveyed to the students that the so-called exact sciences undergo fashions, comings-and-goings of ideas, and, very evidently, a recycling of old ones, often just barely concealed by new terminology. Not only I, but also my research, benefited tremendously from these 10 years of intense summer teaching/learning experience, and I owe it to my Institution, WHOI, with its liberal guidelines such that I could do it at the expense of my participation in its WHOI/MIT Joint Graduate Program. Alex Zehnder persuaded me to transplant some of our MBL-type teaching style to Switzerland at the Kastanienbaum Laboratory on the shore of Lake Lucerne, and later to The Netherlands at the Oceanographic Institute on the island of Texel.

In early 1971 I became involved in the set-up of the Sea Education Association (SEA) at Woods Hole. In return, Cory Cramer, the founder, made the 120-foot schooner *Westward* available to my MBL class almost every summer for one- or two-day sails and sample-taking, an unforgettable experience for the many landlubber students. Some had to experience the unfair struggle: enthusiasm versus seasickness.

Codirected by Martin Dworkin and John Bresnak (1990–1994), molecular studies brought a new dimension to the MBL Microbiology Course, with which I am still loosely associated. Currently Ed Leadbetter and Abigail Salyers are in charge, actively combining cultural, physiological, and biochemical studies with new molecular approaches. Kurt Hanselmann, from the University of Zürich and a course participant from the early 1970s, has now joined it again as a gifted teacher. Because not only the field but also the students keep changing, teaching has to be relearned constantly. I feel that the liberal style practiced in the the MBL courses—with its built-in flexibility, freedom to choose, and updating the curriculum at any moment during all-day activities—will stand as a most effective, if not the most effective, way of teaching, and certainly the most satisfying one. This is true at least for advanced students who are motivated and able to make use of what is offered, including the meeting of many outstanding guest scientists. Unlike Cold Spring Harbor courses, our MBL microbiology course does not emphasize up-to-date technical protocols, except where needed for a particular project, but is designed to help students develop personal views on old or new concepts—currently, for instance, on balancing classical physiological with the newest molecular approaches in understanding the highly complex processes of diverse microbial communities.

SOME REFLECTIONS

Being rooted myself in physiologically and biochemically based, now often called classical, microbiology, I am engrossed by the possibilities promised by powerful new molecular approaches. Through the work of Carl Woese (134) and his coworkers, the first valid microbial phylogenetic systematics was created based on sequences of well-chosen, highly conserved, universally occurring marker genes. The molecular microbiologists' eager ambition to extend these phylogenetic approaches toward microbial ecology (88, 96, 97, 128) intrinsically confirms Thienemann's early notion of ecology as the joint and ultimate goal of all biological disciplines. However, after microbial phylogeny has so long been out of reach, a certain amount of "going overboard" seems unavoidable. Microbial ecology, the interaction of species with the environment and one another, is process-based, and the phylogenetic positioning of a phylotype can only give hints of the organism's actual function in its environment.

Discarding pure culture approaches as too difficult, too slow or unnecessary, or using the term unculturable microbes means throwing the baby out with the bathwater. All those key microbes that make the phylogenetic tree useful appeared to be unculturable until they were cultured, at that point disclosing a wealth of new physiological and biochemical information needed in microbial ecology.

Norman Pace, after presenting an MBL Friday Night Lecture in 1994 on this topic, has enlivened our Woods Hole course with delightful pro and con discussions in this area for several summers. Oligonucleotide hybridization probes and PCR primers are most auspicious tools to select and identify microbes, but they still require much work on the conceptual as well as practical level in order to identify microbial processes qualitatively and quantitatively. The selectivity of PCR amplification of DNA fragments has, for instance, a certain similarity to that of enrichment media. Speaking strictly for biotechnology, Barry Marrs candidly pointed out (at a Grinnell College Symposium in April 1996) that obtaining genes from soil, sediment, or water for a transgenic production of biotechnologically useful enzymes bypasses the cell level and, thereby, any physiological-ecological information. The metabolizing cell is more than the sum of its genes. Pure cultures—as challenging as their isolation may be—and their phenotypic characterization will remain an indispensible counterpart to molecular approaches. A rhetorical question: Where would we stand today if Karl Stetter—or, for that matter, Beijerinck—had forgone isolations and provided us with sequences instead? Winogradski, a genius of microbial ecology, knew around the turn of the century that every one of his soil samples contained thousands of unknown types of microbes, and he wisely concentrated on a few, pioneering the discovery of new and striking microbial processes.

My tales and tacit advocacies for independence or, as it were, for a deferred accountability of taxpayer-supported research may be unrealistic under today's financial constraints that seem to dictate the necessity of setting priorities. Yet, many of the money-saving measures may, in fact, be very costly. When asked, in the late 1980s, to diversify our funding sources, I realized how much time we had been able to spend until that time on research, not minding long days and many working weekends, and how much time was now being lost to preparing multiple proposals and following ever-changing funding policies and priorities. A 70% rejection rate (presently 85% in National Science Foundation's section of Biological Oceanography) means a wasteful spinning of wheels that makes a true competition for quality elusive. A general distrust of scientists, who have been trained to be guided by their own judgment and priorities, and the suggestion that they become activists (15), ignores the fact that the best of them are not often good salesmen. The effect is worse on younger scientists, who are not rewarded for tackling a problem with perseverance, but for hopping from one promising or prioritized subject to another: Funding, not science, is the ultimate goal. Running basic research as a service industry would clearly be at the expense of quality and a waste of national resources, funds as well as talents. If there is an administrative art of allocating limited funds efficiently and equitably, peer-review included, it has not yet been mastered.

The accountability of scientists toward the public is vital. Therefore, I find it embarrassing when calls for exciting news lead to premature announcements of discoveries and claims that remain unresolved—ultimately deceiving rather than serving the public. This newly emerging "soft" science, which was anathema just 20 years ago, is a problem well addressed by Rustum Roy (104). In general, there appears to be a recurring conflict between the somewhat puritan principle that basic research has a right to exist only if it immediately benefits the people and the classical Humboldtian (Wilhelm) viewpoint that support of non-profit research is part of a nation's cultural and educational standards, trusting in long-range benefits and ideally independent of the temporal fluctuations of the economy. Fortunately, this country, unlike any other, has developed some funding stability through a strong tradition and a legislature that encourages private research support from which our work has often benefitted. My critical remarks notwithstanding, I am, as most of my colleagues are, well aware of the privilege of being paid for work that often feels like the pursuit of a hobby.

My ambition to be understood by my nonbiological colleagues was rewarded when I received WHOI's Bigelow Medal in Oceangraphy (1980) and the Cody Award in Ocean Sciences by the Scripps Institution of Oceanography (1992). At the same time, these honors made it difficult for me to comply with the moral dictum *mehr sein als scheinen*, be more than you appear to be. As an incurable individualist, I often had the uncomfortable feeling that our teamwork, necessary as it is, left me with undeserved credit for work that was to a great extent done by others.

My election to the National Academy of Sciences as a Foreign Associate revealed to many of my colleagues and friends the fact that I am not a US citizen. To me it revealed a gracious generosity that I experienced time and again in this country's open and multicultural society and an obliging tact based on a self-assurance that no other nation seems to be able to afford. Yet I feel almost guilty considering those countrymen, one generation older, who had to flee Germany for their lives and embraced the newly won freedom by becoming US citizens enthusiastically. With the luxury to choose, I came under very different circumstances. Although having grown slowly away from the idea of returning to Germany, I remain dependent upon my ancestral cultural heritage that is deeply interwoven with language, literature, history, and so forth, and, not thinking in political terms, I do not want to pretend to be what I really am not.

Remaining an outsider as a first-generation immigrant does have a positive side: the continuously stimulating, almost pioneering existence that comes with the need to make one's home in a socially unfamiliar surrounding, a life-long learning experience. Bilingual daily dealings (speaking German at home) are fun and enrich the usage of both languages. In addition, I can see this society as no born American can. This includes a general appreciation that is difficult to understand by someone who benefitted all his life from the liberties that this country affords, notwithstanding some rough edges of narrow conservatism. Although not being able to vote, there are plenty of opportunities to pay allegiance by "putting your money where your mouth is." Carrying with me the bright as well as the inescapable dark sides of the German history, I hope that it was not only with scientific work but also in my countless nonprofessional interactions with American friends and others around the world that allowed me to do more in my present situation than I would have had in the dignified position of a German professor. Besides, the role of an ambassador in US–German scientific matters appealed to me.

In this capacity, a most gratifying experience was my involvement in the founding of a brand-new research establishment for the very field of my interest, marine microbiology, at Bremen, Germany. In 1989 the Max Planck Society asked me to serve as a member of a founding committee that met several times in Hamburg and Munich from 1989 to 1992. I suggested, successfully, two excellent scientists as joint directors of this new laboratory: Bo Barker Jørgensen from Aarhus, Denmark, and Friedrich Widdel, who had just assumed his first professorship at Munich. I knew both of them through their participation as instructors in the Woods Hole summer course, and I visualized them in terms of, as it were, a Lewis & Clark team, a functional friendship. The opening of the new Institute in rented quarters took place in October 1992, and the dedication of the new building, housing 75 scientists and employees, was celebrated in May of this year (1996). During these starting years, they resolved the enigmatic existence of *Thioploca* mats off the coast of Chile (21) and discovered iron-utilizing microbial photosynthesis (19), as well as the anaerobic hydrocarbon oxidation with a stoichiometric production of sulfide from sulfate (108). I see this Max Planck Institute for Marine Microbiology becoming a most effective research center for students and postdocs from all over the world, as well as an opportunity for colleagues to work during sabbatical leaves, possibly a present-day Naples (see above) for marine microbiology. For US colleagues it could mean a payback for the training of many young German scientists that came over here during the last five decades prior to assuming academic positions back home.

EPILOGUE

I have been fortunate to work as a scientist during a time when the social climate for doing independent basic research was most supportive. As a result, productivity was high and the work mostly pleasurable. My area of interest was rife with opportunities for original work and discoveries. My desire to see a lot of the world was granted, together with a blessed and wonderful family life and a large number of friends of all kinds. There were, of course, disappointments, personal failures, set-backs, goals unreached, and such, but these are not important enough to dwell upon, and they are vastly outweighed by a keen perception of gracious guidance and a goodly portion of undeserved luck. Therefore, coming at this time—just about when I was asked to write this chapter—the detection of a lymphoma and the verdict of a "reduced life expectancy" elicits an unsentimental closing remark to the effect that possibly having to quit while still ahead has many positive aspects. It certainly has no effect on my all-prevailing gratitude for a personally and professionally fulfilled life.

> Visit the *Annual Reviews home page* at
> http://www.annurev.org.

Literature Cited

1. Barker HA, Hungate RE. 1990. Cornelius Bernardus van Niel 1897–1985. *Natl. Acad. Sci. Biogr. Mem.* 59:389–423
2. Bazylinski DA, Frankel RB, Jannasch HW. 1988. Anaerobic magnetite production by a marine magnetotactic bacterium. *Nature* 334:518–19
3. Bazylinski DA, Wirsen CO, Jannasch HW. 1989. Microbial utilization of naturally-occurring hydrocarbons at the Guaymas Basin hydrothermal vent site. *Appl. Environ. Microbiol.* 55:2832–36
4. Beling A, Jannasch HW. 1955. Hydrobiologische Untersuchungen der Fulda unter Anwendung der Membranfiltermethode. *Hydrobiologia* 7:36–51
5. Belkin S, Jannasch HW. 1985. A new extremely thermophilic, sulfur-reducing heterotrophic marine bacterium. *Arch. Microbiol.* 141:181–86
6. Belkin S, Nelson DC, Jannasch HW. 1986. Symbiotic assimilation of CO_2 in two hydrothermal vent animals, the mussel *Bathymodiolus thermophilus* and the tube worm *Riftia pachyptila. Biol. Bull.* 170:110–21
7. Belkin S, Wirsen CO, Jannasch HW. 1986. A new sulfur-reducing, extremely thermophilic eubacterium from a submarine hydrothermal vent. *Appl. Environ. Microbiol.* 51:1180–85
7a. Blöchl E, Rachel R, Burggraf S, Hafenbradl D, Jannasch HW, Stetter KO. 1997. *Pyrolobus fumarii*, gen. and sp. nov., represents a novel group of archaea, extending the upper temperature limit for life to 113°C. *Extremophiles* 1:14–21
8. Broeker WS. 1980. Modelling the carbon system. *Radiocarbon* 22:565–98
9. Bult CJ, White O, Olsen GJ, Zhou L, Fleischmann RD, et al. 1996. Complete genome sequence of the methanogenic archaeon, *Methanococcus jannaschii. Science* 273:1058–72
10. Burggraf S, Jannasch HW, Nicolaus B, Stetter KO. 1990. *Archaeoglobus profundus* sp. nov., represents a new species within the sulfate reducing archaebacteria. *Syst. Appl. Microbiol.* 13:24–28
11. Cavanaugh CM, Gardiner SL, Jones ML, Jannasch HW, Waterbury JB. 1981. Procaryotic cells in the hydrothermal vent tube worm *Riftia pachyptila* Jones: possible chemoautotrophic symbionts. *Science* 213:340–41
12. Ceram CW. 1970. *The March of Archaeology*, p. 64. New York: Knopf. 326 pp.
13. Cuhel RL, Taylor CD, Jannasch HW.

1981. Assimilatory sulfur metabolism in marine microorganisms: characteristics and regulation of sulfate transport in *Pseudomonas halodurans* and *Alteromonas luteo-violaceus. J. Bacteriol.* 147:340–49

14. Cuhel RL, Taylor CD, Jannasch HW. 1981. Assimilatory sulfur metabolism in marine microorganisms: a novel sulfate transport system in *Alteromonas luteoviolaceus. J. Bacteriol.* 147:350–53

15. Daie J. 1996. The activist scientist. *Science* 272:1081

16. DeLong EF. 1992. Novel archaea in coastal marine environments. *Proc. Natl. Acad. Sci. USA* 89:5685–89

17. Distel DL, Felbeck H, Cavanaugh CM. 1994. Evidence for phylogenetic congruence among sulfur-oxidizing chemoautotrophic bacterial endosymbionts and their bivalve hosts. *J. Mol. Evol.* 38:533–42

18. Eberhard C, Wirsen CO, Jannasch HW. 1995. Oxidation of polymetal sulfides by chemolitho-autotrophic bacteria from deep-sea hydrothermal vents. *Geomicrobiol. J.* 13:145–64

19. Ehrenreich A, Widdel F. 1994. Anaerobic oxidation of ferrous iron by purple bacteria, a new type of phototrophic metabolism. *Appl. Environ. Microbiol.* 60:4517–26

20. Fiala G, Stetter KO, Jannasch HW, Langworthy TA, Madon J. 1986. Staphylothermus marinus sp. nov. represents a novel genus of extremely thermophilic submarine heterotrophic archaebacteria growing up to 98°C. *Syst. Appl. Microbiol.* 8:106–13

21. Fossink H, Gallardo VA, Jørgensen BB, Hüttel M, Nielsen LP et al. 1995. Concentration and transport of nitrate by the mat-forming sulphur bacterium *Thioploca. Nature* 374:713–15

22. Goetz FE, Jannasch HW. 1993. Aromatic hydrocarbon degrading bacteria in the petroleum-rich sediments of the Guaymas Basin hydrothermal vent site. *Geomicrobiol. J.* 11:1–18

23. Gorini L. 1960. Antagonism between substrate and repressor in controlling the formation of a biosynthetic enzyme. *Proc. Natl. Acad. Sci. USA* 46:682–90

24. Groeben C. 1984. The Naples Zoological Station and Woods Hole. *Oceanus* 27:60–69

25. Gunderson J, Jørgensen BB, Larsen E, Jannasch HW. 1992. Mats of giant sulfur bacteria in deep-sea sediments due to fluctuating hydrothermal flow. *Nature* 360:454–56

26. Herbert D, Elsworth R, Telling RC. 1956. The continuous culture of bacteria: a theoretical and experimental study. *J. Gen. Microbiol.* 14:601–22

27. Heuss T. 1962. *Anton Dohrn.* Tübingen: Wunderlich Verlag. 448 pp.

28. Hobbie JE, Daley RJ, Jasper S. 1977. Use of nucleopore filters for counting bacteria by epifluorescence microscopy. *Appl. Environ. Microbiol.* 33:1225–28

29. Holland HD, Smith GI, Jannasch HW, Dickson AG, Mianping Z, Tiping D. 1991. Lake Zabuye and the climatic history of the Tibetan Plateau. *Geowissenschaften* 9:37–44

30. Huber R, Kurr M, Jannasch HW, Stetter KO. 1989. A novel group of abyssal methanogenic archaebacteria (*Methanopyrus*) growing at 110°C. *Nature* 342:833–34

31. Huber R, Langworthy TA, König H, Thomm M, Woese CR, et al. 1986. *Thermotoga maritima* sp. nov. represents a new genus of extremely thermophilic eubacteria growing up to 90°C. *Arch. Microbiol.* 144:324–33

32. Hungate RE. 1986. Obituary: C. B. van Niel, 1897–1985. *Photosynth. Res.* 10:139–42

33. Hungate RE, Jannasch HW, Wolfe RS. 1985. Obituary: Cornelius Bernard van Niel. *ASM News* 51:424–25

34. Hutchinson GE. 1965. *The Ecological Theater and the Evolutionary Play.* New Haven: Yale Univ. Press

35. Jannasch HW. 1955. Zur Ökologie der zymogenen planktischen Bakterienflora natürlicher Gewässer. *Arch. Mikrobiol.* 23:146–80

36. Jannasch HW. 1957. Die bakterielle Rotfärbung der Salzseen des Wadi Natrun (Ägypten). *Arch. Hydrobiol.* 53:425–33

37. Jannasch HW. 1960. Denitrification as influenced by photosynthetic oxygen production. *J. Gen. Microbiol.* 23:55–63

38. Jannasch HW. 1963. Bacterial growth at low substrate concentrations. *Arch. Mikrobiol.* 45:323–42

39. Jannasch HW. 1965. Starter populations as determined under steady state conditions. *Biotechnol. Bioeng.* 7:279–83

40. Jannasch HW. 1967. Enrichments of aquatic bacteria in continuous culture. *Arch. Mikrobiol.* 59:165–73

41. Jannasch HW. 1967. Growth of marine bacteria at limiting concentrations of organic carbon in seawater. *Limnol. Oceanogr.* 12:264–71

42. Jannasch HW. 1968. Competitive elimination of Enterobacteriaceae from seawater. *Appl. Microbiol.* 16:1616–18

43. Jannasch HW. 1968. Growth characteristics of heterotrophic bacteria in seawater. *J. Bacteriol.* 95:722–23
44. Jannasch HW. 1969. Current concepts of aquatic microbiology. Baldi Memorial Lecture. *Verh. Int. Ver. Limnol.* 17:25–39
45. Jannasch HW. 1969. Estimations of bacterial growth rates in natural waters. *J. Bacteriol.* 99:156–60
46. Jannasch HW. 1971. Threshold concentrations of carbon sources limiting bacterial growth in seawater. In *Organic Matter in Natural Waters,* ed. DW Hood, pp. 321–28. New York: Pergamon
47. Jannasch HW. 1974. Steady state and the chemostat in ecology. *Limnol. Oceanogr.* 19:717–20
48. Jannasch HW. 1975. Methane oxidation in Lake Kivu (Central Africa). *Limnol. Oceanogr.* 20:861–64
49. Jannasch HW. 1977. Growth kinetics of aquatic bacteria. In *Aquatic Microbiology,* SAB Symp. Ser. 6, ed. JM Shewan, FA Skinner, pp. 55–68. London: Academic
50. Jannasch HW. 1979. Microbial ecology of aquatic low-nutrient habitats. In *Strategy of Life in Extreme Environments,* ed. M Shilo, pp. 243–60. Weinheim: Verlag Chemie. 513 pp.
51. Jannasch HW. 1984. Aspects of measuring bacterial activities in the deep ocean. In *Heterotrophic Activity in the Sea,* ed. J Hobbie, PL Williams, pp. 505–22. New York: Plenum
52. Jannasch HW. 1984. Microbial processes at deep sea hydrothermal vents. In *Hydrothermal Processes at Sea Floor Spreading Centers,* ed. PA Rona, K Bostrom, L Laubier, KL Smith, pp. 677–709. New York: Plenum
53. Jannasch HW. 1985. The chemosynthetic support of life and the microbial diversity at deep sea hydrothermal vents. *Proc. R. Soc. London Ser. B* 225:277–97
54. Jannasch HW. 1989. Litho-autotrophically sustained ecosystems in the deep sea. In *Biology of Autotrophic Bacteria,* ed. HG Schlegel, B Bowien, pp. 147–66. Berlin: Springer-Verlag
55. Jannasch HW. 1995. Microbial interactions with hydrothermal fluids. In *Seafloor Hydrothermal Systems,* ed. SE Humphris, RA Zierenberg, LS Mullineaux, RE Thomson. *Geophys. Monogr.* 91:273–96. Washington, DC: Am. Geophys. Union. Publ.
56. Jannasch HW, Egli T. 1993. Microbial growth kinetics: a historical perspective. *Antonie van Leeuwenhoek* 63:213–24
57. Jannasch HW, Eimhjellen K, Wirsen CO, Farmanfarmaian A. 1971. Microbial degradation of organic matter in the deep-sea. *Science* 171:672–75
58. Jannasch HW, Huber R, Belkin S, Stetter KO. 1988. *Thermotoga neapolitana* sp. nov. of the extremely thermophilic, eubacterial genus *Thermotoga. Arch. Microbiol.* 150:103–4
59. Jannasch HW, Jones GE, 1959. Bacterial populations in seawater as determined by different methods of enumeration. *Limnol. Oceanogr.* 4:128–39
60. Jannasch HW, Mateles RI. 1974. Experimental bacterial ecology studies in continuous culture. *Adv. Microbiol. Physiol.* 11:165–212
61. Jannasch HW, Mottl MJ. 1985. Geomicrobiology of deep sea hydrothermal vents. *Science* 229:717–25
62. Jannasch HW, Nelson DC, Wirsen CO. 1989. Massive natural occurrence of unusual large bacteria (*Beggiatoa* sp.) at a hydrothermal deep-sea vent site. *Nature* 342:834–36
63. Jannasch HW, Wirsen CO. 1977. Retrieval of concentrated and undecompressed microbial populations from the deep sea. *Appl. Environ. Microbiol.* 33:642–46
64. Jannasch HW, Wirsen CO. 1979. Chemosynthetic primary production at East Pacific Ocean floor spreading centers. *BioScience* 29:492–98
65. Jannasch HW, Wirsen CO. 1981. Morphological survey of microbial mats near deep sea thermal vents. *Appl. Environ. Microbiol.* 41:528–38
66. Jannasch HW, Wirsen CO. 1982. Microbial activities in undecompressed and decompressed deep sea water samples. *Appl. Environ. Microbiol.* 43:1116–24
67. Jannasch HW, Wirsen CO. 1984. Variability of pressure adaptation in deep sea bacteria. *Arch. Microbiol.* 139:281–88
68. Jannasch HW, Wirsen CO, Doherty KM. 1996. A pressurized chemostat for the study of marine barophilic and oligotrophic bacteria. *Appl. Environ. Microbiol.* 62:1593–96
69. Jannasch HW, Wirsen CO, Molyneaux SJ. 1991. Chemoautotrophic sulfur-oxidizing bacteria from the Black Sea. *Deep-Sea Res.* 38:1105–20
70. Jannasch HW, Wirsen CO, Molyneaux SJ, Langworthy TA. 1988. Extremely thermophilic fermentative archaebacteria of the genus *Desulfurococcus* from deep-sea hydrothermal vents. *Appl. Environ. Microbiol.* 54:1203–9
71. Jannasch HW, Wirsen CO, Molyneaux SJ, Langworthy TA. 1992. Comparative

physiological studies on hyperthermophilic archaea isolated from deep sea hydrothermal vents with emphasis on *Pyrococcus* Strain GB-D. *Appl. Environ. Microbiol.* 58:3472–81

72. Jannasch HW, Wirsen CO, Nelson DC, Robertson LA. 1985. *Thiomicrospira crunogena* sp. nov., a colourless sulphur oxidizing bacterium from a deep sea hydrothermal vent. *Int. J. Syst. Bacteriol.* 35:422–24

73. Jannasch HW, Wirsen CO, Taylor CD. 1976. Undecompressed microbial populations from the deep sea. *Appl. Environ. Microbiol.* 32:360–67

74. Jannasch HW, Wirsen CO, Taylor CD. 1982. Deep sea bacteria: isolation in the absence of decompression. *Science* 216:1315–17

75. Jannasch HW, Wirsen CO, Winget CL. 1973. A bacteriological, pressure-retaining, deep-sea sampler and culture vessel. *Deep-Sea Res.* 20:661–64

76. Jones WJ, Leigh JA, Meyer F, Woese CR, Wolfe RS. 1983. *Methanococcus jannaschii* sp. nov., an extremely thermophilic methanogen from a submarine hydrothermal vent. *Arch. Microbiol.* 136:254–61

77. Jørgensen BB, Fossing H, Wirsen CO, Jannasch HW. 1991. Sulfide oxidation in the anoxic Black Sea chemocline. *Deep-Sea Res.* 38:1083–103

78. Karl DM. 1986. Determination of in situ microbial biomass, viability, metabolism and growth. In *Bacteria in Nature*, ed. JS Poindexter, ER Leadbetter, pp. 85–176. New York: Plenum

79. Kotre J. 1995. *White Gloves: How To Create Ourselves Through Memory.* New York: Free Press. 276 pp.

80. Kurr M, Huber R, König H, Jannasch HW, Fricke H, et al. 1991. *Methanopyrus kandleri*, gen. and sp. nov. represents a novel group of hyperthermophilic methanogens growing at 110°C. *Arch. Microbiol.* 156:239–47

81. Leigh JA, Mayer F, Wolfe RS. 1981. *Acetogenium kivui*, a new thermophilic hydrogen-oxidizing acetogenic bacterium. *Arch. Microbiol.* 129:275–80

82. McHatton S, Barry J, Jannasch HW, Nelson DC. 1996. Vacuolar nitrate accumulation in autotrophic marine *Beggiatoa*. *Appl. Environ. Microbiol.* 62:954–58

83. Miertsching JA. 1967. *Frozen Ships: The Arctic Diary of Johann Miertsching, 1850–1854.* Transl. LH Neatby. Toronto: Macmillan of Can. 254 pp.

84. Mitchell R, Jannasch HW. 1969. Processes controlling virus inactivation in seawater. *Environ. Sci. Technol.* 3:941–43

85. Mitchell RS, Yankovsky S, Jannasch HW. 1967. Lysis of *Escherichia coli* by marine microorganisms. *Nature* 215:891–93

86. Monod J. 1950. La technique de culture continue; theorie et applications. *Ann. Inst. Pasteur* 79:390–410

87. Murray JW, Jannasch HW, Honjo S, Reeburgh WS, Friederich GE et al. 1989. Unexpected changes in the oxic/anoxic interface in the Black Sea. *Nature* 338:411–41

88. Muyzer G, Teske AP, Wirsen CO, Jannasch HW. 1995. Phylogenetic relationships of *Thiomicrospira* species and their identification in deep-sea hydrothermal vent samples by denaturating gradient gel electrophoresis of 16S rDNA fragments. *Arch. Microbiol.* 164:165–72

89. Nelson DC, Jannasch HW. 1983. Chemoautotrophic growth of a marine *Beggiatoa* in sulfide gradient cultures. *Arch. Microbiol.* 136:262–69

90. Nelson DC, Waterbury JB, Jannasch HW. 1983. Nitrogen fixation and nitrate utilization by marine and freshwater *Beggiatoa*. *Arch. Microbiol.* 133:172–77

91. Nelson DC, Wirsen CO, Jannasch HW. 1989. Characterization of large autotrophic *Beggiatoa* at hydrothermal vents of the Guaymas Basin. *Appl. Environ. Microbiol.* 55:2909–17

92. Neuner A, Jannasch HW, Belkin S, Stetter KO. 1990. *Thermococcus litoralis* sp. nov.: a novel species of extremely thermophilic marine archaebacteria. *Arch. Microbiol.* 153:205–7

93. Novick A, Szilard L. 1950. Description of the chemostat. *Science* 112:715–16

94. Odintsova EV, Jannasch HW, Mamone JA, Langworthy TA. 1996. *Thermothrix azorensis*, nov. spec., a thermophilic, obligate lithoautotrophic, sulfur oxidizing bacterium. *Int. J. Syst. Bacteriol.* 46:422–28

95. Overmann J, Cypionka H, Pfennig N. 1992. An extremely low-light-adapted phototrophic sulfur bacterium from the Black Sea. *Limnol. Oceanogr.* 37:150–55

96. Pace N. 1996. New perspective on the natural microbial world: molecular microbial ecology. *ASM News* 62:463–70

97. Pace NR, Stahl DH, Lane DJ, Olsen GJ. 1986. The use of rRNA sequences to characterize natural microbial populations. *Adv. Microb. Ecol.* 9:1–55

98. Pfeffer W. 1897. *Pflanzenphysiologie.* Leipzig: Engelmann-Verlag. 2nd ed.

99. Pfennig N. 1987. Van Niel remembered. *ASM News* 53:75–77

100. Pfennig N. 1993. Reflections of a microbiologist, or how to learn from the microbes. *Annu. Rev. Microbiol.* 47:1–29

101. Pfennig N, Jannasch HW. 1962. Biologische Grundfragen bei der kontinuierlichen Kultur von Mikroorganismen. *Ergeb. Biol.* 25:93–135

102. Pley Y, Schipka J, Gambacorta A, Jannasch HW, Fricke H, et al. 1991. *Pyrodictium abyssi* sp. nov. represents a novel heterotrophic marine archaeal hyperthermophile growing at 110°C. *Syst. Appl. Microbiol.* 14:255–63

103. Repeta DJ, Simpson DJ, Jørgensen BB, Jannasch HW. 1989. The distribution of bacterio-chlorophylls in the Black Sea: evidence for anaerobic photosynthesis. *Nature* 342:69–72

104. Roy R. 1989. 'Soft cheating' is more harmful to science than cases of outright fraud. *Scientist* 3:14

105. Ruby EG, Jannasch HW. 1982. Physiological characteristics of *Thiomicrospira* sp. isolated from deep sea hydrothermal vents. *J. Bacteriol.* 149:161–65

106. Ruby EG, Jannasch HW, Deuser WG. 1987. Fractionation of stable carbon isotopes during chemoautotrophic growth of sulfur oxidizing bacteria. *Appl. Environ. Microbiol.* 53:1940–43

107. Ruby EG, Wirsen CO, Jannasch HW. 1981. Chemolithotrophic sulfur-oxidizing bacteria from the Galapagos Rift hydrothermal vents. *Appl. Environ. Microbiol.* 42:317–42

108. Rueter P, Rabus R, Wilkes H, Aekersberg F, Rainey F et al. 1994. Anaerobic oxidation of hydrocarbons from crude oil by new types of sulfate-reducing bacteria. *Nature* 372:455–58

109. Russell HL. 1892 Untersuchungen über im Golf von Neapel lebende Bakterien. *Z. Hyg.* 11:165–206

110. Russell HL. 1893. The bacterial flora of the Atlantic Ocean in the vicinity of Woods Hole. *Bot. Gaz.* 18:383–95

111. Schlegel HG, Jannasch HW. 1967. Enrichment cultures. *Annu. Rev. Microbiol.* 21:49–70

112. Schlegel HG, Jannasch HW. 1991. Prokaryotes and their habitats. In *The Prokaryotes,* ed. A Balows, HG Trüper, M Dworkin, W Harder, KH Schleifer, pp. 75–125. Berlin: Springer-Verlag. 2nd ed.

113. Strugger S. 1949. *Fluorenzmikroskopie und Mikrobiologie.* Hannover: Verlag-Schaper. 149 pp.

114. Svetlichny VA, Sokolova TG, Gerhardt M, Kostrikina NA, Zavarzin G. 1991. Anaerobic extremely thermophilic carboxydotrophic bacteria in hydrotherm of Kuril Islands. *Microbiol. Ecol.* 21:1–10

115. Taylor CD. 1979. Growth of a bacterium under a high pressure oxy-helium atmosphere. *Appl. Environ. Microbiol.* 37:42–49

116. Taylor CD. 1987. Solubility properties of oxygen and helium in hyperbaric systems and the influence of high pressure oxygen-helium upon bacterial growth, metabolism and viability. In *Current Perspectives in High Pressure Biology,* ed. HW Jannasch, RE Marquis, AM Zimmerman, pp. 111–28. London: Academic

117. Trüper HG, Jannasch HW. 1968. *Chromatium buderi* nov. spec., eine neue Art der "grossen" Thiorhodaceae. *Arch. Mikrobiol.* 61:363–72

118. Trüper HG, Kelleher JJ, Jannasch HW. 1969. Isolation and characterization of sulfate reducing bacteria from various marine environments. *Arch. Mikrobiol.* 65:208-l7

119. Tuttle JH, Jannasch HW. 1973. Sulfide and thiosulfate oxidizing bacteria in anoxic marine basins. *Mar. Biol.* 20:64–70

120. Tuttle JH, Jannasch HW. 1973. Dissimilatory reduction of inorganic sulfur by facultatively anaerobic marine bacteria. *J. Bacteriol.* 115:732–37

121. Tuttle JH, Jannasch HW. 1979. Microbial dark assimilation of CO_2 in the Cariaco Trench. *Limnol. Oceanogr.* 24:746–53

122. Tuttle JH, Wirsen CO, Jannasch HW. 1983. Microbial activities in emitted hydrothermal waters of the Galapagos Rift vents. *Mar. Biol.* 73:293–99

123. Van Gemerden H, Jannasch HW. 1971. Continuous culture of Thiorhodaceae: sulfide and sulfur limited growth of *Chromatium vinosum. Arch. Microbiol.* 79:345–53

124. Van Niel CB. 1967. The education of a microbiologist: some reflections. *Annu. Rev. Microbiol.* 21:1–30

125. Van Niel CB. 1949. The kinetics of growth of microorganisms. In *The Chemistry and Physiology of Growth,* ed. AK Parpart, pp. 91–105. Princeton, NJ: Princeton Univ. Press

126. Veldkamp H, Jannasch HW. 1972. Mixed culture studies with the chemostat. *J. Appl. Chem. Biotechnol.* 22:105–22

127. von Weitzsäcker CF. 1948. *Die Geschichte der Natur.* Göttingen: Vandenhoeck & Rupprecht. 138 pp.

128. Ward DM, Bateson MM, Weller R, Ruff-Roberts AL. 1992. Ribosomal RNA analysis of microorganisms as they occur in nature. *Adv. Microbiol. Ecol.* 12:219–86

129. Wirsen CO, Jannasch HW. 1975. Activity of marine psychrophilic bacteria at elevated hydrostatic pressures and low temperatures. *Mar. Biol.* 31:201–8

130. Wirsen CO, Jannasch HW. 1976. The decomposition of solid organic materials in the deep-sea. *Environ. Sci. Technol.* 10:880–87

131. Wirsen CO, Jannasch HW. 1978. Physiological and morphological observations of *Thiovulum* sp. *J. Bacteriol.* 136:765–74

132. Wirsen CO, Jannasch HW, Molyneaux SJ. 1993. Chemosynthetic microbial activity at Mid-Atlantic Ridge hydrothermal vent sites. *J. Geophys. Res.* 98:9693–703

133. Wirsen CO, Jannasch HW, Wakeham SG, Canuel EA. 1987. Membrane lipids of a psychrophilic and barophilic deep sea bacterium. *Curr. Microbiol.* 14:319–22

134. Woese CR, Kandler O, Wheelis ML. 1990. Towards a natural system of organisms: proposal for the domains Archaea, Bacteria and Eucarya. *Proc. Natl. Acad. Sci. USA* 87:4576–79

135. Wolfe RS. 1991. My kind of biology. *Annu. Rev. Microbiol.* 45:1–35

136. Yayanos AA, Dietz AS, van Boxtel R. 1979. Isolation of a deep-sea barophilic bacterium and some of its growth characteristics. *Science* 205:808–10

Annu. Rev. Microbiol. 1997. 51:47–72

XENORHABDUS AND *PHOTORHABDUS* SPP.: Bugs That Kill Bugs

Steven Forst
Department of Biological Sciences, University of Wisconsin, Milwaukee, Wisconsin 53201; email: sforst@csd.uwm.edu

Barbara Dowds
Biology Department, Saint Patrick's College, Maynooth, County Kildare, Ireland

Noël Boemare
Laboratoire de Pathologie Comparée, Université Montpellier II, 34095 Montpellier, Cedex 5, France

Erko Stackebrandt
DSMZ, D-38124, Braunschweig, Germany

KEY WORDS: *Xenorhabdus, Photorhabdus,* symbiotic bacteria, virulence, phase variation

---

ABSTRACT

Xenorhabdus and *Photorhabdus* spp. are gram negative gamma proteobacteria that form entomopathogenic symbioses with soil nematodes. They undergo a complex life cycle that involves a symbiotic stage, in which the bacteria are carried in the gut of the nematodes, and a pathogenic stage, in which susceptible insect prey are killed by the combined action of the nematode and the bacteria. Both bacteria produce antibiotics, intracellular protein crystals, and numerous other products. These traits change in phase variants, which arise when the bacteria are maintained under stationary phase conditions in the laboratory. Molecular biological studies suggest that *Xenorhabdus* and *Photorhabdus* spp. may serve as valuable model systems for studying signal transduction and transcriptional and posttranscriptional regulation of gene expression. Such studies also indicate that these bacterial groups, which had been previously considered to be very similar, may actually be quite different at the molecular level.

47

CONTENTS

INTRODUCTION

Xenorhabdus and *Photorhabdus* spp. are motile gram negative bacteria belonging to the family Enterobacteriaceae (3, 15, 17). The general features of the life cycles of these bacteria are quite similar. *Xenorhabdus* and *Photorhabdus* spp. are carried as symbionts in the intestine of the infective juvenile stage of nematodes belonging to the families Steinernematidae and Heterorhabditidae, respectively (1, 2, 50, 77). The nematodes enter the digestive tract of the larval stage of diverse insects and subsequently penetrate into the hemocoel of the host insect. The nematode can also gain access to the hemocoel via the respiratory spiracles or by penetrating directly through the insect cuticle (2, 77). Upon entrance into the hemocoel, the nematodes release the bacteria into the hemolymph. Together, the nematode and the bacteria rapidly kill the insect larva, although in most cases the bacteria alone are highly virulent (2). Within the hemocoel of the larval carcass the bacteria grow to stationary phase conditions while the nematodes develop and sexually reproduce. Nematode reproduction is optimal when the natural symbiont (*Xenorhabdus* or *Photorhabdus* spp.) dominates the microbial flora, suggesting that the bacteria can serve as a food source and/or provide essential nutrients that are required for efficient nematode proliferation (2, 6, 77). During the final stages of development, the nematode and bacteria reassociate and the nematode subsequently develops into its nonfeeding infective juvenile stage. The infective juvenile, carrying the bacteria in its intestinal tract, then emerges from the insect carcass in search of a new insect host.

All of the *Xenorhabdus* isolates studied so far, and almost all of the *Photorhabdus* isolates, have been obtained from nematodes harvested from soil samples. Free-living forms of the bacteria have not yet been isolated from soil or water sources. These findings suggest that the symbiotic association may be essential for the survival of the bacteria in the soil environment. The bacteria in turn are essential for effective killing of the insect host and are required for the nematode to efficiently complete its life cycle (1, 66).

Both *Xenorhabdus* and *Photorhabdus* spp. can be grown as free-living organisms under standard laboratory conditions. As the bacteria enter the stationary phase of their growth cycle, they secrete several extracellular products, including a lipase(s), a phospholipase(s), a protease(s), and several different broad spectrum antibiotics (4, 6, 15, 74). These products are believed to be secreted into the insect hemolymph when the bacteria enter stationary phase conditions. The degradative enzymes break down the macromolecules of the insect cadaver to provide the developing nematode with a nutrient supply, while the antibiotics suppress contamination of the cadaver with other microorganisms. Cytoplasmic inclusion bodies, composed of highly expressed crystalline proteins, are also produced by both bacteria during stationary phase conditions (29, 30) .

Another intriguing property of *Xenorhabdus* and *Photorhabdus* spp. is the formation of phenotypic variant forms (3, 10, 13, 15, 64) that can be isolated at low and variable frequencies during prolonged incubation under stationary phase conditions. The variant forms, or so-called phase II cells, are altered in many properties and are not found as natural symbionts in the nematode. Phase I cells represent the form of the bacteria that naturally associates with the infective juvenile nematode.

Finally, the pathogenic potential of the bacteria/nematode complex has been exploited for its usefulness as a biological pest control agent (67).

While *Xenorhabdus* spp. and *Photorhabdus* spp. are similar in numerous characteristics, they differ in several salient features. *Xenorhabdus* spp. is specifically found associated with entomopathogenic nematodes in the group Steinernematidae, while *Photorhabdus* spp. only associates with the nematode group Heterorhabditidae. A primary property distinguishing *Photorhabdus* from *Xenorhabdus* spp. is the ability of the former to emit light under stationary phase culture conditions and in the infected host insect (78). On the other hand, *Xenorhabdus* is catalase negative, which is an unusual property for bacteria in the Enterobacteriaceae family (15, 46). In this review we present information that has accumulated over the past several years and that supports the contention that these bacteria may represent an example of convergent molecular evolution in which many phenotypic properties required for the symbiotic/pathogenic lifestyle of *Xenorhabdus* and *Photorhabdus* spp. were acquired from disparate genetic origins. Several previous reviews have discussed the biology (1, 2, 6, 50), physiology (73, 74), and taxonomy (6) of both bacterial genera, as well as specific related topics such as bioluminescence in *Photorhabdus* (53, 54). Other topics, such as antibiotic and pigment production, the characterization of intracellular crystalline proteins, and the production of bacteriocins, have also been covered in a recent review (50) and therefore are not discussed in the present chapter.

TAXONOMY AND PHYLOGENY OF *XENORHABDUS* AND *PHOTORHABDUS*

Taxonomic Studies

Xenorhabdus and *Photorhabdus* spp. are facultatively anaerobic gram negative rods, nonsporulating, oxidase negative chemoorganotrophic heterotrophs with respiratory and fermentative metabolisms. They are included in the family of Enterobacteriaceae (59, 87) belonging to group 5 and to sub-group 1 (61). Almost all species of *Xenorhabdus* and *Photorhabdus* are entomopathogenic. However, *Xenorhabdus poinarii* (5) is not pathogenic for *Galleria mellonella* (Lepidoptera) and *Xenorhabdus japonicus* is not pathogenic towards *Spodoptera litura* (76, 93). Production of bacteriocins has been reported for both genera, particularly xenorhabdicin from *Xenorhabdus nematophilus*, that is lytic for closely related bacteria (9, 18, 86). In addition, many strains of *Xenorhabdus* are lysogenic (18), while phage have not yet been identified in *Photorhabdus*. Table 1 summarizes the main phenotypic characteristics of *Xenorhabdus* and *Photorhabdus*.

DNA Relatedness Studies

Strains of each of the individual *Xenorhabdus* species show high DNA similarity values (>60%), while the interspecific values range below 40% (17). This finding confirmed the genomic distinctness of *X. nematophilus, Xenorhabdus bovienii, X. poinarii* and *Xenorhabdus beddingii*, originally defined on the basis of phenotypic data. The DNA relatedness data also show the genomic distinctness of strains of the species formerly identified as *Xenorhabdus luminescens*. These data, together with the significant differences in phenotypic characters and cellular fatty acids (65) that were shown to exist between *X. luminescens* and the other *Xenorhabdus* species, led to the transfer of *X. luminescens* into a new genus, *Photorhabdus*, as *Photorhabdus luminescens* comb. nov. (17).

Analysis of the DNA relatedness data revealed that all the *Photorhabdus* strains studied were congeneric and that the genus *Photorhabdus* was, on the basis of DNA relatedness, more homogeneous than *Xenorhabdus*. Three relatedness groups were recognized in *Photorhabdus* (7). Among the three groups, group I was composed of the symbionts of two or more *Heterorhabditis* species, and its members generally had >60% relatedness. Relatedness between isolates of group II, the symbionts of at least four nematode species, was generally

Table 1 Characteristics of Xenorhabdus[a] and Photorhabdus[a] species

	Nematode Isolation						Sensitive To:											
	Steinernema spp.[b]	Heterorhabditis spp.[c]	Bioluminescence	Catalase	Nitrate-Reductase	Antimicrobial Production	X. nematophilus[d] Antibiotics	X. nematophilus[e] Bacteriocins	Insect Pathogenicity[f]	Protoplasmic Inclusions	Dyes Absorbed	Pigmentation	Swarming[g]	Lecithinase	Lipase	Gelatin Hydrolysis	Urease	Indole	DNase
Xenorhabdus spp.																			
X. nematophilus	+	−	−	−	−	+	−	−	+	+	+	−	+	+	+	+	−	−	+
X. beddingii	+	−	−	−	−	+	−	+	+	+	+	+	+	+	−	+	−	−	+
X. bovienii	+	−	−	−	d	+	−	+	+	+	+	+	+	d	+	+	−	−	+
X. poinarii	+	−	−	−	−	+	−	+	−	+	−	+	−	−	+	+	−	−	−
Photorhabdus spp.																			
P. luminescens I	−	+	+	+	−	+	−	+	+	+	+	+	+	+	+	+	d	d	+
P. luminescens II	−	+	+	+	−	+	−	+	+	+	+	+	nr	+	+	+	d	d	+
P. luminescens III (clinical group)	−	−	+	+	−	−	−	+	nr	−	−	+	nr	−	+	+	−	−	−

Symbols: −, 0–10% positive; d, 26–75% positive; +, 90–100% positive; nr, not reported.

[a]Ref. 15; except when mentioned.

[b]non nematode-symbiotic Enterobacteriaceae (Ref. 14).

[c]non nematode-symbiotic Enterobacteriaceae (Ref. 64a).

[d]for sensitivity of non nematode-symbiotic Enterobacteriaceae (Ref. 18).

[e]Refs. 18 and 86.

[f]Refs. 6 and 15.

[g]Ref. 57; Boemare, unpublished data for Photorhabdus.

>65%, and each member had 70% or greater relatedness to at least one other member of this group. Group III, the clinical strains (47), was the most clearly defined group, with 100% DNA relatedness within the group and <60% relatedness to strains in other groups. It was clear that the clinical strains, which are the only members of the *Xenorhabdus/Photorhabdus* groups to be isolated as saprophytes (i.e. with no nematode host), were appreciably different from the nematode-symbiotic strains.

Phylogenic Studies with 16S rDNA Analyses

In order to determine whether the subclusters that emerged from DNA-DNA hybridization experiments can be recovered by determination of the analysis

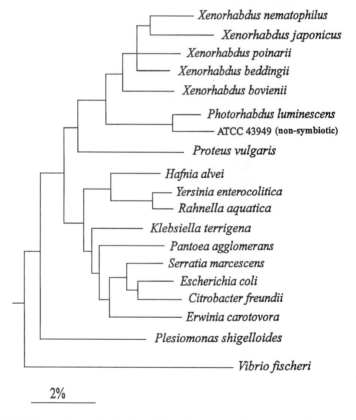

2%

Figure 1 Phylogenetic tree showing the relationships between the type strains of the type species of *Xenorhabdus* and *Photorhabdus*, the nonsymbiotic luminescent strain ATCC 43949, members of the Enterobacteriaceae and *Vibrio fischeri*, inferred from 16S rDNA sequences. The sequence of *Haemophilus influenzae* was used as the root. Analysis was based on more than 1400 nucleotides. Bar represents two nucleotide substitutions per 100 nucleotides.

of the 16S rDNA, this gene was sequenced from 30 representative strains of *Xenorhabdus* and *Photorhabdus* species. These experiments are necessary because, although DNA similarity values around 70% represent the gold standard for species delimitation in the polyphasic approach to classification, the hierarchic intragenus and intergenus relationships cannot be extrapolated from DNA-DNA similarities (83).

16S rDNA sequence was determined for seven strains of *Photorhabdus luminescens* and six strains of *Xenorhabdus* species (79). These sequences were compared to homologous sequences of several species of Enterobacteriaceae. Binary similarity values determined for the *P. luminescens* isolates and previously analyzed strains of this species ranged above 96%, whereas the values were lower by about 2% and 4% when the sequences of the isolates were compared to those of *Xenorhabdus* and Enterobacteriaceae strains, respectively. All *Photorhabdus* strains could be clearly distinguished from strains of *Xenorhabdus* by the occurrence of a CAAG sequence at position 450 (*Escherichia coli* numbering) of the 16S rDNA, whereas strains of *Xenorhabdus* species possess a UC pair at this position. Comparisons of 16S rDNA sequences demonstrate that species of *Photorhabdus* and *Xenorhabdus* form a phylogenetically coherent cluster that branch deeply within the radiation of the family Enterobacteriaceae (Figure 1). It appears that the genus *Photorhabdus* evolved after the main radiation of *Xenorhabdus* species had occurred. This conclusion, originally derived from the analysis of only 7 strains, was verified when the 16S rDNA analysis was extended to include more than 25 strains, including mainly members of *Photorhabdus*. This analysis also verified the existence of several 16S rRNA clusters originally detected by DNA-DNA similarities.

MOLECULAR BIOLOGY

Characterization of Genes of Xenorhabdus and Photorhabdus

Most research on these species concerns the mechanisms and functions of phase variation, pathogenicity, symbiosis, bioluminescence, and adaptation to environmental extremes, and most of the genes characterized thus far are involved in these functions. Cloned genes from *Xenorhabdus* and *Photorhabdus* are easily identified in *E. coli* because they are expressed from their own promoters in *E. coli*, active gene products are made, and secretion signals are recognized in this cloning host.

OUTER MEMBRANE PROTEINS GENES   Outer membrane proteins (Opns) of *X. nematophilus* were shown to be regulated in response to growth phase and growth temperature (69), but not in response to changes in osmolarity (52). The *ompR* and *envZ* two-component signal transduction genes of *X. nematophilus* were shown to constitute a single operon regulated by a σ^{70} promoter. Nucleotide sequence analysis revealed that EnvZ of *X. nematophilus* shared 57% amino acid identity with the cytoplasmic domain of EnvZ of *E. coli*, whereas the OmpR molecules shared 78% amino acid identity (85). OpnP, which is the most highly expressed outer membrane protein of *X. nematophilus*, displayed 55% amino acid sequence identity with OmpF (52, 85). The *opnP* gene has a σ^{70}-type promoter and binding sites for the OmpR and *micF* regulators. Comparison of the EnvZ proteins of *X. nematophilus* and *E. coli* has increased our understanding of the sensing function of EnvZ. The large hydrophilic periplasmic domain of EnvZ (*E. coli*) is missing in EnvZ (*X. nematophilus*), and yet the latter was able to complement an *envZ* mutant of *E. coli* when regulating OmpR in response to changes in osmolarity. This finding indicates that the periplasmic domain of EnvZ may not be essential for sensing osmolarity signals. A strain of *X. nematophilus* in which the *envZ* gene was disrupted displayed significantly reduced levels of several stationary phase–induced Opns as well as OpnP (50, 85; Forst, unpublished data). Thus, it appears that one function of EnvZ in *X. nematophilus* is to regulate the expression of some outer membrane proteins at the stationary phase.

GENES INDUCED AT LOW TEMPERATURE   Clarke & Dowds (27) analyzed exponential phase cells of *P. luminescens* for low temperature specific cytoplasmic proteins. Two major proteins were over-expressed during balanced growth at 9°C relative to 28°C. One was polynucleotide phosphorylase (Pnp) and the other was probably the NusA antiterminator, both of which are cold shock proteins in *E. coli*. Pnp is an mRNA degrading enzyme that has been hypothesized to be less energetically expensive than RNase III, the other major exoribonuclease, because the product generated by Pnp retains the high energy phosphate bond of the ribonucleotide. This suggests the reason for why Pnp is induced at low temperatures. The *pnp* gene of *P. luminescens* is linked to *rpsO* as in *E. coli*, and the predicted amino acid sequence of the proteins revealed 86% identity with the *E. coli* homologues for both Pnp and RpsO. The regions upstream and between the genes are also well conserved, and the sequencing and primer extension studies revealed two possible σ^{70} promoters upstream of the two genes, as well as a cold inducible promoter for *pnp* between the genes. In addition, a putative binding site for the cold inducible transcriptional regulatory protein, CS7.4, was located upstream of the cold inducible promoter, a feature not found in the *E. coli* gene. An RNase III cleavage site was also found upstream of *pnp* and the major 5′ end of the mRNA mapped to this position; this cleavage of *pnp* mRNA by RNase III is the first step in a translational autocontrol mechanism in *E. coli*. *P. luminescens* has been defined as a psychrotroph, and the failure to isolate more cold active mutants by a variety of means (26) suggests that it is indeed well adapted to growth at low temperatures. Clarke & Dowds (27) suggest that one part of this adaptation to low temperatures may be the acquisition of promoters sensitive to low temperature activators such as CS7.4; such a control site is found on the *pnp* gene of *P. luminescens* but is absent from *pnp* in the mesophile *E. coli*.

MALTOSE METABOLISM   *Photorhabdus* spp. can grow on maltose as a sole carbon source, and the genes involved in its utilization are expected to be subject to complex controls by maltose and sugars causing catabolite repression. The *mal* region sequenced from *P. luminescens* has 53% identity with the *malB* region of *E. coli*, and it comprises part of the *malE* and *malK* genes and the regulatory region between them (25a). The predicted amino acid sequence of the protein coding regions reveals 65–75% identity between *E. coli* and *P. luminescens*. Furthermore, the control region has a very similar organization to that in *E. coli*; the *P. luminescens* region has four MalT binding sequences (like *E. coli*) and two binding sites for the cAMP receptor protein (CRP) (*E. coli* has 4 CRP sites). In addition, the U box, a region of unknown function that is conserved in the Enterobacteriaceae and found just upstream of the MalT box closest to the *malE* gene (31), is also found in this position in *P. luminescens*.

LUX GENES   The biochemistry and the physiological regulation of luminescence in *Photorhabdus* and marine bacteria has been reviewed previously (50). The *lux* genes have been isolated from four different strains of *P. luminescens* by screening *E. coli* banks for light emission (53, 84, 88, 90). Five genes are needed for light production: *luxC, D*, and *E* code for the enzymes of the fatty acid reductase complex that produces the long chain aliphatic aldehyde substrate for the luciferase whose two subunits are encoded by the *luxA* and *B* genes. The arrangement and sequence of the *lux* genes is very similar in *P. luminescens* and in the *Vibrio* and *Photobacterium* genera of marine bacteria, which suggests that the genes have a common evolutionary origin. For example, the α subunits of the luciferases of *P. luminescens* and *Vibrio harveyi* have 85% identity, while the β subunits have 60% identity (84). The Vibrionaceae family of bacteria are phylogenetically separate from *P. luminescens* (Figure 1), suggesting that the *lux* genes were acquired by horizontal gene transfer. The transfer seems to have occurred more than once because the chromosomal location of the *lux* genes was found to differ between a strain of *P. luminescens* that was isolated from a human wound and two nematode symbiont strains, though the gene sequences were 85–90% identical (71).

Bacterial bioluminescence results from the luciferase catalyzed simultaneous oxidation of a long chain aldehyde and of $FMNH_2$ by molecular oxygen. $FMNH_2$ is supplied by the reduction of FMN by NAD(P)H-flavin oxidoreductase (Fre). Zenno & Saigo (94) cloned and sequenced the *fre* genes from *P. luminescens* and three *Vibrio* species. The *fre* gene product was a good supplier of $FMNH_2$ to the bioluminescence reaction, indicating that it plays this role in vivo. The *fre* genes were similar in sequence to the *luxG* gene of *Vibrio* spp., but LuxG and Fre appear to constitute two separate groups of flavin associated proteins. The *luxG* gene is either missing from *P. luminescens* or is found at a different position on the chromosome because the sequence diverges at the position where it is found in marine luminous bacteria (71).

EXTRACELLULAR ENZYMES   Phase 1 cells of *P. luminescens* secrete a lipase that hydrolyses Tween 80, *p*-nitrophenylpalmitate, and naphthyl acetate. The *lip*-1 gene was cloned and expressed off its own promoter in *E. coli*, and it codes for a protein of 645 amino acids from which a hydrophobic leader sequence of 24 amino acids is removed during processing in *E. coli*. (89). The leader is presumably a signal for secretion in *P. luminescens*. The lipase gene was originally isolated on a fragment of DNA that also coded for a 32 kDa protein whose N-terminus was 81% identical to the pyridoxal phosphate biosynthetic protein, PdxJ of *E. coli*. Sequencing and primer extension assays revealed two possible lipase promoters, a good σ^{70} sequence starting at -4 and a less conserved promoter starting at -9 whose -10 and -35 regions resemble the corresponding sequences in the *uvrB* and *malEFG* promoters of *E. coli*. The sequence of the structural gene did not show strong homology to other lipases that are, in general, poorly conserved except at the active site. One possible active site was proposed by Wang & Dowds (89), but another site at the N-terminus (amino acids 7–13 of the mature protein) better matches the active site of several lipases, particularly the acyltransferase of *Aeromonas hydrophila* (32, 34).

CRYSTALLINE PROTEIN GENES   Two types of proteinaceous crystals are found in both *X. nematophilus* (29, 30) and *P. luminescens* (12, 20, 45); these account for 55–60% of the total protein of stationary phase cells in the case of *P. luminescens* (20). The proteins of the two species differ in molecular weight and crystalline conformation. The function of the crystals is unknown, but neither has insecticidal activity (20, 30). The crystal proteins of *P. luminescens* do not appear to be a form of nutrient storage for the bacterium (20). It has been hypothesized that they may be a food source for the nematode (12, 29), but there is no direct evidence for this. The *X. nematophilus* crystals contain 26 and 22 kDa proteins respectively (29), while those of *P. luminescens* contain 11.6 and 11.3 kDa proteins (12). The *P. luminescens* genes (coding for the crystal proteins), *cipA* and *cipB*, have been cloned and sequenced, and their products are not similar to any other known proteins but have 25% identity to each other (12). The genes have separate promoters with identical, nearly perfect σ^{70} consensus sequences and are expressed in *E. coli*, resulting in inclusion body formation in the host. The crystals are phase 1 specific, and their role in phase variation was analyzed by inactivating the two genes (see section on phase variation).

Transfer of DNA into Xenorhabdus *and* Photorhabdus

There is considerable applied interest in the genetic engineering of *Xenorhabdus* and *Photorhabdus* spp. Genetic engineering of the bacteria could improve their potential as pest control agents in terms of low temperature activity, greater host specificity, and improved pathogenicity towards certain insects (24). Apart from the need to increase our knowledge of the biological basis of these phenomena, the molecular approaches required to study these bacteria need to be further developed.

Xu et al (91) developed a strain-specific method for transforming *X. nematophilus* with *E. coli* plasmids using the *E. coli* $CaCl_2$-$MgCl_2$ procedure. Xu et al (92) also conjugally transferred four different plasmids into a specific strain of *X. nematophilus*, of which the highest frequency of transfer was 5.8×10^{-2}. Conjugal transfer from *E. coli* into *P. luminescens* was successful in some cases (33, 34) but not others (92). *Photorhabdus* spp. can also be transformed by electroporation with efficiencies of approximately 10^5 transformants/μg of plasmid DNA for strain HF (B Dowds, unpublished data). Francis et al (55) have transduced *X. bovienii* using phage λ as a vector. They did this by first conjugating in a plasmid that encoded the *E. coli* λ receptor protein, LamB. One of the engineered strains expressed LamB on its outer membrane and adsorbed twice as many phage particles as the parent strain. This strain was then transduced with λ :: Tn10 to construct a transposon mutant bank. Another transposon mutant bank in *Xenorhabdus* was also made with an *E. coli* transposon. The delivery system was negative selection plasmid, pHX1— which contains the *Bacillus subtilis* levansucrase gene, *sacB*, a Tet^r gene on the plasmid and Tn5 carrying the Kan^r gene (92). Transformants that were Kan^r but Tet^s and $Sucrose^r$ had lost the plasmid but gained the transposon. More than 250 mutants of *X. nematophilus* were generated in this way, including avirulent strains and variants affected in antibiotic, lipase, lecithinase, or protease production (63, 92). A number of genes of interest: *envZ* (50), *cipA*, *cipB* (12), and *xin* (50) have been insertionally inactivated by allelic exchange between the chromosome and the disrupted cloned gene carried on a plasmid. The extension of this technique to other strains and its application to other cloned genes will increase our knowledge of their functions in key processes.

PATHOGENICITY

Several aspects of the pathogenicity of *Xenorhabdus* and *Photorhabdus* spp. toward insect hosts have been reviewed previously (2, 39, 50). For a more detailed discussion of this topic, the reader is directed to these reviews. In the current section, we discuss recent studies directed at understanding the mechanisms by which *Xenorhabdus* and *Photorhabdus* spp. survive the host defensive response, proliferate in the hemolymph, and kill the larvae.

The initial cellular defensive response to bacterial infection is phagocytosis (35). When a large number of bacteria are present in the hemolymph, then phagocytosis is augmented by nodule formation. Nodules form when the bacteria adhere to and are sequestered by hemocytes. The simultaneous entrapment of many bacteria with large hemocytic aggregates produces nodules that leave the circulation by adhering to tissues (35). The last instar larval stage of the common wax moth, *Galleria mellonella*, has been used as a primary test organism for studying the virulence of *Xenorhabdus* and *Photorhabdus* spp. In most cases the bacteria alone are pathogenic. However, Akhurst has shown that whereas neither *X. poinarii* nor the nematode *Steinernema glaseri* alone resulted in larval mortality, the symbiotic pair was nevertheless highly pathogenic for *G. mellonella* larvae (5). *X. japonica*, which is symbiotically associated with the nematode *Steinernema kushidai*, was found to be nonpathogenic towards *Spodoptera litura* larvae, although other *Xenorhabdus* species that were tested killed this insect host (76, 93). These results indicate that *X. poinarii* and *X. japonica* may be lacking certain pathogenic factors that are present in other species of *Xenorhabdus*. It has been recently discovered that multiple virulence genes in *Shigella*, *Salmonella*, and certain strains of *E. coli* are contained on large segments of DNA (ca. 40 kb), which are referred to as pathogenicity islands (76a). Using various hybridization procedures, the putative pathogenicity islands can be isolated and the genes can be analyzed. This approach could be employed to identify virulence factors in *Xenorhabdus*.

The antibacterial defensive system of the tobacco horn worm, *Manduca sexta*, has been extensively studied (35, 36). This organism is generally resistant to bacterial infection. For example, the LD_{50} of *Pseudomonas aeruginosa* strain P11-1 towards *M. sexta* was shown to be 10^5 bacteria/larvae, whereas less than 100 cells were sufficient to kill *G. mellonella* (36). That *X. nematophilus* is highly pathogenic is emphasized by the fact that as few as 20 bacterial cells are able to kill more than 90% of the *M. sexta* larvae injected with these bacteria (50; S Forst, unpublished data). We have shown that the insect larvae die before bacteria are detectable in the hemolymph. *X. nematophilus* may be sequestered in nodules that temporarily leave the general circulation by adherence of the nodule to fat bodies (36, 85, 50; S Forst, unpublished data). After the insect dies, the number of *X. nematophilus* cells in the hemolymph increases to high levels (10^8/ml). Thus, the pathogenic phase of the bacterial life cycle does not appear to require rapid proliferation in the hemolymph, which suggests that potent insect toxins are produced by the bacteria. Akhurst has recently isolated a gene from *X. nematophilus* that produces a 31 kDa polypeptide that is toxic in *G. mellonella* (50).

One mechanism that *Xenorhabdus* species use to tolerate or evade the humoral defensive response is to inhibit the activation of the insect enzyme, phenoloxidase. The activation of phenoloxidase can occur when bacteria are present in the hemolymph (39). The lipopolysaccharide (LPS) of *Xenorhabdus* species has been shown to prevent the processing of prophenoloxidase into phenoloxidase (40, 41). Phenoloxidase, which is formed by the processing of prophenoloxidase to the active enzyme, converts tyrosine to dihydroxyphenylalanine. The modified phenylalanine binds to the bacterial cell surface, which initiates the process of melanization. The precise role of melanization in the insect immune response remains unclear, although it has been suggested that it functions as an opsonization process promoting adherence to the hemocytes.

The production of several bactericidal proteins (attacins) and peptides (cecropins) is stimulated in many insects following infection with bacteria (25, 48). The ability of *Xenorhabdus* species to grow to high concentrations in the hemolymph suggests that the bacteria may be relatively insensitive to the action of the bactericidal proteins, or it may be able to either inhibit the induction or the function of the bactericidal polypeptides. The presence of the nematode in the hemolymph may also help to inactivate the bactericidal proteins. The insect humoral immune system includes nonspecific hemolymph enzymes such as lysozyme, which is induced upon bacterial infection and can promote bacterial cell wall degradation (35). Other nonspecific antibacterial enzymes (e.g. carbohydrases and proteases) are also present in the hemolymph and can act on and alter the cell surface properties of the invading bacteria (42). *Xenorhabdus* species appear to be generally resistant to the attack by these enzymes in vivo.

Hypovirulent strains of *X. nematophilus* that display reduced pathogenicity towards *G. mellonella* have recently been obtained by a chemical mutagenesis approach (37). Several cell surface properties of the mutant strains were altered. The level of specific outer membrane proteins was also altered in the mutant strains (2). The mutant strains adhered to hemocytes more avidly, and were more efficiently cleared from the hemolymph, than the wild-type cells. These findings were consistent with the idea that the cell surface of *X. nematophilus* possesses anti-hemocytic properties that protect the bacterium from the insect defensive system. Part of the anti-hemocytic process for cells growing in *G. mellonella* may arise from the release of LPS from the bacteria and the subsequent damage that occurs to the hemocytes (40). Since the rate of LPS release was shown to be correlated with the reemergence of the bacteria in the hemolymph, and the injection of purified LPS into *G. mellonella* killed the insects, LPS is considered to be a virulence factor in *X. nematophilus*. Hemocyte lysis was proposed to be stimulated by the lipid A moiety of LPS of *X. nematophilus*. Avirulent mutant strains have also been isolated using a transposon (Tn5) mutagenesis approach (92). These strains were found to possess pleiotropic phenotypes that included a defect in cell motility, an inability to hemolyze sheep red blood cells, and an absence of a highly expressed 32 kDa protein (63). In addition, the Tn5 mutant strains attached to insect hemocytes with greater avidity and the outer membrane protein profile was altered relative to the wild-type cells (63). These approaches should allow identification of genes that contribute to the high level of virulence of *X. nematophilus* toward the insect target.

Recent results by Clarke & Dowds (28) indicate that the lipase activity of *Photorhabdus* spp. strain K122 is a virulence factor towards *G. mellonella*. The authors showed that the sterile extracellular culture media from *E. coli*, which contained a plasmid that encoded the K122 lipase gene, was insecticidal, whereas the extracellular preparations of the control *E. coli* cells were not toxic towards *G. mellonella*. The LPS of *Photorhabdus* was previously shown to damage the hemocytes of *G. mellonella* (41). However, neither purified LPS nor nonviable whole cells were toxic in *G. mellonella* larvae (28). Mutant strains produced by chemical mutagenesis displayed a pleiotropic phenotype in which several cell surface properties were altered (38). The alterations included an increase in surface hydrophobicity and hemocyte attachment, a reduction in the outer membrane protein content, and an increase in total LPS. Unlike the hypovirulent mutant strains of *X. nematophilus*, the insecticidal properties of the *Photorhabdus* strains were enhanced relative to the wild-type strain. These results suggest that the role of the cell surface properties in virulence for *Photorhabdus* may be different than for those found in *Xenorhabdus*. Finally, an extracellular protein complex elaborated by *P. luminescens* was found to be toxic towards *M. sexta*. Injection of nanogram quantities of the toxin was sufficient to kill *M. sexta* larvae (45). Thus, from the data derived from the insecticidal activities studied so far in *P. luminescens* and *X. nematophilus*, it appears that these bacteria make different types of insect toxins. The actual molecular events that cause the demise of the insect host that is infected with *Xenorhabdus* may in fact differ from those that occur during infection with *Photorhabdus*. Resolution of the structure and mechanism of action of the respective insect toxins may shed light on this unresolved issue.

PHASE VARIATION

Bacteria are able to respond to environmental change by a type of genetic instability called phase variation (23, 43, 80). This phenotypic switching occurs in a small proportion of cells in a population, so that a large clonal population is probably never genetically homogeneous. Thus, however sudden the change in growth conditions, there are always some cells that express the phenotype that is needed for survival. Spontaneous phenotypic switching has been reported for synthesis of flagella, fimbriae, toxins, surface antigens, lipopolysaccharide, restriction-modification enzymes, and conjugation pili in a wide range of pathogenic bacteria. The mechanisms of variation differ, but all have in common some type of rearrangement of the genetic material (23, 43, 80). This includes

1. inversion of a fragment of DNA, as in the classical example of flagellar phase variation in *Salmonella typhimurium* (by site-specific recombination);

2. gene conversion using silent copies of genes or transformed genetic material (by homologous recombination);

3. duplication or deletion of tandem repetitive domains within coding regions (by recombination);

4. gain or loss of repeats that affect transcription or translation of a gene (by slipped-strand mispairing during replication); and

5. DNA modification (methylation affecting binding of transcription activators).

These DNA rearrangements affect one or a small number of specific genes and are reversible. High frequency rearrangements through transposition of IS elements or duplication, deletion, or amplification of large specific parts of the chromosome have been observed (e.g. in *Streptomyces* spp.) that result in numerous unstable phenotypes. However it is not known whether this type of instability contributes to the capacity of populations to respond to changes in the environment. Phase variation in *Xenorhabdus* and *Photorhabdus* falls between these two extremes. It coordinately affects a large number of specific characteristics, but unlike in the case for *Neisseria gonorrhoeae*, where different mechanisms govern the switching of different phenotypes, there is evidence suggesting that a common control is exerted on many characteristics.

Characteristics of Xenorhabdus *and* Photorhabdus *Phase Variants*

Strains of *Xenorhabdus* and *Photorhabdus* spp. occur in two forms (3, 8, 13, 15, 50, 54, 64). The first one is the bacterium isolated from the L3 stage, the infective juvenile of the nematode life cycle (dauer larva), and has been named the phase I (primary) variant (15). Phase I is characterized by production of antibiotic molecules; adsorption of dyes (3); production of lipase, phospholipase, and protease (15); and, depending on strains and species, pigmentation or bioluminescence (*Photorhabdus*). Phase I cells are larger than phase II (secondary) cells, and they are pleiomorphic, consist of rods (80–90%) and spheroplasts (10–20%), are motile with peritrichous flagella, and swarm on appropriate agar (57, 58). Protoplasmic paracrystalline inclusions (12, 19, 20, 29, 30, 45), fimbriae (21, 72), and a thick glycocalyx (21) are also phase 1 specific. During the stationary period of in vitro culture or during nematode rearing on an artificial diet, a secondary form, named phase II, appears spontaneously (3, 16). Pure cultures of phase II either lack or have reduced levels of the previously listed properties. Both phases show a similar pathogenic effect, as measured in *G. mellonella* (64a) or *S. litura* (76, 93). However, *X. nematophilus* phase II cells were found to be essentially nonvirulent when tested in the *M. sexta* system (S Forst, unpublished data). Both phases of a given strain of either *Xenorhabdus* or *Photorhabdus* have 100% DNA relatedness (17). Furthermore, restriction digests and Southern Cross analysis of primary and secondary DNA indicate that the organization of the chromosomal DNA is the same in the two phases (8, 8a). The megaplasmid and small plasmid content of the two phases was also found to be identical (34, 68a, 81), and both phases of *Xenorhabdus* contain lysogenic phage (18, 34, 86). In this section, we use the terms phase I cells and primary form interchangeably. Phase II variants also are referred to as secondary form cells.

The question arises as to the significance of phase II variants. This question is rendered more complex by the fact that there is considerable difference between strains in the particular phenotypes that display variation (5, 13, 15, 30, 88). However, it appears to be generally true that phase II provides less suitable nutrient conditions for the nematode (3, 6), particularly for the *Photorhabdus-Heterorhabditis* associations (44). The plasticity of metabolic behavior should probably allow the symbiont to adapt to different niches. It is obvious that the recovery of the symbionts by the L3 nematode escaping the insect cadaver is, for the bacteria, the beginning of starvation, whereas their release into the insect hemolymph provides the most suitable situation for their growth. These opposite environments probably select for the development of phase variation. Stress conditions such as microaerophilic pressure in unshaken broth for both genera (16), or culturing in anaerobic jars for *Xenorhabdus* (3; N Boemare, unpublished data), induce phase variation, and it appears that one cause of secondary formation is prolonged culture time and lack of oxygen. Krasomil-Osterfeld (68) assessed a variety of environmental factors for their ability to induce phase shift of *Photorhabdus* spp. Culturing in low osmolarity liquid medium for 24 h induced shift of almost all phase I cells into the phase II variant. Parallel cultures at high osmolarity revealed no phase shift. Phase shift was reversible after 24 h by increasing the osmolarity, and prolonged subculturing under low salt conditions was needed to produce a stable phase II variant. The author (68) noted that the osmolarity of the nematode gut is the same as the secondary inducing media, whereas that of the insect larva is the same as the primary inducing media. Phase II may be induced by the low salt conditions in the nematode gut and is better adapted to the starvation conditions of this environment, whereas phase I is probably maintained in the nutrient and salt-rich medium of the insect hemocoel. Differences in response to starvation were investigated by Smigielski et al (82), who found that phase II cells recommenced growth within 2–4 h after the addition of nutrients, contrasted with 14 h for phase I cultures. They postulated that the shorter lag period would favor phase II over phase I growth if the bacteria have to compete with other free-living microorganisms outside the insects.

Mechanism of Phase Variation

All available evidence argues against large scale DNA instability. The experiments already mentioned show identical whole genome restriction maps in the two phases, and restriction mapping of some of the phase variant genes (e.g. the lipase, bioluminescence, flagellin, and crystal genes) indicates that the gross structure of the genes is the same in the two forms (12, 58, 88, 89). Thus, any form of large-scale variation in DNA structure between the two phases is unlikely. It would appear that any DNA instability is small, maybe affecting a gene controlling phase variation. In S. typhimurium, flagellar phase variation is mediated by the Hin invertase inverting a piece of DNA. The hin homologue in X. nematophilus (xin) has been cloned and insertionally inactivated in phase I cells (R Akhurst, unpublished data). No differences were found between the xin strain and the wild type with respect to DNA rearrangement or phase variation properties.

Examination of the expression of phase variant genes tells us that different phenotypes can be inactivated in different ways in the secondary phase. The flagellin genes of X. nematophilus are not transcribed in phase II cells (58), whereas lipase and protease protein are made in an inactive form in phase II of P. luminescens (89). In the case of lipase it was shown that an equal amount of lipase protein (as detected with specific antiserum) was made by the two phases, although 6 times more lipase activity was secreted by phase I cells. It was possible to activate phase II lipase with SDS. Protease is also phase regulated, with the activity 10 times greater in phase I than in phase II cells. The protease gene has not been isolated, nor has antiserum been raised against this protein. However, it was shown that protease activity was almost as high in phase II as in phase I extracellular extracts after SDS treatment, indicating that protease is regulated at a post-translational level (B Dowds, unpublished data). The lux genes are transcribed equally in both phases (88) although phase II luminesces very poorly. Hosseini & Nealson (62) have shown that the presence of rifampin, an inhibitor of transcription, results in the early onset of bioluminescence in phase I cells (light is usually emitted during stationary phase). When rifampicin was added to phase II cells, they bioluminesced to the same degree and at the same time as rifampicin-treated phase I cells. Translational inhibitors failed to activate luminescence in phase II, which suggested that the effect occurred at the RNA level. Probably the antibiotic affected transcription of an unstable antisense RNA. It was hypothesized that, in the absence of the putative antisense RNA, the *lux* mRNA would be more abundant, resulting in an increased synthesis of the Lux proteins. Other phase variant characteristics (lipase, protease, pigment, and antibiotics) in the same strain were not activated by rifampicin in phase II cells and are therefore regulated in a different way. In summary, it has been shown that the flagellin gene of *X. nematophilus* is regulated at the level of transcription and that in *P. luminescens*, the *lux* genes are probably post-transcriptionally regulated, whereas the lipase and protease genes are controlled at a post-translational level.

The work described above implies that the different phase variant genes are regulated independently of each other. Data on intermediates and other variants (56, 57, 64), mutants (12, 62; B Dowds, unpublished data), and transformants (54, 58) support this suggestion, but they also indicate that there is probably global control of the phase variant phenotypes because complete phase shift in a given strain is always accompanied by coordinate shift of the complete range of phase variant phenotypes. A number of mutants and transformants with altered phase variation patterns assist us in our efforts to understand the mechanism of this phenomenon:

1. Hosseini & Nealson (62) described a mutant that over-expressed several phase variable characteristics, including light, pigment, and antibiotic production. These products were expressed earlier during the growth cycle than was the primary. The authors did not report whether other phase variant attributes were also over-produced in the mutant.

2. Primary phase *P. luminescens* strains that were disrupted in either the *cipA* or *cipB* gene (12) resulted in cells that exhibited a secondary-like phenotype for a number of characteristics. Intriguingly, the double mutant could not be isolated and is thought to be nonviable, despite the viability of the secondary form that has undetectable amounts of crystal protein.

3. Experiments by Givaudan et al (58) and Frackman & Nealson (54) suggest that the secondary phase produces a repressor of expression of at least two sets of phase variant genes. Complementation experiments showed that motility and flagellin synthesis of phase II cannot be recovered by conjugating the phase I flagellin genes into phase II cells (58). This implies that a repressor that is active in phase II cells may be inhibiting the expression of phase variant genes. When Frackman & Nealson (54) transformed the phase II cells with a high copy number plasmid carrying the *lux* genes of

phase I cells, the resulting strain emitted light to the same degree as the transformed primary cells. This suggests that a repressor in the phase II cells has been sequestered by the high copy number of the genes or by the high concentration of the gene product.

4. K O'Neill & B Dowds (unpublished data) hypothesized that if a repressor does indeed regulate phase variation, then it is expected that it can be inactivated by mutagenesis, thus converting secondary cells into primary cells. They screened mutagenized secondary cultures, initially for primary-like pigmentation. One mutant was identified that is strikingly like the primary for all characteristics except motility, but it could be termed an intermediate in that it has a slightly lower expression of all phenotypes than the primary cell. The isolation of such a mutant indicates that all of the altered characteristics were controlled by a repressor gene in the phase II cells.

It appears that most individual phase variant genes are not directly subject to phase variation (i.e. genetic instability), but rather their expression is regulated by a hypothesized repressor gene. It is proposed that the repressor gene is unstable, its structure differing in the two forms so that it is expressed in phase II cells but not in phase I cells. This small structural difference in the DNA of the two phases could have been missed in the experiments performed thus far. It is clear that various strains have different degrees of instability in "normal" laboratory culture. One extreme example of this is those primary cultures that produce secondary forms so infrequently that it is difficult to isolate a secondary form; at the other extreme are strains that generate secondary variants and other phenotypic variants at a high frequency in all cultures. It is not clear whether the same mechanism of instability applies in all cases.

CELL SURFACE PROPERTIES

The ability of *Xenorhabdus* and *Photorhabdus* spp. to survive the host immune system must depend to a considerable extent on the envelope properties of the bacterium. The properties of the cell surface will also strongly influence the symbiotic interactions between the bacteria and the mucosa of the nematode intestine. Recognition of the importance of cell surface properties in the life cycle of *Xenorhabdus* spp. has stimulated several laboratories to characterize outer membrane proteins (52, 63, 69), to analyze the flagella in phase I and phase II cells (57, 58), and to study the properties of the fimbriae (pilin) protein of this bacterium (11, 72). The cell surface properties of *Photorhabdus* spp. have not been as well studied.

Outer Membrane Proteins of X. nematophilus

The outer membrane of gram negative bacteria functions as a selective diffusion barrier. This specialized membrane contains a limited number of different types of proteins that exist in high copy number (49, 60, 75). Nutrients and other compounds, such as antibiotics, passively diffuse across the outer membrane through water-filled diffusion channels or pores. Channels are formed by porin proteins that associate as homotrimers in the outer membrane. In *E. coli* the regulation and function of two predominant porin proteins, OmpF and OmpC, have been extensively studied (51). The most prominant outer membrane protein of *X. nematophilus*, OpnP, which constitutes more than 50% of the total protein content of the outer membrane, has been recently purified. By reconstituting the purified protein in planar lipid bilayers, the general pore properties of OpnP were shown to be closely similar to the values predicted for the OmpF monomer (52). Nucleotide sequence analysis of *opnP* has revealed that OpnP and OmpF share 50% amino acid identity. Unlike OmpF, OpnP is produced at high constitutive levels and is not repressed by high osmolarity conditions (52). The outer membrane proteins, OpnA, OpnC, OpnD, and OpnT, are also produced by *X. nematophilus* during exponential growth at 30°C (69).

Pulse labeling of *X. nematophilus* with radioactive amino acids during progressive stages of bacterial growth showed that the synthesis of OpnP is dramatically reduced as cultures enter the stationary phase. In contrast, the production of other outer membrane proteins, such as OpnS and OpnB are increased in stationary phase (52, 69). Thus, it appears that one way the cell responds to stationary phase is to induce the synthesis of new outer membrane proteins that may possess pore functions that are optimal for survival under stressful conditions, such as when only a limited nutrient supply is available. Although OpnB is induced during stationary phase growth, it is not produced at elevated temperatures or under higher osmolarity or anaerobic conditions, and it is not produced by the phase II cells (69). This type of regulation suggests that OpnB is involved in altering surface properties of *X. nematophilus* in response to environmental change. The intriguing possibility also exists that Opns, like OpnB, are involved in specific interactions with the nematode.

Analysis of Fimbriae, Flagella, and Glycocalyx

Bacterial cell surface adhesins such as fimbriae mediate the attachment to host tissues. Attachment, in turn, may promote the initial stage of infection in many different microorganisms (70). Fimbriae (pili) of enteric bacteria are generally either rod-like structures of approximately 7-nm diameter (type I) or flexible fibrillae (P pili) of considerably smaller diameter (22). In *X. nematophilus*, fimbriae are thought to be involved in the establishment of the specific association between the bacterium and the nematode gut. The fimbriae of

X. nematophilus (11, 72) are peritrichously arranged, possessing a rigid morphology and are approximately 6–7 nm in diameter. The fimbriae are composed mainly of repeating subunits of a polypeptide of between 16,000 (72) and 17,000 (11) molecular weight. Immunogold labeling experiments have shown that the 17 kDa fimbriae protein is expressed when the bacteria inhabit the gut of the nematode (11). The phase II cells did not produce fimbriae at detectable levels because there was no positive reaction with antisera directed against the 17 kDa protein and the fimbriae were not visible by scanning electron microscopy of negatively stained cells. The apparent absence of fimbriae on the surface of the phase II cells may account, in part, for the poor retention of these cells in the gut of the infective juvenile nematode.

  Xenorhabdus phase I variants displayed a swarming motility when grown on a suitable solid media. Flagellar filaments from strain F1 phase I variants were purified, and the molecular mass of the flagellar structural subunit was estimated to be 36.5 kDa (57). Flagellin from cellular extracts or culture medium of phase II was undetectable with antiserum against the denatured flagellin, using immunoblotting analysis. The lack of flagella in phase II cells appears to be due to a defect in flagellin synthesis.

  The presence of capsular material on the surface of phase I and phase II cells of *Xenorhabdus* and *Photorhabdus* spp. has been determined by transmission electron microscopy (21). The thickness of the glycocalyx layer in the phase I cells of *Xenorhabdus* spp. (142 nm) was approximately threefold greater than that of the phase II cells (49 nm). The same relationship was observed in *Photorhabdus* spp., in which phase I cells possessed a thicker glycocalyx layer (85 nm) than the phase II cells (40 nm). The greater thickness, and the possible chemical differences of the glycocalyx, of phase I cells relative to phase II cells could contribute to the ability of former cell type to adhere to a greater extent to the intestinal cells of the nematode.

CONCLUDING REMARKS

The symbiotic-pathogenic life style of *Xenorhabdus* and *Photorhabdus* spp. provides a system to study biological détente on the microbial level. During the symbiotic stage, the bacteria enter a peaceful coexistence with the symbiotic partner, the nematode. The general aspects of the symbiotic association between *Xenorhabdus* and *Steinernema* nematodes, and between *Photorhabdus* and *Heterorhabditis* nematodes, have been appreciated for several years. The specific details of how the symbiosis is established and maintained remain unclear. Whether common themes exist in the symbiotic process of these different bacteria with their respective nematode hosts is not yet known. A central question is how does the nematode switch from depending on the bacteria as a nutrient source to form a mutualistic association with specific *Xenorhabdus* or *Photorhabdus* species? How do the bacteria and nematode communicate with one another to allow this switch to occur? During the pathogenic phase, the bacteria call upon an arsenal of gene products to survive within the hemolymph and to overcome attacks by the host defensive systems. The extreme pathogenicity of *Xenorhabdus* and *Photorhabdus* spp. toward susceptible insect hosts is illustrated by the fact that a few bacterial cells are able to rapidly inhibit the growth of and to kill the infected larva. Several fascinating questions concerning the pathogenicity of these bacteria include the following: How do they survive the vigorous attack of the insect immune system? What is the molecular nature and mode of action of the insecticidal toxins? Genetic and molecular tools are now available to approach these questions. Of course, a complete understanding of the pathogenic process will require that the cellular and humoral response of the insect immune system be considered. Finally, as more genes from both bacteria are sequenced, we will be able to formulate a clearer picture concerning the intriguing possibility that *Xenorhabdus* and *Photorhabdus* spp. represent a system of lateral gene transfer and convergent molecular evolution.

ACKNOWLEDGMENTS

We express our appreciation to Charles Wimpee for critically reading parts of this manuscript. This work was supported by a grant from The Shaw Scientist Award (The Milwaukee Foundation) to S Forst. B Dowds was supported by grants from the European Community Eclair program (Grant 151) and the European Human Capital and Mobility Institutional Fellowship (Contract 930486).

> **Visit the *Annual Reviews* home page at**
> **http://www.annurev.org.**

Literature Cited

1. Akhurst R. 1993. Bacterial symbionts of entomopathogenic nematodes—the power behind the throne. In *Nematodes and the Biological Control of Insect Pests,* ed. R Bedding, R Akhurst, H Kaya, pp. 127–35. Melbourne, Aust.: CSIRO

2. Akhurst R, Dunphy GB. 1993. Tripartite interactions between symbiotically associated entomopathogenic bacteria, nematodes, and their insect hosts. In *Parasites and Pathogens of Insects,* ed. N Beckage, S Thompson, B Federici, 2:1–23. New York: Academic

3. Akhurst RJ. 1980. Morphological and functional dimorphism in *Xenorhabdus*

spp., bacteria symbiotically associated with the insect pathogenic nematodes *Neoaplectana* and *Heterorhabditis. J. Gen. Microbiol.* 121:303–9

4. Akhurst RJ. 1982. Antibiotic activity of *Xenorhabdus* spp., bacteria symbiotically associated with insect pathogenic nematodes of the families heterorhabditidae and steinernematidae. *J. Gen. Microbiol.* 128:3061–65

5. Akhurst RJ. 1986. *Xenorhabdus nematophilus* subsp. *poinarii*: its interaction with insect pathogenic nematodes. *Syst. Appl. Microbiol.* 8:142–47

6. Akhurst RJ, Boemare NE. 1990. Biology

and taxonomy of *Xenorhabdus*. See Ref. 55a, pp. 79–90

7. Akhurst RJ, Mourant RG, Baud L, Boemare N. 1996. Phenotypic and DNA relatedness study between nematode-symbionts and clinical strains of the genus *Photorhabdus* (*Enterobacteriaceae*). *Int. J. Syst. Bacteriol.* 46:1034–41

8. Akhurst RJ, Smigielski AJ, Mari J, Boemare N, Mourant RG. 1992. Restriction analysis of phase variation in *Xenorhabdus* spp. (*Enterobacteriaceae*), entomopathogenic bacteria associated with nematodes. *Syst. Appl. Microbiol.* 15:469–73

8a. Akhurst RJ, Smigielski AJ. 1994. Is phase variation in *Xenorhabdus nematophilus* mediated by genetic rearrangement? In *Abstr. 6th Int. Coll. Invertebr. Pathol.*, p.1. Montpellier

9. Baghdiguian S, Boyer-Giglio M-H, Thaler J-O, Bonnot G, Boemare NE. 1993. Bacteriocinogenesis in cells of *Xenorhabdus nematophilus* and *Photorhabdus luminescens*: *Enterobacteriaceae* associated with entomopathogenic nematodes. *Biol. Cell.* 79:177–85

10. Bermudes D, Nealson KH, Akhurst RJ. 1993. The genus *Xenorhabdus*: entomopathogenic bacteria for use in teaching general microbiology. *J. Coll. Sci. Teach.* Nov:105–8

11. Binnington K, Brooks L. 1993. Fimbrial attachment of *Xenorhabdus nematophilus* to the intestine of *Steinernema carpocapsae*. In *Nematodes and the Biological Control of Insect Pests*, ed. R Bedding, R Akhurst, H Kaya, pp. 147–55. Melbourne, Australia: CSIRO

12. Bintrim S. 1995. *A Study of the Crystalline Inclusion Proteins of* Photorhabdus luminescens. PhD thesis. Univ. Wis., Madison

13. Bleakley B, Nealson KH. 1988. Characterization of primary and secondary form of *Xenorhabdus luminescens* strain Hm. *FEMS Microbiol. Ecol.* 53:241–50

14. Deleted in proof

15. Boemare NE, Akhurst RJ. 1988. Biochemical and physiological characterization of colony form variants in *Xenorhabdus* spp. (*Enterobacteriaceae*). *J. Gen. Microbiol.* 134:751–61

16. Boemare NE, Akhurst RJ. 1990. Physiology of phase variation in *Xenorhabdus* spp. In *International Colloquium on Invertebrate Pathology and Microbial Control*, ed. DJ Cooper, J Drummond, DE Pinnock, pp. 208–12. Adelaide, Australia: Soc. Invertebr. Path.

17. Boemare NE, Akhurst RJ, Mourant

RG. 1993. DNA Relatedness between *Xenorhabdus* spp. (*Enterobacteriaceae*), symbiotic bacteria of entomopathogenic nematodes, and a proposal to transfer *Xenorhabdus luminescens* to a new genus, *Photorhabdus* gen. nov. *Int. J. Syst. Bacteriol.* 43:249–55

18. Boemare NE, Boyer-Giglio M-H, Thaler J-O, Akhurst RJ, Brehelin M. 1992. Lysogeny and bacteriocinogeny in *Xenorhabdus nematophilus* and other *Xenorhabdus* spp. *Appl. Environ. Microbiol.* 58:3032–37

19. Boemare NE, Louis C, Kuhl G. 1983. Etude ultrastructurale des cristaux chez *Xenorhabdus* spp., bactéries infeodées aux nématodes entomophages Steinernematidae et Heterorhabditidae. *C.R. Soc. Biol.* 177:107–15

20. Bowan D. 1995. *Characterization of a high molecular weight insecticidal protein complex produced by the entomopathogenic bacterium* Photorhabdus luminescens. PhD thesis, Univ. Wis., Madison

21. Brehélin MA, Cherqui L, Drif L, Luciani J, Akhurst R, et al. 1993. Ultrastructural study of surface components of *Xenorhabdus* sp. in different cell phases and culture conditions. *J. Invertebr. Pathol.* 61:188–91

22. Brinton CC. 1965. The structure, function, synthesis and genetic control of bacterial pili and a molecular model for DNA and RNA transport in gram negative bacteria. *Trans. NY Acad. Sci.* 27:1003–54

23. Brunham RC, Plummer FA, Stephens RS. 1993. Bacterial antigenic variation, host immune response, and pathogen-host coevolution. *Infect. Immun.* 61:2273–76

24. Burnell AM, Dowds BCA. 1996. The genetic improvement of entomopathogenic nematodes and their symbiotic bacteria: phenotypic targets, genetic limitations and an assessment of possible hazards. *Biocontrol Sci. Technol.* 6:435–47

25. Carlsson A, Engstrom P, Palva ET, Bennich H. 1991. Attacin, an antibacterial protein of *Hyalophora cecropia*, inhibits synthesis of outer membrane proteins of *Escherichia coli* by interfering with *omp* gene transcription. *Infect. Immun.* 59:3040–45

25a. Clarke D. 1993. *A study into cold activity in the insect pathogenic bacterium,* Xenorhabdus luminescens. PhD thesis. Natl. Univ. Ireland

26. Clarke DJ, Dowds BCA. 1994. Cold adaptation in *Photorhabdus* spp. In *Genetics of Entompathogenic Nematode-Bacterium Complexes*, ed. AM Burnell,

RU Ehlers, JP Masson, pp. 170–77. Brussels: Eur. Commission

27. Clarke DJ, Dowds BCA. 1994. The gene coding for polynucleotide phosphorylase in *Photorhabdus* sp. strain K122 is induced at low temperatures. *J. Bacteriol.* 176:3775–84

28. Clarke DJ, Dowds BCA. 1995. Virulence mechanisms of *Photorhabdus* sp. strain K122 towards wax moth larvae. *J. Invertebr. Pathol.* 66:149–55

29. Couche GA, Gregson RP. 1987. Protein inclusions produced by the entomopathogenic bacterium *Xenorhabdus nematophilus* subsp. *nematophilus.* *J. Bacteriol.* 169:5279–88

30. Couche GA, Lehrbach PR, Forage RG, Cooney DR, Smith DR, et al. 1987. Occurrence of intracellular inclusions and plasmids in *Xenorhabdus* spp. *J. Gen. Microbiol.* 133:967–73

31. Dahl MK, Francoz E, Saurin W, Boos W, Manson MD, et al. 1989. Comparison of sequences from the *malB* regions of *Salmonella typhimurium* and *Enterobacter aerogenes* with *Escherichia coli* K12: a potential new regulatory site in the interoperonic region. *Mol. Gen. Genet.* 218:199–207

32. Derewenda ZS, Sharp AM. 1993. News from the interface: the molecular structures of triglyceride lipases. *Trends Biochem. Sci.* 18:20–25

33. Dowds BCA. 1994. Molecular genetics of *Xenorhabdus* and *Photorhabdus. Proc. 6th Int. Coll. Invertebr. Pathol.* 1:95–100

34. Dowds BCA. 1997. *Photorhabdus* and *Xenorhabdus* and *Photorhabdus*–gene structure and expression, and genetic manipulation. *Symbiosis.* 22:In press

35. Dunn PE. 1987. Biochemical aspects of insect immunology. *Annu. Rev. Entomol.* 31:321–39

36. Dunn PE, Drake DR. 1983. Fate of bacteria injected into naive and immunized larvae of the tobacco hornworm *Manduca sexta. J. Invertebr. Pathol.* 41:77–85

37. Dunphy G. 1993. Interaction of mutants of *Xenorhabdus nematophilus (Enterobacteriaceae)* with antibacterial systems of *Galleria mellonella* larvae (Insecta: Pyralidae). *Can. J. Microbiol.* 40:161–68

38. Dunphy G. 1995. Physicochemical properties and surface components of *Photorhabdus luminescens* influencing bacterial interaction with nonself response systems of nonimmune *Galleria mellonella* larvae. *J. Invertebr. Pathol.* 65:25–34

39. Dunphy GB, Thurston GS. 1990. Insect immunity. See Ref. 55a, pp. 301–23

40. Dunphy GB, Webster JM. 1988. Interac-

tion of *Xenorhabdus nematophilus* subsp. *nematophilus* with the haemolymph of *Galleria mellonella. J. Insect. Physiol.* 30: 883–89

41. Dunphy GB, Webster JM. 1988. Virulence mechanisms of *Heterorhabditis heliothidis* and its bacterial associate, *Xenorhabdus luminescens,* in non-immune larvae of the greater wax moth, *Galleria mellonella. J. Parasitol.* 18:729–37

42. Dunphy GB, Webster JM. 1991. Antihemocytic surface components of *Xenorhabdus nematophilus* var. *dutki* and their modification by serum of nonimmune larvae of *Galleria mellonella. J. Invertebr. Pathol.* 58:40–51

43. Dybvig K. 1993. DNA rearrangements and phenotypic switching in procaryotes. *Mol. Microbiol.* 10:465–71

44. Ehlers R-U, Stoessel S, Wyss U. 1990. The influence of phase variants of *Xenorhabdus* spp. and *Escherichia coli (Enterobacteriaceae)* on the propagation of entomophathogenic nematodes of the genera *Steinernema* and *Heterorhabditis. Rev. Nematol.* 13:417–24

45. Ensign JC, Bowan DJ, Bintrim SB. 1990. Crystalline inclusion proteins and an insecticidal toxin of *Xenorhabdus luminescens* strain of NC-19. *Proc. Abstr. Int. Colloq. Invertebr. Pathol. Microbiol. Control,* 5th, p. 218

46. Farmer JJ III. 1984. Other genera of the family of Enterbacteriaceae. In *Bergey's Manual of Systematic Bacteriology,* ed. NR Krieg, JG Holt, 1:506–516. Baltimore: Williams & Wilkins

47. Farmer JJ III, Jorgensen JH, Grimont PAD, Akhurst RJ, Poinar GO, Jr, et al. 1989. *Xenorhabdus luminescens* (DNA hybridization group 5) from human clinical specimens. *J. Clin. Microbiol.* 27: 1594–600

48. Faye I. 1990. Acquired immunity in insects: The recognition of nonself and the subsequent onset of immune protein genes. *Res. Immunol.* 141:927–32

49. Forst S, Inouye M. 1988. Environmentally regulated gene expression for membrane proteins in *Escherichia coli. Annu. Rev. Cell Biol.* 4:21–42

50. Forst S, Nealson K. 1996. Molecular biology of the symbiotic-pathogenic bacteria *Xenorhabdus* spp. and *Photorhabdus* spp. *Microbiol. Rev.* 60:21–43

51. Forst S, Roberts DL. 1994. Signal transduction by the EnvZ-OmpR phosphotransfer system in bacteria. *Res. Microbiol.* 145:363–74

52. Forst S, Waukau J, Leisman G, Exner M, Hancock R. 1995. Functional and

regulatory analysis of the OmpF-like porin, OpnP, of the symbiotic bacterium *Xenorhabdus nematophilus. Mol. Microbiol.* 18:779–89

53. Frackman S, Anhalt M, Nealson KH. 1990. Cloning, organization, and expression of the bioluminescence genes of *Xenorhabdus luminescens. J. Bacteriol.* 172:5767–73

54. Frackman S, Nealson KH. 1990. The molecular genetics of *Xenorhabdus.* See Ref. 55a, pp. 285–300

55. Francis M, Parker A, Morona R, Thomas C. 1993. Bacteriophage lambda as a delivery vector for the Tn*10*-derived transposons in *Xenorhabdus bovienii. Appl. Environ. Microbiol.* 59:3050–55

55a. Gaugler R, Kaya HK, eds. 1990. *Entomopathogenic Nematodes in Biological Control.* Boca Raton, FL: CRC Press

56. Gerritsen LJM, de Raay G, Smits PH. 1992. Characterization of form variants of *Xenorhabdus luminescens. Appl. Environ. Microbiol.* 58:1975–79

57. Givaudan A, Baghdiguian S, Lanois A, Boemare NE. 1995. Swarming and swimming changes concomitant with phase variation in *Xenorhabdus nematophilus. Appl. Environ. Microbiol.* 61:1408–13

58. Givaudan A, Lanois A, Boemare N. 1996. Cloning and nucleotide sequence of a flagella encoding genetic locus from *Xenorhabdus nematophilus.* Phase variation leads to a different transcription of two flagellar genes (*fliCD). Gene.* 183:243–53

59. Grimont PAD, Steigerwalt AG, Boemare N, Hickman-Brenner FW, Deval C, et al. 1984. Deoxyribonucleic acid relatedness and phenotypic study of the genus *Xenorhabdus. Int. J. System. Bacteriol.* 34:378–88

60. Hancock REW. 1991. Bacterial outer membranes: evolving concepts. *ASM News* 57:175–82

61. Holt JG, Krieg NR, Sneath PA, Staley JT, Williams ST. 1994. *Bergey's Manual of Determinative Bacteriology.* Baltimore: Williams & Wilkins

62. Hosseini PK, Nealson KH. 1995. Symbiotic luminous soil bacteria: unusual regulation for an unusual niche. *Photochem. Photobiol.* 62:633–40

63. Hurlbert RE. 1994. Investigations into the pathogenic mechanisms of the bacterium-nematode complex: The search for virulence determinants of *Xenorhabdus nematophilus* ATCC 19061 could lead to agriculturally useful products. *ASM News* 60:473–89

64. Hurlbert RE, Xu JM, Small CL. 1989.

Colonial and cellular polymorphism in *Xenorhabdus luminescens. Appl. Environ. Microbiol.* 55:1136–43

64a. Jackson TJ, Wang H, Nugent MT, Grifin CT, Burnell AM, Dowds BCA. 1995. Isolation of insect pathogenic bacteria, *Provendencia rettgeri,* from *Heterrhabditis* spp. *J. Appl. Bacteriol.* 78:237–44

65. Janse JD, Smits PH. 1990. Whole cell fatty acid patterns of *Xenorhabdus* species. *Lett. Appl. Microbiol.* 10:131–35

66. Kaya HK, Gaugler R. 1993. Entomopathogenic nematodes. *Annu. Rev. Entomol.* 38:181–206

67. Klein M. 1990. Efficacy against soil-inhabiting insect pests. See Ref 55a, pp. 195–214.

68. Krasomil-Osterfeld KC. 1995. Influence of osmolarity on phase shift in *Photorhabdus luminescens. Appl. Environ. Microbiol.* 61:3748–49

68a. LeClerc MC, Boemare N. 1991. Plasmids and phase variation in *Xenorhabdus* spp. *Appl. Environ. Microbiol.* 57:2597–601

69. Leisman GB, Waukau J, Forst SA. 1995. Characterization and environmental regulation of outer membrane proteins in *Xenorhabdus nematophilus. Appl. Environ. Microbiol.* 61:200–4

70. Massad G, Lockatell CV, Johnson DE, Mobley HLT. 1994. *Proteus mirabilis* fimbrae: construction of an isogenic *pmfA* mutant and analysis of virulence in a CBA mouse model of ascending urinary tract infection. *Infect. Immun.* 62:536–42

71. Meighen EA, Szittner RB. 1992. Multiple repetitive elements and organization of the *lux* operons of luminescent terrestrial bacteria. *J. Bacteriol.* 174:5371–81

72. Moureaux N, Karjalainen T, Givaudan A, Bourlioux P, Boemare N. 1995. Biochemical characterization and agglutinating properties of *Xenorhabdus nematophilus* strain F1 fimbriæ. *Appl. Environ. Microbiol.* 61:2707–12

73. Nealson KH, Schmidt TM, Bleakley B. 1988. Cell to cell signals in plant, animal and microbial symbiosis. In *Luminous Bacteria: Symbionts of Nematodes and Pathogens of Insects,* ed. S Scanerini, pp. 101–13. Berlin: Springer-Verlag

74. Nealson KH, Schmidt TM, Bleakley B. 1990. Physiology and biochemistry of *Xenorhabdus.* See Ref. 55a, pp.271–84

75. Nikaido H. 1994. Porins and specific diffusion channels in bacterial outer membranes. *J. Biol. Chem.* 269:3905–8

76. Nishimura Y, Hagiwara A, Suzuki T, Yamanaka S. 1994. *Xenorhabdus japonicus* sp. nov. associated with the nematode

Steinernema kushidai. World. J. Microbiol. Biotech. 10:207–10

76a. Ochman H, Sonicini FC, Solomon F, Groisman EA. 1996. Identification of a pathogenicity island required for *Salmonella* survival in host cells. *Proc. Natl. Acad. Sci. USA* 93:7800–4

77. Poinar GO Jr. 1990. Biology and taxonomy of Steinermatidae and Heterorhabditidae. See Ref. 55a, pp. 23–62

78. Poinar GO, Thomas G Jr, Haygood M, Nealson KH. 1980. Growth and luminescence of the symbiotic bacteria associated with the terrestrial nematode *Heterorhabditis bacteriophora. Soil Biol. Biochem.* 12:5–10

79. Rainey FA, Ehlers R-U, Stackebrandt E. 1995. Inability of the polyphasic approach to systematics to determine the relatedness of the genera *Xenorhabdus* and *Photorhabdus. Int. J. System. Bacteriol.* 45:379–81

80. Robertson BD, Meyer TF. 1994. Genetic variation in pathogenic bacteria. *Trends Genet.* 8:422–27

81. Smigielski AJ, Akhurst RJ. 1994. Megaplasmids in *Xenorhabdus* and *Photorhabdus* spp., bacterial symbionts of entomopathogenic nematodes (families Steinernematidae and Heterorhabditidae). *J. Invertebr. Pathol.* 64:214–20

82. Smigielski AJ, Akhurst RJ, Boemare NE. 1994. Phase variation in *Xenorhabdus nematophilus* and *Photorhabdus luminescens*: differences in respiratory activity and membrane energization. *Appl. Environ. Microbiol.* 60:120–25

83. Stackebrandt E, Goebel BM. 1994. Taxonomic note: a place for DNA-DNA reassociation and 16S rRNA sequence analysis in the present species definition in bacteriology. *Int. J. System. Bacteriol.* 44:846–49

84. Szittner R, Meighen E. 1990. Nucleotide sequence, expression and properties of luciferase coded by *lux* genes from a terrestrial bacterium. *J. Biol. Chem.* 265:16581–87

85. Tabatabai N. 1995. *Molecular analysis of signal transduction by the two-component regulatory proteins, EnvZ and OmpR, in the symbiotic/pathogenic bacteria* Xenorhabdus nematophilus. PhD thesis. Univ. Wis.-Milwaukee

85a. Tabatabai N, Forst S. 1995. Molecular analysis of the two-component genes,

ompR and *envZ*, in the symbiotic bacterium *Xenorhabdus nematophilus. Mol. Microbiol.* 17:643–52

86. Thaler J-O, Baghdiguian S, Boemare N. 1995. Purification and characterization of Xenorhabdicin, phage tail like bacteriocin, from the lysogenic strain F1 of *Xenorhabdus nematophilus. Appl. Environ. Microbiol.* 61:2049–52

87. Thomas GM, Poinar GO Jr. 1979. *Xenorhabdus* gen. nov., a genus of entomopathogenic, nematophilic bacteria of the family Enterobacteriaceae. *Int. J. System. Bacteriol.* 29:352–60

88. Wang H, Dowds BCA. 1991. Molecular cloning and characterization of the *lux* genes from the secondary form of *Xenorhabdus luminescens*, K 122. *Biochem. Soc. Trans.* 20:68s

89. Wang H, Dowds BCA. 1993. Phase variation in *Xenorhabdus luminescens*: cloning and sequencing of the lipase gene and analysis of its expression in primary and secondary phases of the bacterium. *J. Bacteriol.* 175:1665–73

90. Xi L, Cho K-W, Tu S-C. 1991. Cloning and nucleotide sequences of *lux* genes and characterization of luciferase of *Xenorhabdus luminescens* from a human wound. *J. Bacteriol.* 173:1399–405

91. Xu J, Lohrke S, Hurlbert IM, Hurlbert RE. 1989. Transformation of *Xenorhabdus nematophilus. Appl. Environ. Microbiol.* 55:806–12

92. Xu J, Olsen ME, Kahn ML, Hurlbert RE. 1991. Characterization of Tn5-induced mutants of *Xenorhabdus nematophilus* ATCC 19061. *Appl. Environ. Microbiol.* 57:1173–80

93. Yamanaka S, Hagiwara A, Nishimura Y, Tanabe H, Ishibashi N. 1992. Biochemical and physiological characteristics of *Xenorhabdus* species, symbiotically associated with entomopathogenic nematodes including *Steinernema kushidai* and their pathogenicity against *Spodoptera litura* (Lepidoptera: Noctuidae). *Arch. Microbiol.* 158:387–93

94. Zenno S, Saigo K. 1994. Identification of the genes encoding NAD(P)H-flavin oxidoreductases that are similar in sequence to *Escherichia coli* Fre in four species of luminous bacteria: *Photorhabdus luminescens, Vibrio fischeri, Vibrio harveyi,* and *Vibrio orientalis. J. Bacteriol.* 176:3544–51

Annu. Rev. Microbiol. 1997. 51:73–96

MOLECULAR GENETICS OF SULFUR ASSIMILATION IN FILAMENTOUS FUNGI AND YEAST

George A. Marzluf

Department of Biochemistry and Program in Molecular, Cellular, and Developmental Biology, The Ohio State University, Columbus, Ohio 43210; email: marzluf.1@osu.edu

KEY WORDS: sulfate transport, *Aspergillus*, *Neurospora*, yeast, cysteine

ABSTRACT

The filamentous fungi *Aspergillus nidulans* and *Neurospora crassa* and the yeast *Saccharomyces cerevisiae* each possess a global regulatory circuit that controls the expression of permeases and enzymes that function both in the acquisition of sulfur from the environment and in its assimilation. Control of the structural genes that specify an array of enzymes that catalyze reactions of sulfur metabolism occurs at the transcriptional level and involves both positive-acting and negative-acting regulatory factors. Positive trans-acting regulatory proteins that contain a basic region, leucine zipper–DNA binding domain, are found in *Neurospora* and yeast. Each of these fungi contain a sulfur regulatory protein of the ß-transducin family that acts in a negative fashion to control gene expression. Sulfur regulation in yeast also involves the general DNA binding protein, centromere binding factor I. Sulfate uptake is a highly regulated step and appears to occur in fungi, plants, and mammals via a family of related transporter proteins. Recent developments have provided new insight into the nature and control of the enzymes ATP sulfurylase and APS kinase, which catalyze the early steps of sulfate assimilation, and of the *Aspergillus* enzyme, cysteine synthase, which produces cysteine from O-acetylserine.

CONTENTS

73

0066-4227/97/1001-0073$08.00

INTRODUCTION

In the filamentous fungi and yeasts, entire areas of metabolism are governed by complex regulatory circuits, which control the expression of the permeases and catabolic enzymes that are involved in carbon, nitrogen, phosphorus, and sulfur metabolism and that are involved in amino acid biosynthesis and other major biosynthetic pathways (3, 32, 33). Both positive-acting and negative-acting regulatory genes act in the integrated networks that control the multiple biochemical pathways within each of these major areas of metabolism. Various sulfur compounds, especially cysteine, methionine, and S-adenyosylmethionine are essential for the growth and activities of all cells. Methionine initiates the synthesis of nearly all proteins in all organisms, whereas cysteine plays a critical role in the structure, stability, and catalytic function of many proteins. S-adenosylmethionine plays a pivotal role in methyl group transfer and in polyamine biosynthesis. Furthermore, sulfur is found as an essential component in many other biologically important molecules. Thus, it is not surprising that many organisms possess a complex regulatory circuit that governs the expression of a diverse set of permeases and enzymes that function to acquire and assimilate a steady supply of sulfur. Our most complete understanding of sulfur regulation in eukaryotic organisms has resulted from biochemical and genetic studies with the filamentous fungi, *Aspergillus nidulans*, *Neurospora crassa*, and the yeast *Saccharomyces cerevisiae*. Inorganic sulfate, an important sulfur source, is utilized by a well-defined assimilatory pathway in *Neurospora, Aspergillus*, yeast, and other fungi, as well as in plants. After uptake by the cells, inorganic sulfate is first phosphorylated via adenosine triphosphate (ATP) in two enzymatic steps to generate 3' phosphoadenosine-5' phosphosulfate (PAPS), reduced to sulfite, and then to sulfide, which is condensed with O-acetyl serine to generate cysteine, which also serves as an intermediate in the synthesis of methionine and S-adenosylmethionine. In *Aspergillus* and *Neurospora* this pathway is fully reversible due to the presence of enzymes required for reverse trans-sulfuration; methionine or homocysteine represent excellent sulfur sources for these filamentous fungi (7, 8). Mutations that identify the structural genes that encode each of the enzymes of the sulfur assimilatory pathway and that identify major sulfur regulatory genes have proved invaluable for defining the sulfur regulatory circuits of *Aspergillus, Neurospora*, and *S. cerevisiae* (1, 23, 38, 44). In these model organisms, and presumably in most fungi, a sulfur regulatory circuit operates to insure that the cells maintain an adequate source of sulfur and, conversely, to repress the synthesis of various sulfur catabolic enzymes when the cells possess an adequate internal supply of sulfur. Regulation may occur both at the level of entry of diverse sulfur sources into the assimilatory pathway and as distinct steps within the main pathway itself. In *S. cerevisiae*, S-adenosylmethionine acts as the metabolite responsible for sulfur repression (24, 39). In the case of *N. crassa*, Jacobson and Metzenberg (16) show that exogenous methionine does not repress aryl sulfatase in a leaky serine mutant unless this strain is provided with exogenous serine, when repression is complete. This result provides strong evidence that cysteine, or a compound closely related to or derived from cysteine, represents the true repressing sulfur metabolite. Some important aspects of sulfur metabolism and its genetic and metabolic regulation have been adequately addressed in earlier reviews (33, 34). Thus, our attention here will primarily focus on the significant insights that have resulted from many important and exciting new findings concerning certain assimilatory steps and global sulfur regulation. Major points of similarity as well as distinct differences in the molecular mechanisms underlying sulfur regulation in *A. nidulans, N. crassa*, and *S. cerevisiae* have emerged from these recent studies.

UTILIZATION OF CHOLINE-O-SULFATE AND AROMATIC SULFATE ESTERS

Choline-O-sulfate occurs widely throughout the plant kingdom and is stored in many fungi, where it may serve as an osmoprotectant and as an internal sulfur source; it is synthesized by the transfer of sulfate from 3'-phosphoadenosine-5'-phosphosulfate (PAPS) to choline via the enzyme PAPS-choline sulfotransferase.

N. crassa utilizes choline-O-sulfate as an excellent secondary sulfur source. Exogenous choline-O-sulfate is taken into the cells as an intact molecule via a specific permease and then hydrolyzed by choline sulfatase to yield an internal pool of inorganic sulfate. Expression of choline-O-sulfate permease and choline sulfatase is highly regulated and dependent upon the lifting of sulfur catabolite repression and requires the action of the CYS3 positive regulatory protein.

When primary sulfur sources are not available, *N. crassa, A. nidulans* and other filamentous fungi can utilize tyrosine-O-sulfate and other aromatic sulfate esters as secondary sulfur sources. Tyrosine-O-sulfate is also transported into *Neurospora* cells as an intact molecule. Intracellular tyrosine-O-sulfate is hydrolyzed by aryl sulfatase, yielding internal pools of inorganic sulfate and tyrosine (34). Expression of the aromatic sulfate ester transport system and of aryl sulfatase is dependent upon a functional *cys-3*$^+$ regulatory gene and is strongly repressed by methionine (34). Aryl sulfatase has been extensively studied as a member of the sulfur regulatory circuit because this enzyme is stable, extremely simple to assay, and increases approximately 1000-fold upon sulfur derepression (37). Interestingly, disruption of the aryl sulfatase β gene in mice results in a phenotype resembling mucopolysaccharidosis VI, a human lysosomal genetic disease (6). X-linked chondrodysplasia punctata, a human genetic disease that causes aberrant bone and cartilage development, has been identified with mutations in an arylsulfatase gene (10). It now seems probable that mutation in most, if not all, genes that encode proteins involved in sulfate uptake, its activation, and sulfurylation reactions carried out via PAPS, will cause serious human genetic diseases. *Neurospora ars*$^+$ is a single-copy gene that is transcribed only upon sulfur catabolite derepression to give a 2.3 kb mRNA whose content parallels that of arylsulfatase enzyme activity (40). The *cys-3* mutant lacks any detectable *ars* mRNA and enzyme activity, whereas the sulfur controller *(scon)-1* and *scon-2* control gene mutants each display constitutive expression of both the messenger RNA and enzyme (40). Paietta (40) employed nuclear run-on experiments to demonstrate that the *ars* gene is controlled at the level of transcription.

OTHER SECONDARY SULFUR SOURCES

N. crassa, A. nidulans, S. cerevisiae, and other fungi can utilize various other compounds as sole sulfur sources, which generally requires the de novo synthesis of one or more enzymes or permeases. One particularly intriguing case is the use of an external protein as a sulfur source. Under conditions of sulfur limitation and the availability of an exogenous protein, *Neurospora* and *Aspergillus mycelia* synthesize and secrete an alkaline protease into the growth medium (13, 14). Peptides released from the exogenous protein by the extracellular protease are taken up by the cells and the cysteine and methionine residues they contain fulfil the need for sulfur. This same extracellular protease is expressed in an independent fashion, when the cells have adequate sulfur but are instead limited for nitrogen or for carbon (14). *Neurospora Cys-3* mutants fail to synthesize the extracellular protease in response to sulfur limitation but still express this enzyme when limited for N or C. Thus, the protease structural gene is regulated as a member of three distinct global control circuits that regulate S, N, and C metabolism and thus must be served by a complex promoter control region, which also must respond to environmental pH (2).

EUKARYOTIC SULFATE PERMEASES

In *N. crassa*, two different genes, *cys-13* and *cys-14*, encode distinct sulfate permease species that are readily distinguished genetically and by their significantly different Km values for sulfate uptake (30). These dual sulfate permeases are developmentally regulated; permease I, encoded by *cys-13*, is present mainly in conidiospores, whereas permease II, encoded by *cys-14* is the predominate form in mycelia (30). These two sulfate transport systems are both subject to sulfur catabolic repression and are positively controlled by the *cys-3* regulatory gene; *cys-3* mutants lack all capacity for sulfate transport. Sulfate uptake is energy dependent and does not depend upon assimilation of the sulfate ion by conversion to APS. *Cys-11* mutants of *Neurospora*, which lack ATP sulfurylase, transport sulfate at a wild-type rate (31). In contrast, a mutant of the yeast *S. cerevisiae* that lacks ATP sulfurylase is also deficient in sulfate transport, suggesting the possibility of a structural interaction between the yeast plasma membrane sulfate transporter and the cytosolic ATP sulfurylase (28). On the other hand, the loss of sulfate permease activity occurs in a number of *S. cerevisiae* mutants in the sulfur assimilatory pathway, and may represent a cellular strategy to protect against the toxicity of PAPS and other organic phosphosulfur compounds (49). Yeast mutants that lack the high affinity sulfate permease nevertheless have a normal level of ATP sulfurylase (48). The *N. crassa cys-14*[+] gene, the first eukaryotic sulfate transporter gene to be cloned, encodes a protein of approximately 90 kDa with 12 putative hydrophobic membrane-spanning domains (21). The CYS14 protein is localized within the plasma membrane fraction; its de novo synthesis requires a lifting of sulfur catabolite repression, and its cellular content parallels the kinetics of appearance and turnover of sulfate transport activity (17). These features thus strongly support the conclusion that the CYS14 protein functions as a membrane-bound sulfate transporter.

A number of additional genes that encode H[+]/sulfate cotransporters have recently been identified and together appear to represent a new superfamily of membrane transport proteins. By comparison with the amino acid sequence of CYS14, a soybean nodulin protein and a human colon mucosa protein, DRA, were identified as probable sulfate transport proteins (46). Smith et al (48) obtained a mutant of *S. cerevisiae* that is deficient in sulfate transport and utilized it to isolate the yeast *SUL1* gene, which encodes a high affinity sulfate transporter with homology to the *Neurospora* CYS14 protein. The yeast mutant strain was then employed in elegant work to clone by complementation three sulfate transporter cDNAs derived from the higher plant *Stylosanthes hamata* (47). Two of these, *shst1* and *shst2*, encode closely related but distinct high affinity sulfate permeases that are expressed in the roots, but not in the leaves of the plant, and presumably function in uptake of inorganic sulfate from the environment. The third gene, *shst3*, specifies a low affinity sulfate transporter that is expressed in leaves and apparently is involved in the internal transport of sulfate between cellular or subcellular compartments within the plant (47).

The *sat-1* gene, which specifies a sulfate transport system of rat hepato-cytes, was isolated by its functional expression in *Xenopus laevis* oocytes, and encodes a protein of 703 amino acids that possesses 12 putative membrane spanning domains (4). A novel sulfate transporter, comprised of 739 amino acid residues, has been identified as the product of the gene responsible for the debilitating human genetic disease, diastrophic dysplasia (DTD). The DTD gene was identified by positional cloning and its function was determined by comparison to the known sulfate transporters (15). The severe skeletal abnor-malities associated with DTD appear to be a result of impaired sulfation of proteoglycans in the cartilage matrix because of a deficiency in sulfate uptake (15). A remarkable feature is that the sulfate transporters of the fungi, higher plants, and mammals all appear to possess 12 membrane-spanning domains, show significant homology to each other, and represent a new superfamily that is distinct from other transporters. It is encouraging that basic research with fungi has not only provided a detailed understanding of sulfur metabolism and its genetic regulation, but has also contributed to important studies with higher plants and mammals.

SULFATE ACTIVATION VIA ATP SULFURYLASE

The first step in the assimilation of inorganic sulfate is catalyzed by ATP sul-furylase (ATP: sulfate adenylyltransferase, EC 2.7.7.4), which yields adenosine 5'phosphosulfate (APS), as shown in Figure 1. ATP sulfurylase of *Penicillium chryosgenum* is a hexamer composed of six identical subunits, each of which consists of 572 amino acid residues and contains a single sulfhydryl group that reacts with maleimide (29). The treated enzyme shows a sigmoidal velocity curve and a significant increase in the K_m values for the substrates, APS and

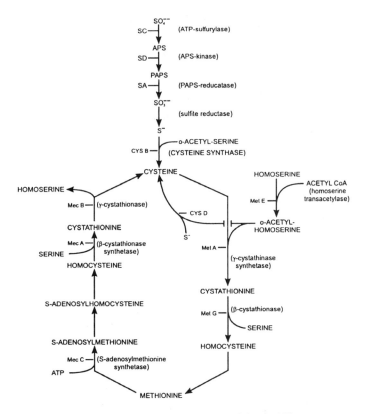

Figure 1 The sulfur assimilatory pathway of *Aspergillus nidulans* and *Neurospora crassa*. The pathway for the assimilation of inorganic sulfate in the biosynthesis of cysteine, methionine, and S-adenosylmethionine is shown. Mutants of *Aspergillus nidulans* that affect specific steps in the pathway are identified. Key enzymes are shown in parentheses.

Mg-ATP (29). ATP sulfurylase from *P. chrysogenum* and other filamentous fungi also shows allosteric inhibition by PAPS, whereas the enzyme from yeast and other eukaryotes shows only a hyperbolic velocity curve and is not affected by PAPS or by treatment with maleimide (29, 45).

ATP sulfurylase from *P. chrysogenum* shows strong homology in the N-terminal 400 amino acids with the enzyme from *S. cerevisiae* and *Arabidopsis thaliana*, with 66% and 28% amino acid identity, respectively. However, the C-terminal 172 amino acids of ATP sulfurylase of *P. chrysogenum* and other filamentous fungi is completely distinct from that of the yeast, plant, and animal enzymes. Cys-508, the residue that reacts with maleimide, is located in the C-terminal region. Of considerable interest is the finding that the C-terminal region of the *P. chrysogenum* enzyme shows very strong homology with APS kinase from several organisms. It appears probable that much of the C-terminal region of *P. chrysogenum* ATP sulfurylase evolved from APS kinase and that this domain provides the allosteric binding site for PAPS (9). It is important to note that the ATP sulfurylase does not have any APS kinase activity. Segel and his colleagues have suggested that this allosteric control, which is unique for ATP sulfurylase of the filamentous fungi, may reflect the branch point position of PAPS that is used to produce large amounts of choline-O-sulfate as well as for the synthesis of the sulfur amino acids (9).

SYNTHESIS OF 3'-PHOSPHOADENOSINE-5'-PHOSPHOSULFATE VIA ADENOSINE-5'-PHOSPHOSULFATE KINASE

3'-Phosphoadenosine-5'-phosphosulfate (PAPS), a pivotal intermediate in the sulfation of many cellular components, is synthesized by phosphorylation of adenosine 5'phosphosulfate (APS), carried out by APS kinase. The *sD* gene of *A. nidulans*, which encodes APS kinase, was isolated from a chromosome-specific cosmid library (DL Clarke, RW Newbert & G Turner, in preparation). The *sD* gene contains one 59 bp intron located at position +8 from the ATG initiation codon. The predicted APS kinase protein contains 206 amino acids and shows a high degree of amino acid identity with the enzyme from *P. chrysogenum* (82%) and from *S. cerevisiae* (75%). The *A. nidulans* APS kinase also shows a similarity to the C-terminal region of ATP sulfurylase, which apparently is regulated by PAPS-mediated feedback inhibition as discussed above. Northern blot analysis revealed that transcription of the *sD* gene occurs in a constitutive fashion and is not subject to methionine-mediated repression (DL Clarke, RW Newbert & G Turner, unpublished data). In sharp contrast, the equivalent yeast gene, MET14 is tightly controlled at the level of transcription by methionine and by the positive regulatory protein Met4 (49).

CYSTEINE SYNTHESIS

The branched pathways that are involved in cysteine and methionine synthesis in *Aspergillus* are shown in Figure 1. The major route for the synthesis of cysteine is the condensation of O-acetylserine with sulfide, catalyzed by cysteine synthase; however, its synthesis also can occur by an alternative pathway: the sulfurylation of O-acetylhomoserine to give homocysteine, that then can be converted to cystathionine and then to cysteine. When the alternative pathway is blocked, *cysB* mutants that encode cysteine synthase, require cysteine for growth. Surprisingly, however, cell extracts of the *cysB* mutant cells possess almost normal levels of cysteine synthase activity. This feature is apparently explained by the presence of distinct isozymes of cysteine synthase in the mitochondria and in the cytoplasm. The loss of the enzyme from either location leads to a loss of cysteine production and consequently of protein synthesis in that compartment. The *cysB* gene has been cloned, and it encodes a protein that possesses a putative N-terminal transit peptide; this suggests a mitochondrial localization (J Topczewski, M Sienko & A Paszewski, Paper submiitted). Expression of *cysB* is constitutive and is not subject to sulfur repression. In addition, the *Aspergillus* structural genes *metG* and *cysD*, which encode cystathionine ß-lyase (GenBank access number U28383), and homocysteine synthase (GenBank access number U19394), respectively, have been isolated and sequenced, and have been shown to be members of a pyridoxal phosphate family of enzymes (A Paszewski, unpublished results).

SULFUR REGULATORY GENES OF *NEUROSPORA CRASSA*

In *Neurospora*, at least three distinct regulatory genes, *scon-1*, *scon-2*, and *cys-3*, control the expression of the structural genes that specify sulfur catabolic enzymes (5, 36, 41). Both *scon-1* and *scon-2* act in the negative control of *cys-3*, because mutation of either leads to constitutive expression of CYS3 and of aryl sulfatase, sulfate permease, and other members of the sulfur circuit. The *scon* genes appear to represent a sequential control network that controls the *cys-3+* regulatory gene, which in turn, directly controls expression of the structural genes.

Interestingly, the 2.6 kb *scon-2+* transcript was present in wild-type strains only during sulfur-limited growth conditions, but was constitutive in *scon-1* mutants strains. Moreover, *scon-2* transcripts were present in a considerably reduced level in the *cys-3* mutant, which suggests a complex feedback loop in which *scon-2* controls *cys-3* that, in turn, along with *scon-1*, also regulates *scon-2*.

THE CYS3 POSITIVE-ACTING REGULATORY PROTEIN

CYS3 is a bZip Protein

The *cys-3+* gene encodes a positive-acting regulatory protein that carries out the final step of what must be a complex control pathway, turning on the expression of *cys-14* and *ars* and presumably each of the other unlinked but coregulated structural genes. *cys-3* null mutants do not express any *cys-14* or *ars* mRNA, and they lack all sulfate permease and aryl sulfatase activity, whereas temperature-sensitive *cys-3* mutants possess these mRNAs and enzyme activities at the permissive temperature but not at the conditional temperature. The CYS3 protein is a member of the large family of bZip DNA-binding proteins that include the yeast GCN4 protein and the mammalian proteins FOS, JUN, CREB, and C/EBP1 (11, 12, 18, 19). The CYS3 DNA-binding domain is bipartite, consisting of a leucine zipper, which is responsible for dimerization, plus an immediate upstream basic region that makes direct contact with DNA (11, 18, 19). Mutant CYS3 monomers that are defective in DNA binding but possess an intact leucine zipper region might be able to dimerize with wild-type CYS3 monomers. Examination of this possibility demonstrated that when cotranslated in vitro, wild-type/mutant hybrid CYS3 dimers do form, but they are incapable of DNA binding: that is, both subunits of the dimer must have a functional basic region for productive DNA binding (19, 42). The question arose as to whether such nonfunctional heterodimers would form in vivo and lead to an inhibitory effect, due to a loss of active wild-type homodimers. Transformation of a functional $cys-3^+$ gene into a host whose resident gene encoded a mutant CYS3 protein, defective for DNA binding but capable of dimer formation, showed significantly reduced ability to turn on the aryl sulfatase gene as compared with the transformation of the same $cys-3^+$ gene into a host that expressed a CYS3 protein lacking the entire bZip region (42). These results almost certainly are due to negative complementation, the formation of inactive hybrid dimers causing a reduced level of functional CYS3 dimers.

In some cases, different bZip proteins form hybrid dimers, e.g. the JUN-FOS heterodimer. In a set of novel experiments, Paietta (43) altered the bZip region of CYS3 in precise ways to modify the specificity of leucine zipper dimerization in order to determine whether any other cellular bZip proteins interact with CYS3 to form heterodimers that are important for sulfur regulation, cell viability, or optimum growth upon minimal medium. These studies very clearly indicated that CYS3 functions only as a homodimeric protein and does not associate with other bZip proteins to form hybrid dimers that are critical for sulfur control or other cellular activities.

CYS3 DNA Binding

The CYS3 protein expressed in *Escherischia coli (E. coli)* binds in vitro to multiple sites upstream of the *cys-14*, *ars*, and *scon-2* genes, as well as to two sites in the 5' promoter region of the *cys-3* gene itself (11, 13, 22, 42). Moreover, the native CYS3 protein in *N. crassa* nuclear extracts binds DNA in gel shift experiments in a similar fashion as does the *E. coli* expressed protein, indicating that CYS3 binds primarily as a single protein species, rather than as a member of a multi-component complex, as observed with the yeast Met4 and Met28 proteins (20, 23; K Coulter & GA Marzluf, unpublished data). The CYS3 binding sites in the promoter regions of the coregulated genes have limited sequence similarities, which initially made it difficult to define the exact requirements for high affinity DNA binding. Recent chemical DNA footprinting and mutagenesis experiments have revealed that the consensus CYS3 DNA binding site is the 10 bp sequence, ATGRYRYCAT, which represents two abutting 5 bp half-sites (26). Substitution of any base of the sequence ATGACGTCAT reduces or entirely abolishes DNA binding; moreover, deletion of a central base to yield a 9 bp overlapping palindromic sequence also eliminates all DNA binding (26). Individual native binding sites differ in their binding affinity for the CYS3 protein and deviate slightly from the consensus sequence (26, 35). The most distal of 3 CYS3 binding sites in the *cys-14* promoter and the most upstream of two sites in the promoter of the *cys-3* gene are actually duplex sites that bind CYS3 with high affinity (11, 26). It remained to be determined whether CYS3 binding sites, detected by in vitro studies, actually were of physiological significance in controlling gene expression in vivo. Analysis of the three CYS3 binding sites in the *cys-14* gene promoter was accomplished by mutating the sites individually and in all possible combinations; constructs containing the manipulated *cys-14*+ genes were then used to transform a *cys-14* null mutant (27). Loss of the two most distal CYS3 binding sites (B and C) results in the loss of all *cys-14* expression. The most distal CYS3 binding site, C, normally located 1.4 kb upstream of the transcriptional initiation point, is necessary and sufficient to mediate a full level of regulated transcriptional activation; this site is able to function equally well when it is located at variable distances upstream of the *cys-14* gene. Site B, located 1 kb upstream, alone supports a moderate degree of *cys-14* expression. Site A, although located in close proximity to the transcription start site, is not required and does not appear to have any role in controlling *cys-14* (27). These results reinforce the concept that DNA binding sites, detected by in vitro assays, may not in fact play any functional role in vivo.

The conventional *cys-3* null and temperature-sensitive mutants examined to date all have amino acid substitutions within the DNA binding domain (12, 35). Several conserved amino acids in the basic region of the CYS3 bzip domain have been changed by site-directed mutagenesis to determine whether they are essential for CYS3 DNA-binding in vitro and for function in vivo. Replacement of basic amino acids within this region by glutamine results in mutant CYS3 proteins that abolish DNA binding or are temperature-sensitive for DNA-binding (19). Furthermore, different substitutions for two uncharged residues within the basic region, serine-113 and phenylalanine-116, demonstrate that they are each important for the specificity and affinity of CYS3 DNA binding (26). When Ser-113 is changed to Ala, the mutant CYS3 protein is functional in vivo but, unlike the wild-type protein, shows a preference for binding sites that possess an A rather than a G at the fourth position; proteins in which Ser-113 is replaced by Cys, Gly, Phe, or Thr are nonfunctional (26). Similarly, mutant CYS3 proteins in which Phe-116 is replaced by Cys, Ser, or Tyr are functional but its substitution by Ile, Asn, Gly, or Val results in the loss of DNA binding in vitro and function in vivo. These results support the concept that the basic region constitutes the DNA contact surface, and that substitution of even a single basic, polar, or uncharged amino acid in this region can alter or abolish CYS3 DNA-binding activity and in vivo function. Differences in the amino acid residues within the basic region of the *N. crassa* CYS3 protein in comparison with the residues in the homologous region of the *S. cerevisiae* Met4 and Met30 bZip proteins apparently must be responsible for the stronger DNA binding activity displayed by CYS3 (Figure 2).

The leucine zipper of CYS3 has been shown to function in dimer formation, and amino acid substitutions in the repeating heptad positions of the zipper prevent dimerization, DNA binding, and function in vivo (18). Interestingly, one asparagine, one tryptophan, and three leucine residues occupy heptad repeat positions of the zipper of the *S. cerevisiae* Met28 protein, and one position in the *N. crassa* CYS3 protein is occupied by methionine (Figure 2). Thus, the zipper motif can actually utilize several of the bulky hydrophobic amino acids: leucine, methionine, tryptophan, isoleucine, and tryptophan. A CYS3 mutant protein with a pure methionine zipper, i.e. methionine in all heptad positions, is functional (18).

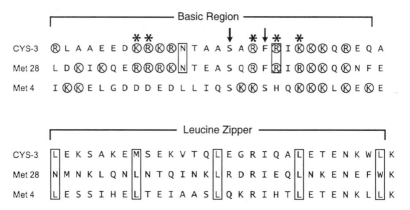

Figure 2   The basic region and leucine zipper (bZip) regions of the *N. crassa* CYS3 and the *S. cerevisiae* Met28 and Met4 proteins are shown. Positively-charged amino acid residues in the basic regions are circled, and those whose substitution has been shown to result in temperature sensitive or complete loss of CYS3 DNA binding are identified with an asterisk (*). The uncharged residues Ser[113] and Phe[116] implicated in CYS3 DNA binding are identified with an arrow. The highly conserved asparagine (N) and arginine (R) residues that occur in most bZip motifs (but not in that of Met4) are boxed for CYS3 and Met28. The leucine zippers of these proteins each immediately follow the basic regions. Heptad repeat positions of the zipper are boxed.

Trans-Activation Domains of CYS3

In addition to its bZip DNA binding domain, the CYS3 protein possesses several distinct regions that are greatly enriched in one or more amino acids, a feature that has been observed with other regulatory proteins. An N-terminal serine/threonine-rich segment of CYS3 is followed by a proline-rich domain, a short acidic region, then the bzip DNA-binding domain, an alanine-rich segment, and a C-terminal serine/threonine-rich region. Any of these regions might be important for proper folding, stability, trans-activation, or for protein-protein interactions with other factors. Different regions of CYS3 were fused to the yeast Gal4 DNA binding domain to identify segments that serve as activation domains in yeast. The N-terminal threonine/serine-rich region, which consists of 28 residues with a net charge of -3 and contains several bulky hydrophobic residues that line up with those in well-characterized activation domains, alone acted as a strong activator. This suggests that the N-terminal threonine/serine-rich region may represent the activation domain that functions in gene expression in *Neurospora* (K Coulter & GA Marzluf, unpublished results).

Control of the cys-3 Regulatory Gene

Transcription of the *cys-3* regulatory gene is itself highly regulated by sulfur catabolite repression and by at least two other unlinked control genes. The *cys-3* gene is expressed at an elevated level when cells are subject to sulfur-limiting conditions, but is only very weakly expressed in cells with high levels of sulfur (12). The products of the two sulfur controller genes, *scon-1*[+] and *scon-2*[+], are both involved in preventing *cys-3* expression during sulfur repression; mutation of either *scon-1* or *scon-2* results in constitutive expression of CYS3 and, consequently, of the entire family of sulfur catabolic enzymes. When the *cys-3* gene is expressed by way of a heterologous promoter in cells subject to repressing levels of sulfur, the *ars* gene is expressed, implying that the presence of CYS3 is sufficient to activate the structural genes (42). This result also suggests that the CYS3 protein itself is not directly antagonized by binding the repressing sulfur metabolite, which would be expected to be present during S-repression.

The CYS3 protein autogenously regulates its own expression in a positive fashion. It appears that when the *cys-3*[+] gene is turned on during S-limited conditions, the resulting CYS3 protein binds at elements in the promoter of its own structural gene and strongly enhances its expression (11, 12, 42). The *cys-3* promoter contains two sites to which CYS3 binds in vitro with high affinity. Consistent with autogenous control, *cys-3* missense mutants that encode a nonfunctional CYS3 protein are greatly deficient in synthesis of both *cys-3* mRNA and CYS3 protein (12, 20, 42). Moreover, a *lac* reporter gene fused to the *cys-3* promoter is only expressed in cells that contain a functional CYS3 protein (42), i.e. expression of *cys-3* is dependent upon its product, which provides compelling evidence for the concept of positive autogenous control (11).

Recent investigations have revealed that both *cys-3* mRNA and CYS3 protein accumulate slowly and are subject to dynamic turnover (Y Tao & GA Marzluf, unpublished results). When *Neurospora* mycelia are transferred from sulfur (S)-repressing to S-derepressing conditions, Northern blot assays show that *cys-3* mRNA is not apparent for approximately two hours, and the CYS3 protein does not accumulate until 3 to 4 hours following the transfer. This slow response may at least partially reflect the time required to deplete an internal cellular sulfur pool. Turnover of both mRNA and the CYS3 protein is much more rapid. When excess methionine is added to S-derepressed cells, the *cys-3* mRNA turns over with a half life of approximately 5 minutes and is completely gone within 20 min; under these same conditions, Western blot experiments show that the pre-existing CYS3 protein also turns over relatively rapidly with a half life of approximately 15 minutes. Interestingly, the CYS3 protein appears to be significantly more stable in cells kept under low S conditions, suggesting control via differential turnover, depending upon the sulfur status of the cells (Y Tao & GA Marzluf, unpublished data).

CYS-3$^+$ Genes from Other Strains or Species

The *cys-3$^+$* genes from the exotic strain Mauriceville of *N. crassa* and of the related species, *N. intermedia* were cloned, sequenced, and characterized (K Coulter & GA Marzluf, unpublished data). In each case, the promoter regions show strong nucleotide sequence conservation, particularly of the binding site for the CYS3 protein. The CYS3 proteins of *N. crassa* Mauriceville and of *N. intermedia* have 98% and 92% amino acid identity, respectively, with that of the standard Oak Ridge wild-type *N. crassa*. All three proteins have the identical length of 236 amino acids. Moreover, the *cys-3$^+$* gene from the Mauriceville strain and from *N. intermedia* transform a *cys-3* mutant strain and show normal expression and regulation of the sulfur catabolic enzymes (K Coulter & GA Marzluf, unpublished data). Thus, it is obvious that the sulfur control circuit of *N. crassa* and of *N. intermedia* operate in an identical or very similar fashion; more distantly related filamentous fungi may also possess a similar mechanism for sulfur regulation.

The Sulfur Controller-2 Gene Encodes a β-Transducin Protein

The sulfur controller-2 (*scon-2*) gene plays a central role in the sulfur regulatory circuit, apparently by controlling the positive activator CYS3. In *scon-2* mutants, *cys-3* and the sulfur-controlled structural genes such as *ars* and *cys-14* are expressed constitutively, even in the presence of high levels of sulfur that repress these genes in *scon-2*$^+$ strains. Thus, during conditions of high sulfur concentrations, unlike wild-type strains, *scon-2* mutants possess sulfate permease and are sensitive to chromate. The *scon-2*$^+$ gene has been isolated from a cosmid library by selection for a clone that transformed *scon-2* mutant cells to chromate resistance (41). The *scon-2*$^+$ gene encodes a protein of 650 amino acids, and its carboxy-terminal half contains six ß-transducin repeats, each consisting of approximately 40 amino acid residues (22). In the WD repeat region and in a distinct amino terminal region, SCON2 shows a high degree of homology with SCONB and MET30, which probably serve a similar function in *Aspergillus* and yeast, respectively (see below). A number of proteins of the ß-transducin family are known to regulate gene expression, and although the exact function of the repeat structures is still unclear, they appear to mediate protein-protein interactions.

Expression of the *scon-2* control gene is also highly regulated; *scon-2* transcription occurs in wild-type cells only during sulfur-derepressing conditions; moreover, full *scon-2* expression requires a functional CYS3 protein; *scon-2* mRNA is not present in *cys-3* mutant cells under S-derepressing or S-repressing conditions. The *scon-2* promoter has several CYS3 binding sites (22). However, the presence of CYS3 is not sufficient to turn on *scon-2* expression. When CYS3 was over-expressed in S-repression conditions by a heterologous promoter, *ars* but not *scon-2* was expressed (42). Significantly, in *scon-1* mutant cells, *scon-2* mRNA is expressed constitutively and is not sulfur regulated. This implies that SCON2 by itself does not prevent *cys-3* expression, since in the *scon-1* mutant *scon-2*, *cys-3*, *ars*, and *cys-14* all are expressed constitutively.

OPERATION OF THE SULFUR CIRCUIT
IN *NEUROSPORA*

We now have considerable insight into the operation of the sulfur regulatory circuit in *Neurospora*, although a number of critical and intriguing questions still must be solved. It is very important that the *scon-1* gene be isolated and the nature of its product fully characterized, including determining whether it is the factor that recognizes the repressing sulfur metabolite. It is also important that any regulatory genes in addition to *cys-3*, *scon-1*, and *scon-2*, which play a role in the control of sulfur metabolism in *Neurospora*, be identified and studied. Finally, the precise mechanism by which *cys-3* expression is regulated must be examined, and any significant protein-protein interactions or protein modification of CYS3 must be identified.

One speculative model, which incorporates all of the information now available, for operation of the sulfur circuit is suggested in Figure 3. One important feature is that the *scon-2* gene is highly expressed during sulfur derepression

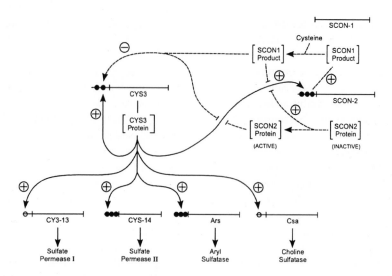

Figure 3 Speculative model for operation of the sulfur regulatory circuit of *Neurospora crassa*. The positive acting *cys-3* gene is itself expressed only when the cells become limited for sulfur. CYS3 is a trans-acting DNA-binding protein which, via autogenous control, increases its own expression and also turns on *cys-14*, *ars*, and other structural genes of the circuit. CYS3 binding sites in the 5′ promoter regions of *ars*, *cys-14*, *cys-3*, and *scon-2* all have been identified. The sulfur-controlled genes, *scon-1* and *scon-2* prevent CYS3 function during conditions of sulfur catabolite repression. The *scon-1* product is postulated to be the factor that recognizes the metabolic repressor, cysteine. The SCON2 ß-transducin protein alone does not inhibit *cys-3* function, but apparently requires modification by or interaction with the unknown product of *scon-1*. An active form of SCON2 or a SCON1-SCON2 complex is postulated to inhibit the transcription of *cys-3* directly (as diagrammed) and/or by binding to the CYS3 protein to prevent its positive autogenous control. Note that CYS3 positively controls *scon-2*, which negatively controls *cys-3*, thus constituting a complex set of feedback regulatory loops. Interactions that have been experimentally demonstrated are shown with solid lines, whereas speculative interactions are shown with dashed lines. Solid circles represent known CYS3 binding sites; open circles represent presumed CYS3 binding sites.

conditions, and is actually turned on by CYS3 and possibly also by the *scon-1* gene product. Thus the SCON2 ß-transducin protein, which plays a central role in preventing *cys-3* expression, is present in the greatest amounts—but presumably in an inactive state—during sulfur limitation, when CYS3 is actively turning on all of the structural genes that encode sulfur catabolic enzymes. This apparent paradox probably represents a complex interacting system of positive and negative feedback loops, so that when the cells suddenly experience an abundant sulfur source, the pool of SCON2 is converted into an active form to prevent CYS3 function, thus rapidly shutting down the operation of the entire sulfur circuit. In this model, it is hypothesized that the *scon-1*[+] gene product is the factor that senses the sulfur repressing metabolite, and under conditions of high levels of sulfur, it acts to prevent *cys-3* function by converting the SCON2 protein into an active form by modifying it or by binding with it. SCON2, or a SCON1-SCON2 heteromultimer, then may bind directly to the CYS3 protein by virtue of the ß-transducin repeat motifs, thus preventing further *cys-3* gene expression by inhibiting its positive autogenous control. Alternatively, SCON2 or a SCON1-SCON2 multimeric protein might bind at an element in the *cys-3* promoter to directly inhibit transcription. Thus, upon establishment of sulfur repression, the cellular content of *cys-3* mRNA and CYS3 protein will rapidly decrease, because both are subject to turnover. Thus the family of structural genes that require CYS3 for expression will be silenced. Once the sulfur circuit is shut down, a much smaller amount of SCON2 may suffice to hold it in a fully quiescent state. When the fungal cells that have enjoyed an abundant supply of sulfur are faced with sulfur limitation, the negative control preventing *cys-3* expression may be lifted due to the conversion of SCON1 and SCON2 into inactive forms. *Cys-3* transcription is then turned on. The resulting CYS3 protein, via positive autogenous control further increases *cys-3* expression, yielding an increasing pool of CYS3 protein that then acts to turn on the entire set of unlinked sulfur-controlled structural genes. Additional and equally speculative models could be proposed that would explain the network of interactions that occur in the sulfur regulatory circuit. The one presented here, however, given the possibility of significant modifications, provides sufficient puzzles to encourage decisive experiments and a major effort to characterize *scon-1* and its role in S-control.

REGULATION OF SULFUR METABOLISM IN *ASPERGILLUS NIDULANS*

Paszewski and his colleagues have made a number of recent and very important contributions to our understanding of the regulation of sulfur metabolism in *Aspergillus nidulans* (25, 38). Certain aspects of sulfur control in *A. nidulans* resemble sulfur regulation in *N. crassa*, but significant differences also occur, as will become apparent. No regulatory gene encoding a positive-acting bZip sulfur regulatory protein homologous to the *N. crassa cys-3* gene or MET4 or MET28 of *S. cerevisiae* has yet been identified, despite major efforts to do so (38). On the other hand, four different sulfur controller genes, *sconA, sconB, sconC,* and *sconD*, have been identified by mutations, each resulting in the loss of methionine repression of sulfur amino acid biosynthetic enzymes, i.e they closely resemble the *scon-1* and *scon-2* mutants of *Neurospora* and MET30 of yeast (38). Mutations in each of the *Aspergillus scon* genes also lead to altered regulation of the folate metabolizing enzymes, apparently due to an increase in the homocysteine pool in the mutant strains (25, 38).

The *sconB* gene has been cloned and sequenced and encodes a protein of the ß-transducin family containing 678 amino acids with seven ß-transducin repeats that span the C-terminal half of the protein (A Paszewski, unpublished data). The SCONB transducin protein of *Aspergillus* is closely related to the *Neurospora* SCON2 protein (74% amino acid identity) and to the *S. cerevisiae* MET30 protein (63% identity). These proteins appear to carry out similar functions in sulfur regulation. This observation is strongly supported by the fact that the *N. crassa scon-2*+ gene complements *Aspergillus sconB* mutations (A Paszewski, unpublished data). However, one important difference is noteworthy: The expression of the *sconB* gene is only weakly repressed by high concentrations of methionine, whereas the *N. crassa scon-2* gene is subject to strong sulfur catabolite repression [(41); A Paszewski, unpublished data]. Similarly, the *sconB* transcript is found in similar amounts in wild-type and in all four *scon* mutants of *Aspergillus*, whereas synthesis of *scon-2* mRNA is constitutive in *scon-1* mutants, but regulated in wild-type and *scon-2 Neurospora* strains (41).

REGULATION OF SULFUR ASSIMILATION IN YEAST

Regulatory Genes

Great advances have recently taken place in the analysis of sulfur regulation in the yeast *S. cerevisiae*. The structural genes that encode the enzymes for the biosynthesis of cysteine, methionine, and S-adenosyl-methionine (SAM) are all repressed when the cells have high concentrations of SAM (Figure 4). At least four different regulatory factors participate in controlling this set of structural genes. One of the yeast regulatory factors, Met30, acts negatively as a transcriptional inhibitor and has five ß-transducin repeats with homology to the *Neurospora* SCON2 protein; MET30 is an essential gene for viability (50). Expression of this entire set of sulfur assmilatory genes (MET3, MET14, MET16, MET5, MET10, and MET25) is dependent upon the yeast Met4 positive-acting protein, which has a bZip motif that has homology with the *N. crassa* CYS3 protein (24, 49). A second bZip protein, Met28, is required for optimal and correctly timed expression of each of the MET structural genes, except for the Met25 gene (23). Met28 only very weakly binds to DNA and lacks intrinsic activation potential; Met28 appears to function by interaction of its leucine zipper with the zipper of Met4 to form Met28-Met4 heterodimers. Different regions of the bZip domain of Met4 and Met28 have strong homology with the *Neurospora* CYS3 protein (Figure 2). The bZip protein CYS3 of *Neurospora* appears to fulfill the functions of the two bZip proteins, Met4 and Met28, of yeast (23). Although certain regions of the sulfur regulatory proteins of *S. cerevisiae* and *N. crassa* are conserved, it appears that the molecular mechanisms

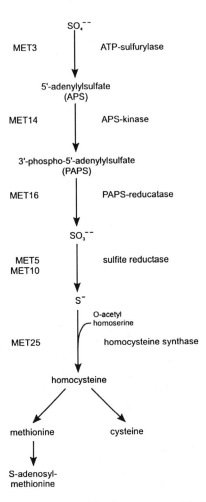

Figure 4   The sulfate assimilatory pathway of *Saccharomyces cerevisiae*. The pathway for the synthesis of homocysteine, methionine, cysteine, and S-adenosylmethionine are shown. The structural genes and the enzymes for the various steps are shown. In yeast, homocysteine is an intermediate in the synthesis of cysteine and of methionine. Expression of all of the MET genes is controlled by the positive-acting Met4 bZip protein, and all, except for MET25, also require the positive-acting Met28 BZip regulatory protein for full expression.

underlying sulfur control in these two lower eukaryotes may differ in significant ways.

A third abundant yeast protein, centromere binding factor I (Cbf1) is required for the normal transcription of most of the MET genes, although it is not required for MET3 expression (49). DNA binding sites for Cbf1, called CDEI elements (for centromere DNA element I), with the sequence 5'-TCACGTGA, are found in the promoter of all of the MET structural genes and in all known yeast centromeres. A *cbf1* null mutant exhibits defects in chromosomal segregation and a nutritional requirement for methionine. Cbf1 has a basic region helix-turn-helix domain and readily binds DNA at the CDEI elements but is not itself capable of gene activation. Rather, it appears to promote the DNA binding of MET4, which is responsible for activation of MET gene expression (24).

Functional Analysis of Met4

A detailed functional analysis of the Met4 protein shows that it contains three important regions in addition to its bZip DNA-binding motif (24). Met4 contains a single activation domain, consisting of approximately 50 amino acids located in its amino terminus. This domain, which is particularly rich in asparagine and is acidic (with 8 acidic and no basic residues), is capable of efficiently activating transcription when fused to the DNA-binding domain of Gal4 (24). A second domain, called the inhibitory region, mediates a significant reduction in gene activation by Met4 when the cellular level of SAM is high. One intriguing possibility is that the Met30 ß-transducin protein, a transcriptional inhibitor of the MET genes, binds at the inhibitory region of Met4 and interferes with the activation domain when the intracellular concentration of SAM is high (50). The two-hybrid technique has demonstrated that the Met4 and Met30 proteins interact in vivo and that the inhibitory domain of Met4 is involved in the interaction (50). A third region of the Met4 protein, the auxiliary domain, is itself devoid of intrinsic transcriptional activation ability, but appears to prevent the negative function of the inhibitory domain during sulfur limiting conditions (24).

DNA Binding by a Heteromeric Protein Complex

Kuras et al (23) have reported a series of elegant studies using extracts from yeast cells that reveal that a heteromeric complex that contains the Cbf1, Met4, and Met28 proteins is bound at the MET16 promoter in gel shifts assays. In fact, neither Met4 nor Met28 alone bind to a detectable level to the MET16 promoter, but only as members of the high molecular weight complex containing Cbf1. It is not clear whether this Cbf1-Met4-Met28 heteromeric complex is present before DNA binding or whether the otherwise weak binding of Met4 and Met28 is stabilized by interaction with Cbf1. Yeast 2 hybrid analysis shows that the bZip domain of Met4 interacts strongly with Cbf1, whereas Met28 and Cbf1 do not display any protein-protein binding. Rather, Met4 and Met28 form heterodimers by virtue of their similar basic leucine zipper domains, and the function of Cbf1 is to tether the Met4 and Met28 factors at promoter elements that contain CDEI elements and adjacent sequences for binding the Met4-Met28 heterodimers (23, 39). Elements that reveal a reasonably conserved consensus sequence <u>CACGTG</u> AAATNA (CDEI element underlined) are found upstream of all six MET structural genes (39). The adjacent sequence (AAATNA) may be significant in distinguishing the CDEI elements that serve as Cbf1 binding sites in the centromeres from those that allow specific binding of the Cbf1-Met4-Met28 complex upstream of the Met genes. Cbf1-dependent changes in the chromatin structure of the MET16 locus are restricted to the immediate region of the CDEI element, indicating that Cbf1 does not act by phasing nucleosomes in this region but, rather, stabilizes the DNA binding of Met4 and Met28 (39).

General Amino Acid Control of MET16

Under certain metabolic conditions, namely during amino acid starvation, the yeast MET16 gene, which encodes PAPS reductase, is expressed in *met4* mutants (39). This Met4-independent transcription of MET16 occurs during tryptophan starvation, elicited by the analog 5-methyl-tryptophan, and requires Gcn4, the transcription factor that mediates general amino acid control. Thus, MET16, whose promoter contains a Gcn4 binding element, can be activated by general control independently of the sulfur-specific activation mediated by Met4 (39).

These various studies provide considerable new information and understanding of the molecular mechanism underlying sulfur regulation in yeast. It will be important to learn whether the synthesis of any of the regulatory proteins, Met4, Met28, and Met30, is strongly controlled, whether they are subject to post-translational modification, and what metabolic conditions might lead to their interconversion between inactive and active states. It will likewise be of interest to determine why the MET30 gene is essential for yeast viability. Another subject of major importance is to identify the factor that is responsible for detecting repressing levels of SAM and analysis of the signal pathway that leads to the precise control of the entire set of sulfur-regulated genes.

CONCLUSION

Although many significant questions remain to be addressed, the recent breakthroughs in our understanding of the operation of the sulfur regulatory circuit in *A. nidulans*, *N. crassa*, and *S. cerevisiae* represent major advances. Regulation occurs primarily at the level of transcription in each of these fungi. Important similarities, but also significant differences, appear in the manner in which sulfur regulation is mediated in these closely related organisms. Positive trans-acting sulfur control factors of the bZip protein family occur in *Neurospora* and yeast, and negative-acting sulfur regulatory proteins of the ß-transducin family are found in all three fungi. The presence of several additional sulfur regulatory genes, identified by mutations, suggests that a multi-component signal transduction system controls sulfur metabolism in *Aspergillus* and in *Neurospora*. Additional work in each of these fungi will be critical in order to compare their mode of sulfur regulation and to illuminate the precise molecular mechanisms and multiple interactions that govern this important area of cellular metabolism.

ACKNOWLEDGMENTS

Research in the author's laboratory has been supported by grant GM-23367 from the National Institutes of Health. I thank Andrzej Paszewski, John Paietta, Yolande Surdin-Kerjan, Irwin Segel, Geoff Turner, Kristin Coulter, and Ying Tao for sharing their ideas and/or unpublished results with me. I would like to acknowledge the pioneering work in sulfur regulation carried out by Robert Metzenberg, whose continuous contributions and insights have been an inspiration to all in the field.

> Visit the *Annual Reviews home page* at
> http://www.annurev.org.

Literature Cited

1. Arst HN. 1968. Genetic analysis of the first steps of sulphate metabolism in Aspergillus nidulans. *Nature* 219:268–70
2. Arst HN. 1996. Regulation of gene expression by pH. In *The Mycota: Biochemistry and Molecular Biology*, ed. R Brambl, GA Marzluf, pp. 235–40. Berlin: Springer-Verlag
3. Arst HN, Scazzocchio C. 1985. Formal genetics and molecular biology of the control of gene expression in *Aspergillus nidulans*. In *Gene Manipulations in Fungi*, ed. J Bennett, L Lasure, pp. 309–43. Orlando: Academic
4. Bissig M, Hagenbuch B, Stieger B, Koller T, Meier PJ. 1994. Functional expression cloning of the canalicular sulfate transport system of rat hepatocytes. *J. Biol. Chem.* 269:3017–21
5. Burton EG, Metzenberg RL. 1972. Novel mutation causing derepression of several enzymes of sulfur metabolism in Neu-

rospora crassa. *J. Bacteriol.* 109:140–50
6. Evers M, Saftig P, Schmidt P, Hafner A, McLoghlin DB, et al. 1996. Targeted disruption of the arylsulfatase β gene results in mice resembling the phenotype of *mucopolysaccharidosis VI*. *Proc. Natl. Acad. Sci. USA* 93:8214–19
7. Flavin M, Slaughter C. 1964. Cystathionine cleavage enzymes of *Neurospora*. *J. Biol. Chem.* 239:2212–19
8. Flavin M, Slaughter C. 1967. The derepression and function of enzymes of reverse trans-sulfuration in *Neurospora*. *Biochim. Biophys. Acta* 132:406–11
9. Foster BA, Thomas SM, Mahr JA, Renosto F, Patel HC, et al. 1994. Cloning and sequencing of ATP sulfurylase from *Penicillium chrysogenum*. *J. Biol. Chem.* 269:19777–86
10. Franco B, Meroni G, Parenti G, Levilliers J, Bernard L, et al. 1995. A cluster of sulfatase genes on Xp22.3: Mutations

in *chondrodysplasia punctata (CDPX)* and implications for warfarin embryopathy. *Cell* 81:15–25

11. Fu YH, Marzluf GA. 1990. Cys-3, the positive-acting sulfur regulatory gene of *Neurospora crassa*, encodes a sequence-specific DNA binding protein. *J. Biol. Chem.* 265:11942–47

12. Fu YH, Paietta JV, Mannix DG, Marzluf GA. 1989. Cys-3, the positive-acting sulfur regulatory gene of *Neurospora crassa*, encodes a protein with a putative leucine zipper DNA-binding element. *Mol. Cell. Biol.* 9:1120–27

13. Hanson MA, Marzluf GA. 1973. Regulation of a sulfur-controlled protease in *Neurospora crassa*. *J. Bacteriol.* 116:785–89

14. Hanson MA, Marzluf GA. 1975. Control of the synthesis of a single enzyme by multiple regulatory circuits in *Neurospora crassa*. *Proc. Natl. Acad. Sci. USA* 72:1240–44

15. Hastbacka J, Chapelle A, Mahtani MM, Clines G, Hamilton BA, et al. 1994. The diastrophic dysplasia gene encodes a novel sulfate transporter: positional cloning by fine-structure linkage disequilibrium mapping. *Cell* 78:1073–87

16. Jacobson ES, Metzenberg RL. 1977. Control of arylsulfatase in a serine auxotroph of *Neurospora*. *J. Bacteriol.* 130:1397–98

17. Jarai G, Marzluf GA. 1991. Sulfate transport in *Neurospora crassa*: Regulation, turnover, and cellular localization of the cys-14 protein. *Biochemistry* 30:4768–73

18. Kanaan M, Fu YH, Marzluf GA. 1992. The DNA-binding domain of the CYS3 regulatory protein of *Neurospora crassa* is bipartite. *Biochemistry* 31:3197–203

19. Kanaan M, Marzluf GA. 1991. Mutational analysis of the DNA-binding domain of the CYS3 regulatory protein of *Neurospora crassa*. *Mol. Cell. Biol.* 11:4356–62

20. Kanaan MN, Marzluf GA. 1993. The positive-acting sulfur regulatory protein CYS3 of *Neurospora crassa*: nuclear localization, autogenous control, and regions required for transcriptional activation. *Mol. Gen. Genet.* 239:334–44

21. Ketter JS, Jarai G, Fu YH, Marzluf GA. 1991. Nucleotide sequence, messenger RNA stability, and DNA recognition elements of Cys–14, the structural gene for sulfate permease II in *Neurospora crassa*. *Biochemistry* 30:1780–87

22. Kumar A, Paietta JV. 1995. The sulfur controller-2 negative regulatory gene of *Neurospora crassa* encodes a protein with β-transducin repeats. *Proc. Natl. Acad. Sci. USA* 92:3343–47

23. Kuras L, Cherest H, Surdin-Kerjan Y, Thomas D. 1996. A heteromeric complex containing the centromere binding factor 1 and two basic leucine zipper factors, Met4 and Met28, mediates the transcription activation of yeast sulfur metabolism. *EMBO J.* 15:2519–29

24. Kuras L, Thomas D. 1995. Functional analysis of Met4, a yeast transcriptional activator responsive to S-adenosylmethionine. *Mol. Cell. Biol.* 15:208–16

25. Lewandowski I, Balinska M, Natorff R, Paszewski A. 1996. Regulation of folate-dependent enzyme levels in *Aspergillus nidulans*: studies with regulatory mutants. *Biochim. Biophys. Acta* 1290:89–94

26. Li Q, Marzluf GA. 1996. Determination of the *Neurospora crassa* CYS3 sulfur regulatory protein consensus DNA-binding site: amino-acid substitutions in the CYS3 bZip domain that alter DNA-binding specificity. *Curr. Genet.* 30:298–304

27. Li Q, Zhou L, Marzluf GA. 1996. Functional in vivo studies of the *Neurospora crassa* cys-14 gene upstream region: importance of CYS3-binding sites for regulated expression. *Mol. Microbiol.* 22:109–17

28. Logan HM, Cathala N, Grignon C, Davidian JC. 1996. Cloning of a cDNA encoded by a member of the *Arabidopsis thaliana* ATP sulfurylase multigene family. *J. Biol. Chem.* 271:12227–33

29. Martin RL, Daley LA, Lovric Z, Wailes LM, Renosto F, et al. 1989. The "regulatory" sulfhydryl group of *Penicillium chrysogenum* ATP sulfurylase. *J. Biol. Chem.* 264:11768–75

30. Marzluf GA. 1970. Genetic and biochemical studies of distinct sulfate permease species in different developmental stages of *Neurospora crassa*. *Arch. Biochem. Biophys.* 138:254–63

31. Marzluf GA. 1974. Uptake and efflux of sulfate in *Neurospora crassa*. *Biochim. Biophys. Acta* 339:374–81

32. Marzluf GA. 1981. Regulation of nitrogen metabolism and gene expression in fungi. *Microbiol. Rev.* 45:437–61

33. Marzluf GA. 1993. Regulation of sulfur and nitrogen metabolism in filamentous fungi. *Annu. Rev. Microbiol.* 47:31–55

34. Marzluf GA. 1994. Genetics and molecular genetics of sulfur assimilation in the fungi. *Adv. Genet.* 31:187–206

35. Marzluf GA, Li Q, Coulter K. 1995. Global regulation of sulfur assimilation in *Neurospora*. *Can. J. Bot.* 73:S167–72

36. Marzluf GA, Metzenberg RL. 1968. Positive control by the CYS3 locus in regulation of sulfur metabolism in *Neurospora*. *J. Mol. Biol.* 33:423–37

37. Metzenberg RL, Parson JW. 1966. Altered repression of some enzymes of sulfur utilization in a temperature-conditional lethal mutant of *Neurospora. Proc. Natl. Acad. Sci. USA* 55:629–35

38. Natorff R, Balinska M, Paszewski A. 1993. At least four regulatory genes control sulphur metabolite repression in *Aspergillus nidulans. Mol. Gen. Genet.* 238:185–92

39. O'Connell KF, Surdin-Kerjan Y, Baker RE. 1995. Role of the *Saccharomyces cerevisiae* general regulatory factor CP1 in methionine biosynthetic gene transcription. *Mol. Cell. Biol.* 15:1879–88

40. Paietta JV. 1989. Molecular cloning and regulatory analysis of the arylsulfatase structural gene of *Neurospora crassa. Mol. Cell. Biol.* 9:3630–37

41. Paietta JV. 1990. Molecular cloning and analysis of the scon-2 negative regulatory gene of *Neurospora crassa. Mol. Cell. Biol.* 10:5207–14

42. Paietta JV. 1992. Production of the CYS3 regulatory, a bZIP DNA-binding protein, is sufficient to induce sulfur gene expression in *Neurospora crassa. Mol. Cell. Biol.* 12:1568–77

43. Paietta JV. 1995. Analysis of CYS3 regulatory function in *Neurospora crassa* by modification of leucine zipper dimerization specificity. *Nucleic Acids Res.* 23:1044–49

44. Paszewski A, Grabski J. 1974. Regulation of S-amino acids biosynthesis in *Aspergillus nidulans. Mol. Gen. Genet.* 132:307–20

45. Renosto F, Martin RL, Wailes LM, Daley LA, Segel IH. 1990. Regulation of inorganic sulfate activation in filamentous fungi. *J. Biol. Chem.* 265:10300–8

46. Sandal NN, Marcker KA. 1994. Similarities between a soybean nodulin, *Neurospora crassa* sulphate permease II and a putative human tumour suppressor. *Trends Biochem. Sci.* 19:19

47. Smith FW, Ealing PM, Hawkesford MJ, Clarkson DT. 1995. Plant members of a family of sulfate transporters reveal functional subtypes. *Proc. Natl. Acad. Sci. USA* 92:9373–77

48. Smith FW, Hawkesford MJ, Prosser IM, Clarkson DT. 1995. Isolation of a cDNA from *Saccharomyces cerevisiae* that encodes a high affinity sulphate transporter at the plasma membrane. *Mol. Gen. Genet.* 247:709–15

49. Thomas D, Jacquemin I, Surdin-Kerjan YH. 1992. MET4, a leucine zipper protein, and centromere-binding factor 1 are both required for transcriptional activation of sulfur metabolism in *Saccharomyces cerevisiae. Mol. Cell. Biol.* 12:1719–27

50. Thomas D, Kuras L, Barbey R, Cherest H, Blaiseau P, et al. 1995. Met30p, a yeast transcriptional inhibitor that responds to S-adenosylmethionine, is an essential protein with WD40 repeats. *Mol. Cell. Biol.* 15:6526–34

Annu. Rev. Microbiol. 1997. 51:97–123

HEMOGLOBIN METABOLISM IN THE MALARIA PARASITE *PLASMODIUM FALCIPARUM*

Susan E. Francis, David J. Sullivan, Jr., and Daniel E. Goldberg

Howard Hughes Medical Institute, Departments of Molecular Microbiology and
Medicine and Barnes-Jewish Hospital, St. Louis, Missouri 63110;
e-mail: goldberg@borcim.wustl.edu

KEY WORDS: protease, plasmepsin, falcipain, heme, quinoline

ABSTRACT

Hemoglobin degradation in intraerythrocytic malaria parasites is a vast process that occurs in an acidic digestive vacuole. Proteases that participate in this catabolic pathway have been defined. Studies of protease biosynthesis have revealed unusual targeting and activation mechanisms. Oxygen radicals and heme are released during proteolysis and must be detoxified by dismutation and polymerization, respectively. The quinoline antimalarials appear to act by preventing sequestration of this toxic heme. Understanding the disposition of hemoglobin has allowed identification of essential processes and metabolic weakpoints that can be exploited to combat this scourge of mankind.

CONTENTS

97

0066-4227/97/1001-0097$08.00

INTRODUCTION

Despite intensive efforts at eradication, malaria remains a major public health problem. The World Health Organization (WHO) estimates that 300–500 million people are afflicted each year (96, 153). Malaria is transmitted to people throughout tropical and subtropical areas, where 40% of the world's population is at risk of infection. Eighty percent of malaria cases worldwide occur in Africa (152), where this disease claims the lives of one to two million children yearly. Vector control was responsible for largely eliminating malaria from North America, but these measures have not been sustainable in most other parts of the world. Partial immunity against malaria is seen in endemic areas and has given rise to the hope that an effective vaccine can be developed. So far, none is available. Thus, the principal means of disease prevention continues to be chemoprophylaxis. Unfortunately, parasite resistance to the widely used quinolines and antifolates is now common (20, 23, 94, 102, 151). Multi-drug-resistant parasites are already a problem, and it is feared that their spread could have disastrous consequences (96). This realization has intensified the efforts to identify new drugs and new drug targets.

Malaria is caused by several species of obligate intracellular protozoa from the genus *Plasmodium. P. falciparum* is the most deadly of these. The parasites have a complicated life cycle that requires a vertebrate host for the asexual cycle and a female Anopheles mosquito for completion of the sexual cycle. During a mosquito blood meal, infectious sporozoites in the mosquito's saliva enter the host bloodstream and invade its hepatocytes. This liver stage is asymptomatic. Parasites proliferate in the hepatocytes, producing thousands of merozoites that are released into the bloodstream. The asexual erythrocytic cycle begins when a single merozoite invades a host red blood cell and is enclosed within a parasitophorous vacuole, separate from the host cell cytoplasm. Three morphologically distinct phases are then observed. The ring stage, lasting approximately 24 h in *P. falciparum*, accounts for about half of the intraerythrocytic cycle, but it is metabolically nondescript (50). It is followed by the trophozoite stage, a very active period during which most of the red blood cell cytoplasm is consumed. Finally, parasites undergo 4–5 rounds of binary divisions during the schizont stage, producing merozoites that burst from the host cell to invade new erythrocytes, beginning another round of infection. The clinical manifestations of malaria result from schizont rupture and additionally, in the case of *P. falciparum*, from trophozoite adherence to endothelial cells (90).

THE ROLE OF HEMOGLOBIN DEGRADATION

Quantitation

Hemoglobin comprises 95% of the cytosolic protein of the red blood cell, where it is present at a concentration of 5 mM. During the intraerythrocytic cycle, the host cell cytoplasm is consumed and an estimated 60–80% of the hemoglobin is degraded (10, 92, 98). A number of lower estimates have also been reported; these most likely reflect assessment of less mature parasites still in the process of hemoglobin consumption. Hemoglobin proteolysis releases heme and generates amino acids. The heme moiety does not appear to be metabolized or recycled (39), but instead, is stored as an inert polymer known as the malaria pigment hemozoin (123).

Utilization

Amino acids derived from globin hydrolysis are incorporated into parasite proteins (124, 140) and also appear to be available for energy metabolism (123). Hemoglobin proteolysis may be essential for survival, because *Plasmodium* has a limited capacity for de novo amino acid synthesis (123, 141). However, hemoglobin degradation alone appears insufficient for the parasite's metabolic needs since it is a poor source of methionine, cysteine, glutamine, and glutamate and contains no isoleucine at all. *P. falciparum* can be maintained in a red blood cell culture system provided that the five amino acids limiting in hemoglobin are present in the medium (36, 48). Thus, the parasite appears able to rely on both hemoglobin proteolysis and exogenous amino acids for growth. Amino acids are readily taken up by parasites from the culture medium (34, 82, 123). This is true for amino acids that are missing from hemoglobin as well as for those that are abundant, and it suggests that the reliance on hemoglobin degradation may reflect a limited availability of certain amino acids in the host serum.

When *Plasmodium* is cultured in a rich medium containing 20 amino acids, the extent of hemoglobin proteolysis is similar to that seen in vivo. Interestingly, more amino acids are generated from hemoglobin breakdown than are used for protein synthesis, resulting in the diffusion of some hemoglobin-derived amino acids into the host cell (160). This raises the possibility that hemoglobin catabolism may have additional functions. One hypothesis is that parasites digest host cell cytosol to prevent premature red cell lysis that might occur if parasite growth were not compensated by reduced host cell volume (50, 160).

Perhaps the strongest argument that hemoglobin degradation is necessary for survival comes from studies with protease inhibitors. When hemoglobin proteolysis is blocked, parasite development is interrupted (48, 76, 107, 109, 113, 115). In addition, when parasites are cultured in a medium lacking most amino acids, they show increased sensitivity to a protease inhibitor known to block hemoglobin proteolysis (48).

HEMOGLOBIN INGESTION

Cytostome

Hemoglobin ingestion during the ring (also known as young trophozoite) stage of development appears to be limited. However, hemozoin can be detected in early stage parasites (4, 98), which suggests that the cellular machinery for ingestion and proteolysis is present. Ultrastructural studies of several species of *Plasmodium* are consistent with this finding (117, 130, 131). These studies reveal that, early in development, small portions of cytoplasm are taken up by micropinocytosis. Vesicles containing a tiny hemozoin crystal can sometimes be seen. As the parasite matures, a larger volume of hemoglobin is ingested by means of a cytostomal system that is formed by invagination of the parasitophorous vacuolar membrane and the parasite plasma membrane. The cytostome is a poorly understood structure spanning these two membranes that separate parasite and erythrocyte cytoplasms (Figure 1). In *P. falciparum*, the cytostome is a large double membrane-enclosed pear-shaped structure (73, 95). In some instances, several cytostomes can be seen in one trophozoite-stage parasite (73, 130). When the cytostomes ingest red cell cytoplasm, double membrane-delimited vesicles are formed by budding. The hemozoin-containing vesicles appear to fuse, forming one or two large, single membrane-enclosed digestive vacuoles that contain a cluster of hemozoin crystals (130) (Figure 1). Variation in cytostome morphology and activity occurs between species (1, 133, 134). In murine plasmodia, a single branching tubular structure appears to fill with hemoglobin and break down into vesicles. Hemozoin accumulates in the vesicles as the parasite matures (132).

Digestive Vacuole

Once the digestive vacuole is formed, it appears to be the primary site of hemoglobin degradation (97). Vesicles budding from the cytostome transport

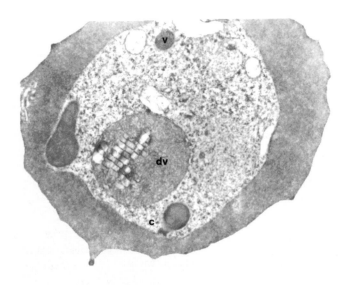

Figure 1   Transmission electron micrograph of a *Plasmodium* falciparum trophozoite inside an erythrocyte. c, cytostome; v, transport vesicle; dv, digestive vacuole. (After Reference 55.)

hemoglobin to the vacuole where their outer membrane fuses with the single membrane–delimited vacuole. Fusion releases a single membrane–enclosed vesicle that can occasionally be seen inside the vacuole (95, 157). The vesicle is then lysed and hemoglobin is hydrolyzed. The agent responsible for vesicular lysis has not been identified, but a phospholipase activity has been hypothesized (51, 71). Whatever the identity of the lytic agent, its activity appears to be highly specific: It is able to distinguish between the vesicular and vacuolar membranes. When parasites are cultured in the presence of chloroquine, unlysed transport vesicles are more abundant (95, 157, 158).

Digestive vacuoles in *P. falciparum* are acidic organelles with a pH estimated at 5.0–5.4 (70, 156). An ATPase activity that could be responsible for the proton gradient was detected in partially purified digestive vacuoles (24), and genes for the proton pump subunits VAP-A and VAP-B have been cloned (66, 67). Both subunits are expressed throughout the intraerythrocytic cycle. Immunolocalization studies of the vacuolar B subunit indicate that it is associated with the digestive vacuole but that it is also distributed over most of the parasite (67). The degradative capacity and acidic pH of digestive vacuoles are characteristic of lysosomes and yeast vacuoles. Acid phosphatase activity typically found in these organelles was associated with hemozoin in the digestive vacuoles of the lower vertebrate parasites *P. berghei* and *P. gallinaceum* (2). In *P. falciparum*, acid phosphatase was found in endocytic vesicles but not in digestive vacuoles (132). β-glucuronidase, β-galactosidase and acid phosphatase activities were absent from digestive vacuoles that were isolated from *P. falciparum* (55). The lack of nonproteolytic acid hydrolases in digestive vacuoles suggests that, during their relatively short intraerythrocytic cycle, parasites may not need to degrade and recycle macromolecules other than hemoglobin and that digestive vacuoles may be specialized for that purpose (55).

HEMOGLOBIN DEGRADATION

Proteases

In *P. falciparum*, some hemoglobin degradation is seen during the ring and early schizont stages of development, but the vast majority of degradation occurs during the 6–12 h trophozoite stage (159). Acid proteases present in parasitized red blood cell extracts or partially purified parasite extracts from different species of *Plasmodium* have been proposed to act as hemoglobinases (3, 9, 58, 59, 74, 75, 119, 125, 144, 145). Some of these were probably derived from the red blood cell or parasite cytoplasm (53, 120, 122). In addition, a trophozoite-stage cysteine protease of 28 kD was observed in cell extracts from *P. falciparum* (112) (see *Falcipain Specificity and Inhibition* section, below).

Enzymes that are present in the digestive vacuole and that are capable of hemoglobin degradation were identified by subcellular fractionation (55). The dense iron-rich hemozoin crystals that are present in digestive vacuoles allowed differential centrifugation and Percoll density gradient separation for purification of these organelles. Marker enzyme analysis and electron microscopy indicated that the digestive vacuoles are not contaminated by other membranes and compartments (55). When the digestive vacuole lysate was added to denatured globin and incubated at acid pH, both aspartic protease and cysteine protease activities were detected. Aspartic proteases account for 60–80%, and cysteine protease(s) account for 20–40% of the globin-degrading activity in purified digestive vacuoles (47, 52, 55). The combination of pepstatin, a specific aspartic protease inhibitor, and E-64, a specific cysteine protease inhibitor, completely blocks this globin digestion, suggesting that aspartic and cysteine proteases are the primary enzymes responsible for globin proteolysis in the vacuole (52).

Order of Action

When native hemoglobin is incubated with digestive vacuole lysate, the disappearance of substrate can be analyzed by SDS-PAGE. Hemoglobin digestion is inhibited when pepstatin is added to such incubations, whereas the effect of E-64 is minimal (52, 55). The data suggest that the process of hemoglobin breakdown is ordered, and that it requires an aspartic protease-mediated initial cleavage, followed by secondary aspartic protease and cysteine protease cleavages. Other enzymes may participate in the breakdown of globin fragments, but this has not been shown. The combined activity of vacuolar hydrolases is expected to produce progressively smaller peptides. Exopeptidases could then generate free amino acids, as is the case in lysosomes and yeast food vacuoles (19, 65, 68). However, in contrast to the activity seen in mammalian and yeast degradative organelles, no exopeptidase activity has been described from *P. falciparum* digestive vacuoles. Neutral pH-requiring aminopeptidases have been identified in trophozoite extract prepared from *P. falciparum* (19, 27, 65, 68) and rodent *Plasmodium* (22, 27). This suggests that either the aminopeptidase has a neutral pH optimum and still functions in the vacuole, that other unidentified peptidases exist in the vacuole, or that the final steps in the hemoglobin proteolysis pathway occur in the cytoplasm. Recent data suggest that vacuolar degradation generates small peptides but no free amino acids, which supports this last mechanism (K Kolakovich et al, unpublished).

Plasmepsin Specificity and Inhibition

Two aspartic proteases, plasmepsins I and II, and one cysteine protease, falcipain, have been purified from digestive vacuoles of *P. falciparum* (52, 54). All three proteases have pH optima near 5, the physiological pH of the vacuole. Plasmepsins I and II were shown by mass spectroscopic analysis to cleave native hemoglobin with remarkable specificity between α33Phe-34Leu (52, 54). This is also the site of initial cleavage by digestive vacuole extracts. The α33–34 bond is in the hinge region of the hemoglobin molecule, the domain responsible for holding the tetramer together when oxygen is bound (101). Cleavage at this site can be envisioned to unravel hemoglobin and facilitate its proteolysis by other degradative enzymes. This region is highly conserved among vertebrate species (35). No homozygous mutations have been mapped to this position (15). Thus, it appears that the parasite has selected a particularly vulnerable site for its initial attack (54).

When the tertiary structure of hemoglobin is relaxed, plasmepsins I and II can cleave at other sites (52, 54). The enzymes have divergent substrate specificities for these secondary cleavages. Plasmepsin I prefers phenylalanine at the P1 position, while plasmepsin II prefers hydrophobic residues on both sides of the scissile bond, especially leucine at the P1' position. It is not entirely clear why the parasite should have two distinct enzymes with apparently similar functions. Interestingly, plasmepsins I and II have completely different patterns of expression during the intraerythrocytic cycle (49). Plasmepsin I is synthesized during the early ring stage, and its synthesis continues through the trophozoite stage, whereas plasmepsin II is almost undetectable in ring stage parasites but is abundant during the trophozoite stage. These observations suggest that only plasmepsin I is involved in the limited hemoglobin proteolysis that is detected during the ring stage. During the trophozoite stage, when most of the host cell cytoplasm is consumed, both plasmepsins, with their differing substrate specificities, may be required to expedite this massive catabolic event.

The plasmepsins also differ by two orders of magnitude in their sensitivity to the peptidomimetic inhibitor, SC-50083 (48). When SC-50083 is added to *P. falciparum* cultures at the late ring stage, hemoglobin degradation is blocked and the parasites are killed with an IC_{50} of 2–5 $\times$ 10^{-6} M. During treatment, parasites have markedly diminished hemozoin production and they appear pyknotic (48). These results indicate that blocking plasmepsin I activity is lethal to the parasite and that, although plasmepsin II is abundant during the trophozoite stage, it cannot compensate for the loss of plasmepsin I activity. The same result might be obtained if, in addition to hemoglobin degradation. plasmepsin I has some other unidentified role that plasmepsin II cannot fill. In that regard, it is interesting to note that plasmepsin I appears to be specifically adapted for hemoglobin proteolysis, and it has been shown to be inactive against a variety of nonglobin substrates (54), which makes other degradative functions in the parasite less likely. To date, no specific plasmepsin II inhibitors have been developed, so it is not known whether that enzyme is also essential. There is evidence that a third vacuolar aspartic protease might exist (146), but the activity has not been purified and cannot be judged to be distinct from plasmepsins I and II at this time.

Falcipain Specificity and Inhibition

A single cysteine protease activity is seen in lysates that are prepared from purified digestive vacuoles (52). The purified enzyme was characterized and was found to be indistinguishable from falcipain, the enzyme identified in trophozoite extract by gelatin substrate PAGE (112). Both proteins are approximately 28 kD and they have substrate specificity similar to cathepsin L. In addition, purified native enzyme has similar kinetics to those obtained with trophozoite extract (47, 115). Consistent with the ordered proteolysis notion, falcipain did not cleave hemoglobin unless this substrate was first denatured by reducing conditions, acid-acetone treatment, or cleavage by plasmepsins (47, 52). These data have suggested the hemoglobin degradation pathway that is shown in Figure 2.

When cysteine protease inhibitors are added to cultures of *P. falciparum* during their hemoglobin-degrading stage, a profound effect on parasite development is seen (37, 107, 109, 112, 113, 115). The digestive vacuoles swell, undigested hemoglobin accumulates, and parasite growth is impaired. The same changes are seen with different classes of cysteine protease inhibitors, indicating that the effect is specific. Hemoglobin accumulation likely occurs

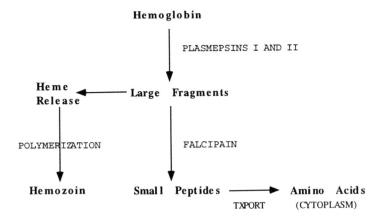

Figure 2   Proposed pathway for hemoglobin degradation in *P. falciparum.*

because of the indirect inhibition of proteolysis (47). It is hypothesized that the plasmepsins generate hemoglobin fragments that cannot be further catabolized without cysteine protease action. Osmotic swelling occurs, vacuolar functions are impaired, and parasite death ensues (47).

Plasmepsin Cloning and Expression

The genes for plasmepsins I and II have been cloned and characterized (30, 48). Their predicted coding regions share 73% amino acid identity. The mature protein sequences are 35% identical to the mammalian aspartic proteases renin and cathepsin D. Both genes predict 51-kD precursor proteins that are cleaved to form 37-kD active enzymes (Figure 3). The plasmepsin pro-pieces (123–124 amino acids) are longer than those of other aspartic proteases, which are 40–50 amino acids in length (137). In addition, 37 (plasmepsin I) or 38 (plasmepsin II) amino acids from the initiator methionine are a hydrophobic stretch that was predicted to be a signal anchor sequence (48).

Recombinant proplasmepsin II has been expressed in *E. coli* (61). Acidification results in the production of a 38-kD protein by autocatalytic cleavage (Figure 3), 12 amino acids upstream of the cleavage site that is deduced from the N-terminal sequence of native plasmepsin II (52, 61). Plasmepsin II does not autoactivate in vivo; however, this fortuitous autocatalytic cleavage produces active plasmepsin II with kinetic properties similar to those of native plasmepsin II (80).

Plasmepsin Structure

Recombinant plasmepsin II has been crystallized, complexed with the inhibitor pepstatin, and its structure solved at 2.7 Å resolution (126). Plasmepsin II has

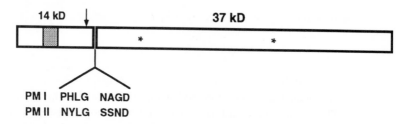

Figure 3 Diagram of proplasmepsin features. The proforms of both enzymes are 51-kD proteins. After a 4-kD cytoplasmic N-terminus, there is a 21 amino acid hydrophobic region (*shaded*). Both are processed into mature, 37-kD enzymes. Amino acid sequences just before and after the cleavage sites determined by N-terminal sequence analysis of mature plasmepsins I and II (PM I and II) are shown. *Asterisks* mark the active site aspartic acids and *the arrow* marks a site used by recombinant plasmepsin II for self-activation.

the bilobal shape that is typical of aspartic proteases. Catalysis is mediated by two aspartic acids found in the active site cleft formed by the two lobes. In contrast to most aspartic proteases, the sequence found at the second active site aspartate is Asp-Ser-Gly, instead of Asp-Thr-Gly. Asp-Ser-Gly is found in some retroviral aspartic proteases (105). Two disulfide bridges typical of mammalian aspartic proteases are found. A loop that forms a flap over the active site in mammalian aspartic proteases is well conserved in plasmepsin II, except that the typical Gly 79 is replaced by Val. This replacement is also found in plasmepsin I. The flap change, in conjunction with other active site substitutions, results in a different conformation of pepstatin in the active site of the plasmodial enzyme. This gives hope that a selective inhibitor can be developed that will target the malaria parasite and not its host. A model of plasmepsin I based on the plasmepsin II structure reveals only a few differences between the two homologs that could help explain the different specificities towards hemoglobin. The plasmepsin I active site appears to be slightly more open and has larger S3 and S3' pockets. The polyproline loop implicated in substrate binding in mammalian renin is quite variant between the two plasmepsins, though it does not make contact with pepstatin in the crystal structure. A hybrid plasmepsin II containing the polyproline loop that is substituted from plasmepsin I has the same kinetic properties and inhibitor sensitivity to SC-50083 as native plasmepsin II (80). This result suggests that this loop is not important for substrate recognition, and that it does not account for different substrate specificities of plasmepsins I and II.

The crystallographic asymmetric unit contained two plasmepsin II molecules in different conformations related by a rotation of 5° between the N- and C-terminal domains (126). This demonstrates an interdomain flexibility that

may explain how the plasmepsins can recognize and cleave their large, proteinaceous substrate hemoglobin.

Falcipain Cloning

A candidate gene for the vacuolar cysteine protease falcipain was cloned and
shown by Northern analysis to be expressed predominantly at the ring stage of
development (114), which is consistent with a proposed hemoglobin-degrading
function during the ring and trophozoite stages. It has not been possible to
confirm that this gene encodes falcipain, because an N-terminal sequence from
falcipain has not been obtained. However, only this single cysteine protease
gene was detectable by Northern analysis of early-stage RNA, making it likely
that gene and protein are a match (47). The deduced gene product encodes a
protein of 56 kD. Cleavage at or near Lys332 would produce a 28-kD enzyme
with 37% amino acid identity to cathepsin L. Homologous genes cloned from
P. vivax (114) and the murine species *P. vinckei* (108) also have exceptionally
long predicted pro-pieces. The predicted mature *P. vivax* and *P. vinckei* enzymes
are 71% and 56% identical to *P. falciparum*, respectively. Nucleotide sequence
encoding the cysteine protease active site was obtained from *P. malaria, P. ovale*,
the primate species *P. rechenowi, P. fragile*, and *P. cynomolgi*, the murine
species *P. berghei*, and the avian species *P. gallinaceum* (110). A number of
residues predicted to be in the active site were conserved among the species
examined.

Hemoglobin Degradation by Other Parasites

Several other parasites are candidates for having hemoglobin degradation systems. *Entameba histolytica*, the causative agent of amebiasis, and hookworm
species ingest erythrocytes, but hemoglobin breakdown has not been characterized. *Babesia*, an intraerythrocytic parasite related to *Plasmodium*, degrades
little if any hemoglobin (72, 97). Unlike *Plasmodium, Babesia* is not contained
within a parasitophorous vacuole. It lacks both cytostomes and digestive vacuoles, and hemozoin has not been detected (72).

Massive quantities of hemoglobin are degraded by blood flukes from the
genus *Schistosoma*, the causative agents of schistosomiasis, a chronic and debilitating disease which afflicts 200 million people in the tropics. Ingested
host erythrocytes are lysed in the gut of the schistosome, and hemoglobin is
degraded as a source of amino acids for parasite metabolism. Adult *S. mansoni*
secrete two forms of cathepsin L (29) as well as cathepsin B (21, 29, 77, 149).
The cathepsin L-like and cathepsin B-like activities have pH optima of 5.2 and
6.2 respectively (29). When cysteine protease inhibitors are added to cultured
S. mansoni, hemoglobin metabolism is blocked and the parasites die (149).

The roles of cathepsin B (Sm 31) and cathepsins L1 and L2 in hemoglobin proteolysis have not been differentiated. Sm32, an asparaginyl endopeptidase, was also identified as a hemoglobinase (31), but it has recently been proposed to participate in hemoglobin metabolism indirectly as a processing enzyme (28). In addition, there is some evidence for aspartic protease participation in hemoglobin degradation (17), but this enzyme has not been isolated and characterized.

Two cathepsin B–like genes (83) and two cathepsin L–like genes that are homologous to the *S. mansoni* genes have been described in *S. japonicum* (32). An aspartic protease gene has also been characterized (12). Both aspartic and cysteine protease activities are present in *S. japonicum* parasite extracts (12, 16, 17, 32), and both activities are proposed to function in hemoglobin proteolysis. Differences in the pH profiles for these enzymes suggest an ordered pathway for schistosome gut hemoglobin degradation in which the cysteine protease activity (pH optimum 5.5) makes the initial cleavage and, after acidification of the bloodmeal has proceeded further, the aspartic protease activity (pH optimum 3.5) is responsible for the bulk of hemoglobin degradation (12). This appears to be the opposite order of enzyme action from that of *P. falciparum.*

INTRACELLULAR TARGETING OF PROTEOLYTIC ENZYMES TO THE DIGESTIVE VACUOLE

Plasmepsin Biosynthesis

The signals that target proteins to the digestive vacuole are unknown. In addition, only a preliminary description of the route taken by vacuolar proteins is available. When observed by immunoEM, plasmepsin I was found to be most abundant in the digestive vacuole, but it was also observed at the parasite cell surface and throughout the hemoglobin ingestion pathway, which suggests that this protein follows an indirect route to the vacuole (48). While no studies of plasmepsin II trafficking are available, the plasmepsins are strikingly similar and may reasonably be predicted to follow the same route to the vacuole.

Biosynthesis studies of plasmepsins I and II indicate that they are processed identically and share several unusual features (49). In contrast to other known aspartic proteases, the plasmepsins are secretory molecules that are synthesized as type II integral membrane proteins. Processing to the soluble mature forms is rapid, occurring with a $t_{1/2}$ of about 20 min. Processing is blocked by lysosomotropic agents and by the proton pump inhibitor bafilomycin A1, which indicates that acid conditions are necessary for maturation. Cleavage of both proplasmepsins is reversibly inhibited by the tripeptide aldehydes ALLN and ALLM. These inhibitors have been shown to block cysteine proteases including calpains (118, 142), as well as the chymotryptic activity of proteasomes (148). However, proplasmepsin cleavage was not inhibited by a variety of other cysteine protease inhibitors or by the proteasome-specific inhibitor lactacystin. This pattern of inhibition suggests that a new class of enzyme may be responsible for the processing of plasmepsins. Similar data were obtained in vitro, using parasite extract to cleave radiolabelled proplasmepsins. Cleavage was shown to occur optimally between pH 4 and 5.5. The same pH profile is characteristic of the plasmepsins and of falcipain (52, 54, 112), and this is consistent with the activation of the plasmepsins in the digestive vacuole. Unlike for many acidic aspartic proteases (137), plasmepsin processing does not appear to be autocatalytic. The activity has been named proplasmepsin processing protease (PPPP) (49).

The following model (Figure 4) can be proposed for plasmepsin trafficking based on the combined immunoEM and biosynthetic data. First, translation of the proplasmepsins by ER-bound ribosomes generates integral membrane proteins with type II topologies. Next, the proplasmepsins probably transit the Golgi apparatus. This is somewhat speculative, because the Golgi is morphologically poorly defined in *Plasmodium* (42, 78) and the plasmepsins are not glycosylated (49). In fact, there has been some suggestion that the Golgi is missing altogether from ring and trophozoite stage parasites (11), as is the case for some stages of *Giardia lamblia* (79). The proplasmepsins then somehow

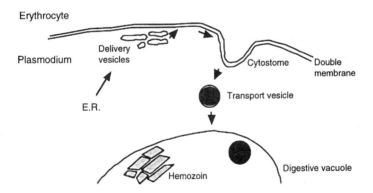

Figure 4   Proposed plasmepsin targeting pathway.   After translation as type II integral membrane proteins in the endoplasmic reticulum, the proplasmepsins transit the secretory pathway to the parasite surface. There they accumulate at the cytostome and are internalized with the substrate hemoglobin. Transport vesicles bring the proplasmepsins to the digestive vacuole, where the vesicle outer membrane fuses and releases the inner membrane–delimited vesicle into the digestive vacuole. Acidic conditions allow PPPP to cleave the proplasmepsins into the lumen, resulting in active mature plasmepsins ready to degrade hemoglobin. (From Reference 48.)

traverse the parasite plasma membrane to the parasitophorous vacuolar membrane and accumulate at the cytostome. This step is also mysterious. Type II orientation of the proplasmepsins could be predicted to result in proplasmepsins that extend into the parasitophorous vacuole rather than into the red blood cell cytoplasm, if *Plasmodium* secretion follows the alternating fusion and vesiculation model that has been proposed (63). At the cytostome, they intersect the hemoglobin ingestion pathway and are transported along with their substrate to the digestive vacuole. Finally, less than half an hour after synthesis, the proplasmepsins arrive at the digestive vacuole and are cleaved by the acid hydrolase PPPP. The proposed vacuolar location of PPPP and the deduced proplasmepsin transit time are based solely on the pH optimum of PPPP.

Utilization of the hemoglobin ingestion pathway for delivery of soluble proteins to the digestive vacuole has been described for the histidine-rich protein HRP II from *P. falciparum* (135), and there are hints that vacuolar membrane proteins may also first traffic to the membranes at the surface of the parasite. Pgh1 and VAP-B are integral membrane proteins which have both been localized to the digestive vacuole of *P. falciparum* (26, 67). In addition to having a vacuolar address, these proteins also appear to be associated with parasite membranes at the parasite cell surface.

Falcipain Biosynthesis

Falcipain biosynthesis has not been characterized. However, as with the plasmepsins, falcipain has an unusually long pro-piece. Falcipain sequences derived from *P. falciparum*, *P. vinckei*, and *P. vivax* predict that a hydrophobic stretch of 20 amino acids, long enough to span the membrane, exists at a point 35 amino acids downstream from the falcipain's deduced initiator methionine (108, 111, 114). It will be interesting to determine whether the hydrophobic stretch that is found in all falcipains that have been cloned, to date, serves as an internal signal sequence or as a signal anchor sequence. The cysteine protease from the related Apicomplexan parasite *Theileria parva* also has a similarly positioned hydrophobic domain (93).

OXIDATIVE STRESS GENERATED BY HEMOGLOBIN DEGRADATION

In the erythrocyte, heme iron is almost entirely in the ferrous (+2) state. Upon degradation in the parasite digestive vacuole, the iron is oxidized to the ferric (+3) state. Electrons liberated by oxidation combine with molecular oxygen to produce reactive oxygen species: superoxide anions (O_2^-), hydroxyl radicals (OH $\cdot$), and hydrogen peroxide (H_2O_2) (7). This appears to be a major source of oxidative stress to the parasite. Oxidative damage to the digestive vacuole is prevented by the action of superoxide dismutase (SOD) which converts superoxide radicals to hydrogen peroxide. Hydrogen peroxide is then cleaved by catalase. Both SOD and catalase are found in the digestive vacuole (43, 44, 47), and they are likely obtained from the host when erythrocyte cytoplasm is ingested. In addition, a parasite-derived SOD gene has been described from *P. falciparum* (13). Its role as a vacuolar antioxidant is uncertain, because the enzyme that is encoded by the cloned gene has not been localized. Reduced glutathione, another erythrocyte antioxidant, is present in red blood cell cytoplasm at 2.2 mM (154) and may also provide some relief from oxidative stress after ingestion.

Reactive oxygen species that are generated by the parasite during hemoglobin degradation have been proposed to be responsible for further damage to the host cell by creating oxidative stress (147). Reactive oxygen species that originate in the parasite have been detected in infected erythrocyte cytoplasm (7). Their role in malarial pathogenesis is not known. Interestingly, conditions which predispose erythrocytes to oxidative stress (like glucose 6-phosphate dehydrogenase deficiency) partially protect the host against infection by *P. falciparum* (64). This is possibly because parasite-derived oxidants overwhelm the host's antioxidant system, resulting in damage to the infected cell and in clearance by the reticuloendothelial system (64, 143).

HEME POLYMERIZATION

Proteolysis of hemoglobin releases the heme moiety. The reactive free heme is detoxified by polymerization into an inert crystalline substance called malaria pigment or hemozoin. This process is unique to *Plasmodium*. The pigment is visible microscopically in stages that are actively degrading hemoglobin, such as trophozoites, schizonts, and gametocytes. Ronald Ross found "granules of black pigment absolutely identical in appearance with the well known and characteristic pigment of the parasite of malaria" in cells outside the stomach of the Anopheles mosquito (116). This enabled him to implicate the mosquito in *Plasmodium* transmission, completing the parasite life cycle. In humans, residual malarial pigment from lysed schizonts accumulates in the reticuloendothelial system, turning the liver and spleen black in cases of chronic or repeated infection (40). Heme polymerization appears to be the target of quinoline antimalarial agents (see *Drugs That May Act on Heme Released During Hemoglobin Degradation* section, below).

Function of Polymer

During hemoglobin degradation, a vast quantity of heme accumulates in the 4 fL digestive vacuole. Simple calculation estimates a vacuolar heme concentration of about 0.4 M! Free micromolar heme can damage cellular metabolism by the inhibition of enzymes (52, 144), the peroxidization of membranes (138), and the production of oxidative free radicals in the acidic environment of the digestive vacuole (7). Lacking the heme oxygenase that vertebrates use for heme catabolism, plasmodial species sequester this toxic by-product into a chemically inert crystal.

An additional function that has been ascribed to hemozoin is the impairment of leukocyte action. Hemozoin is toxic to the phagocytes that scavenge residual hemozoin after rupture of the red blood cell by the parasite. Positive and negative regulation of cytokine secretion has been reported. Monocytes cocultured with pigment-containing parasites, or with crudely isolated pigments, initially release tumor necrosis factor-α and interleukin-1β (104). Protease treatment of isolated pigment abolishes this early production of cytokines. In contrast to the results with crudely isolated pigments, both β-hematin that is chemically synthesized and devoid of protein and monomeric heme inhibit nitric oxide and tumor necrosis factor-α production of lipopolysaccharide-activated macrophages, while microglial cell lines are unaffected (139). Additional experiments showed impairment of monocyte phagocytosis and oxidative burst (121). Sucrose-purified malarial pigment was also found to interfere with macrophage accessory cell function (91). In hypoendemic areas, the degree of pigment-containing neutrophils but not of pigment-laden monocytes predicted fatal outcome with sensitivity and a specificity greater than 73% (103). In hyperendemic areas, pigmented neutrophils or monocytes did not correlate with outcome (89).

Structure of Polymer

Hemozoin is insoluble in DMSO, detergent, and weak base solutions that solubilize monomeric heme (33). Insoluble hemozoin can be converted to free heme in 10 mM NaOH solution (100). Different proteins have been found to be associated with hemozoin, depending on the extraction method utilized to purify the parasite pigment (6, 56). Detergent extraction and proteolysis, which strips the hemozoin of protein, does not change the solubility characteristics (46). Elemental analysis of pigment devoid of protein is identical to that of heme (46, 129). Infrared, ESR, and extended X-ray absorption fine structure spectroscopy demonstrate that hemozoin is a heme polymer with a coordinate bond from the central ferric iron of one heme to the propionate carboxylate group of the next heme (Figure 5). This structure is identical to β-hematin, which is synthesized from heme incubated under supraphysiological salt and temperature conditions (129). Electron microscopy of crystals shows regular fibrils, which provides evidence for hydrogen bonding between straight chains (18). The polymer is also paramagnetic due to high spin iron (45). The

Figure 5   Proposed structure of hemozoin. The propionate group of one heme is linked to the central ferric iron of the next heme in an unusual iron-carboxylate linkage.

degree of branching and the number of monomers in the polymer has yet to be determined.

Process of Polymerization

Polymers of heme do not form spontaneously from free heme or hemoglobin under acidic physiological conditions. Slater & Cerami noted that *P. falciparum* trophozoite lysates could polymerize monomeric heme in a time- and pH-dependent manner. They postulated an enzymatic activity, which they called heme polymerase (128). These results were duplicated with *P. berghei* lysates (25). Ridley and coworkers showed that purified hemozoin or chemically synthesized β-hematin crystals could seed additional incorporation of heme into the polymers (38).

In vitro, *P. falciparum* histidine-rich protein II (HRPII) is capable of binding heme and of initiating polymer formation (135). This protein has been shown to accumulate in the digestive vacuole. The homologous HRP III functions similarly. Clinical *Plasmodium* isolates possess genes for both HRPs II and III. Some laboratory strains lack one or the other. One progeny from a laboratory cross lacks both HRPs II and III, but still produces hemozoin. This clone has a protein in its digestive vacuoles that reacts with anti-HRP II monoclonal antibody (135). The protein has not yet been isolated, nor has it been shown to be able to polymerize heme. Thus, histidine-rich proteins are postulated to mediate heme polymerization in the *Plasmodium* digestive vacuole, but further evidence for an obligate role is needed. These proteins likely participate in initiating the polymerization reaction. Once hemozoin is formed, extension has been shown to proceed in the absence of protein. Polyhistidine, bovine serum albumin and human histidine-rich glycoprotein can all bind heme but are not able to initiate polymer production (135). Nonphysiological lipids can also initiate polymer formation (14) and may well form a micelle that can orient the heme groups appropriately in vitro.

HEMOGLOBIN DEGRADATION AS A TARGET FOR ANTIMALARIAL CHEMOTHERAPY

Drugs That May Act on Heme Released During Hemoglobin Degradation

METHYLENE BLUE   This agent was first reported as a cure for malaria in the 19th century (57) but was not widely used because of toxicity. Derivatization led to the synthesis of primaquine, which is currently used for treatment of *P. vivax* and *P. ovale hypnozoites*. Recently, a series of methylene blue azure analogs were shown to complex with heme in vitro (8). These compounds inhibited *P. falciparum* growth in culture, as well as curing murine malaria caused by *P. vinckei petteri*.

ARTEMISININ   Interaction with parasite-derived heme may also be important for the antimalarial activity of the endoperoxide sesquiterpene lactone, artemisinin, originally known as qinghaosu in ancient Chinese herbal medicine. It has proven effective against both chloroquine and multi-drug-resistant malaria infections (60, 150). When added to cultured parasites, artemesinin has been shown to be associated with hemozoin (62, 84) and to cause damage to digestive vacuoles (81). In vitro, both free iron and heme catalyze the conversion of artemesinin into free radicals (88, 161) which alkylate parasite proteins (5, 155) and which form adducts with heme (84). Unlike chloroquine-heme complexes, artemesinin-heme adducts do not prevent hemozoin formation (4). Instead, artemesinin toxicity appears to be the product of membrane damage that results from protein alkylation (86). More work will be required to determine whether vacuolar heme and vacuolar membranes are involved, but endoperoxide derivatives appear to be promising candidates for future drug development (87).

QUINOLINES   The widely used quinolines, such as chloroquine, quinine, quinidine and mefloquine are thought to act by inhibiting the polymerization of heme released during hemoglobin proteolysis (127), although alternative mechanisms have been proposed (85, 127). Quinoline inhibition is specific for parasites that are actively degrading hemoglobin. Early morphological effects are digestive vacuolar swelling and pigment clumping. Chloroquine and its congeners accumulate in the *Plasmodium* digestive vacuole, where they hyperconcentrate to near millimolar concentration from nanomolar concentration in the plasma (127). This accumulation exceeds that predicted by the Henderson-Hasselbach equation for weak bases. Mammalian lysosomes do not hyperconcentrate and are unaffected at these external quinoline concentrations (69). Recent EM autoradiography and subcellular fractionation has demonstrated that chloroquine binds to hemozoin in the digestive vacuole of *P. falciparum* (136).

In vitro, the polymerization of heme is blocked under approximately physiological conditions by chloroquine, quinine, mefloquine, and related quinolines, regardless of whether extension is mediated by crude trophozoite lysates, seed crystals of hemozoin, or β-hematin (38, 41, 128). Initiation by purified *P. falciparum* HRPs is similarly inhibited (135). From these observations, it is clear that the extension phase of polymer formation is interrupted by the quinolines. Initiation may or may not be inhibited. Quinolines do not bind directly to the HRPs. Nor do they interact with isolated hemozoin. Only in the presence of heme do the quinolines bind measurably to hemozoin (136). It has therefore been proposed that drugs like chloroquine act by forming a complex with heme in the digestive vacuole (127) and incorporating into the growing polymer as a drug-heme complex (136) (Figure 6). This caps further extension, resulting in accumulation of toxic free heme and initiating the irreversible demise of the parasite. These drugs thus appear to turn the organism's own hemoglobin degradation pathway against itself, poisoning the parasite by preventing waste removal.

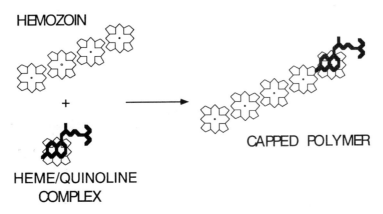

Figure 6 Capping model of quinoline action. Quinoline binds heme to form a drug-heme complex that adds on to the growing hemozoin chain, which prevents further polymerization and promotes toxic monomeric heme accumulation. (From Reference 136.)

Drugs to Block Hemoglobin De'

FALCIPAIN INHIBITORS   The enzymes involved in hemoglobin proteolysis have been shown to be potential new targets for antimalarial chemotherapy. Peptide fluoromethyl ketones are cysteine protease inhibitors that were shown to block both hemoglobin degradation and the growth of *P. falciparum* parasites at nanomolar concentrations (106, 115). The most potent of these inhibitors, benzyloxycarbonyl-Phe-Arg-CH_2F had an IC_{50} of 64 nM. In studies using a murine malaria model, a related fluoromethyl ketone (morpholine urea-Phe-homophenylalanine-CH_2F) cured 80% of *P. vinckei*-infected animals (111), which indicates that targeting hemoglobin catabolism might be practical. Toxicity of fluoromethyl ketones in experimental animals has precluded further drug development (107). Modeling the active site of cloned falcipain (76) permitted a computational search of commercially available molecules, and this led to the identification of oxalic bis [(2-hydroxy-1-naphthylmethylene) hydrazide] as a potential inhibitor (76). A number of chalcone derivatives were prepared and tested for activity against *P. falciparum* and shown to be effective at nanomolar concentrations. The most potent of these, 1-(2,5-dichlorophenyl)-3-(4-quinolinyl)-2-propen-1-one, had an IC_{50} of 200 nM against a chloroquine-resistant strain of *P. falciparum*. Vinyl sulfones were recently shown to be potent irreversible inhibitors of cysteine proteases (99). When tested against cultured *P. falciparum* parasites, Mu-Leu-HPh-VSPh blocked hemoglobin degradation and impaired parasite development at nanomolar concentrations, suggesting that this compound could be the basis of future drug development (107).

PLASMEPSIN INHIBITORS   Plasmepsins I and II are responsible for the initial cleavage of hemoglobin and thus are likely to be good drug targets. In vitro, a specific peptidomimetic inhibitor, SC-50083, was shown to inhibit hemoglobin degradation by plasmepsin I. When added to cultures at the stage during which hemoglobin is degraded, SC-50083 markedly reduces hemozoin formation and kills parasites (48), suggesting that plasmepsin I's hemoglobin-degrading capacity is distinct from that of plasmepsin II, and is essential. The crystal structure of plasmepsin II has been used to generate plasmepsin inhibitors that are active against both plasmepsins (126). These compounds block enzyme activity with Ki values in the picomolar range, but to date, none of these has been shown to be as potent as SC-50083 against cultured parasites. Bioavailability appears to be limited.

PPPP INHIBITORS   Proplasmepsin processing protease (PPPP) can now be added to the list of enzymes that are required for hemoglobin proteolysis (49). It is an acid hydrolase with a novel inhibitor profile that may signal a new class of enzymes. PPPP's role in proplasmepsin maturation makes it central to hemoglobin degradation, and its inhibitors are predicted to be potent antimalarials. The tripeptide aldehydes used to identify PPPP are nonspecific, but they could provide insight into this unusual enzyme's mechanism of cleavage and should be a starting point for new inhibitor design.

CONCLUSIONS

Hemoglobin degradation is a massive process with many requirements and ramifications for the parasite. The vacuole must be maintained at proper ionic strength and pH. The degradative enzymes must be targeted and activated. The peptides and/or amino acids that are generated must be transported out of the digestive vacuole for utilization by the parasite. The active oxygen species that is generated during degradation must be metabolized. The heme moiety that is released must be sequestered. This headquarters of metabolic activity is of crucial importance to the parasite. Interference with any of these processes can be expected to have devastating consequences for the *Plasmodium* organism. There is much critical work to be done for malaria biochemists, cell biologists, and pharmacologists.

ACKNOWLEDGMENTS

Supported by National Institutes of Health (NIH) grants AI-31615 and AI-37977, and by World Health Organization (WHO) grant TDR 950195.

Visit the *Annual Reviews home page* at
http://www.annurev.org.

Literature Cited

1. Aikawa M, Hepler PK, Huff CG, Sprinz H. 1966. The feeding mechanism of avian malarial parasites. *J. Cell Biol.* 28:355–73
2. Aikawa M, Thompson PE. 1971. Localization of acid phosphatase activity in *Plasmodium berghei* and *P. gallinaceum*: an electron microscopic observation. *J. Parasitol.* 57:603–10
3. Aissi E, Charet P, Bouquelet S, Biguet J. 1983. Endoprotease in *Plasmodium yoelii nigeriensis. Comp. Biochem. Physiol.* B7:559–66
4. Asawamahasakda W, Ittarat I, Chang C-C, McElroy P, Meshnick SR. 1994. Effects of antimalarials and protease inhibitors on plasmodial hemozoin production. *Mol. Biochem. Parasitol.* 67:183–91
5. Asawamahasakda W, Ittarat I, Pu Y-M, Ziffer H, Meshnick SR. 1994. Reaction

of antimalarial endoperoxides with specific parasite proteins. *Antimicrob. Agents Chemother.* 38:1854–58
6. Ashong JO, Blench IP, Warhurst DC. 1989. The composition of haemozoin from *Plasmodium falciparum. Trans. R. Soc. Trop. Med. Hyg.* 83:167–72
7. Atamna H, Ginsburg H. 1993. Origin of reactive oxygen species in erythrocytes infected with *Plasmodium falciparum. Mol. Biochem. Parasitol.* 61:231–42
8. Atamna H, Kruliak M, Shalmiev G, Deharo E, Pescarmona G, et al. 1996. Mode of antimalarial effect of methylene blue and some of its analogues in culture and their inhibition of *P. vinckei petteri* and *P. yoelii nigeriensis* in vivo. *Biochem. Pharmacol.* 51:693–700
9. Bailly E, Savel J, Mahouy G,

Jaireguiberry G. 1991. *Plasmodium falciparum*: isolation and characterization of a 55-kDa protease with a cathepsin D-like activity from *P. falciparum*. *Exp. Parasitol.* 72:278–84

10. Ball EG, McKee RW, Anfinsen CB, Cruz WO, Geiman QM. 1948. Studies on malarial parasites: ix. chemical and metabolic changes during growth and multiplication in vivo and in vitro. *J. Biol. Chem.* 175:547–71

11. Banting G, Benting J, Lingelbach K. 1995. A minimalist view of the secretory pathway in *Plasmodium falciparum*. *Trends Cell Biol.* 5:340–43

12. Becker MM, Harrop SA, Dalton JP, Kalinna BH, McManus DP, et al. 1995. Cloning and characterization of the *Schistosoma japonicum* aspartic proteinase involved in hemoglobin degradation. *J. Biol. Chem.* 270:24496–501

13. Becuwe P, Gratapanche S, Fourmeaux M-N, Van Beeumen J, Samyn B, et al. 1996. Characterization of iron-dependent endogenous superoxide dismutase of *Plasmodium falciparum*. *Mol. Biochem. Parasitol.* 76:125–34

14. Bendrat K, Berger BJ, Cerami A. 1995. Haem polymerization in malaria. *Nature* 378:138

15. Beutler E. 1990. Erythrocyte disorders: Anemias related to abnormal globin. In *Hematology*, ed. WJ Williams, E Beutler, AJ Erslev, MA Lichtman, pp. 613–44. New York: McGraw-Hill

16. Bogitsh BJ, Dresden MH. 1983. Fluorescent histochemistry of acid proteases in adult *Schistosoma mansoni* and *Schistosoma japonicum*. *J. Parasitol.* 69:106–10

17. Bogitsh BJ, Kirschner KF, Rotmans JP. 1992. *Schistosoma japonicum*: immunoinhibitory studies on hemoglobin digestion using heterologous antiserum to bovine cathepsin D. *J. Parasitol.* 78:454–59

18. Bohle SD, Conklin BJ, Cox D, Madsen SK, Paulson S, et al. 1994. Structural and spectroscopic studies of beta-hematin (the heme coordination polymer in malaria pigment). In *Inorganic and Organometallic Polymers II: Advanced Materials and Intermediates*, ed. P Wisian-Neilson, HR Allcock, KJ Wynn, pp. 497–515. Washington, DC: Am. Chem. Soc.

19. Bohley P, Seglen PO. 1992. Proteases and proteolysis in the lysosome. *Experientia* 48:151–57

20. Brasseur P, Kouamouo J, Moyou RS, Druilhe P. 1992. Multidrug resistant *falciparum* malaria in Cameroon in 1987–1988. *Am. J. Trop. Med. Hyg.* 46:8–14

21. Chappell CL, Dresden MH. 1986. *Schistosoma mansoni*: proteinase activity of hemoglobinase from the digestive tract of adult worms. *Exp. Parasitol.* 61:160–67

22. Charet P, Aissi E, Maurois P, Bouquelet S, Biguet J. 1980. Aminopeptidase in rodent *Plasmodium*. *Comp. Biochem. Physiol. B* 65:519–24

23. Childs GE, Boudreau EF, Wimonwattratee T, Pang L, Milhous WK. 1991. In vivo and clinical correlates of mefloquine resistance of *P. falciparum* in eastern Thailand. *Am. J. Trop. Med. Hyg.* 44:484–87

24. Choi I, Mego JL. 1988. Purification of *Plasmodium falciparum* digestive vacuoles and partial characterization of the vacuolar membrane ATPase. *Mol. Biochem. Parasitol.* 31:71–78

25. Chou AC, Fitch CD. 1992. Heme polymerase: modulation by chloroquine treatment of a rodent malaria. *Life Sci.* 51:2073–78

26. Cowman AF, Karcz S, Galatis D, Culvenor JG. 1991. A P-glycoprotein homologue of *Plasmodium falciparum* is localized on the digestive vacuole. *J. Cell Biol.* 113:1033–42

27. Curley GP, O'Donovan SM, McNally J, Mullally M, O'Hara H, et al. 1994. Aminopeptidases from *Plasmodium falciparum*, *Plasmodium-chabaudi chabaudi* and *Plasmodium-berghei*. *J. Eukaryot. Microbiol.* 41:119–23

28. Dalton JP, Brindley PJ. 1996. Schistosome asparaginyl endopeptidase Sm32 in hemoglobin digestion. *Parasitol. Today* 12:125

29. Dalton JP, Clough KA, Jones MK, Brindley PJ. 1996. Characterization of the cathepsin-like cysteine proteinases of *Schistosoma mansoni*. *Infect. Immun.* 64:1328–34

30. Dame JB, Reddy GR, Yowell CA, Dunn BM, Kay J, et al. 1994. Sequence, expression and modeled structure of an aspartic proteinase from the human malaria parasite *Plasmodium falciparum*. *Mol. Biol. Parasitol.* 64:177–90

31. Davis AH, Nanduri J, Watson DC. 1987. Cloning and gene expression of *Schistosoma mansoni* protease. *J. Biol. Chem.* 262:12851–55

32. Day SR, Dalton JP, Clough KA, Leonardo L, Tiu WU, et al. 1995. Characterization and cloning of the cathepsin L proteinases of *Schistosoma japonicum*. *Biochem. Biophys. Res. Commun.* 217:1–9

33. Deegan T, Maegraith BG. 1956. Studies on the nature of malarial pigment (hemozoin). *Ann. Trop. Med. Parasitol.* 50:194–222

34. Desjardins RE, Canfield RJ, Haynes JD, Chulay JD. 1979. Quantitative assessment of antimalarial activity in vitro by a semi-automated microdilution technique. *Antimicrob. Agents Chemother.* 16:710–18

35. Dickerson RE, Geis I. 1983. *Hemoglobin,* Ch. 3. Menlo Park, CA: Benjamin/Cummings

36. Divo AA, Geary TG, Davis NL, Jensen JB. 1985. Nutritional requirements of *Plasmodium falciparum* in culture. I. Exogenously supplied dialyzable components necessary for continuous growth. *J. Protozool.* 32:59–64

37. Dluzewski AR, Rangachari K, Wilson RJM, Gratzer WB. 1986. *Plasmodium falciparum*: protease inhibitors and inhibition of invasion. *Exp. Parasitol.* 62:416–22

38. Dorn A, Stoffel R, Matile H, Bubendorf A, Ridley RG. 1995. Malarial haemozoin/beta-haematin supports heme polymerization in the absence of protein. *Nature* 374:269–71

39. Eckman JR, Modler S, Eaton JW, Berger E, Engel RR. 1977. Host heme catabolism in drug-sensitive and drug-resistant malaria. *J. Lab. Clin. Med.* 90:767–70

40. Edington GM. 1967. Pathology of malaria in West Africa. *Br. Med. J.* 1:715–18

41. Egan DC, Adams PA. 1994. Quinoline anti-malarial drugs inhibit spontaneous formation of beta-hematin (malaria pigment). *FEBS Lett.* 352:54–57

42. Elmendorf HG, Haldar K. 1993. Secretory transport in *Plasmodium. Parasitol. Today* 9:98–102

43. Fairfield AS, Abosch AS, Ranz A, Eaton JW, Meshnick SR. 1988. Oxidant defense enzymes of *Plasmodium falciparum. Mol. Biochem. Parisitol.* 30:77–82

44. Fairfield AS, Meshnick SR, Eaton JW. 1983. Malaria parasites adopt host cell superoxide dismutase. *Science* 221:764–66

45. Fairlamb AH, Paul F, Warhurst DC. 1984. A simple magnetic method for the purification of malarial pigment. *Mol. Biochem. Parasitol.* 12:307–12

46. Fitch CD, Kanjananggulpan P. 1987. The state of ferriprotoporphyrin IX in malaria pigment. *J. Biol. Chem.* 262:15552–55

47. Francis SE, Gluzman IY, Oksman A, Banerjee D, Goldberg DE. 1996. Characterization of native falcipain, an enzyme involved in *Plasmodium falciparum* hemoglobin degradation. *Mol. Biochem. Parasitol.* 83:189–200

48. Francis SE, Gluzman IY, Oksman A, Knickerbocker A, Mueller R, et al. 1994. Molecular characterization and inhibition of a *Plasmodium falciparum* aspartic hemoglobinase. *EMBO J.* 13:306–17

49. Francis SE, Banerjee R, Goldberg DE. 1997. Biosynthesis and maturation of the malaria aspartic hemoglobinases plasmepsins I and II. *J. Biol. Chem.* In press

50. Ginsburg H. 1990. Some reflections concerning host erythrocyte-malarial parasite interrelationships. *Blood Cells* 16:225–35

51. Ginsburg H, Krugliak M. 1992. Quinoline-containing antimalarials—mode of action, drug resistance and its reversal. *Biochem. Pharmacol.* 43:63–72

52. Gluzman IY, Francis SE, Oksman A, Smith CE, Duffin KL, et al. 1994. Order and specificity of the *Plasmodium falciparum* hemoglobin degradation pathway. *J. Clin. Invest.* 93:1602–8

53. Goldberg DE. 1992. Plasmodial hemoglobin degradation: An ordered pathway in a specialized organelle. *Infect. Agents Dis.* 1:207–11

54. Goldberg DE, Slater AFG, Beavis R, Chait B, Cerami A, Henderson GB. 1991. Hemoglobin degradation in the human malaria pathogen *Plasmodium falciparum*: A catabolic pathway initiated by a specific aspartic protease. *J. Exp. Med.* 173:961–69

55. Goldberg DE, Slater AFG, Cerami A, Henderson GB. 1990. Hemoglobin degradation in the malaria parasite *Plasmodium falciparum*: An ordered process in a unique organelle. *Proc. Natl. Acad. Sci. USA* 87:2931–35

56. Goldie P, Roth EF Jr, Oppenheim J, Vanderberg JP. 1990. Biochemical characterization of *Plasmodium falciparum* hemozoin. *Am. J. Trop. Med. Hyg.* 43:584–96

57. Guttman P, Ehrlich P. 1891. Uber die wirkung de methylenblau bei malaria. *Klin. Wochenschr.* 28:953–56

58. Gyang FN, Poole B, Trager W. 1983. Peptidases from *Plasmodium falciparum* cultured in vitro. *Mol. Biochem. Parasitol.* 5:263–73

59. Hempelmann E, Wilson RJM. 1980. Endopeptidases from *Plasmodium knowlesi. Parasitology* 80:323–30

60. Hien TT, White NJ. 1993. Qinghaosu. *Lancet* 341:603–8

61. Hill J, Tyas L, Phylip LH, Kay J, Dunn BD, et al. 1994. High level expression and characterisation of plasmepsin II, an aspartic proteinase from *Plasmodium falciparum. FEBS Lett.* 352:155–58

62. Hong U-L, Yang Y-Z, Meshnick SR. 1994. The interaction of artemisinin with

malarial hemozoin. *Mol. Biochem. Parasitol.* 63:121–28

63. Howard RJ, Uni S, Lyon JA, Taylor DW, Daniel W, et al. 1987. Export of *Plasmodium falciparum* proteins to the host eythrocyte membrane: special problems of protein trafficking and topogenesis. In *Host-Parasite Cellular and Molecular Interactions in Protozoal Infections,* ed. KP Chang, D Snary, pp. 281–96. Berlin: Springer-Verlag

64. Hunt NH, Stocker R. 1990. Oxidative stress and the redox status of malaria-infected erythrocytes. *Blood Cells* 16:499–526

65. Jones EW. 1991. Tackling the protease problem in *Saccharomyces cerevisiae. Methods Enzymol.* 194:428–53

66. Karcz SR, Herrmann VR, Cowman AF. 1993. Cloning and characterization of a vacuolar ATPase A subunit homologue from *Plasmodium falciparum. Mol. Biochem. Parasitol.* 58:333–44

67. Karcz SR, Herrmann VR, Trottein F, Cowman AF. 1994. Cloning and characterization of the vacuolar ATPase B subunit from *Plasmodium falciparum. Mol. Biochem. Parasitol.* 65:123–33

68. Klionsky DJ, Herman PK, Emr SD. 1990. The fungal vacuole: composition, function and biogenesis. *Microbiol. Rev.* 54:266–92

69. Krogstad DJ, Schlesinger PH. 1987. Acid-vesicle function, intracellular pathogens, and the action of chloroquine against *Plasmodium falciparum. N. Engl. J. Med.* 317:542–49

70. Krogstad DJ, Schlesinger PH, Gluzman IY. 1985. Antimalarials increase vesicle pH in *Plasmodium falciparum J. Cell Biol.* 101:2302–9

71. Krugliak M, Waldman Z, Ginsburg H. 1987. Gentamicin and amikacin repress the growth of *Plasmodium falciparum* in culture, probably by inhibiting a parasite acid phospholipase. *Life Sci.* 40:1253–57

72. Langreth S. 1976. Feeding mechanisms in extracellular *Babesia microti* and *Plasmodium lophurae. J. Protozool.* 23:215–23

73. Langreth SG, Jensen JB, Reese RT, Trager W. 1978. Fine structure of human malaria in vitro. *J. Protozool.* 25:443–52

74. Levy MR, Chou SC. 1973. Activity and some properties of an acid proteinase from normal and *Plasmodium berghei*-infected red cells. *J. Parasitol.* 59:1064–70

75. Levy MR, Siddiqui WA, Chou SC. 1974. Acid protease activity in *Plasmodium falciparum* and *P. knowlesi* and ghosts of

their respective host red cells. *Nature* 247:546–49

76. Li R, Kenyon GL, Cohen FE, Chen X, Gong B, et al. 1995. In vitro antimalarial activity of chalcones and their derivatives. *J. Protozool.* 38:5031–37

77. Lindquist RN, Senft AW, Petitt M, McKerrow JH. 1986. *Schistosoma mansoni*: purification and characterization of the major acidic proteinase from adult worms. *Exp. Parasitol.* 61:398–404

78. Lingelbach KR. 1993. *Plasmodium falciparum*: A molecular view of protein transport from the parasite into the host erythrocyte. *Exp. Parasitol.* 76:318–27

79. Lujan HD, Marotta A, Mowatt MR, Sciaky N, Lippincott-Schwartz J, et al. 1996. Developmental induction of Golgi structure and function in the primitive eukaryote *Giardia lamblia. J. Biol. Chem.* 270:4612–18

80. Luker KE, Francis SE, Gluzman IY, Goldberg DE. 1996. Kinetic analysis of plasmepsins I and II, aspartic proteases of the *Plasmodium falciparum* digestive vacuole. *Mol. Biochem. Parasitol.* 79:71–78

81. Maeno YT, Toyoshima T, Fujioka H, Ito Y, Meshnick SR. 1993. Morphologic effects of artemisinin in *Plasmodium falciparum. Am. J. Trop. Med. Hyg.* 49:485–91

82. McCormick GJ. 1970. Amino acid transport and incorporation in red blood cells of normal and *Plasmodium knowlesi*-infected rhesus monkeys. *Exp. Parasitol.* 27:143–49

83. Merckelbach A, Hasse S, Dell R, Eschlbeck A, Ruppel A. 1994. cDNA sequences of *Schistosoma japonicum* coding for two cathepsin B-like proteins and sj32. *Trop. Med. Parasitol.* 45:193–98

84. Meshnick SR. 1990. Chloroquine as intercalator: a hypothesis revived. *Parasitol. Today* 6:77–79

85. Meshnick SR. 1994. The mode of action of antimalarial endoperoxides. *Trans. R. Soc. Trop. Med. Hyg.* 88 (Suppl. 1):S31–S32

86. Meshnick SR, Taylor TE, Kamchonwongpaisan S. 1996. Artemisinin and the antimalarial endoperoxides: from herbal remedy to targeted chemotherapy. *Microvas. Rev.* 60:301–15

87. Meshnick SR, Thomas A, Ranz A, Xu CM, Pan HZ. 1991. Artemisinin (qinghaosu): role of intracellular hemin in its mechanism of antimalarial action. *Mol. Biochem. Parasitol.* 49:181–90

88. Meshnick SR, Yang YZ, Lima V, Kuyers F, Kamchonwongpaisan S, et al. 1993. Iron-dependent free radical generation from the antimalarial agent

artemisinin (qinghaosu). *Antimicrob. Agents Chemother.* 37:1108–14

89. Metzger WG, Mordmuller BG, Kremsner PG. 1995. Malaria pigment in leucocytes. *Trans. R. Soc. Trop. Med. Hyg.* 89:637–38

90. Miller LH, Good MF, Milon G. 1994. Malaria pathogenesis. *Science* 264:1878–83

91. Morakote N, Justus DE. 1988. Immunosuppression in malaria: effect of hemozoin produced by *Plasmodium berghei* and *Plasmodium falciparum. Int. Arch. Allergy Appl. Immunol.* 86:28–34

92. Morrison DB, Jeskey HA. 1948. Alterations in some constituents of the monkey erythrocyte infected with *Plasmodium knowlesi* as related to pigment formation. *J. Nat. Malar. Soc.* 7:259–64

93. Nene V, Gobright E, Musoke AJ, Lonsdale-Eccles JD. 1990. A single exon codes for the enzyme domain of a protozoan cysteine protease. *J. Biol. Chem.* 265:18047–50

94. Nosten F, ter Kuile F, Chongsuphajaisiddhi T. 1991. Mefloquine resistant *falciparum* malaria in the Thai-Burmese border. *Lancet* 337:1140–43

95. Olliaro P, Castelli P, Milano F, Filice G, Carosi G. 1989. Ultrastucture of *Plasmodium falciparum* "in vitro". I. Baseline for drug effects evaluation. *Microbiologica* 12:7–14

96. Olliaro P, Cattani J, Wirth D. 1996. Malaria, the submerged disease. *JAMA* 275:230–33

97. Olliaro PL, Goldberg DE. 1995. The *Plasmodium* digestive vacuole: Metabolic headquarters and choice drug target. *Parasitol. Today* 11:294–97

98. Orjih AU, Fitch CD. 1993. Hemozoin production by *Plasmodium falciparum*: variation with strain and exposure to chloroquine. *Biochim. Biophys. Acta* 1157:270–74

99. Palmer JT, Rasnick D, Klaus J, Bromme D. 1995. Vinyl sulfones as mechanism-based cysteine protease inhibitors. *J. Protozool.* 38:3193–96

100. Pandey AV, Tekwani BL. 1996. Identification and quantitation of hemozoin: some additional facts. *Parasitol. Today* 12:370

101. Perutz M. 1987. Molecular anatomy, physiology and pathology of hemoglobin. In *The Molecular Basis of Blood Diseases,* ed. G Stamatoyannopoulis, AW Nienhuis, P Leder, PW Majerus. New York: Saunders

102. Peters W. 1987. *Chemotherapy and Drug Resistance in Malaria.* London: Academic

103. Phu NH, Day N, Diep PT, Ferguson DJP,

White NJ. 1995. Intraleucocytic malaria pigment and prognosis in severe malaria. *Trans. R. Soc. Trop. Med. Hyg.* 89:200–4

104. Pichyangkul S, Saengkral P, Webster HK. 1994. *Plasmodium falciparum* pigment induces monocytes to release high levels of tumor necrosis factor-alpha and interleukin–1beta. *Am. J. Trop. Med. Hyg.* 51:430–35

105. Rao JKM, Erickson JW, Wlodawer A. 1991. Structural and evolutionary relationships between retroviral and eucaryotic aspartic proteinases. *Biochemistry* 30:4663–71

106. Rockett KA, Playfair JHL, Targett GAT, Angliker H, Shaw E. 1990. Inhibition of intraerythrocytic development of *Plasmodium falciparum* by proteinase inhibitors. *FEBS Lett.* 259:257–59

107. Rosenthal P, Olson JE, Lee K, Palmer JT. 1996. Antimalarial effects of vinyl sulfone cysteine proteinase inhibitors. *Antimicrob. Agents Chemother.* 40:1600–3

108. Rosenthal PJ. 1993. A *Plasmodium vinckei* cysteine proteinase shares unique features with its *Plasmodium falciparum* analogue. *Biochem. Biophys. Acta* 1173:91–93

109. Rosenthal PJ. 1995. *Plasmodium falciparum*: effects of proteinase inhibitors on globin hydrolysis by cultured malaria parasites. *Exp. Parasitol.* 80:272–81

110. Rosenthal PJ. 1996. Conservation of key amino acids among the cysteine proteinases of multiple malarial species. *Mol. Biochem. Parasitol.* 75:25–60

111. Rosenthal PJ, Lee GK, Smith RE. 1993. Inhibition of a *Plasmodium vinckei* cysteine proteinase cures murine malaria. *J. Clin. Invest.* 91:1052–56

112. Rosenthal PJ, McKerrow JH, Aikawa M, Nagasawa H, Leech JH. 1988. A malarial cysteine proteinase is necessary for hemoglobin degradation by *Plasmodium falciparum. J. Clin. Invest.* 82:1560–66

113. Rosenthal PJ, McKerow JH, Rasnick D, Leech JH. 1989. *Plasmodium falciparum*: inhibitors of lysosomal cysteine proteinases inhibit a trophozoite proteinase and block parasite development. *Mol. Biochem. Parasitol.* 35:177–84

114. Rosenthal PJ, Nelson RG. 1992. Isolation and characterization of a cysteine protease gene of *Plasmodium falciparum. Mol. Biochem. Parasitol.* 51:143–52

115. Rosenthal PJ, Wollish WS, Palmer JT, Rasnick D. 1991. Antimalarial effects of peptide inhibitors of a *Plasmodium falciparum* cysteine proteinase. *J. Clin. Invest.* 88:1467–72

116. Ross R. 1897. On some peculiar pigmented cells found in two mosquitos fed on malarial blood. *Br. Med. J.* 1:1786–88
117. Rudzinska MA. 1965. Pinocytic uptake and the digestion of hemoglobin in malaria parasites. *J. Protozool.* 12:563–76
118. Sasaki T, Kishi M, Saito M, Katunuma N, Murachi T. 1990. Inhibitory effect of di- and tripeptidyl aldehydes on calpains and cathepsins. *J. Enzym. Inhib.* 3:195–201
119. Sato K, Fukabori Y, Suzuki M. 1987. *Plasmodium berghei*: a study of globinolytic enzyme in erythrocytic parasite. *Zentral Bakteriol. Hyg. Abt.* 264:487–95
120. Scheibel LW, Sherman IW. 1988. Plasmodial metabolism and related organellar function during various stages of the life cycle: proteins, lipids, nucleic acids and vitamins. In *Malaria*, ed. WH Wernsdorfer, I McGregor, pp. 219–52. Edinburgh: Churchill Livingstone
121. Schwarzer E, Turrini F, Ulliers D, Giribaldi G, Ginsburg H, et al. 1992. Impairment of macrophage functions after ingestion of *Plasmodium falciparum*-infected erythrocytes or isolated malarial pigment. *J. Exp. Med.* 176:1033–41
122. Sherman I. 1979. Biochemistry of *Plasmodium. Microbiol. Rev.* 43:453–94
123. Sherman IW. 1977. Amino acid metabolism and protein synthesis in malarial parasites. *Bull. WHO* 55:265–76
124. Sherman IW, Tanigoshi L. 1970. Incorporation of 14C-amino acids by malaria. *Int. J. Biochem.* 1:635–37
125. Sherman IW, Tanigoshi L. 1983. Purification of *Plasmodium lophurae* cathepsin D and its effects on erythrocyte membrane proteins. *Mol. Biochem. Parasitol.* 8:207–26
126. Silva AM, Lee AY, Gulnik SV, Majer P, Collins J, et al. 1996. Structure and inhibition of plasmepsin II, a hemoglobin-degrading enzyme from *Plasmodium falciparum. Proc. Natl. Acad. Sci. USA* 93:10034–39
127. Slater AFG. 1993. Chloroquine: mechanism of drug action and resistance in *Plasmodium falciparum. Pharmacol. Ther.* 57:203–35
128. Slater AFG, Cerami A. 1992. Inhibition by chloroquine of a novel haem polymerase enzyme activity in malaria trophozoites. *Nature* 355:167–69
129. Slater AFG, Swiggard WJ, Orton BR, Flitter WD, Goldberg DE, et al. 1991. An iron-carboxylate bond links the heme units of malaria pigment. *Proc. Natl. Acad. Sci. USA* 88:325–29
130. Slomianny C. 1990. Three-dimensional reconstruction of the feeding process of the malaria parasite. *Blood Cells* 16:369–78
131. Slomianny C, Charet P, Prensier G. 1983. Ultrastructural localization of enzymes involved in the feeding process in *Plasmodium chabaudi* and *Babesia hylomysci. J. Protozool.* 30:376–82
132. Slomianny C, Prensier G. 1990. Cytochemical ultrastructural study of the lysosomal system of different species of malaria parasites. *J. Protozool.* 37:465–70
133. Slomianny C, Prensier G, Charet P. 1985. Ingestion of erythrocytic stroma by *Plasmodium chabaudi* trophozoites: ultrastructural study by serial sectioning and 3-dimensional reconstruction. *Parasitology* 90:579–88
134. Slomianny C, Prensier G, Vivier E. 1982. Ultrastructural study of the feeding process of erythrocytic *P. chabaudi* trophozoite. *Mol. Biochem. Parasitol. Suppl.* 5:695
135. Sullivan DJ Jr, Gluzman IY, Goldberg DE. 1996. *Plasmodium* hemozoin formation mediated by histidine-rich proteins. *Science* 271:219–22
136. Sullivan DJ Jr, Gluzman IY, Russell DG, Goldberg DE. 1996. On the molecular mechanism of chloroquine's antimalarial action. *Proc. Natl. Acad. Sci. USA* 93:11865–70
137. Tang J, Wong RNS. 1987. Evolution in the structure and function of aspartic proteases. *J. Cell. Biochem.* 33:53–63
138. Tappel AL. 1953. The mechanism of oxidation of unsaturated fatty acids catalyzed by hematin compounds. *Arch. Biochem. Biophys.* 44:378–95
139. Taramelli D, Basilico N, Pagani E, Grande R, Monti D, et al. 1995. The heme moiety of malaria pigment (beta-hematin) mediates the inhibition of nitric oxide and tumor necrosis factor-alpha production by lipopolysaccharide-stimulated macrophages. *Exp. Parasitol.* 81:501–11
140. Theakston RDG, Fletcher KA, Maegraith BG. 1970. The use of electron microscope autoradiography for examining the uptake and degradation of haemoglobin by *Plasmodium falciparum. Ann. Trop. Med. Parasit.* 64:63–71
141. Ting IP, Sherman IW. 1966. Carbon dioxide fixation in malaria-I. Kinetic studies in *Plasmodium lophurae. Comp. Biochem. Physiol.* 19:855–69
142. Tsubuki S, Kawasaki H, Saito Y, Miyashita N, Inomata M, et al. 1993. Purification and characterization of a Z-Leu-Leu-MCA degrading protease expected to

regulate neurite formation: A novel catalytic activity in proteasome. *Biochem. Biophys. Res. Commun.* 196:1195–1201

143. Turrini F, Ginsburg H, Bussolino F, Pescarmona F, Serra MV, et al. 1992. Phagocytosis of *Plasmodium*-infected human red blood cells by human monocytes. *Blood* 80:801–8

144. Vander Jagt DL, Hunsaker LA, Campos NM. 1986. Characterization of a hemoglobin-degrading, low molecular weight protease from *Plasmodium falciparum. Mol. Biochem. Parasitol.* 18:389–400

145. Vander Jagt DL, Hunsaker LA, Campos NM. 1987. Comparison of proteases from chloroquine-sensitive and chloroquine-resistant strains of *Plasmodium falciparum. Biochem. Pharmacol.* 36:3285–91

146. Vander Jagt DL, Hunsaker LA, Campos NM, Scaletti JV. 1992. Localization and characterization of hemoglobin-degrading proteinases from the malarial parasite *Plasmodium falciparum. Biochim. Biophys. Acta* 1122:256–64

147. Vennerstrom JL, Easton JW. 1988. Oxidants, oxidant drugs and malaria. *J. Protozool.* 31:1269–77

148. Vinitsky A, Michaud C, Powers JC, Orlowski M. 1992. Inhibition of the chymotrypsin-like activity of the pituitary multicatalytic proteinase complex. *Biochemistry* 31:9421–28

149. Wasilewski MM, Lim KC, Phillips J, McKerrow JH. 1996. Cysteine protease inhibitors block schistosome hemoglobin degradation in vitro and decrease worm burden and egg production in vivo. *Mol. Biochem. Parasitol.* 81:179–89

150. Webster HK, Lehnert EK. 1994. Chemistry of artemisinin: an overview. *Trans. R. Soc. Trop. Med. Hyg.* 88:S27–S29

151. Wellems TE. 1991. Molecular genetics of drug resistance in *Plasmodium falciparum* malaria. *Parasitol. Today* 7:110–12

152. WHO. 1993. Global malaria control. 71:281–84

153. WHO. 1995. *In Control of Tropical Diseases (CTD): Malaria Control.* Geneva, Switzerland: WHO Off. Inf.

154. Williams WJ, Beutler E, Erslev AJ, Lichtman MA. 1990. *Hematology,* p. 319. New York: McGraw-Hill

155. Yang YZ, Asawamahasakda W, Meshnick SR. 1993. Alkylation of human albumin by the antimalarial artemisinin. *Biochem. Pharmacol.* 46:336–39

156. Yayon A, Cabantchik ZI, Ginsburg H. 1984. Identification of the acidic compartment of *Plasmodium falciparum*-infected human erythrocytes as the target of the antimalarial drug chloroquine. *EMBO J.* 3:2695–2700

157. Yayon A, Ginsburg H. 1983. Chloroquine inhibits the degradation of endocytic vesicles in human malaria parasites. *Cell Biol. Int. Rep.* 7:895

158. Yayon A, Timberg R, Friedman S, Ginsburg H. 1984. Effects of chloroquine on the feeding mechanism of the intraerythrocytic human malarial parasite *Plasmodium falciparum. J. Protozool.* 31:367–72

159. Yayon A, Vande Waa JA, Yayon M, Geary TG, Jensen JB. 1983. Stage-dependent effects of chloroquine on *Plasmodium falciparum* in vitro. *J. Protozool.* 30:642–47

160. Zarchin S, Krugliak M, Ginsburg H. 1986. Digestion of the host erythrocyte by malaria parasites is the primary target for quinolone-containing antimalarials. *Biochem. Pharmacol.* 35:2435–42

161. Zhang F, Gosser D, Meshnick SR. 1992. Hemin-catalyzed decomposition of artemisinin (qinghaosu). *Biochem. Pharmacol.* 43:1805–9

Annu. Rev. Microbiol. 1997. 51:125–49

GETTING STARTED: Regulating the Initiation of DNA Replication in Yeast

W. Mark Toone,[1,2] Birgit L. Aerne,[1] Brian A. Morgan,[1,3] and Leland H. Johnston[1,4]

[1]Division of Yeast Genetics, National Institute for Medical Research, The Ridgeway, Mill Hill, London NW7 1AA, UK; [2]Laboratory of Gene Regulation, Imperial Cancer Research Fund, PO Box 123, Lincoln's Inn Fields, London WC2A 3PX; [3]Department of Biochemistry and Genetics, University of Newcastle, Newcastle upon Tyne, NE2 4HH, UK; [4](corresponding author) Division of Yeast Genetics, National Institute for Medical Research, The Ridgeway, Mill Hill, London NW7 1AA, UK; email: l-johnst@nimr.mrc.ac.uk

KEY WORDS: S phase, origins, cell cycle, Start

ABSTRACT

Initiation of DNA replication in yeast appears to operate through a two-step process. The first step occurs at the end of mitosis in the previous cell cycle, where, following the decrease in B cyclin-dependent kinase activity, an extended protein complex called the prereplicative complex (pre-RC) forms over the origin of replication. This complex is dependent on the association of the Cdc6 protein with the Origin Recognition Complex (ORC) and appears concomitantly with the nuclear entry of members of the Mcm family of proteins. The second step is dependent upon the cell passing through a G1 decision point called Start. If the environmental conditions are favorable, and the cells reach a critical size, then there is a rise in G1 cyclin-dependent kinase activity, which leads to the activation of downstream protein kinases; the protein kinases are, in turn, required for triggering initiation from the preformed initiation complexes. These protein kinases, Dbf4-Cdc7 and Clb5/6(B-cyclin)-Cdc28, are thought to phosphorylate targets within the pre-RC. The subsequent rise in B cyclin protein kinase activity following Start not only triggers origin firing, but also inhibits the formation of new pre-RCs, which ensures that there is only one S phase in each cell cycle. The destruction of B-cyclin protein kinase activity at the end of the cell cycle potentiates the formation of new pre-RCs—resetting origins for the next S phase.

"... Or say that the end precedes the beginning."

TS Eliot

0066-4227/97/1001-0125$08.00

CONTENTS

INTRODUCTION

In the eukaryotic cell division cycle, the entire nuclear content of DNA is faithfully replicated within a discrete period of time known as S phase. Due to the large size of eukaryotic chromosomes, replication proceeds by means of two active replication forks that move in opposite directions from multiple chromosomal sites, called origins. The number of origins in a simple eukaryote such as yeast is on the order of 400, whereas in metazoans they may number in the thousands. To maintain the integrity of the genome and the ploidy of the cell, initiation from each of these origins must be regulated such that it occurs only once in each cell cycle. Furthermore, the firing of each origin must be linked with cell cycle controls that dictate whether or not the cell will enter a new round of cell division.

Much of our understanding of the initiation of replication, and the cell cycle in general, has come from studies of yeast. In this review we will focus mainly on work carried out in the budding yeast, *Saccharomyces cerevisiae*, although we include some notable contributions from the fission yeast, *Schizosaccharomyces pombe*. Rather than present a comprehensive review of DNA replication in yeast, we survey recent developments that have led, perhaps for the first time, to a basic understanding of the molecular mechanisms that link cell cycle control to the initiation of DNA synthesis. We begin with a description of the origins of replication themselves, and of the protein complexes that assemble on them to produce a replication-competent state. We then discuss the cell cycle controls that dictate the timing of DNA replication and which, importantly, also ensure that initiation does not recur before the next cell cycle.

YEAST REPLICATION ORIGINS AND ORIGIN BINDING PROTEINS

Origins

In bacterial and viral systems it has been known for some time that short stretches of DNA, called replicators, are bound by initiator proteins and that these DNA-protein complexes act as nucleation sites for the assembly of other complexes that catalyze DNA replication (95). Although they may overlap, a distinction has been drawn between the sequences that constitute the replicator and the actual site where replication begins, which is known as the origin (94). During the late 1970s, a major obstacle to the study of eukaryotic DNA replication was removed by the discovery of yeast replicators or Autonomously Replicating Sequence elements (ARS elements) [see references in (20)]. These sequences allowed plasmids that carry selectable markers to be stably and autonomously maintained in yeast cells. Subsequently, using two-dimensional gel electrophoresis techniques, it was shown that ARSs do, in fact, direct the initiation of DNA replication, and that the origin of replication and the replicator both map to the same region (8, 46). ARSs, therefore, had properties consistent with a role as origins of replication and, furthermore (in some but not all cases), they were found to function as origins when examined in their native chromosomal context (79).

Sequence analysis of ARS elements revealed that they are more A/T rich than is the surrounding chromosomal DNA, and that all ARSs share an 11bp ARS consensus sequence (ACS: 5'-(A/T)TTTAT (A/G)TTT(A/T)-3') (9). The transforming ability and the stability of plasmids that carry ARS elements were then used to identify sequences within the element that were important for ARS function. The ACS and immediate flanking sequences, referred to as domain A, were found to be essential because mutation or deletion of this domain completely abolishes ARS function (99). However, domain A is not sufficient for ARS activity, and flanking sequences—both 5' and 3'—have been shown to contribute to ARS function. Sequences 3' to the T-rich strand of the ACS are called domain B, and sequences 5' to the T-rich strand are referred to as domain C (12, 89) (Figure 1).

The two ARS elements that have been studied in the most detail are *ARS1* and *ARS307*. Using linker substitution analysis, domain B of *ARS1* has been further divided into subdomains B1, B2, and B3, and domain B of *ARS307* has been divided into subdomains B1 and B2 (70, 82, 96). Although there is no obvious sequence similarity between these two ARSs (outside of the ACS), there is functional conservation. For example, the B1 element on *ARS1* will substitute for B1 on *ARS 307*, and vice versa. However, the B element subdomains within a single ARS are not interchangeable, which indicates that they have distinct

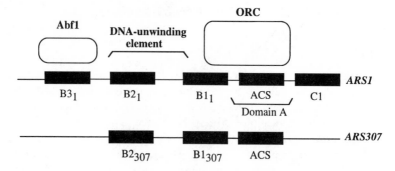

Figure 1   Anatomy of an Autonomously Replicating Sequence (ARS). ARSs were identified based on their ability to allow plasmids to be stably maintained as episomes. Subsequently, it was shown that ARSs do, in fact, function as origins of replication. Mutational analysis of ARSs indicated that they have a modular structure consisting of a highly conserved ARS-consensus sequence (ACS) and, in the case of *ARS1* as well as others, a transcription factor binding domain (B3), separated by an AT-rich sequence. Using linker-scanning substitution analysis, the AT-rich regions of *ARS1* and *ARS307* were further subdivided into discrete functional elements called B1 and B2. The Origin Recognition Complex (ORC) binds to the ACS and to part of the B1 element, and the Abf1 transcription factor binds to the B3 element. The role of the other regions of the ARS in origin function remains conjectural.

roles. The spacing between the B and the A elements, and the orientation of the elements relative to each other have also been shown to be critical for ARS function (23). The functional roles of the A and the B3 domains, as described below, appear to be the specific binding of components of the initiator complex and the Abf1 transcription factor, respectively. The roles of the other elements remain speculative, although they are likely to be involved in such processes as DNA unwinding (89).

As mentioned above, not all ARSs that function as origins on plasmids are active in their chromosomal context (20, 31, 47). Furthermore, although neighboring ARSs do appear to fire synchronously within each S phase, there is an observable temporal pattern of ARS firing along a chromosome (71). These observations imply that sequences that flank the domains described above influence both the frequency and the timing of origin firing. In the case of *ARS501*, on chomosome V, telomere proximity appears to be responsible for firing late in S phase (31). However, for two Chromosome XIV ARSs (*ARS1412* and *ARS1413*), late replication is telomere-independent and is instead dependent on nearby flanking sequences (34). Thus, there are likely to be additional levels of complexity governing the frequency of origin firing and the kinetics of S phase progression that are only just beginning to be uncovered.

ARS Function Requires an Active Transcription Factor

The first identified ARS-binding protein was ARS binding factor 1 (Abf1), which recognizes the B3 element of *ARS1* (23, 89, 93). Abf1 has also been identified as a transcription factor that participates in the regulation of many genes. Abf1 binding sites have been found in some but not all ARSs, although the observation that Abf1 sites can function at a distance may make them difficult to identify in the context of the ARS (100). The Abf1 binding site can be functionally replaced by the binding site for other transcription factors such as Rap1 (69). In these cases, it is the presence of an active transcriptional activation domain per se that appears to be a requirement for replacing Abf1 function. For example, the binding site for the Gal4 transcription factor (a protein involved in galactose utilization which is activated when cells are grown in galactose-containing media) will substitute for the Abf1 binding site within an ARS, but only if galactose is used as a carbon source for growing the cells (69). Exactly how an active transcription factor contributes to the function of the ARS is unknown.

The Origin Recognition Complex

The identification of ARS elements, and in particular the ACS, facilitated the purification of a six-protein subunit complex called the Origin Recognition Complex (ORC). ORC binds to the ACS in an ATP-dependent manner, protecting both the ACS and part of the B1 domain from nuclease digestion (5, 83, 88). Only sequences within the A domain are essential for ORC binding, and mutations in the ACS that reduce ORC binding in vitro have also been found to reduce ARS function in plasmid stability assays (5). Thus, ORC is a good candidate for a eukaryotic initiator. This role has been substantiated by genetic evidence that followed the cloning of the genes which encode the subunits of ORC. Strains carrying viable mutations in *ORC2* or *ORC5* (*orc2-1* or *orc5-1* alleles) show defects in initiation from chromosomal origins and in plasmid maintenance (3, 32, 33, 64, 74). Furthermore, *orc2-1* and *orc5-1* show genetic interactions with mutations in a number of genes known to be involved in the control of DNA synthesis, including *CDC6, CDC14, CDC45, CDC46, CDC54,* and *CDC47* (63) (these are discussed in detail below in the section *Proteins Involved in Formation of Pre-Replicative Complexes*). A genetic interaction was also uncovered between *orc2-1* and a mutation in *CDC7* that encodes a kinase essential for triggering DNA synthesis (39, 63). A mutation in the *ORC3* gene, *orc3-1*, which is lethal in combination with *orc2-1*, confers many of the same phenotypes as do the *orc2* and *orc5* mutations, including a G1–S phase arrest at the nonpermissive temperature and an inability to efficiently maintain plasmids with a single origin of replication (39). *ORC6*, which was cloned using a screen for proteins that interact with the ACS, has also been shown to interact genetically with *CDC6* as well as with genes that encode members of the MCM family of proteins (61); both of these, as discussed below, are presumed to physically associate with origin sequences.

ORC has been found to bind the ACS throughout the cell cycle, demonstrating that, although ORC may be important for DNA replication, it is not alone sufficient for the initiation of DNA synthesis (21, 22). Thus other factors, presumably linked to the cell cycle machinery, must be interacting with ORC to control the initiation of DNA synthesis in S phase. Candidate proteins that interact with ORC to promote replicative competence and their regulation in the cell cycle will be discussed in the following two sections.

Sequence analysis of the ORC protein subunits gives little indication of their likely function, because only Orc1 shows significant similarity to any other proteins. It is most related to the budding yeast Sir3 protein, which has a role in transcriptional silencing (see next section) and which shares 50% identity to Orc1 over a 200 amino acid N-terminal domain (4). Orc1 is also related to two proteins involved in the cell division cycle: Cdc6 from *S. cerevisiae* and Cdc18 from *S. pombe* (4). These proteins are particularly interesting because they play pivotal roles in the control of DNA synthesis. The region of similarity between Cdc18, Cdc6, and Orc1 includes a purine nucleotide binding motif (NTP) which is not present in the Sir3 sequence. The Orc5 subunit also contains a predicted NTP-binding motif and, as mentioned, ORC requires ATP in order to bind the ACS (63). Mutational analysis indicates that the Orc1 NTP-binding motif is essential for cell viability, while inactivation of the Orc5 NTP-binding site results in temperature-sensitive growth (63). The presence of two nucleotide binding sites—only one of which is essential for ORC activity—suggests that ORC may have a function other than just as a platform for further complex formation. Possible roles include unwinding the DNA helix in the region of the ACS, and/or a regulatory role, for example, in preventing reinitiation during S phase.

Homologues of ORC subunits 1, 2, and 5 have recently been cloned from a number of eukaryotes including *Arabidopsis thaliana, Caenorhabditis elegans, S. pombe, Drosophila melanogaster,* and humans (35, 36). In the case of *Drosophila,* Orc2 and Orc5 were found to be part of a multiprotein complex consisting of six subunits with a total molecular mass comparable to that of yeast ORC (36). Furthermore, the *Drosophila ORC2* could suppress some of the phenotypes resulting from an *orc2-1* mutation in yeast (29). In *Xenopus* egg extracts, which can support DNA replication in vitro, immunodepletion of the *Xenopus* Orc1 and Orc2 homologs blocks the initiation of DNA synthesis (11, 87). Thus it is likely that many of the components that regulate DNA synthesis in yeast will be conserved in all eukaryotes.

ORC and Silencing

In addition to their role in DNA replication, the ACS and ORC also play a role in transcriptional silencing. In budding yeast, the mating type is determined by information carried at the *MAT* locus; cells with *MATa* are *a* mating type and cells with *MATα* are α mating type. In addition to the *MAT* locus, cells carry cryptic copies of the *a* and α mating type genes called *HMR* and *HML* respectively. These loci are transcriptionally silent and are used only as a source of mating type information that can be transposed to the *MAT* locus when cells switch their mating type. Four elements in these silent loci (*HMR-E, HMR-I, HML-E,* and *HML-I*) are required for repression. All four elements contain an ACS, and all four of these regions can act as ARS elements on plasmids (59). Only *HMR-E*, however, has been shown to act as an origin of replication in its natural chromosomal context (86). Several lines of evidence link DNA replication and silencing. First, mutations in *HMR-E* that destroy origin function also prevent silencing (72). Second, using a temperature-sensitive allele of *SIR3* (encoding a protein required for silencing) Miller and Nasmyth showed that, following a shift from the nonpermissive to the permissive temperature, silencing was restored, but only if the cells had first passed through S phase (75). Third, loss of the Cdc7 protein kinase activity that is required for the initiation of DNA replication can restore repressing activity to defective silencers, whereas overexpression of *CDC7* can derepress wild-type silencers (2).

Recently, two independent groups devised genetic screens to find defects in silencing that employed strains in which the transcriptional repression of a reporter gene was dependent on the ACS. Both groups identified *ORC2* as a gene required for the maintenance of silencing (32, 74). The isolated *orc2* alleles also showed phenotypes that were consistent with defects in DNA synthesis. Similar alleles of *ORC5* have also been identified (63). Although there is a clear connection between silencing and replication, the mechanism by which the origin recognition complex functions in silencing is still a mystery. (For a cogent discussion of possible models, see Reference 24).

PROMOTING REPLICATION COMPETENCE

Replication Licensing

Experiments performed in the 1970s using HeLa cell fusions demonstrated that cells exist in two distinct states of competence for the initiation of DNA synthesis (84). Thus, fusions of S phase cells with cells in G1 result in unperturbed replication in the S phase nuclei and accelerated DNA replication in the G1 nuclei. In marked contrast, fusion of S phase cells with cells in G2 have no effect on the G2 nuclei. The ability of cells in S phase to prematurely induce DNA synthesis in G1 nuclei suggests that there is a diffusible DNA synthesis–inducing factor in the S phase cells, and that G1 nuclei are in a competent state to respond to this activity. The inability of G2 nuclei to respond to this S phase promoting factor led to the suggestion that nuclei must pass through mitosis before they are competent to begin DNA synthesis (20). Subsequent experiments by Blow and Laskey found that the inability of a G2 nucleus in *Xenopus* egg extracts to re-replicate its DNA could be overcome by permeabilizing the nuclear membrane (6). This dependence of S phase on mitosis and the consequent breakdown of the nuclear membrane led to the formulation of the replication "licensing factor" model. This model proposes that a licensing factor, essential for DNA replication, gains access to the nucleus only during mitosis, when the nuclear membrane is broken down. Then, following initiation of DNA replication, the licensing factor is inactivated, ensuring that replication takes place only once per cell cycle (15, 89).

Recent analysis of complexes formed on yeast replicators suggests that a process analogous to licensing may be occurring. In vivo DNA footprinting analysis shows that the chromatin state of yeast replicators changes in the cell cycle, alternating between a postreplicative state and a prereplicative state, which presumably reflects the binding of distinct complexes (22). The post-replicative complex (post-RC) that can be seen in S phase, G2, and early M phase appears to consist primarily of ORC, which, as discussed above, binds to replicators throughout the cell cycle; the presumed absence of other proteins in the postreplicative complex is based on the observation that the postreplicative footprint is very similar to an in vitro footprint derived solely from purified ORC. In late mitosis an extended footprint appears across the B domain, and this prereplicative complex (pre-RC), which is probably due to the accretion of additional factors onto the ORC, persists until the beginning of the S phase (22) (Figure 2). The temporal appearance and disappearance of the pre-RC suggests that it may represent licensing. If so, what are the components of the prereplicative complex and how do they bring about the intitiation of DNA replication?

Proteins Involved in Formation of Prereplicative Complexes

MCMs   There is now good evidence, in several model systems, that the MCM family of proteins, first identified in budding yeast, are important components of the replication license (14, 58, 67). The budding yeast MCM family is made up of five genes—*MCM2, MCM3, CDC46 (MCM5), CDC47,* and *CDC54*—that were originally isolated as mutants, unable to maintain ARS-containing plasmids (*mcm2/3/5*), or that were isolated as cell cycle mutants, unable to enter S phase (*cdc46/47/54*). These mutants show defects in DNA replication, and they also show genetic interactions with the genes encoding ORC. Thus

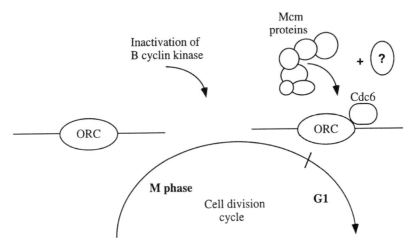

Figure 2   Replication competence is established at the end of mitosis, following the inactivation of the mitotic B-cyclin kinase. The competent state is defined by the appearance of a prereplicative complex at the origin—characterized by an extended nuclease protection pattern. Formation and maintenance of the complex is dependent upon Cdc6 which is synthesized at the end of mitosis. The formation of the pre-RC also coincides with the nuclear localization of the Mcm proteins, which suggests that these proteins contribute to the establishment of the pre-RC. Other proteins, yet to be identified, may also participate in this process.

overexpression of *ORC6* is lethal in *cdc46* strains and mutations in *ORC2* and *ORC5* are synthetically lethal with *cdc46*, *cdc47*, and *cdc54* mutations (61, 63). This genetic data suggests that a complex of Mcm proteins and ORC physically interact. In fact, an Mcm family member from fission yeast, Cdc21, has recently been shown to physically interact with the fission yeast homolog of Orc1 (37). Mcm homologs have been identified in a large number of eukaryotic organisms; sequence comparisons suggest that the protein family falls into six related groups, with each species containing only one member of each group. In *S. cerevisiae*, a sixth family member that is a clear homolog of the fission yeast *mis5* gene has been identified in the budding yeast genome sequencing project (Diffley, personal communication). All of the proteins share a consensus NTP-binding domain, although no specific biochemical function has been ascribed to any of the proteins (15, 57).

One of the problems with the licensing factor model, as it was originally described, is that in yeast and a number of other organisms, nuclear breakdown following mitosis does not occur. In these cases, the licensing factor would either have to be actively imported (or exported) across the nuclear membrane, or alternative mechanisms would have to be invoked, which would limit the activity of the protein(s) to the G1 period. In fact, recent evidence indicates that the mouse P1 (Mcm3) and *Xenopus* Mcm3 proteins are found within nuclei at all times during the cell cycle; however, they are only found tightly bound to chromatin during G1 (58, 68). In budding yeast, the Mcm2, Mcm3, Cdc46, and Cdc47 proteins are localized to the nucleus only during G1, which indicates that they are actively transported across the nuclear membrane (19, 43, 102). The timing of their nuclear entry corresponds to the timing of the switch between the post-RC to the pre-RC, lending credence to the idea that these proteins are involved in the switch from a replication-incompetent state to a replication-competent state.

CDC6   Another protein that has a role in establishing replication competence is Cdc6. Characterization of the original *cdc6* mutant by Hartwell suggested that Cdc6 has a role early in S phase (40). Like the *mcm* mutants, *cdc6* mutants also fail to properly maintain ARS-containing plasmids. This phenotype, in both *cdc6* and *mcm* mutants, can be rescued by providing multiple copies of the ARS element on the plasmid, suggesting that the mutants are defective in replicator utilization (44).

Recently, using a yeast strain in which *CDC6* is expressed from a repressible promoter, Cocker et al (16) showed that expression of *CDC6* was required for the formation of the pre-RC following release from a mitotic block. In the absence of Cdc6, cells exit from mitosis and enter G1 but fail to initiate DNA synthesis. Furthermore, in a *cdc6* temperature-sensitive strain, pre-RCs are thermolabile, which implies that Cdc6 is also required for the mainteinance of pre-RCs. Liang et al (62) have shown that (^{35}S)Methionine-labeled Cdc6, when mixed with affinity-purified ORC from yeast, can be immunoprecipitated using antibodies to ORC, which indicates that there is, in fact, a direct interaction between Cdc6 and ORC. Thus, it is likely that Cdc6 is an essential component of the pre-RC. Genetic experiments which show that increased expression of *CDC6* can suppress the temperature-sensitive phenotype of an *orc5* mutation, and that the phenotype of a *cdc6* mutation is exacerbated by overexpressing *ORC6* (61, 62), are consistent with the above observations. It has been shown that Cdc6 is unstable and therefore must be resynthesized in each cell cycle. Thus, not only is Cdc6 a component of the pre-RC, but the instability of the protein may also provide a means by which pre-RCs can be inactivated following the initiation of DNA synthesis (see below).

Normally, in exponentially growing cells, the *CDC6* gene is cell cycle regulated under the control of the Swi5 protein, with peak expression occurring at the M phase–G1 border. However, after long periods in G0, or even after an α-factor induced G1 arrest, the *CDC6* gene is expressed in G1 under the control of the MBF transcription factor (see seection *Transcriptional Control of S Phase*

Genes, below) (81). The need for two separate mechanisms to control *CDC6* expression may be explained by the observation that prereplicative complexes are lost in stationary phase (22). Thus, as cells reenter the cell cycle from G0, they need to resynthesize components of the prereplicative complex—which, in actively cycling cells, are made at the end of the previous cell cycle.

CDC14   Mutation of *CDC14* leads to a cell cycle block in telophase similar to other late mitotic mutants such as *cdc15, cdc5,* and *dbf2.* However, several lines of evidence suggest that *CDC14* may have a role in S phase. Like *cdc6* and *mcm* mutants, the *cdc14* mutants are deficient in maintaining ARS-containing plasmids, and this deficiency can be suppressed by increasing the number of ARSs on the plasmid (44). *cdc14* Mutations also show a genetic interaction with *orc2-1* mutations, which suggests an interaction with ORC (39, 63). *CDC14* encodes a protein phosphatase (PJ Fitzpatrick and L Johnston, unpublished data) but the molecular role of Cdc14 remains unclear.

CELL CYCLE CONTROL OF S PHASE

From Start to S Phase

Following mitosis and the assembly of the pre-RC, yeast origins are competent to initiate DNA synthesis but fail to do so until the cell passes through a G1 decision point called Start. Start is an event, or the culmination of a series of events, which occurs in late G1 and results in irreversible commitment to a new round of cell division (78). This commitment is independent of subsequent changes in the nutritional environment or the presence of mating pheromones. Following Start, the cells duplicate their spindle pole bodies, they direct growth to the newly forming bud, they activate the G1 transcription machinery, and they initiate DNA synthesis. The execution of Start is dependent upon the activation of the cyclin-dependent protein kinase, Cdc28. In budding yeast, there are nine cyclins that associate with and modulate the activity of Cdc28. The three G1 cyclins (Cln1-3) are required for passage through Start while the six B-type cyclins (Clb1-6) are involved in various aspects of S phase and mitotic progression (78). We will focus on those aspects of G1 regulation that are thought to act on the preformed initiation complex and that lead directly to the initiation of DNA replication (summarized in Figure 3).

Yeast strains lacking all three G1 cyclins (Clns) arrest the cell cycle in G1 with a phenotype similar to that of *cdc28* mutants. Initial analysis of strains carrying pairwise combinations of *cln* null alleles showed that any one of the Clns was sufficient for Start and indicated that the three Cln proteins carried out overlapping or even equivalent functions (85). However, while it is clear that some cyclins can compensate for the loss of others, recent experiments

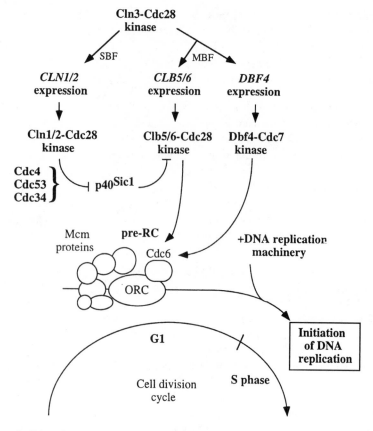

Figure 3 Triggering the initiation of DNA replication in S phase requires not only the formation of the pre-RC, but it also depends on signals resulting from passage through Start. Start culminates in the activation of two kinases, Dbf4-Cdc7 and Clb5/6-Cdc28, which are thought to phosphorylate targets within the pre-RC.

show that the functions of the G1 cyclins are not equivalent (98). *CLN1* and *CLN2* encode highly similar proteins and their expression is cell cycle regulated with peak expression occurring in late G1 (38, 101). *CLN3*, on the other hand, encodes a protein with only weak similarity to other cyclins and is expressed more or less constantly throughout the cell cycle (98). Moreover, once cells have reached a critical size, it is Cln3 that is responsible for activating the late G1 transcription of a whole set of genes including that of *CLN1* and *CLN2*. This antecedent role for Cln3 is based on experiments which show that, unlike Cln1 and Cln2, the ectopic expression of Cln3 leads to an immediate induction of

G1-specific messages. These experiments also show that inactivation of *CLN3* results in a severe delay in the appearance of these mRNAs (25, 98). The Cln3-dependent induction of *CLN1* and *CLN2* transcripts results in an increase in the Cln1/2-Cdc28 kinase activity. It is this activity which is thought to catalyze most or all of the events associated with Start.

Although the initiation of DNA replication is dependent upon activation of the Cln1/2-Cdc28 kinase, it is probably not regulated directly by this kinase activity. Evidence indicates that it is the B-type cyclins, in particular Clb5 and Clb6, which directly activate DNA replication (92). Thus, S phase—but not bud emergence—is delayed by at least 30 minutes in *clb5* and *clb6* double mutants. The fact that S phase occurs at all in these mutants indicates that other B-type cyclins (Clb1-4), expressed later in the cell cycle, have a latent ability to promote S phase. It has been shown by Schwob et al (91) that strains lacking all six B-type cyclins are unable to enter S phase, but that they are able to perform other Start-dependent functions such as budding—in fact, the sextuple mutants form multiple buds. However, Clb5/6 kinase activity peaks at the beginning of S phase, earlier than that of the other B-type cyclins, and therefore, it is likely that Clb5 and Clb6 are usually the relevant cyclins for the activation of DNA replication. It has also been observed that *clb5* mutants have a protracted S phase. This may be the result of fewer origins firing, or it may indicate that the Clb5 kinase has a role in S phase transit as well as in the initiation of S phase (30).

CLB5 and *CLB6* are expressed at the same time as *CLN1* and *CLN2*, and Clb5- and Clb6-Cdc28 complexes can be detected in G1, but these complexes are initially inactive due to inhibition by the Sic1 protein (91). Sic1 was originally identified as a substrate of Cdc28 that associated with the Cdc28 complex. Subsequently, Sic1 was found to be an inhibitor of the Cdc28-Clb protein kinase in vitro (73). As a Clb-kinase inhibitor, Sic1 appears to function at two stages in the cell cycle; first, in promoting the exit from mitosis by contributing to the shut off of the mitotic form of the Clb-Cdc28 kinase and, second, in controlling the correct timing of S phase by regulating the activity of the Clb5/6-Cdc28 kinase (27, 91).

Extraordinarily, the primary function of the Cln1/2-Cdc28 kinase in promoting S phase seems to be the targeted destruction of Sic1. The first indication of this role came from analysis of conditional alleles of three genes, *CDC34*, *CDC4*, and *CDC53*. It turns out that these genes are specifically required for S phase and that, at the restrictive temperature, mutants defective in these genes give phenotypes similar to the *CLB1-6* sextuple delete strain described above (91). Because *CDC34* encodes a ubiquitin-conjugating enzyme, these observations suggest that a protein or proteins have to be destroyed in order to activate the B-type cyclin kinase activity in S phase. Schwob et al (91) showed that in *cdc34* arrested cells there was no Clb2 or Clb5 associated kinase activity. An obvious target of Cdc34 was the Sic1 cyclin-dependent kinase (cdk) inhibitor. In wild-type cells Sic1 was found to accumulate in late mitosis and disappear shortly before S phase (27, 91). Sic1 is phosphorylated at Start in a Cln-dependent manner and the disappearance of Sic1 was found to depend on both Cdc34 and the Cln-Cdc28 kinase (90, 91). Furthermore, in *sic1* mutants DNA replication is uncoupled from other Start regulated events, for example, occurring earlier than budding. The targeted destruction of Sic1 may be the only essential function of the G1 cyclins since deletion of *SIC1* rescues the inviability of the *cln1, cln2, cln3* triple mutant (although the quadruple mutant is far from normal and grows poorly) (90).

So far we have mentioned two ways in which the Clb-Cdc28 kinase, which is required for S phase, is regulated by Start-dependent events; the late G1-specific transcription of *CLB5* and *CLB6* and the targeted destruction of the Clb-Cdc28 kinase inhibitor Sic1. There may also be a third mechanism by which Start triggers activation of the Clb kinase. Inactivation of the mitotic form of the B-cyclin kinase activity is a prerequisite for formation of the prereplicative complex and for exit from mitosis (described below). Amon et al (1) have shown that this B-cyclin proteolysis, which enables the cell to exit mitosis, actually persists after mitosis until the start of the subsequent S phase. Through an unknown mechanism, the Cln-Cdc28 kinase is responsible for shutting off Clb proteolysis. Although this has been shown so far only for Clb2, the regulated proteolysis of B cyclins may be another way that the cell controls the timing of S phase.

A second protein kinase activity which is dependent upon the Cln-Cdc28 kinase, and which is required for S phase entry is Cdc7. Mutations in *CDC7* cause a cell cycle arrest in late G1—after Start, but immediately prior to S phase. When *cdc7* mutants, arrested at the nonpermissive temperature, are returned to the permissive conditions, the cells initiate DNA synthesis even in the absence of protein synthesis; thus, Cdc7 kinase activation is possibly the last regulatory step prior to initiation of DNA replication (45). The Cdc7 protein is present at constant levels throughout the cell cycle, but its kinase activity is periodic, and it peaks at the G1/S boundary (49). This cell cycle regulation of Cdc7 activity may occur via a number of possible mechanisms. First, active Cdc7 kinase requires an interaction with a regulatory subunit encoded by *DBF4* (49, 55). This gene was first identified as a cell cycle mutation with a phenotype essentially identical to a *cdc7* mutation (53, 54). Thus, Dbf4 may activate Cdc7 in a fashion analogous to cdk activation by cyclins. *DBF4* is transcribed in a Start-dependent manner in late G1 (13) (see next section) and the Dbf4 protein is turned over rapidly (Godinho-Ferreira & Diffley, personal communication), thereby allowing for a short burst of Cdc7 activity at the beginning of S phase.

However, the turnover of Dbf4 protein cannot be the only mechanism by which Cdc7 is regulated, because Dbf4 is normally degraded late in the cell cycle— some time after the disappearance of Cdc7 kinase activity (49). In this regard, it is interesting to note that Cdc7 is a phosphoprotein, although it is not yet known to what extent phosphorylation contributes to the activity of the kinase (103).

How do Dbf4-Cdc7 and Clb5/6-Cdc28 kinases facilitate the initiation of DNA synthesis? It is not known if these kinases function on the same or on separate pathways, or whether they share the same target proteins. However, there is evidence that, in both cases, these kinases are targeting the preformed initation complex. The Dbf4 protein was found to interact with replicators in a one-hybrid assay (28), and Cdc7 interacts with the Orc2 subunit both genetically and in a two-hybrid assay (39). Furthermore, Orc2 is a substrate for the Cdc7 kinase in vitro. Physiological targets for the Clb5/6-Cdc28 kinase are lacking, but both Orc2 and Cdc6 contain potential Cdc28 phosphorylation sites. In fission yeast, Cdc2 (a homolog of Cdc28) interacts with spOrc2 (Orp2), and Orp2 in turn has been shown to interact with Cdc18 (a homolog of Cdc6), which raises the possibility that these potential kinase-substrate interactions may be conserved among eukaryotes (60).

Transcriptional Control of S Phase Genes

In budding yeast, many of the genes that are required for control of S phase, for DNA replication itself, and also for postreplication repair are expressed under cell cycle control in late G1. This late G1 transcription is dependent upon the Cln-Cdc28 kinase and, as discussed above, is most likely dependent specifically upon the Cln3-Cdc28 kinase. At present, G1 transcribed genes include a total of approximately 30 genes but, clearly, this number is likely to increase as more S phase genes are characterized (52, 56). Table 1 contains a selection of genes which are regulated by MBF, and which are important for the initiation of DNA synthesis.

The genes are coordinately expressed and contain an element known as the *MluI* cell cycle box or MCB element in their promoter sequences. This element is a six base pair sequence, ACGCGT, although some cell cycle–regulated S phase genes contain only a single ACGCG sequence, which may be the essential core [for more detailed reviews, see References 51, 52, and 56]. These MCB elements by themselves are both necessary and sufficient for the cell cycle regulation (66) (Figure 4).

A transcription factor, initially named DSC1 (for DNA synthesis control), and subsequently MBF (for MCB-binding factor), binds in late G1 to the MCB elements (66). MBF is a heterodimeric factor consisting of Mbp1 and Swi6. Mbp1 is a specific DNA binding protein containing ankyrin repeats, and it is closely related to other G1 transcription factors in both budding and fission yeast

Table 1   Genes controlled by the transcription factor DSC1/MBF that are potentially involved in initiation of S phase

Gene/protein	Motif/function	Interactions shown
CDC6	Purine nucleotide binding motif pre-RC formation/maintenance	*ARS, ORC6*
CDC46/MCM5	Licensing	*ARS, ORC6, CDC45, 47,* and *54*
CLB5	B-type cyclin	*CDC28*
CLB6	B-type cyclin	*CDC28*
DBF4	Cdc7 activation	*ARS, CDC7*
ORC6	ORC subunit	*ARS, CDC6, CDC46*
POL1	Polymerase α-primase subunit	
POL12	Polymerase α-primase subunit	
PRI1	Polymerase α-primase subunit	
PRI2	Polymerase α-primase subunit	
RFA1	Replication factor A subunit	
RFA2	Replication factor A subunit	
RFA3	Replication factor A subunit	

For a more comprehensive list of genes regulated by this transcription factor see reference (97).

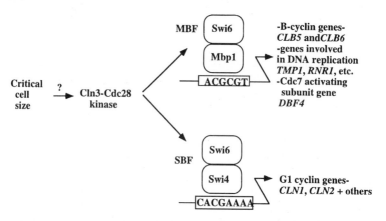

Figure 4   The G1-specific transcription factors in *S. cerevisiae*, MBF and SBF, are shown with their consensus binding sites. Some of the genes that are regulated by these factors are indicated. (For a more complete list, see Table 1 and References 52, 56, and 97.)

(7, 56). In contrast, Swi6 does not bind DNA, specifically, but it associates with Mbp1. Swi6 also associates with another Mbp1-related DNA binding protein, Swi4, to form the transcription factor SBF. SBF, which binds the SCB promoter element, CACGA$_4$, first identified in the promoter of the *HO* gene, was subsequently found to control the expression of genes encoding G1 cyclins such as *CLN1* and *CLN2* (51, 56) as well as genes involved in cell wall biosynthesis (48) (Figure 4). It is important to note that the timing of SBF-regulated gene expression is similar or the same as the timing of MBF-controlled S phase gene expression. This coincidence, as well as the association of Swi6 with two different specific DNA binding proteins, suggests that Swi6 may be responsible for the timing of gene expression in late G1. When Swi6 is deleted, the cell cycle regulation of these genes is either seriously perturbed or it is abolished, depending on the genetic background (26, 65). Furthermore, Swi6, Swi4, or Mbp1 represent either direct or indirect potential targets for the Cln3-Cdc28 kinase. In fact, when Swi4 is overexpressed, cells divide at very small sizes, even in the absence of *CLN3*. The small size implies the advance of Start and a truncated G1. These observations, therefore, are consistent with the notion that the only role of Cln3 in G1 is the activation of the transcription machinery at Start (25). Thus, MBF and SBF, possibly through Swi6, respond to traversal of the cell cycle and are activated at the appropriate time to stimulate expression of relevant genes.

The fact that *SWI6* deletion strains are alive, when the periodic expression of S phase genes is severely perturbed, argues that this regulation has little or no relevance for S phase and its control. Whereas MCB elements occur and DSC1-related transcription factors are found in fission yeast, there are few S phase genes that are periodically expressed in G1 (97). DNA ligase and the DNA polymerase-primase complex, enzymes that are encoded by MCB-regulated S phase genes from budding yeast, have been studied in the cell cycle, and neither is rate limiting. However, *swi6*Δ strains are not normal, they grow slowly, they have a defective cell wall and also a protracted S phase [(48) and our unpublished observations]. Thus, some of the periodic S phase genes may be rate limiting for DNA synthesis. Possible candidates are thymidylate synthase and ribonucleotide reductase, both of which encode enzyme activities that apparently fluctuate in the cell cycle at about the same time as when the genes are expressed. The latter enzyme, ribonucleotide reductase, is also expressed under cell cycle control in all systems so far examined—including fission yeast, where ribonucleotide reductase is the only replication enzyme known to be regulated in this way. Therefore, there may conceivably be a final control over DNA replication at the level of precursor supply.

A summary of the regulatory events initiated by the G1 transcription factors MBF and SBF is as follows: When cells reach a critical size for entry into the cell cycle, the Cln3-Cdc28 kinase activates the G1 transcription machinery. This induces the expression of genes whose products are required for S phase—including *CLN1, CLN2, CLB5, CLB6, DBF4*, and the genes involved in DNA synthesis itself. The Clb5/6 proteins form a complex with Cdc28 that is initially kept inactive by the p40 Sic1 cdk inhibitor. Targeted destruction of Sic1 by the combined actions of the Cln1/2-Cdc28 kinase and the ubiquitin conjugating enzyme Cdc34 leads to activation of the Clb5/6-Cdc28 kinase. A second kinase, Cdc7-Dbf4, is also activated in a Start-dependent manner and together with the Clb5/6-Cdc28 kinase goes on to trigger the initiation of DNA replication (see Figure 3). At this time, the peak expression of the genes needed for S phase could ensure an adequate and stoichiometric supply of the necessary enzymes when they are required.

AFTER THE PARTY, WHO TURNS OUT THE LIGHTS?

As described above, the initiation of DNA replication can now be thought of as a two-step process, characterized by the formation of prereplicative complexes in late M/early G1 and followed by the conversion of these complexes to active replication forks in late G1. This conversion step is a Start-dependent event, and it requires the activity of the B-cyclin kinases that remain active in one form or another until the cell exits mitosis. Why then, in this environment of high B-cyclin kinase activity, do origins not continue to fire throughout the cell cycle? A likely scenario is that the build-up of B-cyclin protein kinase activity in late G1 not only triggers the firing of origins that have formed pre-RCs but also blocks the formation of new pre-RCs—preventing further origin firing until the destruction of the B cyclins in anaphase. Thus, the inactivation of B-cyclin kinase, which is a prerequisite for mitotic exit, may also be required for the switch to the replication-competent state and the initiation of S phase in the subsequent cell cycle (Figure 5).

Budding yeast strains, in which four of the B cyclins (Clb1-4) are inactivated, replicate their DNA once and then arrest in the G2 phase of the cell cycle with high levels of the remaining Clb5/6 B-cyclin kinase activity. Using this strain background, Dahmann et al (18) isolated mutants that were able to undergo a second round of replication without an intervening mitosis. Analysis of these mutants showed that the normally high level of Clb5 kinase activity was reduced, and that it was this decrease in B-cyclin kinase activity that resulted in an extra round of DNA replication. They went on to show that transient ectopic expression of the *SIC1* B-cyclin kinase inhibitor gene in cells that are arrested in G2 (and therefore in a postreplicative state) is sufficient to promote formation of new pre-RCs and to induce rereplication. These experiments are similar to ones that had been performed previously in fission yeast, in which inactivation of the mitotic form of Cdc2 kinase—through either a temperature-sensitive *cdc2* mutation, inactivation of *cdc13* (encoding a B-type cyclin), or through overexpression of *rum1* (a gene encoding an inhibitor of the Cdc13-Cdc2 kinase)—led to endoreduplication, i.e. multiple rounds of DNA replication without an intervening mitosis (10, 17, 41, 76). Thus, there is good evidence that the B-cyclin

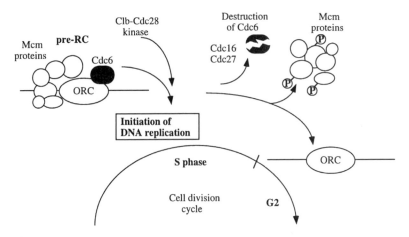

Figure 5   Following Start the (Clb) B-cyclin kinase-dependent conversion of the pre-RC to an active replication fork also inhibits the formation of new pre-RCs, which facilitates a return of origins to the postreplicative state. By analogy with work done in fission yeast, Cdc6 may be a target of the Clb-kinase—resulting in a ubiquitin-mediated proteolysis of Cdc6, which is mediated, in part, by the Cdc16 and Cdc27 proteins. B-cyclin kinase activity also leads to the dissociation of the Mcm complex of proteins from origins. Continued high B-cyclin kinase activity in G2 and M phases prevents reformation of the pre-RCs until the B-cyclins themselves are destroyed at the end of mitosis.

kinase activity that is required for the activation of DNA replication also has a role in preventing the assembly of pre-RCs, and/or in promoting the disassembly of pre-RCs, thereby tying the initiation of DNA replication to the cell cycle oscillations in CDK activity.

A likely target for the B-cyclin kinase that has come out of the fission yeast studies is Cdc18, a protein that shows both structural and functional similarities to budding yeast Cdc6. Overproduction of Rum1, which, as mentioned previously, results in endoreduplication, also results in the concomitant accumulation of Cdc18 protein—probably by increasing the stability of Cdc18 (50). Overproduction of Cdc18, itself, also results in endoreduplication (80). One explanation for these observations is that Cdc18 is a downstream target for the inhibitory B-cyclin kinase activity. Consequently, overexpressing Cdc18 might titrate out the kinase activity such that pre-RC assembly becomes unregulated—resulting in repeated rounds of DNA replication (50, 77). *CDC6* expression, which is required for formation of the pre-RC in budding yeast, occurs in the window of the cell cycle when Clb-kinase activity is low. In fact, the Swi5 transcription factor that is required for *CDC6* expression is negatively regulated by Clb-kinase activity. Thus, B-cyclin kinase activity might negatively regulate the formation of pre-RCs both by targeting the destruction of Cdc6 and by inhibiting its synthesis.

The idea that B cyclins target the destruction of a factor or factors that are involved in the maintenance of pre-RCs is reinforced in recent experiments with budding yeast. *CDC16* and *CDC27* encode components of the 20S complex that is involved in ubiquitin-mediated proteolysis. This complex has been shown to have a role in B cyclin degradation and is often referred to as the anaphase promoting complex (APC). Strains carrying temperature-sensitive alleles of these genes have now been shown, at the nonpermissive temperature, to undergo multiple rounds of DNA replication within a single cell cycle. This is true even though they accumulate high levels of Clb2(B cyclin)-Cdc28 kinase activity. Heichman & Roberts (42) suggest that these mutants are able to bypass the normal inhibitory effects of the Clb-kinase because they are accumulating a positive factor which is normally degraded following initiation of DNA replication. Without the machinery to destroy this factor, the cells behave in a manner that is similar to that of strains with a deficit of B cyclin activity. Based on observations described above for Cdc18 and the role of Cdc6 in maintaining the pre-RC in G1, an obvious target for the Cdc16/27-dependent proteolysis is Cdc6. Conceivably, Clb-Cdc28-dependent phosphorylation of Cdc6 might target it for proteolysis.

FUTURE DIRECTIONS

Although recent work concerning the regulation of S phase, particularly in yeast, answers a number of questions, many more remain. For example, what precisely are the changes to the origin that result in a pre-RC, and how do these changes impart replication competence? What other proteins participate in this process? In G1, the Cln3-Cdc28 kinase activity is thought to sense the nutritional state of the cell, determining, for example, if the cell has reached a critical size for division. It is still unknown how the cell transmits this information to Cln3, and how, in turn, activation of the Cln3-Cdc28 kinase leads to activation of the MBF and SBF transcription factors. The major phosphorylation sites on both Swi4 and Swi6 have been identified and mutated with no real effect on the timing of gene expression. What are the precise targets of the Dbf4-Cdc7 and Clb5/6-Cdc28 kinases, and how do these phosphorylation events potentiate the initiation of DNA synthesis? Finally, the B-cyclin kinase activity leads to the inactivation of the Mcm proteins, as well as to proteolysis of the Cdc6/18 protein. These events appear to be responsible for the inactivation of the pre-RC and for the prevention of reinitiation within the same cell cycle. Although proteolysis, in general, is important for many aspects of cellular metabolism, there is growing evidence that cell cycle–specific proteolysis is a key factor involved in regulating progression through the cell cycle. Determining how the proteolytic machinery is linked to cell cycle controls will be fundamental to our understanding of the cell cycle.

ACKNOWLEDGMENTS

We would like to thank all our colleagues who sent information prior to publication, and to apologize to those whose data we have not discussed due to space limitations and due to the particular emphasis of this review. We would also like to thank John Diffley and Ann Marie Flenniken for critical reading of the manuscript. Birgit Aerne was supported by an EMBO Long Term Fellowship.

> Visit the *Annual Reviews home page* at
> http://www.annurev.org.

Literature Cited

1. Amon A, Irniger S, Nasmyth K. 1994. Closing the cell cycle circle in yeast: G2 cyclin proteolysis initiated at mitosis persists until the activation of G1 cyclins in the next cycle. *Cell* 77:1037–50

2. Axelrod A, Rine J. 1991. A role for *CDC7* in repression of transcription at the silent mating-type locus HMR in *Saccharomyces cerevisiae. Mol. Cell. Biol.* 11:1080–91

3. Bell SP, Kobayashi R, Stillman B. 1993. Yeast origin recognition complex functions in transcription silencing and DNA replication. *Science* 262:1844–49

4. Bell SP, Mitchell J, Leber J, Kobayashi R, Stillman B. 1995. The multidomain structure of Orc1p reveals similarity to regulators of DNA replication and transcriptional silencing. *Cell* 83:563–68

5. Bell SP, Stillman B. 1992. ATP-dependent recognition of eukaryotic origins of DNA replication by a multiprotein complex. *Nature* 357:128–34

6. Blow JJ, Laskey RA. 1988. A role for the nuclear envelope in controlling DNA replication within the cell cycle. *Nature* 332:546–48

7. Breeden L. 1996. Start-specific transcription in yeast. *Curr. Top. Microbiol. Immunol.* 208:95–127

8. Brewer BJ, Fangman WL. 1987. The localization of replication origins on ARS plasmids in *S. cerevisiae. Cell* 51:463–71

9. Broach JR, et al. 1983. Localization and sequence analysis of yeast origins of DNA replication. *Cold Spring Harb. Symp.*

Quant. Biol. 2:1165–73

10. Broek D, Bartlett R, Crawford K, Nurse P. 1991. Involvement of p34^{cdc2} in establishing the dependency of S phase on mitosis. *Nature* 349:388–93

11. Carpenter PB, Mueller PR, Dunphy WG. 1996. Role for a *Xenopus* Orc2-related protein in controlling DNA replication. *Nature* 379:357–60

12. Celniker SE, Sweder K, Srienc F, Bailey JE, Campbell JL. 1984. Deletion mutations affecting autonomously replicating sequence *ARS1* of *Saccharomyces cerevisiae. Mol. Cell. Biol.* 4:2455–66

13. Chapman JW, Johnston LH. 1989. The yeast gene, *DBF4*, essential for entry into S phase is cell cycle regulated. *Exp. Cell. Res.* 180:419–28

14. Chong JP, Mahbubani HM, Khoo CY, Blow JJ. 1995. Purification of an MCM-containing complex as a component of the DNA replication licensing system. *Nature* 375:418–21

15. Chong JPJ, Thommes P, Blow JJ. 1996. The role of MCM/P1 proteins in the licensing of DNA-replication. *Trends Biochem. Sci.* 21:102–6

16. Cocker JH, Piatti S, Santocanale C, Nasmyth K, Diffley JF. 1996. An essential role for the Cdc6 protein in forming the pre-replicative complexes of budding yeast. *Nature* 379:180–82

17. Correa BJ, Nurse P. 1995. p25^{rum1} orders S phase and mitosis by acting as an inhibitor of the p34^{cdc2} mitotic kinase. *Cell* 83:1001–9

18. Dahmann C, Diffley JFX, Nasmyth KA. 1995. S-phase-promoting cyclin-dependent kinases prevent re-replication by inhibiting the transition of replication origins to a pre-replicative state. *Curr. Biol.* 5:1257–69

19. Dalton S, Whitbread L. 1995. Cell cycle-regulated nuclear import and export of Cdc47, a protein essential for initiation of DNA replication in budding yeast. *Proc. Natl. Acad. Sci. USA* 92:2514–18

20. Diffley JFX. 1995. The initiation of DNA-replication in the budding yeast-cell division cycle. *Yeast.* 11:1651–70

21. Diffley JF, Cocker JH. 1992. Protein-DNA interactions at a yeast replication origin. *Nature* 357:169–72

22. Diffley JF, Cocker JH, Dowell SJ, Rowley A. 1994. Two steps in the assembly of complexes at yeast replication origins in vivo. *Cell* 78:303–16

23. Diffley JF, Stillman B. 1988. Purification of a yeast protein that binds to origins of DNA replication and a transcriptional silencer. *Proc. Natl. Acad. Sci. USA* 85:2120–244

24. Dillin A, Rine J. 1995. On the origin of a silencer. *Trends Biochem. Sci.* 20:231–35

25. Dirick L, Bohm T, Nasmyth K. 1995. Roles and regulation of Cln-Cdc28 kinases at the start of the cell cycle of *Saccharomyces cerevisiae. EMBO J.* 14: 4803–13

26. Dirick L, Moll T, Auer H, Nasmyth K. 1992. A central role for *SWI6* in modulating cell cycle Start-specific transcription in yeast. *Nature* 357:508–13

27. Donovan JD, Toyn JH, Johnson AL, Johnston LH. 1994. p40^{SDB25}, a putative CDK inhibitor, has a role in the M/G1 transition in *Saccharomyces cerevisiae. Genes Dev.* 8:1640–53

28. Dowell SJ, Romanowski P, Diffley JF. 1994. Interaction of Dbf4, the Cdc7 protein kinase regulatory subunit, with yeast replication origins in vivo. *Science* 265:1243–46

29. Ehrenhofer-Murray AE, Gossen M, Pak DTS, Botchan MR, Rine J. 1995. Separation of origin recognition complex functions by cross-species complementation. *Science* 270:1671–74

30. Epstein CB, Cross FR. 1992. *CLB5*: a novel B cyclin from budding yeast with a role in S phase. *Genes Dev.* 6:1695–96

31. Ferguson BM, Brewer BJ, Reynolds AE, Fangman WL. 1991. A yeast origin of replication is activated late in S phase. *Cell* 65:507–15

32. Foss M, McNally FJ, Laurenson P, Rine J. 1993. Origin recognition complex (ORC) in transcriptional silencing and DNA replication in *S. cerevisiae. Science* 262:1838–44

33. Fox CA, Loo S, Dillin A, Rine J. 1995. The origin recognition complex has essential functions in transcriptional silencing and chromosomal replication. *Genes Dev.* 9:911–24

34. Friedman KL, et al. 1996. Multiple determinants controlling activation of yeast replication origins late in S phase. *Genes Dev.* 10:1595–1607

35. Gavin KA, Hidaka M, Stillman B 1995. Conserved initiator proteins in eukaryotes. *Science* 270:1667–71

36. Gossen M, Pak DTS, Hansen SK, Acharya JK, Botchan MR. 1995. A *Drosophila* homolog of the yeast origin recognition complex. *Science* 270:1674–77

37. Grallert B, Nurse P. 1996. The *ORC1* homolog *orp1* in fission yeast plays a key role in regulating onset of S phase. *Genes Dev.* 10:2644–54

38. Hadwiger JA, Wittenberg C, Richardson HE, de Barros-Lopes M, Reed SI. 1989. A family of cyclin homologs that control the G1 phase in yeast. *Proc. Natl. Acad. Sci. USA* 86:6255–59

39. Hardy CF. 1996. Characterization of an essential Orc2p-associated factor that plays a role in DNA replication. *Mol. Cell. Biol.* 16:1832–41

40. Hartwell LH. 1976. Sequential function of gene products relative to DNA synthesis in the yeast cell cycle. *J. Mol. Biol.* 104:803–17

41. Hayles J, Fisher D, Woollard A, Nurse P. 1994. Temporal order of S phase and mitosis in fission yeast is determined by the state of the p34^{cdc2}-mitotic B cyclin complex. *Cell* 78:813–22

42. Heichman KA, Roberts JM. 1996. The yeast *CDC16* and *CDC27* genes restrict DNA-replication to once per cell-cycle. *Cell* 85:39–48

43. Hennessy KM, Clark CD, Botstein D. 1990. Subcellular localization of yeast *CDC46* varies with the cell cycle. *Genes Dev.* 4:2252–63

44. Hogan E, Koshland D. 1992. Addition of extra origins of replication to a minichromosome suppresses its mitotic loss in *cdc6* and *cdc14* mutants of *Saccharomyces cerevisiae. Proc. Natl. Acad. Sci. USA* 89:3098–3102

45. Hollingsworth RJ, Sclafani RA. 1990. DNA metabolism gene *CDC7* from yeast

encodes a serine (threonine) protein kinase. *Proc. Natl. Acad. Sci. USA* 87:6272–76

46. Huberman JA, Spotila LD, Nawotka KA, el A-Assouli SM, Davis LR. 1987. The in vivo replication origin of the yeast 2 micron plasmid. *Cell* 51:473–81

47. Huberman JA, Zhu JG, Davis LR, Newlon CS. 1988. Close association of a DNA replication origin and an ARS element on chromosome III of the yeast, *Saccharomyces cerevisiae. Nucleic Acids Res.* 16:6373–84

48. Igual JC, Johnson AL, Johnston LH. 1996. Coordinated regulation of gene-expression by the cell-cycle transcription factor Swi4 and the protein-kinase-C map kinase pathway for yeast-cell integrity. *EMBO J.* 15:5001–13

49. Jackson AL, Pahl PM, Harrison K, Rosamond J, Sclafani RA. 1993. Cell cycle regulation of the yeast Cdc7 protein kinase by association with the Dbf4 protein. *Mol. Cell. Biol.* 13:2899–2908

50. Jallepalli PV, Kelly TJ. 1996. Rum1 and Cdc18 link inhibition of cyclin-dependent kinase to the initiation of DNA replication in *Schizosaccharomyces pombe. Genes Dev.* 10:541–52

51. Johnston LH. 1992. Cell cycle control of gene expression in yeast. *Trends Cell Biol.* 2:353–57

52. Johnston LH, Lowndes NF. 1992. Cell cycle control of DNA synthesis in budding yeast. *Nucleic Acids Res.* 20:2403–10

53. Johnston LH, Thomas AP. 1982. A further two mutants defective in initiation of the S phase in the yeast *Saccharomyces cerevisiae. Mol. Gen. Genet.* 186:445–48

54. Johnston LH, Thomas AP. 1982. The isolation of new DNA synthesis mutants in the yeast *Saccharomyces cerevisiae. Mol. Gen. Genet.* 186:439–44

55. Kitada K, Johnston LH, Sugino T, Sugino A. 1992. Temperature-sensitive *cdc7* mutations of *Saccharomyces cerevisiae* are suppressed by the *DBF4* gene, which is required for the G1/S cell cycle transition. *Genetics* 131:21–29

56. Koch C, Nasmyth K. 1994. Cell cycle regulated transcription in yeast. *Curr. Opin. Cell. Biol.* 6:451–59

57. Koonin EV. 1993. A common set of conserved motifs in a vast variety of putative nucleic acid-dependent ATPases including MCM proteins involved in the initiation of eukaryotic DNA replication. *Nucleic Acids Res.* 21:2541–47

58. Kubota Y, Mimura S, Nishimoto S, Takisawa H, Nojima H. 1995. Identification of the yeast MCM3-related protein as a component of *Xenopus* DNA replication licensing factor. *Cell* 81:601–9

59. Laurenson P, Rine J. 1992. Silencers, silencing, and heritable transcriptional states. *Microbiol. Rev.* 56:543–60

60. Leatherwood J, Lope GA, Russell P. 1996. Interaction of Cdc2 and Cdc18 with a fission yeast Orc2-like protein. *Nature* 379:360–63

61. Li JJ, Herskowit I. 1993. Isolation of *ORC6*, a component of the yeast origin recognition complex by a one-hybrid system. *Science* 262:1870–74

62. Liang C, Weinreich M, Stillman B. 1995. ORC and Cdc6p interact and determine the frequency of initiation of DNA replication in the genome. *Cell* 81:667–76

63. Loo S, et al. 1995. The origin recognition complex in silencing, cell cycle progression, and DNA replication. *Mol. Biol. Cell* 6:741–56

64. Loo S, Rine J. 1994. Silencers and domains of generalied repression. *Science* 264:1768–71

65. Lowndes NF, Johnson AL, Breeden L, Johnston LH. 1992. *SWI6* protein is required for transcription of the periodically expressed DNA synthesis genes in budding yeast. *Nature* 357:505–8

66. Lowndes NF, Johnson AL, Johnston LH. 1991. Coordination of expression of DNA synthesis genes in budding yeast by a cell-cycle regulated transcription factor. *Nature* 350:247–50

67. Madine MA, Khoo CY, Mills AD, Laskey RA. 1995. MCM3 complex required for cell cycle regulation of DNA replication in vertebrate cells. *Nature* 375:421–24

68. Madine MA, Khoo CY, Mills AD, Mushal C, Laskey RA. 1995. The nuclear envelope prevents reinitiation of replication by regulating the binding of MCM3 to chromatin in *Xenopus* egg extracts. *Curr. Biol.* 5:1270–79

69. Marahrens Y, Stillman B. 1992. A yeast chromosomal origin of DNA replication defined by multiple functional elements. *Science* 255:817–23

70. Marahrens Y, Stillman B. 1994. Replicator dominance in a eukaryotic chromosome. *EMBO J.* 13:3395–3400

71. McCarroll RM, Fangman WL. 1988. Time of replication of yeast centromeres and telomeres. *Cell* 54:505–13

72. McNally FJ, Rine J. 1991. A synthetic silencer mediates *SIR*-dependent functions in *Saccharomyces cerevisiae. Mol. Cell. Biol.* 11:5648–59

73. Mendenhall MD. 1993. An inhibitor of p34^{CDC28} protein kinase activity

from *Saccharomyces cerevisiae. Science* 259:216–19

74. Micklem G, Rowley A, Harwood J, Nasmyth K, Diffley JF. 1993. Yeast origin recognition complex is involved in DNA replication and transcriptional silencing. *Nature* 366:87–89

75. Miller AM, Nasmyth KA. 1984. Role of DNA replication in the repression of silent mating type loci in yeast. *Nature* 312:247–51

76. Moreno S, Nurse P. 1994. Regulation of progression through the G1 phase of the cell cycle by the *rum1*+ gene. *Nature* 367:236–342

77. Mui FM, Brown GW, Kelly TJ. 1996. *cdc18*+ regulates initiation of DNA replication in *Schizosaccharomyces pombe. Proc. Natl. Acad. Sci. USA* 93:1566–70

78. Nasmyth K. 1993. Control of the yeast cell cycle by the Cdc28 protein kinase. *Curr. Opin. Cell. Biol.* 5:166–79

79. Newlon CS, Theis JF. 1993. The structure and function of yeast ARS elements. *Curr. Opin. Genet. Dev.* 3:752–58

80. Nishitani H, Nurse P. 1995. p65*cdc18* plays a major role controlling the initiation of DNA replication in fission yeast. *Cell* 83:397–405

81. Piatti S, Lengauer C, Nasmyth K. 1995. Cdc6 is an unstable protein whose de novo synthesis in G1 is important for the onset of S phase and for preventing a 'reductional' anaphase in the budding yeast *Saccharomyces cerevisiae. EMBO J.* 14:3788–99

82. Rao H, Marahrens Y, Stillman B. 1994. Functional conservation of multiple elements in yeast chromosomal replicators. *Mol. Cell. Biol.* 14:7643–51

83. Rao H, Stillman B. 1995. The origin recognition complex interacts with a bipartite DNA binding site within yeast replicators. *Proc. Natl. Acad. Sci. USA* 92:2224–28

84. Rao PN, Johnson RT. 1970. Mammalian cell fusion: studies on the regulation of DNA synthesis and mitosis. *Nature* 225:159–64

85. Richardson HE, Wittenberg C, Cross F, Reed SI. 1989. An essential G1 function for cyclin-like proteins in yeast. *Cell* 59:1127–33

86. Rivier DH, Rine J. 1992. An origin of DNA replication and a transcription silencer require a common element. *Science* 256:659–63

87. Rowles A, et al. 1996. Interaction between the origin recognition complex and the

replication licensing system in *Xenopus. Cell* 87:287–96

88. Rowley A, Cocker JH, Harwood J, Diffley JF. 1995. Initiation complex assembly at budding yeast replication origins begins with the recognition of a bipartite sequence by limiting amounts of the initiator, ORC. *EMBO J. EMBO J.* 14:2631–41

89. Rowley A, Dowell SJ, Diffley JF. 1994. Recent developments in the initiation of chromosomal DNA replication: a complex picture emerges. *Biochem. Biophys. Acta.* 1217:239–56

90. Schneider BL, Yang QH, Futcher AB. 1996. Linkage of replication to start by the cdk inhibitor Sic1. *Science* 272:560–62

91. Schwob E, Bohm T, Mendenhall MD, Nasmyth K. 1994. The B-type cyclin kinase inhibitor p40*SIC1* controls the G1 to S transition in *S. cerevisiae. Cell* 79:233–44

92. Schwob E, Nasmyth K. 1993. *CLB5* and *CLB6*, a new pair of B cyclins involved in DNA replication in *Saccharomyces cerevisiae. Genes Dev.* 7(7A):1160–75

93. Shore D, Nasmyth K. 1987. Purification and cloning of a DNA binding protein from yeast that binds to both silencer and activator elements. *Cell* 51:721–32

94. Stillman B. 1993. DNA replication. Replicator renaissance. *Nature* 366:506–7

95. Stillman B. 1994. Initiation of chromosomal DNA replication in eukaryotes. Lessons from lambda. *J. Biol. Chem.* 269:7047–50

96. Theis JF, Newlon CS. 1994. Domain B of *ARS307* contains two functional elements and contributes to chromosomal replication origin function. *Mol. Cell. Biol.* 14:7652–59

97. Toyn JH, Toone WM, Morgan BA, Johnston LH. 1995. The activation of DNA replication in yeast. *Trends Biochem. Sci.* 20:70–73

98. Tyers M, Tokiwa G, Futcher B. 1993. Comparison of the *Saccharomyces cerevisiae* G1 cyclins: Cln3 may be an upstream activator of Cln1, Cln2 and other cyclins. *EMBO J.* 12:1955–68

99. Van HJ, Newlon CS. 1990. Mutational analysis of the consensus sequence of a replication origin from yeast chromosome III. *Mol. Cell. Biol.* 10:3917–25

100. Walker SS, Francesconi SC, Eisenberg S. 1990. A DNA replication enhancer in *Saccharomyces cerevisiae. Proc. Natl. Acad. Sci. USA* 87:4665–69

101. Wittenberg C, Sugimoto K, Reed SI. 1990. G1-specific cyclins of *S. cerevisiae*: cell cycle periodicity, regulation by mating pheromone, and association with the p34^{CDC28} protein kinase. *Cell* 62:225–37

102. Yan H, Merchant AM, Tye BK. 1993. Cell cycle-regulated nuclear localization of Mcm2 and Mcm3, which are required for the initiation of DNA synthesis at chromosomal replication origins in yeast. *Genes Dev.* 7:2149–60

103. Yoon HJ, Loo S, Campbell JL. 1993. Regulation of *Saccharomyces cerevisiae* CDC7 function during the cell cycle. *Mol. Biol. Cell.* 4:195–208

Annu. Rev. Microbiol. 1997. 51:151–78

RNA VIRUS MUTATIONS AND FITNESS FOR SURVIVAL

E. Domingo
Centro de Biología Molecular "Severo Ochoa" (CSIC-UAM), Universidad Autónoma de Madrid, Cantoblanco, 28049 Madrid, Spain;
e-mail: edomingo@trasto.cbm.uam.es

J. J. Holland
Department of Biology and Molecular Genetics, University of California, San Diego, 9500 Gilman Drive, La Jolla, California 92093-0116

KEY WORDS: virus evolution, quasispecies, polymerase fidelity, phenotype, error threshold

---

ABSTRACT

RNA viruses exploit all known mechanisms of genetic variation to ensure their survival. Distinctive features of RNA virus replication include high mutation rates, high yields, and short replication times. As a consequence, RNA viruses replicate as complex and dynamic mutant swarms, called viral quasispecies. Mutation rates at defined genomic sites are affected by the nucleotide sequence context on the template molecule as well as by environmental factors. In vitro hypermutation reactions offer a means to explore the functional sequence space of nucleic acids and proteins. The evolution of a viral quasispecies is extremely dependent on the population size of the virus that is involved in the infections. Repeated bottleneck events lead to average fitness losses, with viruses that harbor unusual, deleterious mutations. In contrast, large population passages result in rapid fitness gains, much larger than those so far scored for cellular organisms. Fitness gains in one environment often lead to fitness losses in an alternative environment. An important challenge in RNA virus evolution research is the assignment of phenotypic traits to specific mutations. Different constellations of mutations may be associated with a similar biological behavior. In addition, recent evidence suggests the existence of critical thresholds for the expression of phenotypic traits. Epidemiological as well as functional and structural studies suggest that RNA viruses can tolerate restricted types and numbers of mutations during any specific time point during their evolution. Viruses occupy only a tiny portion of their potential sequence space. Such limited tolerance to mutations may open new avenues for combating viral infections.

151

0066-4227/97/1001-0151$08.00

CONTENTS

RNA VIRUS UBIQUITY, EVOLUTION, AND EMERGENCE

New Viruses, New Diseases

RNA viruses are the most abundant molecular parasites infecting humans, animals and plants (138). A comparative analysis of the structures, genetic organization, and replication pathways of RNA viruses indicates that RNA viruses use disparate strategies to ensure their multiplication in cells and their stability as free particles. The possible eradication of some diseases such as poliomyelitis and measles early in next century is likely to be balanced by the emergence or reemergence of new viral pathogens within comparable time periods (112, 135, 139). About 50 new viruses have been recognized as emergent in the past two decades, most of them RNA viruses. They belong to families as diverse as *Arenaviridae, Bunyaviridae, Filoviridae, Flaviviridae, Myxoviridae, Picornaviridae,* and *Retroviridae.* The emergence of new viral pathogens is favored by the genetic plasticity of RNA viruses and by alterations in the environment that have an effect on viral traffic (135). As a result, viruses may come in contact with potential new hosts, thus facilitating host jumping. As examples, changing agricultural practices have led to an increase in the rodent population in areas of the American continents, intensifying contacts between rodents and humans. New arenaviruses associated with human haemorrhagic fevers in South America, or the hantaviruses associated to human respiratory illness in North America, normally exist as endogenous viruses in rodents (112, 119, 135, 139, 143). The emergence of new RNA viral pathogens is but one of several indicators of the variable and adaptable nature of RNA genetic elements. Viroids (100, 159), viroid-like RNAs, satellites (106), and cellular retroelements (161) contribute to a highly dynamic RNA world that is dependent on, and coexists with, a relatively more static cellular DNA world (91).

The Need to Investigate the Population Dynamics of RNA Viruses

The fact that very few viral pathogens can be effectively controlled, either by vaccination or by antiviral therapies, attests to the need to understand the mechanisms by which viruses overcome the pressures that are intended to limit their replication. Current evidence suggests that the origins of such an ability include: the frequent generation of mutations by RNA viruses, the continuous competition among variant genomes, and the selection of those variants which are best adapted to each particular environment (51, 51a, 53, 55, 91). Many questions remain, however, such as: Which factors affect genetic stability and which affect rapid change? What are the mechanisms of long-term persistence versus those of extinction? What mutation pathways are required to reach high fitness values? What is the relevance of viral variability to pathogenesis?

Progress in understanding the implications of high mutation rates and the population dynamics of RNA viruses has already provided a basis for the preference of some types of treatments over others: combination antiretroviral therapy over monotherapy (34, 43, 50) or multicomponent complex vaccines over simple, peptide vaccines (43, 50, 83, 146, 189a, 202). Yet, the broad implications of the complexities of RNA genetics for RNA virus adaptability, viral emergence, and disease control are only beginning to be comprehended.

The genetic and phenotypic variability of RNA viruses allows for an experimental approach to evolutionary phenomena, with attention to molecular detail. The limited complexity of genomes (in terms of the number of encoded proteins), their high mutability, rapid replication rate, and the possibility of testing the effects of several millions–fold differences in the population numbers of replicating genomes, confer great value upon viruses as model systems for understanding molecular evolution. Even though such experimental studies are still in their infancy, they are already the most extensive of research done in any microbial group. This review summarizes the results and implications of such research.

MOLECULAR MECHANISMS OF RNA GENOME VARIATION

The mechanisms of RNA virus variation include mutation, homologous and nonhomologous recombination, and genome segment reassortment, although different virus families exploit these to different extents. Molecular recombination is very active in plant and animal positive strand RNA viruses (107, 108, 182, 187). A number of statistical tests may be needed to distinguish true recombination in the field from the disparate evolution of different genomic segments, because of the existence of variations in the rate of accumulation of mutations (104, 130, 174, 181). Recombination has not been observed for unsegmented negative strand RNA viruses such as rhabdoviruses, in spite of their nonhomologous recombination ability to produce defective-interfering (DI) particles with extensive genomic deletions (86). The ability to undergo molecular recombination may be influenced by the contribution of *cis*-acting versus *trans*-acting factors in the completion of an infectious cycle, and by the processivity of viral replicases. Polymerases with limited processivity have a higher chance of jumping onto another template molecule to generate recombinant progeny by a copy-choice mechanism (102, 107, 108).

Perhaps the feature that most distinguishes RNA genetic elements from cellular DNA is the high mutation rate operating during genome replication. Misinsertion errors during RNA replication and retrotranscription have been estimated to be in the range of 10^{-3} to 10^{-5} substitutions per nucleotide and per round of copying. These misinsertion errors employ a variety of genetic and biochemical approaches (7, 49, 51, 56, 161). The main factor contributing to such high mutation rates is the absence or the low efficiency of proofreading-repair activities that are associated with RNA replicases and transcriptases (74, 185). In addition to biochemical evidence, the crystal structures of reverse transcriptase and RNA polymerases do not reveal the presence of a domain that could be assigned to a 5' to 3' exonucleolytic proofreading activity, such as those found in cellular enzymes like the DNA-dependent DNA polymerase from *Escherichia coli* (96, 103, 185). Also, mismatch repair mechanisms are unlikely to operate on replicating RNA (134) and cannot operate on single-stranded RNA progeny genomes.

Recently, a mutation rate of 2×10^{-5} substitutions per nucleotide copied has been determined for the *Saccharomyces* retrotransposon Ty1 (76). Current evidence suggests that other cellular retroelements share with RNA viruses extensive genetic heterogeneity (22, 161). It has been estimated that as much as 20% of the mammalian cell genome may be composed of retroelements (4) and thus be capable of rapid diversification. Retrotransposition events, combined with suppression of repair activities, could contribute to cellular genetic variation and instability, overtly manifested in cancer development (18, 93, 134). Mutation rates of 10^{-3} per base site per cell generation affecting a specific DNA strand operate in the process of somatic hypermutation of immunoglobulin genes (97, 186). Cells exploit mutant generation, competition, and selection—which are the norm for RNA replicons—when helpful for their adaptation and survival, such as occurs in selective proliferation of B cells, or in the expansion and metastasis of tumor cells within organisms (134, 144, 191).

Mutation and Recombination in RNA Virus Evolution

Mutation rates per nucleotide site in the range of 10^{-3} to 10^{-5} for a 10-Kb genome ensure that each progeny RNA or DNA molecule includes an average of 0.1 to 10 mutations (55, 56, 193, 204, 205). This continuous production of mutants favors adaptability of viruses in the event of environmental changes. Viruses may be able to respond in a nearly deterministic fashion to some selective pressures. Examples are the specific mutations in variant viruses that are resistant to neutralizing antibodies (15, 196) or to some antiviral inhibitors. Other adaptations may require constellations of mutations or genetic jumps via recombination which may take many rounds of replication before they occur. Also, some adaptations may be attained by disparate combinations of mutations and, thus, their pathways of implementation may differ depending on the order of occurrence of mutations (157, 211). Recombination may serve two opposite purposes: exploration of new combinations of genomic regions from different origins, or the rescuing of viable genomes from debilitated parental genomes (107). Recombination occurs at a high frequency in poliovirus when it replicates in vaccinees (133, 211), and it appears to be acting as a major evolutionary force in the recent expansion of the immunodeficiency viruses (168, 180). Western equine encephalitis virus probably originated as a result of a recombination event between a Sindbis virus-like and an Eastern equine encephalitis virus-like genome (81). It may be significant that many emergent viruses belong to groups for which an active recombination potential has been described (112, 135, 139). However, the likely absence of recombination in some highly variable and adaptable viruses, including the highly diverse Rhabdoviruses such as VSV, suggests that recombination may not be a strict requirement for adaptability or for the short-term or long-term evolution of RNA viruses.

Nonhomologous recombination offers a mechanism for the transduction of cellular RNA segments into viral genomes (132). Although the incorporation of cellular sequences often leads to defective viral genomes (86), several DNA and RNA viruses remain functional, or even increase their virulence, upon acquisition of cellular genetic information (132). Genome segment reassortment has been associated with severe pandemics of viral influenza, a topic that has been extensively reviewed (206). It is likely that reassortment plays an active role in the adaptability of viruses with a segmented genome (79).

It is the prevailing view, and some of the evidence will be presented in following sections, that both mutation and recombination (when it operates) occur with high frequencies in the course of RNA genome replication. However, only a minority of viable (selected) mutants and recombinants can actually be detected among the progeny viruses. That is, negative (or purifying) selection is continuously pruning away unfit mutants and recombinants (5, 51).

Hypermutagenesis for the Exploration of Sequence Space

The error-prone nature of RNA replication and retrotranscription is supported by many studies of fidelity measurements using purified enzymes. Such in vitro studies have indicated that fidelity values are dependent on the nucleic acid polymerase, on the sequence context at the template site being copied, and on several environmental parameters, such as the ionic composition of the medium or the relative concentration of nucleoside triphosphate substrates (160, 165, 210, 214). Reverse transcription or DNA amplification using Taq polymerase with biased concentrations of nucleoside triphosphate substrates can result in the synthesis of remarkably hypermutated molecules (126, 198). The ability to accelerate evolution and to explore large portions of sequence space has been exploited to document considerable tolerance of the bacterial enzyme dihydrofolate reductase to accept amino acid substitutions and retain enzyme activity; some active enzymes accumulated up to 20% amino acid replacements (125). Hypermutagenesis and other random mutagenesis procedures may find important applications in the production and analysis of proteins with new properties, and in the generation of highly heterogeneous viral preparations from infectious transcripts of cDNA clones.

Hypermutated viral genomes are found during viral infections in vivo (23, 153, 172, 199). The process often results in a high frequency of one specific type of polymerase misincorporation (biased hypermutation). In addition, some of the modifications seen in RNA genomes have a parallel in editing reactions undergone by some types of cellular RNA (12, 175). One such reaction (A → G hypermutation), thought to mediate hypermutation of measles virus, is catalyzed by a double-stranded RNA adenosine deaminase, an activity induced by interferon. This hypermutation reaction occurs via conversion of adenosine into inosine (6). It is interesting that RNA editing, a process proposed to have evolved in order to slow down the rate of evolution of cellular RNA (12), may often be contributing to large evolutionary jumps of viral RNAs. Hypermutated forms of the measles virus in brain cells are associated with subacute sclerosing panencephalitis, a rare but fatal neurological disease manifested several years after the measles infection (98). Recombination and hypermutation events may yield viral pathogens with new biological properties and highly divergent genomic sequences whose identification may escape standard hybridization or PCR techniques. Given the exhuberant and ubiquitous nature of viral variation, it is quite possible that a number of diseases of unknown etiology may be caused by divergent forms of viruses, still to be identified (91). This is particularly possible for chronic degenerative diseases with slow indolent onset and progression. The rather recent recognition of hepatitis B and C viruses in hepatocellular carcinoma and of human papillomaviruses in cervical carcinoma illustrate this.

FITNESS PEAKS, VALLEYS, AND JUMPS

Theoretical Quasispecies and Real Virus Quasispecies

Quasispecies are complex, dynamic distributions of nonidentical but related replicons. Eigen (61) developed the first theoretical treatment of replication, with limited copying fidelity, as a model for early life forms on earth. Eigen & Schuster (67–69) introduced the concepts of quasispecies, mutant spectrum, and population equilibrium, which are of great importance for the understanding of RNA viruses at the population level (45, 47–49, 51–53, 55, 60, 63–65, 85, 87, 89, 91). The theoretical quasispecies is a steady-state, organized distribution of the error copies of a master sequence. The complexity of the mutant spectrum increases as the fidelity of the replication process decreases (65, 67–69, 189). The most fit master sequence dominates the ensemble but it may or may not coincide with the average or consensus sequence of the mutant distribution. The main departure that quasispecies concepts take from previous models of population biology is the consideration of the wild type not as an individual genome with a defined nucleotide sequence, but as an ensemble of closely related genomes on which selective forces act (61, 64–69).

There are important differences between the theoretical quasispecies concept as originally formulated by Eigen and colleagues, and the real viral quasispecies. A significant difference is that the steady state equilibrium distribution of genomes is very difficult to attain with viral populations that infect host organisms or even cell cultures. Steady state is perturbed by changes in the host microenvironment in which virus replication takes place, as well as by fluctuations in viral population numbers. In spite of such differences, the view of viruses as complex and dynamic distributions in which each individual genome has only a fleeting existence, has been instrumental in the understanding of viral behavior in infected organisms. Not only the absence of a defined wild type viral genome, which is currently recognized, but also the consideration of mutant generation, competition, and selection as an uninterrupted, ongoing chain of events in the life cycle of RNA viruses, have all been adapted from the original description of quasispecies. The links between the theoretical quasi-species and real virus quasispecies have been analyzed in several recent articles (52, 60, 63, 64, 89).

It may be surprising that in spite of several previous models of mutation and selection in populations, it was not until the quasispecies concept had been for-mulated that a link between population biology and experimental virology was established. There are several possible reasons for this. One is the emphasis on mutant generation in the quasispecies concept that was originally intended to represent the evolution of simple macromolecules rather than of complex organisms. Extant RNA viruses share error-prone replication with the putative early replicons that interested Eigen and colleagues (61, 66–69). Another rea-son is the explosion of sequence information regarding cellular and viral genes in the years following the formulation of the quasispecies model. Although se-quence information has also influenced population genetics, the demonstration of extremely high genetic diversity—to the point of invalidating the concept of wild type as a defined genome with a specific nucleotide sequence—became obvious first with RNA viruses (45, 52, 53, 55), and also with simple RNA repli-cons derived from bacteriophage $Q\beta$ RNA (14, 169, 215). In vitro replication, mutation, and selection of simple RNA molecules is an experimental approach that was initiated by S Spiegelman and colleagues three decades ago.

Fitness Loss and Fitness Gain/Quasispecies Optimization by Competitive Selection

Fitness is a complex parameter aimed at describing the replicative adaptabil-ity of an organism to its environment. The concept involves some difficulty concerning which parameters may best be used to adequately reflect fitness (20, 209). For RNA viruses, an experimentally useful approximation to fitness is the relative ability to produce stable infectious progeny in a given environ-ment. This allows a ranking of relative fitness values that can be established by growth-competition experiments in mixed infections in which the initial viruses are genetically or phenotypically distinguishable (7, 24, 88, 123). This procedure has regularly documented that RNA viruses may undergo very large fitness changes, with some passage regimes leading to fitness losses, and others to fitness gains.

Fitness Loss/Muller's Ratchet

Repeated plaque-to plaque transfers of RNA virus clones (genetic bottlenecks) result in average fitness losses of the ensuing populations relative to the parental clones or populations (24, 29, 57, 71). Because of the quasispecies structure of RNA viruses, repeated bottleneck events, such as those mediating serial plaque transfers, are expected to cause progressive deviation of the consensus sequence from the initial one (reviewed in 48). Accumulation of deleterious mutations, when no compensatory mechanisms such as sex or recombination are in operation, was initially proposed by Muller (137), and is known as Muller's ratchet. In serial plaque transfers, the possibility of competitive optimization of the quasispecies is limited to the stage of plaque development. Thus this experimental design results in an accentuation of the deleterious effects of Muller's ratchet. Studies with vesicular stomatitis virus (VSV) have shown that some of the profound fitness losses occurring upon repeated bottlenecking can not be overcome even with intervening trillion-fold amplifications of the viral population (58).

Individual clones from a virus population differ in relative fitness (55, 59). Fitness values approximate a normal distribution with a mean value significantly lower than the average fitness of the parental population (59): The behavior of a quasispecies cannot be explained by the sum of its individual components. The size of a genetic bottleneck (as defined by the number of clonal pools) required to maintain fitness of VSV is dependent on the initial fitness of the clone employed to initiate the successive platings (148). With a low initial fitness clone, pools of five clones result in fitness gain. In contrast, for a virus with high initial fitness, pools of virus from 30 individual plaques are required to maintain the initial fitness value (148).

In a recent study of the types of mutations associated with fitness loss upon repeated plaque transfers of foot-and-mouth disease virus (FMDV), a variety of lesions never seen among natural FMDV isolates, nor in viruses subjected to large population passages, were observed (71). These lesions include substitutions at internal capsid residues and, strikingly, an internal elongation of a polyadenylate tract immediately preceding the second initiation AUG codon. This unusual spectrum of mutants must reflect the occurrence of mutations still compatible with plaque development, but which are selected against during competitive growth of FMDV in vitro or in vivo. A highly debilitated clone, whose genome included a total of six-point mutations, in addition to the elongated internal polyadenylate tract, could not be successfully plated as clones (plaques) beyond transfer 24 (71). In spite of such profound debilitation, a limited number of serial transfers of large populations were sufficient for the clone to regain fitness (Escarmís et al, manuscript in preparation). Thus, as in the case of VSV, very large fitness changes can be transmitted to viral quasispecies by manipulating the population size involved in serial infections. Bottleneck events of different intensities probably occur during the natural life cycle of viruses. Examples are aerosol transmission from an infected host into a susceptible host, activation of a cell harboring a provirus or other quiescent forms of a latent viral genome, or the cyclic transitions from mammalian and arthropod hosts in the arbovirus life cycle (60, 145, 176).

Fitness Gain/Quasispecies Optimization
by Competitive Selection

Fitness gains will generally occur when competitive optimization of viral qua-sispecies is allowed in a constant environment. An analysis of the kinetics of fitness increase of VSV revealed that clones with low initial fitness exhibited biphasic kinetics with larger increases seen at the initial passages, until neutrality was reached. Then fitness gain proceeded exponentially, and reached up to 5000% after 50 passages (147). As a comparison with a cellular organism, gains in *Escherichia coli* amounted to 37% in 2000 generations (114). The two distinct stages of fitness gain seen in the experiments by Novella et al (147) have been modeled with a "mean-field" theory of viral population dynamics operating in a one-dimensional fitness landscape (195).

Fluctuations in the environment in which viral replication takes place may present the virus with conflicting demands for adaptation. Persistent replication of VSV in insect (sandfly) cells profoundly decreased the replicative fitness of the virus in mammalian cells (145). It has been suggested that the slow rates of evolution observed in insect-borne RNA viruses may be due to constraints imposed by the alternation of hosts during the virus life cycles (176). Selection of FMDVs that are resistant to neutralizing polyclonal antibodies result in variant FMDVs with multiple amino acid substitutions in the capsid proteins, and with extremely low fitness when the latter is measured in the absence of the antibodies employed in the selection (15). The process of adaptation to an environment disparate from that of the natural host of a pathogenic virus underlies the classical derivation of attenuated viruses for vaccine production. Knowledge of the types of hosts and cell types in which viral replication may lead to decreases in fitness in the authentic host may help in the design of live-attenuated vaccines.

Competitions initiated with VSV clones of the same initial relative fitness could proceed with the two quasispecies coexisting for many generations until one of them rapidly dominated over the other (28). In these competitions involving VSV clones, both the winners and the losers gained fitness (28, 163). These observations are in agreement with two important concepts of population genetics: the Red Queen hypothesis (197) and the Competitive Exclusion principle (77; reviewed in 48). Recent experiments have documented the remarkable, highly predictable, nonlinear behavior of two competing VSV quasispecies (163). In a series of replicas of the same initial mixtures, the ratio of the two competing viruses remained approximately constant until a critical point was reached from which the viral competitions followed different trajectories. Thus, a nonlinear, nearly deterministic evolutionary behavior can occasionally be seen, in spite of the continuous stochastic generation of mutations in viral quasispecies (163). The molecular basis of such behavior is under investigation. An estimate of relative fitness in vivo has recently been obtained for HIV-1 subpopulations replicating in vivo (80; review in 54).

Population Size and Movements in Fitness Landscapes

In terms of the classical fitness landscape concept of S Wright (212), changes in viral fitness can be viewed as a movement of viral genomes in a rugged and adaptive landscape of peaks, valleys, and pits. Movements through sequence space are generated by mutations and guided toward fitness peaks by selection. Sequence space refers to all possible permutations of sequences for an informational macromolecule (62, 99). For a viral genome of 10 Kb, the total, theoretically possible, sequence space is $4^{10,000}$, a number so immense that it defies our ability to comprehend or even imagine it. In spite of the great connectivity among individual points in sequence space (62), only an extremely tiny fraction of it is actually allowed to functional viruses. Adaptability of RNA viruses can be viewed as an ability to ascend adaptive peaks as the direct consequence of high mutation rates, rapid replication cycles, and large numbers of progeny. Rapid replication cycles with mutation rates of zero would result in evolutionary stasis. For any given genome turnover, the mutation rate determines the ability of a virus to maintain essential information while coping with environmental changes (30, 51–53, 78, 84, 87, 89, 133, 192, 207).

During natural infections, direct competition among equally stable replicating genomes will take place at receptors of, and within, individual cells or even subcellular compartments within a cell. Spatial heterogeneities of quasispecies within an infected organ (173) may result in viral subpopulations with different survival potentials, in spite of their replicating in the same host. Viruses populating organisms have been selected in myriads of replicating quasispecies. When competition occurs among viruses infecting multiple cells, the population size of the virus (viral load) will influence the numbers and types of variants contributing to fitness increases. Upon cytolytic passage of a persistent FMDV in cell culture a unique double substitution, affecting highly conserved positions of a major antigenic loop, becomes dominant as a result of the passage of very large numbers of infectious FMDV. The same double replacement is never seen in the many passages involving a much reduced population size (178). In these series of cytolytic passages, it is observed that phenotypic traits that had been acquired during persistence either reverted or remained invariant, but that the process was accompanied by genetic diversification of the virus (178). The genetic and phenotypic flexibility of an evolving viral quasispecies is dependent on the mutant repertoire which is expanded with population size.

Reversible and Irreversible Phenotypic Change, and Phenotypic Flexibility

Viruses may sometimes cause alterations in host cell functions, facilitating viral persistence (39). In turn, hosts exert important effects on virus evolution (101). The persistence of several RNA viruses in cell culture involves a coevolution of the cells and the resident virus, providing excellent models of the genetic and phenotypic changes of the virus in a continuously changing cellular environment (2, 21, 25, 37, 38, 40, 121, 155). In the case of FMDV, hypervirulent variants, capable of killing more cells more rapidly, become dominant to overcome the resistance of coevolving cells (37, 38, 121). Once these viruses are subjected to cytolytic passage, possible virus variants showing reversion of the virulence trait would kill cells later than their parental counterparts, and could not become dominant. In contrast to virulence, other phenotypic traits associated with FMDV persistence reverted almost deterministically (178). These experiments have allowed a distinction between those phenotypic traits that become irreversibly fixed in the population following an environmental modification, and others that revert, at the return of the virus to the previous environment. Phenotypic reversion occurrs by the fixation of additional mutations rather than by the true reversion of preexisting mutations (178).

The evolution of virulence appears to involve a variety of mechanisms in different viral systems, including mutations in regulatory regions (211) and viral adaptation for utilization of alternative or expanded repertoires of cellular receptors. There is increasing evidence that viruses utilize coreceptors and alternative receptors for their entry into cells, and that this may be an important determinant of virulence. Individuals who are homozygous for genes encoding mutant forms of HIV-1 coreceptors, for example, show resistance to HIV-1 infection (117). Variant polioviruses capable of recognizing altered cell receptors have been selected (21, 31). The fact that residues at antigenic sites may also affect receptor recognition has important implications for the coevolution of antigenicity and host range (21, 82, 141, 183, 208, 211, 213). The likely connection between antigenic variation and modifications of cell tropism becomes even more dramatic in view of the frequent occurrence of antigenic variation in the absence of immune selection (16, 46, 179). Quasispecies are reservoirs of biologically relevant viruses, and the continuous fluctuations in quasispecies distributions within infected organisms favors the stochastic emergence of viruses with altered biological properties (44, 46, 142, 164).

Mutations for Disease

In addition to the presence of phenotypically relevant variants that favor the adaptability of RNA viruses, specific mutations have been associated with the pathogenesis of some RNA viruses. An avirulent (amyocarditic) coxsackievirus B3 evolved to become cardiopathic in selenium or vitamin E–deficient mice (10). Virulent variants differed consistently and precisely from their avirulent parents in six-point mutations at specific base sites scattered along the genome. Once fixed, the cardiopathic phenotype was manifested even in infected animals with adequate selenium levels. These observations establish an important link between nutritional status and viral evolution (9, 115). The generation of mutant coxsackieviruses and their subsequent selection in selenium or vitamin E–deficient mice was probably facilitated by the increased viral load (larger repertoire of multiple mutants in the quasispecies), due to the impaired immune responses that were associated with oxidative stress in these animals. In addition, mutant generation could also be favored by oxidative damage of viral RNA (9, 10, 115). Thus a deficient nutritional status, via impairment of immune functions, can accelerate viral evolution for the emergence of pathogenic variants.

Specific mutations in the lymphocytic choriomeningitis virus (LCMV) genome are associated with tropism for neurons or for cells of the immune system (42 and its references). Some LCMV isolates cause a growth hormone deficiency syndrome in certain strains of newborn mice, although other isolates do not produce such a syndrome. When mixtures of pathogenic and nonpathogenic clones are inoculated at different ratios, it is observed that disease development can be restricted by a large excess of nonpathogenic viral genomes, even though the disease-causing variant remains at low levels in the viral populations (194).

Thresholds for Phenotypic Expression

Suppression of disease potential in LCMV is one of several examples of the phenotypic traits of RNA viruses that are expressed only when the virus manifesting suppression is present above a minimal, critical frequency in the quasispecies. A high fitness VSV clone becomes dominant only when it is present above a critical proportion in the parental uncloned VSV population (36). Live attenuated poliovirus vaccine may include virulent variants in its mutant spectrum, but unless these variants are present above a critical proportion, they cannot induce neurological disease in monkeys (27). Critical population thresholds in viral quasispecies may be an important determinant of viral pathogenesis. The buffering capacity of the mutant swarm with regard to phenotypic expression is, in itself, a prediction of the quasispecies dynamics, in that the fate of all individual variants is conditioned by the mutant spectrum surrounding them (65).

The Error Threshold

Another important prediction of the quasispecies theory is that, at a level below a critical fidelity of copying by the replication machinery, genetic information will be irreversibly lost (61, 65, 189). Such a critical fidelity is termed the error threshold, and its value depends both on the complexity of the information to be maintained and on the superiority of the master sequence over the spectrum of mutants surrounding it. An interesting distinction between the genotypic and the phenotypic error thresholds was established (92) recently, based upon modeling studies with simple replicons, in which the folding of the resulting molecules was taken as the phenotype. This distinction defines a neutral portion of sequence space in which the variants are phenotypically equivalent. Violation of the error threshold with loss of information implies crossing the phenotypic error threshold.

Several experimental results support the concept that RNA viruses replicate with a copying fidelity that is balanced near the error threshold. Mutagenic agents, some acting on infected cells and others acting on the virus particles prior to infection, could increase only two- to threefold the mutant frequency at defined loci of the VSV or poliovirus genomes (90). In a similar experiment, using a retroviral vector, the mutation rate in the presence of 5-azacytidine could be increased only up to 13-fold (154). The difference between the results with a retroviral vector and VSV or poliovirus could either reflect a lower mutation rate for retroviruses than riboviruses (56) or a difference in the complexity of the genomes subjected to mutagenesis. A recent study has further shown that chemical mutagenesis has an adverse effect on the fitness gains and the fitness maintenance of several VSV mutants (113).

Proofreading-repair and postreplicative repair mechanisms necessarily had to evolve to ensure genetic stability of complex DNA genomes (74). Viewing nucleic acid polymerases as modular structures (184), the absence of a proofreading exonucleolytic domain can provide advantages for viral replicases, but its presence was an absolute necessity for cellular DNA polymerases. It is noteworthy that the single site mutation frequencies of DNA-based organisms can be increased hundredsfold or thousandsfold by mutagenesis (33).

OCCUPATION OF SEQUENCE SPACE AND RNA GENOME ADAPTABILITY

A major difference between RNA viruses and cellular organisms lies in the portion of sequence space that they are likely to explore and occupy as a result of replication errors. This distinction requires comparing three relevant parameters for RNA viruses and DNA-based organisms: (a) the genetic heterogeneity of the population, or the average number of mutations in individual genomes, relative to the consensus sequence; (b) the population size; and (c) the genetic complexity. A distinctive feature of RNA viruses, as compared with cellular organisms, is their high ratio of population size to genetic complexity (equated to the genome size). In many viral populations all possible single and double mutants are potentially present (49, 51, 55, 65, 193, 204). In contrast, only a minute fraction of the possible single mutants can be produced in any generation of a cellular organism. This difference underlies the great adaptability of RNA viruses, and validates quasispecies as an adequate descriptor of RNA viruses.

Tolerance and Limitations for the Occupation of Sequence Space by RNA Viruses

The availability of procedures for the amplification and sequencing of viral genomes present in biological specimens has resulted in the production of huge amounts of sequence information, as well as in the definition and establishment of phylogenetic relationships among virus genera and among individual isolates of a genus. From these studies (too numerous to be reviewed here) some general trends are becoming apparent. An observation common to many sequence alignments of related viruses is that of a high frequency of certain specific nucleotide and amino acid substitutions, which are difficult to explain by possible biases derived from codon usage. An abundance of specific replacements is seen even in hypervariable domains such as the V loops of the surface glycoprotein gp120 of HVI-1 (109, 140, 156, 164). Studies with several picornaviruses have amply documented the occurrence of the same subsets of mutations in independent evolutionary lineages or in independent expansions of the same viral clone (15, 26, 32, 35, 120, 178). Even during the course of several decades of viral evolution, variable positions in the capsid proteins of picornaviruses are alternatively occupied by a small subset of all possible amino acids, and a true accumulation of amino acid substitutions is not observed (124). Repeated specific mutation and recombination events between infectious feline leukemia virus (FeLV) and endogenous FeLV-like elements are associated with FeLV-induced lymphomas (171).

Together with repetitive substitutions at some sites, unique mutations in a single genome are frequently seen. When molecular clones are derived after reverse transcription and PCR amplification (RT-PCR) of viral RNA, it cannot be ruled out that such unique mutations may have been introduced during the RT-PCT procedure, due to the limited copying fidelity of the enzymes involved. However, with appropriate control amplifications and sequencing it can often be documented that most of the mutations that are found only once must be present in the viral genomes (142). The shape of quasispecies distributions in natural viral isolates is quite variable; a number of procedures and guidelines are available for quantitating the genetic divergences and the mutation frequencies of collections of related sequences (44, 131, 142, 164, 177).

The abundance of certain types of nucleotide and amino acid substitutions and the absence of many others could be explained by the still limited number of isolates that are sequenced for most viral groups. Alternatively, such biases may most often reflect considerable constraints upon the variation of RNA viruses. Several lines of evidence suggest that RNA virus evolution is constrained by the complexity of viral functions in interaction with cellular functions.

Evidence of Structural and Functional Constraints at the RNA and Protein Levels

It has been amply documented that secondary, and probably tertiary, structures of viral RNAs play important functional roles. This has been well established for some regulatory regions, such as the internal ribosome entry site (IRES) found in several viruses. IRES can accept very limited numbers of nucleotide replacements, and compensatory mutations are often needed to preserve higher order structures and ribosome binding and translation initiation functions (11, 127). Similar constraints have been documented for the pseudoknot structures that are involved in the recognition of specific proteins and in ribosomal frameshifting (158). Analyses of field isolates of FMDV suggest that changes in the number of tandem pseudoknots located within the 5'-untranslated region are allowed, but that compensatory mutations often occur to preserve the folding of each individual pseudoknot (72). If, as in many other biological paradigms, work with RNA bacteriophages anticipates future developments with eukaryotic viruses, secondary and higher order structures in noncoding as well as in coding regions will prove to be important for replicative ability (3, 149, 150). The structural requirements of RNA may contribute to unusual limitations for the silent (synonymous) substitutions that are seen in some RNA viruses as they evolve in cell culture or in vivo (41, 71, 167). Restrictions to the fixation of silent substitutions within open reading frames have frequently been attributed to requirements for codon usage and translational accuracy (19, 73). However, in the case of RNA viruses, the phenotypic involvement of genomic RNA may also contribute to the limitations in the number of silent mutations.

The molecular basis of the restrictions to amino acid substitutions becomes apparent with the establishment of the structure-function relationships of viral particles, the surface viral proteins, or the domains within proteins (116, 183, 200, 201). Several picornaviruses display a canyon or pit on their capsids (170) where the receptor recognition site lies shielded from antibody attack. In contrast, the three-dimensional structure of FMDV reveals a smooth surface without a canyon or pit (1, 111). One of the receptor recognition sites of FMDV, the integrin-binding Arg-Gly-Asp motif is located on a protruding, mobile loop on the capsid (1, 13). This amino acid triplet is directly involved in the interaction with the complementarity determining regions (CDRs) of antibody molecules (200, 201). These structural studies show that the open turn conformation of the Arg-Gly-Asp (RGD) is very similar to the conformation found in the reduced form of another FMDV serotype (118) and in other integrin-binding proteins (95, 105). This open-turn conformation is highly dependent on neighboring residues of the G-H loop of VP1 (200) providing conflicting demands on this loop to keep a functional structure to bind to integrin receptors and also to vary to overcome neutralization of infectivity by antibodies. Thus, the Arg-Gly-Asp sequence is highly conserved in FMDV not because it is not subjected to selective pressure by antibody molecules, but because viruses with replacements at this domain must be generally subjected to negative selection. This observation provides a structural interpretation for the restricted numbers and types of amino acid substitutions in this loop that are found among variant FMDVs (15, 47, 124, 128).

Structural and functional constraints can be documented for many viral proteins, and they are a result of the complexity of biochemical functions to be performed by seemingly simple replicons; perhaps this is a reflection of their modular origin (17, 216). The coordinated activity of functional modules, which are probably borrowed from the cellular world, is linked to a profound dependence of viral replication on cellular functions. In this view, the restrictions to the occupation of sequence space impose a bias on the production of variant phenotypes akin to the influence that developmental constraints have in the evolution of differentiated organisms (129).

Fitness Jumps in Adaptive Landscapes

The "functional freezing" imposed by the complexity of viruses must be counterbalanced by a tolerance to variation that is sufficient to ensure the adaptability and the long-term survival of viruses. The critical observation is that many phenotypic changes that are easily recognized as mediating the adaptability of viruses (antigenic variation, some modifications in host cell tropism, in virulence, etc) often depend on one or a few amino acid substitutions. Several such cases have been recently reviewed (44), and others are continuously being reported. An example is the effect reported of a single codon replacement in brome mosaic virus on compatibility with a new host (75). For statistical considerations these classes of genomes with small numbers of mutations relative to the parental genome are those that are potentially most represented in the quasispecies. Those genomes with even more mutations, however, can be critical for major adaptive changes. These "quasispecies outliers" are critical in order for viruses to find and ascend fitness peaks at different locations of the sequence space. Thus it is the quasispecies mutant swarm, and not individual virus genomes, which are the subject of selection and evolution (64–69).

QUASISPECIES BEHAVIOR PREDICTS FAILURES OF CLASSICAL ANTIVIRAL STRATEGIES AND SUGGESTS IMPROVED STRATEGIES

The dramatic fitness gains attained by replicating viral quasispecies (70, 147) suggest that viruses will be often capable of overcoming selective pressures that are intended to limit their replication. This is an important drawback for disease control, as is reflected in the isolation of vaccine-escape and drug-resistant viral mutants and in the antigenic variation of viral populations (43, 44, 50, 60, 83, 89, 146, 189a, 202). The latter may occur at widely different rates, as exemplified by the generally rapid drift of influenza virus versus the slow drift of measles virus circulating in the human population. Although selection events that compromise disease prevention and therapy have been documented for several pathogenic RNA viruses, they have been most dramatically apparent in the selection of HIV-1 mutants resistant to one or to multiple antiretroviral inhibitors (54, 110, 136, 166; and references therein).

The problem of evolving pathogens overcoming externally applied, inhibitory actions is by no means unique to RNA viruses. Other examples are the selection of multidrug-resistant cellular parasites and cancer cells, reflecting that viruses, cells, and organisms have all evolved a variety of biochemical mechanisms to cope with toxic agents (190). This general problem is aggravated in the case of viral quasispecies, because of the continuous input of mutant genomes during replication that permits adaptability to be reached within short time periods. It is too early to predict whether a suitable combination therapy will be found which will prevent the dominance of multidrug-resistant HIV-1 mutants while maintaining a low viral load throughout the life-span of infected individuals.

In spite of the great success of vaccines for the control of many viral diseases, the great adaptability of RNA replicons leads to the suggestion that antiviral strategies that exploit error catastrophe (the loss of viral genetic information, due to incrreased mutagenesis of the viral genome) should be considered (47, 51, 94). Because of the proximity of RNA viral replication to the error threshold (90, 113, 154), an externally induced decrease in copying fidelity should lead to the loss of viral genetic information (65, 66). A prerequisite for such intervention is that the copying fidelity of viral polymerases could be sensitive to specifically induced structural modifications of the viral enzyme. Some early evidence suggested that mutation rates varied among individual viral clones (162, 188). Recent studies with HIV-1 reverse transcriptase have shown that specific amino acid substitutions can cause a significant modification in the fidelity properties of the enzyme (8, 122, 152, 203). These molecular studies suggest that an induced violation of the error threshold is not unrealistic as a possible antiviral strategy.

CONCLUDING REMARKS

The explosion of information on the genomic sequences of many RNA viruses, the quantitation of genetic heterogeneity within populations, the extreme difficulties in the control of AIDS and chronic active viral hepatitis, and the emergence of many new RNA viral pathogens in just a few decades, have strengthened the concept of a ubiquitous adaptability of RNA genetic elements. Shifts in biological properties and the pathogenic potential of RNA viruses cannot be dissociated from the array of interactions with host cells and organisms (such as immune responses, recognition of alternative receptor molecules, the presence of viral "factories" or "sanctuaries" that may be hidden from external interventions, etc). The evolution of viruses is unavoidably linked to the evolution of their hosts. Human population growth and demographics, habitat destruction, and other more insidious forms of environmental modification are but a few of the influences affecting the selective constraints faced by viruses and by the course of their evolution. The study of viral quasispecies has provided the possibility of linking molecular virology with ecology, and linking evolution with disease control and prevention. The past few years have shown very encouraging trends in multidrug antiviral therapies for AIDS and other RNA viral diseases, and research into novel vaccine approaches has provided significant insights. However, the daunting genetic plasticity and the biological adaptability of RNA viruses will provide continuing, recurring, and quite unexpected challenges for the medical, agricultural, and veterinary sciences. Mankind's confrontation with the RNA pathogens will be endless, for, in the words of the Red Queen "Around here it takes all the running you can do just to stay in the same place" (28, 197).

ACKNOWLEDGMENTS

We are indebted to many colleagues for providing information before publication. We thank D Clarke, E Duarte, IS Novella, J Quer, L Menendez-Arias, C Escarmís, MG Mateu, ME Quiñones-Mateu, N Sevilla, MA Martínez, R Flores, O Levander, M Beck, and S Wain-Hobson for valuable information and discussions, and L Horrillo for careful preparation of the manuscript. Work in Madrid supported by grant DGICYT PB94-0034-C02-01, FIS 95/0034-1, Fundación Rodriguez Pascual, Comunidad Autónoma de Madrid, and Fundación Ramón Areces. Work in La Jolla supported by grant AI-14627 from the National Institute of Allergy and Infectious Diseases. A visit of E Domingo to La Jolla was supported by a fellowship from NATO.

Literature Cited

1. Acharya R, Fry E, Stuart D, Fox G, Rowlands D, et al. 1989. The three-dimensional structure of foot-and-mouth disease virus at 2.9 Å resolution. *Nature* 337:709–16
2. Ahmed R, Canning WM, Kauffman RS, Sharpe AH, Hallum JV, Fields BN. 1981. Role of the host cell in persistent viral infection: coevolution of L cells and reovirus during persistent infection. *Cell* 25:325–32
3. Arora R, Priano C, Jacobson AB, Mills DR. 1996. Cis-acting elements within an RNA coliphage genome: Fold as you please, but fold you must!! *J. Mol. Biol.* 258:433–46
4. Baltimore D. 1985. Retroviruses and retrotransposons: the role of reverse transcription in shaping the eukaryotic genome. *Cell* 40:481–82
5. Banner LR, Lai MMC. 1991. Random nature of coronavirus RNA recombination in the absence of selection pressure. *Virology* 185:441–45
6. Bass BL, Weintraub H, Cattaneo R, Billeter MA. 1989. Biased hypermutation of viral RNA genomes could be due to unwinding/modification of double-stranded RNA. *Cell* 56:331
7. Batschelet E, Domingo E, Weissmann C. 1976. The proportion of revertant and mutant phage in a growing population, as a function of mutation and growth rate. *Gene* 1:27–32
8. Bebenek K, Beard WA, Casas-Finet JR, Kim H-R, Darden TA, et al. 1995. Reduced frameshift fidelity and processivity of HIV-1 reverse transcriptase mutants containing alanine substitutions in helix H of the thumb subdomain. *J. Biol. Chem.* 270:19516–23
9. Beck MA. 1997. Increased virulence of coxsackievirus B3 due to vitamin E or selenium deficiency. *J. Nutr.* 127:966S–70S
10. Beck MA, Shi Q, Morris VC, Levander OA. 1995. Rapid genomic evolution of a non-virulent coxsackievirus B3 in selenium deficient mice results in selection of identical virulent isolates. *Nat. Med.* 1:433–36
11. Belsham GJ, Sonenberg N. 1996. RNA-protein interactions in regulation of picornavirus RNA translation. *Microbiol. Rev.* 60:499–511
12. Benne R. 1996. RNA editing: how a message is changed. *Curr. Opin. Genet. Dev.* 6:221–31
13. Berinstein A, Roivainen M, Hovi T, Mason PW, Baxt B. 1995. Antibodies to the vitronection receptor (integrin $\alpha v \beta 3$) inhibit binding and infection of foot-and-mouth disease virus to cultured cells. *J. Virol.* 69:2664–66
14. Biebricher CK. 1983. Darwinian selection of self-replicating RNA molecules. *Evol. Biol.* 16:1–52
15. Borrego B, Novella IS, Giralt E, Andreu D, Domingo E. 1993. Distinct repertoire of antigenic variants of foot-and-mouth disease virus in the presence or absence of immune selection. *J. Virol.* 67:6071–79
16. Both GW, Shi CH, Kilbourne ED. 1983. Hemagglutinin of swine influenza virus: a single amino acid change pleiotropically affects viral antigenicity and replication. *Proc. Natl. Acad. Sci. USA* 80:6996–7000
17. Botstein D. 1980. A theory of modular evolution for bacteriophages. *Ann. NY Acad. Sci.* 354:484–91
18. Branch P, Hampson R, Karran P. 1995. DNA mismatch binding effects, DNA damage tolerance, and mutator phenotypes in human colorectal carcinoma cell lines. *Cancer Res.* 55:2305–9
19. Britten RJ. 1993. Forbidden synonymous substitutions in coding regions. *Mol. Biol. Evol.* 10:205–20
20. Brodie ED III, Janzen FJ. 1996. On the assignment of fitness values in statistical analyses of selection. *Evolution* 50:437–42
21. Calvez V, Pelletier I, Couderc T, Pavio-Guédo N, Blondel B, et al. 1995. Cell clones cured of persistent poliovirus infection display selective permissivity to the wild-type poliovirus strain Mahoney and partial resistance to the attenuated Sabin 1 strain and Mahoney mutants. *Virology* 212:309–22
22. Casacuberta JM, Vernhettes S, Grandbastien M-A. 1995. Sequence variability within the tobacco retrotransposon Tnt1 population. *EMBO J.* 14:2670–78
23. Cattaneo R, Billeter MA. 1992. Mutations and A/I hypermutations in measles virus persistent infections. *Curr. Top. Microbiol. Immunol.* 176:454–55
24. Chao L. 1990. Fitness of RNA virus decreased by Muller's ratchet. *Nature* 348:454–55
25. Chen W, Baric RS. 1996. Molecular anatomy of mouse hepatitis virus persistence: coevolution of increased host cell

resistance and virus virulence. *J. Virol.* 70:3947–60

26. Chumakov KM, Dragunsky EM, Norwood LP, Douthitt MP, Ran Y, et al. 1994. Consistent selection of mutations in the 5'-untranslated region of oral poliovirus vaccine upon passaging in vitro. *J. Med. Virol.* 43:79–85

27. Chumakov KM, Powers LB, Noonan KE, Roninson IB, Levenbook IS. 1991. Correlation between amount of virus with altered nucleotide sequence and the monkey test for acceptance of oral poliovaccine. *Proc. Natl. Acad. Sci. USA* 88:199–203

28. Clarke DK, Duarte EA, Elena S, Moya A, Domingo E, Holland JJ. 1994. The red queen reigns in the kingdom of RNA viruses. *Proc. Natl. Acad. Sci. USA* 91:4821–24

29. Clarke DK, Duarte EA, Moya A, Elena SF, Domingo E, Holland JJ. 1993. Genetic bottlenecks and population passages cause profound fitness differences in RNA viruses. *J. Virol.* 67:222–28

30. Coffin JM. 1995. HIV population dynamics in vivo: implications for genetic variation, pathogenesis, and therapy. *Science* 267:483–89

31. Colston EM, Racaniello VR. 1995. Poliovirus variants selected on mutant receptor-expressing cells identify capsid residues that expand receptor recognition. *J. Virol.* 69:4823–29

32. Couderc T, Guédo N, Calvez V, Pelletiner I, Hogle J, et al. 1994. Substitutions in the capsids of poliovirus mutants selected in human neuroblastoma cells confer on the Mahoney type 1 strain a phenotype neurovirulent in mice. *J. Virol.* 68:8386–91

33. Cupples CG, Miller JH. 1989. A set of lacZ mutations in *Escherichia coli* that allow rapid detection of each of the six base substitutions. *Proc. Natl. Acad. Sci. USA* 86:5345–49

34. de Jong MD, Boucher CAB, Galasso GJ, Hirsch MS, Kern ER, et al. 1995. Consensus symposium on combined antiviral therapy. *Antivir. Res.* 29:5–29

35. de la Torre JC, Giachetti C, Semler BL, Holland JJ. 1992. High frequency of single-base transitions and extreme frequency of precise multiple-base reversion mutations in poliovirus. *Proc. Natl. Acad. Sci. USA* 89:2531–35

36. de la Torre JC, Holland JJ. 1990. RNA virus quasispecies can suppress vastly superior mutant progeny. *J. Virol.* 64:6278–81

37. de la Torre JC, Martínez-Salas E, Díez

J, Domingo E. 1989. Extensive cell heterogeneity during a persistent infection with foot-and-mouth disease virus. *J. Virol.* 63:69–63

38. de la Torre JC, Martínez-Salas E, Díez J, Villaverde D, Gebauer F, et al. 1988. Coevolution of cells and viruses in a persistent infection of foot-and-mouth disease virus in cell culture. *J. Virol.* 62:2050–58

39. de la Torre JC, Oldstone MBA. 1996. Anatomy of viral persistence: mechanisms of persistence and associated disease. *Adv. Virus Res.* 46:311–43

40. Dermody TS, Nibert ML, Wetzel JD, Tong X, Fields BN. 1993. Cells and viruses with mutations affecting viral entry are selected during persistent infections of L cells with mammalian reoviruses. *J. Virol.* 67:2055–63

41. Díez J, Davila M, Escarmís C, Mateu MG, Domínguez J, et al. 1990. Unique amino acid substitutions in the capsid proteins of foot-and-mouth disease virus from a persistent infection in cell culture. *J. Virol.* 64:5519–28

42. Dockter J, Evans CF, Tishon A, Oldstone MBA. 1996. Competitive selection in vivo by a cell for one variant over another: implications for RNA virus quasispecies in vivo. *J. Virol.* 70:1799–803

43. Domingo E. 1989. RNA virus evolution and the control of viral disease. In *Progress in Drug Research,* ed. E Jucker, 3:93–133. Basel: Birkhauser-Verlag

44. Domingo E. 1996. Biological significance of viral quasispecies. *Viral Hepatitis Rev.* 2:247–61

45. Domingo E, Davila M, Ortín J. 1980. Nucleotide sequence heterogeneity of the RNA from a natural population of foot-and-mouth disease virus. *Gene* 11:333–46

46. Domingo E, Díez J, Martínez MA, Hernandez J, Holguín A, et al. 1993. New observations on antigenic diversification of RNA viruses. Antigen variation is not dependent on immune selection. *J. Gen. Virol.* 74:2039–45

47. Domingo E, Escarmís C, Martínez MA, Martínez-Salas E, Mateu MG. 1992. Foot-and-mouth disease virus populations are quasispecies. *Curr. Top. Microbiol. Immunol.* 176:33–47

48. Domingo E, Escarmís C, Sevilla N, Moya A, Elena SF, et al. 1996. Basic concepts in RNA virus evolution. *FASEB J.* 10:859–64

49. Domingo E, Holland J. 1988. High error rates, population equilibrium, and evolution of RNA replication systems. See Ref. 51a, 3:3–36

50. Domingo E, Holland JJ. 1992. Complications of RNA heterogeneity for the engineering of virus vaccines and antiviral agents. In *Genetic Engineering, Principles and Methods*, ed. JK Setlow, pp. 13–32. New York: Plenum

51. Domingo E, Holland JJ. 1994. Mutation rates and rapid evolution of RNA viruses. See Ref. 135, pp. 161–84. New York: Raven

51a. Domingo E, Holland JJ, Ahlquist P, eds. 1988. *RNA Genetics*, Vols. 2, 3. Boca Raton, FL: CRC Press

52. Domingo E, Holland JJ, Biebricher C, Eigen M. 1995. Quasispecies: the concept and the word. In *Molecular Evolution of the Viruses*, ed. A Gibbs, C Calisher, F García-Arenal, pp. 171–80. Cambridge: Cambridge Univ. Press

53. Domingo E, Martínez-Salas E, Sobrino F, de la Torre JC, Portela A, et al. 1985. The quasispecies (extremely heterogeneous) nature of viral RNA genome populations: biological relevance—a review. *Gene* 40:1–8

54. Domingo E, Menendez-Arias L, Holland JJ. 1997. Virus fitness. *Rev. Med. Virol.* In press

55. Domingo E, Sabo DL, Taniguchi T, Weissmann C. 1978. Nucleotide sequence heterogeneity of an RNA phage population. *Cell* 13:735–44

56. Drake JW. 1993. Rates of spontaneous mutations among RNA viruses. *Proc. Natl. Acad. Sci. USA* 90:4171–75

57. Duarte E, Clarke D, Moya A, Domingo E, Holland JJ. 1992. Rapid fitness losses in mammalian RNA virus clones due to Muller's ratchet. *Proc. Natl. Acad. Sci. USA* 89:6015–19

58. Duarte EA, Clarke DK, Moya A, Elena SF, Domingo E, et al. 1993. Many trillionfold amplification of single RNA virus particles fails to overcome the Muller's ratchet effect. *J. Virol.* 67:3620–23

59. Duarte EA, Novella I, Ledesma S, Clarke DK, Moya A, et al. 1994. Subclonal components of consensus fitness in an RNA virus clone. *J. Virol.* 68:4295–301

60. Duarte E, Novella IS, Weaver SC, Domingo E, Wain-Hobson S, et al. 1994. RNA virus quasispecies: significance for viral disease and epidemiology. *Infect. Agents Dis.* 3:201–14

61. Eigen M. 1971. Self-organization of matter and the evolution of biological macromolecules. *Naturwissenschaften* 58:465–523

62. Eigen M. 1992. *Steps Towards Life.* Oxford: Oxford Univ. Press. 173 pp.

63. Eigen M. 1993. The origin of genetic information: viruses as models. *Gene* 135:37–47

64. Eigen M. 1996. On the nature of viral quasispecies. *Trends Microbiol.* 4:212–14

65. Eigen M, Biebricher C. 1988. Sequence space and quasispecies distribution. See Ref. 51a, 3:211–45.

66. Eigen M, McCaskill J, Schuster P. 1988. Molecular quasi-species. *J. Phys. Chem.* 92:6881–91

67. Eigen M, Schuster P. 1977. The hypercycle: a principle of natural self-organization. Part A: emergence of the hypercycle. *Naturwissenschaften* 64:541–65

68. Eigen M, Schuster P. 1978. The hypercycle: a principle of natural self-organization. Part B: the abstract hypercycle. *Naturwissenschaften* 65:7–41

69. Eigen M, Schuster P. 1978. The hypercycle: a principle of natural self-organization. Part C: the realistic hypercycle. *Naturwissenschaften* 65:341–69

70. Elena SF, Gonzalez-Candelas F, Novella IS, Duarte EA, Clarke DK, et al. 1996. Evolution of fitness in experimental populations of vesicular stomatitis virus. *Genetics* 142:673–79

71. Escarmís C, Davila M, Charpentier N, Bracho A, Moya A. et al. 1996. Genetic lesions associated with Muller's ratchet in an RNA virus. *J. Mol. Biol.* 264:255–67

72. Escarmís C, Dopazo J, Davila M, Palma EL, Domingo E. 1995. Large deletions in the 5'-untranslated region of foot-and-mouth disese virus. *Virus Res.* 35:155–67

73. Eyre-Walker A. 1996. Synonymous codon bias is related to gene length in Escherichia coli: selection for translational accuracy? *Mol. Biol. Evol.* 13:864–72

74. Friedberg EC, Walker GC, Siede W. 1995. *DNA Repair and Mutagenesis.* Washington, DC: Am. Soc. Microbiol. 698 pp.

75. Fujita Y, Mise K, Okuno T, Ahlquist P, Furusawa I. 1996. A single codon change in a conserved motif of a bromovirus movement protein gene confers compatibility with a new host. *Virology* 223:283–91

76. Gabriel A, Willems M, Mules EH, Boeke JD. 1996. Replication infidelity during a single cycle of Ty1 retrotransposition. *Proc. Natl. Acad. Sci. USA* 93:7767–71

77. Gause GF. 1964. *The Struggle for Existence.* New York: Hafner. 163 pp.
78. Gebauer F, de la Torre JC, Gomes I, Mateu MG, Barahona H, et al. 1988. Rapid selection of genetic and antigenic variants of foot-and-mouth disease virus during persistence in cattle. *J. Virol.* 62:2041–49
78a. Deleted in proof
79. Gombold JL, Ramig RF. 1994. Genetics of the rotaviruses. *Curr. Top. Microbiol. Immunol.* 185:129–77
80. Goudsmit J, De Ronde A, Ho D, Perelson AS. 1996. Human immunodeficiency virus fitness in vivo: calculations based on a single zidovudine resistance mutation at codon 215 of reverse transcriptase. *J. Virol.* 70:5662–64
81. Hahn CS, Lustig S, Strauss EG, Strauss JH. 1988. Western equine encephalitis virus is a recombinant virus. *Proc. Natl. Acad. Sci. USA* 85:5997–6001
82. Harber J, Bernhardt G, Lu H-H, Sgro J-Y, Wimmer E. 1995. Canyon rim residues, including antigen determinants, modulate serotype-specific binding of poliovirus to mutants of the poliovirus receptor. *Virology* 214:559–70
83. Haynes BF. 1996. HIV vaccines: where we are and where we are going. *Lancet* 348:933–37
84. Ho DD, Neuman AU, Perelson AS, Chen W, Leonard JM, Markowitz M. 1995. Rapid turnover of plasma virions and CD4 lymphocytes in HIV–1 infection. *Nature* 373:123–26
85. Holland JJ. 1984. Continuum of change in RNA virus genomes. In *Concepts in Viral Pathogenesis,* ed. AL Notkins, MBA Oldstone, pp. 137–43. New York: Springer-Verlag
86. Holland JJ. 1990. Defective viral genomes. In *Virology,* ed. BN Fields, DM Knipe, RM Chanock, MS Hirsch, JL Melnick, TP Monath, B Roizman, pp. 151–65. New York: Raven
87. Holland JJ. 1993. Replication error, quasispecies populations, and extreme evolution rates of RNA viruses. In *Emerging Viruses,* ed. SS Morse, pp. 203–18. Oxford: Oxford Univ. Press
88. Holland JJ, de la Torre JC, Clarke DK, Duarte E. 1991. Quantitation of relative fitness and great adaptability of clonal populations of RNA viruses. *J. Virol.* 65:2960–67
89. Holland JJ, de la Torre JC, Steinhauer D. 1992. RNA virus populations as quasispecies. *Curr. Top. Microbiol. Immunol.* 176:1–20

90. Holland JJ, Domingo E, de la Torre JC, Steinhauer DA. 1990. Mutation frequencies at defined single codon sites in vesicular stomatitis virus and poliovirus can be increased only slightly by chemical mutagenesis. *J. Virol.* 64:3960–62
91. Holland JJ, Spindler K, Horodyski F, Grabau E, Nichol S, et al. 1982. Rapid evolution of RNA genomes. *Science* 215:1577–85
92. Huynen MA, Stadler PF, Fontana W. 1996. Smoothness within ruggedness: the role of neutrality in adaptation. *Proc. Natl. Acad. Sci. USA* 93:397–401
93. Ionov Y, Peinado MA, Malkhosyan S, Shibata D, Perucho M. 1993. Ubiquitous somatic mutations in simple repeated sequences reveal a new mechanism for colonic carcinogenesis. *Nature* 36:558–61
94. Ji J, Hoffmann J-S, Loeb L. 1994. Mutagenicity and pausing of HIV reverse transcriptase during HIV plus-strand DNA synthesis. *Nucleic Acids Res.* 22:47–52
95. Jones EY, Harlos K, Bottonley MJ, Robinson RC, Driscoll PC, et al. 1995. Crystal structure of an integrin-binding fragment of vascular cell adhesion molecule-1 at 1.8 Å resolution. *Nature* 373:539–44
96. Joyce CM, Steitz TA. 1994. Function and structure relationships in DNA polymerases. *Annu. Rev. Biochem.* 63:777–822
97. Kallberg E, Jainandunsing S, Gray D, Leanderson T. 1996. Somatic mutation of immunoglobulin V genes in vitro. *Science* 271:1285–88
98. Katz M. 1995. Clinical spectrum of measles. *Curr. Top. Microbiol. Immunol.* 191:1–12
99. Kauffman SA. 1993. *The Origins of Order. Self-Organization and Selection in Evolution.* New York/Oxford: Oxford Univ. Press. 709 pp.
100. Keese P, Visvader JE, Symons RH. 1988. Sequence variability in plant viroid RNAs. In *RNA Genetics,* ed. E Domingo, JJ Holland, P Ahlquist, 3:71–98. Boca Raton: CRC Press
101. Kilbourne ED. 1994. Host determination of viral evolution: a variable tautology. See Ref. 135, pp. 253–71
102. Kirkegaard K, Baltimore D. 1986. The mechanism of RNA recombination in poliovirus. *Cell* 47:433–43
103. Kohlstaedt LA, Wang J, Rice PA, Friedman JM, Steitz TA. 1993. The structure of HIV-1 reverse transcriptase. In *Re-*

verse Transcriptase, ed. AM Skalka, SP Goff, pp. 223–49. Cold Spring Harbor: Cold Spring Harbor Lab.

104. Korber B, Myers G. 1992. Signature pattern analysis: a method for assessing viral sequence relatedness. *AIDS Res. Hum. Retroviruses* 8:1549–60

105. Krezel AM, Wagner G, Seymour-Ulmer J, Lazarus RA. 1994. Structure of the RGD protein decorsin: conserved motif and distinct function in leech proteins that affect blood clotting. *Science* 264:1944–47

106. Kurath G, Robaglia C. 1995. Genetic variation and evolution of satellite viruses and satellite RNAs. In *Molecular Basis of Virus Evolution,* ed. AJ Gibbs, CA Calisher, F Garcia-Arenal, pp. 385–403. Cambridge Cambridge Univ. Press

107. Lai MMC. 1992. Genetic recombination in RNA viruses. *Curr. Top. Microbiol. Immunol.* 176:21–32

108. Lai MMC. 1995. Recombination and its evolutionary effect of viruses with RNA genomes. See Ref. 78a, pp. 119–32

109. Lamers SL, Sleasman JW, Goodenow MM. 1996. A model for alignment of Env V1 and V2 hypervariable domains from human and simian immunodeficiency viruses. *AIDS Res. Hum. Retroviruses* 12:1169–78

110. Larder BA. 1994. Interactions between drug resistance mutations in human immunodeficiency virus type 1 reverse transcriptase. *J. Gen. Virol.* 75:951–57

111. Lea S, Hernández J, Blakemore W, Brocchi E, Curry S, et al. 1994. The structure and antigenicity of a type C foot-and-mouth disease virus. *Structure* 2:123–39

112. Lederberg J, Shope RE, Oaks SC. 1992. *Emerging Infections. Microbial Threats to Health in the United States.* Washington, DC: Natl. Acad. Press. 294 pp.

113. Lee CH, Gilbertson DL, Novella IS, Huerta R, Domingo E, Holland JJ. 1997. Negative effects of chemical mutagenesis on the adaptive behaviour of vesicular stomatitis virus. *J. Virol.* 71:3636–40

114. Lenski R, Travisano M. 1994. Dynamics of adaptation and diversification: a 10,000-generation experiment with bacterial populations. *Proc. Natl. Acad. Sci. USA* 91:6808–14

115. Levander OA, Beck MA. 1997. Interacting nutritional and infectious etiologies of Keshan disease: insights from coxsackievirus B-induced myocarditis in mice deficient in selenium or Vitamin E. *Biol. Trace Element Res.* 56:5–22

116. Liljas L. 1996. Viruses. *Curr. Opin. Struct. Biol.* 6:151–56

117. Liu R, Paxton WA, Choe S, Ceradini D, Martin SR, et al. 1996. Homozygous defect in HIV-1 coreceptor accounts for resistance of some multiply-exposed individuals to HIV-1 infection. *Cell* 86:367–77

118. Logan D, Abu-Ghazaleh R, Blakemor W, Curry S, Jackson T, et al. 1993. Structure of a major immunogenic site on foot-and-mouth disease virus. *Nature* 362:566–68

119. Lopez N, Padula P, Rossi C, Lazaro ME, Franze-Fernandez MT. 1996. Genetic identification of a new hantavirus causing severe pulmonary syndrome in Argentina. *Virology* 220:223–26

120. Lu Z, Rezapkin GV, Douthitt MP, Ran Y, Asher DM, et al. 1996. Limited genetic changes in the Sabin 1 strain poliovirus occurring in the central nervous system of monkeys. *J. Gen. Virol.* 77:273–80

121. Martín-Hernandez AM, Carrillo EC, Sevilla N, Domingo E. 1994. Rapid cell variation can determine the establishment of a persistent viral infection. *Proc. Natl. Acad. Sci. USA* 91:3705–9

122. Martín-Hernandez AM, Domingo E, Menéndez-Arias L. 1996. Human immunodeficiency virus type 1 reverse transcriptase: role of Tyr-115 in deoxynucleotide binding and misinsertion fidelity of DNA synthesis. *EMBO J.* 15:4434–42

123. Martínez MA, Carrillo C, González-Candelas F, Moya A, Domingo E, et al. 1991. Fitness alteration of foot-and-mouth disease virus mutants: measurement of adaptability of viral quasispecies. *J. Virol.* 65:3954–57

124. Martínez MA, Dopazo J, Hernandez J, Mateu MG, Sobrino F, et al. 1992. Evolution of the capsid protein genes of foot-and-mouth disease virus: antigenic variation without accumulation of amino acid substitutions over six decades. *J. Virol.* 66:3557–65

125. Martínez MA, Pezo V, Marlière P, Wain-Hobson S. 1996. Exploring the functional robustness of an enzyme by in vitro evolution. *EMBO J.* 15:1203–10

126. Martinez MA, Vartanian J-P, Wain-Hobson S. 1994. Hypermutagenesis of RNA using human immunodeficiency virus type 1 reverse transcriptase and biased dNTP concentration. *Proc. Natl. Acad. Sci. USA* 91:11787–91

127. Martínez-Salas E, Regalado MP, Domingo E. 1996. Identification of an essential region for internal initiation of

translation in the aphthoviruses IRES, and implications for viral evolution. *J. Virol.* 70:992–98

128. Mateu MG. 1995. Antibody recognition of picornaviruses and escape from neutralization: a structural view. *Virus Res.* 38:1–24

129. Maynard Smith J, Burian R, Kauffman S, Alberch P, Campbell J, et al. 1985. Developmental constraints and evolution. *Quart. Rev. Biol.* 60:265–87

130. McClure MA. 1997. The complexities of viral genome analysis: the primate lentiviruses. *Curr. Opin. Genet. Dev.* 6: 749–56

131. Meyerhans A, Cheynier R, Albert J, Seth M, Kwok S, et al. 1989. Temporal fluctuations in HIV quasispecies in vivo are not reflected by sequential isolations. *Cell* 58:901–10

132. Meyers G, Tantz N, Theil H-J. 1995. Cellular sequences in viral genomes. See Ref. 78a, pp. 91–102

133. Minor PD. 1992. The molecular biology of poliovaccines. *J. Gen. Virol.* 73:3065–77

134. Modrich P, Lahue R. 1996. Mismatch repair in replication fidelity, genetic recombination, and cancer biology. *Annu. Rev. Biochem.* 65:101–33

135. Morse SS. 1994. The viruses of the future? Emerging viruses and evolution. In *The Evolutionary Biology of Viruses,* ed. SS Morse, pp. 325–35. New York: Raven

136. Moyle GJ. 1996. Use of viral resistance patterns to antiretroviral drugs in optimising selection of drug combinations and sequences. *Drugs* 52:168–85

137. Muller HJ. 1964. The relation of recombination to mutational advance. *Mutat. Res.* 1:2–9

138. Murphy FA. 1996. Virus taxonomy. In *Fields Virology,* ed. BN Fields, DM Knipe, PM Howley, RM Chanock, JL Melnick, et al, pp. 15–57. Philadelphia: Lippincott-Raven

139. Murphy FA, Nathanson N. 1994. The emergence of new virus diseases: an overview. *Semin. Virol.* 5:87–102

140. Myers GM, Korber B, Berzofsky JA, Smith RF, Paulakis GN. 1995. *A compilation and analysis of nucleic acid and amino acid sequences.* Theor. Biol. Biophys. Group, Los Alamos Natl. Lab., Los Alamos, NM

141. Naeve CW, Hinshaw VS, Webster RG. 1984. Mutations in the influenza hemagglutinin receptor-binding site can change the biological properties of an influenza virus. *J. Virol.* 51:567–69

142. Nájera I, Holguín A, Quiñones-Mateu ME, Muñoz-Fernández MA, Nájera R, et al. 1995. The pol gene quasispecies of human immunodeficiency virus. Mutations associated with drug resistance in virus from patients undergoing no drug therapy. *J. Virol.* 69:23–31

143. Nichol ST, Spiropoulou CF, Morzunov S, Rollin PE, Ksiazek TG, et al. 1993. Genetic identification of a hantavirus associated with an outbreak of acute respiratory illness. *Science* 262:914–17

144. Nicolson GL. 1987. Tumor cell instability, diversification, and progression to the metastatic phenotype: from oncogene to oncofetal expression. *Cancer Res.* 47:1473–87

145. Novella IS, Clarke DK, Quer J, Duarte EA, Lee CH, et al. 1995. Extreme fitness differences in mammalian and insect hosts after continuous replication of vesicular stomatitis virus in sandfly cells. *J. Virol.* 69:6805–9

146. Novella IS, Domingo E, Holland JJ. 1995. Rapid viral quasispecies evolution: implications for vaccine and drug strategies. *Mol. Med. Today* 1:248–53

147. Novella IS, Duarte EA, Elena SF, Moya A, Domingo E, et al. 1995. Exponential increases of RNA virus fitness during large populations transmissions. *Proc. Natl. Acad. Sci. USA* 92:5841–44

148. Novella IS, Elena SF, Moya A, Domingo E, Holland JJ. 1995. Size of genetic bottlenecks leading to virus fitness loss is determined by mean initial population fitness. *J. Virol.* 69:2869–72

149. Olsthoorn RCL, Licis N, van Duin J. 1994. Leeway and constraints in the forced evolution of a regulatory RNA helix. *EMBO J.* 13:2660–68

150. Olsthoorn RCL, van Duin J. 1996. Random removal of inserts from an RNA genome: selection against single-stranded RNA. *J. Virol.* 70:729–36

151. Ortín J, Nájera R, López C, Dávila M, Domingo E. 1980. Genetic variability of Hong Kong (H3N2) influenza viruses: spontaneous mutations and their location in the viral genome. *Gene* 11:319–31

152. Pandey VN, Kaushik N, Rege N, Sarafianos SG, Yadav PNS, et al. 1996. Role of methionine 184 of human immunodeficiency virus type 1 reverse transcriptase in the polymerase function and fidelity of DNA synthesis. *Biochemistry* 35:2168–79

153. Pathak VK, Temin HM. 1990. Broad spectrum of in vivo forward mutations,

hypermutations, and mutational hotspots in a retroviral shuttle vector after a single replication cycle: substitutions, frameshifts and hypermutations. *Proc. Natl. Acad. Sci. USA* 87:6019–23

154. Pathak VK, Temin HM. 1992. 5-Azacytidine and RNA secondary structure increase the retrovirus mutation rate. *J. Virol.* 66:3093–100

155. Pelletier I, Couderc T, Borzakian S, Wyckoff E, Crainic R, et al. 1991. Characterization of persistent poliovirus mutants selected in human neuroblastoma cells. *Virology* 180:729–37

156. Penny MA, Thomas SJ, Douglas NW, Ranjbar S, Homes H, et al. 1996. Env gene sequences of primary HIV Type 1 isolates of subtypes B, C, D, E, and F obtained from the World Health Organization Network for HIV isolation and characterization. *AIDS Res. Hum. Retroviruses* 12:741–47

157. Pilipenko EV, Gmyl AP, Maslova SV, Svitkin YV, Sinyakov AN, et al. 1992. Prokaryotic-like cis elements in the cap-independent internal initiation of translation on picornavirus RNA. *Cell* 68:119–31

158. Pleij CWA. 1994. RNA pseudoknots. *Curr. Opin. Struct. Biol.* 4:337–44

159. Polivka H, Staub U, Gross HJ. 1996. Variation of viroid profiles in individual grapevine plants: novel grapevine yellow speckle viroid 1 mutants show alterations of hairpin I. *J. Gen. Virol.* 77:155–61

160. Pop MP, Biebricher CK. 1996. Kinetic analysis of pausing and fidelity of human immunodeficiency virus type 1 reverse transcription. *Biochemistry* 35:5045–62

161. Preston BD. 1996. Error-prone retrotransposition: rime of the ancient mutators. *Proc. Natl. Acad. Sci. USA* 93:7427–31

162. Pringle CR, Devine V, Wilkie M, Preston CM, Dolan A, McGeoch DJ. 1981. Enhanced mutability associated with a temperature sensitive mutant of vesicular stomatitis virus. *J. Virol.* 39:377–89

163. Quer J, Huerta R, Novella IS, Tsimring L, Domingo E, et al. 1996. Reproducible nonlinear population dynamics and critical points during replicative competition of RNA virus quasispecies. *J. Mol. Biol.* 264:465–71

164. Quiñones-Mateu ME, Holguín A, Dopazo J, Nájera I, Domingo E. 1996. Point mutant frequencies in the pol gene of human immunodeficiency virus type 1 are two- to three-fold lower that those

of env. *AIDS Res. Hum. Retroviruses* 12:1117–28

165. Ricchetti M, Buc H. 1990. Reverse transcriptases and genomic variability: the accuracy of DNA replication is enzyme specific and sequence dependent. *EMBO J.* 9:1583–93

166. Richman DD. 1994. Resistance, drug failure, and disease progression. *AIDS Res. Hum. Retroviruses* 10:901–5

167. Rima BK, Earle JAP, Baczko K, Rota PA, Bellini WJ. 1994. Measles virus strain variations. *Curr. Top. Microbiol. Immunol.* 191:65–83

168. Robertson DK, Sharp PM, McCutchan FE, Hahn BH. 1995. Recombination in HIV-1. *Nature* 374:124–26

169. Rohde N, Draum H, Biebricher CK. 1995. The mutant distribution of an RNA species replicated by Qβ replicase. *J. Mol. Biol.* 249:754–62

170. Rossmann MG. 1989. The canyon hypothesis. *J. Biol. Chem.* 333:426–31

171. Roy-Burman P. 1996. Endogenous env elements: partners in generation of pathogenic feline leukemia viruses. *Virus Genes* 11:147–61

172. Rueda P, Garcia Barreno B, Melero JA. 1994. Loss of conserved cysteine residues in the attachment glycoprotein of two human respiratory syncytial virus escape mutants that contain multiple A-G substitutions (hypermutations). *Virology* 198:653–62

173. Sala M, Zambruno G, Vartanian JP, Marconi A, Bertazzoni U, et al. 1994. Spatial discontinuities in human immunodeficiency virus type 1 quasispecies derived from epidermal langerhans cells of a patient with AIDS and evidence for double infection. *J. Virol.* 68:5280–83

174. Sawyer S. 1989. Statistical test for detecting gene conversion. *Mol. Biol. Evol.* 6:526–38

175. Scott J. 1995. A place in the world for RNA editing. *Cell* 81:833–36

176. Scott TW, Weaver SC, Mallampali VL. 1994. Evolution of mosquito-borne viruses. See Ref. 135, pp. 293–324

177. Seillier-Moiseiwitsch F, Margolin BH, Swanstrom R. 1994. Genetic variability of the human immunodeficiency virus: statistical and biological issues. *Annu. Rev. Genet.* 28:559–96

178. Sevilla N, Domingo E. 1996. Evolution of a persistent aphthovirus in cytolytic infections: partial reversion of phenotypic traits accompanied by genetic diversification. *J. Virol.* 70:6617–24

179. Sevilla N, Verdaguer N, Domingo E. 1996. Antigenically profound amino

acid substitutions occur during large population passages of foot-and-mouth disease virus. *Virology* 225:400–5

180. Sharp PM, Robertson DL, Gao F, Hahn BH. 1994. Origins and diversity of human immunodeficiency viruses. *AIDS* 8:S27–S42

181. Siepel AC, Halpern AL, Macker C, Korber BTM. 1995. A computer program designed to screen rapidly for HIV type 1 intersubtype recombinant sequences. *AIDS Res. Hum. Retroviruses* 11:1413–16

182. Simon AE, Bujarski JJ. 1994. RNA-RNA recombination and evolution in virus-infected plants. *Annu. Rev. Phytopathol.* 32:337–62

183. Skehel JJ, Wiley DC. 1988. Antigenic variation in influenza virus hemagglutinions. See Ref. 49, pp. 139–46

184. Sousa R. 1996. Structural and mechanistic relationships between nucleic acid polymerases. *Trends Biochem. Sci.* 21:186–90

185. Steinhauer D, Domingo E, Holland JJ. 1992. Lack of evidence for proofreading mechanisms associated with an RNA virus polymerase. *Gene* 122:281–88

186. Storb U. 1996. The molecular basis of somatic hypermutation of immunoglobulin genes. *Curr. Opin. Immunol.* 8:206–14

187. Strauss JH. 1993. Recombination in the evolution of RNA viruses. In *Emerging Viruses*, ed. SS Morse, pp. 241–51. Oxford: Oxford Univ. Press

188. Suarez P, Valcarcel J, Ortín J. 1992. Heterogeneity of the mutation rates of influenza A viruses: isolation of mutator mutants. *J. Virol.* 66:2491–94

189. Swetina J, Schuster P. 1982. Self-replication with error—a model for polynucleotide replication. *Biophys. Chem.* 16:329–45

189a. Taboga O, Tami C, Carrillo E, Nunez JI, Rodriguez A, et al. 1997. A large-scale evaluation of peptide vaccines against foot-and-mouth disease: lack of solid protection in cattle and isolation of escape mutants. *J. Virol.* 71:2606–14

190. Taylor M, Feyereisen R. 1996. Molecular biology of resistance to toxicants. *Mol. Biol. Evol.* 13:719–34

191. Temin HM. 1988. Evolution of cancer genes as a mutation-driven process. *Cancer Res.* 48:1697–1701

192. Temin HM. 1989. Is HIV unique or merely different? *J. AIDS* 2:1–9

193. Temin HM. 1993. The high rate of retrovirus variation results in rapid evolution. In *Emerging Viruses*, ed. SS Morse, pp. 219–25. Oxford: Oxford Univ. Press

194. Teng MN, Oldstone MBA, de la Torre JC. 1996. Suppression of lymphocytic choriomeningitis virus-induced growth hormone deficiency syndrome by disease-negative virus variants. *Virology* 223:113–19

195. Tsimring L, Levine H, Kessler DA. 1996. RNA virus evolution via a fitness-space model. *Phys. Rev. Lett.* 76:4440–43

196. VandePol SB, Lefrançois L, Holland JJ. 1986. Sequences of the major antibody binding epitopes of the Indiana serotype of vesicular stomatitis virus. *Virology* 148:312–25

197. van Valen L. 1973. A new evolutionary law. *Evol. Theory* 1:1–30

198. Vartanian J-P, Henry M, Wain-Hobson S. 1996. Hypermutagenic PCR involving all four transitions and a sizeable proportion of transversions. *Nucleic Acids Res.* 24:2627–31

199. Vartanian J-P, Meyerhans A, Asjö B, Wain-Hobson S. 1991. Selection, recombination and G/A hypermutation of human immunodeficiency virus type 1 genomes. *J. Virol.* 65:1779–88

200. Verdaguer N, Mateu MG, Andreu D, Giralt E, Domingo E, et al. 1995. Structure of the major antigenic loop of foot-and-mouth disease virus complexed with a neutralizing antibody: direct involvement of the Arg-Gly-Asp motif in the interaction. *EMBO J.* 14:1690–96

201. Verdaguer N, Mateu MG, Bravo J, Domingo E, Fita I. 1996. Induced pocket to accommodate the cell attachment Arg-Gly-Asp motif to a neutralizing antibody against foot-and-mounth disease virus. *J. Mol. Biol.* 256:364–76

202. Wagner R, Deml L, Fitzon T, Wolf H. 1995. HIV assembly: target for antiviral therapy and basis of a rationally designed candidate vaccine. In *Vaccines 95*, ed. RM Chanock, F Brown, HS Ginsberg, E Norrby, pp. 347–56. Cold Spring Harbor, NY: Cold Spring Harbor Lab.

203. Wainberg MA, Drosopoulos WC, Salomon H, Hsu M, Borkow G, et al. Enhanced fidelity of 3TC-selected mutant HIV-1 reverse transcriptase. *Science* 271:1282–85

204. Wain-Hobson S. 1994. Is antigenic variation of HIV important for AIDS and what might be expected in the future? See ref. 135, pp. 185–209

205. Wain-Hobson s. 1996. Running the gamut of retroviral variation. *Trends Microbiol.* 4:135–41

206. Webster RG, Bean WJ, Gorman OT, Chambers TM, Kawaoka Y. 1995. Evolution and ecology of influenza A viruses. *Microbiol. Rev.* 56:152–79

207. Wei X, Ghoosh SK, Taylor ME, Johnson VA, Emini EA, et al. 1995. Viral dynamics in human immunodeficiency virus type 1 infection. *Nature* 373:117–22

208. Weis W, Brown J, Cusack S, Paulson JE, Skehel JJ, Wiley DC. 1988. The structure of the influenza virus haemagglutinin complexed with its receptor, sialic acid. *Nature* 333:14587–90

209. Williams GC. 1992. *Natural Selection. Domains, Levels and Challenges.* Oxford: Oxford Univ. Press. 208 pp.

210. Williams KJ, Loeb LA. 1992. Retroviral reverse transcriptases: error frequencies and mutagenesis. *Curr. Top. Microbiol. Immunol.* 176:165–80

211. Wimmer E, Hellen CUT, Cao X. 1993. Genetics of poliovirus. *Annu. Rev. Genet.* 27:353–436

212. Wright S. 1931. Evolution in Mendelian populations. *Genetics* 16:97–159

213. Yewdell JW, Caton AJ, Gerhard W. 1986. Selection of influenza A virus adsorptive mutants by growth in the presence of a mixture of monoclonal antihemagglutinin antibodies. *J. Virol.* 57:623–28

214. Yu H, Goodman MF. 1992. Comparison of HIV-1 and avian myeloblastosis virus reverse transcriptase fidelity on RNA and DNA templates. *J. Biol. Chem.* 267:10888–96

215. Zamora H, Luce R, Biebricher CK. 1995. Design of artificial short-chained RNA species that are replicated by Q replicase. *Biochemistry* 34:1261–66

216. Zimmern D. 1988. Evolution of RNA viruses. See Ref. 51a, 2:211–40

Annu. Rev. Microbiol. 1997. 51:179–202
Copyright © 1997 by Annual Reviews Inc. All rights reserved

MAKING AND BREAKING DISULFIDE BONDS

S. Raina
Centre Médical Universitaire, Département de Biochimie Médicale, 1 Rue
Michel-Servet, 1211 Genève 4, Switzerland; e-mail: Satish.Raina@medecine.unige.ch

D. Missiakas
Centre National de Recherche Scientifique, LIDSM-CBBM, 31 Chemin
Joseph-Aiguier, 13402 Marseille Cedex 20, France;
e-mail: missiaka@ibsm.cnrs-mrs.fr

KEY WORDS: protein folding, redox proteins, disulfide isomerase, DTT, periplasm

ABSTRACT

It is now well established that protein folding requires the assistance of folding
helpers in vivo. The formation or isomerization of disulfide bonds in proteins
is a slow process requiring catalysis. In nascent polypeptide chains the cysteine
residues are in the thiol form. The formation of the disulfide bonds usually
occurs simultaneously with the folding of the polypeptide, which means in the
endoplasmic reticulum of eukaryotes or in the periplasm of Gram-negative bacteria. In prokaryotes, the existence of redox proteins involved in the formation of
disulfide bonds containing proteins has recently been revealed in the periplasm.
The discovery of these redox proteins through various genetic approaches will be
summarized, as well as the most recent insights regarding their biochemical and
biological activities.

CONTENTS

179

0066-4227/97/1001-0179$08.00

INTRODUCTION

This review summarizes current knowledge about redox proteins in *Escherichia coli* (*E. coli*) and their involvement in the folding of transported proteins containing disulfide bonds. In *E. coli*, such redox proteins are located in the membranes or in the periplasm and their active site is always facing the periplasm. Most of these proteins are called Dsb's for disulfide bond formation (7). Each of them carries in its primary structure an active site containing two vicinal cysteines, Cys-X-X-Cys, and they can exist either in the reduced form, Dsb-$(SH)_2$ with a dithiol, or in the oxidized form Dsb-S_2 with an intramolecular disulfide bond established between both half cysteine residues.

Thioredoxin is one of the best characterized redox proteins, containing two redox-active half-cysteine residues in an exposed active center. It is a small ubiquitous protein, present in the cytoplasm of *E. coli*, and evolutionarily conserved in both eukaryotes and prokaryotes. Thioredoxin serves as a model for our understanding of *E. coli* Dsb proteins and eukaryotic Protein Disulfide Isomerase (PDI). Thioredoxin was first discovered as a hydrogen donor for ribonucleotide reductase (50) but was later recognized as a general protein disulfide reductant (37, 69). It is an effective reductant of insulin in animal cells (36), and this activity has been used in vitro for the preliminary biochemical characterization of many Dsb redox proteins. Thioredoxin-$(SH)_2$ and DsbA/DsbC-$(SH)_2$ are all much faster at reducing insulin disulfides in vitro than is dithiothreitol (DTT) (7, 36, 57).

The known *E. coli* redox proteins, such as thioredoxin, the four Dsb proteins, and the three glutaredoxins, lack an overall sequence homology. Nevertheless, they all contain the active site Cys-X-X-Cys and they behave either as general protein disulfide reductants or as oxidants. Therefore, they all belong to what can be referred to as the thioredoxin super-family (Figure 1) (Table 1). In this review, we will discuss the discovery of Dsb proteins and their roles as folding

DsbA	Cys-Pro-His-Cys
DsbB	Cys-Val-Leu-Cys
DsbC	Cys-Gly-Tyr-Cys
DsbD	Cys-Val-Ala-Cys
DsbE	Cys-Pro-Thr-Cys
Thioredoxin	Cys-Gly-Pro-Cys
PDI	Cys-Gly-His-Cys

Figure 1   The thiol/disulfide active sites of members of the thioredoxin super-family.

helpers. A particular emphasis is placed on the progress achieved in our understanding of the chemistry of thiol:disulfide exchanges during protein folding, by comparing thioredoxin, and the various Dsb and PDI in their respective environments and cellular locations.

FOLDING OF DISULFIDE-BOND–CONTAINING PROTEINS

Disulfide Bonds Stabilize the Folded Conformation of Proteins

Two cysteine residues which are adjacent in the three-dimensional structure of a protein can be oxidized to form a disulfide bond. Such covalent bonds usually contribute to the stabilization of the folded conformation (14, 20). No denaturant is required to unfold such proteins: Unfolding will occur simply upon the addition of reducing agents (14). Proteins such as β-lactamase or α-lactalbumin are rare examples that can fold with and without the disulfide bonds. Nevertheless, even in such cases, disulfide bonds define a particular structure. For example, the three disulfides of α-lactalbumin cannot be paired randomly without greatly destabilizing the protein (21).

Where Does the Oxidation of Cysteines Occur In Vivo?

The formation of disulfide bonds requires an oxidative environment. The intracellular proteins seldom contain disulfide bridges because intracellular compartments, such as the bacterial cytoplasm or the eukaryotic cytosol, are essentially reductive. Most of the disulfide-bond–containing proteins are secreted proteins.

Table 1 The various redox proteins of *E. coli*

Gene name	Genetic map (min)	Protein name	MW of monomer (kDa)	Cellular location	Function (pH7, 30°C)	Redox potential
dsbA	87	DsbA	21	periplasm	oxidant	−0.089 V (45) ([a]K_{ox} 80 μM; 39)
dsbB	26	DsbB	20	inner membrane	oxidant	ND
dsbC	62	DsbC	2 × 23	periplasm	disulfide isomerase	[a]K_{ox} 200 μM (40)
dsbD	94	DsbD	50	inner membrane	reductant	
dsbE	46	DsbE	20	periplasm	reductant	
trxA	86	Thioredoxin	12	cytoplasm	reductant	−0.270 V (46) ([a]K_{ox} 10 M; 53)
trxB	21	Thioredoxin reductase	2 × 70	cytoplasm	reductant	
grxA	19	Glutaredoxin 1	10	cytoplasm	reductant	
grxB		Glutaredoxin 2	27	cytoplasm	reductant	
grxC		Glutaredoxin 3	10	cytoplasm	reductant	
gor	77	Glutaredoxin reductase	48	cytoplasm	reductant	
gshA	58	Glutathione synthetase		cytoplasm	synthesizes the tripeptide Glutathione	

[a]K_{ox} are equilibrium constants between oxidized and reduced forms of the proteins, i.e. disulfide bond formation as measured by interchange with glutathione.

In fact, the translocation of proteins across membranes often requires a fully unfolded conformation (66). Hence, disulfide formation should be prevented until the proteins reach their final destination, where they ultimately aquire their native and active structure.

The formation of disulfide bridges is largely prevented in the cytosol of eukaryotes because of the cytosol's reduced environment, which has a ratio of reduced/oxidized glutathione (GSH/GSSG) estimated to be between 30/1 to 100/1 (25) [the critical factor for disulfide bond stability is the ratio $(GSH)^2/(GSSG)$]. In the endoplasmic reticulum (ER), a different ratio—of about 3/1—is maintained between the two species GSH/GSSG (39), albeit by a so far unknown mechanism. Such a ratio favors the thermodynamic formation of disulfide bonds that are trapped within folded, or almost folded, proteins. In other words, because the ER is more oxidizing than the cytosol, the disulfide bonds have more stability in the ER and, consequently, give more stability to a protein conformation.

A redox potential of about -165 ± 5 mV, determined by a total glutathione concentration of about 2 mM and a GSH/GSSG ratio 2/1 (39), has been predicted to be the optimum redox environment for refolding a disulfide containing protein such as RNase A. Such conditions are approximated in the ER and are in agreement with the redox state of protein disulfide isomerase (PDI)—with a standard redox potential of -0.111 V (35) [this value is higher when measured with methods involving a faster alkylation reaction to block the free thiol groups (see Reference 18)].

In Gram negative bacteria, most of the disulfide-bond–containing proteins are also secreted proteins such as periplasmic, outer membrane, or extracellular proteins. It has long been known that the cytoplasm provides mostly a reducing environment with the thioredoxin/thioredoxin reductase and glutaredoxin/glutathione/glutaredoxin reductase systems (Figure 2) (Table 1). It was assumed that the oxidation between two cysteines of secreted proteins occurs in an uncatalyzed manner and in the periplasm, where a favorable redox environment can be maintained, presumably because diffusion of many small oxidizing compounds could easily occur through the porous outer membrane. It is only recently that thiol:disulfide oxido-reductases have been identified as key players in protein folding in the periplasm of Gram negative bacteria.

Role of the Eukaryotic Protein Disulfide Isomerase

An enzyme able to reactivate reduced and denatured ribonuclease A (RNAse A) in the presence of GSSG was isolated from microsomes (31) about 30 years ago. The ability of reduced PDI to catalyze the reformation of the correct intramolecular disulfide bonds in scrambled RNAse A was the reason why such an enzyme was named protein disulfide isomerase (PDI) (75). In addition

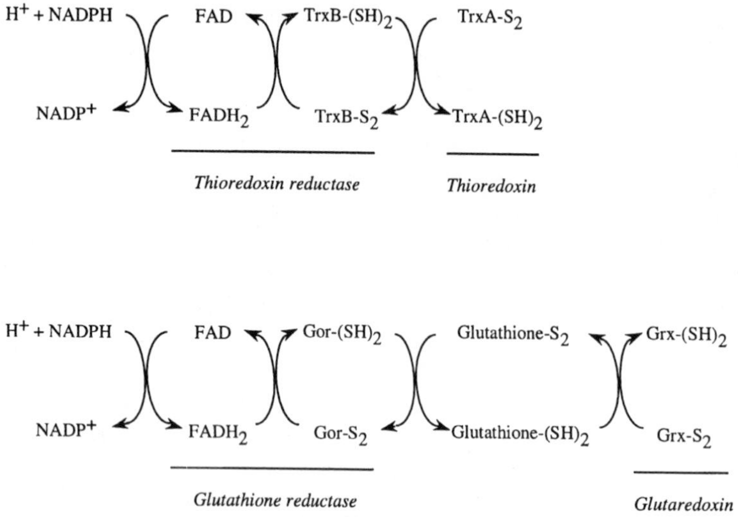

Figure 2   The thioredoxin and glutaredoxin cytoplasmic systems. TrxA: thioredoxin; TrxB: thioredoxin reductase; Grx: glutaredoxin; Gor: glutathione reductase.

to this isomerase activity, PDI was found to be a very good oxidant, able to introduce disulfide bonds into a fully reduced protein (30). PDI was later found to be an abundant enzyme in the lumen of the ER; its abundance correlates with the amounts of disulfide-bond–containing proteins that are synthesized (26). PDI is a modular protein and carries two active sites that closely resemble that of the redox protein thioredoxin. The latest domain composition of the protein, as determined by protein engineering, limited proteolysis, and NMR analysis, is presented in Figure 3.

THE GENETIC BASIS OF ISOLATION OF PDI-LIKE ACTIVITIES IN PROKARYOTES

Because PDI was found to carry out redox reactions in the appropriate cellular compartment in eukaryotes, it was tempting to try to isolate a similar activity from the bacterial periplasm. Such a search implied that non-catalyzed oxidation, such as the oxygen-dependent oxidation of disulfide-bond–containing proteins, was to be ruled out. The first attempts were based on the biochemical purification of a thiol:disulfide oxido-reductase activity from *E. coli* osmotic shock fractions (8). Although not fully successful, these experiments gave the first evidence for the presence of redox proteins in the periplasm.

(A)

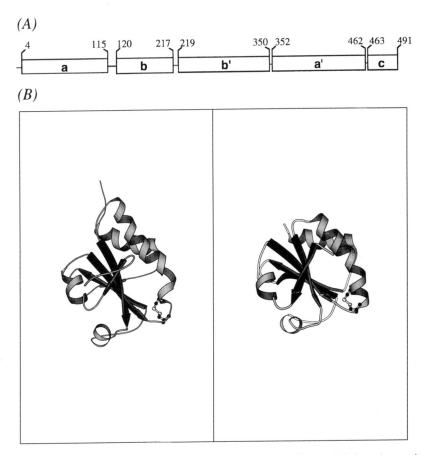

Figure 3   The eukaryotic enzyme PDI. (*A*) Domain organization of human PDI. Domains a and a′ are thioredoxin-like domains. Residue numbers are indicated for each domain boundary, as determined by a combination of protein engineering, limited proteolysis and NMR analysis (44). The c domain at the carboxyl terminus is a 24-residue acidic segment in which over half residues are Glu or Asp, followed by a typical KDEL sequence for retention in the ER. (*B*) Recent NMR structure analyses suggest that the a domain contains the thioredoxin fold (44). Left side: human PDI a domain (J Kemmink, NJ Darby, K Dijkstra, M Nilges, TE Creightton; personal communication); right side: *E. coli* thioredoxin.

Isolation of the dsbA and dsbB Genes in Escherichia coli

It was only three years later that a first gene encoding a redox activity was described in *E. coli* by two different groups.

In the first genetic screen, mutants were isolated that impaired the proper insertion of inner membrane proteins (7). To facilitate screening, a hybrid between the membrane MalF protein and β-galactosidase had been used. β-galactosidase is only active in the cytoplasm, and mutants exhibiting a Lac[+] phenotype reflected the inability of the hybrid to integrate in the membrane. Such a mutant candidate was mapped into a previously unknown gene and designated as *dsbA* (after the deduced function encoded by this gene) (7). In fact, in a *dsbA* mutant, β-galactosidase probably remains reduced and, as a consequence, unfolded, allowing the hybrid MalF-β-gal to slide back to the cytoplasm. The same genetic approach identified yet another gene, *dsbB*, which also encodes a redox protein (6). Mutations in the *dsbB* gene were independently isolated because they conferred a motility defect to *E. coli*, presumably due to the accumulation of reduced P-ring protein forming the flagellar basal body (17) (Table 1).

A second independent genetic study, carried out at about the same time, identified mutations in a *ppfA* (periplasmic protein formation) gene resulting in a loss of alkaline phosphatase (AP) activity (43). Loss of AP activity was correlated with a defect in AP folding. AP is a periplasmic enzyme containing two disulfide bonds in its native conformation. In the *ppfA* mutant, disulfide bond formation in AP is impaired, leading to its misfolding. The *ppfA* gene was shown to be identical to *dsbA* (43).

There are many other cases in which mutations in *dsbA* or *dsbB* have been isolated because they affected the proper folding of specific transported proteins, such as acid phosphatase (11) or enterotoxin I and II (61, 80). Finally, the *Bacteroides fragilis* metallo-β-lactamase CcrA was found to be best expressed in *E. coli* strains carrying mutations either in *dsbA* or *dsbB* (3).

A Global Genetic Approach Identifies Multiple Thiol:Disulfide Oxido-Reductases in Escherichia coli

Various genes encoding redox proteins were identified by using a rather simple, but more direct, genetic screen based on the ability or inability of *E. coli* cells to cope with changes in the redox environment. Wild-type bacteria were found to grow normally on agar plates containing dithiothreitol (DTT) concentrations of up to 7 mM. The possibility that DTT could be neutralized by re-oxidation, and by a general disulfide oxidant protein was examined. Multicopy libraries were used to transform *E. coli*, and candidates that could improve growth in the presence of DTT were selected on media containing sub-lethal DTT amounts (56). Cloning of the wild-type *dsbB* gene was found to allow *E. coli* to grow normally on such reductive media. In a second approach, null mutants unable to cope with low concentrations of DTT in the medium (5 mM and lesser) were isolated. Lack of any putative redox protein should lead to bacteria that are hypersensitive to the redox environment. Multiple DTT-sensitive mutants were isolated based on this hypothesis, and were mapped to various unlinked genes (56). The most interesting mutants mapped to the known *trxA, trxB, dsbA* and *dsbB* genes, but also to some other genes, including the *dsbC, dsbD* and *dsbE* genes.

Later findings revealed that the hypersensitivity to DTT exhibited by *dsbA* mutants could be alleviated in two ways: either by selectively overexpressing *dsbC* from a multicopy library (57) or by mutations blocking either *dsbD* (59) or *dsbE* (Table 1). DsbC was also independently identified in *Erwinia chrysanthemi* by Shevchik and colleagues (72), who used *E. chrysanthemi* DNA libraries to complement an *E. coli dsbA* mutant in order to clone the *dsbA* homologue.

Homologues of the dsb Genes in Other Bacteria

Homologues of *dsbA* have been identified in many other bacteria. Some of them have been isolated from chromosomal DNA libraries and selected by complementing an *E. coli* strain that lacks its own functional *dsbA* gene. This is the case for the *dsbA* homologues found in *E. chrysanthemi* (71). Other *dsbA* homologues have been found to complement the *E. coli dsbA* mutation by restoring its motility defect and/or AP activity. This is the case for the *Azotobacter vinelandii dsbA* homologue (accession number L76098) or the *bdb* gene from *Bacillus brevis* (41).

Mutations in *dsbA* have also been isolated because they block cellular pathways or activities that depend solely on the presence of one disulfide-bond–containing protein. The biogenesis of the toxin-coregulated colonization pilus (TCP) was found to require the *tcpG* gene product, a protein 40% identical to *E. coli*'s DsbA in *Vibrio cholerae* (63). An independent study showed that production of enterotoxin B subunit (EtxB), a disulfide-bond–containing protein, was highly decreased in *tcpG* mutants, thereby leading to an attenuated virulence (82). The Por protein of *Haemophilus influenza* was proposed to be a periplasmic redox protein that participates in the competence-induced DNA uptake process. Por was shown to be 45% identical to the *E. coli* DsbA protein and *por* mutants could block DNA uptake, presumably, by impairing the folding of disulfide-bond–containing outer membrane proteins (74).

Sequence homologues of *dsbB* can be found in *H. influenza* (U32726), *Shigella flexneri* (accession number D38254), *Pseudomonas aeruginosa* (accession number M30145) or *Vibrio alginolyticus* (accession number D83728).

THE BIOCHEMICAL AND BIOLOGICAL ACTIVITIES OF THE DSB PROTEINS

The Chemistry of Thiol:Disulfide Exchange Reactions In Vitro

Biochemical characterization of the thiol:disulfide exchange abilities of the known Dsb proteins is clearest for DsbA and DsbC, primarily because both of them are soluble proteins, whereas DsbB and DsbD are membrane proteins. Despite the lack of sequence homology, all these redox proteins contain an active site that is similar to thioredoxin (Figure 1; Table 1). The resolution of DsbA three-dimensional structure (53) has revealed that despite its lack of overall sequence homology, DsbA contains a domain that is superimposable on the thioredoxin fold. This is also the case with glutaredoxin which, despite its low degree of sequence homology, exhibits a thioredoxin fold (79). From secondary structure prediction, DsbC is likely to contain such a fold as well (27). These findings further sustain the existence of a thioredoxin super-family (Figure 1).

Redox proteins such as thioredoxin contain a reversible disulfide bridge that forms between the two sulfur atoms of the cysteines in the active site. The first N-terminal cysteine in the motif is exposed and reacts with the thiol groups of other proteins by forming a mixed disulfide species (Figure 4). The chemistry of such thiol:disulfide reversible active sites varies among all members of the thioredoxin superfamily in terms of stability and reactivity. The stability of the disulfide bond in thioredoxin-S_2 confers its property as a good hydrogen donor to the protein. The presence of the disulfide bond greatly stabilizes the overall conformation of thioredoxin-S_2 (37). In PDI, DsbA, or DsbC this disulfide bond is very unstable and, therefore, the reduced form $Dsb(SH)_2$ is favored (52, 78, 83, 85). Hence such redox proteins are quite good at transfering their disulfide bonds to other proteins. Although both DsbA and DsbC active sites are nearly equally unstable, the reactivity of the cysteines differs considerably for each active site. Conclusive evidence, summarized below, has shown that DsbA is a very potent disulfide oxidant (78, 83), whereas DsbC possesses a better isomerase activity (85) (Figure 5).

BIOCHEMICAL PROPERTIES OF DsbA   Purified DsbA protein has been shown to catalyze: *(a)* the reduction of insulin in the presence of DTT (7); *(b)* the regeneration of active RNAse A from the scrambled form in the presence of reduced glutathione (2) or DTT (81); and *(c)* the oxidative refolding of RNAse A and alkaline phosphatase in the presence of oxidized glutathione (1, 2). The standard redox potential, determined by fluorescence emission spectra (77), is found to be -0.089 V at pH 7, 30°C, whereas it is -0.11 V for PDI (35), and -0.27 V for thioredoxin (49) (Table 1). Therefore, DsbA should be a much better oxidant than thioredoxin.

The redox properties of DsbA have been explained by the fact that its reduced form, $DsbA-(SH)_2$, is more stable than its oxidized form $DsbA-S_2$ (78, 83). Using two variants of the DsbA protein (Cys[30]-Pro-His-Ser[33]) and (Ser[30]-Pro-His-Cys[33]), it has been proven that the formation of mixed disulfide with other thiol substrates (glutathione, proteins) is initiated by the most N-terminal, most highly accessible Cys residue Cys[30] (84). The mixed disulfide is an obligatory

A

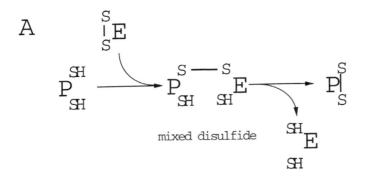

B

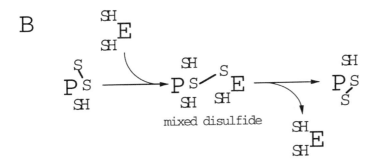

C

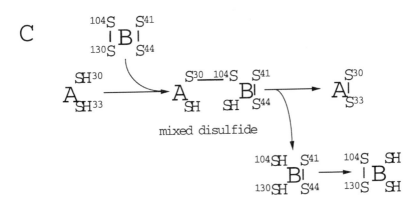

Figure 4 Formation of mixed disulfide species between a polypeptide (P) and a thiol:disulfide oxido-reductase (E). (*A*) Oxidation of a polypeptide, the enzyme being the direct oxidant. In case of E being DsbA, the mixed disulfide reacts with the protein's (P) free thiol group 10^3-fold faster than any other mixed disulfide would do. (*B*) Isomerization reaction. (E could be either PDI or DsbC-(SH)$_2$. (*C*) Reoxidation of reduced DsbA-(SH)$_2$ by oxidized DsbB-S$_2$. The letters *A* and *B* represent DsbA and DsbB respectively. DsbB contains two disulfide bonds between Cys41:Cys44 (the thioredoxin-like motif) and Cys104:Cys130. The mixed disulfide species is very transient but can be stabilized when Cys33 of DsbA and Cys130 of DsbB are substituted for Ala (34, 46, 45).

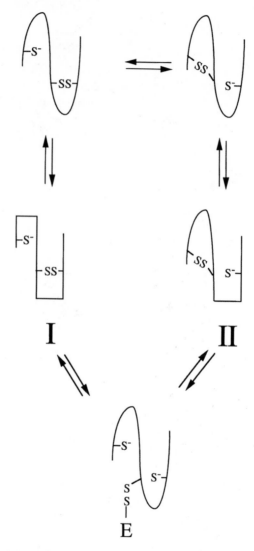

Figure 5   Effect of DsbC and PDI redox proteins on the folding of BPTI. There are three native disulfide bridges in folded BPTI (5–55, 30–51, 14–38)$_N$. The numbers indicate the position of the cysteines in the sequence and N a native-like structure. DsbA can rapidly introduce disulfide bond into unfolded reduced BPTI but not in species I and II (85). Species I can be both intermediates (30–51, 14–38)$_N$ or (5–55, 14–38)$_N$ with a native-like structure. Species II can be both (30–51, 5–14)$_N$ or (30–51, 5–38)$_N$ with a non-native disulfide bond. Rearrangement of species I and II can be catalyzed both by PDI and DsbC (E for enzymes), albeit to a lesser extent.

and transient intermediate of all thiol:disulfide exchange reactions (Figure 4). A mixed disulfide, formed with DsbA, reacts with an external thiol at a rate 10^3-fold faster than a normal mixed disulfide does (83, 84) (Figure 4A). Such a property reflects the high oxidizing capacity of DsbA. Yet, DsbA appears to oxidize unfolded proteins much more rapidly than small thiol compounds would do, e.g. oxidized DTT or oxidized glutathione (23, 78). This last property has been attributed to the presence of a binding site in DsbA for unfolded polypeptides (18, 23).

The two amino acid residues present in the di-thiol active site sequence are not conserved among the known redox proteins (Figure 1); these residues seem to govern some of the redox properties of such proteins. The thiol group of Cys^{30}, the accessible thiol in the redox active site, ionizes with a very low pKa value of 3.5 (60), the pKa of a normal cysteine being about 8.7. Such a low pKa is critical for the high oxidizing capacity of DsbA. This pKa originates in part from the location of the accessible cysteine Cys^{30} near an α helix (53); the dipole of this α helix stabilizes the negative charge of the thiolate anion (47). His^{32}, within DsbA's motif Cys^{30}-Pro-His-Cys^{33}, also contributes to lowering this pKa value (32). The relative instabilities of the disulfide bond and the folded conformation of DsbA-S_2 are therefore linked, simply by mass action. The opposite situation is found for thioredoxin, which has a stable disulfide bond and a high pKa value of its active site thiol group (73).

BIOCHEMICAL PROPERTIES OF DsbC   Like DsbA, DsbC is a periplasmic protein, but it is found as a dimer (57, 85). The disulfide bond present in the active site of the oxidized DsbC-S_2 protein is nearly as unstable as the one found in DsbA-S_2 (85). The equilibrium constant for its formation by interchange with glutathione is 200 μM, which is rather close to the value of 80 μM found for DsbA (83), but much lower than the 10 M equilibrium constant value measured for thioredoxin (38), or the values of up to 10^5 M that are measured for normal stabilizing bonds (15). Again, only the most N-terminal cysteine residue in the active site of DsbC has an accessible thiol (85).

The putative ability of DsbC to rapidly transfer its disulfide bond in vitro has been studied using reduced, unfolded bovine pancreatic trypsin inhibitor (BPTI) (which contains three disulfides in the native state) as the substrate. The disulfide coupled folding of reduced BPTI occurs when defined–folding intermediates form (Figure 5). The early–folding intermediates can be rapidly oxidized, by the simple formation of disulfide bridges. Comparative analyses, using either DsbA-S_2 or DsbC-S_2 in equimolar amounts to the substrate, have led to the suggestion that DsbA is more efficient at transferring its disulfide bond than is DsbC (83, 85). DsbA-S_2 was found to react extremely fast with cysteine residues in unfolded proteins such as TEM 1 β-lactamase or RNase

T1, suggesting, again, an early role for DsbA-S$_2$ during protein folding (23, 24). The ability of PDI to reshuffle wrongly paired disulfides in scrambled intermediates (Figures 4B, 5) has prompted Zapun and colleagues (85) to compare the ability of DsbA-(SH)$_2$ and DsbC-(SH)$_2$ to do the same, i.e. to catalyze disulfide rearrangements (85). DsbC-(SH)$_2$ appeared to be very efficient in such reactions. In addition, like PDI, DsbC seems to carry an unfoldase activity, which permits the direct formation of the missing disulfide bridge in quasinative species of BPTI (Figure 5). The dimeric form of DsbC could be the clue to this PDI-like behavior, because PDI contains two redox active sites in its amino acid sequence.

DsbA and DsbB Constitute a Redox Couple Involved in the Formation of Disulfide Bonds

The pleiotropic phenotype observed in vivo, with bacteria-carrying mutations in either *dsbA* or *dsbB*, can be attributed to defects of the disulfide-coupled folding of transported proteins. In most cases, *dsbA* and *dsbB* mutations confer the same phenotypes. For example, the non-motile phenotype observed with *dsbB* mutant bacteria can also be observed with a *dsbA* mutation (17), and the lack of AP activity occurs both with *dsbA* or *dsbB* mutants (6, 56). The active DsbA-S$_2$ protein, after transferring its disulfide bond to exported proteins, should accumulate in the inactive reduced form: DsbA-(SH)$_2$. Because both *dsbA* and *dsbB* mutants exhibit similar defects, the possibility that DsbA and DsbB function in the same pathway could be considered. Likewise, DsbB-S$_2$ could be responsible for re-oxidizing DsbA-(SH)$_2$ in vivo. The observed accumulation of DsbA-(SH)$_2$ in *dsbB* mutant bacteria provided the first evidence in support of the proposed model (6).

While DsbA is a periplasmic protein with only two cysteines, DsbB is located in the inner membrane and contains six cysteines, two of which form the Cys[41]-Val-Leu-Cys[44] motif (Figure 1). Topological and site directed mutagenesis experiments have shown that Cys[41], Cys[44], Cys[104] and Cys[130] were all essential for DsbB activity and that they all face the periplasm (42). Because of the thiol:disulfide chemistry, an interaction between DsbA and DsbB implies the formation of a mixed disulfide, cross-linked via the cysteine of their active sites (Figure 4C). Such mixed disulfides are very transient species, but in the case of glutaredoxin and glutathione, a stable mixed disulfide could be isolated, provided that only the reactive, first cysteine of each di-thiol motif was present (Figure 4C) (13). Similarly a mixed disulfide between DsbA-Cys[30]:Cys[104]-DsbB could be readily trapped in vivo using various mutant DsbA and DsbB proteins (34, 46). Cys[104] was shown to form a reversible disulfide bond with Cys[130], which is the one responsible for recycling DsbA (45). The Cys[104]:Cys[130] linkage is re-oxidized intra-molecularly by the N terminally located Cys[41]-Val-Leu-Cys[44] thioredoxin-like motif of the DsbB protein (45).

One quite interesting finding is that rat PDI, when produced in the periplasm, can complement several phenotypes that are exhibited by *dsbA* mutants, in a DsbB-dependent manner (62). Hence, reoxidation of eukaryotic PDI, which is glutathione-dependent in the ER, proceeds in the periplasm by direct interaction with the bacterial redox proteins.

Disulfide Isomerase Activity in the Periplasm: The DsbC Protein

The *dsbC* gene was first isolated as a multicopy suppressor of a *dsbA* null mutation. Overproduction of DsbC allows growth on DTT-containing media, AP activity, and motility of *dsbA* mutant cells (57, 72). A *dsbC* null mutant more severely impairs the folding of multiple-disulfide-bond–containing proteins such as alkaline phosphatase and penicillin binding protein 4 (57) than the folding of single disulfide-bond–containing proteins. Given its ability to act as a disulfide isomerase in vitro (85), it is not so surprising that multiple-disulfide–containing proteins should be the preferred substrates of DsbC. More recently, it has been shown that the folding of mouse urokinase, which contains 12 disulfide bonds (67) or BPTI (G Georgiou, personal communication) is severely affected in the periplasm of *dsbC* mutant bacteria, again arguing for the importance of DsbC isomerase activity in vivo. Finally, tissue plasminogen activator, which contains 35 disulfide bonds and does not fold properly in *E. coli* periplasm, can efficiently fold upon overexpression of either DsbC or rat PDI (G Georgiou, personal communication).

Folding defects observed with *dsbC* mutants are not always as dramatic as in a *dsbA* null mutant, but they are severe enough to trigger a stress response owing to the misfolded polypeptides that accumulate in the cell envelope (65, 58). This stress response is reminiscent of that induced by protein misfolding in the cytoplasm, known as the heat shock response (28), or in the endoplasmic reticulum (29).

In vivo, DsbC functions rather independently from the DsbA/DsbB system (57). The DsbC redox active site is not recycled by the inner membrane DsbB protein (57). Rather, to function as an isomerase (Figure 4B), DsbC should be maintained in its reduced state. In vivo, DsbC is present in both forms: approximately 50% DsbC-S_2 and 50% DsbC-$(SH)_2$. Hence the amount of DsbC-S_2, when expressed from a plasmid, is probably sufficient to complement a *dsbA* mutation (57, 72).

DsbD and DsbE are Two Thiol:Disulfide Reductases

THE *DsbD* GENE PRODUCT: AN INNER MEMBRANE THIOL:DISULFIDE OXIDO-RE-DUCTASE   Mutations in the *dsbD* gene were isolated because they can revert some of the phenotypic defects observed in a *dsbA* mutant. Closer analyses have shown that this suppression effect is *dsbC*[+] dependent. Bacteria carrying a mutation in *dsbD* accumulate the DsbA-S$_2$ and DsbC-S$_2$ species, and excessive disulfide oxidation occurs (59). Presumably, in the double *dsbA dsbD* null, the accumulation of DsbC-S$_2$ is sufficient to supply the cell with enough active redox protein in order to oxidize and properly fold most, if not all, disulfide bond containing proteins. For all these reasons, DsbD was proposed to act as a thiol:disulfide reductase (Figure 1). DsbD is also referred to as CutA2 (22) or DipZ (16). The *dsbD* transcription start site is indeed part of the *cutA* locus. The *cutA* locus has been implicated in copper homeostasis, since such mutant bacteria are sensitive to Cu^{2+} ions (22). Disulfide isomerase protein-like domain (DipZ) was discovered because it is involved in the biogenesis of periplasmic *c*-type cytochromes (16) (see *Role of the Dsb Proteins in c-Type Cytochrome's Biogenesis*, below).

The DsbD protein is located in the inner membrane and contains a periplasmic protruding 16 kDa C-terminal domain (55, 59). This C-terminal part is closely related to eukaryotic PDIs with about 40–45% sequence identity. There are two Cys-X-X-Cys motifs in DsbD. The Cys^{403}-Val-Ala-Cys^{406} motif located in the C-terminal PDI-like domain has been proposed to be the active redox site, based on the finding that the secreted, truncated, C-terminal domain can partly restore the phenotypic defects associated with *dsbD* mutant bacteria (59). Recent biochemical analyses have confirmed that DsbD redox properties, at least those carried by the soluble C-terminal domain, are much like those of thioredoxin (English et al, unpublished).

DsbD: A POTENT REDUCTANT FOR DsbC-(SH$_2$)?   To function as an isomerase, the thiol group of Cys 98, the most N-terminal Cys in the active site of DsbC, should be free and accessible (Figure 4). DsbD could maintain Cys 98 of DsbC in its thiolate form. Such an interaction is supported by various genetic evidence:

1. A *dsbD* null mutation suppresses only the phenotypic defects associated with *dsbA* or *dsbB* mutant backgrounds and not those of *dsbC* mutants (59).

2. Folding of multiple-disulfide–containing proteins in *E. coli*'s periplasm, such as human alkaline phosphatase (9) or BPTI (G Georgiou, personal communication), is particularly affected in *dsbD* mutants, when DsbC-S$_2$ accumulates (59).

3. The suppression effect of a *dsbD* null mutation in the *dsbA* mutant strain is solely dependent on the presence of a wild-type DsbC protein, which accumulates in its oxidized form DsbC-S$_2$ (59, 67). A direct DsbC-DsbD interaction in vivo still needs to be demonstrated.

A PERIPLASMIC THIOL:REDUCTASE: THE DsbE PROTEIN   Overexpressing yet another protein, designated as DsbE, from *E. coli* chromosomal libraries, could suppress some phenotypic defects associated with *dsbD* null mutants. In particular, random oxidation of proteins due to the accumulation of oxidized DsbA and DsbC was avoided, largely by overproducing DsbE (D Missiakas & S Raina, unpublished). DsbE is a soluble periplasmic protein, with Cys-Pro-Thr-Cys as its active site (Figure 1). Biochemical data suggest that DsbE, like DsbD, has redox properties closer to thioredoxin than either DsbA or DsbC have (English et al, unpublished.)

Role of the Dsb Proteins in c-Type Cytochromes' Biogenesis

E. coli periplasm contains at least five *c*-type cytochromes that are synthesized during anaerobic growth. *c*-Type cytochromes are electron transfer proteins of the respiratory chains that use either nitrite, nitrate or trimethylamine-*N*-oxide as electron acceptors (33, 40). They differ from other cytochromes by their heme moiety, which is covalently liganded via the sulfur atoms of two cysteines. Although no evidence for a heme lyase has yet been demonstrated, many mutants that impair apo-protein maturation have been isolated. Some of them have been shown to map in genes that code for thioredoxin-like proteins, like HelX in *Rhodobacter capsulatus* (10) and TlpA in *Bradyrhizobium japonicum* (51). In *E. coli*, the *dsbD* and *dsbA* genes have also been shown to be essential for formate-dependent nitrite reductase (Nrf) activity and *c*-type cytochrome biogenesis (22, 54: *dsbD* is referred to as *dipZ* in these studdies). Both DsbD and DsbE, which have been shown to behave as thiol:disulfide reductases in vitro (English et al, unpublished), may be good candidates for maintaining the cysteine residues of the apo-cytochrome in a reduced state, thereby allowing the proper covalent linkage with the haem moiety.

Synthesis of *c*-type cytochromes appears to require a functional *dsbB* gene as well (55). It is more surprising that periplasmic nitrate reductase activity is found to be defective in *dsbB* mutants, but not in *dsbA* mutant bacteria (55). This could be the first example where DsbA and DsbB work independently and oxidize distinct substrates.

A MODEL FOR DISULFIDE BOND FORMATION IN VIVO

Prevention of Disulfide Bond Oxidation in the Cytoplasm

The formation of a disulfide bond within a protein introduces a three-dimensional constraint in the protein's structure. Exported proteins have to be maintained in an extended conformation to be translocated across the inner membrane. This implies that disulfide formation has to be prevented in the cytoplasm, either because of the particular redox environment that exists, or because of the presence of specific redox proteins.

REDOX CONDITIONS IN THE CYTOPLASM: THIOTEDOXIN AND GLUTAREDOXINS
At first examination, one could assume that the redox conditions in *E. coli* cytoplasm cannot favor cysteine oxidation. The glutathione concentration (5 mM) (48) and the GSH/GSSG ratio, which ranges from 50/1 to 200/1 (39), provide a redox environment that is in agreement with the estimated redox potential of *E. coli* thioredoxin (about −0.27 V) and can prevent, at least as demonstrated in vitro, disulfide bond formation in many proteins (39, 70, 76). In addition, there are many redox proteins that have the properties of thiol:disulfide reductases, one thioredoxin (68) and three glutaredoxins: These include glutaredoxin 1, glutaredoxin 2 and glutaredoxin 3 (4) (Table 1). Thioredoxin (TrxA) is kept reduced by the flavoenzyme thioredoxin reductase (TrxB), which uses NADPH (Figure 2). TrxB is specific for thioredoxin and is unable to reduce any of the three glutaredoxins. Glutaredoxins are reduced by glutathione, which in turn is reduced by another NADPH-dependent flavoenzyme, glutathione reductase (Figure 2). In vitro, either of the two redox proteins (thioredoxin and glutaredoxin 1) have been shown to reduce disulfide-bond–containing polypeptides. Nevertheless, this is not a direct proof that such systems are responsible for maintaining the newly synthesized polypeptides in a reduced state.

IS TrxB PLAYING A SPECIFIC ROLE?   Genetic experiments have shed new light regarding the systems that could prevent the disulfide oxidation of cytoplasmic proteins. First, mutations in the *trxB* and *trxA* genes can be isolated because they render *E. coli* cells abnormally sensitive to changes in the redox environment, a phenotype reminiscent of the one exhibited by *dsb* mutants (56). Second, mutations in the *trxB* genes were selected because they allow the proper folding of export-defective versions of alkaline phosphatase or mouse urokinase (19). Folding occurs because, in the absence of proper thioredoxin reductase activity, disulfide bonds were formed in these two proteins. However, thioredoxin was found to be unnecessary for disulfide bond formation in these two cases (19).

More recently, it has been shown that *E. coli* mutants that are missing parts of both the thioredoxin and glutaredoxin systems allow substantially more disulfide bond formation in the cytoplasm than do mutants missing components of only one system (WA Prinz et al, personal communication.)

All these results suggest that TrxB contributes very efficiently as a disulfide reductant, but that the glutaredoxins [even though less efficient reductants of disulfides than thioredoxin in vitro (5, 38)] could also play a role in preventing protein disulfide oxidation in the cytoplasm.

A Sequential Pathway for Disulfide Bond Formation in the Periplasm

A model for disulfide formation in *E. coli* is presented in Figure 6. It suggests that transported proteins that contain disulfide bonds in their final native

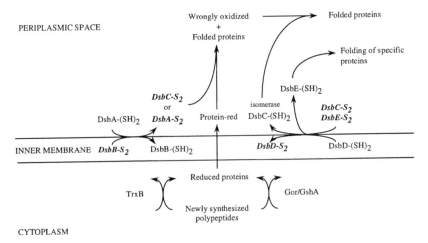

Figure 6   Model of disulfide formation catalyzed by the various Dsb enzymes in the periplasm of *E. coli*.

conformation are maintained as reduced species in the cytoplasm. After translocation through the inner membrane, they are still unfolded, and are probably rapidly oxidized by DsbA-S$_2$ (the role of DsbC-S$_2$ can not fully be ruled out, even though DsbC seems to favor more structured proteins as substrates). DsbA-(SH)$_2$ is in turn re-oxidized by DsbB. What re-oxidizes DsbB is still a matter of speculation. Because DsbB is located in the inner membrane, the possibility that it may be coupled to the electron transfer system is rather attractive, but nevertheless not proven. Finally, proteins that are wrongly oxidized could undergo a reshuffling and aquire the right disulfide pairings in a DsbC-(SH)$_2$-dependent manner. The DsbC-(SH)$_2$ pool could be maintained by the presence of the DsbD reductant in the inner membrane.

Are All Proteins Oxidized in the Periplasm Prior to Secretion into the Extra-cellular Medium?

Because all the redox proteins involved in the disulfide-coupled folding of transported proteins are present in the periplasm, it was tempting to suggest that all the proteins could acquire their three-dimensional structure before their translocation across the outer membrane. In fact, it has been shown for pullulanase, a cell surface protein, that mutations in *dsbA* impair both the folding and the translocation across the outer membrane (64).

A rather similar result was obtained in the case of secretion of cellulase to the cell exterior of *E. chrysanthemi*. Cellulase contains a disulfide bond formed in the periplasm. Adding DTT to *E. chrysanthemi* cultures leads to immediate arrest of cellulase secretion, which accumulates in the periplasm (12). Here again, periplasmic disulfide bond formation is shown to be an obligatory step for secretion competence (12).

CONCLUDING REMARKS

It is now clear that the oxidation of transported proteins that contain disulfide bonds in their native structure occurs in the periplasm and is catalyzed by various redox proteins. As proteins emerge from the inner membrane, unfolded and reduced, they encounter the appropriate oxidizing environment and oxidation occurs very rapidly. *E. coli* and other Gram-negative bacteria synthesize two redox proteins, DsbA and DsbB, which act as strong thiol:disulfide oxidants. Surprisingly enough, the periplasm also contains thiol:disulfide reductases such as DsbD and DsbE, which are needed for maintaining a proper redox state for the disulfide isomerase DsbC and for *c*-type cytochrome maturation.

Because oxidation in vivo occurs very rapidly, it would be interesting to know what is the exact location of the various redox proteins. One can imagine that DsbA could interact with the inner membrane secretion machinery. Also, the regulation of *dsb* genes has not yet been fully addressed. It is not known whether aerobic or anaerobic growth can affect the transcription of any of the *dsb* genes. It has been shown in the case of *dsbB* that transcription of the gene is increased when cells are subjected to various environmental stresses (56).

Many oxidizing compounds can probably diffuse into the periplasm, because the outer membrane is porous. It is tempting to speculate that fluctuations in the redox environment could be controlled in the periplasm in order to avoid nonspecific thiol:disulfide oxidation. In this case, some specific redox protein could be devoted to the maintainance of the proper redox conditions. It also remains to be understood how the membrane-located DsbB and DsbD proteins get recycled. Although much progress has been made during the last six years, the final picture of the various steps leading to disulfide-coupled protein folding in the periplasm still needs more work before it can be fully deciphered.

ACKNOWLEDGMENTS

We thank TE Creighton for carefully reading the manuscript and for many suggestions. We are grateful to J Beckwith and K Ito for sharing unpublished information. S Raina is supported by grants from the Fond National Scientifique Suisse (FN31-42429-94), and D Missiakas is supported by grants from the Centre National de Recherche Scientifique (3A9027).

Literature Cited

1. Akiyama Y, Ito K. 1993. Folding and assembly of bacterial alkaline phosphatase in vitro and in vivo. *J. Biol. Chem.* 268:8146–50
2. Akiyama Y, Kamitani S, Kusukawa N, Ito K. 1992. In vitro catalysis of oxidative folding of disulfide-bonded proteins by the *Escherichia coli dsbA (ppfA)* gene product. *J. Biol. Chem.* 267:22440–45
3. Alksne LE, Keeney D, Rasmussen BA. 1995. A mutation in either *dsbA* or *dsbB*, a gene encoding a component of a periplasmic disulfide bond-catalyzing system, is required for high-level expression of the *Bacteroides fragilis* metallo-beta-lactamase, CcrA, in *Escherichia coli*. *J. Bacteriol.* 177:462–64
4. Åslund F, Ehn B, Miranda-Vizuet A, Pueyo C, Holmgren A. 1994. Two additional glutaredoxins exist in *Escherichia coli*: glutaredoxin 3 is a hydrogen donor for ribonucleotide reductase in a thioredoxin/glutaredoxin 1 double mutant. *Proc. Natl. Acad. Sci. USA* 91:9813–17
5. Åslund F, Nordstand K, Berndt KD, Nikkola M, Bergman T, et al. 1996. Glutaredoxin-3 from *Escherichia coli*. Amino acid sequence, ^{1}H and ^{15}N NMR assignments, and structural analysis. *J. Biol. Chem.* 271:6736–45
6. Bardwell JCA, Lee J-O, Jander G, Martin N, Belin D, et al. 1993. A pathway for disulfide bond formation in vivo. *Proc. Natl. Acad. Sci. USA* 90:1038–42
7. Bardwell JCA, McGovern K, Beckwith J. 1991. Identification of a protein required for disulfide bond formation in vivo. *Cell* 67:581–89
8. Barth PT, Bust C, Hawkins HC, Freedman RB. 1988. Protein disulfide isomerase activity in bacterial osmotic shock preparations. *Biochem. Soc. Trans.* 16:57
9. Beck R, Crooke H, Jarsch M, Cole J, Burtscher H. 1994. Mutation in *dipZ* leads to reduced production of active human placental alkaline phosphatase in *Escherichia coli*. *FEMS Microbiol. Lett.* 124:209–14
10. Beckman DL, Kranz RG. 1993. Cytochromes c biogenesis in a photosynthetic bacterium requires a periplasmic thioredoxin-like protein. *Proc. Natl. Acad. Sci. USA* 90:2179–83
11. Belin P, Boquet PL. 1993. A second gene involved in the formation of disulfide bonds in proteins localized in *Escherichia coli* periplasmic space. *C. R. Acad. Sci. III* 316:469–73
12. Bortoli-German I, Brun E, Py B, Chippaux M, Barras F. 1994. Periplasmic disulphide bond formation is essential for cellulase secretion by the plant pathogen *Erwinia chrysanthemi*. *Mol. Microbiol.* 11:545–53
13. Bushweller JH, Åslund F, Wütrich K, Holmgren A. 1992. Structural and functional characterization of the mutant *Escherichia coli* glutaredoxine (C14S) and its mixed disulfide with glutathione. *Biochemistry* 31:9288–93
14. Creighton TE. 1986. Disulfide bonds as probes of protein folding pathways. *Methods Enzymol.* 131:83–106
15. Creighton TE, Goldenberg J. 1984. Kinetic role of a meta-stable native-like two-disulphide species in the folding transition of bovine pancreatic trypsin inhibitor. *J. Mol. Biol.* 179:497–526
16. Crooke H, Cole J. 1995. The biogenesis of c-type cytochromes in *Escherichia coli* requires a membrane-bound protein, DipZ, with a protein disulphide isomerase-like domain. *Mol. Microbiol.* 15:1139–50
17. Dailey FE, Berg HC. 1993. Mutants in disulfide bond formation that disrupt flagellar assembly in *Escherichia coli*. *Proc. Natl. Acad. Sci. USA* 90:1043–47
18. Darby NJ, Creighton TE. 1995. Characterization of the active site cysteine residues of the thioredoxin-like domains of protein disulfide isomerase. *Biochemistry* 34:16770–80
19. Derman AI, Prinz WA, Belin D, Beckwith J. 1993. Mutations that allow disulfide bond formation in the cytoplasm of *Escherichia coli*. *Science* 262:1744–47
20. Doig AJ, Williams DH. 1991. Is the hydrophobic effect stabilizing or destabilizing in protein? The contribution of disulphide bonds to protein stability. *J. Mol. Biol.* 217:389–98
21. Ewbank JJ, Creighton TE. 1993. Pathway of disulfide-coupled unfolding and refolding of bovine α-lactalbumin. *Biochemistry* 32:3677–93
22. Fong S-T, Camakaris J, Lee BTO. 1995. Molecular genetics of a chromosomal locus involved in copper tolerance in *Escherichia coli* K-12. *Mol. Microbiol.* 15:1127–37
23. Frech C, Wunderlich M, Glockshuber R, Schmid FX. 1996. Preferential binding of an unfolded protein to DsbA. *EMBO J.* 15:392–98
24. Frech C, Wunderlich M, Glockshuber R, Schmid FX. 1996. Competition beween DsbA-mediated oxidation and

conformational folding of RTEM1 β-lactamase. *Biochemistry* 35:11386–95

25. Freedman RB. 1989. Protein disulfide isomerase: multiple roles in the modification of nascent secretory proteins. *Cell* 57:1069–72

26. Freedman RB. 1984. Native disulphide bond formation in protein biosynthesis: evidence for the role of protein disulphide isomerase. *Trends Biochem. Sci.* 9:438–41

27. Frishman D. 1996. DsbC protein: a new member of the thioredoxin fold-containing family. *Biochem. Biophys. Res. Commun.* 219:686–89

28. Georgopoulos C, Liberek K, Zylicz M, Ang D. 1994. Properties of the heat shock proteins of *Escherichia coli* and the autoregulation of the heat shock response. In *The Biology of Heat Shock Proteins and Molecular Chaperones*, ed. R Morimoto, A Tissières, C Georgopoulos, pp. 209–49. Cold Spring Harbor, NY: Cold Spring Harbor Lab.

29. Gething M-J, Blond-Elguindi S, Mori K, Sambrook J. 1994. Structure, function, and regulation of the endoplasmic reticulum chaperone, BiP. See Ref. 28, pp. 111–35

30. Givol D, Goldberger RF, Anfinsen CB. 1964. Oxidation and disulfide interchange in the reactivation of reduced ribonuclease. *J. Biol. Chem.* 239:3114–16

31. Goldberger RF, Epstein CJ, Anfinsen CB. 1963. Acceleration of reactivation of reduced bovine pancreatic ribonuclease by a microsomal system from rat liver. *J. Biol. Chem.* 238:628–35

32. Grauschopf U, Winther JR, Korber P, Zander T, Dallinger P, et al. 1995. Why is DsbA such an oxidizing disulfide catalyst? *Cell* 83:947–55

33. Grove J, Tanapongpipat S, Thomas G, Griffiths L, Crooke H, et al. 1996. *Escherichia coli* K-12 genes essential for the synthesis of c-type cytochromes and a third nitrate reductase located in the periplasm. *Mol. Microbiol.* 19:467–81

34. Guilhot C, Jander G, Martin NL, Beckwith J. 1995. Evidence that the pathway of disulfide bond formation in *Escherichia coli* involves interactions between the cysteines of DsbB and DsbA. *Proc. Natl. Acad. Sci. USA* 92:9895–99

35. Hawkins HC, DeNardi M, Freedman RB. 1991. Redox properties and cross-linking of the dithiol/disulfide active sites of mammalian protein disulfide isomerase. *Biochem. J.* 275:349–53

36. Holmgren A. 1979. Thioredoxin catalyzes the reduction of insuline disulfides by dithiothreitol and dihydrolipoamide. *J. Biol. Chem.* 254:9113–19

37. Holmgren A. 1985. Thioredoxin. *Annu. Rev. Biochem.* 54:237–71

38. Holmgren A. 1989. Thioredoxin and glutaredoxin systems. *J. Biol. Chem.* 264:13963–66

39. Hwang C, Sinskey AJ, Lodish HF. 1992. Oxidized redox state of glutathione in the endoplasmic reticulum. *Science* 257:1496–502

40. Iobbi-Nivol C, Crooke H, Griffiths L, Grove J, Hussain H, et al. 1994. A reassessment of the range of c-type cytochromes synthesized by *Escherichia coli* K-12. *FEMS Microbiol. Lett.* 119:89–94

41. Ishihara T, Tomita H, Hasegawa Y, Tsukagoshi N, Yamagata H, et al. 1995. Cloning and characterization of the gene for a protein thiol-disulfide oxidoreductase in *Bacillus brevis*. *J. Bacteriol.* 177:745–49

42. Jander G, Martin NL, Beckwith J. 1994. Two cysteines in each periplasmic domain of the membrane protein DsbB are required for its function in protein disulfide bond formation. *EMBO J.* 13:5121–27

43. Kamitani S, Akiyama Y, Ito K. 1992. Identification and characterization of an *Escherichia coli* gene required for the formation of correctly folded alkaline phosphatase, a periplasmic enzyme. *EMBO J.* 11:57–62

44. Kemmink J, Darby NJ, Dijkstra K, Nilges M, Creighton TE. 1996. Structure determination of the N-terminal thioredoxin-like domain of protein disulfide isomerase using multidimensional heteronuclear $^{13}C/^{15}N$ NMR spectroscopy. *Biochemistry* 35:7684–91

45. Kishigami S, Ito K. 1996. Roles of cysteine residues of DsbB in its activity to reoxidize DsbA, the protein disulphide bond catalyst of *Escherichia coli*. *Genes Cells* 1:201–8

46. Kishigami S, Kanaya E, Kikuchi M, Ito K. 1995. DsbA-DsbB interaction through their active site cysteines. *J. Biol. Chem.* 270:17072–74

47. Kortemme T, Creighton TE. 1995. Ionization of cysteine residues at the termini of model α-helical peptides. Relevance to unusual thiol pKa values in proteins of the thioredoxin family. *J. Mol. Biol.* 253:799–812

48. Kosower NS, Kosower EM. 1978. The glutathione status of cells. *Int. Rev. Cytol.* 54:109–60

49. Krause G, Lundström J, Barea JL, de la Cuesta C, Holmgren A. 1991. Mimicking the active site of protein disulfide-isomerase by substitution of proline 34 in *Escherichia coli* thioredoxin. *J. Biol. Chem.* 266:9494–500

50. Laurent TC, Moore EC, Reichard P. 1964. Enzymatic synthesis of deoxyribonucleotides: IV. Isolation and characterization of thioredoxin, the hydrogen donor from *Escherichia coli B. J. Biol. Chem.* 239:3436–44

51. Loferer H, Bott M, Hennecke H. 1993. *Bradyrhizobium japonicum* TlpA, a novel membrane-anchored thioredoxin-like protein involved in the biogenesis of cytochrome aa3 and development of symbiosis. *EMBO J.* 12:3373–83

52. Lundström J, Holmgren A. 1993. Determination of the reduction-oxidation potential of the thioredoxin-like domains of protein disulfide-isomerase from the equilibrium with glutathione and thioredoxin. *Biochemistry* 32:6649–55

53. Martin JL, Bardwell JCA, Kuriyan J. 1993. Crystal structure of the DsbA protein required for disulphide bond formation in vivo. *Nature* 365:464–68

54. Metheringham R, Griffiths L, Crooke H, Forsythe S, Cole J. 1995. An essential role for DsbA in cytochrome c synthesis and formate-dependent nitrite reduction by *Escherichia coli* K-12. *Arch. Microbiol.* 164:301–7

55. Metheringham R, Tyson K, Crooke H, Missiakas D, Raina S, et al. 1996. Effects of mutations in genes for proteins involved in disulfide bond formation in the periplasm on the activities of anaerobically induced electron transfer chains in *Escherichia coli* K-12. *Mol. Gen. Genet.* 253:95–102

56. Missiakas D, Georgopoulos C, Raina S. 1993. Identification and characterization of the *Escherichia coli* gene *dsbB*, whose product is involved in the formation of disulfide bonds in vivo. *Proc. Natl. Acad. Sci. USA* 90:7084–88

57. Missiakas D, Georgopoulos C, Raina S. 1994. The *Escherichia coli dsbC (xprA)* gene encodes a periplasmic protein involved in disulfide bond formation. *EMBO J.* 13:2013–20

58. Missiakas D, Raina S, Georgopoulos C. 1996. Heat shock regulation. In *Regulation of Gene Expression in Escherichia coli*, ed. ECC Lin, SA Lynch, pp. 481–501. Austin, TX: Landes

59. Missiakas D, Schwager F, Raina S. 1995. Identification and characterization of a new disulfide-isomerase like protein (DsbD) in *Escherichia coli*. *EMBO J.* 14:3415–24

60. Nelson JW, Creighton TE. 1994. Reactivity and ionization of the active site cysteine residues of DsbA, a protein required for disulfide bond formation in vivo. *Biochemistry* 33:5974–83

61. Okamoto K, Baba T, Yamanaka H, Akashi N, Fujii Y. 1995. Disulfide bond formation and secretion of *Escherichia coli* heat-stable enterotoxin II. *J. Bacteriol.* 177:4579–86

62. Ostermeier M, De Sutter K, Georgiou G. 1996. Eukaryotic protein disulfide isomerase complements *Escherichia coli dsbA* mutants and increases the yield of a heterologous secreted protein with disulfide bonds. *J. Biol. Chem.* 271:10616–22

63. Peek JA, Taylor RK. 1992. Characterization of a periplasmic thiol:disulfide interchange protein required for the functional maturation of secreted virulence factors of *Vibrio cholerae*. *Proc. Natl. Acad. Sci. USA* 89:6210–14

64. Pugsley AP. 1992. Translocation of a folded protein across the outer membrane in *Escherichia coli*. *Proc. Natl. Acad. Sci. USA* 89:12058–62

65. Raina S, Missiakas D, Georgopoulos C. 1995. The *rpoE* gene encoding the $\sigma^E (\sigma^{24})$ heat shock sigma factor of *Escherichia coli*. *EMBO J.* 14:1043–55

66. Randall LL, Hardy SJ. 1995. High selectivity with low specificity: how SecB has solved the paradox of chaperone binding. *Trends Biochem. Sci.* 20:65–69

67. Rietsch A, Belin D, Martin N, Beckwith J. 1996. An in vivo pathway for disulfide bond isomerization in *Escherichia coli*. *Proc. Natl. Acad. Sci. USA* 93:13048–53

68. Russel M. 1995. Thioredoxin genetics. *Methods Enzymol.* 252:264–74

69. Russel M, Model P. 1986. The role of thioredoxin in filamentous phage assembly. Construction, isolation, and characterization of mutant thioredoxins. *J. Biol. Chem.* 261:14997–5005

70. Saxena VP, Wetlaufer DB. 1970. Formation of three-dimensional structure in proteins. I. Rapid nonenzymatic reactivation of reduced lysozyme. *Biochemistry* 9:5015–23

71. Shevchik VE, Bortoli-German I, Robert-Baudouy J, Robinet S, Barras F, et al. 1995. Differential effect of *dsbA* and *dsbC* mutations on extracellular enzymes in *Erwinia chrysanthemi*. *Mol. Microbiol.* 166:745–53

72. Shevchik VE, Condemine G, Robert-Baudouy J. 1994. Characterization of DsbC, a periplasmic protein of *Erwinia chrysanthemi* and *Escherichia coli* with disulfide isomerase activity. *EMBO J.* 13:2007–12

73. Takahashi T, Creighton TE. 1996. On the reactivity and ionization of the active site cysteine residues of *Escherichia coli* thioredoxin. *Biochemistry* 35:8342–53

74. Tomb J-F. 1992. A periplasmic protein

disulfide oxidoreductase is required for transformation of *Haemophilus influenzae* Rd. *Proc. Natl. Acad. Sci. USA* 89:10252–56

75. Venetianer P, Straub FB. 1963. The enzymic reactivation of reduced ribonuclease. *Biochem. Biophys. Acta* 67:166–68
76. Walker KW, Gilbert HF. 1994. Effect of redox environment on the *in vitro* and *in vivo* folding of RTEM-1 β-lactamase and *Escherichia coli* alkaline phosphatase. *J. Biol. Chem.* 269:28487–93
77. Wunderlich M, Glockshuber R. 1993. Redox properties of protein disulfide isomerase (DsbA) from *Escherichia coli*. *Protein Sci.* 2:717–26
78. Wunderlich M, Otto A, Seckler R, Glockshuber R. 1993. Bacterial protein disulfide isomerase: efficient catalysis of oxidative protein folding at acidic pH. *Biochemistry* 32:12251–56
79. Xia TH, Bushweller JH, Sodano P, Billeter M, Björnberg O, et al. 1992. NMR structure of oxidized *Escherichia coli* glutaredoxin: comparison with reduced *E. coli* glutaredoxin and functionally related proteins. *Protein Sci.* 1:310–21
80. Yamanaka H, Kameyama M, Baba T, Fujii Y, Okamoto K. 1994. Maturation pathway

of *Escherichia coli* heat-stable enterotoxin I: requirement of DsbA for disulfide bond formation. *J. Bacteriol.* 176:2906–13

81. Yu J, McLaughlin S, Freedman RB, Hirst TR. 1993. Cloning and active site mutagenesis of *Vibrio cholerae* DsbA, a periplasmic enzyme that catalyzes disulfide bond formation. *J. Biol. Chem.* 268:4326–30
82. Yu J, Webb H, Hirst TR. 1992. A homologue of the *Escherichia coli* DsbA protein involved in disulphide bond formation is required for enterotoxin biogenesis in *Vibrio cholerae*. *Mol. Microbiol.* 6:1949–58
83. Zapun A, Bardwell JCA, Creighton TE. 1993. The reactive and destabilizing disulfide bond of DsbA, a protein required for protein disulfide bond formation *in vivo*. *Biochemistry* 32:5083–92
84. Zapun A, Cooper L, Creighton TE. 1994. Replacement of the active-site cysteine residues of DsbA, a protein required for disulfide bond formation in vivo. *Biochemistry* 33:1907–14
85. Zapun A, Missiakas D, Raina S, Creighton TE. 1995. Structural and functional characterization of DsbC, a protein involved in disulfide bond formation in *Escherichia coli*. *Biochemistry* 34:5075–89

Annu. Rev. Microbiol. 1997. 51:203–24

AGAINST ALL ODDS: The Survival Strategies of *Deinococcus radiodurans*

J. R. Battista
Department of Microbiology, 508 Life Sciences Building, Louisiana State University, Baton Rouge, Louisiana 70803; jbattis@unix1.sncc.lsu.edu

KEY WORDS: DNA repair, ionizing radiation resistance, UV resistance, regulatory checkpoints, interchromosomal recombination

ABSTRACT
Bacteria of the genus *Deinococcus* exhibit an extraordinary ability to withstand the lethal and mutagenic effects of DNA damaging agents—particularly the effects of ionizing radiation. These bacteria are the most DNA damage–tolerant organisms ever identified. Relatively little is known about the biochemical basis for this phenomenon; however, available evidence indicates that efficient repair of DNA damage is, in large part, responsible for the deinococci's radioresistance. Obviously, an explanation of the deinococci's DNA damage tolerance cannot be developed solely on the basis of the DNA repair strategies of more radiosensitive organisms. The deinococci's capacity to survive DNA damage suggests that (*a*) they employ repair mechanisms that are fundamentally different from other prokaryotes, or that (*b*) they have the ability to potentiate the effectiveness of the conventional complement of DNA repair proteins. An argument is made for the latter alternative.

CONTENTS

203

0066-4227/97/1001-0203$08.00

INTRODUCTION

The *Deinococcaceae* are distinguished by their extraordinary ability to tolerate the lethal effects of DNA damaging agents, particularly those of ionizing
radiation. Although the physiological basis of the deinococci's extreme ra-
diotolerance has never been adequately explained, it is clear that irradiated
cells are not passively protected from the damaging effects of the incident radiation. Instead, available evidence (such as is presented in Figure 1) argues
that the deinococci do suffer massive DNA damage following irradiation, and
that extensive DNA repair is necessary if these cells are to survive such expo-
sures. To generate panel (*A*) of Figure 1, a *D. radiodurans* R1 culture, grown
to mid-log phase, was exposed to 3000 Gy γ radiation, a dose that introduces
approximately 110 double strand breaks (dsbs) into the chromosome of each

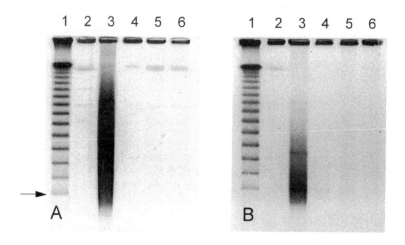

Figure 1   The ability of *D. radiodurans* R1 (*panel A*) and IRS41 (*panel B*) to survive the accumulation of chromosomal DNA double strand breaks following exposure to 3000 Gy γ radiation. For
each panel: *Lane 1* is a lambda ladder size standard, *Lane 2* is chromosomal DNA prepared from
an untreated culture, *lane 3* is chromosomal DNA prepared from a culture immediately after irradiation, and *lanes 4–6* are chromosomal DNA prepared from a culture 3, 6, and 9 h post-irradiation,
respectively.

cell in the population.[1] At 0, 3, 6, and 9 h post-irradiation, aliquots of the irradiated culture were removed, and the chromosomal DNA was isolated for analysis. Pulsed field gel electrophoresis (PFGE) was used to provide a visual record of the cell's recovery from the dsbs. Fragmentation of the chromosome is obvious immediately after irradiation. The band of chromosomal DNA present in unirradiated cell preparations (*lane 2*) is gone, replaced by a broad smear of lower molecular weight material (*lane 3*). Within three hours, however, this smear has disappeared and the chromosome has reformed (*lane 4*). Remarkably, *D. radiodurans* survives this degree of damage without loss of viability and without evidence of induced mutation.

Interest *Deinococcus radiodurans* (*D. radiodurans*) R1 was first isolated in 1956 (3) from tins of meat that had been given what was believed to have been a sterilizing dose of γ radiation. Although this organism and its relatives have been studied over the past 40 years, our knowledge of the survival strategies employed by *D. radiodurans* following DNA damage lacks the biochemical and mechanistic detail that characterizes our knowledge of the better known prokaryotic and eukaryotic organisms [e.g. *Escherichia coli* (*E. coli*) and *Saccharomyces cerevisiae* (*S. cerevisiae*)]. The absence of tractable genetics stifled the study of *D. radiodurans* for many years. *D. radiodurans* was substantially more difficult to work with and, as a consequence, few laboratories made the effort to study this organism.

Interest in *D. radiodurans* appears to be growing, as is exemplified by the number of review articles published in recent years on this organism and its DNA repair capabilities. Between 1956 and 1992, only one review on the radiobiology of *D. radiodurans* was published (52); since 1992, five such reviews have appeared (5, 46–48, 65). For this reason, it is my intent to shorten the more conventional descriptions of *D. radiodurans*' DNA damage tolerance and the enzymology of DNA repair and to focus on aspects of deinococcal biology that appear to enhance the capabilities of this organism's DNA repair proteins.

SUMMARY DESCRIPTION OF *DEINOCOCCUS RADIODURANS*

General Characteristics

D. radiodurans is a pigmented, nonsporeforming, nonmotile, spherical bacterium that ranges from 1.5 to 3.5 μm in diameter (59, 60). Colonies are convex, smooth, and vary from pink to red. *D. radiodurans* is gram positive but has a complex cell envelope similar to that of gram negative organisms. A

---

[1]Calculation of this value assumes that the *D. radiodurans* genome consists of 3×10^6 bp (22) and that one double-strand break forms 10 Gy^{-1} 5×10^9 daltons^{-1} of double-stranded DNA (9).

thick peptidoglycan layer and outer membrane are present, and some strains exhibit a paracrystalline S layer. Cells are chemoorganotrophic with respiratory metabolism and are typically grown with aeration in TGY broth (0.5% tryptone, 0.1% glucose, 0.3% yeast extract). Under optimal conditions, the generation time of wild-type *D. radiodurans* R1 is approximately 80 min. Defined media have been described (61, 62), but their use results in slow and erratic growth. Cells divide alternately in two planes, generating pairs and tetrads in liquid culture. The optimal growth temperature is 30°C, but growth remains strong to 37°C. Growth ceases at temperatures below 4°C and above 45°C.

DNA Damage Tolerance

D. radiodurans cultures exhibit unusually high resistance to many DNA damaging agents. Most studies of this organism have focused on its tolerance of DNA damage induced by ionizing radiation, ultraviolet (UV) light, and cross-linking agents. *D. radiodurans'* earlier reported resistance to nitrous acid, hydroxylamine, N-methyl-N'-nitro-N-nitrosoguanidine (MNNG), and 4-nitroquinoline-N-oxide (66, 67) has not been further characterized by new research for 20 years. *D. radiodurans* is one of the most ionizing radiation–resistant organisms ever identified; exponential phase cultures routinely survive exposure to 15,000 Gy γ radiation (34, 59). The typical γ radiation survival curve (Figure 2) for *D. radiodurans* R1 exhibits a shoulder of resistance to approximately 5,000 Gy (57), in which there is no loss of viability. For purposes of comparison, the survival of an *E. coli* B/r culture is also plotted in Figure 2. At

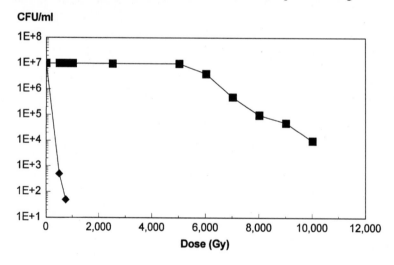

Figure 2   Representative survival curve for *D. radiodurans* R1 (*squares*) and for *E. coli* B/r (*diamonds*) following exposure to γ radiation.

CFU/ml

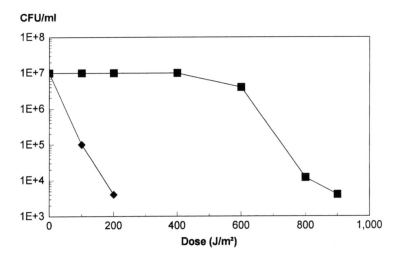

Figure 3   Representative survival curve for *D. radiodurans* R1 (*squares*) and for *E. coli* B/r (*diamonds*) following exposure to UV radiation.

doses above 5,000 Gy, however, the survival of *D. radiodurans* declines rapidly; the D_{37} dose[2] is approximately 6,000 Gy for exponential cultures. The D_{37} dose is 30 Gy for cultures of *E. coli* B/r exposed to ionizing radiation (67). In terms of DNA damage, 6,000 Gy γ radiation will induce approximately 200 DNA double strand breaks (9), over 3,000 single strand breaks (7), and greater than 1,000 sites of base damage per *D. radiodurans* genome (65). Although there are reports of deinococcal strains that survive as much as 50,000 Gy (4), in the author's experience viability falls to undetectable levels when exponential phase cultures of wild-type *D. radiodurans* are exposed to doses of 18,000 Gy or higher.

Wild-type *D. radiodurans* is also extremely resistant to UV radiation, surviving doses as high as 1000 J/m^2 (57). UV survival curves of *D. radiodurans* and *E. coli* B/r are compared in Figure 3. *E. coli* exhibits an exponential decline in viability, whereas *D. radiodurans* has a shoulder of resistance that extends to 500 J/m^2. The D_{37} dose for UV light is approximately 30 J/m^2 for *E. coli* B/r (67) and between 550 and 600 J/m^2 for exponentially growing cultures of *D. radiodurans* R1. The doses of UV light that *D. radiodurans* will tolerate cause an enormous amount of base damage. At 500 J/m^2, for example, it is estimated (6, 71) that 5000 thymine-containing pyrimidine dimers form an average of one lesion for every 600 base pairs.

---

[2]This is the dose of radiation required to reduce the number of individuals in the irradiated population to 37%, i.e. the dose that, on average, is required to inactivate a single colony-forming unit of the irradiated population.

Wild-type *D. radiodurans* is more resistant to the cross-linking agent mitomycin C than are most vegetative bacteria, as it can survive incubation in the presence of 20 μg/ml mitomycin for 10 min at 30°C without loss of viability. After 40 min of incubation at this concentration, 1% of the culture still remains viable (35). Kitayama (35) has reported that, given cultures that are treated for 10 min at this dose, greater than 90% of the isolated chromosomal DNA exists as fragments of nondenaturable double-stranded DNA, with an average molecular mass of 2×10^7 Da. This indicates that at least 100 mitomycin C-induced crosslinks form per genome at this dose.

Phylogeny and Habitat

Four species make up the genus *Deinococcus: D. radiodurans, D. proteolyticus, D. radiopugnans,* and *D. radiophilus. D. radiodurans* is the type species for the genus (8, 59, 60). The deinococci were originally classified as members of the genus *Micrococcus.* However, chemotaxonomic studies of the *Micrococcaceae* suggest that this classification is incorrect, and subsequent phylogenetic analysis of deinococcal 16S and 5S rRNA sequences confirm that the deinococci are not related to the *Micrococcaceae* (33, 70, 73). The genus *Deinococcus* is specifically related to the gram negative genera *Deinobacter* ($S_{AB} = 0.58$ to 0.68) and *Thermus* ($S_{AB} = 0.22$ to 0.29). *Deinobacter* strains are radiation-resistant rod-shaped organisms that exhibit chemotaxonomic characteristics very similar to those of *Deinococcus.* In contrast, the thermophiles of the genus *Thermus* do not exhibit any obvious phenotypic relationship to *Deinococcus.*

The natural habitat of the deinococci has not been defined, even though members of this family have been isolated from a variety of locations worldwide. The environments from which these isolations have been made vary greatly in terms of their overall ecology and geographic distribution. This variability has undoubtedly contributed to the uncertainty concerning the deinococci's habitat. Two ecological studies have suggested that the deinococci are widely distributed soil organisms (8, 59). It is consistent with this conclusion that most deinococcal isolates have been recovered from environments that are rich in organic nutrients, including soil (8, 59, 60), animal feces (34), processed meats (14, 21), and sewage (34).

Deinococcal strains have also been recovered from more unforgiving environments. Successful isolation of deinococci from dried foods (41, 45), room dust (10), medical instruments (10), and textiles (38) suggests that this family has the capacity to survive in dry, nutrient-poor surroundings, which greatly expands the number of niches that this organism could exploit. These isolations are significant because they suggest that the deinococci could have evolved, much as the spore-forming organisms did, to survive periods of prolonged environmental stress. It is noteworthy that *D. radiodurans* is exceptionally resistant to desiccation: According to one anecdotal account (59), it can survive for six years in a desiccator with 10% viability.

Mattimore & Battista (43) have recently provided evidence of a connection between the ionizing radiation resistance of *D. radiodurans* and its desiccation resistance. They evaluated the ability of 41 ionizing radiation-sensitive (IRS) strains of *D. radiodurans* to survive six weeks in a desiccator and demonstrated that every IRS strain was sensitive to desiccation. In addition, they established that, during dehydration *D. radiodurans* accumulates DNA damage—including accumulating DNA double-strand breaks. It appears that *D. radiodurans* is an organism that has adapted to dehydration and that its DNA repair capability is one manifestation of that evolutionary adaptation.

The Genetics of D. Radiodurans

D. radiodurans is multigenomic (28, 32), with stationary phase cells that each carry an estimated four genome equivalents. Each chromosome is a single covalently closed circular molecule that contains approximately 3×10^6 base pairs (22). The base composition of all deinococcal species is characterized by a high GC content, ranging from 65 to 71 mol%. The genome of *D. radiodurans* R1 is currently being sequenced, and publication of that sequence is expected by November, 1997 (O White, The Institute for Genomic Research, personal communication).

Stable naturally occurring plasmids have been found in all species examined to date (63). Most are larger than 20 kb and are present in low copy number. These plasmids have not been extensively studied, and their function in cellular metabolism is unknown.

The phylogenetic isolation of *D. radiodurans* makes genetic studies of this organism difficult. In general, *D. radiodurans* genes are not expressed in other bacteria, and the genes of other bacteria are not expressed in *D. radiodurans* (39, 64). Presumably, those structural features that signal the initiation of transcription in *D. radiodurans* evolved differently than they did in other prokaryotes. As a consequence, attempts at defining the function of a *D. radiodurans* gene product by interspecies complementation are effectively blocked unless hybrids are made—either by fusing the *D. radiodurans* gene to an appropriate promoter, or by fusing a foreign gene to a *D. radiodurans* promoter. Once transcribed, however, heterologous sequences are translated. Shuttle plasmids have been generated that are capable of replicating in *D. radiodurans* and *E. coli* by inserting an *E. coli* plasmid into a naturally occurring *D. radiodurans* plasmid, and by removing sequences that are not needed for plasmid replication (63). Expression of promoterless *cat* and *tet* genes from these vectors in *D. radiodurans* requires that an uncharacterized segment of *D. radiodurans* DNA (presumably containing a recognized promoter) be located immediately upstream of the drug resistance gene. Minton and colleagues have exploited this finding and developed methods that utilize this putative promoter to create selectable, insertional mutations in the *pol* (25) and *rec* (23) genes of *D. radiodurans* and to express *E. coli* genes in *D. radiodurans*.

Transposon mutagenesis has never been reported in *D. radiodurans*, probably because the genes required for transposition are not expressed in this organism. *D. radiodurans* is, however, mutable when treated with simple alkylating agents such as MNNG, and most mutant strains are produced by chemical mutagenesis (51, 54, 69). Typically, cultures are treated with a mutagen, diluted into fresh media, and allowed to grow (usually overnight) before identification of the desired mutant phenotype is attempted. The outgrowth of mutagenized cells is necessary because *D. radiodurans* is multigenomic, and recessive mutations need the opportunity to segregate and form cells homozygous for that mutation before the mutant phenotype can be expressed.

The study of *D. radiodurans* has been hampered by a lack of genetic methods appropriate for use with this organism. Many of the tools that prokaryotic geneticists take for granted are not available when studying *D. radiodurans*. There is no evidence of conjugation, and no phage capable of infecting *D. radiodurans* has been identified (52, 65). Fortunately, *D. radiodurans* is relatively easy to manipulate using natural transformation, and experimental methods exploiting this property have been developed. *D. radiodurans* is fully competent throughout exponential growth (68). This species readily takes up and incorporates transforming DNA into its chromosomes, with marker-specific efficiencies from 0.01 to 3.0%, when cells are transformed in liquid culture in the presence of calcium (68). *D. radiodurans* strains are the only members of the deinococci that are naturally transformable. It is largely for this reason that *D. radiodurans* is the only species routinely used when studying the extreme radioresistance of the *Deinococcaceae*. The ability to transform *D. radiodurans* with high efficiency allows for relatively rapid isolation of DNA fragments that carry genes involved in DNA damage resistance. A genomic library generated from the wild-type strain can be efficiently screened for clones that are capable of restoring the wild-type phenotype to mutant strains. As part of the transformation process, the mutant allele is replaced by the wild-type sequence in a recombinational event. Moseley and coworkers (2) were the first to use this technique when they cloned the wild-type gene necessary for *D. radiodurans*' mitomycin resistance from a cosmid library. They transformed individual cosmids into *D. radiodurans* 302, a mitomycin-sensitive strain, and they plated transformants onto media containing mitomycin C. The restoration of mitomycin resistance identified the cosmid that carries the wild-type sequence that corresponds to the allele responsible for mitomycin C sensitivity. Once cosmids capable of restoring mitomycin resistance were identified, the gene of interest was isolated by transforming restriction fragments, obtained from that cosmid, into strain 302.

The efficiency of transformation remained high until fragment size dropped below 1.2 kb.

The large window of competence found in strains of *D. radiodurans* also permits the investigator to use in situ or "dot" transformation when manipulating these strains (69). In this protocol, exponential phase *D. radiodurans* cells are simply plated onto a rich medium, dotted with transforming DNA, and allowed to grow into a lawn. The lawn is replica plated and selective pressure is applied. Successful transformants appear within the area where the transforming DNA was dotted. Dot transformation has the advantage of avoiding most of the tedium associated with conducting the large number of transformations needed to screen a genomic library.

The genetic methods available for the study of *D. radiodurans* are, at present, relatively primitive in comparison with that of many other prokaryotes. This should change rapidly, however, when the sequence of the R1 genome becomes available.

DNA DAMAGE–SENSITIVE STRAINS OF *D. RADIODURANS*

Screens for mitomycin C (54), UV (51), and ionizing radiation (69) sensitive strains of *D. radiodurans* following chemical mutagenesis have been described. Table 1 lists characterized genotypes that exhibit damage-sensitive phenotypes.

Mutational inactivation of the *pol* or *rec* loci of *D. radiodurans* R1 results in strains that are sensitive to all three agents. The *pol* and *rec* gene products of *D. radiodurans* have been shown to be homologues of *E. coli*'s DNA polymerase I (25) and RecA protein (23), respectively. The deinococcal proteins appear to play roles similar to those that their *E. coli* counterparts play in DNA repair.

The *uvrA* gene product also plays a part in the cell's tolerance to these three types of DNA damage, but its contribution to UV and ionizing radiation resistance is redundant and, therefore, not essential in wild-type *D. radiodurans*.

Table 1   *D. radiodurans* strains sensitive to DNA damage

Mitomycin C	Ultraviolet radiation	Ionizing radiation
pol	*pol*	*pol*
recA	*recA*	*recA*
uvrA	*uvrA, uvsC*	*uvrA, irrB*
mtcC	*uvrA, uvsD*	*irrI*
mtcD	*uvrA, uvsE*	
mtcE	*irrI*	
mtcF		
mtcG		

A *uvrA* strain is quite sensitive to mitomycin C, however (54). The first *uvrA* strains were derived from *D. radiodurans* R1, and the loci inactivated were designated *mtcA* and *mtcB* (54). Clones of these "loci" were recently sequenced and found to be two parts of a single gene that encodes a protein that is 60% identical to the UvrA proteins of *E. coli* and *Micrococcus luteus* (1). The *uvrA* strains of *D. radiodurans* only become sensitive to UV light when a second locus, designated *uvs*, is also inactivated (20, 56). Three *uvs* loci—*uvsC*, *uvsD*, and *uvsE*—have been identified (20). In a wild-type background *uvs* mutants exhibit near wild-type levels of UV resistance, indicating that the *uvrA* and *uvs* gene products encode functionally redundant proteins (18, 19). In addition, *uvs* mutants exhibit wild-type resistance to mitomycin C and γ radiation. The *uvrA uvs* strain is sensitive to mitomycin C, but it is nearly as resistant to γ radiation as is wild-type *D. radiodurans* (20, 26).

The *uvrA* mutation can influence ionizing radiation resistance in *D. radiodurans* as well. Udupa et al (69) screened 45,000 MNNG-treated colonies of the *uvrA* strain *D. radiodurans* 302 for ionizing radiation sensitivity, and they identified 49 putative ionizing radiation–sensitive (IRS) strains in this screen. All of the IRS strains are, therefore, double mutants. Subsequent characterization of one of these strains, IRS18, showed that ionizing radiation resistance was largely restored when the strain was transformed to mitomycin C resistance with the appropriate *uvrA*+ sequence. In other words, the *irrB* mutant was only ionizing radiation sensitive when the background was also *uvrA*, indicating that the *uvrA* gene product can, when necessary, contribute significantly to ionizing radiation resistance.

In addition to the *pol, recA*, and *uvrA* mutants, there are six other mutations that will render *D. radiodurans* more sensitive to at least one of the DNA damaging agents listed in Table 1. Inactivation of any of five independent loci, designated *mtcC* through *mtcG*, enhances *D. radiodurans'* sensitivity to mitomycin C–induced DNA damage (36). The functions of these *mtc* gene products have not been described. Inactivation of the *irrI* locus (69) results in a significant loss of resistance to ionizing radiation, and the *irrI uvrA* double mutant and the *irrI* mutant are equally sensitive to ionizing radiation. The *irrI* mutations also reduce *D. radiodurans* resistance to ultraviolet radiation, but this effect is only observed in *uvrA*+ backgrounds. The function of the IrrI protein is unknown, but there is preliminary evidence that IrrI may be a regulatory protein.

THE ENZYMOLOGY OF DNA REPAIR IN *D. RADIODURANS*

Only a handful of studies in the last 40 years has attempted to detail the biochemistry of any process associated with DNA repair in *D. radiodurans*. As

Table 2   Characterized DNA repair proteins of *D. radiodurans*

Repair process	Repair protein	Associated loci	Comments	Reference
Nucleotide excision repair	Endonuclease α	*uvrA*	Shares 60% amino acid sequence identity with the UvrA protein of *E. coli*	(1)
Base excision repair	Endonuclease β	*uvsC, uvsD, uvsE*	Appears to be a pyrimidine-dimer DNA glycosylase	(19, 26)
	Uracil DNA glycosylase	?	Activity detected in extracts of *D. radiodurans,*	(42)
	Thymine glycol glycosylase	?	Activity detected in extracts of *D. radiodurans*	(58)
	AP endonuclease	?	Activity detected in extracts of *D. radiodurans*	(42)
	Deoxyribophospho-diesterase	?	Activity detected in extracts of *D. radiodurans*	(58)
Recombinational repair	RecA	*rec*	Shares 56% amino acid sequence identity with RecA protein of *E. coli*	(23)
Accessory proteins	DNA polymerase I	*pol*	Shares 51% amino acid sequence identity with DNA Polymerase I protein of *E. coli*	(24)

a consequence, our knowledge of these processes is rudimentary. Three types of DNA repair have been described: (*a*) nucleotide excision repair, (*b*) base excision repair, and (*c*) recombinational repair. A list of those *D. radiodurans* proteins that have been at least partially characterized, and that appear to function in DNA repair, is presented in Table 2. Since a detailed description of the biochemistry of *D. radiodurans* DNA repair proteins has been published in two recent reviews (5, 46), it will not be repeated here.

Nucleotide excision repair is mediated by the activity of a protein identified as endonuclease α (56). Endonuclease α recognizes a broad range of DNA damage, and it incises the DNA at or near the site of that damage. It appears to be a functional homologue of the *uvrABC* endonuclease of *E. coli* (1). Mutational inactivation of the deinococcal *uvrA* locus results in loss of endonuclease α activity. Given that a UvrA homologue has been identified as part of endonuclease α, it is not unreasonable to expect that *D. radiodurans* expresses homologues of the *uvrB* and *uvrC* gene products of *E. coli* as well. There is no direct evidence, however, that these homologues are present in *D. radiodurans*.

Base excision repair has also been detected in *D. radiodurans*. There have been reports describing an apurinic/apyrimidinic (AP) endonuclease activity (42), a uracil-N-glycosylase activity (42), a DNA deoxyribophophodiesterase (dRPase) activity (58), and a thymine glycol glycosylase activity (58) in cell extracts of *D. radiodurans*. The genes that encode the proteins responsible for these activities have not been identified. Thus, the influence of these repair activities on DNA damage resistance has not been assessed.

Endonuclease β is a pyrimidine-dimer DNA glycosylase (PD DNA glycosylase) analogous to the PD DNA glycosylases isolated from *Micrococcus luteus* or bacteriophage T4 (26). Expression of the *denV* gene of T4 in *uvrA uvs* and *uvs* strains of *D. radiodurans* partially restores UV resistance to these strains, which indicates that endonuclease β has specificity for pyrimidine dimers. This protein appears to be involved exclusively in the repair of UV-induced DNA damage. Three mutations, designated *uvsC*, *uvsD*, and *uvsE*, inactivate endonuclease β (20), which renders *uvrA* strains of *D. radiodurans* UV sensitive. Strains with wild-type *uvs* resist crosslinking agents and ionizing radiation.

Mutational inactivation of the *rec* locus of *D. radiodurans* R1 will result in strains that are sensitive to all forms of DNA damage, presumably because this mutant cannot carry out homologous recombination. The *rec* gene has been cloned and sequenced, and its gene product has been shown to be a homologue to *E. coli*'s RecA protein, which shares 56% identity with *E. coli*'s RecA (23). Despite this similarity, the deinococcal RecA will not restore recombinational repair to an *E. coli recA* null mutant, which distinguishes the deinococcal gene product from the majority of prokaryotic RecA homologues. Evidence has been presented that suggests that when the deinococcal RecA is expressed in *E. coli*, even at low levels, it is toxic (23). The reasons for the toxicity are unknown.

Gutman et al (25) have cloned and sequenced a gene that encodes a DNA polymerase and that is necessary for DNA damage resistance in *D. radiodurans*. It appears that the *pol* gene product of *D. radiodurans* and *E. coli*'s DNA polymerase I have similar, if not identical, functions. The *D. radiodurans' pol* gene product shares 51.1% identity with DNA polymerase I of *E. coli*. In addition, a *pol* strain of *D. radiodurans* can be restored to DNA damage resistance by the intracellular expression of a clone of E. coli's DNA polymerase I (24).

An evaluation of the enzymology of DNA repair in *D. radiodurans* reveals that each deinococcal DNA repair protein that has been identified, to date, has a functional homologue in other prokaryotes. Further, with the possible exception of the RecA homologue, nothing exceptional has been reported concerning the catalytic abilities of this group of proteins. Although Table 2 defines an admittedly short list of repair proteins, the similarity between the deinococcal proteins and those found in *E. coli* is striking, which suggests that *D. radiodurans* may use the same complement of DNA repair proteins as does this much more DNA damage–sensitive organism. If this is true, it seems unlikely that an exhaustive characterization of the enzymology of deinococcal DNA repair will provide insight into this organism's extreme DNA damage tolerance. Instead, it is perhaps more pertinent to ask whether the context in which these proteins are utilized in *D. radiodurans* is somehow different from the context in which they are used in more radiosensitive organisms.

D. RADIODURANS' STRATEGIES FOR SURVIVAL

Redundant Genetic Information

As stated earlier, *D. radiodurans* is multigenomic (28, 32). Stationary phase cells contain four copies of their chromosome, and actively dividing cells contain from four to ten copies. Since multiple copies of the genome provide the cell with a reservoir of genetic information, it is expected that DNA repair processes will be more efficient in organisms with higher chromosome multiplicity. Mortimer (50) demonstrated in 1958 that diploid and tetraploid forms of *S. cerevisiae* are more radiation resistant than are haploid strains. From these results, it was assumed that the additional genetic information "protected" the polyploid cells from radiation-induced lethality, because the presence of multiple genomes offered the possibility of restoring the DNA sequence that had been damaged or lost during irradiation through homologous recombination. Similar results were obtained by Krasin & Hutchinson (37) in their studies of *E. coli*. They showed that *E. coli* cultures grown in minimal media were more radiation sensitive than were exponential phase cultures grown in rich media. When growing under optimal conditions, the *E. coli* chromosome replicates faster than the cell septates, which results in multigenomic cells. In contrast, in minimal media the cell replicates slowly, and individual cells are haploid. Thus, in *E. coli* and *S. cerevisiae*, chromosome multiplicity confers enhanced radiation resistance. Clearly, just as the materials necessary to build a house don't assemble themselves, the availability of redundant genetic information is of little use if the mechanisms are not in place to take advantage of that information. Simply being multigenomic cannot, in and of itself, provide additional protection. As discussed below, *D. radiodurans* exhibits extremely efficient recombinational repair and it is here that the availability of redundant genetic information appears to play a critical role in cell survival.

It should be noted that, at this time, there is no physical evidence that redundant genetic information is necessary for the extreme radioresistance of *D. radiodurans*. Attempts at reducing chromosomal copies to less than four have been unsuccessful (32). Defined media have been used to vary chromosome multiplicity between 5 and 10 copies in *D. radiodurans*, but these studies fail to demonstrate any correlation between the number of copies and the radioresistance of *D. radiodurans* (32).

Interchromosomal Recombination

When an exponential phase culture of *D. radiodurans* is exposed to 5000 Gy γ radiation, more than 150 DNA double-stranded breaks (dsbs) are introduced into the chromosome (9). Given the number of dsbs generated, it is remarkable that this level of damage is repaired without loss of genetic integrity. Cellular exonucleases, acting at strand breaks, should rapidly destroy overhanging ends, with a concomitant loss of sequence information. Even so, *D. radiodurans* is apparently able to reassemble an intact chromosome from the fragments produced following γ radiation in a way that conserves the linear continuity of its genome. This ability undoubtedly requires chromosome multiplicity. Because strand breaks are generated randomly, the probability of losing genetic information at the same site on every chromosome is very low at sublethal doses of radiation. Although each chromosome will suffer strand breaks, the distribution of that damage will, on average, be different from chromosome to chromosome. In principle, the total complement of fragments that remain after sublethal damage and repair should be sufficient to form an intact chromosome, provided that the cell has a way of mediating this reassembly.

Daly & Minton (11–13, 48) have suggested that *D. radiodurans* uses interchromosomal recombination to take advantage of the additional information present in multiple chromosomal copies. They have provided physical evidence that interchromosomal recombination occurs, reporting that γ irradiation induces as many as 600 crossovers per four-chromosome nucleoid (11). Approximately one third of these crossovers were identified as nonreciprocal, which suggests that *D. radiodurans* is, at least in part, restoring its fragmented chromosomes by piecing available sequences together.

The idea that interchromosomal recombination reassembles an entire chromosome from a myriad of small, randomly generated fragments implies that *D. radiodurans* is able to efficiently bring homologous regions of those fragments together by an unprecedented mechanism. In an elegant theoretical discussion of this concept, Minton & Daly (48) point out that a search for homology among hundreds of fragments is a logistical nightmare—a problem that is perhaps most easily explained by the following analogy. Assume that the fragments of each *D. radiodurans* chromosome are pieces of a jigsaw puzzle scattered on a table and, because *D. radiodurans* is multigenomic, there are four identical and intermixed sets of puzzle pieces. Putting the puzzle together requires that each piece be examined individually and fit to other compatible pieces. Since there is no obvious way of determining whether a piece has already been used, solving the puzzle will require an extensive and frequently futile search for compatibility. The *D. radiodurans* cell must, therefore, not only have the means for assembling an intact chromosome from component parts, but it must also have a mechanism for avoiding these potentially fruitless homology searches. Without such a mechanism, it is difficult to envision how the process of chromosome restitution could be completed within a reasonable time frame.

Minton & Daly (48) have suggested that the problem of futile searches for homology can be overcome by physically eliminating the need for such searches.

They have proposed that there is a pre-existing alignment between homologous regions on the different *D. radiodurans* chromosomes, and that this alignment facilitates the association of homologous sequences after irradiation. However, there is no experimental evidence to support this hypothesis.

Coordinate Regulation of DNA Repair Functions

When microorganisms are treated with DNA damaging agents, such as UV light, the lesions that are introduced will stop DNA replication by blocking movement of the polymerase along the template strand (27, 49). While this phenomenon might contribute to the inhibition of DNA replication that is observed following DNA damage in *D. radiodurans*, it does not appear to be the primary reason for the inhibition. Numerous studies (15, 40, 55) have shown that the DNA damage–induced delay in *D. radiodurans'* chromosomal DNA replication is dose dependent, and that the length of the delay always exceeds the time required to repair the DNA damage that caused the inhibition. These observations suggest that the process of DNA replication in *D. radiodurans* is sensitive to the level of intracellular DNA damage, and that the cell has a mechanism for sensing the completion of DNA repair (i.e. the level of DNA damage) and relaying that information to the replication machinery. In other words, these observations suggest that there is a regulatory checkpoint in *D. radiodurans* that controls chromosome replication and, as a consequence, controls cell division.

Mattimore et al (44) have provided genetic evidence supporting the existence of a DNA damage checkpoint in *D. radiodurans*. This group identified three mutant strains of *D. radiodurans*, designated SLR2, SLR4, and SLR5, that exhibit a slow recovery phenotype following exposure to ionizing radiation. The SLR mutants are as resistant to γ radiation as the wild-type organism is, but after irradiation they require from 48 to 72 h longer than the wild type does to form colonies on agar. These strains do not express a growth defect, however, because they have generation times identical to wild-type *D. radiodurans*. While they are obviously able to repair the DNA damage that is introduced following irradiation, the SLR strains appear to be defective in control of the reinitiation of DNA repair following irradiation. The SLR strains have been divided into two classes (NC Shank & JR Battista, unpublished observations), based on the rate at which they repair ionizing radiation–induced DNA damage. Figure 4 illustrates how the SLR strains recover from exposure to 5000 Gy γ radiation, a dose at which all three strains exhibit 100% viability. Using the restitution of DNA dsbs as an endpoint, recovery was followed using PFGE. Samples of each culture were removed immediately after irradiation and at 24 h intervals thereafter for the next 72 h. Restoration of chromosomal DNA was not observed in SLR2 until 48 h after irradiation, whereas the chromosomal DNA of R1, SLR4, and SLR5 had reassembled within 24 h. (SLR4 and SLR5 recover with

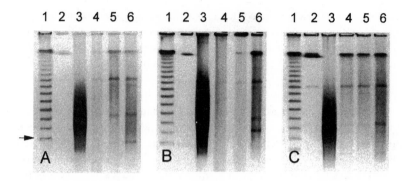

Figure 4 The ability of *D. radiodurans* R1 (*panel A*), SLR2 (*panel B*), and SLR5 (*panel C*) to survive the accumulation of chromosomal DNA double strand breaks following exposure to 5000 Gy γ radiation. For each panel: *Lane 1* is a lambda ladder size standard, *lane 2* is chromosomal DNA prepared from an untreated culture, *lane 3* is chromosomal DNA prepared from a culture immediately after irradiation, and *lanes 4–6* are chromosomal DNA prepared from a culture at 24, 48, and 72 h post-irradiation, respectively.

the same kinetics. Only SLR5 is represented in Figure 4.) In fact, for these strains, recovery is complete in less than 12 h (data not shown). SLR2 mends dsbs slowly relative to wild type, and this obvious delay in DNA repair correlates with the delay in colony formation (cell division?) noted by Mattimore et al (44). It appears that SLR2 cells have the ability to suspend DNA replication until repairs are complete, even though that delay is 36 h longer than what would normally occur following 5000 Gy γ radiation.

If there is a mechanism that is both sensitive to DNA damage and capable of restricting DNA replication, it must be assumed that there is also a mechanism for alleviating that inhibition, once DNA damage has been removed. The phenotypes observed in SLR4 and SLR5 may reflect a breakdown in the cell's ability to "realize" that DNA repair is complete. In contrast to SLR2, SLR4 and SLR5 do not exhibit a delay in double strand break repair. Once repairs are finished, however, these strains seem unable to reinitiate DNA synthesis and/or cell division. This defect could be in the strain's ability to detect removal of the signal that triggers the inhibition of DNA replication, or, alternatively, the strain may not be able to clear that signal from the cell with the same kinetics as the wild-type organism.

One of the most pronounced physiological effects observed following the administration of ionizing radiation to *D. radiodurans* is an immediate and extensive breakdown of chromosomal DNA (40, 53, 71). This degradation seems to be caused by the introduction of single and double strand breaks into the chromosome that serve as substrates for exonucleases present in the cell. For most species, the loss of genetic information that accompanies this breakdown is considered one of the lethal consequences of ionizing radiation–induced DNA damage. In *D. radiodurans*, however, the DNA degradation observed appears to be part of the DNA repair process. There is evidence of degradation within 5 min of the administration of 5000 Gy γ radiation (72).

The duration of postirradiation DNA degradation is determined by the dose of radiation administered: The larger the dose, the longer the degradation continues and the greater the loss of chromosomal DNA (40, 53, 71). The rate of degradation is independent of dose, with an estimated 0.1% of the total genomic DNA lost per minute. The degradative process, once started, must be "turned off" by an uncharacterized inhibitory protein that seems to be induced by DNA damage (29). The termination of DNA degradation requires protein synthesis post-irradiation. Administration of either chloramphenicol (16, 17) or actinomycin D (16) to cultures prior to irradiation results in extensive loss of chromosomal DNA, and it ultimately results in cell death. The observation that the extent of DNA degradation is governed by the dose of radiation administered indicates that this process is regulated, and that this regulation keys on the level of intracellular DNA damage. It therefore appears that the signal(s) that control DNA replication and the process of DNA degradation (and DNA repair?) are interrelated. DNA replication is held in check until DNA repair is complete, and DNA repair ends before DNA replication is reinitiated, suggesting that these phenomena are coordinately regulated. If this is true, it is not unreasonable to expect that the same signal affects each process. The nature of the proposed signal is unknown.

The protein that inhibits DNA degradation appears to be a component of the cell's mechanism for sensing DNA damage. As noted, this protein is responsible for stopping DNA degradation. Presumably, it is sensitive to DNA damage and, when damage drops to an "acceptable" level, it relays that information to the proteins catalyzing the degradation. The recently described IrrI protein of *D. radiodurans* (69) is a candidate for this inhibitor of DNA degradation. *irrI* strains are extremely sensitive to ionizing radiation, exhibiting a dramatic reduction in survival following doses that are sublethal to the wild-type organism. As illustrated in Figure 1 (*panel B*), the chromosome of the *irrI* strain, IRS41, is completely destroyed by 3000 Gy γ radiation and fails to recover (MD Manuel & JR Battista, unpublished observations). Only 0.1% of an IRS41 culture will survive this dose. In contrast, the wild-type *D. radiodurans* restores its genome in less than 3 h (Figure 1, *panel A*), and exhibits no loss of viability at this dose. Following irradiation, an *irrI* mutant exhibits the same unrestricted DNA degradation that typifies wild-type cultures that are pre-treated with chloramphenicol, which suggests that IrrI is the protein needed to stop DNA degradation.

DNA Damage Export

UV and γ radiation–induced DNA degradation are accompanied by export of DNA damage. Initially, the products formed are oligonucleotides approximately 2000 bp long, and a mixture of damaged and undamaged nucleotides and nucleosides (72). These products are found in the cytoplasm and the surrounding growth medium, which indicates that *D. radiodurans* exports the degradation products once they are formed (6, 30, 31). Release of degradation products continues throughout this stage in the cell's recovery, ending shortly after degradation ends. While the release of DNA damage has been exploited for many years as a convenient means of following the repair of base damage in *D. radiodurans*, the relationship of this phenomenon to DNA damage resistance has never been explored. The removal of damaged nucleotides from the intracellular nucleotide pool, and their subsequent conversion to nucleosides, could represent a survival strategy. Two possibilities seem likely: 1. Moving damaged nucleotides outside the cell might protect the organism from elevated levels of mutagenesis by preventing the reincorporation of damaged bases into the genome during DNA synthesis subsequent to repair. 2. Removal of nucleotides from the cell is part of the signal that coordinates the DNA repair functions described above.

CONCLUDING REMARKS

Despite 40 years of investigation, our knowledge of *D. radiodurans*' extraordinary ability to survive DNA damage is still limited. Even so, evidence is accumulating that indicates that *D. radiodurans* uses a set of proteins not unlike those of other prokaryotes to carry out DNA repair functions. The picture that emerges suggests that *D. radiodurans* influences the effectiveness of an "ordinary" complement of DNA repair proteins by changing the conditions in which they operate. Many of the unintended consequences of the DNA repair process observed in other organisms seems to be avoided. Redundant genetic information is always available and through interchromosomal recombination a new genome can be built, from fragments if necessary, without loss of genetic integrity. In effect, this system seems to prevent the loss of genetic information through deletion during DNA repair. Also, the process of DNA repair appears to be coordinately regulated with the cell cycle, providing adequate time for repair and averting a potentially deleterious attempt to replicate a damaged genome. Finally, damaged nucleotides are sequestered by expulsion from the cell, which prevents their reincorporation during DNA synthesis subsequent to their removal during repair. In the author's estimation, appreciating how this combination of characteristics influences the activity of DNA repair proteins is key to understanding the extreme DNA damage tolerance of *D. radiodurans*.

ACKNOWLEDGMENTS

The author is grateful to Misty D. Manuel for critically evaluating this manuscript. The unpublished work from the author's laboratory was supported by National Science Foundation grant MCB-9418594.

> **Visit the *Annual Reviews home page* at
> http://www.annurev.org.**

Literature Cited

1. Agostini HJ, Carroll JD, Minton KW. 1996. Identification and characterization of *uvrA*, a DNA repair gene of *Deinococcus radiodurans. J. Bacteriol.* 178:6759–65

2. Al-Bakri GH, Mackay MW, Whittaker PA, Moseley BEB. 1985. Cloning of the DNA repair genes *mtcA, mtcB, uvsC, uvsD, uvsE* and the *leuB* gene from *Deinococcus radiodurans. Gene* 33:305–11

3. Anderson AW, Nordon HC, Cain RF, Parrish G, Duggan D. 1956. Studies on a radio-resistant micrococcus. I. Isolation, morphology, cultural characteristics, and resistance to gamma radiation. *Food Technol.* 10:575–78

4. Auda H, Emborg C. 1973. Studies on post-irradiation DNA degradation in *Micrococcus radiodurans*, Strain R^{II}5. *Radiat. Res.* 53:273–80

5. Battista JR. 1997. DNA repair in *Deinococcus radiodurans*. In *DNA Damage and Repair, Vol. I: DNA Repair in Prokaryotes and Lower Eukaryotes*, ed. JA Nickoloff, M Hoekstra. Totowa, NJ: Humana. In press

6. Boling ME, Setlow JK. 1966. The resistance of *Micrococcus radiodurans* to ultraviolet radiation III. A repair mechanism. *Biochim. Biophys. Acta* 123:26–33

7. Bonura T, Town CD, Smith KC, Kaplan HS. 1975. The influence of oxygen on the yield of DNA double-strand breaks in X-irradiated *Escherichia coli* K12. *Radiat. Res.* 63:567–77

8. Brooks BW, Murray RGE. 1981. Nomenclature for "*Micrococcus radiodurans*" and other radiation-resistant cocci: Deinococcaceae fam. nov. and *Deinococcus* gen. nov., including five species. *Int. J. Syst. Bacteriol.* 31:353–60

9. Burrell AD, Feldschreiber P, Dean CJ. 1971. DNA membrane association and the repair of double breaks in X-irradiated *Micrococcus radiodurans. Biochim. Biophys. Acta* 247:38–53

10. Christensen EA, Kristensen H. 1981. Radiation-resistance of micro-organisms from air in clean premises. *Acta Pathol. Microbiol. Scand. Sect. B.* 89:293–301

11. Daly MJ, Minton KW. 1995. Interchromosomal recombination in the extremely radioresistant bacterium *Deinococcus radiodurans. J. Bacteriol.* 177:5495–505

12. Daly MJ, Ouyang L, Fuchs P, Minton KW. 1994. In vivo damage and *recA*-dependent repair of plasmid and chromosomal DNA in the radiation-resistant bacterium. *J. Bacteriol.* 176:3608–17

13. Daly MJ, Ouyang L, Minton KW. 1994. Interplasmidic recombination following irradiation of the radioresistant bacterium *Deinococcus radiodurans. J. Bacteriol.* 176:7506–15

14. Davis NS, Silverman GJ, Mausurosky EB. 1963. Radiation resistant, pigmented coccus isolated from haddock tissue. *J. Bacteriol.* 86:294–98

15. Dean CJ, Feldschreiber P, Lett JT. 1966. Repair of X-ray damage to the deoxyribonucleic acid in *Micrococcus radiodurans. Nature* 209:49–52

16. Dean CJ, Little JG, Serianni RW. 1970. The control of post irradiation DNA breakdown in *Micrococcus radiodurans. Biochem. Biophys. Res. Commun.* 39:126–34

17. Driedger AA, James AP, Grayston MJ. 1970. Cell survival and X-ray-induced DNA degradation in *Micrococcus radiodurans. Radiat. Res.* 44:835–45

18. Evans DM, Moseley BEB. 1988. *Deinococcus radiodurans* UV endonuclease β DNA incisions do not generate photoreversible thymine residues. *Mutat. Res.* 207:117–19

19. Evans DM, Moseley BEB. 1985. Identification and initial characterization of a pyrimidine dimer UV endonuclease

(UV endonuclease β) from *Deinococcus radiodurans:* a DNA-repair enzyme that requires manganese ions. *Mutat. Res.* 145:119–28

20. Evans DM, Moseley BEB. 1983. Roles of the *uvsC, uvsD, uvsE,* and *mtcA* genes in the two pyrimidine dimer excision repair pathways of *Deinococcus radiodurans. J. Bacteriol.* 156:576–83

21. Grant IR, Patterson MF. 1989. A novel radiation resistant *Deinobacter* sp. isolated from irradiated pork. *Lett. Appl. Microbiol.* 8:21–24

22. Grimsley JK, Masters CI, Clark EP, Minton KW. 1991. Analysis by pulsed-field gel electrophoresis of DNA double-strand breakage and repair in *Deinococcus radiodurans* and a radiosensitive mutant. *Int. J. Radiat. Biol.* 60:613–26

23. Gutman PD, Carroll JD, Masters CI, Minton KW. 1994. Sequencing, targeted mutagenesis and expression of a *recA* gene required for extreme radioresistance of *Deinococcus radiodurans. Gene* 141:31–37

24. Gutman PD, Fuchs P, Minton KW. 1994. Restoration of the DNA damage resistance of *Deinococcus radiodurans* DNA polymerase mutants by *Escherichia coli* DNA polymerase I and Klenow fragment. *Mutat. Res.* 314:87–97

25. Gutman PD, Fuchs P, Ouyang L, Minton KW. 1993. Identification, sequencing, and targeted mutagenesis of a DNA polymerase gene required for the extreme radioresistance of *Deinococcus radiodurans. J. Bacteriol.* 175:3581–90

26. Gutman PD, Yao H, Minton KW. 1991. Partial complementation of the UV sensitivity of *Deinococcus radiodurans* excision repair mutants by the cloned *denV* gene of bacteriophage T4. *Mutat. Res.* 254:207–15

27. Hall JD, Mount DW. 1981. Mechanisms of DNA replication and mutagenesis in ultraviolet-irradiated bacteria and mammalian cells. *Prog. Nucleic Acid Res. Mol. Biol.* 30:53–65

28. Hansen MT. 1978. Multiplicity of genome equivalents in the radiation-resistant bacterium *Micrococcus radiodurans. J. Bacteriol.* 134:71–75

29. Hansen MT. 1982. Rescue of mitomycin C- or psoralen-inactivated *Micrococcus radiodurans* by additional exposure to radiation or alkylating agents. *J. Bacteriol.* 152:976–82

30. Hariharan PV, Cerutti PA. 1972. Formation and repair of γ-ray-induced thymine damage in *Micrococcus radiodurans. J. Mol. Biol.* 66:65–81

31. Hariharan PV, Cerutti PA. 1971. Repair of γ-ray-induced thymine damage in *Micrococcus radiodurans. Nature New Biol.* 229:247–49

32. Harsojo S, Kitayama S, Matsuyama A. 1981. Genome multiplicity and radiation resistance in *Micrococcus radiodurans. J. Biochem.* 90:877–80

33. Hensel R, Demharter W, Kandler O, Kroppenstedt RM, Stackebrandt E. 1986. Chemotaxonomic and molecular-genetic studies of the genus *Thermus:* Evidence for a phylogenetic relationship of *Thermus aquaticus* and *Thermus ruber* to the genus *Deinococcus. Int. J. Syst. Bacteriol.* 36:444–53

34. Ito H, Watanabe H, Takeshia M, Iizuka H. 1983. Isolation and identification of radiation-resistant cocci belonging to the genus *Deinococcus* from sewage sludges and animal feeds. *Agric. Biol. Chem.* 47:1239–47

35. Kitayama S. 1982. Adaptive repair of cross-links in DNA of *Micrococcus radiodurans. Biochim. Biophys. Acta* 697:381–84

36. Kitayama S, Asaka S, Totsuka K. 1983. DNA double-strand breakage and removal of cross-links in *Deinococcus radiodurans. J. Bacteriol.* 155:1200–7

37. Krasin F, Hutchinson F. 1977. Repair of DNA double-strand breaks in *Escherichia coli,* which requires *recA* function and the presence of a duplicate genome. *J. Mol. Biol.* 116:81–98

38. Kristensen H, Christensen EA. 1981. Radiation-resistant microorganisms isolated from textiles. *Acta Pathol. Microbiol. Scand. Sect. B* 89:303–9

39. Lennon E, Minton KW. 1990. Gene fusions with *lacZ* by duplication insertion in the radioresistant bacterium *Deinococcus radiodurans. J. Bacteriol.* 172:2955–61

40. Lett JT, Feldschreiber P, Little JG, Steele K, Dean CJ. 1967. The repair of X-ray damage to the deoxyribonucleic acid in *Micrococcus radiodurans:* A study of the excision process. *Proc. R. Soc. London, Ser. B* 167:184–201

41. Lewis NF. 1971. Studies on a radio-resistant coccus isolated from Bombay Duck (*Harpodon nehereus*). *J. Gen. Microbiol.* 66:29–35

42. Masters CI, Moseley BEB, Minton KW. 1991. AP endonuclease and uracil DNA glycosylase activities in *Deinococcus radiodurans. Mutat. Res.* 254:263–72

43. Mattimore V, Battista JR. 1995. Radioresistance of *Deinococcus radiodurans:* Functions necessary to survive ionizing radiation are also necessary to

survive prolonged desiccation. *J. Bacteriol.* 178:633–37

44. Mattimore V, Udupa KS, Berne GA, Battista JR. 1995. Genetic characterization of forty ionizing radiation-sensitive strains of *Deinococcus radiodurans*: Linkage information from transformation. *J. Bacteriol.* 177:5232–37

45. Maxcy RB, Rowley DB. 1978. Radiation-resistant vegetative bacteria in a proposed system of radappertization of meats. In *Food Preservation by Irradiation*, 1:347–59. Vienna: International Atomic Energy Agency

46. Minton KW. 1994. DNA repair in the extremely radioresistant bacterium *Deinococcus radiodurans*. *Mol. Microbiol.* 13:9–15

47. Minton KW. 1996. Repair of ionizing-radiation damage in the radiation resistant bacterium *Deinococcus radiodurans*. *Mutat. Res.* 363:1–7

48. Minton KW, Daly MJ. 1995. A model for repair of radiation-induced DNA double strand breaks in the extreme radiophile *Deinococcus radiodurans*. *BioEssays* 17:457–64

49. Moore P, Strauss BS. 1979. Sites of inhibition of in vitro DNA synthesis in carcinogen- and UV-treated DNA. *Nature* 278:664–66

50. Mortimer RK. 1958. Radiobiological and genetic studies on a polyploid series (haploid to hexaploid) of *Saccharomyces cerevisiae*. *Radiat. Res.* 66:158–69

51. Moseley BEB. 1967. The isolation and some properties of radiation sensitive mutants of *Micrococcus radiodurans*. *J. Gen. Microbiol.* 49:293–300

52. Moseley BEB. 1983. Photobiology and radiobiology of *Micrococcus* (*Deinococcus*) *radiodurans*. *Photochem. Photobiol. Rev.* 7:223–75

53. Moseley BEB 1967. The repair of DNA in *Micrococcus radiodurans* following ultraviolet irradiation. *J. Gen. Microbiol.* 48:4–24

54. Moseley BEB, Copland HJR. 1978. Four mutants of *Micrococcus radiodurans* defective in the ability to repair DNA damaged by mitomycin-C, two of which have wild-type resistance to ultraviolet radiation. *Mol. Gen. Genet.* 160:331–37

55. Moseley BEB, Copland HJR. 1976. The rate of recombination repair and its relationship to the radiation-induced delay in DNA synthesis in *Micrococcus radiodurans*. *J. Gen. Microbiol.* 93:251–58

56. Moseley BEB, Evans DM. 1983. Isolation and properties of strains of *Micrococcus* (*Deinococcus*) *radiodurans* un-

able to excise ultraviolet light-induced pyrimidine dimers from DNA: Evidence of two excision pathways. *J. Gen. Microbiol.* 129:2437–45

57. Moseley BEB, Mattingly A. 1971. Repair of irradiated transforming deoxyribonucleic acid in wild type and a radiation-sensitive mutant of *Micrococcus radiodurans*. *J. Bacteriol.* 105:976–83

58. Mun C, Del Rowe J, Sandigursky M, Minton KW, Franklin WA. 1994. DNA deoxyribophosphodiesterase and an activity that cleaves DNA containing thymine glycol adducts in *Deinococcus radiodurans*. *Radiat. Res.* 138:282–85

59. Murray RGE. 1992. The family *Deinococcaceae*. In *The Prokaryotes*, ed. A Ballows, HG Truper, M Dworkin, W Harder, KH Schleifer 4:3732–44. New York: Springer-Verlag

60. Murray RGE. 1986. Genus 1. *Deinococcus* Brooks and Murray 1981. In *Bergey's Manual of Systematic Bacteriology*, ed. PHA Sneath, NS Mair, ME Sharpe, JG Holt, 2:1035–43. Baltimore: Williams & Wilkins

61. Raj HJ, Duryee FL, Deeny AM, Wang CH, Anderson AW, Elliker PW. 1960. Utilization of carbohydrates and amino acids by *Micrococcus radiodurans*. *Can. J. Microbiol.* 6:289–98

62. Shapiro A, DiLello D, Loudis MC, Keller DE, Hutner SH. 1977. Minimal requirements in defined media for improved growth of some radio-resistant pink tetracocci. *Appl. Env. Microbiol.* 33:1129–33

63. Smith MD, Abrahamson R, Minton KW. 1989. Shuttle plasmids constructed by the transformation of an *Escherichia coli* cloning vector into two *Deinococcus radiodurans* plasmids. *Plasmid* 22:132–42

64. Smith MD, Masters CI, Lennon E, McNeil LB, Minton KW. 1991. Gene expression in *Deinococcus radiodurans*. *Gene* 98:45–52

65. Smith MD, Masters CI, Moseley BEB. 1992. Molecular biology of radiation resistant bacteria. In *Molecular Biology and Biotechnology of Extremophiles*, ed. RA Herbert, RJ Sharp, 258–80. New York: Chapman & Hall

66. Sweet DM, Moseley BEB. 1974. Accurate repair of ultraviolet-induced damage in *Micrococcus radiodurans*. *Mutat. Res.* 23:311–18

67. Sweet DM, Moseley BEB. 1976. The resistance of *Micrococcus radiodurans* to killing and mutation by agents which damage DNA. *Mutat. Res.* 34:175–86

68. Tigari S, Moseley BEB. 1980. Transformation in *Micrococcus radiodurans*:

Measurement of various parameters and evidence for multiple, independently segregating genomes per cell. *J. Gen. Microbiol.* 119:287–96

69. Udupa KS, O'Cain PA, Mattimore V, Battista JR. 1994. Novel ionizing radiation-sensitive mutants of *Deinococcus radiodurans. J. Bacteriol.* 176:7439–46

70. Van den Eynde H, Van de Peer Y, Vandenabeele H, Van Bogaert M, De Wachter R. 1990. 5S rRNA sequences of myxobacteria and radioresistant bacteria and implications for eubacterial evolution. *Int. J. Syst. Bacteriol.* 40:399–404

71. Varghese AJ, Day RS. 1970. Excision of cytosine-thymine adduct from the DNA of ultraviolet-irradiated *Micrococcus radiodurans. Photochem. Photobiol.* 11:511–17

72. Vukovic-Nagy B, Fox BW, Fox M. 1974. The release of DNA fragments after X-irradiation of *Micrococcus radiodurans. Int. J. Radiat. Biol.* 25:329–37

73. Weisburg WG, Giovannoni SJ, Woese CR. 1989. The *Deinococcus-Thermus* phylum and the effect of rRNA composition on phylogenetic tree construction. *Syst. Appl. Microbiol.* 11:128–34

Annu. Rev. Microbiol. 1997. 51:225–55

GENETICS OF THE ROTAVIRUSES

Robert F. Ramig

Division of Molecular Virology, Baylor College of Medicine, Houston, Texas 77030;
email: rramig@bcm.tmc.edu

KEY WORDS: rotavirus genetics, rotavirus mutants, reassortment of genome segments, reassortment mapping of the genome, potential for reverse genetics

---

ABSTRACT

Genetic analyses have contributed significantly to our understanding of the biology of the rotaviruses. The distinguishing feature of the virus is a genome consisting of 11 segments of double-stranded RNA. The segmented nature of the genome allows reassortment of genome segments during mixed infections, which is the major distinguishing feature of rotavirus genetics. Reassortment has been a powerful tool for mapping viral mutations and other determinants of biological phenotypes to specific genome segments. However, more detailed genetic analysis of rotaviruses is currently limited by the inability to perform reverse genetics. Development of a reverse genetic system will facilitate analysis of the molecular mechanisms involved in various genetic, biochemical, and biological phenomena of the virus.

CONTENTS

225

0066-4227/97/1001-0225$08.00

INTRODUCTION

Rotaviruses are classified as a genus within the virus family Reoviridae. They share common properties with other members of the Reoviridae, including a multilayered capsid structure, a virion-associated RNA-dependent RNA polymerase, a segmented genome consisting of double-stranded RNA, and a cytoplasmic life cycle.

Rotaviruses are an agent of gastroenteritis and are the major viral etiologic agent of severe diarrheal disease in children and in the young of other mammalian and avian species (3). The genus is divided into a number of groups between which antigenicity is distinct (8). Most studies have been performed with group A rotaviruses, which constitute the focus of this review.

Numerous books and reviews have recently considered various aspects of the rotaviruses (19, 22, 23, 28, 48, 50, 79, and others), including reviews of rotavirus genetics (33, 83). The literature cited here was chosen for illustrative purposes and is, therefore, a selective list.

THE VIRUS AND ITS LIFE CYCLE

The rotavirion is a complex structure consisting of three concentric proteinaceous capsid layers that surround a double-stranded RNA (dsRNA) genome made up of 11 segments (23). The outermost layer of the mature virion, or triple-layered particle, consists of viral proteins VP4 and VP7. Removal of the outer capsid proteins produces the double-layered particle with VP6 on its surface and activates the virion-associated RNA-dependent RNA polymerase, or transcriptase. The mRNAs produced by the transcriptase are complete end-to-end copies of each genome segment, with a 5'-terminal cap 1 structure and lacking a 3'-terminal poly(A). Removal of the intermediate capsid layer produces the viral core, or single-layered particle, which displays VP2 on its surface and contains an inactive transcriptase. Small amounts of VP1 and VP3, together with the 11 segments of dsRNA, reside within the core particle. The dsRNA segments range in size from 667 to 3302 base pairs, and they generally contain a single open reading frame, with relatively short 5'- and 3'-terminal noncoding regions. Each of the 11 segments shares short conserved 5'- and 3'-terminal sequences. The positive-strand of each genomic duplex is identical to the mRNA derived from it. In addition to the six structural proteins, rotaviruses encode five nonstructural proteins, four of which possess RNA-binding activity (71).

The rotavirus replication cycle is cytoplasmic (23). Figure 1 shows a simplified life cycle, in which the features critical to understanding genetics are emphasized. Upon entry of the virion, the outer capsid is removed, activating the

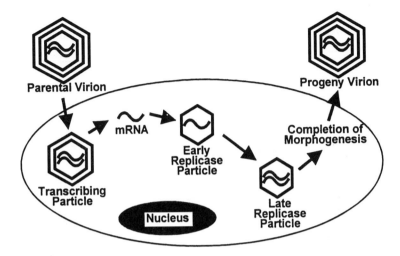

Figure 1   Essential elements of the rotavirus life cycle. Note that polymerase activities and dsRNA are always particle-associated and that mRNA is the only viral nucleic acid that is free for genetic interaction. For simplicity, only one of the 11 genome segments is shown.

virion-associated transcriptase. The resulting mRNAs function either as messengers for translation into protein or as templates for replication of progeny genomes. After sufficient viral protein is translated, subviral particles are formed that contain a mixture of structural and nonstructural proteins, together with one copy of each of the 11 template RNAs. After particle formation, or during the process, an RNA-dependent RNA polymerase activity (replicase) synthesizes the negative-strand on the positive-strand template. During the replication process, VP6 is added, so that particles containing a full complement of dsRNA also display VP6 on their surface. Subsequent steps of viral morphogenesis occur and progeny virions are released. Several features of the life cycle are significant in genetic terms: (*a*) mRNA is the only viral nucleic acid ever free in the cytoplasm of the infected cell. Thus, genetic interactions must occur at the level of mRNA. (*b*) Parental and progeny dsRNA genomes are always particle-associated; free dsRNA is never found. (*c*) All polymerase activity, transcriptase or replicase, is also particle-associated.

THE GENETIC SYSTEM

A Reassorting Genetic System

The segmented nature of the genome suggests that, like other segmented genome viruses [reovirus (80), influenza virus (69)], rotaviruses undergo recombination

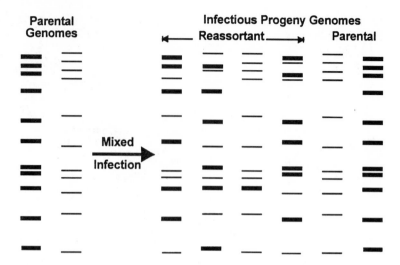

Figure 2   Schematic representation of reassortment in rotavirus. (*left*) The genomes of two parental virus strains as they appear upon electrophoretic separation. (*right*) Electrophoretically separated genomes derived from infectious progeny. Note that electrophoretic mobility can be used to assign parental origin of each genome segment.

by a mechanism of reassortment. A corollary of recombination by a mechanism of reassortment is that recombination frequencies are expected to be high or undetectably low, with no continuum between low and high frequencies (26, 92). Physical evidence for reassortment (59) and a high frequency of recombination (75) have both been reported for rotaviruses.

Physical demonstration of reassortment requires that the genome segments of the two parental viruses be distinguishable. The most commonly used differentiation between parental origin of genome segments is electrophoretic mobility (Figure 2) (86), although other means of identifying parental genes, such as hybridization (27), are sometimes used. Following mixed infection of two virus strains with distinguishable genomes, progeny virus is plaque-purified, and the parental origin of genome segments is determined for the resulting progeny clones. All progeny have genome segment constellations that are (*a*) parental or (*b*) reassortant mosaics containing segments derived from both parents (Figure 2).

The 11 genome segments of the parental virus strains can theoretically reassort into 2^{11} different possible genome constellations, if reassortment is random. No study has identified all possible reassortant progeny from a cross, presumably because insufficient progeny have been examined. Selection for fitness seems likely to limit the viable constellations. However, reassortment of genome segments among viable progeny generally appears to be random during infection of cultured cells (76) and animals (30). Reassortment may be somewhat restricted in certain cases that involve parental viruses that are distantly related (see *Genetic Distance and Reassortment,* below) (96). Reassortment can also be restricted by the imposition of selective pressure during the growth of progeny virus, such as selection against temperature-sensitive (*ts*) mutants present in one or both parents (29, 31).

Random reassortment of genome segments in rotaviruses, like reassortment of chromosomes in cells, is expected to occur at high frequency, because two alternative parental cognates for each segment are available to be packaged. In studies where progeny were analyzed without selection, reassortants dominated over wild-type parentals among the progeny and segment segregation appeared to be relatively random (30). However, in cases where two temperature-sensitive (*ts*) mutants were crossed, and the frequency of wild-type (*ts*$^+$) reassortant progeny was determined, recombination was either high or undetectably low (75). In the case in which temperature was used to select *ts*$^+$ reassortants, the frequency of *ts*$^+$ reassortants did not approach the expected 50%. Rather, the frequency of *ts*$^+$ progeny was significantly lower, being in the range of 1%–20% of the progeny. The lower-than-expected frequency of *ts*$^+$ progeny has been ascribed to interference of the mutant parental virus with wild-type reassortants in the reovirus system, where the phenomenon was systematically examined (12). It seems likely that a similar explanation applies in the case of rotaviruses, because the ts mutants generally have strong interference phenotypes (76).

The high frequency of reassortment observed in crosses of some *ts* mutant pairs is interpreted to indicate that the two *ts* mutations reside on noncognate segments that allow reassortment of constellations of segments that contain no *ts* mutations. Conversely, when no *ts*$^+$ progeny are observed, the *ts* mutations of the mutant pair must reside on cognate segments of the parental viruses so that reassortment cannot generate *ts*$^+$ constellations of genome segments. The failure to detect frequencies of *ts*$^+$ progeny between zero and the high value obtained in the case of reassortment indicates that intramolecular recombination (e.g. recombination mediated by mechanisms of breakage and reformation of covalent bonds or copy-choice strand-switching) does not occur at detectable levels in rotaviruses (33).

Reassortment is a powerful tool and is used to map rotavirus mutations and phenotypes to specific genome segments (see *Reassortment as a Tool to Identify Gene Functtion,* below).

Genetic Markers

Genetic analyses require genetic markers with phenotypes that can be readily identified in the progeny of mixed infections. A wide variety of phenotypes have served as genetic markers in rotaviruses.

CONDITIONAL-LETHAL MUTATIONS   Two types of conditional-lethal mutations, temperature-sensitive (*ts*) and cold-adapted (*ca*), have been isolated from rotaviruses. Although *ca* mutants of rotavirus have been isolated (38), they have not been used for genetic studies. In contrast, *ts* mutants have been used extensively for genetic studies. Attempts to isolate suppressor-sensitive (amber) rotavirus mutants, using genetically engineered mammalian cells that express suppressor tRNAs (88), were unsuccessful because of the extreme instability of the mutants (RF Ramig, unpublished observations).

Temperature-sensitive mutants are quite useful because the mutant phenotype is conditional. The mutants replicate well at the permissive condition ($\sim 31°C$) but fail to replicate at the nonpermissive condition ($\sim 39°C$). This allows propagation of mutants at the permissive condition and the analysis of the mutation at the nonpermissive conditions where the mutant phenotype is expressed. *ts* mutants of rotavirus generally meet the criteria of good mutations in that they (*a*) have phenotypic expression with complete penetrance, (*b*) are easily detected and scored, (*c*) result from single mutations, and (*d*) are stable. The small number of genes in a rotavirus makes it likely that all genes are essential and subject to *ts* lesions. Saturation of the 11 genome segments with mutations is expected to lead to mutants that can be divided into 11 groups by reassortment.

Temperature-sensitive mutants were isolated from the Simian rotavirus SA11 (36, 75, 77), the Rhesus rotavirus RRV (46), and the bovine rotavirus UK (25, 37), following mutagenesis. These mutant collections were divided into groups by scoring reassortment in mixed infection (Table 1); pairs of mutants yielding *ts*[+] progeny lay on different, reassorting segments and were assigned to different reassortment groups, whereas pairs of mutants yielding no *ts*[+] progeny lay on cognate segments and were assigned to the same reassortment group. The set of SA11 *ts* mutants was the most complete, identifying 10 of the 11 reassortment groups expected (76). Preliminary reconciliation of the SA11 and RRV mutants isolated by Greenberg (37) and Kalica (46), against the SA11 collection isolated by Ramig (75, 77) indicated that all of the mutants in the former collections contained single, double, and triple mutations in groups identified by the collection of Ramig. The expected eleventh reassortment group of mutants was not identified (RF Ramig, unpublished observations). The UK mutant collections have not been reconciled against each other or against the SA11 and RRV mutant collections, but the UK mutants of McCrae et al (25) represented nine reassortment groups.

NEUTRALIZATION-RESISTANT MUTANTS   Spontaneous mutations, or those induced by generalized mutagenesis, are randomly distributed over the genome. The use of monoclonal antibodies (MAbs) allows selection of MAb-resistant

Table 1   Grouping of rotavirus SA11 *ts* mutants by reassortment[a]

Mutant	Percentage ts^+ reassortants when crossed with:[b]								
	*ts*234	*ts*339	*ts*606	*ts*778	*ts*868	*ts*971	*ts*975	*ts*1025	*ts*1400
*ts*146	8.70	7.22	4.41	0.02	10.85	0.03	18.65	0.03	2.66
*ts*234	—	0.00	25.90	6.31	0.00	23.41	25.13	14.07	1.21
*ts*339		—	16.03	15.71	0.00	10.79	4.79	6.20	1.50
*ts*606			—	1.95	3.63	2.46	1.64	1.66	0.68
*ts*778				—	6.08	0.00	16.00	0.00	5.81
*ts*868					—	7.65	20.77	25.60	1.44
*ts*971						—	6.02	0.00	3.83
*ts*975							—	4.70	1.48
*ts*1025								—	4.10

[a]Modified from Ramig (75) with permission.

[b]On the basis of the results shown, the mutants were divided into five reassortment groups. Group A: *ts*146, *ts*778, *ts*971, *ts*1025; group B: *ts*234, *ts*339, *ts*868; group C: *ts*606; group D: *ts*975; group E: *ts*1400.

(MAR) mutations whose location is limited to genes encoding the neutralizing proteins VP4 and VP7 (20, 89). MAR mutants have been isolated from numerous rotavirus strains, and many of them have been mapped within the VP4 or VP7 open reading frame by sequence analysis. Although sequenced MAR mutants offer the possibility of studying mutants with precisely known mutations, the mutations present in MAR mutants may be distant from the site of antibody binding because of the conformational nature of many epitopes (20). In cases where multiple, independently isolated MAR mutants have been obtained with the same MAb, mutations are found in a small region of the target gene rather than at a single site, reflecting the fact that an epitope consists of several amino acids. Thus, MAR mutants must be considered regional rather than point mutations. If this distinction is kept in mind, MAR mutants can be useful.

GENETICALLY ENGINEERED MUTANTS   The absence of a reverse genetic system for the rotaviruses has precluded the engineering of specific mutations into infectious virus. However, many mutations have been engineered into the various rotavirus genes using standard molecular genetic methods for the production of deletion mutations, site-directed mutations, and mutations of other types (5). These mutations have been studied primarily in expression systems and have led to the understanding of numerous functional domains within the individual proteins. However, the understanding of the functions of these mutations and their phenotypes, in the context of a viral infection, awaits the development of a reverse genetic system for rotaviruses.

POLYMORPHISMS   Polymorphisms in the electrophoretic migration rate of genomic dsRNA segments (47) can be used as genetic markers (Figure 2). The polymorphic genome profiles of different viral strains are often referred to as electropherotypes, and they provide a means of determining the parental origin of each genome segment in reassortant viruses. Genome segment polymorphisms have been used extensively as genetic markers in experiments to physically map specific phenotypes by correlating genome segment segregation with a given phenotype.

GENETIC REARRANGEMENTS   Rotavirus genome segments undergo rearrangement, particularly under conditions of serial high-multiplicity passage in tissue culture, and rearrangements have been found in animals and normal or immunodeficient children (19). Rearrangements were identified when rotavirus isolates were found that contained extra genome segments (73). Viable viruses containing genome segment rearrangements could be plaque purified and could reassort the rearranged segments during mixed infection, indicating that rearranged segments contain functional genes and were not defective, interfering RNAs (45). Northern hybridization and sequencing of rearranged segments indicated that they are concatemers of sequences within single genome segments (2, 73), and they occur most often in segments that encode nonstructural proteins (19). No rearrangement has been identified with a mosaic of sequences from two or more different segments (19), although two have been identified that contain sequences of unknown origin (58). Rearrangements have been identified that (a) contain a normal open reading frame (ORF) followed by an extended 3'-noncoding region (NCR), (b) an extended ORF with duplicated amino acid sequence and normal 3'-NCR, and (c) a truncated ORF with an extended 3'-NCR (Figure 3). The most common type of rearrangement seems to be one which contains a normal ORF with an extended 3'-NCR (19). The mechanism of formation of rearranged segments is unknown, but the data are consistent with a copy-choice model of polymerase that jumps during transcription or replication. Viruses with rearranged genome segments have been exploited to demonstrate that rotavirions are capable of encapsidating an additional 10% over the normal complement of base pairs (61).

BIOLOGICAL PHENOTYPES AS GENETIC MARKERS   Several biological phenotypes associated with rotaviruses have been used as genetic markers in reassortment experiments designed to map the determinants of those phenotypes. Among the phenotypes used in this way are serological determinants (37), hemagglutination (46), growth properties (37), plaque size (13), protection from challenge (67), and virulence (9, 66). Other phenotypes associated with specific virus strains have the potential for similar use in mapping studies.

A) Rearrangement with conserved ORF

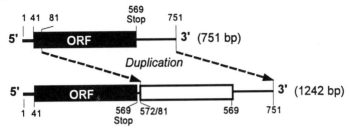

B) Rearrangement with extended ORF

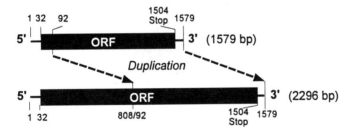

C) Rearrangement with point mutation & truncated ORF

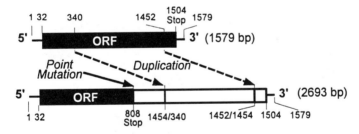

Figure 3 Structures of rearranged rotavirus genome segments. *Lines* indicate noncoding regions (NCR), *solid bars* show open reading frames (ORF), and *open bars* show untranslated regions derived from ORF sequences that were duplicated. *Position numbers* indicate nucleotide numbers in normal genes. The lengths are of the parental and rearranged segments are indicated in base pairs. (*A*) Rearrangement of genome segment 10 from a human rotavirus, showing conservation of the ORF with generation of an extended 3'-NCR (2). (*B*) Rearrangement of segment 5 from a bovine rotavirus with an in-frame duplication generating an extended ORF (91). (*C*) Rearrangement of segment 5 from a bovine rotavirus with a duplication and a point mutation generating a truncated ORF (44). (Modified from indicated references with permission.)

Genetic Interactions in Mixed Infection

The majority of studies to examine the genetic interactions that occur between rotaviruses during mixed infection have been performed using *ts* mutants as genetic markers because of the ease of scoring the *ts* phenotype. In classical viral genetics, the most informative tests have been the complementation test and the recombination test (i.e. special case of reassortment in segmented viruses). The genetic characteristics of rotaviruses determined using these tests are described below.

COMPLEMENTATION   Complementation tests are generally used to group viral mutants into groups that have lesions in the same viral function. In this test, cells are mixedly infected with the two *ts* mutants to be tested, and the infection is allowed to proceed to completion at the nonpermissive condition (NPC; $\sim$39°C) where mutant phenotype is expressed. Each *ts* mutant is also grown in single infection at NPC as a control. The total yield of progeny virus resulting from the mixed and control infections is determined by plaque assay at the permissive condition (PC; 31°C). An increase in yield of the mixed infection, relative to the sum of the two single parent controls, indicates that each of the viruses is able to complement the function that is defective in the other. Often, complementation occurs to significant levels, so that complementing mixed infections yield greater than five-fold more progeny virus than the single control infections. Complementation tests should be corrected for the formation of ts^+ revertants or reassortants during the mixed infection at NPC, because these viruses would have a selective advantage and could contribute significantly to the yield of the mixed infection. This correction is made by determining, at NPC, the yield of ts^+ and subtracting this value from the yield determined at PC. The standard complementation index (C.I.) is determined by the following equation:

$$\text{C.I.} = \left[(\text{Yield A} \times \text{B}_{\text{NPC}})^{\text{PC}} - (\text{Yield A} \times \text{B}_{\text{NPC}})^{\text{NPC}}\right]$$
$$/\left[(\text{Yield A}_{\text{NPC}})^{\text{PC}} + (\text{Yield B}_{\text{NPC}})^{\text{PC}}\right]$$

(subscript indicates condition of infection and superscript indicates condition at which yield is measured). A C.I. greater than 1.0 indicates complementation, although a C.I. greater than 5 is generally considered significant.

When pairs of rotavirus *ts* mutants were subjected to complementation tests, the indices obtained were quite low (often less than 1.0) and irreproducible (75). The complementation indices obtained did not allow the mutants to be placed into self-consistent groups and suggested that complementation was not a useful assay with rotavirus *ts* mutants.

Interference   In the reovirus system, similar low and irreproducible complementation indices were obtained, and these were correlated with the ability of *ts* mutants to interfere with the growth of other mutants or wild-type during mixed infection (12). When rotavirus *ts* mutants were tested for their ability to interfere with the growth of wild-type virus in mixed infection at either PC or NPC, all of the *ts* mutants were found to interfere strongly with the growth of wild-type in either condition (76, 77). Interference with the growth of wild-type was taken as an indication that the *ts* mutants also interfered with the growth of each other and that the interference phenotype of the *ts* mutants prevented the use of complementation tests for functional grouping of the mutants. The mechanism of interference between *ts* mutants is not understood.

REASSORTMENT   In rotaviruses, reassortment tests have provided most of our knowledge of the interactions between mutant viruses during mixed infections.

Quantitative reassortment tests   Reassortment between pairs of rotavirus mutants is analyzed as follows (in the case of *ts* mutants). Cells are mixedly infected with the two *ts* mutants to be tested, and the infection is allowed to proceed to completion at PC. As a control, each *ts* mutant is grown alone under identical conditions. The yield of the mixed infection (and the control parental infections) is then assayed for the presence of *ts*+ reassortants by plaque assay at NPC. In addition to *ts*+ reassortants, progeny with *ts*+ phenotype can result from reversion of the *ts* mutation. To correct for reversion, the yields of the single control infections are measured at NPC to quantitate the reversion of each of the mutants alone. The total yield of the mixed infection is determined by plaque assay at PC. The percentage of *ts*+ reassortants is generally calculated from the following equation, which corrects for the contribution of revertants to the yield:

$$\% \, ts^+ = \left\{ (\text{Yield A} \times \text{B}_{\text{PC}})^{\text{NPC}} - \left[(\text{Yield A}_{\text{PC}})^{\text{NPC}} + (\text{Yield B}_{\text{PC}})^{\text{NPC}} \right] \right\}$$
$$/ (\text{Yield A} \times \text{B}_{\text{PC}})^{\text{PC}}$$

(subscripts indicate condition of the infection and superscripts indicate the condition under which the yield is quantitated). Occasionally, the recombination frequency is calculated without correction for reversion during the test. In this case, the values obtained are termed reassortment indices and are subject to significant inaccuracy because of reversion.

With DNA genome viruses, recombination has been shown to be reciprocal when measured on a population basis, so that the numerator of the formula is multiplied by two to account for the unscored double mutant recombinant class. This correction is not made when working with segmented genome viruses, because the reassortment mechanism has not been demonstrated to be reciprocal, even on a population basis.

Stability of the reassortant genotype   When working with conditional mutants (*ts*), wild-type phenotype progeny could arise from a number of phenomena other than reassortment or reversion. Primary among these phenomena are aggregates of complementing mutants or particles containing heterozygous genomes. Neither phenomenon could be demonstrated to contribute significantly to the wild-type yield (76). In other viral systems, wild phenotype progeny could also result from the generation of suppressor mutations (84). However, suppression of mutant phenotype could not be demonstrated with rotavirus *ts* mutations (RF Ramig, unpublished observations). Furthermore, reassortants can be passaged sequentially from plaque-to-plaque many times without alteration of the reassortant segment constellation, indicating that they are genetically stable and that wild phenotype resulted from the reassortment event.

All-or-none reassortment   As noted above, reassortment of genome segments was demonstrated as the mechanism of reassortment in rotaviruses (59). Mixed infections between *ts* mutants were characterized by either high frequencies of ts^+ reassortants in the yield ($>1.0\%$) or undetectably low frequencies of ts^+ reassortants ($<0.2\%$) (Table 1), a result termed all-or-none recombination (25, 75) and predicted for recombination by a mechanism of reassortment. The all-or-none nature of reassortment was subsequently demonstrated to result from the ability of *ts* mutations on noncognate segments to reassort ts^+ constellations of genome segments, whereas *ts* mutations on cognate segments were unable to generate constellations of genome segments with ts^+ phenotypes (29, 31). Thus, reassortment tests divide mutants into groups that reside on individual genome segments. Ordering of mutations within genome segments is not possible by recombination tests, as intramolecular recombination has not been documented in rotaviruses.

Mapping viral phenotypes by reassortment   Phenotypes associated with specific viral strains can be mapped to genome segments and their encoded proteins through the application of reassortment in a segregation analysis. For example, *ts* mutations have been mapped to genome segments by crossing two mutants belonging to different reassortment groups, selecting ts^+ reassortant progeny, and analyzing them for the cosegregation of ts^+ phenotype with specific genome segment constellations. Specifically, selection for the ts^+ phenotype demands that the segment(s) bearing the *ts* lesion(s) be excluded from ts^+ reassortants (29, 31; Figure 4). The locations of the *ts* mutants, obtained by reassortment and segregation analysis, are shown in Table 2, along with the RNA synthesis phenotypes of the mutant groups.

Reassortment has also been used to map determinants of phenotypes that cannot be easily selected in vitro. In this case, two wild-type parental viruses

Table 2   The map locations and phenotypes of rotavirus SA11 *ts* mutations

ts Mutant group	Prototype mutant[a]	Genome segment[b]	Mutant protein product[c]	Infected cell RNA synthesis phenotype[d]		Virion transcriptase phenotype[e]
				ssRNA	dsRNA	
A	*ts*A(778)	4	VP4	+	+	+
B	*ts*B(339)	3	VP3	−	−	+
C	*ts*C(606)	1	VP1	−	−	+
D	*ts*D(975)	?	?	+	+	+
E	*ts*E(1400)	8	NSP2	−	−	+
F	*ts*F(2124)	2	VP2	+	+	+
G	*ts*G(2130)	6	VP6	+	−	+
H	*ts*H(2384)	?	?	+	+	+
I	*ts*I(2403)	?	?	+	+	+
J	*ts*J(2131)	?	?	+	+	+

[a]Data from (75, 77).
[b]Data from (29, 31).
[c]Coding assignments from (23).
[d]Phenotype of mutant at nonpermissive temperature (15).
[e]Phenotype of mutant at nonpermissive temperature (Crook, Chen, & Ramig, unpublished observations).

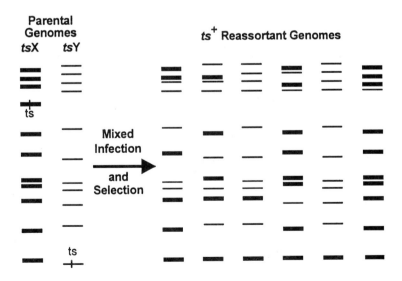

Figure 4   Reassortment mapping of rotavirus *ts* mutants to genome segments. Note that the parental segments bearing *ts* lesions are excluded from *ts*⁺ reassortant progeny.

Table 3 Protein coding assignments and functions/phenotypes associated with the rotavirus genome segments

Genome segment[a]	Encoded protein[a]	Function/phenotype	References
1	VP1	RNA Polymerase	(99)
2	VP2	Nonspecific RNA binding	(7)
3	VP3	Guanylyltransferase	(55)
4	VP4	Hemagglutination	(46)
		Neutralization Antigen	(67)
		Protease-enhanced infectivity	(24)
		Virulence	(67)
		Cell attachment	(18)
5	NSP1	Specific viral mRNA binding	(43)
6	VP6	Subgroup antigen	(37)
7	NSP3	3′-specific viral mRNA binding	(74)
8	NSP2	Nonspecific RNA binding	(51)
9	VP7	RER integral membrane glycoprotein	(21)
		Neutralization antigen	(67)
10	NSP4	RER integral membrane glycoprotein	(21)
		Virion morphogenesis	(1)
11	NSP5/6	RNA binding?	(60)

[a]Coding assignments shown for rotavirus SA11.

differing in the phenotype of interest are crossed. Random progeny are chosen and screened for reassortants. The reassortant progeny are then used in an appropriate assay to demonstrate the phenotype under test. The data are analyzed to determine the segment of the parent that expresses the phenotype of interest that is present in all reassortant progeny expressing the same phenotype (or excluded from all reassortant progeny that do not express the phenotype). This type of mapping is laborious because the progeny that have the phenotype of interest often cannot be specifically selected. An example of this type of mapping is found in the mapping of virulence determinants to genome segment 4 (66). A summary of phenotypes mapped to the various rotavirus genes and gene products is shown in Table 3.

Reassortment kinetics   A kinetic analysis of reassortment between two rotavirus *ts* mutants revealed that reassortment frequency was maximal at the earliest times after infection that reassortment could be detected (76). Maximal reassortment at early times indicates that each progeny genome undergoes only a single round of reassortment, an observation consistent with removal of reassortant genomes from the mating pool by morphogenesis (Figure 1). This does not imply that reassortment occurs only early in the infectious cycle, but rather that the number of reassortant progeny increases in parallel with the yield late in infection. Demonstration of reassortment late in the infectious cycle came from the analysis of asynchronous mixed infections. A second, superinfecting virus, added up to 24 h after infection with the first virus, contributed genome segments to reassortants formed at a time when the replication cycle of the first virus was essentially complete (78).

Reassortment and multiplicity of infection   The frequency of reassortant progeny increases in parallel with multiplicity of infection (m.o.i.) until an m.o.i. of 2.5–5.0 pfu/cell of each parent is reached (76). This correlates with the fact that an m.o.i. in the range of 2.5–5.0 is required to infect the majority of the cells with at least one pfu of each parental virus. Because the m.o.i. is calculated from the infectious titer of the parental viruses, this result also suggests that the noninfectious virus particles, which are generally present in excess in viral stocks, do not contribute significantly to genetic interactions in the mixedly infected cell (76). Trypsin activation of viruses prior to mixed infection also increases the frequency of reassortment (75). This probably reflects the fact that trypsin treatment increases the effective m.o.i., because the titer was determined prior to trypsin activation of the parental viruses.

Reassortment at nonpermissive condition   Temperature-sensitive mutants do not grow significantly at NPC. However, those mutants may be able to undergo abortive replication or low levels of complete replication at NPC. When mixed infections were performed at NPC with SA11 *ts* mutants representing all 10 reassortment groups (defined at PC), high frequencies of *ts*$^+$ reassortants were observed for most mutant pairs (76). Yields of these infections were low, and reassortant frequencies were higher than observed at PC, indicating that the mutants remained restricted at the NPC and that the *ts*$^+$ reassortants had a significant growth advantage. However, no reassortment was noted with the *ts*G/*ts*J mutant pair, although each of these mutants reassorted with representatives of the other groups at NPC. This result suggested that reassortment was restricted in the *ts*G/*ts*J mixed infection at NPC, possibly because the product containing the *ts*G lesion (VP6) could not interact with the product containing the *ts*J lesion (not mapped) in a manner consistent with formation of viable reassortants at high temperature.

Reassortment and the host   The host has been shown to affect the constellations of genome segments present in reassortants isolated following mixed infection. Mixed infections between bovine and human rotavirus strains in RMK cells (secondary Rhesus kidney) yielded a high frequency of reassortants when reassortants were selected on either MA104 or BSC-1 cells (both African green monkey kidney cells; 35). Comparison of reassortants isolated on MA104 and BSC-1 revealed that the cell line used to isolate the reassortants influenced the result. Statistical analysis revealed significant differences in the isolation frequency of different reassortant constellations from the two cell lines and showed that different constellations were favored in one cell line over the other. This observation emphasizes the importance of the indicator cell line, especially when data from different experiments are to be compared.

Reassortment has also been demonstrated during experimental infection of mice (30, 82). In this case, the segregation of genome segments in a large population of reassortants derived from several mixedly infected animals was not random, and significant differences in constellations isolated were noted from animal to animal. Furthermore, the immune status of the animal was shown to affect reassortment (32). In animals immune to both virus strains, viral replication was effectively ablated as was reassortment. However, in the case of animals immune to a virus serologically distinct from either of the infecting viruses, reassortment occurred but was restricted about three-fold relative to nonimmune mice. These observations may be relevant because rotavirus infections are widespread and most animals have neutralizing antibodies by a relatively early age.

Reassortment and phenotypic mixing Phenotypic mixing has been shown to occur between viruses in infections carried out at the permissive condition, suggesting that various combinations of viral protein from heterologous parents can functionally interact to form infectious virions. In one study of mixed infection between antigenically distinguishable parental viruses, 40–75% of the progeny virus were neutralization mosaics, which indicates that the neutralization antigens of the parents had mixed in individual viral particles (93). The possibility of phenotypic mixing of viral proteins, other than those of the outer capsid, has not been examined. The presence of phenotypically mixed particles in yields of mixed infections suggests that progeny virus should be passaged once at low multiplicity of infection to rectify genotype and phenotype prior to analysis of phenotypes present in the progeny. Without such low multiplicity passage, the phenotype of an individual virus particle may not reflect its genetic potential and may lead to spurious results.

OTHER GENETIC STUDIES

Reassortment as a Tool to Identify Gene Function

Reassortment has been useful in studying the basic genetics of the rotaviruses, and is, in addition, the primary tool for identification of functions of the viral genes. The use of reassortment to map gene function is conceptually similar to the mapping of *ts* mutations, but there are significant differences in approach. Reassortants are most easily selected if they result from the cross of two parents

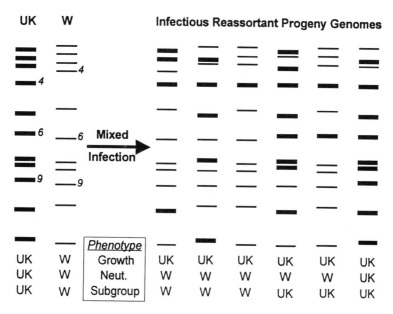

Figure 5 Simultaneous mapping of three rotavirus phenotypes, using reassortants derived from a single cross. Note that growth in culture segregates with UK segment 4 (ratio 6UK:0W), neutralization segregates with segment 9 (ratio 1UK:5W), and subgroup specificity segregates with segment 6 (ratio 3UK:3W). Relevant segments are numbered in parental lanes. [Based on data from Greenberg et al (37).]

with selectable phenotypes. However, because most conditional mutations were generated following mutagenesis, the possibility exists that nonconditional mutations also exist in the mutant. Because these nonconditional mutations have the potential to affect the expression of phenotypes of interest, crosses to map functions generally involve two wild-type parental viruses that differ in the phenotype of interest. Once reassortants are identified and characterized for the constellation of parental genes, the expression of the phenotype of interest is scored for each reassortant. Cosegregation of a parental genome segment with a phenotype demonstrates that the phenotype is encoded in the segment with which it segregates (Figure 5). In some cases, expression of a phenotype segregates with multiple genome segments, or other factors arise that confound the analysis (see *Limitations of Reassortment Analysis*, below).

An early and elegant use of reassortment to map the genome segment encoding a viral function was the use of a single cross to map viral determinants of growth restriction in cultured cells, neutralization specificity, and subgroup specificity (37). Bovine rotavirus strain UK (grows in cultured cells, neutralization by UK antiserum, subgroup 1) was crossed with human rotavirus strain W (does not grow in cultured cells, neutralization by W antiserum, subgroup 2). Progeny were picked and reassortants were identified by polyacrylamide gel electrophoresis. Each of the progeny reassortants was then tested by neutralization and enzyme-linked immunosorbent assay (ELISA) for subgroup determination. The fact that all reassortants grew in the cell line used for selection of plaques indicated that they all had obtained the UK gene that allowed growth in cultured cells. This type of segregation analysis is shown in Figure 5.

Numerous other rotavirus phenotypes have been mapped by reassortment and segregation analysis, including: (a) determinants of two neutralization specificities (41), (b) determinants of virulence and pathogenesis (9, 39, 66), (c) determinants of passive protection from challenge (67), (d) determinants of active protection from challenge (40), (e) VP3 as the product encoded within genome segment 3 (56), (f) VP4 as the viral hemagglutinin (46), (g) the determinant of neutralization by a monoclonal antibody with unusual properties (100), and (h) determinants of viral growth phenotype in cultured human liver cells (81). The wide array of phenotypes mapped by reassortment illustrates the utility of the method.

Reassortment and Vaccination

Reassortment has been used to generate vaccine candidate strains for use in a modified "Jennerian" approach to protection of children from rotavirus disease (49). In this approach, relevant neutralizing antigen-encoding genome segments from a human rotavirus are moved by reassortment onto a recipient genetic background that consists of an animal virus of known avirulence and immunogenicity in humans. In the reassortant, the antigen of a virulent virus is presented to the host in the context of the remainder of the genes derived from an avirulent virus, leading to the development of a protective immune response without a disease response to the vaccine virus. Human neutralization antigen-encoding genes have been moved by reassortment onto recipient backgrounds of bovine strains UK or WC3 and rhesus strain RRV for the production of vaccine candidate strains.

Genetic Distance and Reassortment

Natural isolates of rotaviruses have tended to maintain certain combinations of phenotypes. For example, human rotaviruses of subgroup 2 have tended to be serotype 1 and have long electropherotypes (SG2/ST1/long), whereas human viruses of subgroup 1 have tended to be serotype 2 and have short electropherotypes (SG1/ST2/short). Viruses that could be reassortants between these two phenotypic combinations have been isolated relatively rarely (65), which suggests that the viruses are relatively genetically isolated and do not undergo frequent reassortment in nature. When human SG1/ST2/short viruses were crossed with human SG2/ST1/long viruses in cultured cells, reassortant progeny were easily isolated (94), although reassortant frequency was significantly lower in this case than when two SG2/ST1/long viruses were crossed (95). This suggested that reassortment between SG1/ST2/short and SG2/ST1/long viruses is somewhat restricted in cultured cells. In both the intracellular RNA pool and in individual reassortants, segments from the SG2/ST1/long virus predominated, and when the progeny of the cross were passaged as a population, viruses containing mostly segments from the SG2/ST1/long parent came to dominate. These data were interpreted as indicating that SG1/ST2/short and SG2/ST1/long viruses evolve in relative genetic isolation, and the dominance of one over the other in mixed infection reflected evolution of sets of gene products that functioned optimally as homologous sets. Subsequently, the relative genetic distance between viruses that maintain the phenotype constellations has been demonstrated through nucleic acid hybridization (64), and the distantly related groups defined by hybridization are called genogroups.

Mammalian and avian group A rotaviruses are also distantly related by hybridization (62), and the isolation of apparent mammalian/avian reassortants has not been reported—although apparent avian viruses have been isolated from mammals (10). Attempts to generate reassortants between avian and mammalian viruses in the laboratory were unsuccessful (HB Greenberg, personal communication), although the identification of an avian/mammalian reassortant as a laboratory contaminant (53) suggests that reassortment is not impossible between these related, but genetically distant, viruses.

No evidence of reassortment between rotaviruses that represent different serological groups (between which there is no serological cross reaction) has been presented. A single report that failed to detect reassortment between group A and group B rotaviruses has been published (98). Investigations of reassortment between rotavirus and reovirus (different genera within the family Reoviridae) proved negative, even when both parental viruses contained *ts* mutations so that extremely rare reassortants could be selected (42; RF Ramig, unpublished data).

Limitations of Reassortment Analysis

Reassortment and segregation analyses have been used to map viral phenotypes to genome segments and their encoded proteins, and they have generally resulted in simple mapping of the phenotype to a single segment. However, a phenotype has been mapped to a single segment in one study and to multiple segments in another study. For example, virulence in mice was mapped to segment 4 in a study using reassortants derived from simian strain SA11 and bovine strain NCDV (66), but virulence in pigs was mapped to multiple genome segments using reassortants derived from porcine strain SB-1A and human strain DS-1 (39). Despite obvious variables in virus strains and hosts between the two experiments, these results called into question the general applicability of results obtained from reassortment analysis.

The following data suggest that the two viruses chosen as parents for the generation of reassortants can affect the mapping of functions by reassortment. A variant of the simian virus SA11 (SA11-4F) contained an unusual form of segment 4 and the encoded VP4. SA11-4F had the phenotype of making unusually large plaques, growing to unusually high titer, and synthesizing VP4 that was unusually susceptible to trypsin cleavage (11). These phenotypes had previously been mapped to segment 4 using different parental virus strains (46). To confirm that segment 4 was indeed responsible for the unusual phenotypes of the variant, SA11-4F was used to generate reassortants with the bovine strain B223, which makes small plaques, grows to low titer, and has a VP4 that is relatively resistant to trypsin (13). Analysis of the phenotypes of reassortants revealed that the formation of large plaques, the growth to high titer, and the trypsin sensitivity of VP4, which were expected to segregate with segment 4 and VP4 of SA11-4F, actually segregated with an unexpected pattern in reassortants (Figure 6, *top*). Specifically, when the outer capsid proteins VP4 and VP7 (encoded in segment 9) were of homologous parental origin, the expected phenotypic expression was seen in reassortants; but when VP4 was from SA11-4F and VP7 was from B223, unexpected phenotypes were observed. To further test the phenotypic expression of SA11-4F segment 4, that segment was reassorted onto a different genetic background by reassortment between reassortant 144 and SA11-Cl3. Reassortant 144 was used as the segment 4 donor in this cross because SA11-4F and SA11-Cl3 share electrophoretic mobility of all segments except 4. In several independently isolated reassortants containing SA11-4F segment 4 on a genetic background derived from SA11-Cl3, the phenotypes associated with SA11-4F segment 4 segregated with that segment alone (Figure 6, *bottom*). Clearly, in this case, transfer of the large plaque, growth to high titer, and trypsin sensitivity of VP4 phenotypes of SA11-4F to a B223 recipient genetic background required transfer of two segments (segments 4 and 9), whereas transfer of the same phenotypes to an SA11-Cl3 recipient genetic background required transfer of only one segment (segment 4).

The basis for unexpected segregation of phenotypes, depending on recipient genetic background, was not understood, but interactions between VP4 and VP7 in the structure of the virion were postulated to be involved (13). Further evidence in support of this model came from examination of neutralization and ELISA reactivity of the same group of reassortants with a VP4-specific neutralizing monoclonal antibody, 2G4. Reassortants gave the expected pattern of neutralization and ELISA reactivity when segments 4 and 9 (VP4 and VP7)

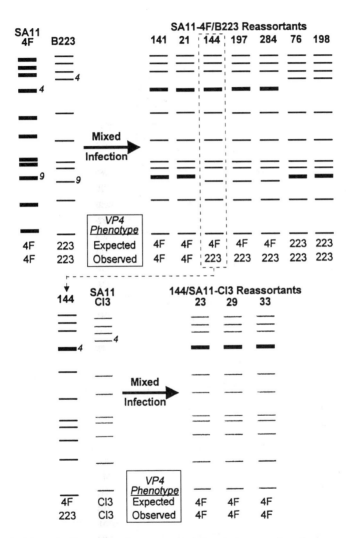

Figure 6 Schematic diagram showing nonexpected phenotypic expression of genome segment 4 of SA11–4F, when reassorted onto a B223 recipient genetic background (*top*); and expected phenotypic expression of genome segment 4 of SA11–4F, when reassorted from reassortant 144 onto an SA11-Cl3 recipient genetic background (*bottom*). Relevant segments are numbered in parental lanes. [Figure based on data from Chen et al (13).]

were of homologous parental origin, but gave unexpected neutralization and ELISA reactivity if segments 4 and 9 were of heterologous parental origin. However, if ELISA was performed using soluble proteins derived from infected cell lysates of the reassortants, antibody 2G4 ELISA reactivity segregated with segment 4 alone as expected (14). The expected reactivity of unassembled, soluble VP4, together with the unexpected reactivity of virion VP4 when VP4 and VP7 were of heterologous parental origin, indicated that the unexpected results were the result of interaction of VP4 with a heterologous VP7 in the virion. Thus, within the constraints of viral structure imposed by the requirement of infectivity, changes in protein conformation induced by protein:protein interactions can affect phenotypic expression.

Similar unexpected results from reassortment experiments have been documented in other publications (54, 57). However, in neither of these cases was the segment in question reassorted to a second recipient genetic background to confirm that the unexpected result was recipient background specific.

One must conclude that, whereas reassortment is a very useful technique for analysis of rotaviruses, the results must be interpreted with caution and that for results to be considered generally applicable, they should be repeated using different parental virus strains for the reassortment analysis.

EVOLUTION OF ROTAVIRUSES

Rotaviruses are characterized by a high degree of diversity. Natural isolates of virus show wide diversity of electropherotype and serology, so that viruses with identical electropherotype and serotype are not isolated exclusively, even within outbreaks or from individuals within those outbreaks (85). Three primary sources for this diversity have been proposed.

Mutation is generally considered the primal source of genetic diversity. As RNA genome viruses, rotaviruses have been assumed to have high mutation rates because RNA replication is an error-prone process. The rate of mutation in rotavirus was recently examined using a porcine virus with a rearrangement of segment 11 that had a conserved ORF (Figure 3A). In this case, 328 base pairs of the ORF were duplicated as part of the extended 3′-nontranslated region and were, therefore, presumably not subject to selection. Sequence analysis of the nontranslated, duplicated region of segment 11 from the parental virus, and a virus separated from the parent by 8 plaque-to-plaque purifications (31 rounds of replication assumed), revealed a mutation rate of $<5 \times 10^{-5}$ mutations per nucleotide per replication (4). This rate of mutation suggests that the average rotavirus genome differs from its parental genome by at least one mutation, and it is consistent with the expected high mutation rate.

Reassortment clearly contributes to the diversity of rotaviruses, because mixed infections in vitro and in vivo are characterized by high frequencies of reassortment (33). Contribution of reassortment to natural diversity is suggested by the identification of apparently mixedly infected individuals (87) and the isolation of apparently reassortant viruses (41). In addition to a simple shuffling of genome segments, reassortment may alter the phenotypic expression of genes in reassortants (14). Reassortment-mediated alterations in protein:protein interactions were shown to affect serology in reassortants (14, 54), suggesting that reassortment, as well as mutation, can contribute to evolution of new antigenic types of the virus. Clearly, any mutation or rearrangement conferring selective advantage on the virus containing it theoretically could be rapidly moved throughout the virus population by reassortment.

Finally, rearrangement of rotavirus genome segments has the potential to contribute to evolution of new rotavirus genes or functions. Changes in protein structure that are mediated by rearrangement could introduce new and potentially useful domains within rearranged proteins, and these domains could gain function through the accumulation of mutations. Indeed, rearranged proteins with altered function have been described (90). In addition, duplicated sequences that are not part of functional ORFs have the potential to evolve new functional products. The recent demonstration of high mutation rates in these extended, nontranslated regions (4) suggests that these regions could mutate to function.

Thus, it seems likely that (*a*) mutation, (*b*) reassortment, and (*c*) rearrangement are the three major forces that contribute to rotavirus diversity and drive the evolution of the virus (19).

FUTURE PROSPECTS

Genetic analysis is likely to continue to play an important role in our quest to understand the molecular mechanisms that characterize the rotavirus life cycle and the pathogenesis of viral infection. Currently, genetic studies of the Reoviridae, including the rotaviruses, are hampered by the inability to rescue RNA copies of viral cDNAs into infectious virions (reverse genetics), so that specific, site-directed mutations cannot be studied in the context of the infected cell or host. The ability to perform reverse genetic studies with rotaviruses will play a major role in the elucidation of molecular mechanisms involved in genetic, biochemical, and biological phenomena that are identified by classical techniques.

Potential for Reverse Genetics

The recent development of reverse genetic techniques for a number of RNA genome viruses has allowed rapid advances in understanding many facets of their replication, genetics, and biology (6). The paradigmatic models for rotaviruses are the reverse genetic systems devised for the segmented genome;

(−)-sense influenza virus (70); the dsRNA, three-segmented genome phage $\phi6$ (68); and the dsRNA, two-segmented genome birnaviruses (63). Although no reverse genetic system currently exists for any member of the Reoviridae, including rotaviruses, recent progress indicates that we may be on the threshold of such a system.

A template-dependent, in vitro replication system has been developed that synthesizes complete rotavirus dsRNA on (+)-strand template RNAs (17). This system utilizes a replicase enzyme that is particulate and resembles the core particle of the virion in terms of protein content and structure (17, 99). If the system can be manipulated so that the newly synthesized dsRNA is packaged into the particulate replicase, the resultant particles can be transcapsidated to yield transcriptionally active particles that are infectious (16), in a manner that is analogous to the formation of transcriptionally active nucleocapsids in the influenza system. The potential application of the in vitro replication system to the rescue problem is being actively pursued.

An alternative approach to reverse genetics, the expression of rotavirus RNA from plasmids in cells infected with helper virus, has been reported with promising preliminary results (34). However, no additional applications of this approach have been reported.

Regardless of which approach ultimately yields a reverse genetic system, such a system will greatly enhance the numbers and types of genetic and biochemical studies that can be performed with rotaviruses. Many genetically engineered mutations in rotavirus proteins have been generated and studied in expression systems. The application of reverse genetics will allow for the study of these and other mutations in the context of the infected cell and/or host and for the examination of the molecular mechanisms involved in viral replication and pathogenesis.

CIS-ACTING ELEMENTS ON mRNA TEMPLATES   Examination of template mRNAs containing deletions, truncations, and site-directed mutations in the replication system has allowed for the localization of cis-acting elements that are required for replication (72, 97). These studies, examining segment 9 (VP7) of OSU and segment 8 (NSP2) of SA11, identified a tripartite cis-acting replication signal on the respective templates. The size and locations of these signals were virtually identical on the two templates, suggesting that similar size and location signals can be expected on all 11 rotavirus mRNAs (Figure 7). The minimal promoter of (−)-strand synthesis is necessary and sufficient to confer replication competence on a foreign RNA, if it is present at the 3'-terminus of that template. Immediately upstream of the minimal promoter lies the 3'-enhancing sequence of some 25–30 nucleotides that significantly increases the activity of the minimal promoter. At the 5'-terminus of the template in the

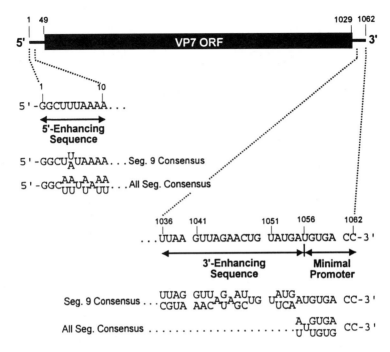

Figure 7 Location of the components of the tripartite *cis*-acting replication signals mapped on genome segment 9 template RNA from virus strain OSU (97). For comparison, the consensus sequence for segments 9 and for all segments are shown. Virtually the same locations of sequences were identified as the tripartitie *cis*-acting replication signal on the genome segment 8 template from virus strain SA11 (72).

region of nucleotides 1–10, is the 5'-enhancing sequence that significantly increases template activity of templates with both the complete 3'-enhancing sequence and the minimal promoter. Interestingly, the minimal promoter and the 5'-enhancing sequence are identical to the regions at the termini of the template that are conserved on all rotavirus mRNAs and genome segments. The 3'-enhancing sequence contains the only segment-specific sequence within the *cis*-acting replication signal. Deciphering the function of each of the *cis*-acting signals is an active area of research that may provide information relevant to the development of a reverse genetic system.

Assortment Signals in Rotaviruses

One of the long-standing problems in rotaviruses, and other viruses with segmented genomes, is the mechanism by which segments are chosen from among the pool of intracellular RNAs for packaging and replication, so that a virus contains at least one copy of each genome segment. Clearly, the process is not random, because rotaviruses with particle to plaque-forming unit ratios of less than 5 have been documented (11). The identification of *cis*-acting replication signals on template RNAs suggests that an answer to this question may be imminent. In the tripartite *cis*-acting signal identified on two of the genome segments, only one portion of the signal, the 3′-enhancer sequence, lies in segment-specific sequences (72, 97). This signal may provide the specificity necessary for assortment of RNAs during genome replication and viral morphogenesis. Although the function of the 3′-enhancer sequence as the assortment signal remains to be demonstrated, the segment-specific nature of this sequence provides a model that can be tested.

Does Intramolecular Recombination Occur in Rotaviruses?

Genetic evidence does not directly support the notion of intramolecular recombination between markers in cognate genome segments. However, the description of the structure of rearranged genome segments suggests that the polymerase is capable of template release and reassociation with continued synthesis, as documented in other RNA viruses that recombine by a mechanism of copy-choice (52). Indeed, mechanisms for polymerase jumping during (+)-strand or (−)-strand RNA synthesis have been proposed for rotavirus (19). Application of modern methods of molecular biology should allow determination of whether copy-choice recombination occurs in rotaviruses and if it occurs during (+)-strand or (−)-strand RNA synthesis. This question merits examination, because intramolecular recombination, even if rare, could contribute to the development of new viral phenotypes. Of particular interest for vaccine development strategies is the possibility that copy-choice recombination could generate neutralization antigens with new, previously unknown combinations of neutralizing epitopes that represent new serological types.

CONCLUSIONS

Clearly, genetic studies have played a major role in the development of our understanding of the structure, replication, and pathogenesis of the rotaviruses during the two and a half decades since their initial description. Reassortment has been the most useful genetic tool; thus the emphasis on reassortment in this review. Reassortment is likely to remain an important tool in the future. However, development of a reverse genetic system will enhance progress, because it will allow for the examination of specific mutations in the context of viral infection of tissue culture cells or the vertebrate host. Although much has been accomplished, more remains to be done.

ACKNOWLEDGMENTS

I thank my colleagues in the dsRNA virus field for helpful discussions over the years and for sharing unpublished results. I am indebted to Keith Barnhart and Jeff Lawton for their critical review of the manuscript during its preparation. The studies by RF Ramig, cited here, were supported by grants from the National Institutes of Health, the National Science Foundation, and the World Health Organization.

> Visit the *Annual Reviews home page* at
> http://www.annurev.org.

Literature Cited

1. Au KS, Chan WK, Burns JW, Estes MK. 1989. Receptor activity of rotavirus non-structural glycoprotein NS28. *J. Virol.* 63:4553–62
2. Ballard A, McCrae MA, Desselberger U. 1992. Nucleotide sequences of normal and rearranged RNA segments 10 of human rotaviruses. *J. Gen. Virol.* 73:633–38
3. Bern C, Glass RI. 1994. Impact of diarrheal diseases worldwide. In *Viral Infections of the Gastrointestinal Tract*, ed. AZ Kapikian, pp. 1–26. New York: Marcel Dekker. 2nd ed.
4. Blackhall J, Fuentes A, Magnusson G. 1996. Genetic stability of a porcine rotavirus RNA segment during repeated plaque isolation. *Virology* 225:181–90
5. Both GW, Bellamy AR, Mitchell DB. 1994. Rotavirus protein structure and function. In *Rotaviruses*, ed. RF Ramig, pp. 67–105. Berlin: Springer-Verlag
6. Boyer JC, Haenni A-L. 1994. Infectious transcripts and cDNA clones of RNA viruses. *Virology* 198:415–26
7. Boyle JF, Holmes KV. 1986. RNA-binding proteins of bovine rotavirus. *J. Virol.* 58:561–68
8. Bridger J. 1994. Nongroup A rotaviruses. In *Viral Infections of the Gastrointestinal Tract*, ed. AZ Kapikian, pp. 369–407. New York: Marcel Dekker. 2nd ed.
9. Broome RL, Vo PT, Ward RL, Clark HF, Greenberg HB. 1993. Murine rotavirus genes encoding outer capsid proteins VP4 and VP7 are not major determinants of host range restriction and virulence. *J. Virol.* 67:2448–55
10. Brüssow H, Nakagomi O, Minamoto N, Eichhorn W. 1992. Rotavirus 993/83, isolated from calf faeces, closely resembles an avian rotavirus. *J. Gen. Virol.* 73:1873–75

11. Burns JW, Chen D, Estes MK, Ramig RF. 1989. Biological and immunological characterization of a simian rotavirus SA11 variant with an altered genome segment 4. *Virology* 169:427–35
12. Chakraborty PR, Ahmed R, Fields BN. 1979. Genetics of reovirus: the relationship of interference to complementation and reassortment of temperature-sensitive mutants at nonpermissive temperature. *Virology* 94:119–27
13. Chen D, Burns JW, Estes MK, Ramig RF. 1989. Phenotypes of rotavirus reassortants depend upon the recipient genetic background. *Proc. Natl. Acad. Sci. USA* 86:3743–47
14. Chen D, Estes MK, Ramig RF. 1992. Specific interactions between rotavirus outer capsid proteins VP4 and VP7 determine expression of a cross-reactive, neutralizing VP4-specific epitope. *J. Virol.* 66:432–39
15. Chen D, Gombold JL, Ramig RF. 1990. Intracellular RNA synthesis directed by temperature-sensitive mutants of simian rotavirus SA11. *Virology* 178:143–51
16. Chen D, Ramig RF. 1993. Rescue of infectivity by sequential *in vitro* transcapsidation of rotavirus core particles with inner capsid and outer capsid proteins. *Virology* 194:743–51
17. Chen D, Zeng CQ-Y, Wentz MJ, Gorziglia M, Estes MK, Ramig RF. 1994. Template-dependent, in vitro replication of rotavirus RNA. *J. Virol.* 68:7030–39
18. Crawford SE, Labbé M, Cohen J, Burroughs MH, Zhou Y-J, Estes MK. 1994. Characterization of virus-like particles produced by the expression of rotavirus capsid proteins in insect cells. *J. Virol.* 68:5945–52

19. Desselberger U. 1996. Genome rearrangements of rotaviruses. *Adv. Virus Res.* 46:69–95

20. Dyall-Smith ML, Lazdins I, Tregear GW, Holmes IH. 1986. Location of the major antigenic sites involved in rotavirus serotype-specific neutralization. *Proc. Natl. Acad. Sci. USA* 83:3465–68

21. Ericson BL, Graham DY, Mason BB, Hanssen HH, Estes MK. 1983. Two types of glycoprotein precursors are produced by the simian rotavirus SA11. *Virology* 127:320–32

22. Estes MK. 1996. Advances in molecular biology: Impact on rotavirus vaccine development. *J. Infect. Dis.* 174:S37–46

23. Estes MK. 1996. Rotaviruses and their replication. In *Fields Virology*, ed. BN Fields, DM Knipe, PM Howley, pp. 1625–56. Philadelphia: Lippincott-Raven. 3rd ed.

24. Estes MK, Graham DY, Mason BB. 1981. Proteolytic enhancement of rotavirus infectivity: molecular mechanisms. *J. Virol.* 39:879–88

25. Faulkner-Valle GP, Clayton AV, McCrae MA. 1982. Molecular biology of rotaviruses. III. Isolation and characterization of temperature-sensitive mutants of bovine rotavirus. *J. Virol.* 42:669–77

26. Fields BN. 1971. Temperature-sensitive mutants of reovirus—features of genetic recombination. *Virology* 46:142–48

27. Flores J, Greenberg HB, Myslinski J, Kalica AR, Wyatt RG, et al. 1982. Use of transcription probes for genotyping rotavirus reassortants. *Virology* 121:288–95

28. Franco MA, Feng NG, Greenberg HB. 1996. Molecular determinants of immunity and pathogenicity of rotavirus infection in the mouse model. *J. Infect. Dis.* 174:S47–50

29. Gombold JL, Estes MK, Ramig RF. 1985. Assignment of simian rotavirus SA11 temperature-sensitive mutant groups B and E to genome segments. *Virology* 143:309–20

30. Gombold JL, Ramig RF. 1986. Analysis of reassortment of genome segments in mice mixedly infected with rotaviruses SA11 and RRV. *J. Virol.* 57:110–16

31. Gombold JL, Ramig RF. 1987. Assignment of simian rotavirus SA11 temperature-sensitive mutant groups A, C, F, and G to genome segments. *Virology* 161:463–73

32. Gombold JL, Ramig RF. 1989. Passive immunity modulates genetic reassortment between rotaviruses in mixedly infected mice. *J. Virol.* 63:4525–32

33. Gombold JL, Ramig RF. 1994. Genetics of the rotaviruses. In *Rotaviruses*, ed. RF Ramig, pp. 129–78. Berlin: Springer-Verlag

34. Gorziglia MI, Collins PL. 1992. Intracellular amplification and expression of a synthetic analog of rotavirus genomic RNA bearing a foreign marker gene: Mapping *cis*-acting nucleotides in the 3'-noncoding region. *Proc. Natl. Acad. Sci. USA* 89:5784–88

35. Graham A, Kudesia G, Allen AM, Desselberger U. 1987. Reassortment of human rotavirus possessing genome rearrangements with bovine rotavirus: evidence for host cell selection. *J. Gen. Virol.* 68:115–22

36. Greenberg HB, Kalica AR, Wyatt RG, Jones RW, Kapikian AZ, Chanock RM. 1981. Rescue of noncultivatable human rotavirus by gene reassortment during mixed infection with ts mutants of a cultivatable bovine rotavirus. *Proc. Natl. Acad. Sci. USA* 78:420–24

37. Greenberg HB, Flores J, Kalica AR, Wyatt RG, Jones R. 1983. Gene coding assignments for growth restriction, neutralization and subgroup specificities of the W and DS-1 strains of human rotavirus. *J. Gen. Virol.* 64:313–20

38. Hoshino Y, Kapikian AZ, Chanock RM. 1994. Selection of cold-adapted mutants of human rotaviruses that exhibit various degrees of growth restriction in vitro. *J. Virol.* 68:7598–602

39. Hoshino Y, Saif LJ, Kang S-Y, Sereno MM, Chen W-K, Kapikian AZ. 1995. Identification of group A rotavirus genes associated with virulence of a porcine rotavirus and host range restriction of a human rotavirus in the gnotobiotic piglet model. *Virology* 209:274–80

40. Hoshino Y, Saif LJ, Sereno MM, Chanock RM, Kapikian AZ. 1988. Infection immunity of piglets to either VP3 or VP7 outer capsid protein confers resistance to challenge with a virulent rotavirus bearing the corresponding antigen. *J. Virol.* 62:744–48

41. Hoshino Y, Sereno MM, Midthun K, Flores J, Kapikian AZ, Chanock RM. 1985. Independent segregation of two antigenic specificities (VP3 and VP7) involved in neutralization of rotavirus infectivity. *Proc. Natl. Acad. Sci. USA* 82:8701–4

42. Hrdy DB. 1982. Investigation of genetic interaction between simian rotavirus SA11 and reovirus. *Microbiologica* 5:207–13

43. Hua J, Chen X, Patton JT. 1994. Deletion mapping of the rotavirus metalloprotein NS53 (NSP1): The conserved cysteine-rich region is essential for virus-specific RNA binding. *J. Virol.* 68:3990–4000

44. Hua J, Patton JT. 1994. The carboxyl-half of the rotavirus nonstructural protein NS53

(NSP1) is not required for virus replication. *Virology* 198:567–76

45. Hundley F, Biryahwaho B, Gow M, Desselberger U. 1985. Genome rearrangements of bovine rotavirus after serial passage at high multiplicity of infection. *Virology* 143:88–103

46. Kalica AR, Flores J, Greenberg HB. 1983. Identification of the rotaviral gene that codes for hemagglutination and protease-enhanced plaque formation. *Virology* 125:194–205

47. Kalica AR, Sereno MM, Wyatt RG, Mebus CA, Chanock RM, Kapikian AZ. 1978. Comparison of human and animal rotavirus strains by gel electrophoresis of viral RNA. *Virology* 87:247–55

48. Kapikian AZ. 1994. *Viral Infections of the Gastrointestinal Tract.* New York: Marcel-Dekker. 785 pp.

49. Kapikian AZ. 1994. Jennerian and modified jennerian approach to vaccination against rotavirus diarrhea in infants and young children: an introduction. In *Viral Infections of the Gastrointestinal Tract,* ed. AZ Kapikian, pp. 409–17. New York: Marcel-Dekker

50. Kapikian AZ, Chanock R. 1996. Rotaviruses. In *Fields Virology,* ed. BN Fields, DM Knipe, PM Howley, pp. 1657–709. Philadelphia: Lippincott-Raven. 3rd ed.

51. Kattoura MD, Clapp LL, Patton JT. 1992. The rotavirus nonstructural protein, NS35, possesses RNA-binding activity *in vitro* and *in vivo. Virology* 191:698–708

52. Kirkegaard K, Baltimore D. 1986. The mechanism of RNA recombination in poliovirus. *Cell* 47:433–41

53. Kool DA, Matsui SM, Greenberg HB, Holmes IH. 1992. Isolation and characterization of a novel reassortant between avian Ty-1 and simian RRV rotaviruses. *J. Virol.* 66:6836–39

54. Lazdins I, Coulson BS, Kirkwood C, Dyall-Smith M, Masendycz PJ, Sonza S, Holmes IH. 1995. Rotavirus antigenicity is affected by the genetic context and glycosylation of VP7. *Virology* 209:80–89

55. Liu M, Mattion NM, Estes MK. 1992. Rotavirus VP3 expressed in insect cells possesses guanylyltransferase activity. *Virology* 188:77–84

56. Liu M, Offit PA, Estes MK. 1988. Identification of the simian rotavirus SA11 genome segment 3 product. *Virology* 163:26–32

57. Ludert JE, Feng NG, Yu JH, Broome RL, Hoshino Y, Greenberg HB. 1996. Genetic mapping indicates that VP4 is the rotavirus cell attachment protein in vitro and in vivo. *J. Virol.* 70:487–93

58. Matsui SM, Mackow ER, Matsuno S, Paul PS, Greenberg HB. 1990. Sequence analysis of gene 11 equivalents from "short" and "super short" strains of rotavirus. *J. Virol.* 64:120–24

59. Matsuno S, Hasegawa A, Kalica AR, Kono R. 1980. Isolation of a recombinant between simian and bovine rotaviruses. *J. Gen. Virol.* 48:253–56

60. Mattion NM, Mitchell DB, Both GW, Estes MK. 1991. Expression of rotavirus proteins encoded by alternative open reading frames of genome segment 11. *Virology* 181:295–304

61. McIntyre M, Rosenbaum V, Rappold W, Desselberger M, Wood D, Desselberger U. 1987. Biophysical characterization of rotavirus particles containing rearranged genomes. *J. Gen. Virol.* 68:2961–66

62. Minamoto N, Nakagomi O, Sugiyama M, Kinjo T. 1991. Characterization of an avian rotavirus strain by neutralization and molecular hybridization assays. *Res. Virol.* 142:271–75

63. Mundt E, Vakharia VN. 1996. Synthetic transcripts of double-stranded Birnavirus genome are infectious. *Proc. Natl. Acad. Sci. USA* 93:11131–36

64. Nakagomi O, Nakagomi T, Akatani K, Ikegami N. 1989. Identification of rotavirus genogroups by RNA-RNA hybridization. *Mol. Cell Probes* 3:251–61

65. Nakagomi O, Nakagomi T, Hoshino Y, Flores J, Kapikian AZ. 1987. Genetic analysis of a human rotavirus that belongs to subgroup I but has an RNA pattern typical of subgroup II human rotaviruses. *J. Clin. Microbiol.* 25:1159–64

66. Offit PA, Blavat G, Greenberg HB, Clark HF. 1986b. Molecular basis of rotavirus virulence: role of gene segment 4. *J. Virol.* 57:46–49

67. Offit PA, Clark HF, Blavat G, Greenberg HB. 1986a. Reassortant rotaviruses containing structural proteins vp3 and vp7 from different parents induce antibodies protective against each parental serotype. *J. Virol.* 60:491–96

68. Olkkonen VM, Gottlieb P, Strassman J, Qiao XY, Bamford DH, Mindich L. 1990. In vitro assembly of infectious nucleocapsids of bacteriophage phi 6: formation of a recombinant double-stranded RNA virus. *Proc. Natl. Acad. Sci. USA* 87:9173–77

69. Palese P. 1977. The genes of influenza virus. *Cell* 10:1–10

70. Palese P. 1995. Genetic engineering of infectious negative-strand RNA viruses. *Trends in Microbiology* 3:123–25

71. Patton JT. 1995. Structure and function of the rotavirus RNA-binding proteins. *J.*

Gen. Virol. 76:2633–44

72. Patton JT, Wentz M, Jiang XB, Ramig RF. 1996. *cis*-Acting signals that promote genome replication in rotavirus mRNA. *J. Virol.* 70:3961–71

73. Pedley S, Hundley F, Chrystie I, McCrae MA, Desselberger U. 1984. The genomes of rotaviruses isolated from chronically infected immunodeficient children. *J. Gen. Virol.* 65:1141–50

74. Poncet D, Aponte C, Cohen J. 1993. Rotavirus protein NSP3 (NS34) is bound to the 3′ end consensus sequence of viral mRNAs in infected cells. *J. Virol.* 67:3159–65

75. Ramig RF. 1982. Isolation and genetic characterization of temperature-sensitive mutants of simian rotavirus SA11. *Virology* 120:93–105

76. Ramig RF. 1983a. Factors that affect genetic interaction during mixed infection with temperature-sensitive mutants of simian rotavirus SA11. *Virology* 127:91–99

77. Ramig RF. 1983b. Isolation and genetic characterization of temperature-sensitive mutants that define five additional recombination groups in simian rotavirus SA11. *Virology* 130:464–73

78. Ramig RF. 1990. Superinfecting rotaviruses are not excluded from genetic interactions during asynchronous mixed infections in vitro. *Virology* 176:308–10

79. Ramig RF, ed. 1994a. *Rotaviruses. Curr. Top. Microbiol. Immunol.* Vol. 185. Berlin: Springer-Verlag. 380 pp.

80. Ramig RF, Fields BN. 1983. Genetics of reoviruses. In *The Reoviridae,* ed. WK Joklik, pp. 197–228. New York: Plenum

81. Ramig RF, Galle KL. 1990. Rotavirus genome segment 4 determines viral replication phenotype in cultured liver cells (HepG2). *J. Virol.* 64:1044–49

82. Ramig RF, Gombold JL. 1991. Rotavirus temperature-sensitive mutants are genetically stable and participate in reassortment during mixed infection of mice. *Virology* 182:468–74

83. Ramig RF, Ward RL. 1991. Genomic segment reassortment in rotaviruses and other reoviridae. *Adv. Virus Res.* 39:164–207

84. Ramig RF, White RM, Fields BN. 1977. Suppression of the temperature-sensitive phenotype of a mutant of reovirus type 3. *Science* 195:406–7

85. Rodger SM, Bishop RF, Birch C, McLean B, Holmes IH. 1981. Molecular epidemiology of human rotaviruses in Melbourne, Australia, from 1973 to 1979, as determined by electrophoresis of genome ribonucleic acid. *J. Clin. Microbiol.* 13:272–78

86. Rodger SM, Holmes IH. 1979. Comparison of the genomes of simian, bovine, and human rotaviruses by gel electrophoresis and detection of genomic variation among bovine isolates. *J. Virol.* 30:839–46

87. Rodriguez WJ, Kim HW, Brandt CD, Gardner MK, Parrott RH. 1983. Use of electrophoresis of RNA from human rotavirus to establish the identity of strains involved in outbreaks in a tertiary care nursery. *J. Infect. Dis.* 148:34–40

88. Sedivy JM, Capone JP, RajBhandary UL, Sharp PA. 1987. An inducible mammalian amber suppressor: propagation of a poliovirus mutant. *Cell* 50:379–89

89. Shaw RD, Vo PT, Offit PA, Coulson BS, Greenberg HB. 1986. Antigenic mapping of the surface proteins of rhesus rotavirus. *Virology* 155:434–51

90. Shen S, Burke B, Desselberger U. 1994. Rearrangement of the VP6 gene of a group A rotavirus in combination with a point mutation affecting trimer stability. *J. Virol.* 68:1682–88

91. Tian Y, Tarlow O, Ballard A, Desselberger U, McCrae MA. 1993. Genomic concatemerization/deletion in rotaviruses: A new mechanism for generating rapid genetic change of potential epidemiological importance. *J. Virol.* 67:6625–32

92. Tobita K. 1971. Genetic recombination between influenza viruses Ao-NWS and A2-Hong Kong. *Arch. Ges. Virusforsch.* 34:119–30

93. Ward RL, Knowlton DR, Greenberg HB. 1988. Phenotypic mixing during coinfection of cells with two strains of human rotavirus. *J. Virol.* 62:4358–61

94. Ward RL, Knowlton DR. 1989. Genotypic selection following coinfection of cultured cells with subgroup 1 and subgroup 2 human rotaviruses. *J. Gen. Virol.* 70:1691–99

95. Ward RL, Knowlton DR, Hurst PF. 1988b. Reassortant formation and selection following coinfection of cultured cells with subgroup 2 human rotaviruses. *J. Gen. Virol.* 69:149–62

96. Ward RL, Nakagomi O, Knowlton DR, McNeal MM, Nakagomi T, Huda N, Clemens JD, Sack DA. 1991. Formation and selection of intergenogroup reassortants during cell culture adaptation of rotaviruses from dually infected subjects. *J. Virol.* 65:2699–2701

97. Wentz MJ, Patton JT, Ramig RF. 1996. The 3′-terminal consensus sequence of rotavirus mRNA is the minimal promoter of negative-strand RNA synthesis. *J. Virol.* 70:7833–41

98. Yolken R, Arango-Jaramillo S, Eiden J,

Vonderfecht S. 1988. Lack of genomic reassortment following infection of infant rats with group A and group B rotaviruses. *J. Infect. Dis.* 158:1120–23

99. Zeng CQ-Y, Wentz MJ, Cohen J, Estes MK, Ramig RF. 1996. Characterization and replicase activity of double-layered and single-layered rotavirus-like particles expressed from baculovirus recombinants. *J. Virol.* 70:2736–42

100. Zheng SL, Woode GN, Melendy DR, Ramig RF. 1989. Comparative studies of the antigenic polypeptide species VP4, VP6, and VP7 of three strains of bovine rotavirus. *J. Clin. Microbiol.* 27:1939–45

Annu. Rev. Microbiol. 1997. 51:257–83

INTRACELLULAR ANTIBODIES (INTRABODIES) FOR GENE THERAPY OF INFECTIOUS DISEASES

Isaac J. Rondon and Wayne A. Marasco

Division of Human Retrovirology, Dana-Farber Cancer Institute, Harvard Medical School, 44 Binney Street, Boston, Massachusetts 02115; email: isaac_rondon@dfci.harvard.edu; wayne_marasco@dfci.harvard.edu

KEY WORDS: single-chain antibodies, intrabodies, intracellular immunization, HIV-1, AIDS, gene therapy

---

ABSTRACT

Intracellular antibodies (intrabodies) represent a new class of neutralizing molecules with a potential use in gene therapy. Intrabodies are engineered single-chain antibodies in which the variable domain of the heavy chain is joined to the variable domain of the light chain through a peptide linker, preserving the affinity of the parent antibody. Intrabodies are expressed inside cells and directed to different subcellular compartments where they can exert their function more effectively. The effects of intrabodies have been investigated using structural, regulatory, and enzymatic proteins of the human immunodeficiency virus (HIV-1) as targets. These intrabodies have demonstrated their versatility by controlling early as well as late events of the viral life cycle. In this article, we review studies of the use of intrabodies as research tools and therapeutic agents against HIV-1.

CONTENTS

0066-4227/97/1001-0257$08.00

INTRODUCTION

Molecular genetic techniques are powerful tools for understanding many infectious diseases. Scientists have accumulated a great wealth of knowledge on the basic mechanisms that control gene expression and protein synthesis, as well as the molecular anatomy of many viruses and their possible mechanisms of pathogenesis. This knowledge has motivated scientists to develop alternative therapies with a genetic-based strategy using DNA itself as a therapeutic agent. The successful delivery and expression of nucleic acids in cells holds promise for the treatment of numerous infectious diseases.

Gene therapy is a new form of molecular medicine that has gained special interest among Acquired Immunodeficiency Syndrome (AIDS) researchers, since conventional therapies have shown limited success. Alteration of the host cell could potentially confer permanent suppression of viral replication, after infection, or could provide protection against viral infection. When considering Human immunodeficiency virus (HIV-1) infection, different innovative strategies have been investigated to control the progress of this virus to infectious disease. Among these strategies are immune reconstitution, nucleic acid–based therapeutic vaccines, and intracellular immunization. Immune reconstitution or adoptive cell therapy for HIV-1 infection involves ex vivo expansion of selected, and sometimes genetically modified, T cell populations (e.g. CD8 cytotoxic T lymphocytes (CTL) and CD4 lymphocytes), followed by their reinfusion into the HIV-1-infected patient (125). Nucleic acid–based therapeutic vaccines involve direct delivery of HIV-1 genes (e.g. env) to mimic viral infection; the expression of viral proteins encoded by these nucleic acids elicits both cellular and humoral responses (149). Lastly, intracellular immunization transfers a therapeutic gene into target cells to render them resistant to viral replication. The resistant cells will then limit the spread of the virus in the patient (3). Different modalities of intracellular immunization have been developed, broadly divided into two categories: RNA-based and protein-based suppressors. The RNA-based suppressors include antisense RNA, ribozymes (130), and RNA decoys (141). Antisense RNA and ribozymes hybridize to target viral transcripts to inactivate them. RNA decoys interact and sequester regulatory proteins (e.g.

Tat, Rev) that are essential for virus replication. The RNA-based suppressors are limited to the cytoplasmic compartment. The protein-based suppressors include transdominant mutant proteins (91), suicide molecules (14), and intracellular antibodies (96). Transdominant mutant proteins are altered viral proteins that compete with the native viral protein. Suicide molecules, unlike other approaches, do not protect the cells but rather destroy the infected cell. Intracellular antibodies, hereafter referred to as "intrabodies," are the focus of this review. Intrabodies are synthesized by the cell and directed to a particular cellular compartment to inactivate, in a highly specific manner, a target molecule.

Early studies showed that the cDNA of an antibody (heavy and light chains) against yeast alcohol dehydrogenase I (ADH I) can be expressed in the cytoplasm of *Saccharomyces cerevisiae* and could neutralize the enzyme in vivo (24). Further studies verified that the assembly of an antibody can take place in the reducing environment of a mammalian cytoplasm (7). Protein engineering of antibody binding sites demonstrated the feasibility of joining heavy-chain and light-chain variable domains through a synthetic linker, maintaining binding specificity and affinity of the parent molecule. This designed antibody is termed single-chain antibody (sFv) and it was first produced in *Escherichia coli* (10, 65). With these advances in the field and further improvements, intracellularly expressed sFvs were shown to inhibit functions in mammalian cells (96).

The constant development and refinement of tools used by immunologists and antibody engineers has made it possible to recreate the humoral immune system in vitro (64, 100, 152), thus allowing the isolation of high-affinity and highly specific antibodies against almost any target, including nonantigenic molecules (58). Likewise, basic cell biology studies have uncovered a body of information regarding intracellular protein trafficking signals (122). These signals can be used to direct intrabodies to subcellular compartments where neutralization of the target can be more effective. Combining the fast-growing field of gene transfer (137) with intrabody technology holds great promise in the future for genetically treating a number of infections and other diseases.

This review concentrates on the expression of intrabodies against HIV-1 proteins with relevance to future applications in gene therapy. It does not include intrabodies against the envelope of flaviviruses (68), oncoproteins (5, 8, 9, 57), growth factor receptors (123), nor antibodies expressed in plants (61, 71, 142).

DEFINING TERMS

An antibody or immunoglobulin of the type "G" (IgG) is a bivalent molecule composed of four chains, two heavy chains and two light chains. The heavy chains have four domains, one variable domain (V_H) and three constant domains (C_H1, C_H2, C_H3); the light chains have one variable domain (V_L) and one constant domain (C_L). The monovalent antigen binding fragment or "Fab" is composed of light chain (V_L–C_L) and Fd fragment (heavy chain V_H–C_H1), and the Fab construct is also used as an intrabody. The "Fv" portion of an antibody consists of the variable heavy chain (V_H) and the variable light-chain (V_L) domains. The Fv is the smallest fragment that maintains the binding specificity and affinity of the whole antibody (55). However, the nature of the noncovalent association between the V_H and V_L makes this fragment relatively unstable. The recombinant sFv (intrabody) represents a stabilized Fv fragment in which the V_H and V_L are joined by a flexible polypeptide linker (10, 65). An alternative method to stabilize the V_H–V_L domains uses an interchain disulfide bond; these constructs are termed dsFvs (16). Intrabodies of the form sFv have also been constructed fused to domains, such as the human constant kappa (hCk), or fused to peptide signals, such as the simian virus 40 (SV40) nuclear localization signal (73, 106). Figure 1 is a schematic representation of these molecules.

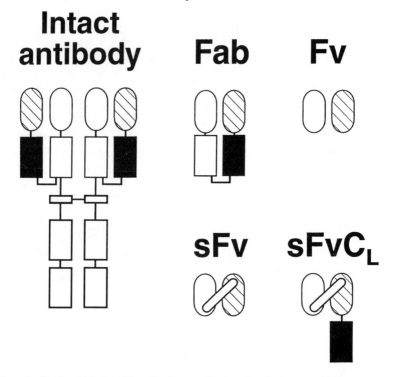

Figure 1 Structural representation of an intact antibody and antibody fragments. Abbreviations: V_H, heavy-chain variable domain; C_H, heavy-chain constant domain; V_L, light-chain variable domain; C_L, light-chain constant domains; Fv, variable-region fragment; sFv, single-chain variable-region fragment; sFvC$_L$, single-chain variable-region fragment with C_L domain.

STARTING MATERIAL FOR INTRABODY CONSTRUCTION

To obtain the variable domain sequences to construct an intrabody, two sources have commonly been used: hybridomas and antibody libraries. Hybridomas producing either murine monoclonal (9, 40, 124) or human monoclonal antibodies (96) have been the source of the V_H and V_L cDNA for that particular antibody. Primer sequences, such as those described by Ørum et al (113), or Zhou et al (154), can be used to amplify the immunoglobulin chains. These immunoglobulin chains can be used to construct sFvs (81). Some problems, however, have been encountered when multiple light-chain transcripts are present in the hybridoma fusion partner (25, 41). Antibody libraries have been built using cDNA from mouse spleens (31), human peripheral blood lymphocytes (99), and bone marrow from HIV-1-infected donors (22). These antibody fragments are expressed on the surface of phage (bacterial viruses), which are able to infect and multiply in *E. coli*. The antibody genes, cloned into a phage vector, are fused to a coat protein of the phage; this process, therefore, allows the display of the antibody. The genetically engineered phage will express and carry the genotype of the antibody. Phage-expressing antibodies can bind to a particular antigen and be separated from nonbinding phage, permitting a selection step. The bound phage can then be eluted and used to infect healthy *E. coli*, thus enriching that population of phage. The enriched binding phage can be used in a new round of selection. After three or four rounds of selection, the antigen binding phage can be isolated and characterized.

Modifications to the antibody phage display technology have allowed the cloning of antibody fragments with fine specificity and high affinity (1, 67, 98, 132, 146).

INTRABODIES AGAINST THE HUMAN IMMUNODEFICIENCY VIRUS

HIV-1, a member of the lentivirus subfamily (retrovirus family), is the etiologic agent of AIDS (118). AIDS develops after a long incubation period of the virus, coupled with one or more of several other infections, resulting in immunologic disorders and neurological disease (85). The HIV-1 genome, like that of other retroviruses, contains *gag* (group antigen), *pol* (polymerase), and *env* (envelope) structural genes. HIV-1 also encodes three regulatory proteins—Rev (regulator of virion protein), Tat (transactivator) and Nef (negative regulator factor), and three proteins believed to be involved in virus maturation and release—Vif (virion infectivity factor), Vpu (viral protein U), and Vpr (viral protein R) (144, 150). Figure 2 is a schematic representation of the HIV-1 virion.

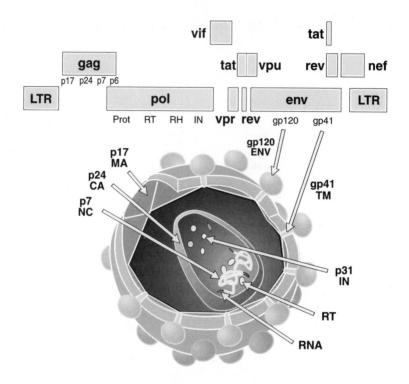

Figure 2   Schematic representation of the HIV-1 virion.

As an anti-HIV-1 treatment strategy, anti-HIV-1 intrabodies are synthesized by the cell and targeted to specific cellular compartments where they bind to their target HIV-1 protein and inhibit its function. These intrabodies target structural as well as regulatory proteins essential in the life cycle of the virus.

Envelope Glycoprotein

The HIV-1 envelope (Env) protein is synthesized as a gp160 precursor and cleaved into mature gp120/41 proteins in the Golgi complex (33). Once transported to the plasma membrane, gp120/41 proteins are incorporated into virions. These virions can infect CD4$^+$ cells through the interaction of gp120 with the CD4 receptor (34) and the co-receptor (Fusin, CCR5, CCR3, or CCR2B) (29, 35, 38, 39, 51). Likewise, the interaction between gp120 on the surface of an HIV-1-infected cell with the CD4 molecule and the co-receptor on an uninfected cell mediates the formation of syncytia, which results in the death of the fused cells (139).

The human monoclonal antibody, F105, which competes with CD4 for binding to gp120, was chosen as a candidate in constructing an intrabody targeted to gp120 (95, 119). The strategy was to design an sFv intrabody directed to the secretory system. In theory, the intrabody could be directed to different points within the secretory pathway. However, the endoplasmic reticulum (ER) is the first compartment for newly made secretory proteins and the natural site of antibody assembly. The ER is also the residence of molecular chaperones, such as BiP and GRP94, which assist in the correct folding of immunoglobulin molecules (103, 104). In addition, peptide signals required for the ER retention of soluble proteins are well characterized and consist of the carboxy terminal tetrapeptide Lys-Asp-Glu-Lev (KDEL) (107). Taking into consideration these facts, the sFv intrabody was directed to the lumen of the ER, where it could bind gp160 precursor and prevent its transport to the cell surface (see Figure 3). Two different sFv 105 intrabodies were designed, differing only by the presence of a C-terminal KDEL sequence, to retain the intrabody in the ER. Unexpectedly, the sFv 105, designed for secretion, was retained in the ER, whereas the sFv 105 KDEL intrabody yielded an unstable protein that was rapidly degraded if not coexpressed with HIV-1 envelope (96). Further experiments demonstrated

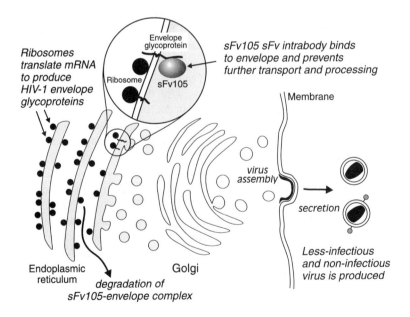

Figure 3 Inhibition of gp160 precursor processing in sFv 105 expressing cells leads to inhibition of envelope transport and the release of less infectious and noninfectious virions.

coprecipitation of sFv 105 and Env, inhibition of proteolytic processing of the Env precursor, a marked reduction of cytopathic effects due to a decrease in the envelope-mediated syncytium formation, and reduced infectivity of HIV-1 particles released by sFv 105 expressing cells.

The sFv 105 was also used to construct an inducible expression vector. The sFv 105 gene was placed under the control of the HIV-1 long terminal repeat (LTR) promoter, which could be induced after HIV-1 infection, or in the presence of exogenous Tat protein, which is taken up by the cells. Stably transfected cells exhibited resistance to the virus-mediated syncytium formation and a decreased ability to support HIV-1 production. These stably transfected cells also showed a reduced surface gp120 expression and normal surface CD4 levels, in contrast to cells transfected with a control vector in which surface CD4 was down-regulated following HIV-1 infection [association of gp160–CD4 in the lumen of the ER causes down-regulation of surfaces of CD4 (63)]. Cell surface phenotype, replication rate, morphology, and response to mitogenic stimulation of transformed cells were normal. These results demonstrated that intracellular protein-protein interactions could be inhibited by the sFv intrabodies.

To examine the reason for sFv 105 retention in the ER, protein-protein interaction with resident proteins of the ER was considered. Newly synthesized immunoglobulin (Ig) chains associate with several ER proteins during their maturation. The best characterized is BiP. BiP appears to interact predominantly with the C_H1 domain of the H chain and the V_L domain of the L chain (60, 77, 102). Normal Ig H or L chain expressed in the absence of the appropriate counterpart L or H chains also binds to BiP and can be retained in the ER. These normal proteins are not exported until the correct partner is also present (13). Interestingly, a single point mutation in the framework of a V_L, similar to a point mutation found in the F105 V_L framework, caused the retention of the L chain in the ER. However, the defect in secretion of the L chain was not due to misfolding of the L chain, as the L chain assembled into functional antibodies when coexpressed with a H chain (45). Considering these facts, sFv 105 ER retention was assessed by immunoprecipitation using a rat monoclonal antibody made against the ER chaperone protein, BiP (GRP 78) (13). As expected, sFv 105 was coimmunoprecipitated with BiP using anti-BiP monoclonal antibody (96). From this experiment, it was concluded that retention of sFv 105 was probably a consequence of association with BiP; however, binding to BiP did not impair the specific binding activity of sFv 105. Variable secretion in eukaryotic cells of ER-directed sFv intrabodies that lack a KDEL sequence has also been reported by other researchers (5, 124).

A novel strategy was also investigated whereby the antibody F105 was constructed in the format of Fab intrabody (Fab105). In this study, Fab105 directed to the ER exerted an effect intra- and extracellularly. In transduced cells infected with HIV-1, the nascent Fab105 binds intracellularly to the HIV-1 Env protein gp160 and inhibits its processing and incorporation into HIV-1 virions, whereas the secreted Fab105 neutralized cell-free HIV-1 virions (28). In addition, the nascent Fab105 also neutralized intracellular envelope mutants that escaped neutralization by extracellular F105 antibody. This finding suggested that the high concentration of Fab105 in the secretory vesicles could compensate for decreases in affinity that resulted from point mutations in the HIV-1 Env.

In summary, the use of vectors to transduce anti-Env intrabodies into CD4$^+$ peripheral blood T lymphocytes or CD34$^+$ stem cell in a clinical gene therapy setting should result in the production of virions that are markedly less infectious. As a consequence, the spread of HIV-1 in infected patients may be reduced. This gene therapy approach may provide a therapeutic benefit for HIV-1-infected patients by several different mechanisms. The interaction of gp120–CD4 either on the cell surface or intracellularly is crucial not only for virus infection, but also for many of the postulated mechanisms of CD4$^+$ depletion and functional abnormalities such as syncytium formation, induction of apoptosis or single-cell lysis, disruption of CD4-mediated cell signaling, and inappropriate immune responses (96, 115). To further these ideas, conventional gene therapy vectors containing either the sFv105 or the Fab105 have been used to transduce human lymphocytes.

The murine leukemia virus (MuLV) vector containing the sFv105 gene was used to transduce peripheral blood mononuclear cells (PBMC) of uninfected and infected patients. Transduced PBMCs challenged with HIV-1 showed cytoprotection as well as inhibition of HIV-1 replication (70, 120). Based on these results, a clinical gene therapy protocol was approved by the United States Federal Recombinant DNA Advisory Committee (RAC) in June, 1995. This study will evaluate the safety and efficacy of intracellular antibody gene therapy in asymptomatic HIV-1-infected patients, by reinfusing autologous CD4$^+$ T cells that have been transduced ex vivo with a MuLV vector that expresses the sFv105 intrabody.

Similarly, the adeno-associated virus (AAV) vector containing the Fab105 gene was used to transduce human lymphocytes. The Fab105 was initially built into a bicistronic expression vector, under the control of the cytomegalovirus immediate early (CMVIE) promoter. This vector allows near stoichiometric co-expression of the H and L chain of a Fab fragment (84, 94). An internal ribosomal entry site (IRES) corresponding to the 5' untranslated region (UTR) of encephalomyocarditis virus (EMCV) (44, 46) was used to obtain CAP-independent ribosomal binding and same level translation of the L chain (pCMV-Fab-IRES) (Figure 4). The Fab105 expression cassette was then cloned into an AAV shuttle vector, and encapsidated recombinant AAV-Fab105 vectors were produced. AAV is a suitable vector for HIV-1 gene therapy since AAV

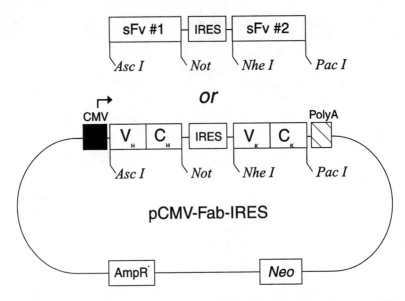

Figure 4 Bicistronic expression vector pCMV-Fab-IRES. An internal ribosomal entry site (IRES), the 5' UTR of the encephalomyocarditis virus (EMCV), has been cloned into the vector. Unique restriction enzyme sites allow the cloning of either heavy and light chains of Fab intrabodies or two sFv intrabodies.

incorporates the therapeutic gene(s) into replicating cells as well as nonreplicating cells (80, 109), thereby eliminating the potential problem of activating HIV-1 during the necessary stimulation step required for murine retroviral gene transfer (126). The Fab105 intrabody genes were shown to be transduced into human lymphocytes using recombinant AAV virus, and infection with several primary isolates from HIV-1 patients was effectively blocked in transduced T-lymphocytes (26).

Tat

Tat is a regulatory RNA-binding protein essential in the life cycle of HIV-1. HIV-1 encodes two forms of Tat, encoded by one (72 amino acid) or two (86 amino acid) exons. Tat protein activates transcription by association with a stable RNA hairpin [transactivating region (TAR)] that forms in the 5'-untranslated leader of nascent viral transcripts and promotes efficient elongation of transcription in vivo (82, 92, 97). This Tat-mediated stimulation of transcription involves an interaction between Tat and Tat-binding cellular proteins, some of which may be transcription factors themselves (79, 111), as well as cellular proteins bound to the TAR region (54, 97) (Figure 5). In addition, Tat protein has other roles.

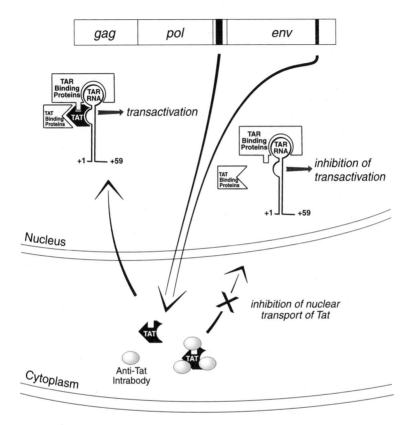

Figure 5 Strategy for inhibition of Tat-mediated transactivation of viral gene expression by anti-Tat sFv intrabodies. *Left side,* Tat binding to TAR RNA and other cellular factors are required for transactivation. *Right side,* anti-Tat sFv binding to Tat may inhibit transactivation by inhibition of Tat nuclear import.

Tat can easily be taken up by cells growing in tissue culture; it can enter the nucleus; and it can transactivate cellular genes such as tumor necrosis factor (TNF) α, TNF β, and interleukin 2 (Il-2) (20, 131, 151). Some of these cytokines in turn can activate HIV-1 transcription through their effect on nuclear factor kB (NF-kB) (21). NF-kB binding to the enhancer elements in the HIV-1 LTR can cause TAR-independent activation of viral transcripts (11, 87). Thus, Tat is likely to have both direct and indirect effects in the pathogenesis of HIV-1.

The anti-HIV-1 activity of anti-Tat sFv intrabodies has been investigated (106). In these studies, two monoclonal antibodies with different specificities were used to construct sFv intrabodies. Their specificity was to either exon

1 or exon 2 of the Tat protein. Each intrabody was directed to either the cytoplasm or the nuclear compartment. In addition, for each parent anti-Tat sFv, an additional anti-Tat sFv derivative (sFv-fusion protein) was constructed that had the entire human domain Ck fused in frame at the carboxy terminus of the sFv cassette (sFvTat–Ck). The addition of the Ck to the intrabody appears to promote multimeric forms (101). For nuclear targeting, the sFvTat and sFvTat–Ck were further modified to contain the carboxy-terminal SV40 nuclear localization signal (sFvTat-SV40 and sFvTat–Ck-SV40), which directs heterologous proteins into the nucleus (74) (Figure 5).

When the various anti-Tat intrabodies were assessed, the following results were obtained. First, the anti-Tat sFv specific to exon 1 domain (sFvTat1) blocked Tat-mediated transactivation of the HIV-1 LTR (using a CAT reporter gene), as well as the cytoplasm-nuclear transport of Tat in mammalian cells. Therefore, stably transfected lymphocytes expressing this anti-Tat1 sFv were resistant to HIV-1 infection, whereas the anti-Tat sFv specific to the exon 2 domain (sFvTat2) showed only a moderate effect. The differential effects of exon 1- and exon 2-specific anti-Tat sFvs may be due to difference in affinity. However, a more likely explanation is supported by previous works on Tat mutagenesis where a region of exon 1 of Tat was shown to be required for Tat-mediated trans-activation and exon 2 was dispensable (59, 128). Second, nuclear targeting of the anti-Tat sFv was not required to inhibit the function of Tat (which acts in the nucleus). This result demonstrates that the inhibitory effect of the intrabody sFvTat1 is obtained by sequestering Tat in the cytoplasm, away from the nucleus where Tat would exert its transactivation function. Third, the sFvTat1–Ck intrabody had greater inhibitory activity than the sFvTat had in transfected CD4$^+$ T lymphocytic cells. This result shows the beneficial effect obtained by the addition of the Ck domain that promotes multimeric forms (101). Lastly, the biological effect shown by the different intrabodies expressed in the cytoplasm and in the nuclear compartments of eukaryotic cells demonstrates that proper folding of intrabodies occurs in the reducing environment of the cytoplasm.

In gene transfer studies, a MuLV vector containing the sFv Tat1–Ck was used to transduce CD4$^+$ T cells derived from HIV-1–infected individuals (120). This MuLV vector contains a β-galactosidase gene with a nuclear localization signal to allow for determination of transduction efficiencies. In these studies, expression of sFv Tat1–Ck intrabody was more effective at stably inhibiting HIV-1 replication in transduced cells from HIV-1-infected patients than was sFv 105 in analogous studies.

Rev

Rev is another regulatory RNA-binding protein essential in the life cycle of HIV-1. Rev is encoded by multiply spliced viral mRNA and appears to function efficiently only in a multimeric form (112). The Rev protein interacts with a short RNA sequence in the envelope gene [Rev Responsive Element (RRE)] to control the export of late mRNA from the nucleus; this allows expression of virus structural proteins and late regulatory proteins (91, 93, 139). In the absence of the Rev protein, only spliced mRNAs can be transported to the cytoplasm, allowing the expression of only the early regulatory proteins (Figure 6).

The HIV-1 inhibitory activity of anti-Rev sFv intrabodies has been investigated (40, 42, 153). In these studies, the investigators showed that cytoplasmic

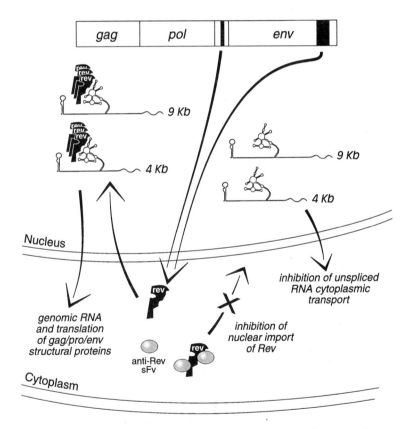

Figure 6   Strategy for inhibition of Rev-mediated cytoplasmic transport of intron containing viral mRNAs. *Left side*, Rev binding to the *cis*-acting RRE sequence in unspliced and singly spliced viral mRNA leads to cytoplasmic RNA transport, viral structural protein synthesis, virus assembly, and production of infectious virions. *Right side*, anti-Rev sFv binding to any of several critical epitopes on Rev may inhibit nuclear import and/or inhibit Rev interactions in the nucleus with cellular factors and/or RRE containing viral mRNA.

expression of a sFv intrabody against Rev in CD4$^+$ HeLa cell lines inhibits nuclear transport of Rev, syncytia formation, viral mRNA processing, and HIV-1 production (40). The specific effect of anti-Rev sFv expression on Rev function was demonstrated; there were no alterations in HIV-1 internalization, reverse transcription, or initial transcription of multiply spliced viral mRNAs in sFv-immunized cells, compared to controls (42). In addition, HIV-1 infection was inhibited in human PBMCs, transduced with retroviral vectors expressing the anti-Rev sFv intrabodies (43). These findings suggest that the anti-Rev sFv intrabody works in a manner similar to the anti-Tat sFv (106), by sequestering Rev in the cytoplasm and preventing its nuclear translocation and possibly its proposed nucleo-cytoplasmic shuttling activities (52, 53, 75, 76, 105). As a result, subsequent interaction of Rev with cellular Rev-binding proteins (52, 53, 89, 129) or with the RRE containing viral mRNA and associated RRE-binding proteins (116, 135) may be inhibited (Figure 6).

To further understand the molecular mechanism(s) underlying the effects of intracellular anti-Rev sFvs on HIV-1, two anti-Rev sFvs that specifically bind to differing epitopes of Rev were cloned (153). One sFv (D8) binds to the Rev activation domain, and the second sFv (D10) binds to the distal C terminus of Rev in the nonactivation region. Although the anti-Rev sFv D8 had an extracellular binding affinity significantly lower than that of the anti-Rev sFv D10, the sFv D8 demonstrated more potent activity in inhibiting virus production in human T-cell lines and PBMCs than did sFv D10. These results indicate that extracellular binding affinities of an sFv for a target viral protein cannot be used to predict its intracellular activity. The important feature might be the epitope targeted.

Reverse Transcriptase

The HIV-1 enzymes protease, reverse transcriptase (RT), and integrase (IN) are encoded by the pol gene. They are the processed product of a Gag-Pol polyprotein precursor, produced by ribosome frameshifting (66, 145). HIV-1 RT and other viral proteins are required in the early stages of HIV-1 infection. RT specifically reverse transcribes the viral ribonucleic acid (RNA) to DNA, needed for integration into the host chromosome (Figure 7).

The anti-HIV-1 activity of anti-RT intrabodies has been studied by two different groups (90, 133). One group reported the inhibition of RT function in CD4$^+$ T cell lines expressing a cytoplasmic Fab intrabody against HIV-1 RT (90). Their results showed reduced levels of both unintegrated and integrated viral DNA in cells stably expressing anti-RT Fab intrabodies challenged with HIV-1. The investigators used two separate plasmids to express the light chain and the Fd fragment. Interestingly, this anti-RT monoclonal antibody, from which the Fab was derived, was not neutralizing in assays of enzyme activity.

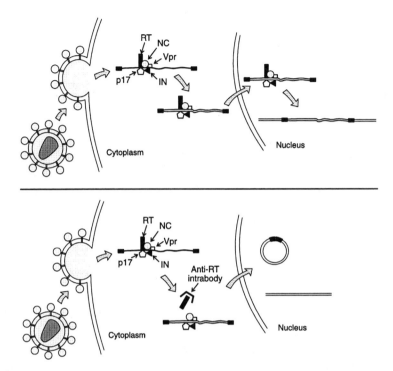

Figure 7   *Top panel*, representation of normal integration of HIV-1 provirus. *Bottom panel*, representation of the same process in the presence of an anti-RT intrabody. There is no reverse transcription; thus, there is no viral integration.

The authors proposed that intracellular binding of anti-RT Fabs may sterically hinder movement of the enzyme along the RNA template or otherwise disrupt RT secondary structure.

Although most of the inhibitory effect on HIV-1 replication was observed in cell lines expressing the Fab, some inhibition was also reported for cells expressing the Fd fragment only. This observation supports previous antibody studies indicating that most of the binding specificity of an antibody is determined by the V_H domain (148), specifically by its complementary determining region 3 (CDR h3) (2, 30, 72).

A second group reported the construction of an anti-RT sFv intrabody from a murine hybridoma producing anti-RT IgG (133). In vitro studies demonstrated that this anti-RT IgG inhibits DNA polymerase activity of HIV-1 RT. Cytoplasmic expression of anti-RT sFv in T-lymphocytic cells specifically neutralizes the RT activity in the preintegration stage, affecting the reverse transcription process and thereby decreasing HIV-1 propagation and HIV-1-induced cyto-pathic effects. These data also show that intrabodies can gain access to viral proteins of the HIV-1 preintegration complex (Figure 7).

Integrase

The HIV-1 enzyme IN is encoded by the pol gene and its sequence is similar to that of other retroviruses (69). HIV-1 DNA integration occurs through a defined set of DNA cutting and joining reactions (23). The HIV-1 integration function is required in vivo for viral replication (32). Site directed mutagenesis abrogating IN function results in unintegrated proviral DNA with intracellular accumulation of circular viral DNA (19, 47). HIV-1 IN can be divided into several functional domains: a zinc-finger-like domain (N-terminal), a central catalytic domain, and a nonspecific DNA-binding domain (C-terminal) (48).

The anti-HIV-1 inhibitory activity of anti-IN sFv intrabodies have been inves-tigated (86). A panel of five anti-HIV-1 IN hybridomas was used to construct anti-IN sFv intrabodies. These intrabodies recognize different domains within IN. The intracellular expression of only those sFvs that bind to IN catalytic and C-terminal domains resulted in resistance to productive HIV-1 infection. The inhibition of HIV-1 replication was observed with sFvs localized in either the cytoplasmic or nuclear compartment of the cell. The nuclear localization signal (NLS) selected in this study was the HIV-1 Tat NLS (136), which was placed at the 3' end of the interchain linker. The expression of anti-IN sFvs in human T-lymphocytic cells and PBMCs appears to neutralize IN activity prior to integration and thus decreases HIV-1 replication (Figure 8).

Matrix

The core of an HIV-1 virion contains four nucleocapsid proteins, p24 (capsid), p17 (matrix), p7 (nucleocapsid), and p6. Each of these proteins is proteolyti-cally cleaved from a 55-kd Gag polyprotein precursor by the HIV-1 protease (78, 134). The HIV-1 matrix (MA) comes from the amino-terminus of p55 and is myristylated at its N terminus. MA is believed to be involved in two critical stages of the viral life cycle; it is required for both nuclear import of the viral preintegration complex and for particle assembly. MA contains a NLS that ap-pears to play a role in the nuclear transport of the viral preintegration complex, especially in nondividing cells (18, 147). MA also plays a critical role in parti-cle formation, and a myristylated residue and charged N-terminal amino acids of MA direct the Gag polyprotein precursor to the plasma membrane. This targeting is essential for the proper assembly of viral particles (140, 155) and for their release into the extracellular space (17, 56). The anti-HIV-1 activity of anti-MA Fab intrabodies has been investigated (84). The binding site for the 3H7 monoclonal antibody, used to clone the Fab, had been previously epitope

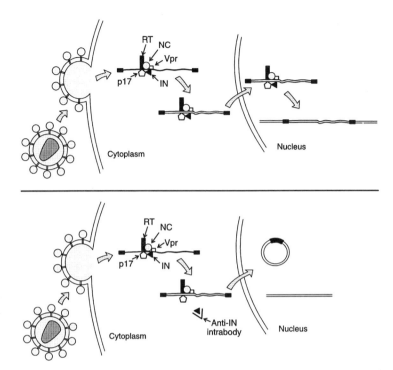

Figure 8   Top panel, representation of normal integration of HIV-1 provirus into the host chromo-some. *Bottom panel*, representation of the same process in the presence of an anti-IN intrabody. There is accumulation of circular viral DNA and no viral integration.

mapped to the amino acid sequence KKAQQAAADT (residues 113–122) near the carboxy terminus of MA (110). Co-expression of the Fd and light chain was optimized using the pCMV-Fab-IRES vector (Figure 4). CD4$^+$ Jurkat cells stably expressing cytoplasmic anti-MA Fab and challenged with HIV-1 showed a marked delay of virus infection over a period of 25 days compared to control. This inhibition of infection suggests that the anti-MA Fab intrabody interferes with the integration of HIV-1 provirus. To examine the role of the anti-MA in-trabody during particle formation, the infectivity of the virus particles released from Jurkat cells expressing anti-MA Fab and infected with HIV-1 were exam-ined. These experiments demonstrated a significant decrease in the infectivity of the virions released from these HIV-1-infected Fab intrabody expressing cells. Further experiments need to be addressed to delineate the mechanism(s) of inhibition of anti-MA Fab intrabodies observed in infection studies or virion infectivity studies.

Nucleocapsid

The inner core of most retroviruses is formed by a shell of capsid protein molecules (p24) surrounding the dimeric RNA genome in close association with approximately 2000 molecules of nucleocapsid protein (NC), 20–50 molecules of reverse transcriptase and integrase (27), as well as tRNAs, 5S RNA, 7S RNA, and ribosomal RNAs (37). The NC protein contains two highly conserved zinc finger motifs (6). In vitro studies have shown that the NC is required for genomic RNA dimerization, for correct encapsidation, for annealing of the tRNA primer to its complementary binding site (PBS) on the genomic RNA, and for the minus-strand DNA transfer during reverse transcription (117, 121). These observations indicate that the NC plays an essential role in the afferent as well as the efferent arms of the HIV-1 life cycle.

Preliminary reports show the anti-HIV-1 activity of anti-NC sFv intrabodies (127). An sFv intrabody has been cloned from a murine hybridoma made against one of the zinc fingers of HIV-1 NC (62). Jurkat cells stably expressing cytoplasmic sFv-NC show resistance to single round HIV-1 infection. In addition, infectious particle formation is substantially reduced in cells expressing sFv-NC intrabodies. These data indicate that sFv-NC intrabodies affect afferent and efferent arms of the HIV-1 life cycle. However, further studies are needed to delineate the mechanism(s) of HIV-1 inhibition.

COMBINATION TARGETS

Cancer treatment studies conducted over many years have concluded that combination chemotherapy is often more effective than are drugs used individually. In theory, drugs for combination chemotherapy are chosen because they have different mechanisms of action, are effective individually, and have qualitatively different toxicities so that each drug can be given at or near its individual maximum tolerated dose (49). Likewise, the idea of using a multitargeted approach for AIDS treatment has gained considerable ground in the past few years because of the failures of many single agents to halt the spread of HIV-1 (83).

In the case of intrabodies, studies presented in this review demonstrate that a variety of HIV-1 proteins are sensitive to neutralization by intrabodies and inhibit HIV-1 replication. This HIV-1 inhibition by intrabodies can be obtained at different stages of the viral life cycle. Therefore, the combination of intrabodies blocking HIV-1 replication at different points of the viral life cycle may be useful for the gene therapy of HIV-1 infection and AIDS. A bicistronic expression vector such as pCMV-Fab-IRES (Figure 4), capable of expressing two different genes simultaneously, would allow different combinations of intrabodies to be tested for their combined or synergistic inhibition of HIV-1 replication.

INTRABODIES AGAINST CELLULAR PROTEINS INVOLVED IN HIV-1 PATHOGENESIS

Cellular proteins involved in HIV-1 entry or pathogenesis may be excellent targets for intrabody based-gene therapy to treat HIV-1 infection and AIDS. Cellular targets would be less prone to somatic mutations that occur with high frequency in HIV-1 owing to the low fidelity of HIV-1 RT. Somatic mutations in the HIV-1 genome could lead to escape mutants, allowing the viral protein to be unaffected by the intrabody.

Among the cellular targets involved in HIV-1 entry are the HIV-1 co-receptors, such as fusin, CCR5, CCR3, and CCR2B (29, 35, 38, 39, 51). These co-receptors are needed for entry of different strains of HIV-1. However, a homozygous defect in CCR5 appears to have no obvious phenotypic defect in the affected individuals (88).

Intrabodies are well suited to cause "phenotypic knock-out" of viral and cellular proteins, such as the alpha subunit of the Il-2 receptors from trafficking through the secretory system (124). Several examples to support this statement have been reported (5, 36, 57, 123). Therefore, cell surface expression of the HIV-1 co-receptor CCR5 could be abrogated with an anti-CCR5 intrabody directed to the ER.

Other cellular targets of intrabodies may also be exploited for HIV-1 gene therapy. For example, HIV-1 apoptosis is probably mediated through the combined effect of viral as well as cellular proteins (114). Candidate cellular proteins could include Grb3-3 (50), interleukin-1β-converting enzyme (ICE) and related molecules (12, 108), as well as other cellular proteins that mediate apoptosis (4, 143).

SUMMARY

Intrabodies provide a powerful research tool that could potentially have many applications in the field of therapeutic molecules for gene therapy of infectious diseases. The specificity and affinity that can be obtained in intrabodies make these molecules effective modulators of viral or cellular targets. In principle, intrabodies can block or stabilize molecular interactions, modulate enzyme function, or even sequester targets in a particular cellular compartment. Some of the reasons that make intrabodies good effector molecules reside in the ability to direct them to different cellular compartments and to bind different epitopes on a target. Using HIV-1 as an example, intrabodies demonstrate broad versatility in their ability to inhibit different stages of the viral life cycle by targeting structural, regulatory, and enzymatic proteins of the virus.

Several potential problems may limit the use of intrabodies in the clinical setting for the treatment of HIV-1 and AIDS. One problem, also shared by other gene-based therapies, is the efficiency of gene delivery. The development of a gene-transfer system is needed that is capable of delivering the intrabody gene into a sufficient number of a specific cell type. The cell type, probably $CD4^+$ and differentiated $CD34^+$, will have to expand in the host and allow for long-term immunologic reconstitution. A second problem that is only shared with some of the gene-based therapies is the immunogenicity that the intrabody could elicit. Although intrabodies are expected to cause a cytotoxic T-cell response in humans, cells transduced with the prokaryotic neomycin-resistance gene have remained in the circulation of cancer patients for over 18 months (15). In addition, the use of human intrabodies should minimize the issue of immune recognition, but this will have to be determined in a human clinical trial. A third potential problem, previously mentioned, is the development of HIV-1 escape mutants. A way to minimize the development of HIV-1 escape mutants would be choosing conserved epitopes needed for virus survival or cellular targets involved in virus replication. An alternative approach would be the use of combination targets, two intrabodies targeting different stages of the viral life cycle. Lastly, the expression of the intrabody gene will need to be sustained and at a sufficient level to achieve the desired effect. Different factors play a role in this last issue, such as vector choices for insertion into the chromosomes, the promoter used to direct expression, half-life of the intrabody, cell type, and other factors. Only the persistence and perseverance of scientists addressing pertinent issues will answer these questions.

ACKNOWLEDGMENTS

We thank Dr. Carla Pumphrey and Dr. Bijan Etemad-Moghadam for helpful comments, and Ms. Yvette McLaughlin and Ms. Amy Emmert for preparation of this manuscript.

> **Visit the *Annual Reviews home page* at**
> **http://www.annurev.org.**

Literature Cited

1. Akamatsu Y, Cole MS, Tso JY, Tsurushita N. 1993. Construction of a human Ig combinatorial library from genomic V segments and synthetic CDR3 fragments. *J. Immunol.* 151:4651–59
2. Amit AG, Mariuzza RA, Phillips SEV, Poljak RJ. 1986. Three-dimensional structure of an antigen-antibody complex at 2. 8 Å resolution. *Science* 233:747–53
3. Baltimore D. 1988. Intracellular immunization. *Nature* 335:395–96
4. Barr PJ, Tomei LD. 1994. Apoptosis and its role in human disease. *Bio/Technology* 12:487–93
5. Beerli RR, Wels W, Hynes NE. 1994. Intracellular expression of single chain

antibodies reverts ErbB-2 transformation. *J. Biol. Chem.* 269:23931–36

6. Bess JW Jr, Powell PJ, Issaq HJ, Schumack LJ, Grimes MK, et al. 1992. Tightly bound zinc in human immunodeficiency virus Type 1. Human T-cell leukemia virus Type 1, and other retroviruses. *J. Virol.* 66:840–47

7. Biocca S, Neuberger MS, Cattaneo A. 1990. Expression and targeting of intracellular antibodies in mammalian cells. *EMBO J.* 9:101–8

8. Biocca S, Pierandrei-Amaldi P, Campion N, Cattaneo A. 1994. Intracellular immunization with cytosolic recombinant antibodies. *Bio/Technology* 12:396–99

9. Biocca S, Pierandrei-Amaldi P, Cattaneo A. 1993. Intracellular expression of anti-p21[ras] single chain Fv fragments inhibits meiotic maturation of xenopus oocytes. *Biochem. Biophys. Res. Commun.* 197:422–27

10. Bird RE, Hardman KD, Jacobson JW, Johnson S, Kaufman BM, et al. 1988. Single-chain antigen-binding proteins. *Science* 242:423–26

11. Biswas DK, Salas TR, Wang F, Ahlers CM, Dezube BJ, Pardee AB. 1995. A Tat-induced auto-up-regulatory loop for superactivation of the human immunodeficiency virus type 1 promoter. *J. Virol.* 69:7437–44

12. Boldin MP, Goncharov TM, Goltsev YV, Wallach D. 1996. Involvement of MACH, a novel MORT1/FADD-interacting protease, in Fas/APO-1- and TNF receptor-induced cell death. *Cell* 85:803–15

13. Bole DG, Hendershot LM, Kearney JF. 1986. Posttranslational association of immunoglobulin heavy chain binding protein with nascent heavy chains in nonsecreting and secreting hybridomas. *J. Cell Biol.* 102:1558–66

14. Brady HJM, Miles CG, Pennington DJ, Dzierzak EA. 1994. Specific ablation of human immunodeficiency virus Tat-expressing cells by conditionally toxic retroviruses. *Proc. Natl. Acad. Sci. USA* 91:365–69

15. Brenner MK, Rill DR, Holladay MS, Heslop HE, Moen RC, et al. 1993. Gene marking to determine whether autologous marrow infusion restores long-term haemopoiesis in cancer patients. *Lancet* 342:1134–37

16. Brinkman U, Reiter Y, Jung SH, Lee B, Pastan I. 1993. A recombinant immunotoxin comparing a disulfide-stabilized Fv fragment. *Proc. Natl. Acad. Sci. USA* 90:7538–42

17. Bryant M, Ratner L. 1990. Myristoylation-dependent replication and assembly of human immunodeficiency virus 1. *Proc. Natl. Acad. Sci. USA* 87:523–27

18. Bukrinsky MI, Haggerty S, Dempsey MP, Sharova N, Adzhubei A, et al. 1993. A nuclear localization signal within HIV-1 matrix protein that governs infection of non-dividing cells. *Nature* 365:666–69

19. Bukrinsky M, Sharova N, Stevenson M. 1993. Human immunodeficiency virus Type 1 2-LTR circles reside in a nucleoprotein complex which is different from the preintegration complex. *J. Virol.* 67:6863–65

20. Buonaguro L, Barillari G, Chang HK, Bohan CA, Kao V, et al. 1992. Effects of the human immunodeficiency virus Type 1 Tat protein on the expression of inflammatory cytokines. *J. Virol.* 66:7159–67

21. Buonaguro L, Buonaguro FM, Giraldo G, Ensoli B. 1994. The human immunodeficiency virus type 1 Tat protein transactivates tumor necrosis factor beta gene expression through a TAR-like structure. *J. Virol.* 68:2677–82

22. Burton DR, Barbas CF III, Persson MAA, Koenig S, Chanock RM, Lerner RA. 1991. A large array of human monoclonal antibodies to type 1 human immunodeficiency virus from combinatorial libraries of asymptomatic seropositive individuals. *Proc. Natl. Acad. Sci. USA* 88:10134–37

23. Bushman FD, Fujiwara T, Craigie R. 1990. Retroviral DNA integration directed by HIV integration protein in vitro. *Science* 249:1555–58

24. Carlson JR. 1988. A new means of inducibly inactivating a cellular protein. *Mol. Cell. Biol.* 8:2638–46

25. Carroll WL, Mendel E, Levy S. 1988. Hybridoma fusion cell lines contain an aberrant Kappa transcript. *Mol. Immunol.* 25:991–95

26. Chen J, Yang Q, Yang A-G, Marasco WA, Chen S-Y. 1996. Intra- and extracellular immunization against HIV-1 infection with lymphocytes transduced with an AAV vector expressing a human anti-gp120 antibody gene. *Hum. Gene Ther.* 7:1515–25

27. Chen M, Garon C, Papas T. 1980. Native ribonucleoprotein is an efficient transcriptional complex of avia myeloblastosis virus. *Proc. Natl. Acad. Sci. USA* 77:1296–300

28. Chen SY, Khouri Y, Bagley J, Marasco W. 1994. Combined intra- and extra-cellular immunization against human immunodeficiency virus type 1 infection with a

human anti-gp120 antibody. *Proc. Natl. Acad. Sci. USA* 91:5932–36

29. Choe H, Farzan M, Sun Y, Sullivan N, Rollins B, et al. 1996. The β-chemokine receptors CCR3 and CCR5 facilitate infection by primary HIV-1 isolates. *Cell* 85:1135–48

30. Chothia C, Lesk AM, Tramontano A, Levitt M, Smith-Gill SJ, et al. 1989. Conformations of immunoglobulin hypervariable regions. *Nature* 342:877–83

31. Clackson T, Hoogenboom HR, Griffiths AD, Winter A. 1991. Making antibody fragments using phage display libraries. *Nature* 352:624–28

32. Clavel F, Hoggan MD, Willey RL, Strebel K, Martin MA, Repaske R. 1989. Genetic recombination of human immunodeficiency virus. *J. Virol.* 63:1455–59

33. Crise B, Rose JK. 1992. Human immunodeficiency virus Type 1 glycoprotein precursor retains a CD4–p56^{lck} complex in the endoplasmic reticulum. *J. Virol.* 66:2296–301

34. Dalgleish AG, Beverleym PC, Clapham PR, Crawford DH, Greaves MF, Weiss RA. 1984. The CD4 (T4) antigen is an essential component of the receptor for the AIDS retrovirus. *Nature* 312:763–67

35. Deng HK, Liu R, Ellmeier W, Choe S, Unutmaz D, et al. 1996. Identification of a major co-receptor for primary isolates of HIV-1. *Nature* 381:661–66

36. Deshane J, Loechel F, Conry RM, Siegal GP, King CR, Curiel DT. 1994. Intracellular single-chain antibody directed against erbB2 down-regulates cell surface erbB2 and exhibits 1 selective antiproliferative effect in erbB2 overexpressing cancer cell lines. *Gene Ther.* 1:332–37

37. Dickson C, Eisenman R, Fan H, Hunter E, Reich N. 1985. Protein biosynthesis and assembly. See Ref. 150, pp. 513–648

38. Doranz BJ, Rucker J, Yi Y, Smyth RJ, Samson M, et al. 1996. A dual-tropic primary HIV-1 isolate that uses fusin and the β-chemokine receptors CKR-5, CKR-3, and CKR-2b as fusion cofactors. *Cell* 85:1149–58

39. Dragic T, Litwin V, Allaway GP, Martin SR, Huang Y, et al. 1996. HIV-1 entry into CD4$^+$ cells is mediated by the chemokine receptor CC-CKR-5. *Nature* 381:667–73

40. Duan L, Bagasra O, Laughlin MA, Oakes JW, Pomerantz RJ. 1994. Potent inhibition of human immunodeficiency virus type 1 replication by an intracellular anti-Rev single-chain antibody. *Proc. Natl. Acad. Sci. USA* 91:5075–79

41. Duan L, Pomerantz RJ. 1994. Elimination of endogenous aberrant kappa chain transcripts from sp2/0–derived hybridoma cells by specific ribozyme cleavage: utility in genetic therapy of HIV-1 infections. *Nucleic Acids Res.* 22:5433–38

42. Duan L, Zhang H, Oakes JW, Bagasra O, Pomerantz RJ. 1994. Molecular and virological effects of intracellular anti-rev single-chain variable fragments on the expression of various human immunodeficiency virus-1 strains. *Hum. Gene Ther.* 5:1315–24

43. Duan L, Zhu M, Bagasra O, Pomerantz RJ. 1995. Intracellular immunization against HIV-1 infection of human T lymphocytes: utility of anti-rev single-chain variable fragments. *Hum. Gene Ther.* 6:1561–73

44. Duke GM, Hoffman MA, Palmenberg AC. 1992. Sequence and structural elements that contribute to efficient encephalomyocarditis virus RNA translation. *J. Virol.* 66:1602–9

45. Dul JL, Argon Y. 1990. A single amino acid substitution in the variable region of the light chain specifically blocks immunoglobulin secretion. *Proc. Natl. Acad. Sci. USA.* 87:8135–39

46. Elroy-Stein O, Fuerst TR, Moss B. 1989. Cap-independent translation of mRNA conferred by encephalomyocarditis virus 5′ sequence improves the performance of the vaccinia virus/bacteriophage T7 hybrid expression system. *Proc. Natl. Acad. Sci. USA* 86:6126–30

47. Engelman A, Englund G, Orenstein JM, Martin MA, Craigie R. 1995. Multiple effects of mutations in human immunodeficiency virus Type 1 integrase on viral replication. *J. Virol.* 69:2729–36

48. Engelman A, Hickman AB, Craigie R. 1994. The core and carboxyl-terminal domains of the integrase protein of human immunodeficiency virus Type 1 each contribute to nonspecific DNA binding. *J. Virol.* 68:5911–17

49. Erlichman C, Kerr IG. 1985. Antineoplastic drugs. In *Principles of Medical Pharmacology*, ed. H Kalant, WHE Roschlau, EM Sellers, 4:699–710

50. Fath I, Schweighoffer F, Rey I, Multon M-C, Boiziau J, et al. 1994. Cloning of a Grb2 isoform with apoptotic properties. *Science* 264:971–74

51. Feng Y, Broder CC, Kennedy PE, Berger EA. 1996. HIV-1 entry cofactor: functional cDNA cloning of a seven-transmembrane, G protein-coupled receptor. *Science* 272:872–77

52. Frankhauser C, Izaurralde E, Adachi Y, Wingfield P, Laemmli UK. 1991. Specific complex of human immunodeficiency

virus type 1 Rev and nucleolar B23 proteins: dissociation by the Rev response element. *Mol. Cell. Biol.* 11:567–75

53. Fritz CC, Zapp ML, Green MR. 1995. A human nucleoporin-like protein that specifically interacts with HIV Rev. *Nature* 376:530–33

54. Gatignol A, Buckler-White A, Beckhout B, Jeang K-T. 1991. Characterization of a human TAR RNA-binding protein that activates the HIV-1 LTR. *Science* 251:1597–600

55. Glockshuber R, Malia M, Pfitzinger I, Pluckthun A. 1990. A comparison of strategies to stabilize immunoglobulin Fv-fragments. *Biochemistry* 29(6):1362–67

56. Gottlinger HG, Sodroski JG, Haseltine WA. 1989. Role of capsid precursor processing and myristylation in morphogenesis and infectivity of HIV-1. *Proc. Natl. Acad. Sci. USA* 86:5781–85

57. Graus-Porta D, Beerli RR, Hynes NE. 1995. Single-chain antibody-mediated intracellular retention of ErbB-2 impairs Neu differentiation factor and epidermal growth factor signaling. *Mol. Cell. Biol.* 15(3):1181–91

58. Griffiths AD, Malmqvist M, Marks JD, Bye JM, et al. 1993. Human anti-self antibodies with high specificity from phage display libraries. *EMBO J.* 12:725–34

59. Hauber J, Malim MH, Cullen BR. 1989. Mutational analysis of the conserved basic domain of human immunodeficiency virus tat protein. *J. Virol.* 63(3):1181–87

60. Hendershot L, Bole D, Köhler G, Kearney JF. 1987. Assembly and secretion of heavy chains that do not associate postranslationally with immunoglobulin heavy chain-binding protein. *J. Cell Biol.* 104:761–67

61. Hiatt A, Cafferkey R, Bowdish K. 1989. Production of antibodies in transgenic plants. *Nature* 342:76–78

62. Hinkula J, Rosen J, Sundqvist V-A, Stigbrand T, Wahren B. 1990. Epitope mapping of the HIV-1 Gag region with monoclonal antibodies. *Mol. Immunol.* 27:395–403

63. Hoxie JA, Alpers JD, Rackowski JL, Huebner K, Haggarty BS, et al. 1986. Alterations in T4 (CD4) protein and MRNA synthesis in cells infected with HIV. *Science* 234:1123–27

64. Huse WD, Sastry L, Iverson SA, Kang AS, Alting-Mees M, et al. 1989. Generation of a large combinatorial library of the immunoglobulin repertoire in phage lambda. *Res. Article* 246:1275–81

65. Huston JS, Levinson D, Mudgett-Hunter M, Tai M-S, et al. 1988. Protein engineering of antibody binding sites: recovery of specific activity in an anti-digoxin single-chain Fv analogue produced in *Escherichia coli. Proc. Natl. Acad. Sci. USA* 85:5879–83

66. Jacks T, Power MD, Masiarz FR, Luciw PA, Barr PJ, Varmus HE. 1988. Characterization of ribosomal frameshifting in HIV-1 gag-pol expression. *Nature* 331:280–83

67. Jespers LS, Roberts A, Mahler SM, Winter G, Hoogenboom HR. 1994. Guiding the selection of human antibodies from phage display repertoires to a single epitope of an antigen. *Bio/Technology* 12:899–903

68. Jiang W, Venugopal K, Gould EA. 1995. Intracellular interference of tick-borne flavivirus infection by using a single-chain antibody fragment delivered by recombinant sindbis virus. *J. Virol.* 69:1044–49

69. Johnson MS, McClure MA, Feng D-F, Gray J, Doolittle RF. 1986. Computer analysis of retroviral pol genes: assignment of enzymatic functions to specific sequences and homologies with nonviral enzymes. *Proc. Natl. Acad. Sci. USA* 83:7648–52

70. Jones SD, Porter-Brooks J, Eberhardt B, Chen S-Y, Mhashilkar A, et al. 1995. Gene therapy for HIV using intracellular antibodies. *J. Cell Biol.* 21A:395

71. Julian K-CM, Hiatt A, Hein M, Vine ND, Wang F, et al. 1995. Generation and assembly of secretory antibodies in plants. *Science* 268:716–19

72. Kabat EA, Wu TT. 1991. Identical V region amino acid sequences and segments of sequences in antibodies of different specificities. *J. Immunol.* 147:1709–19

73. Kalderon D, Richardson WD, Markham AF, Smith AE. 1994. Sequence requirements for nuclear localisation of SV40 large-T antigen. *Cell* 311:33–38

74. Kalderon D, Roberts BL, Richardson WD, Smith AE. 1984. A short amino acid sequence able to specify nuclear location. *Cell* 39:499–509

75. Kalland K-H, Szilvay AM, Brokstad KA, Sætrevik W, Haukenes G. 1994. The human immunodeficiency virus type 1 Rev protein shuttles between the cytoplasm and nuclear compartments. *Mol. Cell. Biol.* 14:7436–44

76. Kalland K-H, Szilvay AM, Langhoff E, Kaukenes G. 1994. Subcellular distribution of human immunodeficiency virus type 1 Rev and colocalization of Rev with

RNA splicing factors in a speckled pattern in the nucleoplasm. *J. Virol.* 68:1475–85

77. Kaloff CR, Haas IG. 1995. Coordination of immunoglobulin chain folding and immunoglobulin chain assembly is essential for the formation of functional IgG. *Immunity* 6:629–37

78. Kaplan AH, Swanstrom R. 1991. Gag proteins are processed in two cellular compartments. *Proc. Natl. Acad. Sci. USA* 88:4528–32

79. Kashanchi F, Piras G, Rodonovich MF, Duvall JF, Fattaey A, et al. 1994. Direct interaction of human TFIID with the HIV-1 transactivator tat. *Nature* 367:295–99

80. Kotin RM. 1994. Prospects for the use of adeno-associated virus as a vector for human gene therapy. *Hum. Gene Ther.* 5:793–801

81. Lake DF, Bernstein RM, Schluter SF, Marchalonis JJ. 1995. Generation of diverse single-chain proteins using a universal $(Gly_4–Ser)_3$ encoding oligonucleotide. *Bio/Technology* 19:700–2

82. Laspia MF, Rice AP, Mathews MB. 1989. HIV-1 Tat protein increases transcriptional initiation and stabilizes elongation. *Cell* 59:282–92

83. Lederman MM. 1995. Host-directed and immune-based therapies for human immunodeficiency virus infection. *Ann. Intern. Med.* 122(3):218–22

84. Levin R, Mhashilkar AM, Dorfman T, Bukovsky A, Zani C, et al. 1997. Inhibition of early and late events of the HIV-1 replication cycle by cytoplasmic Fab intrabodies against the matrix protein. *Mol. Med.* 2:96–110

85. Levy JA. 1989. HIV and the pathogenesis of AIDS. *JAMA* 261:2997–3006

86. Levy-Mintz P, Duan L, Zhang H, Hu B, Dornadula G, et al. 1996. Intra-cellular expression of single-chain variable fragments to inhibit early stages of the viral life cycle by targeting human immunodeficiency virus Type 1 integrase. *J. Virol.* 70:8821–32

87. Liu J, Perkins ND, Schmid RM, Nabel GJ. 1992. Specific NF-kB subunits act in concert with Tat to stimulate human immunodeficiency virus type 1 transcription. *J. Virol.* 66:3883–87

88. Liu R, Paxton WA, Choe S, Ceradini D, Martin SR, et al. 1996. Homozygous defect in HIV-1 coreceptor accounts for resistance of some multiply-exposed individuals to HIV-1 infection. *Cell* 86:367–77

89. Luo Y, Yu H, Peterlin BM. 1994. Cellular protein modulates effects of human immunodeficiency virus type 1 rev. *J. Virol.* 68:3850–56

90. Maciejewski JP, Weichold FF, Young NS, Cara A, Zella D, et al. 1995. Intracellular expression of antibody fragments directed against HIV reverse transcriptase prevents HIV infection in vitro. *Nat. Mag.* 1:667–73

91. Malim MH, Bohnlein S, Hauber J, Cullen BR. 1989. Functional dissection of the HIV-1 Rev transactivator-derivation of a trans-dominant repressor of Rev function. *Cell* 58:205

92. Malim MH, Hauber J, Fenrick R, Cullen BR. 1988. Immunodeficiency virus rev trans-activator modulates the expression of the viral regulatory genes. *Nature* 335:181–83

93. Malim MH, Tiley LS, McCarn DF, Rusche JR, Hauber J, Cullen BR. 1990. HIV-1 structural gene expression requires binding of the Rev *trans*-activator to its RNA target sequence. *Cell* 60:675–83

94. Marasco WA. 1995. Intracellular antibodies (intrabodies) as research reagents and therapeutic molecules for gene therapy. *Immunotechnology* 1:1–19

95. Marasco WA, Bagley J, Zani C, Posner M, Cavacini L, et al. 1992. Characterization of the cDNA of a broadly reactive neutralizing human anti-gp120 monoclonal antibody. *J. Clin. Invest.* 90:1467–78

96. Marasco WA, Haseltine WA, Chen SY. 1993. Design, intracellular expression, and activity of a human anti-human immunodeficiency virus type 1 gp120 single-chain antibody. *Proc. Natl. Acad. Sci. USA* 90:7889–93

97. Marciniak RA, Garcia-Blanco MA, Sharp PA. 1990. Identification and characterization of a HeLa nuclear protein that specifically binds to the trans-activation-response (TAR) element of human immunodeficiency virus. *Proc. Natl. Acad. Sci. USA* 87:3624–38

98. Marks JD, Griffiths AD, Malmqvist M, Clackson TP, Bye JM, Winter G. 1992. By-passing immunization: building high affinity human antibodies by chain shuffling. *Bio/Technology* 10:779–83

99. Marks JD, Hoogenboom HR, Bonnert TP, McCafferty J, Griffiths AD, Winter G. 1991. By-passing immunization human antibodies from V-gene libraries displayed on phage. *J. Mol. Biol.* 222:581–97

100. McCafferty J, Griffiths AD, Winter G, Chiswell DJ. 1990. Phage antibodies: filamentous phage displaying antibody variable domains. *Lett. Nat.* 348:552–54

101. McGregor DP, Molloy PE, Cunningham C, Harris WJ. 1994. Spontaneous assembly of bivalent single chain antibody fragments in *Escherichia coli. Mol. Immunol.* 31:219–26

102. Melnick J, Argon Y. 1995. Molecular chaperones and the biosynthesis of antigen receptors. *Immunol. Today* 16:243–50

103. Melnick J, Aviel S, Argon Y. 1992. The endoplasmic reticulum stress protein GRP94, in addition to BiP, associates with unassembled immunoglobulin chains. *J. Biol. Chem.* 267:21303–6

104. Melnick J, Dul JL, Argon Y. 1994. Sequential interaction of the chaperones BiP and GRP94 with immunoglobulin chains in the endoplasmic reticulum. *Nature* 370:373–75

105. Meyer BB, Malim MH. 1994. The HIV-1 Rev *trans*-activator shuttles between the nucleus and the cytoplasm. *Genes Dev.* 8:1538–47

106. Mhashilkar MA, Bagley J, Chen SY, Szilvay AM, Helland DG, Marasco WA. 1995. Inhibition of HIV-1 Tat-mediated LTR transactivation and HIV-1 infection by anti-Tat single chain intrabodies. *EMBO J.* 14:1542–51

107. Munroe S, Pelham HRB. 1987. A C-terminal signal prevents secretion of luminal ER proteins. *Cell* 48:899–97

108. Muzio M, Chinnaiyan AM, Kischkel FC, O'Rourke K, Shevchenko A, et al. 1996. FLICE, a novel FADD-homologous ICE/CED-3–like protease, is recruited to the CD95 (Fas/APO-1) death-inducing signaling complex. *Cell* 85:817–27

109. Muzyczka N. 1992. Use of adeno-associated virus as a general transduction vector for mammalian cells. *Curr. Top. Microbiol. Immunol.* 158:97–129

110. Niedrig M, Hinkula J, Weigelt W, L'Age-Stehr J, Pauli G, et al. 1989. Epitope mapping of monoclonal antibodies against human immunodeficiency virus type 1 structural proteins by using peptides. *J. Virol.* 63:3525–28

111. Ohana B, Moore PA, Ruben SR, Southgate CD, Green MR, Rosen CA. 1993. The type 1 human immunodeficiency virus Tat binding protein is a transcriptional activator belonging to an additional family of evolutionarily conserved genes. *Proc. Natl. Acad. Sci. USA* 90:138–42

112. Olsen HS, Cochrane AW, Dillon PJ, Nalin CM, Rosen CA. 1990. Interaction of the human immunodeficiency virus type 1 *rev* protein with structured region in *env* mRNA is dependent upon multimer formation mediated through a basic stretch of amino acids. *Genes Dev.* 4:1357–65

113. Ørum H, Andersen PS, Øster A, Johansen LK, Riise E, et al. 1993. Efficient method for constructing comprehensive murine Fab antibody libraries displayed on phage. *Nucleic Acids Res.* 21:4491–98

114. Pantaleo G, Fauci AS. 1995. Apoptosis in HIV infection. *Nat. Med.* 1:118–20

115. Pantaleo G, Graziosi C, Fauci AS. 1993. New concepts in the immunopathogenesis of human immunodeficiency virus infection. *N. Engl. J. Med.* 328(5):327–35

116. Park H, Davies MV, Langland JO, Chang H-W, Nam YS, et al. 1994. TAR RNA-binding protein is an inhibitor of the interferon-induced protein kinase PKR. *Proc. Natl. Acad. Sci. USA* 91:4713–17

117. Peliska JA, Balasubramanian S, Giedroc DP, Benkovic SJ. 1994. Recombinant HIV-1 nucleocapsid protein accelerates HIV-1 reverse transcriptase catalyzed strand transfer reactions and modulates RNase H activity. *Biochemistry* 33: 13817–23

118. Popovic M, Sarngadharan MG, Read E, Gallo RC. 1984. Detection, isolation, and continuous production of cytopathic retroviruses (HTLV-III) from patients with AIDS and pre-AIDS. *Science* 224:497–500

119. Posner MR, Hideshima T, Cannon T, Mukherjee M, Mayer KH, Byrn RA. 1991. An IgG human monoclonal antibody that reacts with HIV-1/gp120, inhibits virus binding to cells, and neutralizes infection. *J. Immunol.* 146:4325–32

120. Poznansky MC, Foxall R, Ramstedt U, Coker R, Mhashilkar A, et al. 1996. Intracellular antibodies against Tat and gp120 inhibit the replication of HIV-1 in T-cells from HIV-infected individuals. *Abstr. Int. AIDS Conf., 10th, Vancouver*

121. Prats AC, Sarih L, Gabus C, Litvak S, Keith G, Darlix JL. 1988. Small finger protein of avian and murine retroviruses has nucleic acid annealing activity and positions the replication primer tRNA onto genomic RNA. *EMBO J.* 7:1777–83

122. Pugsley AP. 1989. In *Protein Targeting*, San Diego: Academic

123. Richardson JH, Marasco WA. 1995. Intracellular antibodies: development and therapeutic potential. *Trends Biotechol.* 13:306–10

124. Richardson JH, Sodroski JG, Waldmann TA, Marasco WA. 1995. Phenotypic knockout of the high-affinity human interleukin 2 receptor by intracellular single-

chain antibodies against the subunit of the receptor. *Proc. Natl. Acad. Sci. USA* 92:3137–41

125. Roberts MR, Qin L, Zhang D, Smith DH, Tran A-C, et al. 1994. Targeting of human immunodeficiency virus-infected cells by CD8$^+$ T lymphocytes armed with universal T-cell receptors. *Blood* 84:2878–89

126. Roe TY, Reynolds TC, Yu G, Brown PO. 1993. Integration of murine leukemia virus DNA depends on mitosis. *EMBO J.* 12:2099–108

127. Rondon IJ, Marasco WA. 1996. Inhibition of HIV-1 replication by a sFv intrabody against the nucleocapsid, NCp7. *IBC Annu. Int. Conf. Antibody Eng., 7th. Coronado, CA* (Abstr.) Dec. 2–4

128. Ruben S, Perkins A, Purcell R, Joung K, Sai R, et al. 1989. Structural and functional characterization of human immunodeficiency virus tat protein. *J. Virol.* 63(1):1–8

129. Ruhl M, Himmelspach M, Bahr GM, Hammerschmid F, Jacksche H, et al. 1993. Eukaryotic initiation factor 5A is a cellular target of the human immunodeficiency virus type 1 Rev activation domain mediating *trans*-activation. *J. Cell Biol.* 123:1309–20

130. Sarver N, Cantin EM, Chang PS, Zaia JA, Ladne PA. 1990. Ribozymes as potential anti-HIV-1 therapeutic agents. *Science* 247:1222–25

131. Sastry KJ, Reddy RHR, Pandita R, Totpal K, Aggarwall BB. 1990. HIV-1 Tat gene induces tumor necrosis factor-β (Lymphotoxin) in a human B-lympho-blastoid cell line. *J. Biol. Chem.* 265:20091–93

132. Schier R, Bye J, Apell G, McCall A, Adams GP, et al. 1996. Isolation of high-affinity monomeric human anti-c-*erb*B-2 single chain Fv using affinity-driven selection. *J. Mol. Biol.* 255:28–43

133. Shaheen F, Duan L, Zhu M, Bagasra O, Pomerantz RJ. 1996. Targeting human immunodeficiency virus Type 1 reverse transcriptase by intracellular expression of single-chain variable fragments to inhibit early stages of the viral life cycle. *J. Virol.* 70:3392–400

134. Sheng N, Erickson-Viitranen S. 1994. Cleveage of p15 in vitro by HIV-1 protease is RNA dependent. *J. Virol.* 68:6207–14

135. Shukla RR, Kimmel PL, Kumar A. 1994. Human immunodeficiency virus type 1 Rev-responsive element RNA binds to host cell-specific proteins. *J. Virol.* 68:2224–29

136. Siomi H, Shida H, Maki M, Hatanaka M. 1990. Effects of a highly basic region of

human immunodeficiency virus Tat protein on nucleolar localization. *J. Virol.* 64:1803–7

137. Smith AE. 1995. Viral vectors in gene therapy. *Annu. Rev. Microbiol.* 49:807–38

138. Sodroski J, Goh WC, Rosen C, Dayton A, Terwilliger E, Haseltine WA. 1986. A second post-transcriptional *trans*-activator gene required for HTLV-III replication. *Nature* 321:412–17

139. Sodroski J, Goh WC, Rosen C, Tartar A, Portetelle D, et al. 1986. Replication and cytopathic potential of HTLV-III/LAV with sor gene deletions. *Science* 231:1549–53

140. Spearman P, Wang J-J, Vander Heyden N, Ratner L. 1994. Identification of human immunodeficiency virus type 1 gag protein domains essential to membrane binding and particle assembly. *J. Virol.* 68:3232–42

141. Sullenger BA, Gallardo HF, Ungers GE, Gilboa E. 1991. Analysis of trans-acting response decoy RNA-mediated inhibition of human immunodeficiency virus type 1 transactivation. *J. Virol.* 65(12):6811–16

142. Taviadoraki P, Benvenuto E, Trinca S, De Martinis D, Cattaneo A, Galeffi P. 1993. Transgenic plants expressing a functional single-chain Fv antibody are specifically protected from virus attack. *Nature* 366:469–72

143. Thompson CB. 1995. Apoptosis in the pathogenesis and treatment of disease. *Science* 267:1456–62

144. Varmus H. 1988. Retroviruses. *Science* 240:1427–35

145. Varmus HE, Swanstrom R. 1985. Replication of retroviruses. See Ref. 150, pp. 75–134 (Suppl.)

146. Vaughan TJ, Williams AJ, Pritchard K, Osbourn JK, Pope AR, et al. 1996. Human antibodies with sub-nanomolar affinities isolated from a large non-immunized phage display library. *Nat. Biotech.* 14:309

147. von Schwedler U, Kornbluth RS, Trono D. 1994. The nuclear localization signal of the matrix protein of human immunodeficiency virus type 1 allows the establishment of infection in macrophages and quiescent T lymphocytes. *Proc. Natl. Acad. Sci. USA* 91:6992–96

148. Ward ES, Güssow D, Griffiths AD, Jones PT, Winter G. 1989. Binding activities of a repertoire of single immunoglobulin variable domains secreted from *Escherichia coli. Nature* 341:544–46

149. Wang B, Ugen KE, Srikantan V, Agadjanyan MG, Dang K, et al. 1993. Gene inoculation generates immune responses

against human immunodeficiency virus type 1. *Proc. Natl. Acad. Sci. USA* 90: 4156–60

150. Weiss R, Teich N, Varmus H, Coffin J, eds. 1985. *RNA Tumor Virus.* Cold Spring Harbor, NY: Cold Spring Harbor Lab.

151. Westendorp MO, Li-Weber M, Frank RW, Krammer PH. 1994. Human immunodeficiency virus type 1 Tat unregulates interleukin-2 secretion in activated T cells. *J. Virol.* 68:4177–85

152. Winter G, Milstein C. 1991. Man-made antibodies. *Nature* 349:293–99

153. Wu Y, Duan L, Zhu M, Hu B, Kubota S, et al. 1996. Binding of intracellular anti-

Rev single chain variable fragments to different epitopes of human immunodeficiency virus type 1 Rev: variations in viral inhibition. *J. Virol.* 70:3290–97

154. Zhou H, Fisher RJ, Papas TS. 1994. Optimization of primer sequences for mouse scFv repertoire display library construction. *Nucleic Acids Res.* 22:888–89

155. Zhou W, Parent LJ, Wills JW, Resh MD. 1994. Identification of a membrane-binding domain within the amino terminal region of human immunodeficiency virus type 1 gag protein which interacts with acidic phospholipids. *J. Virol.* 68:2556–69

\

Annu. Rev. Microbiol. 1997. 51:285–310

MICROBIAL ALDOLASES AND TRANSKETOLASES: New Biocatalytic Approaches to Simple and Complex Sugars

Shuichi Takayama, Glenn J. McGarvey, and Chi-Huey Wong

Department of Chemistry, The Scripps Research Institute, 10550 North Torrey Pines Road, La Jolla, California 92037; e-mail: wong@scripps.edu

KEY WORDS: recombinant enzymes, carbohydrate recognition, enzyme inhibitors, multienzyme synthesis, stereoselective synthesis

ABSTRACT

Enzymes have become exceedingly valuable tools in organic synthesis as the reactions they catalyze generally proceed under mild conditions and in high stereo- and regioselectivity. Advances in microbiology and genetic engineering have greatly increased the availability of various enzymes. One of the most useful applications of enzyme-catalyzed chemical transformations is in the synthesis of water-soluble, polyfunctional organic molecules such as carbohydrates. As the pivotal roles that carbohydrates play in biological processes become more evident, access to these compounds becomes increasingly important. This review gives a brief overview of the use of aldolases and transketolases in the synthesis of sugars, sugar analogs, and related compounds.

CONTENTS

GENERAL INTRODUCTION

Although fermentation dates the use of biocatalytic transformations as centuries old, it has only been in recent decades that chemists have begun to rationally exploit the synthetic potential of such methodology. Increased understanding of the in vivo role of enzymes as the biological catalysts necessary for the proper regulation of crucial life-sustaining chemical transformations is largely responsible for the advances in this area. Among the earlier successes in exploiting the synthetic potential of enzymes was the application of various hydrolytic enzymes, such as lipases, esterases, and amidases. Recent advances have extended the field of enzymatic synthesis to many other complex types of reactions such as oxidoreductions, carbon-carbon bond-forming reactions, glycosylation reactions, and various group transfer reactions that are difficult or impossible to execute with conventional synthetic methods.

Microbiology has been invaluable to the development of enzymatic synthesis. Many synthetically useful enzymes are, in fact, of microbial origin, and the ease by which microorganisms can be grown provides an attractive source for large quantities of these enzymes. Microorganisms also provide the means for cloning and overexpressing enzymes of both microbial and nonmicrobial origins. Their enormous variety and ability to adapt to extreme environments makes them an excellent source of new enzymes, many possessing unique properties. The value of microorganisms is further augmented through the use of site directed or random mutagenesis to allow the production of enzymes with new, previously unobserved properties.

One of the most attractive applications of enzyme-catalyzed chemical transformations is in the synthesis of water-soluble, polyfunctional organic molecules. A prominent example, which has no competition from other techniques, is found in the use of deoxyribonucleic acid ligases (DNA ligases) and restriction enzymes to chemically manipulate DNA. Another area in which enzymes have proven especially useful is in carbohydrate synthesis. Conventional synthetic approaches to these molecules generally require multiple protection/deprotections sequences and face difficulties in stereocontrol. The high regio- and stereoselectivity and mild reaction conditions found in enzyme-catalyzed reactions provide new opportunities to tackle many synthetic problems encountered in the synthesis of carbohydrate species. With increasing evidence that carbohydrates play key roles in many important biological processes, including immune response, inflammation, metastasis, and other cell-recognition events, access to these compounds becomes correspondingly more important. Consequently, synthetic methods that provide access to natural structures, as well as carbohydrate analogs, carbohydrate probes, and glycoprocessing enzyme inhibitors, are invaluable.

This review gives a brief overview of the use of aldolases and transketolases to synthesize sugars, sugar analogs, and related compounds. Aldolases and transketolases (TK) are enzymes that catalyze the reversible transfer of carbonyl-group–containing donor units to acceptor aldehydes. Although these enzymes are quite specific for the donor component, a wide range of aldehydes can be used as acceptors, which thus provides access to a variety of natural and unnatural sugar structures. These enzyme-catalyzed aldol condensations are stereospecific and provide products in high optical purity. In combination with other enzymatic and chemical transformations, the aldolase- and transketolase-catalyzed reactions provide a powerful and convenient route to sugars, sugar analogs, glycoprocessing enzyme inhibitors, and carbohydrate probes.

ALDOL ADDITION REACTIONS WITH ALDOLASES

Introduction

Aldol condensations have emerged as an important method of carbon-carbon bond formation in organic synthesis (44, 44a, 50a, 66a, 68a, 79a). These methods have become increasingly valuable as recent efforts have improved the stereoselection of these reactions. Complementing these chemical methods is use of aldolases that catalytically promote this reaction, generally in a highly stereoselective manner. Early in this century, a class of enzymes was recognized that catalyzed the reversible formation of hexoses from their three carbon components via an aldol condensation (61). Originally named zymohexase, the enzymes capable of catalyzing an aldol condensation are now known as aldolases. Several of these aldolases have been explored for use as catalysts in organic synthesis and have been reviewed recently (23, 39, 90, 94). In this section, the different types of aldolases and their recent applications in the synthesis of sugars and sugar-related compounds are discussed.

Most of the over 30 aldolases identified and isolated so far catalyze the reversible stereospecific addition of a ketone donor to an aldehyde acceptor. Mechanistically, two distinct classes can be recognized (Figure 1) (10a, 23a, 48): Type I aldolases form a Schiff-base intermediate in the active site with the donor substrate, which subsequently adds stereospecifically to the acceptor; and type II aldolases use a Zn^{2+} cofactor, which acts as a Lewis acid in the active site. Type I aldolases are primarily found in animals and higher plants, whereas type II aldolases are predominantly encountered in bacteria and fungi. In general, type II aldolases are more stable than type I aldolases. Both types of enzymes are rather specific for the donor substrate but exhibit a more relaxed specificity for the acceptor substrate. The stereoselectivity of the aldol reaction is, in

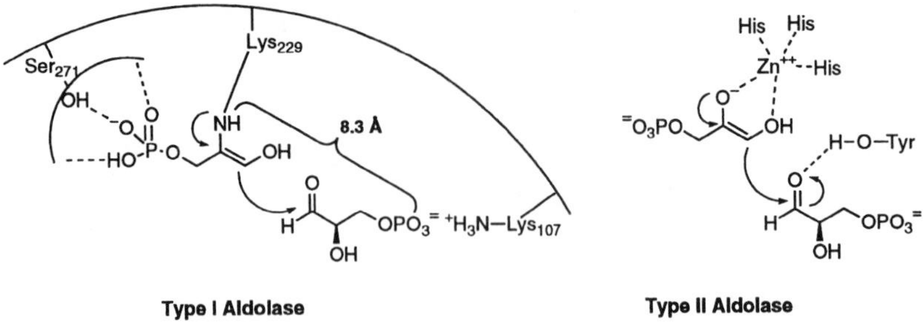

Type I Aldolase   **Type II Aldolase**

Figure 1   Mechanism of type I and type II aldolases.

general, controlled by the enzyme and does not depend on the structure or stereochemistry of the substrate; this characteristic thus makes the stereochemistry of the products highly predictable.

The aldolases that have been investigated for use in synthesis can be divided into four main groups, based on the structure of the donor substrate: (*a*) the dihydroxyacetone phosphate (DHAP)-dependent aldolases, (*b*) the pyruvate- or phosphoenol pyruvate (PEP)-dependent aldolases, (*c*) 2-deoxyribose-5-phosphate aldolase (DERA, the lone member of the acetaldehyde-dependent aldolases), and (*d*) the glycine-dependent aldolases. In the following sections, these four groups are discussed in more detail.

Apart from these main groups, several other aldolases are known, but they have not been investigated very much in terms of their use in synthesis or their substrate specificity. These enzymes are mentioned in other reviews (39, 94) and will not be discussed here.

DHAP-Dependent Aldolases

GENERAL BACKGROUND   In vivo, four DHAP-dependent aldolases catalyze the reversible asymmetric aldol addition of DHAP to D-glyceraldehyde 3-phosphate (G3P) or L-lactaldehyde, and each reaction generates one unique product whose stereochemistry at C-3 and C-4 is complementary to the others (Figure 2).

To date, fructose 1,6-diphosphate aldolase (FDP aldolase) has been the most extensively studied. Both type I and type II enzymes are known and they have been isolated from several mammalian and selected microbial sources. For synthesis, the type I enzyme from rabbit muscle (RAMA) has been used most often and so is included in this review even though it is not of microbial origin. This enzyme is not particularly air-sensitive, despite the presence of an active site thiol group. The half-life of the free enzyme, which is ca 2 days in aqueous solution at pH 7.0 (7, 84), can be lengthened by immobilization or by enclosure in a dialysis membrane. Recently, the monomeric type I FDP aldolase from *Staphylococcus carnosus* was shown to be much more stable and was also used

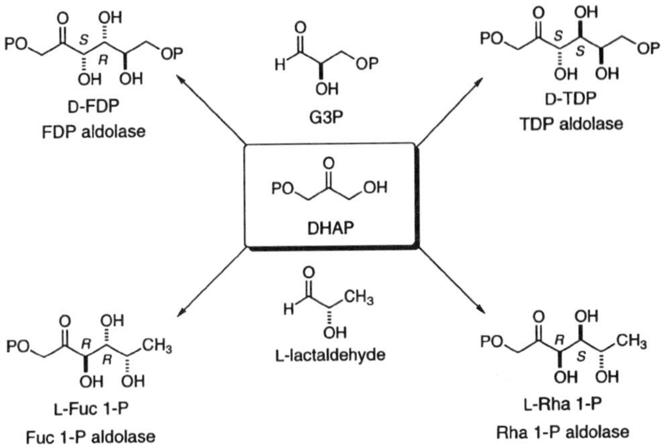

Figure 2   Product stereochemistries generated by the four complementary DHAP aldolases.

in synthesis (13). Type II FDP aldolases were found to be even more stable. For example, the enzyme from *Esherichia coli* has no thiol group in the active site and has a half-life of ca 60 days in 0.3 mM Zn^{2+} at pH 7.0 (84). Despite the small degree of homology in the primary sequences between the enzymes from rabbit muscle (type I) and *E. coli* (type II), they possess almost the same substrate specificity (45).

Fuculose 1-phosphate aldolase (Fuc 1-P aldolase) and rhamnulose 1-phosphate aldolase (Rha 1-P aldolase) have only been found as type II enzymes in several microorganisms (15, 16, 79). They have been cloned and overexpressed and subsequently purified (32, 36, 66). Tagatose 1,6-diphosphate aldolase (TDP aldolase), a type I aldolase involved in the galactose metabolism of *Streptococci*, has also been isolated from several sources (21) and has been recently cloned and overexpressed (36).

All four types of DHAP-dependent aldolases have been explored for synthetic application. Extensive studies have demonstrated that these enzymes, while being quite specific for the donor substrate DHAP, accept a wide range of aldehydes as the acceptor substrates. A detailed description of the substrate specificity of these enzymes has appeared recently (90). In general, unhindered aliphatic aldehydes, the α-heteroatom substituted aldehydes, including monosaccharides and their derivatives, are suitable acceptors. With FDP aldolase, phosphorylated aldehydes react more rapidly than their unphosphorylated analogs as a consequence of their resemblance to its natural substrate, G3P. In general, aromatic aldehydes, sterically hindered aliphatic, and α,β-unsaturated aldehydes are not substrates. The specificity for the donor substrate is much more stringent. Among many DHAP analogs tested, only compounds **1** and **2** were found to be weak substrates for rabbit muscle aldolase (RAMA) (ca 10% of the activity of DHAP) (7, 9) (Figure 3). Recently, compounds **3**, **4**, and **5** were also reported to be accepted by aldolase as DHAP analogs (31).

The aldol reactions with unnatural substrates catalyzed by these aldolases generate ketose phosphates as the initial products. The charged phosphate group facilitates the product separation and can be easily removed by enzymatic hydrolysis with acid phosphatase (EC.3.1.3.2). The stereochemistry of the vicinal diols produced usually follows closely that of the natural substrates. The configurations at C-3 are invariably conserved; however, the stereoselectivity at C-4 can be less consistent.

Resolution may be observed when racemic aldehydes are used as the acceptor. The diastereoselectivity exhibited by aldolases depends on the reaction conditions. In experiments wherein the reaction is stopped before reaching equilibrium, the diastereomer reflecting the kinetic preference of the enzyme is obtained. Alternatively, the reaction may be allowed to equilibrate to afford the more energetically favored product. This thermodynamic approach is especially useful when the aldol products can cyclize to form pyranoses, in which case the products with the fewest 1,3-diaxial interactions will predominate as a result of the reversible nature of the aldol reaction. This result is illustrated by the aldol condensation of a series of 2- or 3-substituted 3-hydroxypropanals with DHAP catalyzed by FDP aldolase (Figure 4) (7, 25).

SYNTHETIC APPLICATIONS   Owing to their broad substrate tolerance and defined product stereochemistry, DHAP-dependent aldolases have been employed to synthesize a large number of monosaccharides from simple precursors. These include ^{13}C-labeled sugars (75, 92, 93), heterosubstituted sugars, deoxysugars, fluorosugars (88), and high-carbon sugars (8, 24, 91). Many of these monosaccharides were subsequently converted into other interesting derivatives, such as azasugars, thiosugars, and carbocyclic compounds.

Mono- and disaccharides   More than 100 aldehydes have been used as the acceptor substrates for DHAP-dependent aldolases to prepare monosaccharides.

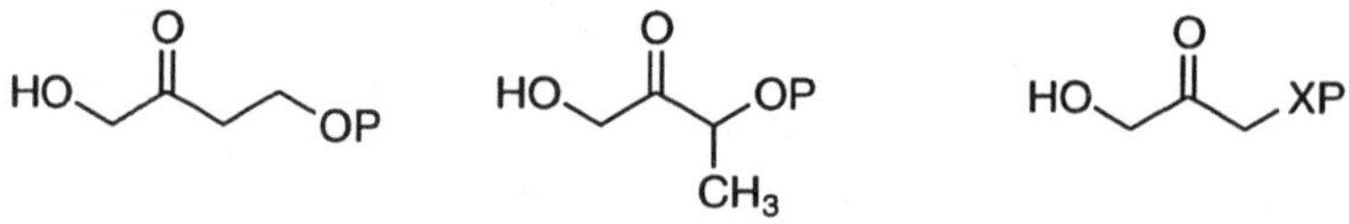

1   **2**   **3**: $X = CH_2$
                              4: $X = NH$
                              5: $X = S$

Figure 3   DHAP analogs that are substrates for RAMA.

DHAP-dependent aldolases also catalyze the condensation of pentose and hexose phosphates with DHAP, consequently extending the sugar chain by three carbons while introducing two new stereogenic centers. This process provides a new route to novel high-carbon sugars that are difficult to obtain from either chemical syntheses or natural sources. A number of these compounds have been synthesized, including analogs of sialic acid and KDO (e.g. **6, 7**) (Figure 5) (8). When an appropriate dialdehyde such as **8** is the substrate, *C*-disaccharide mimetics such as **9**, where two sugars are linked by carbon instead of oxygen, can be prepared by enzymatic tandem aldol reactions (Figure 5) (29).

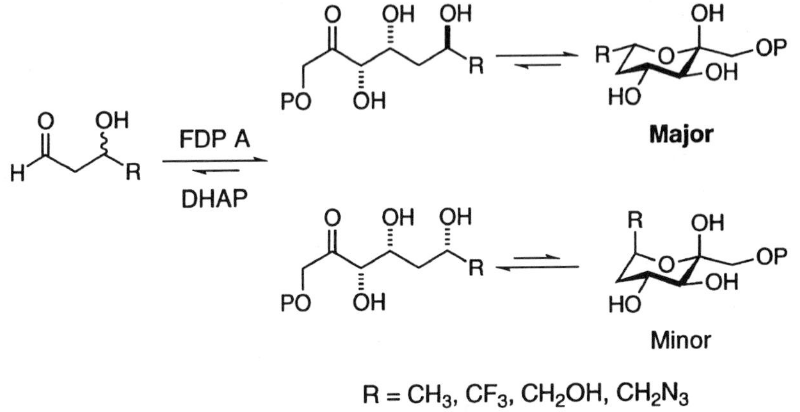

$$R = CH_3, CF_3, CH_2OH, CH_2N_3$$

Figure 4 Diastereoselectivity exhibited by FDP aldolase for acceptor aldehydes under thermodynamic conditions.

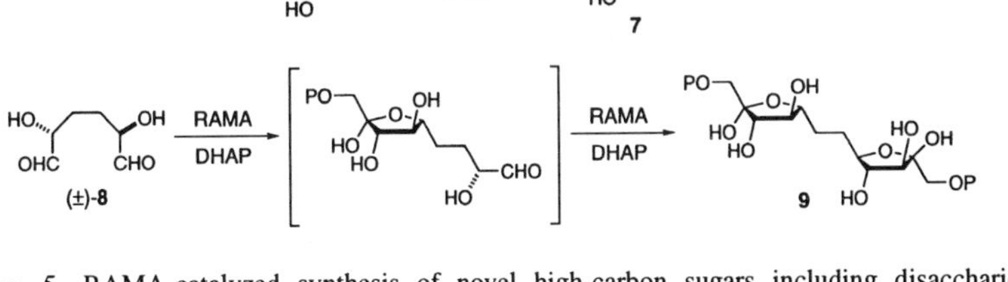

Figure 5 RAMA-catalyzed synthesis of novel high-carbon sugars including disaccharide mimetics.

As illustrated by these examples, DHAP-dependent aldolases generate several types of ketoses. However, most of the important naturally occurring carbohydrates are aldoses. One strategy to convert aldolase-generated ketoses to aldoses is by the application of isomerases. Glucose isomerase [GI, or xylose isomerase (EC 5.3.1.5)] catalyzes the isomerization of fructose to glucose and is used in the food industry to produce high-fructose corn syrup. GI also accepts fructose analogs that are modified at the 3, 5, and 6 positions. Various FDP aldolase products can be isomerized to a mixture of the ketose and aldose products, then separated. A series of 6-substituted glucoses (**10–14**) were synthesized using this FDP aldolase/GI methodology (Figure 6) (24, 25). Fuc isomerase (Fuc I, EC 5.3.1.3) and Rha isomerase (Rha I; EC 5.3.1.14) have also been cloned and overexpressed (30). They have been used in combination with Fuc-1-P aldolase and Rha aldolase, respectively, to synthesize various L-aldoses and their derivatives (2, 87).

Another method to convert ketoses to aldoses is through the so-called "inversion strategy" (12). Monoprotected dialdehydes were used as substrates for aldolase to generate protected aldehyde ketoses, which were then stereoselectively reduced, either enzymatically (19, 91) or chemically (12). Subsequent deprotection of the aldehyde afforded the aldoses.

Aza- and thiosugars   Azasugars have become increasingly important targets owing to their potential value as enzyme inhibitors and therapeutic agents (28, 49). The direct extension of the aldolase strategy has led to one of the most effective and practical routes to synthesize azasugars (58). Ketoses containing a suitable amine synthon, such as an azido group, can be generated by an enzymatic aldol condensation. Unmasking of the amine function, followed by reductive amination by a palladium-mediated hydrogenation, results in the production of azasugars. Based on this general strategy, a variety of 5-, 6-, and 7-membered ring azasugars have been synthesized (Figure 7) (63, 67, 90). These heterocycles are potent inhibitors of glycosidases and also have been used as key components in the synthesis of glycosyltransferase inhibitors (Figure 7) (67).

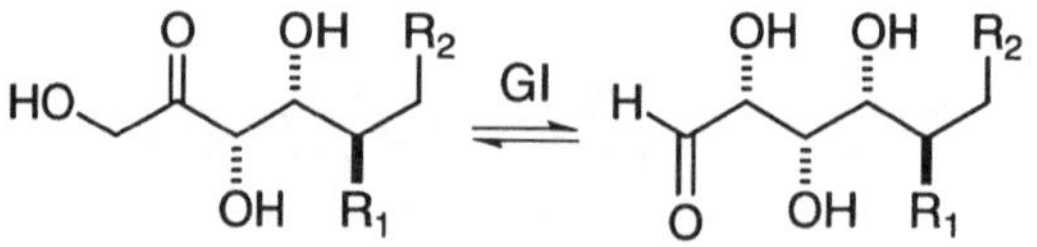

	R_1	R_2
10:	H	OH
11:	OH	H
12:	OH	OCH_3
13:	OH	F
14:	OH	N_3

Figure 6   Enzymatic isomerization of ketoses and aldoses using glucose isomerase.

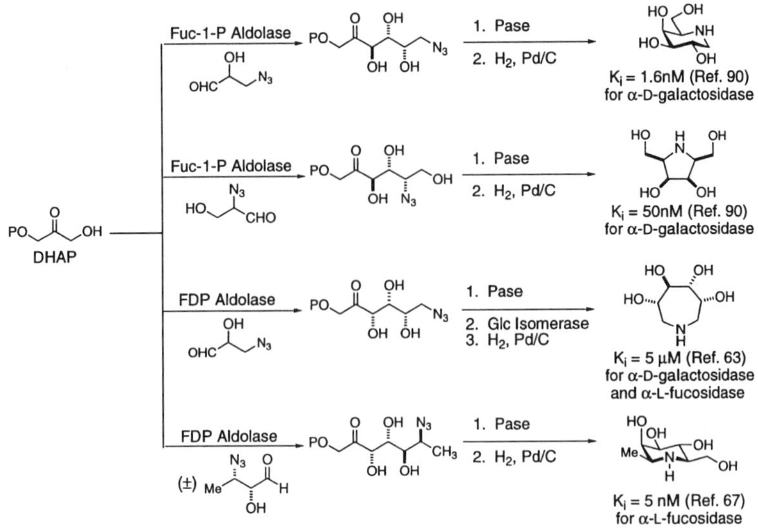

Figure 7   Chemoenzymatic preparation and biological activities of some 5-, 6-, and 7-membered ring azasugars.

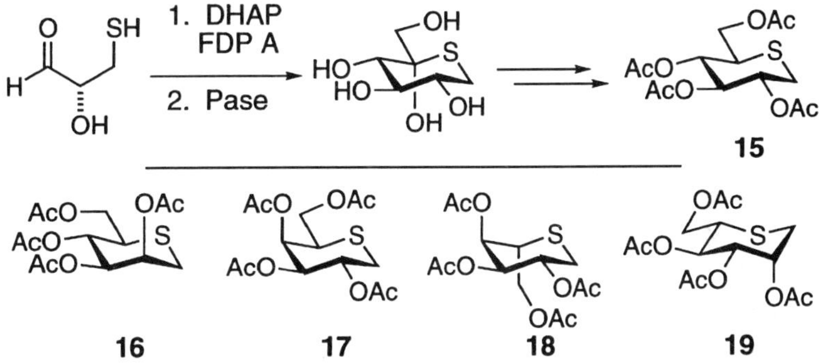

Figure 8   Chemoenzymatic preparation of deoxythiosugars.

Similar to the synthesis of azasugars, a series of deoxythiosugars were prepared by aldol condensation of thioaldehydes and DHAP, followed by reduction of the resulting thioketoses (17) (Figure 8). RAMA-catalyzed aldol condensation followed by dephosphorylation gave the corresponding thioketose (27), which was then acetylated and reduced to the 1-deoxy-5-thio-D-glucopyranose peracetate **15** (17). Similarly, deoxythiosugars **16–19** were obtained by using different combinations of aldehyde acceptors and aldolases (Figure 8) (17).

Other chemoenzymatic applications   The products from the FDP aldolase-catalyzed reaction have also been used to synthesize a variety of other compounds. These include the α–keto acid sugar 3-deoxy-D-*arabino*-heptulosonic acid (DAHP) (**20**) (80), the beetle pheromone (+)-exo-brevicomin (**21**) (74), homo-C-nucleoside **22** (57), C-glycoside **23** (65, 73), spirocycle **24** (59), and cyclitol **25** (Figure 9). The aldolase was used in the key step to establish the desired chirality in the target molecules. Although the synthesis of all the compounds in Figure 9 is based on aldol reactions catalyzed by FDP aldolase, stereoisomeric derivatives can be obtained by using other DHAP-dependent aldolases.

Cyclitols are an interesting class of biologically active compounds and the use of aldolases provides a chemoenzymatic strategy for their synthesis. An example is the synthesis of nitrocyclitols **27** and **28**, which was accomplished by FDP aldolase-catalyzed reaction with nitroaldehyde **26**, followed by a nonenzymatic intramolecular nitro-aldol reaction (Figure 10) (18).

A one-pot synthesis of the cyclitol **29** was reported involving a FDP aldolase-catalyzed reaction between a phosphonate aldehyde and DHAP (43). The aldol product cyclized in situ via an intramolecular Horner-Wadsworth-Emmons olefination to give the polyhydroxylated cyclopentene **29** (Figure 11). Using these approaches, different functionalized cyclitols may become easily accessible.

Pyruvate and Phosphoenolpyruvate-Dependent Aldolases

Pyruvate-dependent aldolases have catabolic functions in vivo, whereas their counterparts employing phosphoenolpyruvate as the donor substrate are involved in the biosynthesis of ketoacids. Both classes of enzymes can be used

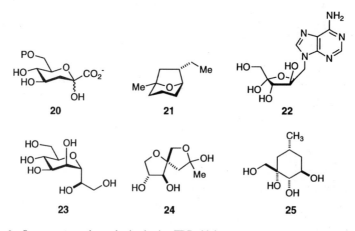

Figure 9   Some compounds synthesized using FDP-aldolase.

to prepare similar ketoacid products in vitro, and they are discussed jointly in the following section.

N-ACETYLNEURAMINIC ACID ALDOLASE AND NeuAc SYNTHASE   In vivo, the enzyme *N*-acetylneuraminic acid (NeuAc) aldolase (EC 4.1.3.3), also named sialic acid aldolase, catalyzes the reversible aldol reaction of *N*-acetyl-D-mannosamine [ManNAc (**30**)] and pyruvate (Figure 12) (20).

With an equilibrium constant of 12.7 M^{-1} the retro-aldol reaction is favored, because the enzyme has a catabolic function in vivo. When the aldolase is

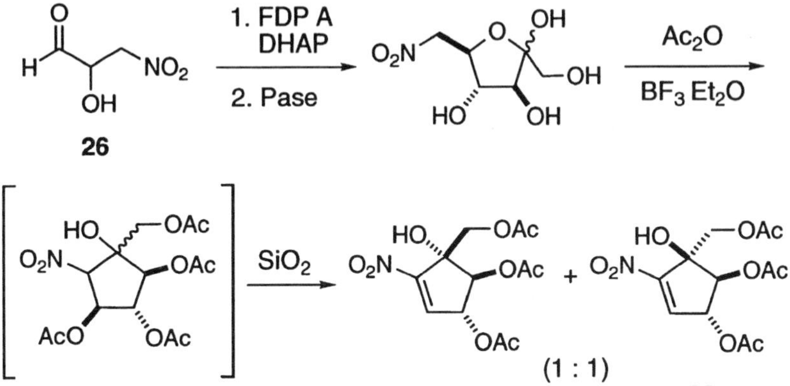

Figure 10 Chemoenzymatic preparation of nitrocyclitols.

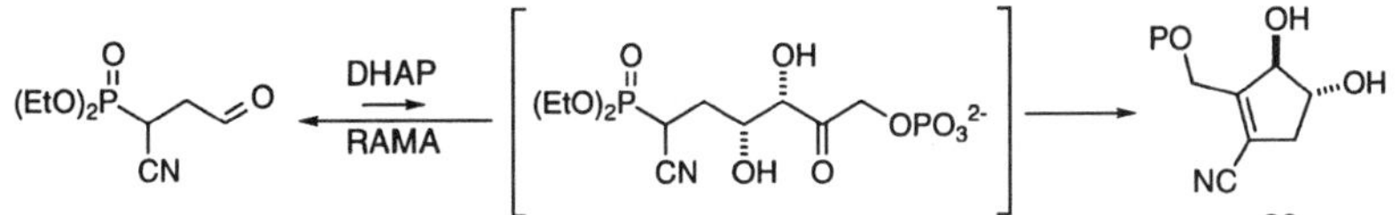

Figure 11 Chemoenzymatic preparation of a cyclopentene by a one-pot enzymatic aldol condensation-chemical cyclization sequence.

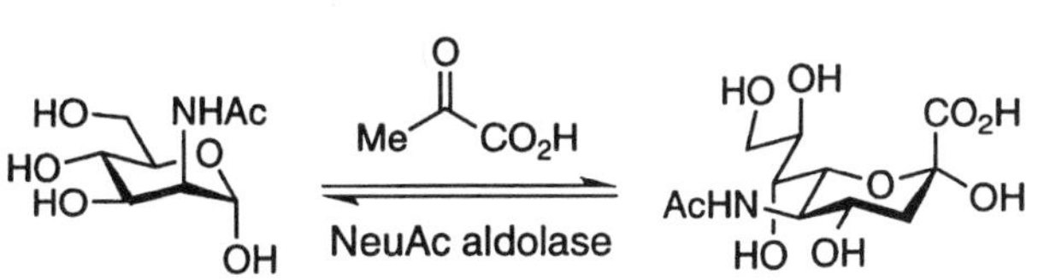

Figure 12 Aldol addition reaction catalyzed in vivo by NeuAc aldolase.

used for synthetic purposes, an equilibrium favoring the condensation product is usually induced by providing pyruvate in excess (81). NeuAc aldolase has been isolated from bacteria and mammals, and in both cases, it is a Schiff base–forming, type I aldolase. The enzyme is active between pH 6 and 9, has a temperature optimum of 37°C, and is stable under oxygen (10, 81). The enzymes from *Clostridia* and *E. coli* are now commercially available, and the enzyme from *E. coli* has been cloned and overexpressed (1).

Synthetic studies have shown that a high conversion (>80%) of ManNAc (**30**) to NeuAc (**31**) was achieved using the isolated enzyme, although several equivalents of pyruvate were required. Alternatively, the need to employ an excess of pyruvate can be avoided by coupling the NeuAc synthesis to an irreversible second step. For example, NeuAc aldolase reactions were coupled with a sialyltransferase reaction to produce sialylsaccharides (50).

With regard to its substrate specificity, NeuAc aldolase is specific for pyruvate as the donor substrate. However, the enzyme is remarkably flexible to a variety of acceptor substrates, including hexoses, pentoses, and tetroses, in both D- and L-forms. To date, over 60 aldoses have been demonstrated to be accepted by the enzyme (39). While substitutions at C(4), C(5), and C(6) of ManNAc (**30**) are allowed, the C(2) substituent is critical both in terms of size and stereochemistry and a free hydroxyl at C(3) is a prerequisite for successful aldol condensation (33).

In contrast to the strict enzymatic control of the stereochemical course usually encountered with aldolases, the stereochemical outcome of NeuAc aldolase-catalyzed reactions depends on the structure of the substrate. With most substrates [e.g. the natural substrate D-ManNAc (**30**)], the acceptor carbonyl group is attacked from the *si* face, which results in the formation of a new stereogenic center with an *S* configuration (Figure 13). However, with some substrates (e.g. L-mannose, D-altrose, L-xylose), the stereochemical course of the reaction can be reversed with the enamine attacking the carbonyl component from the *re* face, which leads to an *R* configuration for the C(4) center of the resulting sialic acid (Figure 13). The stereochemical outcome of these anomalous reactions appears to be under thermodynamic control since the resulting *R* configured

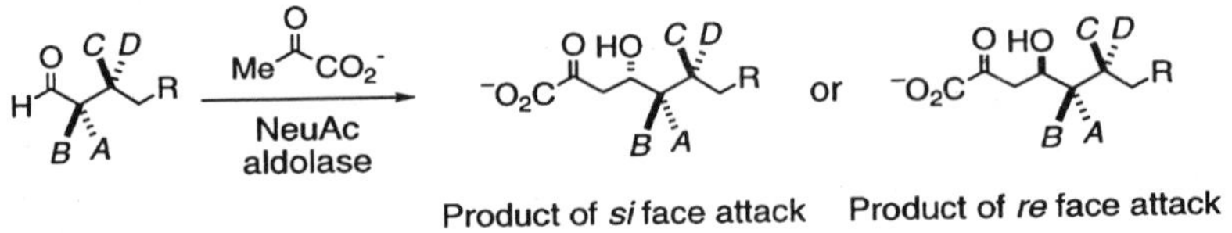

Figure 13   Product stereochemistry obtained by attack of the acceptor carbonyl on the *si* or *re* faces.

product is the more stable species with the hydroxy group at C(4) in the equatorial position.

Several biologically interesting L-sugars, including L-NeuAc (**32**), L-KDO (**34**), and L-KDN (**33**), were synthesized by taking advantage of the stereochemical preferences of the enzyme (Figure 14) (4, 33, 53).

Because of the biological importance of sialic acids and the commercial availability of NeuAc aldolase, various NeuAc-derivatives were synthesized enzymatically. For example, an efficient synthesis of 9-*O*-acetyl-NeuAc was developed by regioselective irreversible acetylation of ManNAc catalyzed by subtilisin followed by NeuAc aldolase-catalyzed condensation of the resulting 6-*O*-acetyl-ManNAc with pyruvate (51). This two-step enzymatic synthesis provided 9-*O*-acetyl-NeuAc (**36**) in ca 80% yield. A similar procedure was applied in the preparation of 9-*O*-lactyl-NeuAc (**37**) (51, 56) and the fluorescent derivative of sialic acid **38** (Figure 15) (34).

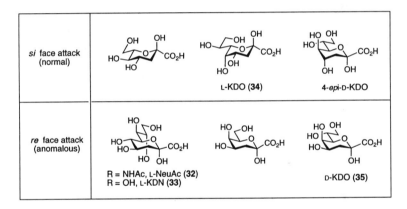

Figure 14   Biologically interesting sugars obtained using NeuAc aldolase.

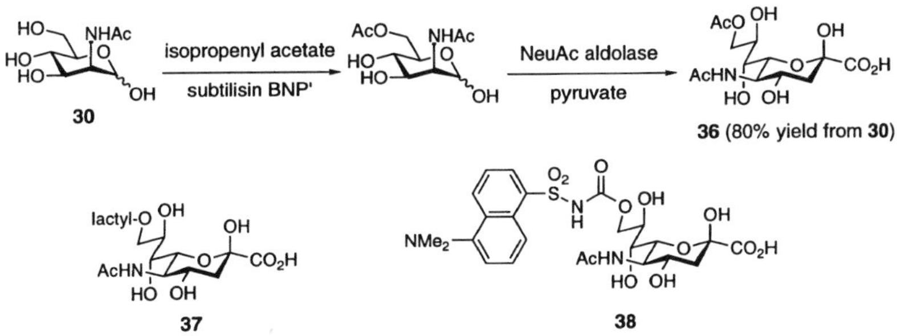

Figure 15   Synthesis of 9-*O*-derivatized NeuAc by combined use of subtilisin and NeuAc aldolase.

Azasugars were also prepared with NeuAc aldolase. Condensation of pyruvate with *N*-Cbz-D-mannosamine (**39**), followed by an intramolecular reductive amination of the deprotected aldolase product, gave the pyrrolidine **40**, which was further converted to 3-(hydroxymethyl)-6-epicastanospermine (**41**) (Figure 16) (96).

The synthesis of NeuAc (**31**) in vivo is accomplished by NeuAc synthetase (EC 4.1.3.19) through the irreversible condensation of PEP and *N*-acetylmannnosamine (**30**) (Figure 17) (10). Although this enzyme has yet to be isolated and characterized, it has promising synthetic potential, as this reaction is irreversible.

3-DEOXY-D-*MANNO*-2-OCTULOSONATE ALDOLASE AND 3-DEOXY-D-*MANNO*-2-OCTULOSONATE 8-PHOSPHATE SYNTHETASE 3-Deoxy-D-*manno*-2-octulosonate aldolase (EC 4.1.2.23), also named 2-keto-3-deoxyoctanoate (KDO) aldolase, is the enzyme responsible for degrading KDO in vivo. Since KDO (**35**) and its activated form, cytidine 5′-monophosphate-KDO (CMP-KDO), are key intermediates in the biosynthesis of the outer membrane lipopolysaccharide (LPS) of gram-negative bacteria (68), analogs of KDO may inhibit LPS

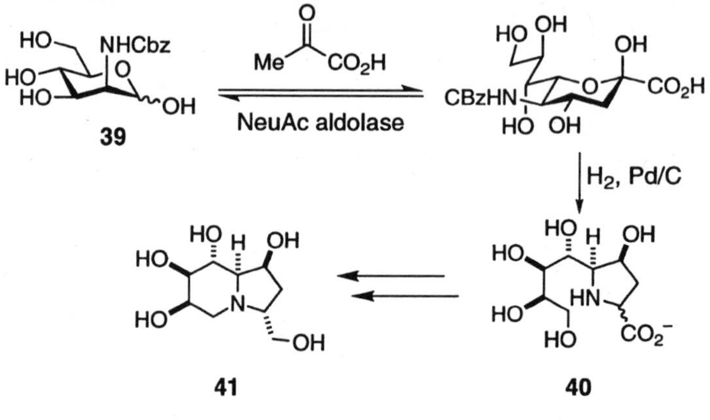

Figure 16 Synthesis of 3-(hydroxymethyl)-6-epicastanospermine using NeuAc aldolase.

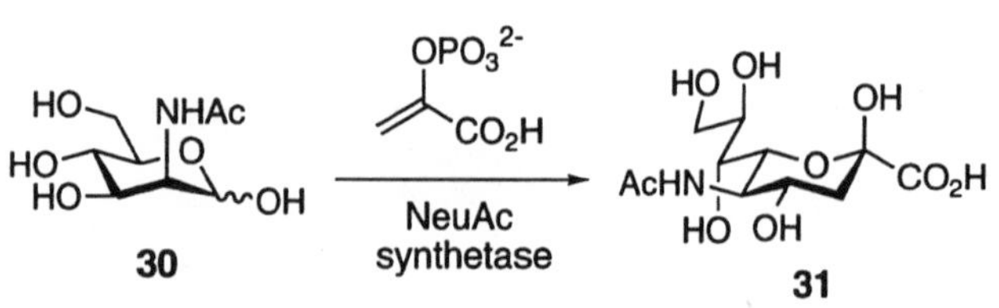

Figure 17 Aldol addition reaction catalyzed in vivo by NeuAc synthetase.

biosynthesis or interact with the LPS binding protein. KDO aldolase catalyzes the reversible condensation of pyruvate with D-arabinose (41) to form KDO with an equilibrium constant in the cleavage direction of $0.77 \ M^{-1}$ (Figure 18).

The enzyme from *E. coli* (38), *Aerobacter cloacae* (37), and *Aureobacterium barkerei*, strain KDO-37-2 have been described. The KDO aldolase from *A. barkerei*, strain KDO-37-2 accepts the widest variety of aldose substrates, including hexoses, pentoses, tetroses, and trioses, with pentoses and tetroses providing the best substrates (78). The enzyme was found to be specific for substrates having an *R*-configuration at C-3, whereas the stereochemical requirements at C-2 were less stringent. Under kinetic control, the C-2-*S*-configuration is favored while the C-2-*R*-configuration is favored under thermodynamic conditions (Figure 18). 3-Deoxy-D-*manno*-2-octulosonate 8-phosphate synthetase (EC 4.1.2.16), also named phospho-2-keto-3-deoxyoctanoate (KDO 8-P) synthetase, participates in the biosynthesis of bacterial lipopolysaccharides by catalyzing the irreversible aldol reaction of PEP and D-arabinose 5-phosphate (42) to give KDO 8-P (43) (Figure 19) (69, 70).

The enzyme has been isolated from *E. coli* B(6) and *Pseudomonas aeruginosa* (54). The *E. coli* enzyme has been cloned and overexpressed in *E. coli* and *Salmonella typhimurium* (85). Little is known about the substrate specificity of KDO 8-P synthetase, but initial studies suggest that this enzyme is highly specific for its natural substrates. A preparative scale synthesis of KDO 8-P itself has been reported, which involved the use of KDO 8-P synthetase (6).

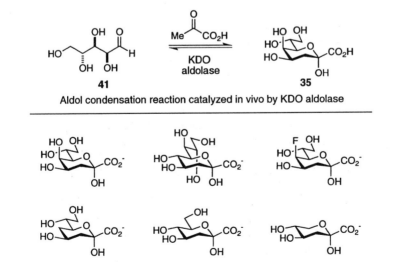

Aldol condensation reaction catalyzed in vivo by KDO aldolase

Figure 18   Compounds synthesized on a preparative scale using KDO aldolase.

2-KETO-3-DEOXY-6-PHOSPHOGLUCONATE ALDOLASE   Unlike most other pyruvate aldolases, the 2-keto-3-deoxy-6-phosphogluconate (KDPG) aldolases accept short-chain, noncarbohydrate electrophilic aldehydes as substrates. The enzyme stability, pH dependence, tolerance of cosolvent, and substrate specificity of the enzymes from *Pseudomonas putida, E. coli,* and *Zymomonas mobilis* have been described, and compounds **44**, **45**, and **46** have been prepared on a preparative scale using these enzymes (Figure 20) (76).

OTHER PYRUVATE AND PHOSPHOENOLPYRUVATE-DEPENDENT ALDOLASES   Although not extensively investigated for synthetic exploitation to date, a variety of other ketoacid-producing aldolases appear promising for future applications. These enzymes include 3-deoxy-D-arabino-2-heptulosonic acid 7-phosphate (DAHP) synthetase (EC 4.1.2.15), 2-keto-4-hydroxyglutarate (KHG) aldolase (EC 4.1.2.31) and 2-keto-3-deoxy-D-gluconate (KDG) aldolase (EC 4.1.2.20).

DAHP synthetase has been employed for the straightforward synthesis of DAHP (**20**) on a preparative-scale (Figure 21) (71). This multienzyme system contains transketolase, which generates D-erythrose 4-phosphate (**48**) from fructose 6-phosphate (Fru 6-P) (**47**) in the presence of D-ribose 5-phosphate. Fru 6-P, in turn, was generated from D-fructose and adenosine 5'-triphosphate (ATP) through a hexokinase-catalyzed reaction in the presence of an ATP regeneration system.

KHG aldolase displays a somewhat atypical substrate specificity, when compared to other aldolases, in that various pyruvate derivatives are accepted

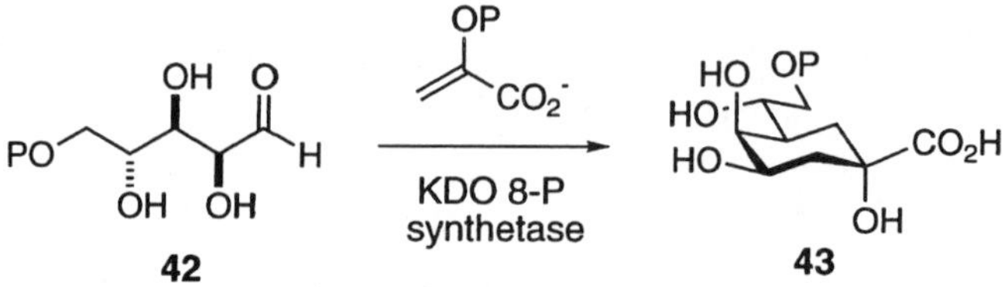

Figure 19   Aldol addition reaction catalyzed in vivo by KDO 8-P synthetase.

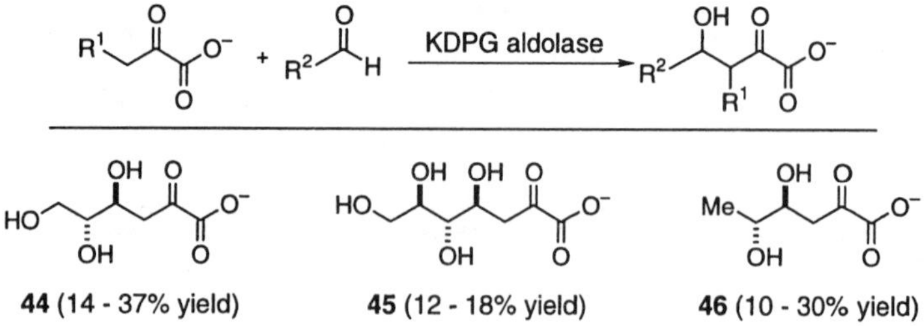

Figure 20   Compounds prepared on a preparative scale using KDPG aldolase.

as donor substrates by this enzyme (35).   Using KDG aldolase from *Aspergillus niger*, 2-keto-3-deoxy-D-gluconate was synthesized on a preparative scale (3).

2-Deoxyribose-5-Phosphate Aldolase

In vivo 2-deoxyribose-5-phosphate aldolase (DERA, EC 4.1.2.4) catalyzes the reversible aldol reaction of acetaldehyde and D-glyceraldehyde-3-phosphate (G3P) to form 2-deoxyribose-5-phosphate (**49**) (Figure 22). DERA is a type I aldolase that has been isolated from animal tissues (60) and microorganisms (47). It has been overexpressed in *E. coli*, allowing large quantities of the enzyme to be easily obtained (5, 14, 89). Under reaction conditions with a temperature of 25°C and pH 7.5, the enzyme is fairly stable, retaining 70% of its activity after 10 days. In addition to acetaldehyde, DERA is also able to accept propanal, acetone, and fluoroacetone as donor substrates, albeit at much slower rates. The acceptor substrates have very few structural requirements, and a variety of products including 2-deoxy-sugar analogs, thiosugars, azasugars, and disaccharide mimetics have been prepared (89).   2-Hydroxyaldehydes appear to react the fastest, with D-isomers being preferred over L-isomers, making selective formation of products from a racemic mixture of acceptor substrates possible. The stereochemistry of the newly generated chiral center is strictly determined by the enzyme, generally giving products with the (*S*)-configuration.

An interesting feature of DERA is that when acetaldehyde is used as the donor, the products from the DERA-catalyzed reaction are also aldehydes and are capable of becoming acceptor substrates for a second aldol condensation. Such sequential aldol reactions were first observed for DERA itself (40). When

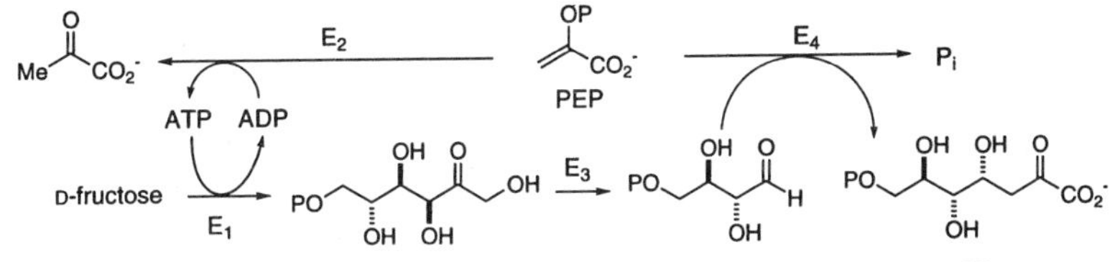

E_1) Hexokinase; E_2) Pyruvate kinase; E_3) Transketolase + D-ribose 5-P; E_4) DAHP synthetase

Figure 21   Multienzyme synthesis of DAHP.

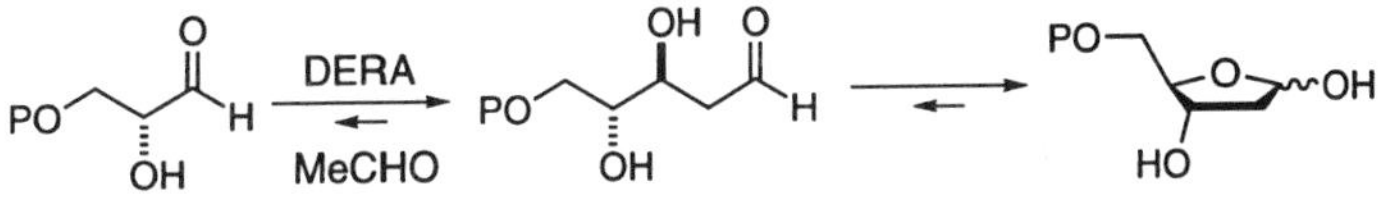

Figure 22   Aldol addition reaction catalyzed in vivo by DERA.

α-substituted acetaldehydes were used that did not incorporate functionality that will cyclize to a hemiacetal after the first aldol reaction, the products condensed with a second molecule of acetaldehyde to form 2,4-dideoxyhexoses **50** (Figure 23). These products cyclize to a stable hemiacetal **51**, thus stopping the oligomerization after two sequential aldol reactions. The products can be converted to derivatives of the lactone moiety of mevinic acids, which are active as cholesterol-lowering agents. This two-step synthesis provides the shortest route available to these chiral lactone functionalities **52** using extremely simple starting materials. The best substrate for these DERA-catalyzed sequential reactions appears to be succinic semialdehyde (R = CH_2CH_2COOH), which has a carboxylic acid group, mimicking the phosphate group in the natural substrate G3P (89).

Similar one-pot sequential aldol reactions were performed by combining DERA with FDP-aldolase (Figure 24) (41,42). The products of these reactions are 5-deoxyketoses **53** with three substituents in the axial positions. Due to the formation of these thermodynamically unfavorable products and the long

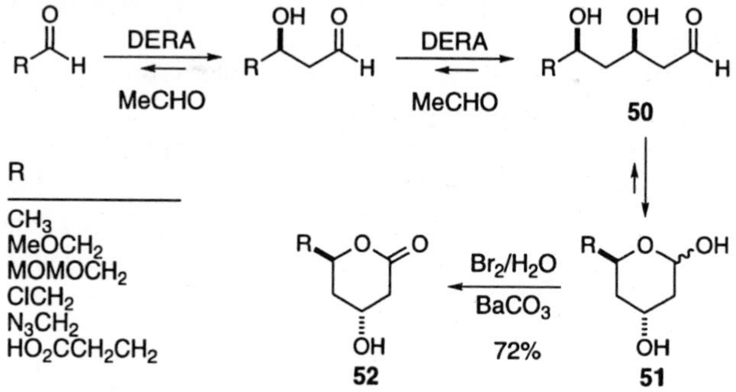
DERA
MeCHO
DERA
MeCHO
50
R
CH3
MeOCH2
MOMOCH2
ClCH2
N3CH2
HO2CCH2CH2
Br2/H2O
BaCO3
72%
52
51

Figure 23 Sequential aldol reactions catalyzed by DERA.

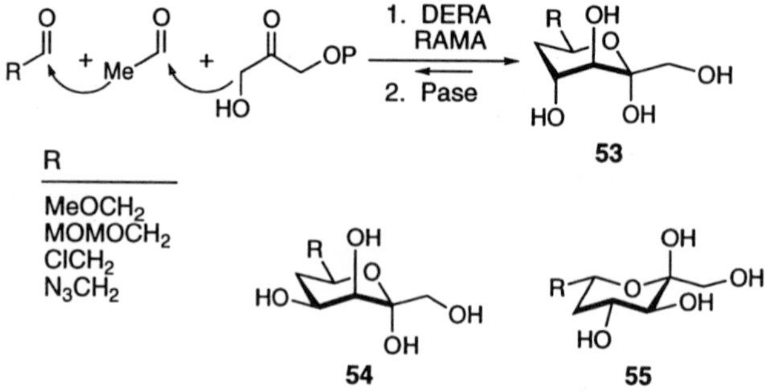
1. DERA
RAMA
2. Pase
53
R
MeOCH2
MOMOCH2
ClCH2
N3CH2
54
55

Figure 24 Sequential aldol reactions catalyzed by DERA and RAMA.

reaction times for these reactions, some inversion of the expected stereochemistry of both DERA and FDP-aldolase was observed, leading to products of type **54** and **55**. Combination of DERA and NeuAc-aldolase gave sialic acid derivatives of type **56** (Figure 25) (42).

Glycine-Dependent Aldolases

The glycine-dependent aldolases are pyridoxal-5-phosphate-dependent enzymes that catalyze the reversible aldol reaction of glycine with an aldehyde acceptor to form a β-hydroxy-α–amino acid (72, 77). Two types of glycine-dependent aldolases have been found: the serine hydroxymethyltransferases (SHMT) and the threonine aldolases. Only a few examples are known of the use of these aldolases in bond-forming reactions. Recently, the 4-hydroxy L-allothreonine derivative **57**, used as a key intermediate in the synthesis of potent sialyl Lewis[x] mimetics such as **58**, was prepared using threonine aldolase (82, 83, 86, 95). The hydroxyl groups of this unnatural amino acid are designed to mimic some of the essential carbohydrate hydroxyl groups in sialyl Lewis[x], a carbohydrate ligand that mediates some important cell adhesion events in the body (Figure 26).

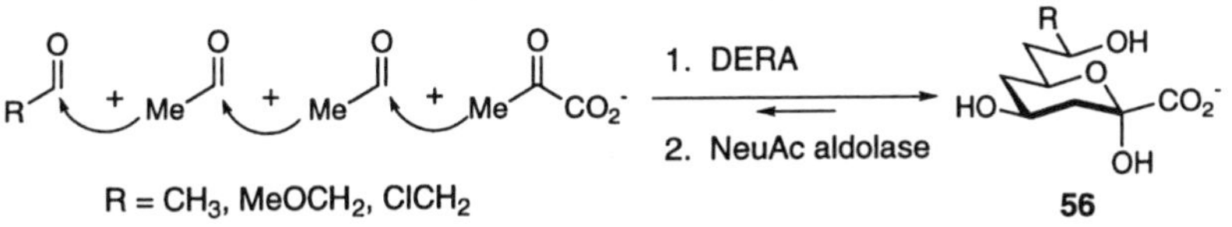

R = CH₃, MeOCH₂, ClCH₂

Figure 25   Sequential aldol reactions catalyzed by DERA and NeuAc aldolase.

Sialyl Lewis X (E,P Selectin Ligand)

20 times better binding to E-selectin
compared to sialyl Lewis[x] (Ref. 86)

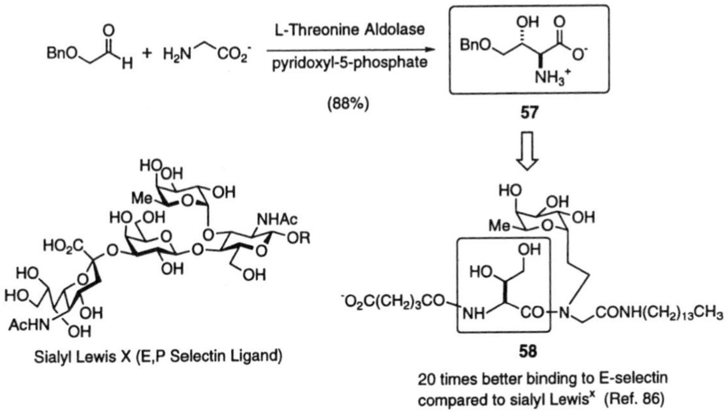

Figure 26   L-Threonine aldolase-catalyzed synthesis of an unnatural amino acid used as a key component of potent sialyl Lewis[x] mimetics.

KETOL AND ALDOL TRANSFER REACTIONS

Transketolases

One of the enzymes in the pentose phosphate pathway is transketolase (TK), which reversibly transfers the C1-C2 ketol unit from D-xylulose 5-phosphate to D-ribose 5-phosphate to generate D-sedoheptulose 7-phosphate and G3P (Figure 27). The enzyme relies on two cofactors for activity: thiamine pyrophosphate (TPP) and Mg^{2+}. TK from baker's yeast is commercially available, and the enzyme from *E. coli* has been overexpressed and prepared on a large scale (55). While several ketose phosphates besides D-xylulose 5-phosphate also act as ketol donors, the synthetically most useful donor is β-hydroxypyruvic acid (HPA) (11, 46).

Though reactions with aldose acceptors, such as HPA, are less readily catalyzed compared to the transfer reactions with D-xylulose 5-phosphate, the former are rendered irreversible through loss of carbon dioxide following transfer of the ketol unit by TK. A wide range of aldehydes are ketol acceptors, including

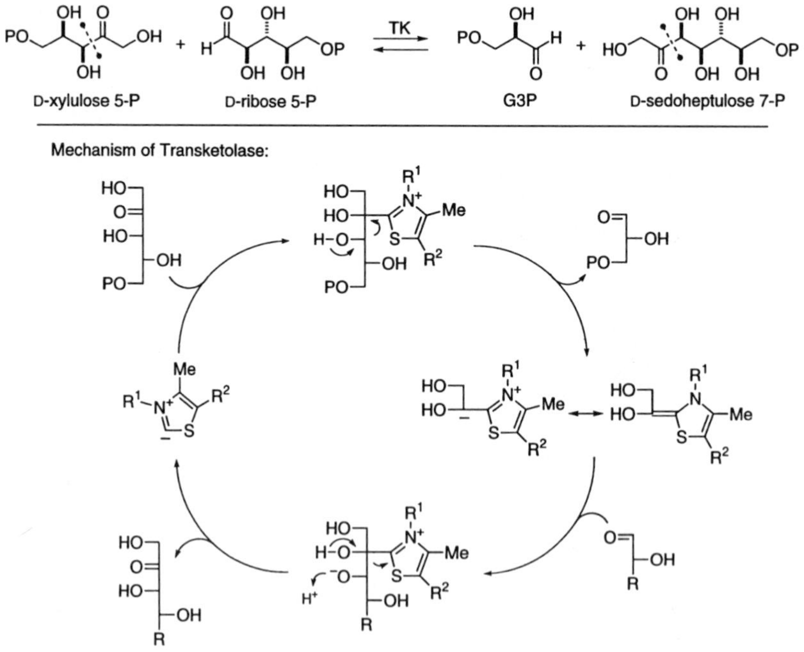

Figure 27   Ketol transfer reaction in the oxidative pentose phosphate pathway catalyzed by TK.

aliphatic, α,β-unsaturated, aromatic and heterocyclic aldehydes (22, 52, 64). β-Hydroxy aldehydes epimeric at C-2 can be efficiently resolved by TK since only the D-enantiomers are substrates and give D-*threo* products (26, 52). Beyond the C-2 position, the enzyme appears to have no stereochemical preference. A few synthetic examples employing TK are illustrated in Figure 28 (11, 46, 52, 97).

Transaldolases

Like transketolase, transaldolase (TA) is an enzyme in the oxidative pentose phosphate pathway. TA, which operates through a Schiff-base intermediate, catalyzes the transfer of the C1–C3 aldol unit from D-sedoheptulose 7-phosphate to G3P to produce D-Fru 6-P and D-erythrose 4-phosphate (Figure 29). Although it is commercially available, TA has rarely been used in organic synthesis, and no detailed substrate specificity study has yet been performed. In one application, TA was used in the synthesis of D-fructose from starch (Figure 30) (62). The aldol moiety was transferred from Fru 6-P to D-glycerol in the final step of this multienzyme synthesis of D-fructose.

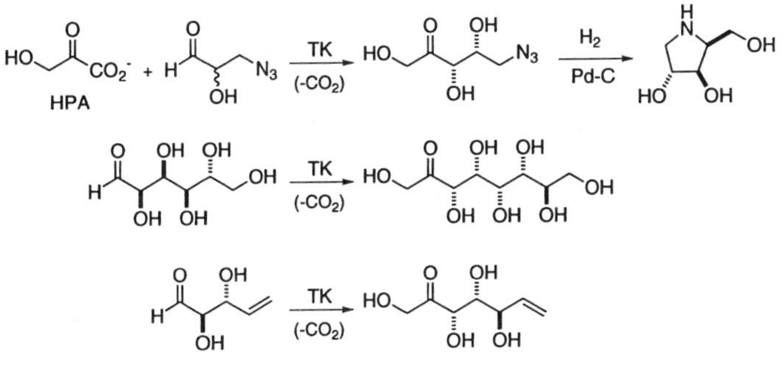

Figure 28   Examples of synthesis using TK.

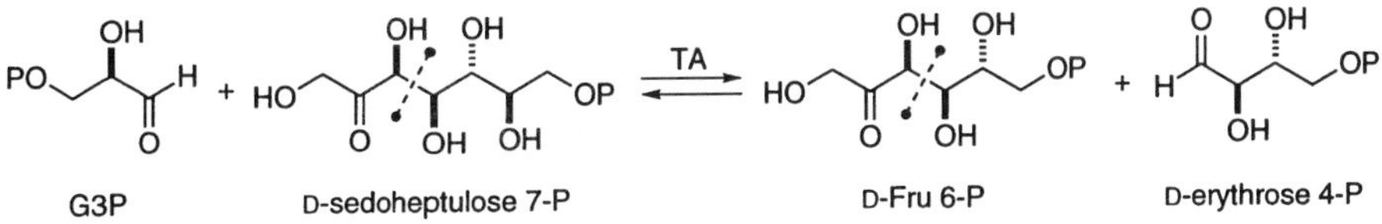

Figure 29   Aldol transfer reaction in the oxidative pentose phosphate pathway catalyzed by TA.

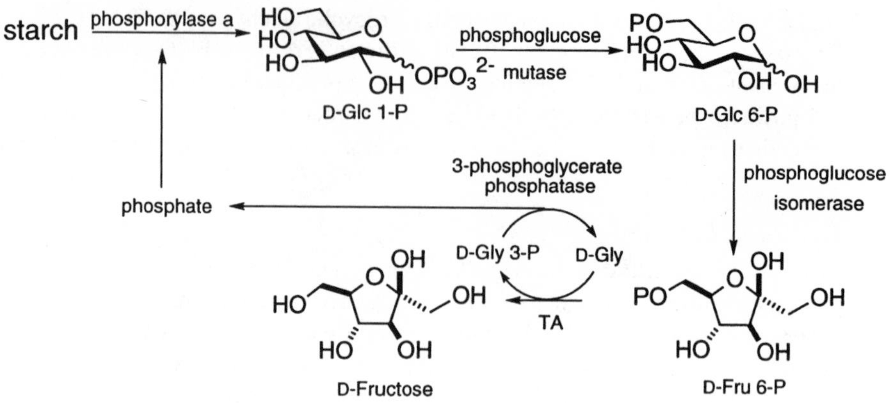

Figure 30   Multienzyme synthesis of D-fructose from starch.

CONCLUDING REMARKS

This review has briefly described the use of aldolases and transketolases for the synthesis of sugars and related compounds. The increased accessibility of these compounds promises to lead to a better understanding of important processes in glycobiology such as receptor-mediated recognition and glycoprotein processing, which can lead to the subsequent development of new therapeutics and diagnostics. Enzymatic synthesis is becoming increasingly indispensable as biologists and chemists strive to understand the molecular basis of living systems. The remarkable successes achieved in the area of enzymatic synthesis are a tribute to the close interaction between microbiology and chemistry. As our knowledge of the molecular basis of biological systems grows, the prospects for such interdisciplinary studies in the future are limitless.

> Visit the *Annual Reviews home page* at
> http://www.annurev.org.

Literature Cited

1. Aisaka K, Tamura S, Arai Y, Uwajima T. 1987. Hyperproduction of *N*-acetyl-neuraminate lyase by the gene-cloned strain of *Escherichia coli. Biotechnol. Lett.* 9:633–37
2. Aljarín R, Garcia-Junceda E, Wong C-H. 1995. A short enzymic synthesis of L-glucose from dihydroxyacetone phosphate and L-glyceraldehyde. *J. Org. Chem.* 60:4294–95

3. Augé C, Delest V. 1993. Microbiological aldolizations. Synthesis of 2-keto-3-deoxy-D-gluconate. *Tetrahedron Asymmetry* 4:1165–68
4. Augé C, Gautheron C, David S. 1990. Sialyl aldolase in organic synthesis: from the trout egg acid, 3-deoxy-D-glycero-D-galacto-2-nonulosonic acid (KDN), to branched-chain higher ketoses as possible new chirons. *Tetrahedron* 46:201–14

5. Barbas CFI, Wang Y-F, Wong C-H. 1990. Deoxyribose-5-phosphate aldolase as a synthetic catalyst. *J. Am. Chem. Soc.* 112:2013–14

6. Bednarski MD, Crans DC, DiCosimo R, Simon ES, Stein PD, Whitesides GM. 1988. Synthesis of 3-deoxy-D-manno-2-octulosonate-8-phosphate (KDO-8-P) from D-arabinose: generation of D-arabinose-5-phosphate using hexokinase. *Tetrahedron Lett.* 29:427–30

7. Bednarski MD, Simon ES, Bischofberger N, Fessner W-D, Kim M-J, et al. 1989. Rabbit muscle aldolase as a catalyst in organic synthesis. *J. Am. Chem. Soc.* 111:627–35

8. Bednarski MD, Waldmann HJ, Whitesides GM. 1986. Aldolase-catalyzed synthesis of complex C8 and C9 monosaccharides. *Tetrahedron Lett.* 27:5807–10

9. Bischofberger N, Waldmann H, Saito T, Simon ES, Lees W, et al. 1988. Synthesis of analogs of 1,3-dihydroxyacetone phosphate and glyceraldehyde 3-phosphate for use in studies of fructose-1,6-diphosphate aldolase. *J. Org. Chem.* 53:3457–65

10. Blacklow RS, Warren L. 1962. Biosynthesis of sialic acids by *Neisseria meningitidis. J. Biol. Chem.* 237:3520–26

10a. Blom N, Sygusch J. 1997. Product binding and role of the C-terminal region in class I D-fructose 1,6-bisphosphate aldolase. *Nature Struct. Biol.* 4:36–9

11. Bolte J, Demuynck C, Samaki H. 1987. Utilization of enzymes in organic chemistry: transketolase catalyzed synthesis of ketoses. *Tetrahedron Lett.* 28:5525–28

12. Borysenko CW, Spaltenstein A, Straub JA, Whitesides GM. 1989. The synthesis of aldose sugars from half-protected dialdehydes using rabbit muscle aldolase. *J. Am. Chem. Soc.* 111:9275–76

13. Brockamp HP, Kula MR. 1990. *Staphylococcus carnosus* aldolase as catalyst for enzymic aldol reactions. *Tetrahedron Lett.* 31:7123–26

14. Chen L, Dumas DP, Wong C-H. 1992. Deoxyribose 5-phosphate aldolase as a catalyst in asymmetric aldol condensation. *J. Am. Chem. Soc.* 114:741–48

15. Chiu T-H, Feingold DS. 1969. L-Rhamnulose-1-phosphate aldolase from *Escherichia coli.* Crystallization and properties. *Biochemistry* 8:98–108

16. Chiu T-H, Otto R, Power J, Feingold DS. 1966. An improved preparation of L-rhamnulose 1-phosphate. *Biochim. Biophys. Acta* 127:249–51

17. Chou W-C, Chen L, Fang J-M, Wong C-H. 1994. A new route to deoxythiosugars based on aldolases. *J. Am. Chem. Soc.* 116:6191–94

18. Chou W-C, Fotsch C, Wong C-H. 1995. Synthesis of nitrocyclitols based on enzymic aldol reaction and intramolecular nitroaldol reaction. *J. Org. Chem.* 60:2916–17

19. Christensen U, Tuchsen E, Andersen B. 1975. Initial velocity and product inhibition studies on L-iditol:NAD oxidoreductase. *Acta Chem. Scand. B* 29:81–87

20. Comb DG, Roseman S. 1960. The sialic acids. *J. Biol. Chem.* 235:2529–37

21. Crow VL, Thomas TD. 1982. D-Tagatose 1,6-diphosphate aldolase from lactic *Streptococci*: purification, properties, and use in measuring intracellular tagatose 1,6-diphosphate. *J. Bacteriol.* 151:600–8

22. Demuynck C, Bolte J, Hecquet L, Dalmas V. 1991. Enzyme-catalyzed synthesis of carbohydrates: synthetic potential of transketolase. *Tetrahedron Lett.* 32:5085–88

23. Drauz K, Waldmann H. 1995. Formation of carbon-carbon bonds. In *Enzyme Catalysis in Organic Synthesis,* B4:547–593. New York: VCH Publ.

23a. Dreyer MK, Schulz GE. 1996. Catalytic mechanism of the metal-dependent fuculose aldolase from *Escherichia coli* as derived from the structure. *J. Mol. Biol.* 259:458–66

24. Durrwachter JR, Drueckhammer DG, Nozaki K, Sweers HM, Wong C-H. 1986. Enzymic aldol condensation/isomerization as a route to unusual sugar derivatives. *J. Am. Chem. Soc.* 108:7812–18

25. Durrwachter JR, Wong C-H. 1988. Fructose 1,6-diphosphate aldolase-catalyzed stereoselective synthesis of C-alkyl and N-containing sugars: thermodynamically controlled C-C bond formations. *J. Org. Chem.* 53:4175–81

26. Effenberger F, Null V, Ziegler T. 1992. Enzyme catalyzed reactions. 12. Preparation of optically pure L-2-hydroxy aldehydes with yeast transketolase. *Tetrahedron Lett.* 33:5157-60

27. Effenberger F, Straub A, Null V. 1992. Enzyme-catalyzed reactions. 14. Stereoselective synthesis of thiosugars from achiral starting compounds by enzymes. *Liebigs Ann. Chem.* pp. 1297–301

28. Elbein AD. 1987. Inhibitors of the biosynthesis and processing of N-linked oligosaccharide chains. *Annu. Rev. Biochem.* 56:497–534

29. Eyrisch O, Fessner W-D. 1995. Disaccharide mimetics by enzymatic tandem

aldol reaction. *Angew. Chem. Int. Ed. Engl.* 34:1639–41

30. Fessner W-D, Badia J, Eyrisch O, Schneider A, Sinerius G. 1992. Enzymes in organic synthesis. 5. Enzymic syntheses of rare ketose 1-phosphates. *Tetrahedron Lett.* 33:5231–34

31. Fessner W-D, Sinerius G. 1994. Enzymes in organic synthesis. 7. Synthesis of dihydroxyacetone phosphate (and isosteric analogs) by enzymic oxidation: sugars from glycerol. *Angew. Chem. Int. Ed. Engl.* 33:209–12

32. Fessner W-D, Sinerius G, Schneider A, Dreyer M, Schulz GE, et al. 1991. Enzymes in organic synthesis. Part 1. Diastereoselective, enzymatic aldol addition with L-rhamnulose- and L-fuculose-1-phosphate aldolases from *E. coli. Angew. Chem. Int. Ed. Engl.* 30:555–58

33. Fitz W, Schwark J-R, Wong C-H. 1995. Aldotetroses and C(3)-modified aldohexoses as substrates for *N*-acetylneuraminic acid aldolase: a model for the explanation of the normal and the inversed stereoselectivity. *J. Org. Chem.* 60:3663–70

34. Fitz W, Wong C-H. 1994. Combined use of subtilisin and *N*-acetylneuraminic acid aldolase for the synthesis of a fluorescent sialic acid. *J. Org. Chem.* 59:8279–80

35. Floyd NC, Liebster MH, Turner NJ. 1992. A simple strategy for obtaining both enantiomers from an aldolase reaction: preparation of L- and D-4-hydroxy-2-ketoglutarate. *J. Chem. Soc. Perkin Trans. I*, pp. 1085–86

36. Garcia-Junceda E, Shen G-J, Sugai T, Wong C-H. 1995. A new strategy for the cloning, overexpression and one step purification of three DHAP-dependent aldolases: rhamnulose-1-phosphate aldolase, fuculose-1-phosphate aldolase and tagatose-1,6-diphosphate aldolase. *Bioorg. Med. Chem.* 3:945–53

37. Ghalambor MA, Heath EC. 1966. 3-Deoxyoctulosonic acid aldolase. *Methods Enzymol.* 9:534–38

38. Ghalambor MA, Heath EC. 1966. The biosynthesis of cell wall lipopolysaccharide in *Escherichia coli. J. Biol. Chem.* 241:3222–27

39. Gijsen HJM, Qiao L, Fitz W, Wong C-H. 1996. Recent advances in the chemoenzymatic synthesis of carbohydrates and carbohydrate mimetics. *Chem. Rev.* 96:443–73

40. Gijsen HJM, Wong C-H. 1994. Unprecedented asymmetric aldol reactions with three aldehyde substrates catalyzed by 2-deoxyribose-5-phosphate aldolase. *J. Am. Chem. Soc.* 116:8422–23

41. Gijsen HJM, Wong C-H. 1995. Sequential one-pot aldol reactions catalyzed by 2-deoxyribose-5-phosphate aldolase and fructose-1,6-diphosphate aldolase. *J. Am. Chem. Soc.* 117:2947–48

42. Gijsen HJM, Wong C-H. 1995. Sequential three- and four-substrate aldol reactions catalyzed by aldolases. *J. Am. Chem. Soc.* 117:7585–91

43. Gijsen HJM, Wong C-H. 1995. Synthesis of a cyclitol via a tandem enzymatic aldol-intramolecular Horner-Wadsworth-Emmons reaction. *Tetrahedron Lett.* 36:7057–60

44. Heathcock CH. 1991. The aldol reaction: acid and general base catalysis. See Ref. 79a, pp. 133–179

44a. Heathcock CH. 1991. The aldol reaction: Group I and Group II enolates. See Ref. 79a, pp. 181–238

45. Henderson I, Garcia-Junceda E, Liu KK-C, Chen Y-L, Shen G-J, Wong C-H. 1994. Cloning, overexpression and isolation of the Type II FDP aldolase from *E. coli* for specificity study and synthetic application. *Bioorg. Med. Chem.* 2:837–43

46. Hobbs GR, Lilly MD, Turner NJ, Ward JM, Willets AJ, Woodley JM. 1993. Enzyme-catalyzed carbon-carbon bond formation:use of transketolase from *Escherichia coli. J. Chem. Soc. Perkin Trans. I*, pp. 165–66

47. Hoffee PA. 1968. 2-Deoxyribose-5-phosphate aldolase of *Salmonella typhimurium*: purification and properties. *Arch. Biochem. Biophys.* 126:795–802

48. Horecker BL, Tsolas O, Lai C-Y. 1975. *The Enzymes*. New York: Academic

49. Hughes AB, Rudge AJ. 1994. Deoxynojirimycin: synthesis and biological activity. *Nat. Prod. Rep.* pp. 135–62

50. Ichikawa Y, Liu JL-C, Shen G-J, Wong C-H. 1991. A highly efficient multienzyme system for the one-step synthesis of a sialyl trisaccharide: in situ generation of sialic acid and N-acetyllactosamine coupled with regeneration of UDP-glucose, UDP-galactose and CMP-sialic acid. *J. Am. Chem. Soc.* 113:6300–2

50a. Kim BW, Williams SF, Masamune S. 1991. The aldol reaction: group III enolates. See Ref. 79a, pp. 239–75

51. Kim M-J, Hennen WJ, Sweers HM, Wong C-H. 1988. Enzymes in carbohydrate synthesis: N-acetylneuraminic acid aldolase catalyzed reactions and preparation of N-acetyl-2-deoxy-D-neuraminic acid derivatives. *J. Am. Chem. Soc.* 110:6481–86

52. Kobori Y, Myles DC, Whitesides GM. 1992. Substrate specificity and carbo-

hydrate synthesis using transketolase. *J. Org. Chem.* 57:5899–907

53. Kragl U, Güdde A, Wandrey C, Lubin N, Augé C. 1994. New synthetic applications of sialic acid aldolase, a useful catalyst for KDO synthesis. Relation between substrate conformation and enzyme stereoselectivity. *J. Chem. Soc. Perkin Trans. I*, pp.119–24

54. Levin DH, Racker E. 1959. Condensation of arabino 5-phosphate and phosphoenol pyruvate by 2-keto-3-deoxy-8-phosphooctonic acid synthetase. *J. Biol. Chem.* 234:2532–39

55. Lilly MD, Chauhan R, French C, Gyamerah M, Hobbs GR, et al. 1996. Carboncarbon bond synthesis: The impact of rDNA technology on the production and use of *E. coli* transketolase. *Ann. NY Acad. Sci.* 782:513–25

56. Liu JL-C, Shen G-J, Ichikawa Y, Rutan JF, Zapata G, et al. 1992. Overproduction of CMP-sialic acid synthetase for organic synthesis. *J. Am. Chem. Soc.* 114:3901–10

57. Liu KK-C, Wong C-H. 1992. A new strategy for the synthesis of nucleoside analogs based on enzyme-catalyzed aldol reactions. *J. Org. Chem.* 57:4789–91

58. Look GC, Fotsch CH, Wong C-H. 1993. A new strategy for the synthesis of nucleoside analogs based on enzyme-catalyzed aldol reactions. *Acc. Chem. Res.* 26:182–90

59. Maliakel BP, Schmidt W. 1993. Total synthesis of sphydrofuran. *J. Carbohydr. Chem.* 12:415–24

60. McGeown MG, Malpress FH. 1952. Synthesis of deoxyribose in animal tissue. *Nature* 170:575–76

61. Meyerhof O, Lohmann K. 1934. Enzymic equilibrium reaction between hexosediphosphate and dihydroxyacetonephosphate. *Biochem. Z.* 271:89–110

62. Moradian A, Benner SA. 1992. A biomimetic biotechnological process for converting starch to fructose: thermodynamic and evolutionary considerations in applied enzymology. *J. Am. Chem. Soc.* 114:6980–87

63. Morís-Varas F, Qian XN, Wong C-H. 1996. Enzymatic/chemical synthesis and biological evaluation of seven-membered iminocyclitols. *J. Am. Chem. Soc.* 118:7647–52

64. Morris KG, Smith MEB, Turner NJ, Lilly MD, Mitra RK, Woodley JM. 1996. Transketolase from *Escherichia coli*: a practical procedure for using the biocatalyst for asymmetric carbon-carbon bond synthesis. *Tetrahedron: Asymmetry* 7:2185–88

65. Nicotra F, Panza L, Russo G, Verani A. 1993. A biomimetic biotechnological process for converting starch to fructose: thermodynamic and evolutionary considerations in applied enzymology. *Tetrahedron: Asymmetry* 4:1203–4

66. Ozaki A, Toone EJ, von der Osten CH, Sinskey AJ, Whitesides GM. 1990. Overproduction and substrate specificity of a bacterial fuculose-1-phosphate aldolase: a new enzymic catalyst for stereocontrolled aldol condensation. *J. Am. Chem. Soc.* 112:4970–71

66a. Paterson I. 1991. The aldol reaction: transition metal enolates. See Ref. 79a, pp. 301–19

67. Qiao L, Murray B, Shimazaki M, Schultz J, Wong C-H. 1996. Synergistic inhibition of human α-1,3-fucosyltransferase V. *J. Am. Chem. Soc.* 118:7653–62

68. Raetz CRH, Dowhan W. 1990. Biosynthesis and function of phospholipids in *Escherichia coli*. *J. Biol. Chem.* 265:1235–38

68a. Rathke MW, Weipert P. 1991. Zinc enolates: the Reformatsky and Blaise reactions. See Ref. 79a, pp. 277–99

69. Ray PH. 1980. Purification and characterization of 3-deoxy-D-manno-octulosonate 8-phosphate synthetase from *Escherichia coli*. *J. Bacteriol.* 141:635–44

70. Ray PH. 1982. 3-Deoxy-D-manno-octulosonate-8-phosphate (KDO-8-P) synthase. *Methods Enzymol.* 83:525–30

71. Reimer LM, Conley DL, Pompliano DL, Frost JW. 1986. Construction of an enzyme-targeted organophosphonate using immobilized enzyme and whole-cell synthesis. *J. Am. Chem. Soc.* 108:8010–15

72. Schirch L. 1982. Serine hydroxymethyltransferase. *Adv. Enzymol.* 53:83–112

73. Schmidt W, Whitesides GM. 1990. A new approach to cyclitols based on rabbit-muscle aldolase (RAMA). *J. Am. Chem. Soc.* 112:9670–71

74. Schultz M, Waldmann H, Vogt W, Kunz H. 1990. Stereospecific carbon-carbon bond formation with rabbit muscle aldolase. A chemoenzymic synthesis of (+)-exo-brevicomin. *Tetrahedron Lett.* 31:867–68

75. Serianni AS, Cadman E, Pierce J, Hayes ML, Barker R. 1982. Enzymic synthesis of carbon-13-enriched aldoses, ketoses, and their phosphate esters. *Methods Enzymol.* 89:83–92

76. Shelton MC, Cotterill IC, Novak STA, Poonawala RM, Sudarshan S,

Toone EJ. 1996. 2-Keto-3-deoxy-6-phosphonogluconate aldolase as catalysts for stereocontrolled carbon-carbon bond formation. *J. Am. Chem. Soc.* 118:2117–25

77. Stover P, Zamora M, Shostak K, Gautam-Basak M, Schirch V. 1992. *Escherichia coli* serine hydroxymethyltransferase. The role of histidine 228 in determining reaction specificity. *J. Biol. Chem.* 267:17679–87

78. Sugai T, Shen G-J, Ichikawa Y, Wong C-H. 1993. Synthesis of 3-deoxy-D-manno-2-octulosonic acid (KDO) and its analogs based on KDO aldolase-catalyzed reactions. *J. Am. Chem. Soc.* 115:413–21

79. Takagi Y. 1966. L-Rhamnulose 1-phosphate aldolase. *Methods Enzymol.* 9:542–45

79a. Trost BM, Flemming I, Heathcock CH, eds. 1991. *Comprehensive Organic Synthesis*, Vol. 2. Oxford: Pergamon

80. Turner NJ, Whitesides GM. 1989. A combined chemical-enzymic synthesis of 3-deoxy-D-arabino-heptulosonic acid 7-phosphate. *J. Am. Chem. Soc.* 111:624–27

81. Uchida Y, Tsukada Y, Sugimori T. 1984. Purification and properties of N-acetylneuraminate lyase from *Escherichia coli*. *J. Biochem.* 96:507–22

82. Uchiyama T, Vassilev VP, Kajimoto T, Wong W, Huang H, et al. 1995. Design and synthesis of sialyl Lewis X mimetics. *J. Am. Chem. Soc.* 117:5395–96

83. Vassilev VP, Uchiyama T, Kajimoto T, Wong C-H. 1995. An efficient chemoenzymic syntheses of α-amino-β-hydroxy-γ-butyrolactone. *Tetrahedron Lett.* 36:5063–64

84. Von der Osten CH, Sinskey AJ, Barbas CF III, Pederson RL, Wang Y-F, Wong C-H. 1989. Use of a recombinant bacterial fructose-1,6-diphosphate aldolase in aldol reactions: preparative syntheses of 1-deoxynojirimycin,1-deoxymannojirimycin, 1,4-dideoxy-1,4-imino-D-arabinitol, and fagomine. *J. Am. Chem. Soc.* 111:3924–27

85. Woisetschlager M, Hogenauer G. 1986. Cloning and characterization of the gene encoding 3-deoxy-D-manno-octulosonate 8-phosphate synthetase from *Escherichia coli*. *J. Bacteriol.* 168:437–39

86. Woltering T, Weitz-Schmidt G, Wong C-H. 1996. C-Fucopeptides as selective antagonists: attachment of lipid moieties enhances the activity. *Tetrahedron Lett.* 37:9033–36

87. Wong C-H, Alajarín R, Morís-Varas F, Blanco O, Garcia-Junceda E. 1995. Enzymatic synthesis of L-fucose and analogs. *J. Org. Chem.* 60:7360–63

88. Wong C-H, Druekhammer DG, Sweers HM. 1988. Chemoenzymatic synthesis of flourosugars. In *Fluorinated Carbohydrates: Chemical and Biochemical Aspects*, ed. NF Taylor, pp. 29–42. Washington, DC: Am. Chem. Soc.

89. Wong C-H, Garcia-Junceda E, Chen L, Blanco O, Gijsen HJM, Steensma DH. 1995. Recombinant 2-deoxyribose-5-phosphate aldolase in organic synthesis: use of sequential two-substrate and three-substrate aldol reactions. *J. Am. Chem. Soc.* 117:3333–39

90. Wong C-H, Halcomb RL, Ichikawa Y, Kajimoto T. 1995. Enzymes in organic synthesis. *Angew. Chem. Int. Ed. Engl.* 34:412–32, 521–46

91. Wong C-H, Mazenod FP, Whitesides GM. 1983. Chemical and enzymatic syntheses of 6-deoxyhexoses. Conversion to 2,5-dimethyl-4-hydroxy-2,3-dihydrofuran-3-one (Furaneol) and analogs. *J. Org. Chem.* 48:3493–97

92. Wong C-H, Shen G-J, Pederson RL, Wang Y-F, Hennen WJ. 1991. Enzymic catalysis in organic synthesis. *Methods Enzymol.* 202:591–620

93. Wong C-H, Whitesides GM. 1983. Synthesis of sugars by aldolase-catalyzed condensation reactions. *J. Org. Chem.* 48:3199–205

94. Wong C-H, Whitesides GM. 1994. C-C bond formation. In *Enzymes in Synthetic Organic Chemistry*, pp. 195–251. Oxford: Pergamon

95. Wu S-H, Shimazaki M, Lin C-C, Qiao L, Moree WJ, et al. 1996. Synthesis of fucopeptides as sialyl Lewis[x] mimetics. *Angew. Chem. Int. Ed. Engl.* 35:88–89

96. Zhou P, Salleh HM, Honek JF. 1993. Facile chemoenzymic synthesis of 3-(hydroxymethyl)-6-epicastanospermine. *J. Org. Chem.* 58:264–66

97. Ziegler T, Straub A, Effenberger F. 1988. Enzyme-catalyzed synthesis of 1-deoxymannojirimycin, 1-deoxynojirimycin and 1,4-dideoxy-1,4-imino-D-arabinitol. *Angew. Chem. Int. Ed. Engl.* 27:716–17

Annu. Rev. Microbiol. 1997. 51:311–40
Copyright © 1997 by Annual Reviews Inc. All rights reserved

INTERACTION OF ANTIGENS AND ANTIBODIES AT MUCOSAL SURFACES

Michael E. Lamm

Institute of Pathology, Case Western Reserve University, Cleveland, Ohio 44106;
email: mel6@po.cwru.edu

KEY WORDS: immune defense, immune barrier, IgA, infections, disposition of antigens

---

ABSTRACT

Infections often involve the mucosal surfaces of the body, which form a boundary with the outside world. This review focuses on immunoglobulin A (IgA) antibodies because IgA is the principal mucosal antibody class. IgA is synthesized by local plasma cells and has a specific polymeric immunoglobulin receptor-mediated transport mechanism for entry into the secretions. By serving as an external barrier capable of inhibiting attachment of microbes to the luminal surface of the mucosal epithelial lining, IgA antibodies form the first line of immune defense. In addition to this traditional mode of extracellular antibody function, recent evidence suggests that IgA antibodies can also function in a nontraditional fashion by neutralizing viruses intracellularly, if a virus is infecting an epithelial cell through which specific IgA antibody is passing on its way to the secretions. IgA antibodies are also envisaged as providing an internal mucosal barrier beneath the mucosal lining. Antigens intercepted by IgA antibodies here can potentially be ferried through the epithelium and thereby excreted. In addition to the polymeric immunoglobulin receptor on mucosal epithelial cells, IgA antibodies can bind to receptors on a variety of leukocytes and have been shown, in some experimental systems, to be capable of activating the alternative complement pathway, making IgA antibodies potential participants in inflammatory reactions. Although the relationship of IgA antibodies to inflammation is not entirely clear, the bias presented is that IgA is basically noninflammatory, perhaps even anti-inflammatory. According to this view, the major role of the Fc portion of IgA antibodies is to transport IgA across mucosal epithelial cells and not, as in the case of the other classes of antibody, to activate secondary phenomena of the kind that contribute to inflammation. Because of IgA's key role as an initial barrier to infection, much

311

current research in mucosal immunology is directed toward developing new vectors and adjuvants that can provide improved approaches to mucosal vaccines. Finally, because of advances in monoclonal antibody technology, topical application of antibodies to mucosal surfaces has significant potential for preventing and treating infections.

CONTENTS

INTRODUCTION

Host defense at mucous membranes is mediated by both nonspecific and immunologically specific mechanisms. Although the nonspecific mechanisms can vary from one anatomic location to another, in general they include the following: In the secretions of mucous membranes, the mucus itself acts as a physicochemical barrier and as a viscous matrix in which microbes can be trapped. Mucosal secretions can include acid, digestive enzymes, bile, lysozyme, and lactoferrin, all of which can inhibit microbes. Peristaltic contraction of smooth muscle can help propel microbes out of the intestinal, urinary, and respiratory tracts; ciliary action can also do the same for the respiratory tract. By keeping microbes moving, these physical factors also decrease the time available for adherence to the epithelial lining. For microbes that have managed to pass these hurdles, the epithelial cells that line mucous membranes, together with the glycocalyx covering their apical surface and the tight junctions between neighboring cells, act as a physical barrier against penetration.

The immunologically specific mechanisms of mucosal defense against microbes are both cellular (mediated by T lymphocytes) and humoral (mediated by antibodies), with the latter as the subject of this review. First, some features of the antimicrobial actions of antibodies in general will be mentioned briefly.

For example, specific antibodies can bind to microbes and to their toxins. Binding can be to a site that is required for adherence to a specific receptor on host epithelial cells, thus inhibiting attachment and uptake. Even when binding is not directly to a specific microbial adhesin, the relatively bulky antibody may interfere with the adhesin for steric reasons, or an antibody may bind to a nonadhesin site that interferes with an important microbial function. Antibodies may also bind to flagella and inhibit bacterial motility. In the case of a toxin, antibodies may bind directly to the portion of the molecule that mediates the toxic effect. Moreover, the effects of binding by a single antibody combining site in an Fab region can be enhanced by multipoint binding when multiple combining sites on the same antibody molecule are engaged—ranging from two in an immunoglobulin G (IgG) antibody, to four in dimeric immunoglobulin A (IgA), and, potentially, to ten in pentameric immunoglobulin M (IgM). Because IgA and IgM are abundant in mucosal tissues and secretions, mucosal antibodies tend to be good agglutinators. Collectively, the inhibition of microbial motility and the agglutination of microbes within viscous mucous secretions, in concert with peristaltic and ciliary action, are highly efficient at preventing the attachment of microbes to epithelial surfaces and at promoting their removal.

When antibodies combine with microbes in internal body fluids, important secondary antimicrobial phenomena, mediated by antibody Fc regions, can come into play. Thus, complement may be activated and lyse microbes, and microbes may be opsonized via complement receptors and Fc receptors on phagocytic cells. In contrast, in the mucosal secretions there is no compelling evidence for the presence of a biologically active complement system, nor are significant numbers of phagocytes normally present. Although both complement components and leukocytes can be found in mucosal fluids, in the presence of inflammation (with its attendant increase in permeability and chemotactic factors), the significance of phenomena such as complement activation, opsonization, and antibody-dependent cellular cytotoxicity in mucosal defense is still not clear.

Because the mucosal surfaces of the body provide an extensive interface that physically separates the interior of the body from the outside world, the mucosal immune system is constantly interacting with the environment. This is especially true in the intestinal tract, which contains an endogenous microbial flora and carries the components of the daily food intake, and also in the respiratory tract, which is exposed to inhaled foreign substances. These elements of normal homeostasis account for much of the bulk of the mucosa-associated lymphoid tissue. Superimposed on the normal activity are periodic encounters with microbial agents of disease in the nasopharynx and in the gastrointestinal, respiratory, urinary, and genital tracts. Because of this extensive exposure to foreign substances, it is not surprising that the majority of the body's lymphoid tissue is distributed along mucous membranes, especially in the intestine. In fact, although immunologists have traditionally tended to focus on the systemic lymphoid tissue (lymph nodes, spleen, bone marrow, and thymus), there are probably two to three times as many lymphoid cells in mucosa-associated lymphoid tissue. This compartment includes lymphocytes in the mucosal epithelial layer and cells dispersed through the loose connective tissue of the lamina propria beneath the epithelium, as well as aggregates of densely packed lymphoid cells, such as the intestinal Peyer's patches. Lymph nodes that drain mucous membranes, e.g. the mesenteric nodes, combine features of both mucosa-associated and systemic lymphoid tissue.

Not only is there a difference in bulk between the mucosal and the systemic lymphoid compartments, with respect to antibodies, there is also a major difference in what they produce. Mucosal plasma cells tend to make antibodies of the IgA class, whereas the systemic lymphoid tissue is more geared to IgG production. These differences are readily observed by immunofluorescence analysis (the large majority of mucosal plasma cells make IgA, which is only a minor product of systemic plasma cells, where IgG cells represent the predominant type) and by measurement of immunoglobulin content: IgA predominates in the mucosal secretions, whereas IgG predominates in serum. Because of the great bulk of the mucosa-associated lymphoid tissue, IgA is by far the major class of antibody synthesized by the body as a whole. Most of this IgA is synthesized locally within mucous membranes and is destined to enter the mucosal secretions. In addition to abundant local synthesis, another factor accounting for its predominance in mucosal secretions is a receptor-mediated mechanism for selectively transporting IgA across the epithelial cells that line mucous membranes.

Because IgA is so dominant in mucosal tissues and secretions, this review will emphasize the interactions between IgA antibodies and antigens and their significance for host defense against infectious agents. The major actions of IgA in this context are summarized in Table 1. (For a comprehensive consideration

Table 1   Mechanisms of immunoglobulin A–mediated mucosal defense[a]

Provides an immune exclusion barrier in secretions against microbes, toxins, and other antigens
Binds, via its carbohydrate moieties, to lectin-like bacterial adhesins, thereby blocking bacterial adhesion to receptors on epithelial cells
Synergizes with nonspecific antimicrobial substances in secretions
Neutralizes viruses within epithelial cells
Excretes antigens from subepithelial compartment across epithelium into secretions
Promotes phagocytosis
Promotes antibody-dependent cellular cytotoxicity
Activates complement (alternative pathway)

[a]The first mechanism is well established; the others are less so and in some cases are controversial or may not be of major significance in vivo.

of the multiple aspects of mucosal immunity, the reader is referred to Reference 115; other general or focused recent reviews include References 10, 69, 75, 95, 103, 109, 138.)

IGA, THE PRINCIPAL MUCOSAL ANTIBODY

Transport of IgA into Mucosal Secretions

IgA from the plasma cells that underlie the mucosal epithelium is largely secreted as dimers containing two conventional immunoglobulin subunits, each with two H and two L polypeptide chains, and an intersubunit J chain. These IgA dimers are able to bind to the polymeric immunoglobulin receptor (pIgR) on the basolateral surface of the epithelial cells that line the mucous membrane (103). The pIgR is a transmembrane protein with an external IgA-binding domain, an intramembranous domain, and a cytoplasmic domain that contains cellular sorting signals. After synthesis and processing in the endoplasmic reticulum, the Golgi and trans-Golgi network, these signals direct the pIgR to the basolateral surface of the epithelial cell, where it has an opportunity to bind the dimeric IgA that is secreted by the local plasma cells. The IgA-pIgR complex is endocytosed, and other sorting signals cause it to be transcytosed in vesicles to the apical surface, probably via apical recycling endosomes (4). At the apical surface, proteolytic cleavage splits the external domain of the pIgR (now termed the secretory component) from its membrane spanning domain; this releases the dimeric IgA-secretory component complex, known as secretory or exocrine IgA, into the mucosal secretions. Shortly before secretion, the IgA dimer and the pIgR are thought to be disulfide-bonded to one another (22).

Immune Barrier Function of IgA

Once in the secretions, secretory IgA can bind to antigens, including microbes, and it can help prevent them from attaching to or penetrating the mucosal surface (1, 45, 53, 118, 140, 141, 147, 150, 158). Exactly how this is accomplished in the case of microbes is not fully understood; the mechanisms are undoubtedly many. The abilities to agglutinate microbes, to interfere with the action of their flagella, rendering them less motile, and to block interactions between microbial adhesins and their receptors on the apical surface of mucosal epithelial cells are thought to be important (93, 141, 158). Also, because bacterial adhesins may function as lectins, the oligosaccharides on the constant regions of IgA molecules may bind, thus competing with carbohydrate moieties on epithelial cells that act as receptors (161). The antibacterial effects of IgA may be potentiated by such nonspecific antibacterial substances as lactoferrin, lactoperoxidase, and lysozyme (72). In addition to binding to intact organisms, IgA antibodies can block the actions of bacterial products (29, 34, 49). IgA antibodies against viruses may prevent their binding to cells, as well as their internalization and replication (6, 118, 119).

The immune barrier function of IgA has been inferred from correlations between the content of specific IgA antibodies in mucosal secretions and the resistance to infection by a variety of microbes, especially viruses (23, 43, 48, 62, 63, 68, 76, 80, 99, 106, 117, 131, 135, 137, 151). Such correlations have been made in clinical studies, after deliberate immunization or recovery from infection, and in experiments with animals. A particularly noteworthy clinical example is infection with poliovirus, in which oral vaccination (Sabin vaccine) was developed in light of the route of natural infection, whereby virus enters the body via the intestine, spreads to the regional lymph nodes, and reaches the central nervous system via the bloodstream. Later studies showed that oral vaccination leads to the production of secretory IgA antibodies that block the primary mucosal infection (114, 117). This prominent secretory IgA antibody response represents a principal difference from the largely systemic antibody response induced by parenteral immunization (Salk vaccine). The data on the antibody response to oral vaccination for poliomyelitis thus provided strong clinical support for the concept of IgA as an immune exclusion barrier in the mucosal secretions. Another example of the barrier function is the passive protection against infection provided to suckling infants by mother's milk, where there is also good evidence that the principal mediator is IgA antibody (29, 50, 132). Although the benefits of milk antibodies may not be apparent or significant in economically developed areas of the world, they could be considerable in less developed countries.

After either deliberate immunization or infection, antibody responses are heterogeneous, never being restricted to the IgA class. Although it is possible to fractionate the different immunoglobulin classes from a heterogeneous mixture, it is difficult to determine that an individual immunoglobulin class is rendered pure. Nevertheless, a number of studies of passive protection by purified antibody fractions have also supported the concept that the most important class is IgA (8, 113, 145, 146). A different approach to the same question was to show in mice that infection-induced immunity to influenza virus could be abrogated by instilling antibodies to mouse IgA into the respiratory tract (130). This experiment, under relatively natural conditions, provided important evidence that secretory IgA is, in fact, the main mediator of resistance to infectious challenge by influenza virus.

A particularly powerful means of investigating the protective properties and potential of IgA antibodies, or any class of antibody for that matter, is offered by monoclonal antibody technology. To make IgA monoclonal antibodies, standard procedures can be modified in several ways, with the goal of targeting gut-associated lymphoid tissue for immunization before taking cells for fusion.

This strategy dramatically increases the percentage of resulting hybridomas that make IgA. To evaluate their host defense properties and potential, IgA monoclonal antibodies of desired specificity can be introduced passively into immunologically naive animals to test their ability to provide resistance to infectious challenge. In such studies, IgA antibodies have been delivered in two ways. One way is the direct introduction to a mucosal surface, e.g. by instilling IgA antibodies into the respiratory or gastrointestinal tracts. The other route to the secretions is indirect, e.g. by systemic injection or via so-called backpack tumors in which hybridoma cells are injected subcutaneously into the back of a mouse; as the tumor enlarges, the IgA antibody it secretes appears in both serum and mucosal secretions (159). In a number of studies in recent years, it has been amply demonstrated that immunologically naive animals containing passive IgA monoclonal antibody as their only specific immune reactant can be protected against a variety of viruses (Sendai, influenza, respiratory syncytial) and bacteria (*Salmonella typhimurium, Shigella flexneri, Vibrio cholerae, Chlamydia trachomatis, Helicobacter felis*) that afflict different mucosae (5, 9, 25, 35, 90, 91, 100, 124, 129, 155, 157, 159). These studies with monoclonal antibodies collectively illustrate the protective potential of an IgA response. Under natural conditions a mucosal immune response would be polyclonal—with antibodies directed to multiple microbial antigenic determinants, rather than to the single determinant of a monoclonal antibody—and, consequently, should be even more effective.

The work cited validates the concept that IgA antibody provides an effective exclusion barrier in the mucosal secretions. Moreover, because most human infections involve mucous membranes, as either the locus of clinical disease or the portal of entry, these studies provide added incentive for developing immunization procedures that are designed to induce strong mucosal antibody responses. Under natural conditions, infections tend to be initiated by relatively small inocula of a pathogen, whether inhaled into the respiratory tract or ingested; therefore, even small amounts of pre-existing secretory IgA antibody can be effective in preventing disease.

Intraepithelial Cell Neutralization of Virus by IgA Antibody

It is generally believed that antibodies function in defense against pathogens by combining with them extracellularly. This holds for all classes of antibody, in all locations. Once an intracellular pathogen has entered a cell, it is considered to be inaccessible to the actions of antibodies. Cell-mediated immunity, which is ineffective against extracellular microbes, is thought, instead, to be integral to the immune system's ability to combat infections by intracellular pathogens. These considerations also underlie the classical concept that, although pre-existing antibody may be effective at preventing infections by intracellular pathogens (if it can react before the microbe has entered a cell), antibody offers little in promoting recovery from infection once microbes are sheltered intracellularly.

Recent evidence, however, suggests that the particular case of infection of mucous membranes by intracellular pathogens such as viruses may offer an exception to the general rules just mentioned. Because of its mandatory transepithelial route to the secretions, IgA antibody, in transit through an epithelial cell, could conceivably contact the antigens of a virus that is infecting the same cell. Opportunities and probabilities for such intracellular interactions between IgA antibodies and viral components would be determined by the transcytotic pathway of the IgA, its specificity for particular components, and the life cycle of the individual virus within the mucosal epithelial cell. These parameters would collectively determine the probability of a given IgA antibody and a viral component actually meeting intracellularly.

The polarized cell monolayer system that has contributed much to the understanding of the cell biology of transepithelial transport of IgA (104) has also offered an opportunity to investigate the possibility of intraepithelial cell neutralization of viruses by IgA antibodies. Experiments to date have used the following general protocol. Viruses that can infect epithelial cells from the apical surface are added from above to a polarized monolayer in a two-compartment tissue culture chamber. IgA monoclonal antibodies to a viral antigen are added from below and, because the epithelial cells express the pIgR at the basolateral surface, the IgA is internalized and transcytosed. This setup mimics the situation in vivo, in which the apical surface of a mucosal epithelial cell faces a lumen that is accessible to viruses, and the basolateral surface faces the lamina propria with its abundant IgA secreting plasma cells. In vitro, the epithelial cells are polarized (as they are in vivo), with individual cells attached to their neighbors by tight junctions. Therefore, the IgA antibody coming from below cannot diffuse between cells to cross the monolayer. It can only enter the fluid above the epithelial cells (the upper compartment in vitro, the luminal secretions in vivo) by epithelial transcytosis. Intersection of IgA antibody and viral antigen intracellularly can be assessed in a number of ways, including microscopy and measurement of virus in cell lysates and fluid above the epithelial monolayer.

Such experiments have been done with Sendai virus, a parainfluenza virus in the paramyxovirus group, and influenza virus, an orthomyxovirus (88, 89). In both cases, double-label immunofluorescence analysis showed that IgA antibody and virus envelope antigen (HN in the case of Sendai virus, HA in the case of influenza) colocalized within the epithelial cells. This was taken as evidence of antibody and viral antigen having encountered one another intracellularly. Such colocalization did not occur with IgA antibody of irrelevant specificity. When virus titers were measured in cell lysates and apical culture fluids in these experiments, it was observed that specific IgA antibody inhibited virus production, whereas irrelevant IgA did not; nor did IgG antibodies to the same envelope antigens inhibit virus production, which is explainable by the fact that IgG is not internalized by the pIgR. These results were interpreted as reflecting intracellular neutralization of virus by IgA antibody against envelope protein. A limited number of experiments have suggested that IgA antibodies against internal viral proteins are much less effective (MB Mazanec, personal communication).

Recent data in animal model systems have provided in vivo evidence for intracellular virus neutralization by IgA antibody. IgA monoclonal antibodies to outer and inner capsid proteins of rotavirus were delivered systemically and evaluated for their ability to prevent or cure enteritis in mice (13). IgA antibodies to the inner capsid protein VP6 were highly effective, in both respects, whereas IgA antibodies to the outer capsid protein VP4 were ineffective. These in vivo results were opposite to those obtained in an in vitro neutralization assay, in which antibodies to the external capsid protein were neutralizing and antibodies to the inner capsid protein were not. The results for the in vitro neutralization assay are to be expected because antibodies to proteins displayed on the surface of the virus (such as VP4) can combine with intact virus, whereas antibodies to proteins that are not on the surface (such as VP6) cannot. The observation that IgA antibodies that were incapable of neutralizing virus in a plaque reduction assay (or when presented to the luminal aspect of the intestinal mucosa in vivo) were highly effective at neutralizing virus in vivo when given systemically argues for neutralization of rotavirus inside intestinal epithelial cells during transport of the IgA in the basal to apical direction.

Data from another system also support intracellular neutralization in vivo. In this approach, the ability of the rodent hepatocyte to transport polymeric IgA via the pIgR from blood to bile was exploited, to show that systemically injected IgA antibody to mouse hepatitis virus E_2 surface–spike protein could reduce the virus content of liver cells from infected mice, whereas IgG antibody could not (D Huang & MB Mazanec, personal communication).

In both the in vitro and the in vivo experiments on intracellular neutralization, IgA antibodies to different components of a given virus were found to have different degrees of effectiveness. Conceivably, this could reflect reactivities to epitopes that are intrinsically more or less neutralizing. More likely, however, they reflect different intracellular pathways for synthesis, transport, and packaging of individual viral proteins. Certainly, the various proteins of a given virus can follow different routes during synthesis, transport, and assembly, and viruses can vary from one another in these respects. In theory, intracellular neutralization of virus by IgA antibody could be mediated at a number of stages in a virus life cycle, including the point of entry of virus into the cell, uncoating, replication, transit of newly synthesized components through different subcellular compartments, and assembly and release. The major elements of IgA's transcellular route through epithelial cells have been characterized as follows: receptor mediated endocytosis into basolateral endosomes, vesicular transport to apical endosomes, vesicular transit to the apical cell surface, and release (4). A key factor determining the potential of a transcytosing IgA antibody to mediate intraepithelial cell neutralization would appear to be its chances for encountering a given viral protein. For an internal viral protein, synthesized on free cytoplasmic ribosomes, such chances should be minimal. On the other hand, a viral envelope protein that follows the secretory pathway (vesicular transport from rough endoplasmic reticulum to Golgi, to trans-Golgi network, to the cell surface) is likely to have a greater potential for intersecting transcytosing IgA antibody. One possibility is that post-Golgi vesicles containing viral envelope protein might join apical endosomes containing IgA. Such intracellular communication would be consistent with a current view regarding trafficking within epithelial cells (112). The exact site of intracellular neutralization of virus by IgA antibody, however, has not been determined in any instance. Preliminary double-label immunoelectron microscopy data indicate that IgA antibody and viral envelope protein do meet within vesicles (H Fujioka & MB Mazanec, personal communication).

The two major host defense functions of IgA discussed thus far, namely an immune exclusion barrier in the lumen and intraepithelial cell neutralization, would serve different roles in resistance to infection. Pre-existing IgA antibody in the lumen, as could arise from previous infection or deliberate immunization, can prevent infection. As mentioned earlier, comparatively small amounts of pre-existing antibody should suffice because, under natural conditions, the infectious inoculum is usually small. However, once mucosal infection has occurred in a nonimmune host, the resulting immune response can be expected to include IgA antibodies. Although the antibodies that then reach the luminal secretions might help to limit cell-to-cell spread of viruses that are released from apical cell surfaces, intracellular neutralization would seem to offer significant potential to assist in recovery. Functioning in this fashion, IgA antibodies should be able to synergize with the cell-mediated immune response that would also have been initiated by the infection. If these speculations are correct, then in the special case of virus infections of mucous membranes, immune mechanisms of recovery would not necessarily be limited to the cellular arm of the immune response, as has long been thought.

Excretory Immune System

In addition to serving host defense roles in the luminal secretions and in the epithelium that lines mucous membranes, IgA can also potentially function in an immune elimination role. This possibility follows from the fact that the Fab regions that bind antigen are separate from the Fc regions that bind to the pIgR. Thus, epithelial cells can transport not only free IgA, but also IgA antibodies that are bound to antigen (65, 66). The transcellular route for IgA immune complexes and for free IgA appears to be identical, in that essentially everything that is internalized by the pIgR is transported intact into the secretions, with no major diversion to an intracellular degradation pathway (65). Such a mechanism for excreting antigens could be physiologically significant, because antigens are likely to be present continuously in the lamina propria of mucous membranes— especially in the intestinal tract, where there is an abundant microbial flora and also components of food. Proteins are known to be absorbed in a relatively intact condition, to a small extent (31, 59, 122). Infections caused by a replicating pathogen that has entered a mucous membrane would periodically provide additional sources of antigens. The IgA from local plasma cells that secrete antibody to the antigens that are derived from food and microbes would combine with antigens that reached the lamina propria, to give rise to IgA-containing immune complexes. Although some of these complexes could be drained to the liver and systemically via blood and lymph, a more direct and efficient means of removal is offered by pIgR-mediated transcytosis across the adjacent mucosal epithelium.

Overall, such an IgA-mediated excretory immune system should be beneficial. In special cases, however, there could be adverse consequences. In this context, it has been suggested that mucosal IgA antibodies can introduce microbial pathogens into epithelial cells, to the detriment of the host. For example, after an initial pharyngitis, Epstein-Barr virus spreads to the regional lymph nodes, which constitute a reservoir of virus. After the initial infection, there is often a brisk IgA antibody response. Thus, if virus is released from chronically infected lymphoid cells and is complexed by IgA antibody, then the antibody-virus complexes could be endocytosed into epithelial cells via the pIgR. This would then provide a mechanism to account for the association of Epstein-Barr virus infection with the later complication of nasopharyngeal carcinoma, which is postulated to arise from IgA-mediated infection of epithelial cells and their subsequent transformation (136).

Figure 1 shows the normal transport route of polymeric IgA across epithelial cells, as well as the three anatomic levels where IgA can potentially function in mucosal defense: in the lamina propria, within a lining epithelial cell, and in the lumen.

IgA Proteases

In humans, there are two subclasses of IgA, IgA1 and IgA2, which are present in different proportions in different secretions, as well as in the immune responses

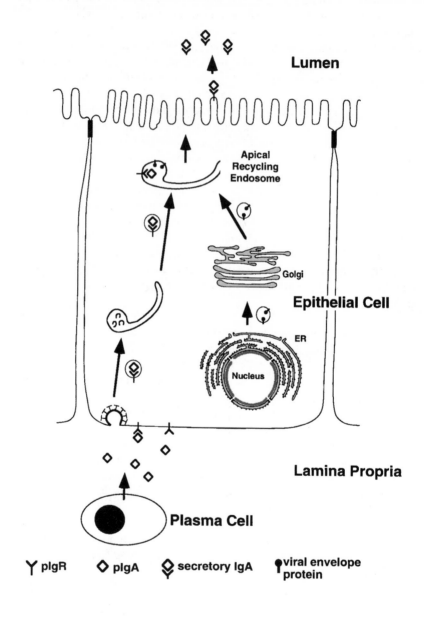

Lumen

Apical
Recycling
Endosome

Golgi

Epithelial Cell

ER

Nucleus

Lamina Propria

Plasma Cell

Y pIgR   ◇ pIgA   ⬙ secretory IgA   ●viral envelope
                                              protein

to different antigens. IgA proteases are extracellular enzymes, produced by certain pathogenic and commensal bacteria, that are able to cleave the hinge region of the IgA1 subclass with great specificity to produce Fab and Fc fragments (71, 126, 127). Cleavage occurs after proline residues that are available in IgA1 but not in IgA2, which has a deletion in the hinge region that makes it resistant to these enzymes. Because the enzymes are produced by limited groups of pathogens that colonize mucous membranes—including *Neisseria meningitidis*, *Streptococcus pneumoniae*, and *Haemophilus influenzae* (the three main agents of bacterial meningitis) and *Neisseria gonorrhoeae*—they have been envisioned as virulence factors that promote infection by degrading antibacterial IgA1 antibodies. The degraded antibodies could be directed not only to the bacteria that produce the enzyme but to other kinds of bacteria as well. The resulting univalent Fab fragments could still bind to the bacteria, but they would be nonagglutinating and less inhibitory than intact IgA antibodies would be (53, 85). Indeed, in binding to bacteria, the Fab fragments produced from cleaved IgA1 antibodies might be capable of blocking the antibacterial activities of the resistant IgA2 antibodies, as well as the antibodies of other classes.

Fc_α Receptors

The antimicrobial actions of antibodies in general can involve interactions with cellular receptors that bind to the Fc portion of immunoglobulins. In the case of IgA, such receptors, referred to as Fc_α receptors, have been described on different kinds of leukocytes (2, 70, 92, 101, 102). Within mucous membranes, they could help to mediate the phagocytosis of immune complexes, including microbes, that contain IgA. In experimental systems, however, IgA antibodies have been shown both to promote and to inhibit phagocytosis (69, 72). There is also evidence that IgA antibodies may be able to mediate antibody-dependent cellular cytotoxicity (134, 143).

←───────────────────────────────────────

Figure 1   The different levels at which IgA antibodies function in host defense in relation to mucosal epithelium. Plasma cells in the lamina propria of a mucous membrane secrete polymeric IgA (pIgA), which binds to its receptor, pIgR, on the basolateral surface of an epithelial cell. The complex is endocytosed and transcytosed in vesicles to the apical surface, where the pIgR is cleaved, releasing secretory IgA into the lumen. Antigens in the lamina propria can be bound by IgA antibody and excreted in immune complexes through the epithelium by the same route as free IgA. Virus infecting an epithelial cell can potentially be neutralized intracellularly if transcytosing IgA antibody is able to bind a viral component and interfere with virus synthesis or assembly. In the hypothetical scheme shown, viral envelope protein that was synthesized in the rough endoplasmic reticulum (ER) and processed in the Golgi is transported in a post-Golgi vesicle to an apical recycling endosome that contains transcytosing antiviral IgA, which then disrupts virus assembly. A final level of action for free secretory IgA antibody in the lumen is to bind antigens, thus providing an immune exclusion barrier that inhibits access to the epithelium.

Under normal conditions, mucosal secretions contain many fewer leukocytes than do mucosal tissues, and although the number of leukocytes in secretions can increase in the presence of inflammation, the role of leukocyte Fc_α receptors in this compartment may be a limited one. In aggregate, the functional significance of leukocyte Fc_α receptors in mucous membranes and their secretions is far from clear.

Some streptococci also produce surface proteins that bind with great avidity to the Fc region of IgA (3, 81). The significance of this phenomenon is uncertain.

Anti-Inflammatory Properties of IgA

The Fc portions of antibodies in general have evolved to mediate secondary functions that come into play after the Fab portions have combined with antigen. IgG and IgM antibodies can activate complement efficiently, and immunoglobulin E (IgE) antibodies on the surface of tissue mast cells and blood basophils can trigger the release of histamine and other mediators of inflammation. In the case of IgA, the Fc portion of polymeric IgA is responsible for binding to the pIgR, thus initiating the transepithelial transport mechanism. This is probably the main functional role of its Fc portion. It is not at all clear that the Fc portion of IgA mediates significant secondary phenomena analogous to those mediated by the other classes of immunoglobulin (72). IgA cannot activate the classical complement pathway, and it is inefficient, at best, at activating the alternative pathway (100, 110, 111, 123). Certainly IgA is less efficient than IgG. Therefore, to the extent that IgA competes successfully with IgG for binding sites on the same antigen when both are present, IgA will decrease the overall activation of complement. As an example, when IgA and IgG antibodies bind to the same bacteria, IgA interferes with the ability of the IgG to activate complement (61). In addition, although mucosal tissues proper have a functional complement system, the mucosal secretions probably do not. Finally, as just mentioned, although Fc_α receptors can be detected on leukocytes, their biological significance is not well understood.

Overall, the major functions of IgA, especially in mucous membranes and their secretions, would not seem to require the kinds of secondary phenomena that are mediated by antibodies elsewhere. Instead, the ability to be transported across epithelia and to cross-link antigens, including microbes, appears to account for the major host defense functions of IgA. It even seems advantageous to the host that IgA is not pro-inflammatory. Indeed, it may make sense to consider IgA an anti-inflammatory immunoglobulin (Figure 2). This argument follows from the extensive exposure of mucous membranes to foreign matter, especially in the intestinal tract, with its indigenous microflora and its exposure to the contents of food. If IgA were efficient at activating potent secondary phenomena, the body would seem to be at excessive risk for a state of chronic

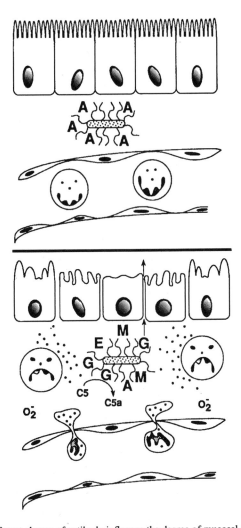

Figure 2   The different classes of antibody influence the degree of mucosal inflammation. *(Top)* IgA antibodies combine with antigen, in this case a bacterium, in mucosal lamina propria between the epithelium above and a blood vessel containing leukocytes below. IgA tends not to incite inflammation. *(Bottom)* Most of the antibacterial antibodies are pro-inflammatory (IgG, IgE, IgM). They are potent activators of complement and leukocytes. As a result, inflammation ensues, with production of phlogistic mediators (O_2^-, C5a, etc), increased intercellular permeability, chemotaxis and degranulation of leukocytes, phagocytosis, antibody-dependent cellular cytotoxicity, and damage to bystander cells. Usually, IgA comprises the majority of the antibody in mucosal lamina propria and secretions; hence, mucosal tissues tend to remain quiescent, even in the presence of antigen-antibody interactions.

mucosal inflammation because the interactions of IgA antibodies with an enormous array of antigens are both extensive and constant. In this context, whether the mucosal immune system is involved in inflammatory diseases of mucous membranes, such as regional enteritis and ulcerative colitis, is not known.

IgA Deficiency

Selective IgA deficiency is the most common immunodeficiency. Because of the impairment in immune barrier function, individuals with this deficiency tend to absorb increased amounts of foreign proteins, to which they then make systemic (IgG) antibody. Following a meal, they may, therefore, have increased levels of circulating immune complexes (30). Although IgA-deficient people can be subject to excessive respiratory infections, intestinal disorders, allergies, and autoimmune diseases, other affected individuals appear perfectly normal (55). Because of the extent of IgA production in normal individuals, especially in mucous membranes, and the role envisioned for IgA in host defense, it is surprising that overt illness is not more prevalent among IgA-deficient persons. Perhaps the ability to resist infections in the absence of IgA antibodies simply reflects the heterogeneity and redundancy of immune defenses, which helps to minimize the phenotypic expression of isolated defects. Thus, the ability of other immunoglobulins, especially IgM, to substitute for IgA may afford reasonably normal mucosal defense in its absence (11,97). Another potentially significant factor is that most of the data on IgA deficiency are based on studies in economically developed countries, where standards of public health and nutrition are high. It is conceivable that in less developed areas of the world, where infections, particularly diarrheal diseases, are more prevalent, deficiencies in IgA would be more of a liability, as well as less likely to be diagnosed (19).

OTHER CLASSES OF IMMUNOGLOBULIN

Although IgA is certainly the major mucosal immunoglobulin, the other classes of antibody also have opportunities to interact with antigens in relation to mucous membranes. IgM, for example, shares important features with mucosal IgA and can be considered the primordial mucosal antibody. IgM, like IgA, is polymeric (i.e. it is pentameric, whereas mucosal IgA, though mainly dimeric, can also be present as higher polymers) and can bind to the pIgR, which, likewise, affords IgM selective transport into mucosal secretions, although less efficiently than is the case with IgA (ME Lamm & JK Robinson, unpublished data). A major difference from IgA is IgM's greater potency as an activator of complement. In patients with selective IgA deficiency, mucosal IgM may substitute and provide a measure of mucosal defense (11).

IgG, the principal class of systemic antibody, is secreted by a small minority of mucosal plasma cells. Although there is no selective transport mechanism for IgG to reach mucosal secretions, local inflammation or damage to the epithelial lining can present opportunities for IgG to enter by diffusion. Thus, in selected instances, the secondary effects that are mediated by IgG complexed to antigen might come into play. These could include the activation of complement and binding to receptors on phagocytes. In addition, immune complexes that form in mucous membranes with multivalent antigens may include monomeric antibodies like IgG in the same complex with polymeric IgA. In this way, monomeric antibodies may modulate the properties of mucosal immune complexes and may even be transported in piggyback fashion through epithelial cells via the usual pIgR-mediated transport mechanism (66).

Despite its extremely low concentration in extracellular fluid, IgE is a significant mucosal immunoglobulin (107). IgE-producing cells are more abundant in mucosal lymphoid than they are in systemic lymphoid tissue (142), and mast cells, which are numerous in mucosal connective tissue, have high affinity receptors for IgE. When cell surface IgE is aggregated by multivalent antigen, potent inflammatory mediators—including histamine, derivatives of arachidonic acid, and cytokines—are liberated. Thus, antigens impinging on mucous membranes can readily trigger allergic inflammation in individuals who make excessive amounts of IgE antibodies. This phenomenon can underlie such common illnesses as allergic rhinitis, food allergies, and asthma.

Although there is a well-established association between IgE antibodies and infections by macroparasites, the significance of IgE responses and immediate hypersensitivity reactions in relation to defense against parasites is still not clear (15, 73, 86, 107). It has been proposed that the inflammatory mediators that are released as a result of antigen–IgE antibody reactions may damage parasites directly or may cause contractions of smooth muscle that help expel parasites, or that the IgE antibodies that are bound to leukocytes may promote phagocytosis and antibody-dependent cell-mediated cytotoxic killing. These attractive ideas, however, have been difficult to prove in humans, and experimental models of infection with different parasites have yielded inconsistent results, which suggests that conclusions drawn from studies of a particular parasite may not be generalizable to other parasites (73, 86).

PASSIVE ANTIBODY FOR PREVENTION AND TREATMENT OF MUCOSAL INFECTIONS

The increasing prevalence of antibiotic-resistant bacterial pathogens and the rising number of immunocompromised patients have provided a stimulus to reconsider the use of antibodies, in the form of monoclonal antibodies, to combat clinical infections (18). The development of monoclonal antibodies for a wide variety of pathogens as part of the therapeutic armamentarium may now be feasible—owing to advances in monoclonal antibody technology that potentially enable the generation and production of large quantities of human or humanized antibodies in cells of diverse origin, including bacteria, insects, and plants (14, 16, 84, 160, 162). The rationale for increased clinical use of passive antibodies presented by Casadevall & Scharff (18) is based on an emphasis on systemic delivery for therapeutic purposes.

Several experimental studies employing passively administered monoclonal IgA antibodies against infectious pathogens were discussed above (see "Immune Barrier Function of IgA"), in the context of demonstrating and elucidating the mucosal host–defense functions of IgA. The success of IgA antibodies in these studies suggests that it could be worthwhile, under certain circumstances, to give passive antibody not only to treat but also to prevent respiratory and enteric infections in humans. The obvious route of administration would be topical; oral, in the case of the gastrointestinal tract; or by drops, sprays, or inhalants for the respiratory tract. Topical antibodies could be effective against intraluminal pathogens as well as against viruses that spread laterally after release from the luminal surface of infected mucosal epithelial cells. To the extent that systemically administered dimeric IgA antibodies are able to gain access to mucosal epithelium, where they could be internalized by the pIgR-mediated transport mechanism, they might also be able to inhibit viruses by intracellular effects.

Eventually, as suggested in Reference 18, passive therapy with monoclonal antibodies could be broadly useful. In the interim, such measures could well be applicable in particular situations, such as when selected populations are at risk—e.g. in an institutional environment, where a particular infection is prevalent during an acute spread in a hospital or school setting, or for reproductive health (24).

Human or humanized antibodies offer advantages over murine antibodies for avoiding adverse immune reactions to the foreign protein of the nonhuman antibody, if given systemically. Such hazards would be of less concern if the antibodies were applied topically to a mucosal surface because mucosal, especially intestinal, exposure to foreign protein tends to induce systemic immune tolerance—so-called oral (or more generally mucosal) tolerance—rather than overt immunity (152). Consequently, if human(ized) antibodies are not available, use of murine monoclonal antibodies topically might be feasible.

Studies of passive immunity by individual antibody classes, in the context of mucosal immunity, have generally focused on IgA. Just because these antibodies normally predominate in mucosal secretions, however, it does not necessarily follow that for passive immunity of mucosal surfaces, IgA antibodies would be more effective than other classes. Polymeric IgA and pentameric IgM do have theoretical advantages, in that they contain more than the two antibody combining sites per molecule that are present in IgG. This additional valence affords more potential for cross-linking microbes. In practice, this may or may not be significant. In fact, a study of passive immunity to respiratory virus infection in mice suggested that IgG and monomeric and polymeric IgA antibodies had comparable benefits (90), and another study on a bacterial gastroenteritis found that IgA and IgG antibodies appeared comparable (9). On the other hand, in a study of passive protection of the intestinal tract against bacterial colonization, polymeric IgA antibody was more effective than monomeric antibody was (77), and in studies of antibacterial and antiviral protection of the nasopharynx and intestine, IgG was less effective than secretory IgA was (8, 113). In a study of children with rotaviral gastroenteritis, pooled human serum immunoglobulin, mostly IgG, given orally, promoted recovery (51). It would therefore be worthwhile to do thorough comparative studies of the efficacy of the different classes of antibodies for delivering passive protection to the respiratory, gastrointestinal, and genital tracts. This is a complicated issue because results could vary with different pathogens and different antigenic determinants on the same pathogen. Separate studies should be done to evaluate prophylaxis and recovery because the same antibody class may not be most effective in both situations.

In addition to intact antibodies, univalent and bivalent fragments can be effective and they deserve attention (20, 28, 128, 163). The success achieved with antibody fragments illustrates that the cross-linking of antigens and Fc-mediated antibody effector functions may not always be required.

Another important consideration is the degree of added benefit from the presence of secretory component in the IgA (and IgM) found naturally in secretions. Studies confined to the products of plasma cells (H and L chains in all classes and J chain in IgM and polymeric IgA—but not secretory component, which is derived from the pIgR produced by the mucosal epithelial cells through which IgA and IgM enter secretions) may not provide complete information. Secretory component protects against proteolytic degradation of secretory IgA (12, 82)—a factor that may be important for passive immunization of the gastrointestinal tract, where there are abundant digestive enzymes, but that is less likely to be important for the respiratory tract.

For comparative studies of secretory IgA versus serum-type IgA (i.e. IgA with and without bound secretory component) it will be necessary to develop efficient methods for preparing IgA antibodies that contain secretory component. For example, cells could be genetically engineered to produce secretory IgA, polymeric IgA antibodies could be passed through epithelial cells that make pIgR, or polymeric IgA antibodies and secretory component could be combined in solution. Even so, the added benefit from inclusion of secretory component would have to be considerable to outweigh the extra inconvenience and cost.

Finally, rigorous comparisons of the degree of passive protection offered by different classes of antibody would require that the antigenic determinants that are recognized by the antibodies and the binding affinities be consistent. In principle, this is achievable with monoclonal antibodies by selecting class switch variants or by genetic engineering.

Although it would be of great interest to have rigorous comparative data on the relative efficacies of different classes of antibody for passive prophylaxis and therapy, an effective and practical advance probably could be achieved by simply using cocktails of monoclonal antibodies of different classes (and antigenic specificities).

MUCOSAL VACCINATION

In the early part of the century, there was interest in developing oral vaccination procedures for enteric infections. Subsequently, vaccination procedures increasingly focused on delivery through the skin, which targets the systemic immune system. However, the success of oral vaccination for poliomyelitis, which was based on knowledge that the primary infection involved the intestine, together with a burgeoning appreciation of the large bulk and wide distribution of mucosa-associated lymphoid tissue and the compartmentalization of the IgA antibody response, has led to renewed interest in exploiting the mucosal immune system for purposes of vaccination against the many infections that occur in or on mucous membranes or that use mucous membranes as portals of entry. Although an in-depth consideration of current approaches to mucosal immunization is beyond the scope of this review, some general considerations and examples are mentioned.

The origin and differentiation of the mucosal plasma cells that secrete IgA are relevant to the specificity of the antibodies that are produced at a given site (local immunity) and therefore are also relevant to strategies for mucosal immunization that are designed to yield antibodies of a desired specificity at a particular mucous membrane that is at risk for a particular infection. Mucosal antibody responses are most efficiently stimulated when antigen impinges on mucosal epithelium overlying organized lymphoid tissue. The lymphoid follicle–associated epithelium contains numerous M cells, which have an extended but attenuated cytoplasm that forms a pocket that surrounds underlying lymphocytes in an umbrella-like fashion. The M cells are able to transport relatively intact macromolecules, including certain viruses and bacteria, from the lumen to the lymphocytes in the underlying pocket (44, 108, 120, 121). Such transport may be facilitated when IgA antibody is bound to the antigen

(156). M-cell transport thus provides a mechanism for focusing antigens on organized mucosal lymphoid tissue, where mucosal antibody responses are initiated (26, 58). Under the influence of cytokines that are characteristic of so-called Th2 type T cell responses, such as IL-4, IL-5, IL-6, IL-10, and TGF$_\beta$ that are preferentially produced in mucosa-associated lymphoid tissue, local B cells proliferate and differentiate in an IgA direction (38, 46, 153). At the same time, via the lymph flow, they leave the mucosa for the draining lymph nodes. After exiting these lymph nodes via the efferent lymph, the B cells eventually reach the bloodstream, by which they return to mucous membranes, displaying a preference for their site of origin (125). Thus, B cells that were originally stimulated in intestinal Peyer's patches and that migrated through mesenteric lymph nodes tend to return to the intestinal mucosa—not to the organized lymphoid tissue where their precursors originally met antigen but rather to the diffuse lymphoid tissue of the lamina propria. On arriving at the lamina propria via the blood circulation, many of the B cells differentiate into IgA-secreting plasma cells. Despite the tendency for B cells that initially interacted with antigen in one site, such as the intestine, to return preferentially to that site, there is also abundant evidence that the various mucous membranes of the body and their associated lymphoid tissue comprise an interrelated, integrated system (33, 94, 133, 154). Thus, when cells are initially stimulated by antigen in organized lymphoid tissue at a particular location, some of their descendants settle in the lamina propria of a distant mucous membrane so that immunization at one mucosal site may yield an antibody response at another mucosal site. Therefore, although there are advantages to direct immunization of the site where a particular antibody response is desired, it is also possible for IgA antibody responses at that site to be induced by immunization of mucosa-associated lymphoid tissue elsewhere, or alternatively, IgA antibody responses can be induced by a regimen of primary immunization and boosting that includes both a distant site and the site where the infection in question occurs (7, 21, 27, 57, 64, 67, 79, 99, 105, 106, 139, 148). There is great interest in learning the most efficient and practical routes of immunization for inducing mucosal antibody responses in different anatomical sites in order to prevent a variety of infections (96, 139, 149). In addition to the more traditional avenues of the intestinal and respiratory tracts, rectal and genital (40, 54, 57, 74, 78, 79) immunization protocols are also being investigated. Still another approach is to convert a peripheral immunization of lymphoid tissue not draining a mucous membrane into a mucosal type of antibody response (37).

There is also much interest in developing new and effective means for packaging antigens for mucosal immunization. For example, antigens can be administered in aerosols to the respiratory tract and can be given orally inside biodegradable microspheres that penetrate M cells to reach gut-associated lymphoid tissue with increased efficiency (42, 64, 98, 116). Antigens for oral delivery have also been combined with liposomes (52). Through genetic engineering, it is possible to develop attenuated bacteria and viruses (and also plants) that express protective antigens or epitopes of either the parent pathogen or another pathogen (7, 17, 32, 36, 57, 60, 87, 105). If the vector is able to colonize a mucosal surface, it can provide a particularly effective means of stimulating a mucosal immune response. Attenuated microbes can also be constructed that (a) retain the ability of the original pathogen to invade mucous membranes, affording better access to mucosa-associated lymphoid tissue, but (b) have only a limited capacity to replicate, and are therefore unable to cause disease. Another promising approach is DNA vaccines delivered mucosally (47).

In addition to new vaccines, there is considerable interest in adjuvants that will boost mucosal immune responses and that will be clinically acceptable. For example, cholera toxin and the related *Escherichia coli* heat-labile enterotoxin are highly effective mucosal adjuvants in experimental systems, but they cannot be used in native form in humans; however, it may become possible to create nontoxic variants that retain adjuvant activity (39, 40, 41, 56, 60, 83, 144).

CONCLUSION

It has long been recognized that most infections involve mucous membranes, either as the locus of disease or as the site of invasion and perhaps initial replication. It is also well known that antibodies, in general, are highly effective against extracellular pathogens and against the extracellular phases of intracellular pathogens. Thus, pre-existing antibodies are capable of preventing infections with great efficiency. Increasing insights into the role of IgA as the major mucosal antibody and into the natural pathways and requirements for eliciting its production, together with rapid progress in developing new vectors and adjuvants to stimulate the formation of mucosal antibodies in greater amounts and over longer periods of time, should make it possible to create a new generation of vaccines. In addition, topical administration of monoclonal antibodies to mucous membranes offers significant potential for treating and, in selected circumstances, for preventing infections.

ACKNOWLEDGMENTS

The author thanks Drs. Steven Czinn, Steven Emancipator, Yung Huang, Edward Medof, John Nedrud and Sanjay Pimplikar for helpful comments. Drs. Pimplikar and Emancipator assisted with Figures 1 and 2. Research in the author's laboratory was supported by NIH grants AI-26449 and AI-36359.

Literature Cited

1. Abraham SN, Beachey EH. 1985. Host defenses against adhesion of bacteria to mucosal surfaces. In *Advances in Host Defense Mechanisms,* ed. JF Gallin, AS Fauci, 4:63–88. New York: Raven
2. Abu-Ghazaleh RI, Fujisawa T, Mestecky J, Kyle RA, Gleich GJ. 1989. IgA-induced eosinophil degranulation. *J. Immunol.* 142:2393–400
3. Åkerström B, Lindqvist A, Vander Maelen C, Grubb A, Lindahl G, Vaerman J-P. 1994. Interaction between streptococcal protein ARP and different molecular forms of human immunoglobulin A. *Molec. Immunol.* 31:393–400
4. Apodaca G, Katz L, Mostov KE. 1994. Receptor-mediated transcytosis of IgA in MDCK cells is via apical recycling endosomes. *J. Cell Biol.* 125:67–86
5. Apter FM, Michetti P, Winner LS III, Mack JA, Mekalanos JJ, Neutra MR. 1993. Analysis of the roles of antipolysaccharide and anti-cholera toxin immunoglobulin A (IgA) antibodies in protection against *Vibrio cholerae* and cholera toxin by use of monoclonal IgA antibodies *in vivo*. *Infect. Immun.* 61:5279–85
6. Armstrong SJ, Dimmock NJ. 1992. Neutralization of influenza virus by low concentrations of hemaglutinin-specific polymeric immunoglobulin A inhibits viral fusion activity, but activation of the ribonucleoprotein is also inhibited. *J. Virol.* 66:3823–32
7. Bender BS, Rowe CA, Taylor SF, Wyatt LS, Moss B, Small PA Jr. 1996. Oral immunization with a replication-deficient recombinant vaccinia virus protects mice against influenza. *J. Virol.* 70:6418–24
8. Bessen D, Fischetti VA. 1988. Passive acquired mucosal immunity to group A streptococci by secretory immunoglobulin A. *J. Exp. Med.* 167:1945–50
9. Blanchard TG, Czinn SJ, Maurer R, Thomas WD, Soman G, Nedrud JG. 1995. Urease-specific monoclonal antibodies prevent *Helicobacter felis* infection in mice. *Infect. Immun.* 63:1394–99
10. Brandtzaeg P. 1995. Molecular and cellular aspects of the secretory immunoglobulin system. *Acta Path. Microbiol. Immunol. Scand.* 103:1–19

11. Brandtzaeg P, Karlsson G, Hansson G, Petruson B, Björkander J, Hanson LA. 1987. The clinical condition of IgA-deficient patients is related to the proportion of IgD- and IgM-producing cells in their nasal mucosa. *Clin. Exp. Immunol.* 67:626–36
12. Brown WR, Newcomb RW, Ishizaka K. 1970. Proteolytic degradation of exocrine and serum immunoglobulins. *J. Clin. Invest.* 49:1374–80
13. Burns JW, Siadat-Pajouh M, Krishnaney AA, Greenberg HB. 1996. Protective effect of rotavirus VP6-specific IgA monoclonal antibodies that lack neutralizing activity. *Science* 272:104–7
14. Burton DR, Barbas CF. 1994. Human antibodies from combinatorial libraries. *Adv. Immunol.* 57:191–280
15. Capron M, Capron A. 1994. Immunoglobulin E and effector cells in schistosomiasis. *Science* 264:1876–77
16. Carayannopoulos L, Max EE, Capra JD. 1994. Recombinant human IgA expressed in insect cells. *Proc. Natl. Acad. Sci. USA* 91:8348–52
17. Cárdenas L, Clements JD. 1992. Oral immunization using live attenuated *Salmonella* spp. as carriers of foreign antigens. *Clin. Microbiol. Rev.* 5:328–42
18. Casadevall A, Scharff MD. 1995. Return to the past: the case for antibody-based therapies in infectious diseases. *Clin. Infect. Dis.* 21:150–61
19. Castrignano SB, Carlsson B, Carneiro-Sampaio MS, Söderström T, Hanson LA. 1993. IgA and IgG subclass deficiency in a poor population in a developing country. *Scand. J. Immunol.* 37:509–14
20. Chanock RM, Crowe JE Jr, Murphy BR, Burton DR. 1993. Human monoclonal antibody Fab fragments cloned from combinatorial libraries: potential usefulness in prevention and/or treatment of major human viral diseases. *Infect. Agents Dis.* 2:118–31
21. Chen KS, Quinnan GV Jr. 1989. Efficacy of inactivated influenza vaccine delivered by oral administration. *Curr. Top. Microbiol. Immunol.* 146:101–6
22. Chintalacharuvu KR, Tavill AS, Louis LN, Vaerman J-P, Lamm ME, Kaetzel CS. 1994. Disulfide bond formation between

dimeric immunoglobulin A and the polymeric immunoglobulin receptor during hepatic transcytosis. *Hepatology* 19:162–73

23. Clements ML, Betts RF, Tierney EL, Murphy BR. 1986. Serum and nasal wash antibodies associated with resistance to experimental challenge with influenza A wild-type virus. *J. Clin. Microbiol.* 24:157–60

24. Cone RA, Whaley KJ. 1994. Monoclonal antibodies for reproductive health: Part I. Preventing sexual transmission of disease and pregnancy with topically applied antibodies. *Am. J. Reprod. Immunol.* 32:114–31

25. Cotter TW, Meng Q, Shen Z-L, Zhang Y-X, Su H, Caldwell HD. 1995. Protective efficacy of major outer membrane protein-specific immunoglobulin A (IgA) and IgG monoclonal antibodies in a murine model of *Chlamydia trachomatis* genital tract infection. *Infect. Immun.* 63:4704–14

26. Craig SW, Cebra JJ. 1975. Rabbit Peyer's patches, appendix, and popliteal lymph node B-lymphocytes: a comparative analysis of their membrane immunoglobulin components and plasma cell precursor potential. *J. Immunol.* 114:492–502

27. Cripps AW, Dunkley ML, Clancy RL. 1994. Mucosal and systemic immunizations with killed *Pseudomonas aeruginosa* protect against acute respiratory infection in rats. *Infect. Immun.* 62:1427–36

28. Crowe JE, Murphy BR, Chanock RM, Williamson RA, Barbas CF III, Burton DR. 1994. Recombinant human respiratory syncytial virus (RSV) monoclonal antibody Fab is effective therapeutically when introduced directly into the lungs of RSV-infected mice. *Proc. Natl. Acad. Sci. USA* 91:1386–90

29. Cruz JR, Gil L, Cano P, Caceres P, Pareja G. 1988. Breastmilk anti-*Escherichia coli* heat-labile toxin IgA antibodies protect against toxin-induced infantile diarrhea. *Acta Paediatr. Scand.* 77:658–62

30. Cunningham-Rundles C, Brandeis WE, Good RA, Day NK. 1979. Bovine antigens and the formation of circulating immune complexes in selective immunoglobulin A deficiency. *J. Clin. Invest.* 64:272–79

31. Curtis GH, Gall DG. 1992. Macromolecular transport by rat gastric mucosa. *Am. J. Physiol.* 262:G1033–40

32. Curtiss R III, Kelly SM, Gulig PA, Nakayama K. 1989. Selective delivery of antigens by recombinant bacteria. *Curr. Top. Microbiol. Immunol.* 146:35–49

33. Czerkinsky C, Prince SJ, Michalek SM, Jackson S, Russell MW, et al. 1987. IgA antibody producing cells in peripheral blood after antigen ingestion: evidence for a common mucosal immune system in humans. *Proc. Natl. Acad. Sci. USA* 84:2449–53

34. Czerkinsky C, Svennerholm A-M, Quiding M, Jonsson R, Holmgren J. 1991. Antibody-producing cells in peripheral blood and salivary glands after oral cholera vaccination of humans. *Infect. Immun.* 59:996–1001

35. Czinn SJ, Cai A, Nedrud JG. 1993. Protection by germ-free mice from infection by *Helicobacter felis* after active oral or passive IgA immunization. *Vaccine* 11:637–42

36. Davis NL, Brown KW, Johnston RE. 1996. A viral vaccine vector that expresses foreign genes in lymph nodes and protects against mucosal challenge. *J. Virol.* 70:3781–87

37. Daynes RA, Enioutina EY, Butler S, Mu H-H, McGee ZA, Araneo BA. 1996. Induction of common mucosal immunity by hormonally immunomodulated peripheral immunization. *Infect. Immun.* 64:1100–9

38. Defrance T, Vanbervliet B, Briere F, Durand I, Rousset R, Banchereau J. 1992. Interleukin 10 and transforming growth factor β cooperate to induce anti-CD40-activated naive human B cells to secrete immunoglobulin A. *J. Exp. Med.* 175:671–82

39. Dickinson BL, Clements JD. 1995. Dissociation of *Escherichia coli* heat-labile enterotoxin adjuvanticity from ADP-ribosyltransferase activity. *Infect. Immun.* 63:1617–23

40. Di Tommaso A, Saletti G, Pizza M, Rappuoli R, Dougan G, et al. 1996. Induction of antigen-specific antibodies in vaginal secretions by using a nontoxic mutant of heat-labile enterotoxin as a mucosal adjuvant. *Infect. Immun.* 64:974–79

41. Douce G, Trucotte C, Cropley I, Roberts M, Pizza M, et al. 1995. Mutants of *Escherichia coli* heat labile toxin lacking ADP-ribosyltransferase activity act as nontoxic, mucosal adjuvants. *Proc. Natl. Acad. Sci. USA* 92:1644–48

42. Eldridge JH, Staas JK, Meulbroek JA, McGhee JR, Tice TR, Gilley RM. 1991. Biodegradable microspheres as a vaccine delivery system. *Mol. Immunol.* 28:287–94

43. Feng N, Burns JW, Bracy L, Greenberg HB. 1994. Comparison of mucosal and systemic humoral immune responses and

subsequent protection in mice orally inoculated with a homologous or a heterologous rotavirus. *J. Virol.* 68:7766–73

44. Frey A, Giannasca KT, Weltzin R, Giannasca PJ, Reggio H, et al. 1996. Role of the glycocalyx in regulating access of microparticles to apical plasma membranes of intestinal epithelial cells: implications for microbial attachment and oral vaccine targeting. *J. Exp. Med.* 184:1045–59

45. Fubara ES, Freter R. 1973. Protection against enteric bacterial infection by secretory IgA antibodies. *J. Immunol.* 111:395–403

46. Fujihashi K, Kweon M-N, Kiyono H, VanCott JL, van Ginkel F, et al. 1997. A T cell/B cell/epithelial cell internet for mucosal inflammation and immunity. *Springer Sem. Immunopath.* 18:477–94

47. Fynan EF, Webster RG, Fuller DH, Haynes JR, Santoro JC, Robinson HL. 1993. DNA vaccines: protective immunizations by parenteral, mucosal, and gene-gun inoculations. *Proc. Natl. Acad. Sci. USA* 90:11478–82

48. Gerber JD, Ingersoll JD, Gast AM, Christianson KK, Selzer NL, et al. 1990. Protection against feline infectious peritonitis by intranasal inoculation of a temperature-sensitive FIPV vaccine. *Vaccine* 8:536–42

49. Gilbert JV, Plaut AG, Longmaid B, Lamm ME. 1983. Inhibition of microbial IgA proteases by human secretory IgA and serum. *Molec. Immunol.* 20:1039–49

50. Glass RE, Svennerholm AM, Stoll BJ, Khan MR, Hossein KMB, et al. 1983. Protection against cholera in breast-fed children by antibodies in breast-milk. *N. Engl. J. Med.* 308:1389–92

51. Guarino A, Canani RB, Russo S, Albano F, Canani MB, et al. 1994. Oral immunoglobulins for treatment of acute rotaviral gastroenteritis. *Pediatrics* 93:12–16

52. Guzman CA, Molinari G, Fountain MW, Rohde M, Timmis KN, Walker MJ. 1993. Antibody responses in the serum and respiratory tract of mice following oral vaccination with liposomes coated with filamentous hemagglutinin and pertussis toxoid. *Infect. Immun.* 61:573–79

53. Hajishengallis G, Nikolova E, Russell MW. 1992. Inhibition of *Streptococcus mutans* adherence to saliva-coated hydroxyapatite by human secretory immunoglobulin A antibodies to the cell surface protein antigen I/II: reversal by IgA1 protease cleavage. *Infect. Immun.* 60:5057–64

54. Haneberg B, Kendall D, Amerongen HM,

Apter FM, Kraehenbuhl J-P, Neutra MR. 1994. Induction of specific immunoglobulin A in the small intestine, colon-rectum, and vagina measured by a new method for collection of secretions from local mucosal surfaces. *Infect. Immun.* 62:15–23

55. Hanson LA, Björkander J, Carlsson B, Roberton D, Söderström T. 1988. The heterogeneity of IgA deficiency. *J. Clin. Immunol.* 8:159–62

56. Holmgren J, Lycke N, Czerkinsky C. 1993. Cholera toxin and cholera B subunit as oral-mucosal adjuvant and antigen vector systems. *Vaccine* 11:1179–84

57. Hopkins S, Kraehenbuhl J-P, Schödel F, Potts A, Peterson D, et al. 1995. A recombinant *Salmonella typhimurium* vaccine induces local immunity by four different routes of immunization. *Infect. Immun.* 63:3279–86

58. Husband AJ, Gowans JL. 1978. The origin and antigen-dependent distribution of IgA-containing cells in the intestine. *J. Exp. Med.* 148:1146–60

59. Husby S, Jensenius JC, Svehag SE. 1985. Passage of undegraded dietary antigen into the blood of healthy adults: quantitation, estimation of size distribution, and relation of uptake to levels of specific antibodies. *Scand. J. Immunol.* 22:83–92

60. Jagusztyn-Krynicka EK, Clark-Curtiss JE, Curtiss III R. 1993. *Escherichia coli* heat-labile toxin subunit B fusions with *Streptococcus sobrinus* antigens expressed by *Salmonella typhimurium* oral vaccine strains: importance of the linker for antigenicity and biological activities of the hybrid proteins. *Infect. Immun.* 61:1004–15

61. Jarvis GA, Griffiss JM. 1991. IgA1 blockade of IgG-initiated lysis of *Neisseria meningitidis* is a function of antigen-binding fragment binding to the polysaccharide capsule. *J. Immunol.* 147:1962–67

62. Jertborn M, Svennerholm AM, Holmgren J. 1986. Saliva, breast milk, and serum antibody responses as indirect measures of intestinal immunity after oral cholera vaccination or natural disease. *J. Clin. Microbiol.* 24:203–9

63. Johnson PR, Feldman S, Thompson JM, Mahoney JD, Wright P. 1986. Immunity to influenza A virus infection in young children: a comparison of natural infection, live cold-adapted vaccine, and inactivated vaccine. *J. Infect. Dis.* 154:121–27

64. Jones DH, McBride BW, Thornton C, O'Hagan DT, Robinson A, Farrar GH.

1996. Orally administered microencapsulated *Bordetella pertussis* fimbriae protect mice from *B. pertussis* respiratory infection. *Infect. Immun.* 64:489–94

65. Kaetzel CS, Robinson JK, Chintalacharuvu KR, Vaerman J-P, Lamm ME. 1991. The polymeric immunoglobulin receptor (secretory component) mediates transport of immune complexes across epithelial cells: a local defense function of IgA. *Proc. Natl. Acad. Sci. USA* 88:8796–800

66. Kaetzel CS, Robinson JK, Lamm ME. 1994. Epithelial transcytosis of monomeric IgA and IgG cross-linked through antigen to polymeric IgA. A role for monomeric antibodies in the mucosal immune system. *J. Immunol.* 152:72–76

67. Kanesaki T, Murphy BR, Collins PL, Ogra PL. 1991. Effectiveness of enteric immunization in the development of secretory immunoglobulin A response and the outcome of infection with respiratory syncytial virus. *J. Virol.* 65:657–63

68. Keren DF, McDonald RA, Carey JL. 1988. Combined parenteral and oral immunization results in an enhanced mucosal immunoglobulin A response to *Shigella flexneri*. *Infect. Immun.* 56:910–15

69. Kerr MA. 1990. The structure and function of human IgA. *Biochem. J.* 271:285–96

70. Kerr MA, Mazengera RL, Stewart WW. 1990. Structure and function of immunoglobulin A receptors on phagocytic cells. *Biochem. Soc. Trans.* 18:215–17

71. Kilian M, Reinholdt J, Lomholt H, Poulsen K, Frandsen EVG. 1996. Biological significance of IgA1 proteases in bacterial colonization and pathogenesis: critical evaluation of experimental evidence. *Acta Path. Microbiol. Immunol. Scand.* 104:321–38

72. Kilian M, Russell MW. 1994. Function of mucosal immunoglobulins. In *Handbook of Mucosal Immunology*, ed. PL Ogra, J Mestecky, ME Lamm, W Strober, JR McGhee, J Bienenstock, pp. 127–37. San Diego: Academic

73. King CL, Nutman TB. 1993. Cytokines and immediate hypersensitivity in protective immunity to Helminth infections. In *Infectious Agents and Disease*, 2:103–8. New York: Raven

74. Kiyono H, Miller CJ, Lu Y, Lehner T, Cranage M, et al. 1995. The common mucosal immune system for the reproductive tract: basic principles applied toward an AIDS vaccine. *Adv. Drug Delivery Rev.* 18:23–51

75. Kraehenbuhl J-P, Neutra MR. 1992. Molecular and cellular basis of immune protection of mucosal surfaces. *Physiol. Rev.* 72:853–79

76. Kris RM, Asofsky R, Evans CB, Small PA Jr. 1985. Protection and recovery in influenza virus-infected mice immunosuppressed with anti-IgM. *J. Immunol.* 134:1230–35

77. Lee CK, Weltzin R, Soman G, Georgakopoulos KM, Houle DM, Monath TP. 1994. Oral administration of polymeric immunoglobulin A prevents colonization with *Vibrio cholerae* in neonatal mice. *Infect. Immun.* 62:887–91

78. Lehner T, Brookes R, Panagiotidi C, Tao L, Klavinskis LS, et al. 1993. T- and B-cell functions and epitope expression in nonhuman primates immunized with simian immunodeficiency virus antigen by the rectal route. *Proc. Natl. Acad. Sci. USA* 90:8638–42

79. Lehner T, Tao L, Panagiotidi C, Klavinskis LS, Brookes R, et al. 1994. Mucosal model of genital immunization in male rhesus macaques with a recombinant simian immunodeficiency virus p27 antigen. *J. Virol.* 68:1624–32

80. Liew FY, Russell SM, Appleyard G, Brand CM, Beale J. 1984. Cross protection in mice infected with influenza A virus by the respiratory route is correlated with local IgA rather than serum antibody or cytotoxic T cell reactivity. *Eur. J. Immunol.* 14:350–56

81. Lindahl G, Akerström B, Frithz E, Hedén L-O, Stenberg L. 1990. Protein Arp, the immunoglobulin A receptor of group A streptococci. In *Bacterial Immunoglobulin-Binding Proteins. Microbiology, Chemistry, and Biology,* ed. MDP Boyle, 1:193–200. San Diego: Academic

82. Lindh E. 1975. Increased resistance of immunoglobulin A dimers to proteolytic degradation after binding of secretory component. *J. Immunol.* 14:284–86

83. Lycke N, Tsuji T, Holmgren J. 1992. The adjuvant effect of *Vibrio cholerae* and *E. coli* heat labile enterotoxins is linked to their ADP-ribosyltransferase activity. *Eur. J. Immunol.* 22:2277–81

84. Ma JK-C, Hiatt A, Hein M, Vine ND, Wang F, et al. 1995. Generation and assembly of secretory antibodies in plants. *Science* 268:716–19

85. Ma JK-C, Hunjan M, Smith R, Kelly C, Lehner T. 1990. An investigation into the mechanism of protection by local passive immunization with monoclonal antibodies against *Streptococcus mutans*. *Infect. Immun.* 58:3407–14

86. Maizels RM, Bundy DAP, Selkirk ME, Smith DF, Anderson RM. 1993. Immunological modulation and evasion by helminth parasites in human populations. *Nature* 365:797–805

87. Mason HS, Lam DM-K, Arntzen CJ. 1992. Expression of hepatitis B surface antigen in transgenic plants. *Proc. Natl. Acad. Sci. USA* 89:11745–49

88. Mazanec MB, Coudret CL, Fletcher DR. 1995. Intracellular neutralization of influenza virus by IgA anti-HA monoclonal antibodies. *J. Virol.* 69:1339-43

89. Mazanec MB, Kaetzel CS, Lamm ME, Fletcher D, Nedrud JG. 1992. Intracellular neutralization of virus by immunoglobulin A antibodies. *Proc. Natl. Acad. Sci. USA* 89:6901–05

90. Mazanec MB, Lamm ME, Lyn D, Portner A, Nedrud JG. 1992. Comparison of IgA versus IgG monoclonal antibodies for passive immunization of the murine respiratory tract. *Virus. Res.* 23:1–12

91. Mazanec MB, Nedrud JG, Lamm ME. 1987. Immunoglobulin A monoclonal antibodies protect against Sendai virus. *J. Virol.* 61:2624–26

92. Mazengera RL, Kerr MA. 1990. The specificity of the IgA receptor purified from human neutrophils. *Biochem. J.* 272:159–65

93. McCormick BA, Stocker BAD, Laux DC, Cohen PS. 1988. Roles of motility, chemotaxis and penetration through and growth in intestinal mucus in the ability of a virulent strain of *S. typhimurium* to colonize the large intestine of the streptomycin-treated mouse. *Infect. Immun.* 50:2209–17

94. McDermott MR, Bienenstock J. 1979. Evidence for a common mucosal immunologic system. I. Migration of B immunoblasts into intestinal, respiratory and genital tissues. *J. Immunol.* 122:1892–98

95. McGhee JR, Mestecky J. 1990. In defense of mucosal surfaces. *Infect. Dis. Clin. N. Am.* 4:315–41

96. McGhee JR, Mestecky J, Dertzbaugh MT, Eldridge JH, Hirasawa M, Kiyono H. 1992. The mucosal immune system: from fundamental concepts to vaccine development. *Vaccine* 10:75–88

97. Mellander L, Björkander J, Carlsson B, Hanson LA. 1986. Secretory antibodies in IgA deficient and immunosuppressed individuals. *J. Clin. Immunol.* 6:284–91

98. Mestecky J, Moldoveanu Z, Novak M, Huang W-Q, Gilley RM, et al. 1994. Biodegradable microspheres for the delivery of oral vaccines. *J. Control. Rel.* 28:131–41

99. Michalek SM, McGhee JR, Babb JL. 1978. Effective immunity to dental caries: dose-dependent studies of secretory immunity by oral administration of *Streptococcus mutans* to rats. *Infect. Immun.* 19:217–24

100. Michetti P, Mahan MJ, Slauch JM, Mekalanos JJ, Neutra MR. 1992. Monoclonal secretory immunoglobulin A protects mice against oral challenge with the invasive pathogen *Salmonella typhimurium. Infect. Immun.* 60:1786–92

101. Monteiro RC, Cooper MD, Kubagawa H. 1992. Molecular heterogeneity of Fc_α receptors detected by receptor-specific monoclonal antibodies. *J. Immunol.* 148:1764–70

102. Monteiro RC, Hostoffer RW, Cooper MD, Bonner JR, Gartland GL, Kubagawa H. 1993. Definition of IgA receptors on eosinophils and their enhanced expression in allergic individuals. *J. Clin. Invest.* 92:1681–85

103. Mostov KE. 1994. Transepithelial transport of immunoglobulins. *Annu. Rev. Immunol.* 12:63–84

104. Mostov KE, Deitcher DL. 1986. Polymeric immunoglobulin receptor expressed in MDCK cells transcytoses IgA. *Cell* 46:613–21

105. Muster T, Ferko B, Klima A, Purtscher M, Trkola A, et al. 1995. Mucosal model of immunization against human immunodeficiency virus type 1 with a chimeric influenza virus. *J. Virol.* 69:6678–86

106. Nedrud JG, Liang X, Hague N, Lamm ME. 1987. Combined oral/nasal immunization protects mice from Sendai virus infection. *J. Immunol.* 139:3484–92

107. Negrão-Corrêa D, Adams LS, Bell RG. 1996. Intestinal transport and catabolism of IgE. A major blood-independent pathway of IgE dissemination during a *Trichinella spiralis* infection of rats. *J. Immunol.* 157:4037–44

108. Neutra MR, Frey A, Kraehenbuhl J-P. 1996. Epithelial M cells: gateways for mucosal infection and immunization. *Cell* 86:345–48

109. Neutra MR, Pringault E, Kraehenbuhl J-P. 1996. Antigen sampling across epithelial barriers and induction of mucosal immune responses. *Annu. Rev. Immunol.* 14:275–300

110. Nikolova EB, Tomana M, Russell MW. 1994. The role of the carbohydrate chains in complement (C3) fixation by solid-phase-bound human IgA. *Immunology* 82:321–27

111. Nikolova EB, Tomana M, Russell MW. 1994. All forms of human IgA antibodies

bound to antigen interfere with complement (C3) fixation induced by IgG or by antigen alone. *Scand. J. Immunol.* 39: 275–80

112. Odorizzi G, Pearse A, Domingo D, Trowbridge IS, Hopkins CR. 1996. Apical and basolateral endosomes of MDCK cells are interconnected and contain a polarized sorting mechanism. *J. Cell Biol.* 135:139–52

113. Offit PA, Clark HF. 1985. Protection against rotavirus-induced gastroenteritis in a murine model by passively acquired gastrointestinal but not circulating antibodies. *J. Virol.* 54:58–64

114. Ogra PL, Karzon DT, Righthand F, McGillivray M. 1968. Immunoglobulin response in serum and secretions after immunization with live and inactivated poliovaccine and natural infection. *N. Engl. J. Med.* 279:893–900

115. Ogra PL, Mestecky J, Lamm ME, Strober W, McGhee JR, Bienenstock J, eds. 1994. *Handbook of Mucosal Immunology.* San Diego: Academic. 766 pp.

116. O'Hagan DT, McGee JP, Holmgren J, Mowat AM, Donachie AM, et al. 1993. Biodegradable microparticles for oral immunization. *Vaccine* 11:149–54

117. Onorato IM, Modlin JF, McBean AM, Thomas ML, Losonsky GA, Bernier RH. 1991. Mucosal immunity induced by enhanced-potency inactivated and oral polio vaccines. *J. Infect. Dis.* 163:1–6

118. Outlaw MC, Dimmock NJ. 1990. Mechanisms of neutralization of influenza virus on mouse tracheal epithelial cells by mouse monoclonal polymeric IgA and polyclonal IgM directed against the viral haemagglutinin. *J. Gen. Virol.* 71:69–76

119. Outlaw MC, Dimmock NJ. 1991. Insights into neutralization of animal viruses gained from study of influenza virus. *Epidemiol. Infect.* 106:205–20

120. Owen RL. 1994. M cells: entryways of opportunity for enteropathogens. *J. Exp. Med.* 180:7–9

121. Owen RL, Ermak TH. 1992. Structural specializations for antigen uptake and processing in the digestive tract. *Springer Semin. Immunopathol.* 12:139–52

122. Paganelli R, Lennsky RJ, Atherton DS. 1981. Detection of specific antigen within circulating immune complexes. Validation of the assay and its application to food antigen-antibody complexes found in healthy and food-allergic subjects. *Clin. Exp. Immunol.* 46:44–53

123. Pfaffenbach G, Lamm ME, Gigli I. 1982. Activation of the guinea pig alternative complement pathway by mouse IgA immune complexes. *J. Exp. Med.* 155:231–47

124. Phalipon A, Kaufmann M, Michetti P, Cavaillon J-M, Huerre M, et al. 1995. Monoclonal immunoglobulin A antibody directed against serotype-specific epitope of *Shigella flexneri* lipopolysaccharide protects against murine experimental shigellosis. *J. Exp. Med.* 182:769–78

125. Phillips-Quagliata JM, Lamm ME. 1994. Lymphocyte homing to mucosal effector sites. See Ref. 115, pp. 225–39

126. Plaut AG, Bachovchin WW. 1994. IgA-specific prolyl endopeptidases (serine-type). *Methods Enzymol.* 244:137–51

127. Plaut AG, Wright A. 1995. IgA-specific prolyl endopeptidases (metallotype). *Methods Enzymol.* 248:634–42

128. Prince GA, Hemming VG, Horswood RL, Baron PA, Murphy BR, Chanock RM. 1990. Mechanism of antibody-mediated viral clearance in immunotherapy of respiratory syncytial virus infection of cotton rats. *J. Virol.* 64:3091–92

129. Renegar KB, Small PA Jr. 1991. Passive transfer of local immunity to influenza virus infection by IgA antibody. *J. Immunol.* 146:1972–78

130. Renegar KB, Small PA Jr. 1991. Immunoglobulin A mediation of murine nasal anti-influenza virus immunity. *J. Virol.* 65:2146–48

131. Reuman PD, Bernstein DI, Keely SP, Sherwood JR, Young EC, Schiff GM. 1990. Influenza-specific ELISA IgA and IgG predict severity of influenza disease in subjects pre-screened with hemagglutination inhibition. *Antiviral Res.* 13:103–10

132. Ruiz-Palacios GM, Calva JJ, Pickering LK. 1990. Protection of breastfed infants against *Campylobacter* diarrhoea by antibodies in human milk. *J. Pediatr.* 116:707–13

133. Russell MW, Moldoveanu Z, White PL, Sibert GJ, Mestecky J, Michalek SM. 1996. Salivary, nasal, genital, and systemic antibody responses in monkeys immunized intranasally with a bacterial protein antigen and the cholera toxin B subunit. *Infect. Immun.* 64:1272–83

134. Sestini P, Nencioni L, Villa L, Boraschi D, Tagliabue A. 1988. IgA-driven antibacterial activity against *Streptococcus pneumoniae* by mouse lung lymphocytes. *Am. Rev. Respir. Dis.* 137:138–43

135. Shroff KE, Meslin K, Cebra JJ. 1995. Commensal enteric bacteria engender a self-limiting humoral mucosal immune response while permanently colonizing the gut. *Infect. Immun.* 63:3904–13

136. Sixbey JW, Yao Q-Y. 1992. Immunoglobulin A-induced shift of Epstein-Barr virus tissue tropism. *Science* 255:1578–80

137. Smith CB, Bellanti JA, Chanock RM. 1967. Immunoglobulins in serum and nasal secretions following infection with type 1 parainfluenza virus and injection of inactivated vaccines. *J. Immunol.* 99:133–41

138. Staats HF, Jackson RJ, Marinaro M, Takahashi I, Kiyono H, McGhee JR. 1994. Mucosal immunity to infection with implications for vaccine development. *Curr. Opin. Immunol.* 6:572–83

139. Staats HF, Nichols WG, Palker TJ. 1996. Mucosal immunity to HIV-1. Systemic and vaginal antibody responses after intranasal immunization with the HIV-1 C4/V3 peptide T1SP10 MN(A). *J. Immunol.* 157:462–72

140. Stokes CR, Soothill JF, Turner MW. 1975. Immune exclusion is a function of IgA. *Nature* 255:745–46

141. Svanborg-Eden C, Svennerholm A-M. 1978. Secretory immunoglobulin A and G antibodies prevent adhesion of *Escherichia coli* to human urinary tract epithelial cells. *Infect. Immun.* 22:790–97

142. Tada T, Ishizaka K. 1970. Distribution of γE-forming cells in lymphoid tissues of the human and monkey. *J. Immunol.* 104:377–87

143. Tagliabue A, Nencioni L, Villa L, Keren DF, Lowell GH, Boraschi D. 1983. Antibody-dependent cell-mediated antibacterial activity of intestinal lymphocytes with secretory IgA. *Nature* 306:184–86

144. Takahashi I, Marinaro M, Kiyono H, Jackson RJ, Nakagawa I, et al. 1996. Mechanisms for mucosal immunogenicity and adjuvancy of *Escherichia coli* labile enterotoxin. *J. Infect. Dis.* 173:627–35

145. Tamura S-I, Funato H, Hirabayashi Y, Kikuta K, Suzuki Y, et al. 1990. Functional role of respiratory tract haemagglutinin-specific IgA antibodies in protection against influenza. *Vaccine* 8:479–85

146. Tamura S, Funato H, Hirabayashi Y, Suzuki Y, Nagamine T, et al. 1991. Cross-protection against influenza A virus infection by passively transferred respiratory tract IgA antibodies to different hemagglutinin molecules. *Eur. J. Immunol.* 21:1337–44

147. Taylor HP, Dimmock NJ. 1985. Mechanism of neutralization of influenza virus by secretory IgA is different from that of monomeric IgA or IgG. *J. Exp. Med.* 161:198–209

148. Waldman RH, Bergmann KC, Stone J, Howard S, Chiodo V, et al. 1987. Age-dependent antibody response in mice and humans following oral influenza immunization. *J. Clin. Immunol.* 7:327–32

149. Walker RI. 1994. New strategies for using mucosal vaccination to achieve more effective immunization. *Vaccine* 12:387–400

150. Walker WA, Isselbacher KJ, Bloch, KJ. 1972. Intestinal uptake of macromolecules: effect of oral immunization. *Science* 177:608–10

151. Watt PJ, Robinson BS, Pringle CR, Tyrrell DAJ. 1990. Determinants of susceptibility to challenge and the antibody response of adult volunteers given experimental respiratory syncytial virus vaccines. *Vaccine* 8:231–36

152. Weiner HL, Friedman A, Miller A, Khoury SJ, Al-Sabbagh A, et al. 1993. Oral tolerance: immunological mechanisms and treatment of murine and human organ specific autoimmune diseases by oral administration of autoantigens. *Annu. Rev. Immunol.* 12:809–37

153. Weinstein PD, Cebra JJ. 1991. The preference for switching to IgA expression by Peyer's patch germinal center B cells is likely due to the intrinsic influence of their microenvironment. *J. Immunol.* 147:4126–35

154. Weisz-Carrington P, Roux ME, McWilliams M, Phillips-Quagliata JM, Lamm ME. 1979. Organ and isotype distribution of plasma cells producing specific antibody after oral immunization: evidence for a generalized secretory immune system. *J. Immunol.* 123:1705–8

155. Weltzin R, Hsu SA, Mittler ES, Georgakopoulos K, Monath TP. 1994. Intranasal monoclonal immunoglobulin A against respiratory syncytial virus protects against upper and lower respiratory tract infections in mice. *Antimicrob. Agents Chemother.* 38:2785–91

156. Weltzin R, Lucia-Jandris P, Michetti P, Fields BN, Kraehenbuhl J, Neutra MR. 1989. Binding and transepithelial transport of immunoglobulins by intestinal M cells: demonstration using monoclonal IgA antibodies against enteric viral proteins. *J. Cell Biol.* 108:1673–85

157. Weltzin R, Traina-Dorge V, Soike K, Zhang J-Y, Mack P, et al. 1996. Intranasal monoclonal IgA antibody to respiratory syncytial virus protects rhesus monkeys against upper and lower respiratory tract infection. *J. Infect. Dis.* 174:256–61

158. Williams RC, Gibbons RJ. 1972. Inhi-

bition of bacterial adherence by secretory immunoglobulin A: a mechanism of antigen disposal. *Science* 177:697–99

159. Winner III L, Mack J, Weltzin R, Mekalanos JJ, Kraehenbuhl J-P, Neutra M. 1991. New model for analysis of mucosal immunity: Intestinal secretion of specific monoclonal immunoglobulin A from hybridoma tumors protects against *Vibrio cholerae* infection. *Infect. Immun.* 59:977–82

160. Winter G, Griffiths AD, Hawkins RE, Hoogenboom HR. 1994. Making antibodies by phage display technology. *Annu. Rev. Immunol.* 12:433–55

161. Wold A, Endo T, Kobata A, Mestecky J, Tomana JM, et al. 1990. Secretory immunoglobulin A carries oligosaccharide receptors for the *E. coli* type 1 fimbrial lectin. *Infect. Immun.* 58:3073–77

162. Wright A, Shin S-U, Morrison SL. 1992. Genetically engineered antibodies: progress and prospects. *Crit. Rev. Immunol.* 12:125–68

163. Zeitlin L, Whaley KJ, Sanna PP, Moench TR, Bastidas R, et al. 1996. Topically applied human recombinant monoclonal IgG_1 antibody and its Fab and $F(ab')_2$ fragments protect mice from vaginal transmission of HSV-2. *Virology* 225:213–15

Annu. Rev. Rev. Microbiol. 1997. 51:341–73

TRANSCRIPTIONAL CONTROL OF THE *PSEUDOMONAS* TOL PLASMID CATABOLIC OPERONS IS ACHIEVED THROUGH AN INTERPLAY OF HOST FACTORS AND PLASMID-ENCODED REGULATORS

Juan L. Ramos and Silvia Marqués
Consejo Superior de Investigaciones Científicas, Department of Biochemistry and Molecular and Cellular Biology of Plants, Profesor Albareda 1, 18008 Granada, Spain; email: jlramos@eez.csic.es

Kenneth N. Timmis
Gesellschaft für Biotechnologische Forschung MBH, National Center for Biotechnology, Division of Microbiology, Mascheroder Weg 1, D-38125 Braunschweig, Germany

KEY WORDS: DNA bending proteins, sigma factors, XylS family, NtrC family, catabolite repression, promoters

---

ABSTRACT

The *xyl* genes of *Pseudomonas putida* TOL plasmid that specify catabolism of toluene and xylenes are organized in four transcriptional units: the upper-operon *xylUWCAMBN* for conversion of toluene/xylenes into benzoate/alkylbenzoates; the *meta*-operon *xylXYZLTEGFJQKIH*, which encodes the enzymes for further conversion of these compounds into Krebs cycle intermediates; and *xylS* and *xylR*, which are involved in transcriptional control. The XylS and XylR proteins are members of the XylS/AraC and NtrC families, respectively, of transcriptional regulators. The *xylS* gene is constitutively expressed at a low level from the Ps2 promoter. The XylS protein is activated by interaction with alkylbenzoates, and this active form stimulates transcription from Pm by σ^{70}- or σ^{S}-containing RNA polymerase (the *meta* loop). The *xylR* gene is also expressed constitutively. The XylR protein, which in the absence of effectors binds in a nonactive form to target DNA sequences, is activated by aromatic hydrocarbons and ATP; it subsequently

341

undergoes multimerization and structural changes that result in stimulation of transcription from Pu of the upper operon. This latter process is assisted by the IHF protein and mediated by σ^{54}-containing RNA polymerase. Once activated, the XylR protein also stimulates transcription from the Ps1 promoter of $xylS$ without interfering with expression from Ps2. This process is assisted by the HU protein and is mediated by σ^{54}-containing RNA polymerase. As a consequence of hyperexpression of the $xylS$ gene, the XylS protein is hyperproduced and stimulates transcription from Pm even in the absence of effectors (the cascade loop). The two σ^{54}-dependent promoters are additionally subject to global (catabolite repression) control.

CONTENTS

INTRODUCTION

Metabolism of Aromatic Hydrocarbons Specified by TOL Plasmids

Plasmids that specify the degradation of toluene and xylenes have been found in bacteria isolated in different locations around the world, from Japan to Wales (5). Although these TOL plasmids vary in size, incompatibility group, genetic organization of catabolic genes, and other characteristics, they encode pathways that are very similar (74, 116). The archetypal TOL plasmid is designated pWW0 (130).

Pseudomonas putida bacteria containing the pWW0 plasmid grows on toluene, *m*- and *p*-xylene, *m*-ethyltoluene, and 1,2,4-trimethylbenzene (79, 134).

The genetic information for coordinated metabolism of these hydrocarbons is contained within the *xyl* operons on the TOL plasmid. Figure 1 shows the sequence of reactions involved in the oxidation of toluene and related hydrocarbons to Krebs cycle intermediates. The methyl group at carbon 1 in the aromatic ring of these compounds is sequentially oxidized to yield the corresponding benzyl alcohol, benzaldehyde, and benzoate (or the respective substituted compounds). This set of reactions is called the upper pathway. The aromatic carboxylic acids are then further metabolized through the *meta*-cleavage pathway, in which the benzoate, which can be substituted with an alkyl group at positions *meta, para,* or both, is then oxidized and decarboxylated to produce the corresponding catechol. These compounds undergo *meta* fission to yield 2-hydroxymuconic acid semialdehyde or the corresponding alkyl derivatives. Metabolism of the semialdehydes occurs via a branched pathway that rejoins at a common intermediate, 2-oxopent-4-enoate. The semialdehyde produced from *m*-toluate is hydrolyzed, whereas the semialdehyde from benzoate and *p*-methylbenzoate is metabolized via the oxalocrotonate branch, which involves at least three enzymatic steps (49). 2-Oxopent-4-enoate is further converted in 2-oxo-4-hydroxypentonate, which eventually renders Krebs cycle intermediates. For a more detailed description of the characteristics of the pathway enzymes, the reader is referred to earlier reviews and original papers (5, 52, 95, 106, 115, 123).

Tn4651 and Tn4653, the Toluene Transposons of TOL Plasmid pWW0

pWW0, a detailed physical map of which has been reported (25), is about 117 kb in size and belongs to the plasmid IncP9 incompatibility group. Unlike other TOL plasmids, pWW0 is self-transmissible, and on solid agar plates the frequency of conjugational transfer between different pseudomonads can be as high as 10^{-1} to 1 transconjugant per recipient cell (96, 114). pWW0 can also be transferred to enterobacteriaceae, although at much lower frequencies, ranging from 10^{-5} to 10^{-7} transconjugants per recipient cell (8, 71, 96, 114). The constitutive expression of pili genes in pWW0 may account for the high transmissibility of this plasmid (10). Replication of the plasmid is thermosensitive (96). Catabolic operons borne on the TOL plasmid pWW0 can also be integrated into the host chromosome because the toluene catabolic genes for these pathways are located on nested transposons Tn4651 (56 kb in size) and Tn4653 (70 kb in size) (119). These transposons are closely related to the Tn1721 family of class II transposons (44). Transposition of these elements has been shown to occur by cointegrate formation and resolution, mechanisms characteristic of class II elements (126–128). Transposition of Tn4653 and Tn4651 requires its own *tnpA* gene product. The resolution of cointegrates formed by

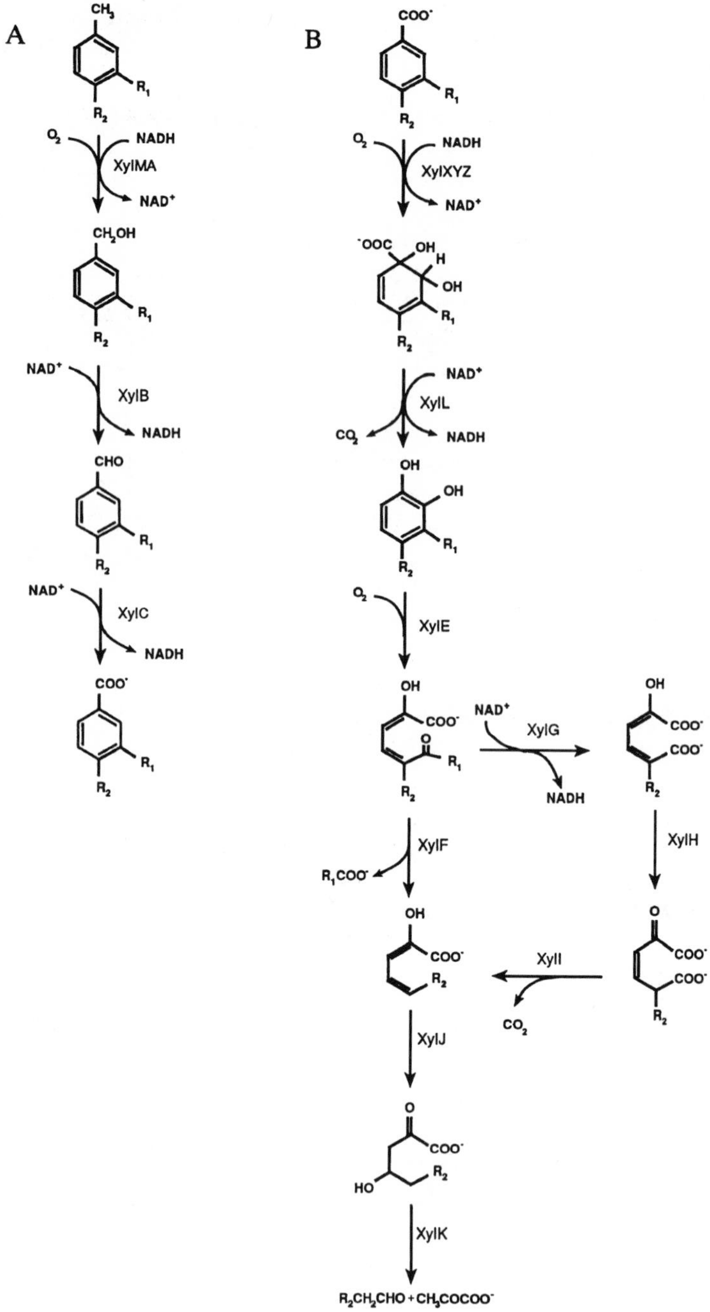
A
B
XylMA
XylB
XylC
XylXYZ
XylL
XylE
XylG
XylF
XylH
XylI
XylJ
XylK
O2
NADH
NAD+
CO2
R1COO-
R2CH2CHO + CH3COCOO-

Tn*4651* or Tn*4653* transposition requires three additional factors, all encoded by the smaller element, Tn*4651*. These factors are the *cis*-acting *res* region and the *trans*-acting *tnpS* and *tnpT* gene products.

The origin of the *xyl* genes within the Tn*4651* element remains unresolved, although it should be noted that the catabolic operons are readily deleted by homologous recombination between two 1.4-kb direct repeat sequences, located at the boundaries of the region (7, 90). In enterobacteria, such as *Erwinia* spp., and in pseudomonads growing on benzoate, the variants that lack catabolic genes arise at high frequencies, often yielding deletion plasmids that are similar to the 79-kbp plasmid pWW0-8 (7, 30, 31, 114).

THE CURRENT MODEL OF TRANSCRIPTIONAL CONTROL OF THE TOL PLASMID pWW0 CATABOLIC PATHWAYS

The genetic information for the upper pathway (genes *xylUWCMABN*) is located in the upper operon (47, 51, 131). Somewhat surprisingly, the *xylU*, *xylW*, and *xylN* gene products are not required for growth on toluene/xylenes (47, 131). The information for the *meta*-cleavage pathway is encoded by a separate operon (36, 48). This operon comprises 13 genes, *xylXYZLTEGFJQKIH* (50), extending over 11 kb, and it is thus one of the largest operons in prokaryotes. In addition to the catabolic operons, two regulatory genes, *xylS* and *xylR*, located 3′ from the meta operon, are involved in transcriptional control (37, 60–62, 133). These regulatory genes are transcribed from physically close, but functionally divergent, promoters (Figure 2) (66, 109). In this review, the promoters for the upper operon, the *meta* operon, and the *xylS* and *xylR* genes are referred to as Pu, Pm, and Ps and Pr, respectively.

Figure 2 depicts the current model for the control of expression of the catabolic pathway operons of TOL plasmid pWW0 (85). This model explains the induction of enzymes of each operon in *P. putida* (pWW0) cells growing exponentially on xylenes or toluates.

The genetic control of the catabolic operons reflects their biochemical organization, i.e. there are two regulatory loops: one that operates when cells are growing on toluates (the *meta* loop) and another complex system (the cascade

---

Figure 1   Degradation of toluene and related hydrocarbons as specified by the TOL plasmid pWW0. (*A*) Oxidation of toluene ($R_1 = R_2 = H$) and xylenes (R_1 or $R_2 = CH_3$; $R_1 = R_2 = CH_3$) to benzoate and toluates. (*B*) Oxidation of benzoate/toluates to Krebs cycle intermediates. Gene coding for the enzymes involved in each step of the pathways is indicated. Further details on biochemical reactions are found in References 5, 52.

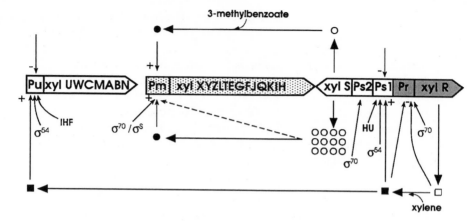

Figure 2 Regulatory circuits controlling expression from the TOL plasmid pWW0. *Squares*: XylR; *circles*: XylS; *open symbols*: transcriptional regulator forms unable to stimulate transcription; *closed symbols*: forms able to stimulate transcription. +: stimulation of transcription; −: inhibition of transcription. The regulatory loops are explained in the text.

loop) that operates when cells are growing on xylenes and ensures that both the upper and *meta* pathways are coordinately expressed.

The *meta* loop functions as follows: In cells growing in the absence of effectors, e.g. on glucose or glycerol, the *xylS* gene is expressed at a low level from an σ^{70}-dependent promoter (Ps2); and small quantities of XylS protein, in an inactive form (XylS$_i$), are produced. When a benzoate effector is added to the culture medium, the XylS$_i$ protein interacts with the effector and becomes activated (XylS$_a$) to stimulate transcription from the *meta*-pathway operon promoter Pm. Maintainance of high levels of expression from Pm throughout growth phases seems to involve the use of σ^{70} in the early exponential phase and of σ^S in the late exponential phase or stationary phase.

The cascade loop functions as follows:

1. The master regulator that is involved in transcriptional control of the catabolic pathways in cells growing on xylenes is *xylR*, which is transcribed from two σ^{70}-dependent tandem promoters. The XylR protein controls its own synthesis.

2. The inactive protein (XylR$_i$) becomes activated (XylR$_a$) when it binds toluene/xylenes to stimulate transcription from the upper pathway operon Pu promoter. This process requires σ^{54}-containing RNA polymerase and the DNA-bending protein integration host factor (IHF).

3. Activated XylR also stimulates expression of the *xylS* gene by inducing transcription from a second σ^{54}-dependent promoter (Ps1), a process that may be assisted by the chromatin-associated DNA-bending protein HU.

4. As a result, in cells growing on xylenes, the *xylS* gene is expressed from two promoters: the *xylS* Ps2 promoter, which is constitutive and XylR-independent, and the Ps1 promoter, which is σ^{54}- and XylR-dependent.

5. When the *xylS* gene is overexpressed, the XylS protein is hyperproduced; as a consequence, presumably a certain amount of the XylS protein becomes active, because under these conditions, transcription from Pm is achieved even in the absence of *meta* pathway effectors.

The expression of TOL plasmid operons is integrated in overall metabolic control in *Pseudomonas* spp.: Both the upper pathway operon promoter and the *xylS* Ps1 promoter are subject to catabolite repression, and as a consequence of the latter, the expression of the *meta*-cleavage operon is also subject to moderate catabolite repression.

THE *META* LOOP

This section will analyze how the regulatory protein XylS is expressed; how this protein, a member of XylS/AraC family of regulators, becomes activated by effector binding; and how the active regulator stimulates transcription from the Pm promoter.

xylS *Gene Expression in Cells Growing on Alkylbenzoates*

Growth of *P. putida* (pWW0) on alkylbenzoates requires expression of the *meta*-pathway operon, which is mediated by the XylS protein that is activated by the binding of a benzoate effector (36, 37, 40, 61, 63, 84, 112, 133). The *xylS* gene is expressed at low constitutive levels from a σ^{70}-dependent promoter called Ps2, and these levels are not significantly affected by the presence or absence of an XylS effector (39). The transcription initiation point of the mRNA that originates from Ps2 maps 9 bp upstream of the proposed first ATG of the *xylS* gene; the transcript contains a ribosome-binding site sequence 5′-GTGA-3′.

Putative −10 (TAAAAT) and −35 (TGTACA) sequences, similar to those recognized by σ^{70}-containing RNA polymerase, were identified at a point shortly upstream from the transcription start site. The amount of XylS protein that is made when the *xylS* gene is expressed from the Ps2 promoter is sufficient to stimulate high transcription levels from Pm in the presence of *m*-methylbenzoate. This is shown by the finding that near-maximal mRNA levels of *meta*-cleavage pathway were reached in the *P. putida* (pWW0) cells growing on this aromatic carboxylic acid (39).

The XylS *Regulator*

The XylS protein is 321 amino acid residues long (65, 89, 121). It belongs to a family of regulators that comprises at least 90 different proteins that are involved in the transcription stimulation of several cell processes, e.g. carbon metabolism, pathogenesis, and the response to alkylating agents in bacteria. This set of regulators constitutes the so-called XylS/AraC family (38, 41, 111). On the basis of protein alignments, two domains can be defined in these proteins. At the C-terminal end, these regulators exhibit a highly conserved stretch of about 100 amino acids that seems to be involved in DNA binding and is probably also involved in interactions with RNA polymerase (11–13, 16, 38, 87). The remainder of the protein is not conserved and may be involved in interactions with effectors and dimerization (34, 81, 110).

Inspection of the polypeptide sequence by an appropriate algorithm for predicting secondary structures suggests the presence of six α-helices that are highly conserved in the C-terminal region. Within this conserved region, an α-helix-turn-α-helix motif comprising residues 228–251 of XylS is probably involved in DNA binding. This is indicated in several ways: (*a*) The motif is homologous with the DNA-binding domain of lambda Cro and other regulators that bind DNA (89); (*b*) mutations within this motif result in mutants mediating high basal transcription stimulation from Pm in the absence of effectors (i.e. Ser229→Leu; Phe248→Cys) (38, 111, 135); (*c*) the corresponding aligned amino acids of AraC have been shown to be involved in protein-DNA interactions (11, 12); and (*d*) downstream of this α-helix-turn-α-helix DNA binding motif, another helix (residues 256–274 in XylS) is thought to participate in the correct organization of the DNA binding motif (122). Mutations within this third α-helix resulted in mutants with notably decreased transcriptional activity, e.g. Pro256→Leu or Ser (38). Although the α-helix-turn-α-helix DNA binding motif is conserved among the members of the family, some sequence variation has also been observed (JL Ramos, unpublished data), which probably reflects the fact that each regulator recognizes different promoter sequences.

The motif defined by helices α_5 (residues 281–289) and α_6 (residues 294–305) within the conserved C-terminal end is the most conserved region in the proteins of the family, with certain residues showing almost 90% identity (38). This may reflect a feature shared by all these regulators, e.g. this region may be involved in interactions with RNA polymerase, although no direct evidence for this hypothesis is available. Outside of the α_6 helix, a proline residue (Pro309 in XylS) is also conserved in more than 90% of the family members. Substitution of Pro309 by Ser, Ala, or Glu in XylS results in mutant regulators unable to stimulate transcription from Pm (MT Gallegos, M Manzanera, S Marqués, JL Ramos, unpublished observations).

Low levels of XylS protein are made in its inactive form when the the *xylS* gene is expressed from the Ps2 promoter. However, upon the addition of a *meta*-cleavage substrate, expression from the Pm promoter occurs immediately, which suggests that the regulator becomes active after effector binding (84). Interactions between effector molecules and the regulator have been studied by analyzing XylS-dependent transcriptional activation from the Pm promoter, which is fused to a promoterless *lacZ* gene in the presence of more than 50 different benzoate analogues. These studies reveal that substituted benzoates are XylS effectors, although not all positions in the planar benzoate molecule are equivalent. For example, position 3 is highly permissible ($-CH_3$, $-C_2H_5$, and $-OCH_3$ groups and F, Cl, Br, and I atoms are permissible substituents), whereas positions 2 and 4 pose some restrictions to substituents ($-CH_3$ and F and Cl atoms are allowed, whereas $-C_2H_5$ and I atoms are not) (112). Although disubstitutions involving positions 2 and 3, and 3 and 4, are permissible, other combinations are usually nonpermissible, which suggests that interactions between the effector and the regulator are nonsymmetrical. Ramos et al (110, 113) and Michán et al (92) isolated and sequenced a series of mutant regulators that are able to recognize substituted benzoate effectors that are not recognized by the wild-type regulator. Mutations were found to be clustered at positions 37–45, 88–92, 151–155, and around residues 256 and 288. These findings suggest, but do not prove, that the recognition pocket for XylS effectors may be composed of two or more noncontiguous segments of its primary sequence. Arg-41 seems to be a critical residue for interaction(s) with effectors, as changes in this position result in many different phenotypes. For example, XylSArg41Gly is a mutant regulator whose ability to recognize *o*- and *p*-methylbenzoate was lost, although it remained activatable by *m*-methylbenzoate. Substitution of Arg41 by Leu resulted in a mutant that was unable to respond to benzoate effectors (92).

XylS mutants such as XylSArg41Cys, XylSPro37Gly XylSGly44Ser, XylSSer229Ile, XylSAsp274Val, and XylSAsp274Glu mediated transcription from Pm in the absence of effectors (92, 135). These results support the hypothesis that XylS exists in vivo in a dynamic equilibrium between an inactive and an active form, with respect to transcriptional stimulation. Therefore, transition from the inactive to the active form may be mediated by effector binding.

How the interaction between benzoates and XylS leads to an active regulator is not yet understood. One possibility is that after effector binding, an intramolecular signal is transmitted from the N-terminal end to the C-terminal end, resulting in a conformational change that activates the regulator and promotes transcription. Other possibilities are that binding of the effector either favors oligomerization (with the oligomer as the active form of XylS), or it increases affinity for target sequences, thus facilitating DNA occupancy by XylS at the Pm promoter. Regardless of the mechanism, the effector-binding pocket and the DNA-binding motif are not independent domains, as shown by the intramolecular dominance of C-terminal mutations over N-terminal ones, and a reversal of this dominance in double mutants that were constructed in vitro (91).

Overproduction of XylS via a natural cascade regulatory system (66, 84, 109) (see below, the cascade loop), or after expression from strong promoters (67, 89, 121), leads to the stimulation of transcription from the Pm promoter in the absence of effectors. This finding further supports the idea that XylS may exist in an equilibrium between an inactive and an active form so that, if the total amount of XylS protein is increased in the cell, part of the XylS population becomes active, from a transcriptional point of view (85).

The σ^{70}/σ^{S} Promoter of the Meta-Cleavage Pathway Operon

Early studies located a main transcription initiation point 29 bp upstream from the first ATG of the *xylX* gene (63, 88) (Figure 2). This promoter was subsequently shown to be the only one driving the expression of the 11-kbp long *meta*-cleavage pathway operon (48, 86).

Stimulation of transcription from the Pm promoter requires a DNA sequence extending to about 80 bp upstream of the transcription initiation point (76, 109, 120). In the architecture of the Pm promoter, two regions can be distinguished on the basis of genetic data: the XylS interaction region, which extends from about −40 to −80 bp, and the downstream RNA polymerase recognition sequences, which exhibit atypical −35 and −10 DNA sequences. The DNA sequence between nucleotides −35 and −80 was first deduced to include the XylS binding site by Mermod et al (88), who created a "new" weak constitutive promoter, upstream of Pm, with the single mutation C-47→T. It was observed that expression from this new promoter was repressed in the presence of XylS and an effector, which suggests that, in this mutant promoter, XylS and RNA polymerase compete for the same binding sites. Kessler et al (75, 76) generated deletions and point mutations in the Pm region and analyzed transcription from the mutant promoters that are mediated by XylS activated by *m*-methylbenzoate. Their results suggested that the XylS binding region in Pm was organized as two homologous 15-bp tandemly repeated motifs (5′-TGCAAPuAAPuPyGGNTA-3′), separated by 6 bp and partially overlapping the −35 region (Figure 3). These results narrowed down the XylS recognition sequences to the nucleotides that span positions −35 to −72 and led Kessler et al (76) to suggest that the XylS-responsive element was arranged as an imperfect direct repeat of the motif, approximately 36 bp in length, and involving three and a half DNA helix turns.

Recently, Gallegos et al (40) readdressed the organization of XylS binding sites in the Pm promoter. They also generated a series of 5′ sequential deletions

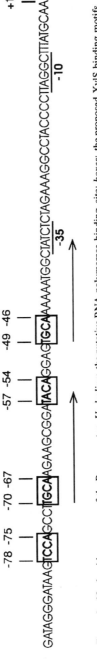

Figure 3   Nucleotide sequence of the Pm promoter. *Underline*: the putative RNA polymerase-binding site; *boxes*: the proposed XylS binding motifs, as deduced from genetic experiments (40); *arrows*: the direct repeats, deduced from in vitro experiments with N-XylS bound to affinity sepharose beads (73); *numbers*: distance in basepairs upstream from the transcription start site (+1).

in the promoter and analyzed transcription from the resulting mutant promoters that are mediated by the wild-type XylS protein and by mutant XylS regulators that are constitutive. It was found that promoters deleted up to -60 could be activated by constitutive XylS mutants (but not by the wild-type regulator) and that extension of the deletion to -51 prevented transcription. This led to the suggestion that shorter sequences than those proposed by Kessler et al (76) might suffice for XylS activation of transcription. On the basis of sequence analyses, Gallegos et al (40) proposed that the XylS binding site may be represented by the motif T(C or A)CAN$_4$TGCA, which appears twice, such that the exact location of the RNA polymerase binding-site proximal motif is between -46 and -57 and the distal motif is between -67 and -78. The $-46/-57$ proximal site appears to constitute the minimum sequence required for transcription stimulation, as XylSSer299Ile—a mutant regulator that mediates transcription from wild-type Pm in the absence of effectors—was able to promote transcription from this promoter in the presence of m-methylbenzoate, whereas in the absence of effectors it did not mediate transcriptional activity. The wild-type protein requires both motifs and an effector to stimulate transcription from Pm, although deletion of the furthest submotif TCCA has little effect on expression from Pm. Point mutations inside the proposed motifs (i.e. C-68→G and C-47→G) decreased the activity to different degrees, whereas mutations outside the motifs had little or no effect (40, 76).

Therefore, in view of the available genetic information, it is proposed that the XylS recognition site consists of two submotifs (TNCA) separated by 4 bp, which are simultaneously direct and inverted repeats. In this motif, the 3′ submotif is TGCA, which suggests that it is the primary recognition element for XylS, with the remaining 5′ submotif contributing to overall affinity.

Direct examination of XylS:Pm interactions through in vitro studies with purified activator have long been hampered by the difficulty of obtaining XylS protein that is active in vitro (21). Recently, Kaldalu et al (73) reported the immunopurification of a functionally active XylS protein bearing a hemagglutinin epitope that is fused at its N terminus (N-XylS). This N-XylS variant, bound to affinity sepharose beads, could specifically bind and retain a DNA fragment bearing the proposed XylS binding region in Pm. A set of footprinting experiments, using matrix-bound N-XylS, indicated that N-XylS binds along one side of the DNA, covering four helix turns (from -28 to -72) and making base-specific contacts in four adjacent major groove regions on the same helix face. This footprinting extended beyond the site defined by genetic means, and it is not known whether it represents a unique characteristic of the N-XylS variant used in these studies or whether it reflects oligomerization of N-XylS after recognition of a primary binding site. More in vitro studies with purified RNA polymerase and XylS are needed in order to determine whether the binding sites for each protein overlap. Results obtained by Kaldalu et al (73) suggest that an effector is not required to obtain the footprint with N-XylS, although the degree of protection is increased in the presence of the effector m-methylbenzoate, with no alteration in the protection pattern. This, together with the observation that overproduction of the regulator is sufficient to activate Pm in vivo in the absence of effector (66, 67, 109, 120), supports the hypothesis that effectors increase the cellular concentration of XylS in its active conformation ($XylS_a$ may exhibit higher affinity for its target DNA sequence) at the DNA target site.

The fact that the proposed XylS binding sites are very close to, or partially overlap, the RNA polymerase binding region suggests that the regulator probably contacts RNA polymerase to promote transcription, as is the case in a number of positively regulated σ^{70}-dependent promoters (18). Several single mutants of the α subunit of RNA polymerase are known to produce a protein that is unable to transcribe a number of activatable promoters, and direct contact between the α subunit and the activator protein has been suggested (46, 70). XylS activation of Pm has been monitored in vivo in *Escherichia coli* backgrounds with different mutant α subunits of RNA polymerase that are not responsive to certain transcriptional activators. None of the mutants tested had any significant effect on XylS-mediated transcription from Pm (77, 83).

In summary, it seems that Pm exhibits two sites recognized by XylS. The proximal motif from -46 to -57 seems to be adjacent to, or to overlap, the RNA polymerase site; its role may be to favor interaction between the regulator and RNA polymerase. This interaction may be facilitated by a track of six As, located between -46 and -41, that curves the DNA molecule (38), thereby facilitating these interactions. The second XylS binding motif further upstream, from -67 to -78, may increase DNA occupancy and thus enhance transcription from Pm. At present there is no indication of how many XylS monomers participate in interactions with Pm, nor is it known whether the XylS monomers interact with each other.

In an *xylS*-minus background of *P. putida*, some activation from the Pm promoter was seen in response to the addition of benzoate. This activity may be due to cross-talk regulation mediated by a chromosomal regulator, e.g. the BenR protein (72, 77).

Initially, Pm was assumed to be a σ^{70}-dependent promoter. However, its DNA sequence does not display the organization and consensus expected for promoters transcribed by the RNA polymerase with σ^{70}: the -10 box is TTAGGC rather than the more canonical TATAAT. Sequence deviation from the consensus in the -35 region was not unexpected, as this feature occurs in many positively regulated σ^{70}-dependent promoters (18). Recent experiments that have analyzed transcription from Pm in different *E. coli* RNA polymerase backgrounds revealed that in a strain lacking the sigma-S factor (σ^S), high levels of transcription from Pm were obtained only in the early logarithmic phase; transcription decreased sharply thereafter (83). This contrasts with the situation in the σ^S proficient strain, in which high levels of transcription were obtained throughout the growth curve. In all cases, the transcription initiation point was the same. This led Marqués et al (83) to propose that Pm belongs to the family of promoters whose transcription may be driven by RNA polymerase with either σ^{70} or σ^S (4, 125). They suggested that in the early exponential phase, transcription was mediated by σ^{70} and later by σ^S, although it remained dependent on XylS. A requirement for σ^S is not exhibited by the mutant promoter Pm5U (C-47→G) in combination with the constitutive XylSGly44Ser mutant, suggesting that σ^S participates directly in XylS-dependent transcription from Pm (83).

THE CASCADE LOOP

The cascade loop of the TOL plasmid catabolic operons organizes the coordinated expression of the two operons, such that a balance in assimilation of aromatic hydrocarbons and gene expression is achieved. This section will describe how the master regulator XylR is expressed, how this protein becomes activated by effector binding, and how the active regulator stimulates transcription from cognate promoters.

xylR Gene Expression

The XylR protein is expressed from two tandem promoters, one distal (Pr1) and one proximal (Pr2) (64). Expression from these promoters is high regardless of the growth phase, although expression from Pr1 is stronger than that from Pr2 (64, 84). In an XylR-minus background, expression from Pr1 and Pr2 is about 2- to 10-fold higher than in the isogenic XylR-plus background, which suggests that XylR controls its own synthesis (2, 43, 64; S Marqués, unpublished findings).

The basis for the different strengths of Pr1 and Pr2 promoters has not been analyzed in detail. Overlap between these promoters may explain this feature: The location of T-7 in Pr1 corresponds with T-35 in Pr2 (Figure 4). Therefore the location of RNA polymerase in Pr1 and the initiation of transcription could well interfere with recognition of the Pr2 promoter. Such transcriptional occlusion by upstream promoters has been noted in other systems (33).

Holtel et al (56) found, by in vitro footprinting, that IHF binds to the Pr region, which protects the zone corresponding to −8/−27 of Pr1 from DNase digestion. This is the zone that shows no significant overlap with the Pr2 region (Figure 4). In spite of this in vitro finding, in vivo expression from the *xylR* promoters in IHF⁻ and IHF⁺ *E. coli* backgrounds, determined by using Pr::'*lacZ*

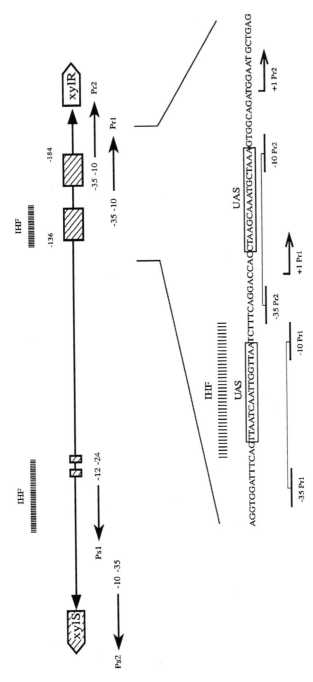

Figure 4   Organization of the *xylR-xylS* intergenic region. The Ps1 promoter includes UAS sequences recognized by σ^{54}-containing RNA polymerase (*small boxes*), and two potential IHF binding sites (*shaded bars*). Also indicated are the $-12/-24$ sequences recognized by σ^{54}-containing RNA polymerase (*large boxes*) for the master activator of the system, XylR, are consensus σ^{70}-containing RNA polymerase-binding sites ($-10/-35$) and transcription initiation points ($+1$) for the *xylR* gene Pr tandem promoters Pr1 and Pr2, and for *xylS* gene Ps2 promoter. The nucleotide sequence of the Pr promoter region overlapping the Ps1 UAS is shown in detail.

fusions, was found not to be affected. Therefore this IHF binding site seems to be irrelevant for transcriptional control of Pr promoters.

It should be noted that the *xylR* gene is transcribed divergently from *xylS* gene and that the target enhancer sequences for XylR protein (UAS) in the *xylS* gene Ps1 promoter overlap the *xylR* promoters: The target inverted repeats of XylR in Ps1 that are located between -139 and -154 (UAS1) overlap the RNA polymerase recognition site of Pr1 between -11 and -26. The XylR target in UAS2, which is between -169 and -184, overlaps Pr1 between $+5$ and $+20$, and overlaps Pr2 between -24 and -19 (Figure 4) (43, 56, 103). Given that nonactivated XylR protein seems to be persistently bound to target sequences (1), XylR most likely binds to a common site in order to inhibit the XylR promoters and to activate the *xylS* Ps1 promoter in the presence of effectors. It is consistent with this proposal that the XylR mutant XylRAsp135Asn that constitutively activates *xylS* gene expression shows an increased level of autoregulation (20), which suggests increased affinity for target sequences. The extent of autoregulation of XylR was slightly higher in the presence of toluene and related compounds (2, 56, 64). This suggests that XylR–target sequence contacts may be altered upon the binding of effector or that structural changes that are induced by the activated XylR may somehow reduce the binding of the RNA polymerase-σ^{70} complex in the Pr promoters.

In conclusion, a fine balance between at least two DNA binding proteins (XylR and the RNA polymerase-σ^{70} complex), interacting with overlapping DNA target sequences, and involved in transcription from divergent promoters, accounts for the modulation of XylR expression.

The XylR *Regulator*

The master regulator of transcriptional control of TOL plasmid catabolic pathways is the XylR protein, which stimulates transcription from Pu and Ps1 upon exposure of the cells to an effector. The XylR protein is 566 amino acid residues long and belongs to the NtrC/NifA family of regulators (68). Sequence alignment of the regulators of this family has revealed that they exhibit four domains, three of which are highly conserved among members of the family (80). The carboxy-terminal D domain contains an α-helix-turn-α-helix DNA binding motif. The most conserved central domain C, is involved in ATP hydrolysis and probably contacts the σ^{54} factor (80). The B domain or Q-linker is a short hydrophilic region that probably serves as a linker between the C and A domains. The nonhomologous N-terminal A domain has been implicated in signal reception, either via a sensory protein as in the NtrC/NifB pair or via interaction with an effector (93).

XylR is one of the few examples in this family in which the primary environmental signal (the presence of *m*-xylene or another effector in the medium) directly affects its transcriptional activity. In this respect, XylR behaves like the closely related phenol/cresol-responsive activator DmpR (117) and the less closely related formate-activated FhlA protein (57).

XylR is activated by a wide variety of aromatics with different functional groups, e.g. toluene, xylenes, benzyl alcohol, p-chlorobenzaldehyde, alkylbenzyl alcohols, m-aminotoluene, and o- and p-nitrotoluene, but not m-nitrotoluene (2, 19). Different substituents on the aromatic ring differentially affect XylR-mediated transcription stimulation from Pu. With toluene and xylenes, transcription from Pu follows a hyperbolic saturation curve in response to increasing effector concentration. In contrast, with o-nitrotoluene the saturation curve shows a sigmoidal shape at low concentrations, which suggests cooperative interactions, whereas higher concentrations of this aromatic inhibit transcription from Pu (R Salto, JL Ramos, unpublished observations). m-Nitrotoluene behaves as an inhibitor for XylR (19) and competitively inhibits o-nitrotoluene stimulation of transcription from Pu (R Salto, JL Ramos, unpublished). From these features and the fact that XylR-mediated transcription from Pu can be reproduced in the heterologous host E. coli, it was concluded that the XylR protein is directly activated through interaction with effectors, rather than through a cascade of events that could involve the phosphorylation of this regulator, as others have described for members of this family (9, 129). This hypothesis was confirmed when XylR mutants were selected for their ability to recognize substituted toluenes that were not recognized by the wild-type protein (i.e. m-nitrotoluene), or when they were selected for their ability to activate transcription in the absence of toluene (19–20), and by replacement of the receptor module of DmpR by that of XylR and demonstration that the resulting chimeric protein responded to toluene instead of cresols (35).

Delgado & Ramos (19) isolated one XylR mutant that is activated by m-nitrotoluene and that exhibits the single amino acid change Glu172→Lys. Unlike the wild-type XylR, the mutant XylRGlu172Lys lost the ability to respond to m-aminotoluene. It is therefore possible that the substituent at position 3 on the aromatic ring interacts with amino acid 172 in XylR. Activation of transcription from Pu by this mutant regulator was still dependent on the IHF protein and σ^{54}, which suggests that the Glu172→Lys change specifically affected effector recognition.

Replacement of aspartic acid 135 by either asparagine, or glutamine, or of proline 85 with serine, resulted in XylR mutants that mediated constitutive expression from Pu, i.e. they stimulated transcription in the absence of effectors (20; A Delgado, R Salto, JL Ramos, unpublished data). The high basal levels of transcription mediated by these constitutive mutant regulators increased further in response to certain effectors. For example, XylRAsp135Asn increased transcription from Pu in response to o- and m-nitrotoluene. The mutant XylRPro85Ser also responded to *o*-nitrotoluene; however, with *m*-nitrotoluene the basal level was significantly diminished, which confirmed that *m*-nitrotoluene behaved as an inhibitor of XylR (20).

The importance of residue 135 in the activation of XylR was further supported by the finding that substitution of aspartic acid 135 by glutamic acid resulted in a mutant regulator that did not respond to effectors (R Salto, JL Ramos, unpublished findings), as well as by the fact that the constitutive character conferred on XylR by the change Asp135→Asn was abolished in the XylRAsp135Asn, Glu172Lys double mutant, because the latter lost the constitutive character (20). This loss suggests that the binding of the effector to XylR induces conformational changes that are transmitted to the DNA-binding and the central domains of the protein, thereby stimulating transcription; and that an alteration in this signal transmission chain may have important consequences for transcriptional activity of the regulator.

Fernández et al (35) and Pérez-Martín & de Lorenzo (102) defined subdomains within domain A after analyzing nested or internal deletions throughout the entire A domain of XylR. All mutants were able to shut down the expression of the Pr promoters of the *xylR* gene in vivo, which indicated that they were able to bind DNA. Deletion of the 150 residues closest to the N-terminal resulted in regulators that were unable to stimulate transcription, with or without an effector. Deletion of 30 additional amino acids (XylRΔ180) partially restored the ability of the protein to promote transcription, although the protein was unresponsive to *m*-xylene and hence its activity was constitutive. Deletion of 30 additional amino acids (XylRΔ210, also called XylRΔA), which entirely removed the A domain, resulted in a mutant regulator that promoted the highest level of transcription (100, 102–104). This suggests that the N-terminal domain of the nonactivated form of XylR inhibits activity of the regulator and that effector binding or severe deletion of this domain relieves this intramolecular repression and permits productive interactions with the RNA polymerase-σ^{54} complex, located at the downstream promoter element (19, 35, 100–104). A remarkable finding was that internal deletions Δ160/210 and Δ190/210, within the domain A, generated mutant XylR regulators that, although able to activate transcription in the absence of inducer, maintained a significant degree of responsiveness to *m*-xylene. This indicates that the first two thirds of the A domain are important for effector binding and that the third proximal to the Q-linker may be involved in intramolecular repression (100, 102).

The B-domain or Q-linker in members of the NtrC family serves as a bridge between the A domain and the central domain; however its specific role(s) is(are) unclear. In the NtrC and NifA regulators, the Q-linker tolerates a range of mutations without a significant change in activity (132), whereas point mutations in the hinge region of XylR (residues 211–223) resulted in a constitutive phenotype (35).

The central domain of XylR (residues 224–472) contains a putative nucleotide binding site (93) that may endow the protein with ATPase activity. Pérez-Martín & de Lorenzo (104, 105) showed that XylRΔA possessed an intrinsic ATP-cleavage activity like that of other activators of the σ^{54}-dependent family (6, 9, 57, 82, 118, 129). The ATPase activity of XylR increased when XylRΔA was incubated in the presence of DNA carrying the UAS of Pu. This stimulatory effect required integrity of the complete UAS region. Measurements of the rate of ATP hydrolysis indicated that adding Pu enhancer DNA to the reaction increased the V_{max} apparent 10-fold, whereas the affinity of XylRΔA for ATP did not change significantly (104, 105). Pérez-Martín and de Lorenzo (104) generated a mutant in which glycine 268 within the nucleotide-binding domain was replaced by asparagine, which resulted in an XylR protein that could not bind ATP and hence had no ATPase activity. Binding of ATP (or ATPγS, a nonhydrolyzable analog) to XylRΔA caused a significant conformational change in the activator; suggesting that ATP is an allosteric effector of the protein independently of any subsequent hydrolysis (104, 105).

XylRΔA binds DNA in vitro, in accordance with predictions based on genetic data showing that the A domain does not affect the DNA binding ability of XylR (35). DNA binding is not, therefore, a regulated step. This is in agreement with the early observation of Abril et al (1) that XylR is always bound to DNA. Because binding to DNA is not regulated (22, 23, 105), the step(s) most likely to be inhibited by the A domain seem(s) to be either ATP binding or ATPase activity, or both (105).

The role of the D domain (residues 515–558) as a DNA binding domain has not been addressed yet in detail, but on the basis of the homology of NtrC and XylR, it seems reasonable to assume that residues 534–553 constitute the actual DNA binding domain of this regulator (68).

Architecture of the Pu and Ps1 Promoters

Pu and Ps1, the two promoters regulated by XylR, belong to the σ^{54}-dependent family; they include two regions of high homology, at $-12/-24$ and at $-120/-190$, and they also include a region rich in A and T, between -40 and -80.

THE σ^{54} RECOGNITION SITE   The $-12/-24$ region of Pu and Ps1 contains the sequence 5′-TGGCN$_7$TTGCTA-3′, which is typical of σ^{54}-dependent promoters (59, 66, 109). Early work showed that in an *E. coli* σ^{54}-deficient background, no stimulation of transcription from Pu and Ps occurred (24, 109). *P. putida* σ^{54}-deficient mutants bearing the TOL plasmid pWW0 were unable to grow on toluene, and when grown with glucose as a C-source and in the presence of toluene, the upper and *meta*-cleavage pathway enzymes were not induced (69, 78). These results confirmed that these two promoters belong to the class of σ^{54} promoters described in other gram-negative bacteria (80).

SITES FOR ACCESSORY PROTEINS INVOLVED IN DNA BENDING   The $-40/-70$ region in both Pu and Ps is rich in A and T. The $-55/-67$ segment of Pu showed good homology to the consensus motif for the binding of integration host factor (IHF) protein (53), a DNA-bending protein whose presence in *Pseudomonas* spp. was recently confirmed (14). In vitro DNase I footprinting assays have shown that *E. coli* IHF specifically binds to Pu, and that it protects this motif (22). It was also shown that in an *E. coli* IHF⁻ background, effector-activated XylR-dependent stimulation of transcription from Pu was only about 25% of the stimulation obtained in an IHF⁺ background (1, 22). Substitution of the IHF DNA-binding sequences of Pu by intrinsically curved DNA sequences gave mutant promoters that displayed 40 to 100% of the activity of Pu and that were independent of IHF (99). It was therefore proposed that IHF coregulates expression from Pu by bending the DNA rather than by interacting directly with RNA polymerase. Residual activity from the Pu promoter in an IHF⁻ *E. coli* background seems to be dependent on the chromatin-associated DNA-bending protein HU, because in an *E. coli* IHF⁻, HU⁻ background, the Pu promoter was completely silent (101). This observation has been confirmed by in vitro assays (J Pérez-Martín & V de Lorenzo, unpublished data). An IHF mutant of *P. putida* was constructed by replacing the wild-type *ihfA* gene with a knockout derivative. Calb et al (14) showed that IHF is essential for the expression of the Pu promoter of the upper pathway operon and that this IHF mutant is unable to efficiently degrade benzyl alcohol.

The Ps1 promoter contains two putative IHF binding sites at $-8/-27$ and at $-137/-156$ (54, 56). Both sites were clearly protected, in vitro, from DNase I digestion (54). However, in an IHF⁻ background, transcription from Ps1 was as high as that in an IHF⁺ background (56), which suggests that, in vivo, these sequences do not play a significant role with IHF. On the other hand, Pérez-Martín & de Lorenzo (101) showed, in vivo, that the chromatin-associated DNA-bending protein HU is probably involved in transcription stimulation from the Ps1 promoter because expression from Ps1 was significantly diminished in *E. coli* HU⁻ backgrounds. The HU protein may be guided by a static DNA bend at the intervening region of Ps1 (42, 101).

THE TARGET SITES FOR XYLR BINDING: ENHANCER-LIKE SEQUENCES   Nested deletions in Pu (1, 60) and in Ps1 (43, 53, 56) defined the $-120/-190$ region of Pu and Ps1 as the target for the XylR regulator. In vivo and in vitro footprinting experiments with the Pu promoter region revealed that XylR recognizes the 5′-TTGATCAAATC-3′ motif—which appears twice, around -160 (UAS1)

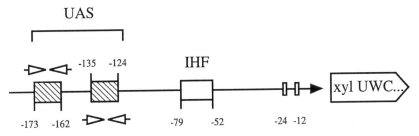

Figure 5 Scheme of the Pu promoter showing location of the relevant DNA sequences. (*Dashed boxes*): XylR-target sequences; (*arrows*): inverted repeat sequences; (*small boxes*): σ⁵⁴-containing RNA polymerase-binding site; (*open box*): IHF-binding site. Numbers indicate distance in bp from the transcription point.

and around −130 (UAS2)—in inverted orientation in both promoters (1, 22) (Figures 4, 5). In vitro DNase I footprinting patterns of Pu and Ps1 confirmed that the truncated XylRΔA protein interacts with target DNA clustered within the two UAS of Pu and Ps1; and hydroxyl radical footprinting revealed a pattern of six discontinuous patches of protected bases in both strands, extending from −125 to −180 in Pu and from −135 to −190 in Ps1 (103).

Alignment of the DNA sequences of the upstream regions of Pu and Ps1, based exclusively on DNase I protection and hydroxyl radical nicking patterns, gave rise to a potential consensus sequence: 5′-TTGATCAATTGATCAA-3′ (103). The consensus sequences can be defined either as a tandem repeat of the 5′-TTGATCAA-3′ half-site (which is itself an inverted repeat sequence) or as a twofold rotation of the same half-site, which gives rise to an inverted repeat. The symmetric pattern of interactions between XylRΔA and each half-site suggest that the target sequences are recognized as an inverted repeat and not as a tandem array (103). On the basis of the proposed consensus sequence, the two XylR binding sites of Pu are thought to span −172 to −157 for the distal site and −142 to −127 for the proximal site. In Ps1, the sites are located at −184 to −169 (distal) and −154 to −139 (proximal), i.e. one whole helix-turn farther from the binding site of RNA polymerase than the corresponding distance in Pu (103). In vivo experiments have shown that XylR is always bound to the regulatory sequences in Pu and that changes occur in the methylation protection pattern of Gs after effector binding, which suggests that XylR makes different contacts with DNA once it is activated by an effector (1). It has been observed that offsetting the two binding sites by 5 bp considerably decreased promoter activity in vivo, which suggests that protein-protein interactions between XylR dimers bound to each site were required for transcriptional activity (1, 3, 103, 105).

Delgado et al (20) showed that, in contrast with the wild-type protein, XylRAsp135Asn was still able to activate transcription from ΔPu that lacks the most distal UAS. Pérez-Martín & de Lorenzo (103) showed that replacement of the proximal site by a heterologous sequence virtually obviated significant activity of the mutant Pu promoter. These results suggest that the proximal UAS plays a key role in positioning the nucleoprotein complex that is involved in transcription initiation.

The physical interactions between XylR and target sequences were studied in vitro by Pérez-Martín & de Lorenzo (105), who showed that cooperative interactions involving the XylRΔA and the two upstream binding sites of the enhancer were dependent on ATP binding but not on ATP hydrolysis. In the absence of ATP, high concentrations of XylRΔA were needed for UAS occupation; sites were occupied in an apparently preferential order, the proximal one occupied first. The addition of ATP or ATPγS altered this pattern remarkably, so that both sites appeared to be simultaneously occupied, even at low concentrations of protein. This difference seems to be due to the formation of XylR multimers, detected with cross-linking agents under various conditions. Limited proteolysis of activated XylR in the presence of ATP, but in the absence of target DNA, also revealed that XylR underwent a conformational change upon ATP binding. ATP hydrolysis is, therefore, preceded by multimerization of XylR at the enhancer, an event that is itself triggered by this allosteric effect of ATP binding to the protein.

CRITICAL ROLE OF THE POSITION OF ENHANCER SITES IN PU AND PS1 PROMOTERS   The mechanism of transcription of the RNA polymerase-σ^{54} complex is similar to activation of σ^{70}, in the sense that the activator needs to make contact with the RNA polymerase that is bound at the promoter site (18). In the σ^{54} promoter, it has been proposed that the intervening DNA sequences loop out to accomplish such contacts (80). Evidence for DNA looping out in the activation of transcription from Pu arose from the fact that activation of XylR by an effector changed the in vivo footprinting pattern of methylation outside the specific protected region, rendering Gs at -92, -70, and -54 and rendering T at -45 hyperreactive (1). This suggests that the activation of XylR markedly distorts the $-50/-90$ region.

Planar projections of binding sites in Pu and Ps1 showed that σ^{54} and XylR binding sites are located on the same DNA face, whereas the IHF binding site in Pu (22) is located on the opposite face. This suggests that XylR must be positioned stereospecifically with respect to the bound RNA polymerase-σ^{54} complex, a hypothesis that is supported by the fact that the introduction, in Pu, of a half-helix turn between the IHF site and the XylR-binding site, or between IHF and RNA polymerase-σ^{54} complex sites, decreased transcription activity by more than 90% (3). In contrast, the introduction of one full helix-turn on either side of the IHF site did not significantly affect transcription stimulation (1, 3, 60). The upstream activator sequences in most σ^{54}-regulated promoters can be moved more than 1 kb away without losing their capability to mediate transcription stimulation. Inouye et al (60) and Gomada et al (43) have shown that XylR-binding sequences can be moved between 200 bp and 1 kb in the Pu promoter, and about 1.2 kb in the Ps1 promoter, without loss of promoter function.

Physiological Consequences of Exposure of P. putida (pWW0) to m-Xylene: A Cascade of Events Leading to Coordinate Expression of Upper and Meta Pathways

After binding an effector, the XylR protein binds ATP and oligomerizes and undergoes a series of conformational changes that lead to a state with enhanced ATPase activity that is active, from a transcriptional point of view (104, 105). The resulting protein, called $XylR_a$, activates transcription from the Pu promoter and, as a consequence, triggers the synthesis of upper pathway enzymes that transform m-xylene into m-methylbenzoate. In addition, $XylR_a$ stimulated the expression of $xylS$ from the σ^{54}-dependent Ps1 promoter without affecting expression of the $xylS$ regulator from the constitutive σ^{70} Ps2 promoter (39). When the $xylS$ gene is expressed from these two promoters, the XylS protein is hyperproduced; this, in turn, induces expression of the $meta$ pathway, even in the absence of $meta$-pathway substrates, which allows the coordinate induction of the upper and the $meta$-cleavage pathways (109). As a result, the aromatic hydrocarbon substrate is metabolized, without accumulation of intermediate products or perturbation of the carbon flux.

INTEGRATION OF TOLUENE METABOLISM IN CELL METABOLISM

Catabolite Repression

The TOL plasmid provides P. putida with the advantage of a broad spectrum of growth substrates. However, expression of the catabolic operons in response to the presence of an aromatic compound requires de novo synthesis of at least 14 proteins and a total of 18 new peptides. The energy cost of this process needs to be compensated by the advantage of using the aromatic compound as a carbon and as an energy source. The expression of a complex catabolic pathway may, nonetheless, be a burden to the organism when alternative carbon sources are available. The expression of the TOL pathways, for example, has been found to be tightly regulated, depending on the carbon sources available for growth (27, 28, 55, 58, 84).

Earlier studies showed that when the TOL catabolic pathways were induced in bacteria growing in a rich medium, such as Luria broth, the response to the addition of an effector was slow and included a relatively long lag period before a steady state was reached (2, 84). This phenomenon is also seen with cells growing in defined minimal medium on substrate mixtures, e.g. a TOL pathway substrate and glucose. This suggests that the TOL catabolic pathways are subject to catabolite repression (28, 29, 55, 58, 134).

Several observations have now demonstrated that Pu and Ps1, the two σ^{54}-dependent promoters regulated by XylR, are subject to catabolite repression:

1. High concentrations of glucose (1.6% wt/vol) impaired the utilization of benzyl alcohol by *P. putida* (pWW0) growing on minimal medium. Inhibition was caused by repression of mRNA synthesis from Pu and Ps1 promoters (55).

2. Effector-mediated induction of both promoters was strongly repressed during the first hours of growth in rich medium, in contrast with the immediate full expression observed in minimal medium that contains the aromatic compound as the only carbon source (84). Addition of yeast extract to cells that exhibited full expression of the Pu and Ps1 promoter, while these cells were growing on minimal medium in the presence of *m*-methylbenzyl alcohol, led to complete repression of mRNA synthesis from both promoters (84).

3. Pu and Ps1 promoters were completely repressed in bacteria that were growing in a continuous culture at near-maximum growth rates, in the presence of nonlimiting concentrations of succinate and in the presence of the nonmetabolizable inducer *o*-xylene. In contrast, full expression of Pu and Ps1 was observed at low growth rates, when no excess of succinate could be detected in the medium (28). Repression was not a consequence of the growth rate itself: When phosphate or sulfate limitation was used, instead of carbon, to obtain a low growth rate, and when excess succinate was present, the Pu promoter was totally repressed (27, 28).

These experiments show that a number of compounds, related to primary carbon metabolism and able to provide energy to the cell, can override the specific regulation of both promoters. This is also true for the aromatic inducer itself, as long as it can be metabolized. Duetz (27) has shown that *m*-xylene acts not only as a pathway inducer, but also as a repressor under N-, P-, and S-limiting conditions in which anabolic processes limit growth. He proposed that, under these circumstances, cells accumulate a "repressing metabolite" or exhibit an "energy excess state," which switches off the upper pathway operon.

As a result of the catabolite repression of the Ps1 promoter (28) in continuous cultures that are growing at the maximum growth rate in the presence of an upper pathway effector, the XylS protein is not hyperproduced. Consequently, the expression of the *meta* pathway under these conditions (the cascade loop) is also under catabolite repression control. However, effector-activated XylS-dependent expression of the *meta*-cleavage pathway (the *meta* loop) is not subject to catabolite repression. The same was true for the σ^{70}-dependent promoters of *xylR* (28).

Unlike for Enterobacteriaceae, catabolite repression in pseudomonads does not occur according to the classical system involving cAMP (55). Instead, more global regulatory systems, responding to metabolic conditions related to the energy status of the cell, seem to play a role. Such systems may well involve changes in the level and activity of the sigma factor σ^{54} (15, 26). This possibility was raised by the finding that overproduction of σ^{54} protein partially alleviated the lack of expression from chromosomally inserted Pu::'*lacZ* fusions in *P. putida* cells growing on LB-rich medium (15).

Catabolite repression has also been described in other catabolic routes for aromatic compounds in *Pseudomonas* spp., including the *phl* operon for the degradation of (methyl)phenol (94) and the *dmp* operon of pVI150 for the degradation of phenol (124). Both operons are under the control of σ^{54}-dependent promoters. The σ^{54}-dependent promoter of the *dctA* gene of *Rhizobium* spp. for succinate uptake is also known to be susceptible to catabolite repression; this was shown by the finding that DctD-dependent expression from this promoter was repressed in the presence of glucose (25). In these two cases the corresponding activator protein seems to play a key role in overcoming negative regulation: overproduction of DmpR bypassed the catabolite repressive effect of glucose (124), and activation of transcription from the *dctA* gene by a mutant DctD regulator that lacks the A domain was not subject to catabolite repression (25).

Elucidation of the mechanism of catabolite repression of the *xyl* system in *Pseudomonas* spp. will require mutant strains that escape catabolite repression control.

PRESENT AND FUTURE PERSPECTIVES

The catabolic activities encoded by the *P. putida* TOL plasmid were detected more than 25 years ago (97, 98), and were confirmed to be plasmid-encoded by Worsey & Williams in 1975 (134), shortly after the first description of catabolic pathways borne by plasmids (17, 32). Since then, both the structural organization of the catabolic genes and their regulation have been extensively studied. One of the main conclusions of these studies is that the expression of the TOL plasmid pWW0 catabolic pathways involves the participation of plasmid regulators and a number of host accessory regulatory elements that are not exclusive to this pathway; these elements are themselves subject to fine-tuning in response to the physiological status of the cell. This sophisticated interplay between regulatory elements is probably an important requirement for tight control of the level of expression of this energetically expensive pathway. The regulation of the TOL catabolic pathways can thus be considered an illuminating model of the regulatory networks in prokaryotes.

During the last five years, much knowledge has accumulated on the regulatory proteins XylS and XylR, particularly with regards to their mechanism of activation of transcription. However, one critical feature, i.e. the activation of these regulators via effector binding, is a completely unexplored topic.

The σ^{54}-dependent promoters and the Pm promoter for the *meta*-pathway operon have been analyzed in detail. However, an area that deserves particular attention is the 300-bp sequence that contains the *xylS* and *xylR* promoters. The constitutive pair of overlapping promoters for the *xylR* gene is divergent from the tandem promoters for the *xylS* gene, one of which is constitutive and the other of which is regulated. In this region, at least one regulator (XylR), two accessory proteins (IHF and HU), and RNA polymerase with either σ^{70} or σ^{54} each have binding sites. How these proteins interact with one another and with their target DNA sequences is an interesting area that deserves further analysis.

Studies of *E. coli* have produced much knowledge on the regulation of the TOL pathways, owing to the fact that the regulatory system of the TOL catabolic pathways can be reproduced accurately in this enterobacterium. However, the fact that many accessory proteins play a role in the system constituted in *E. coli* requires confirmation in the natural host *Pseudomonas* spp., whereas other host factors are expected to influence the expression of the TOL pathways. This is particularly relevant for studies on the integration of toluene degradation and central metabolism because the global regulatory circuits in *E. coli* and *Pseudomonas* spp. seem to operate through different global sensors. In this regard, one fascinating area of study is the subject of regulatory "cross-talk" (23), which has major implications for the operation and the integration of genetic circuits.

It is also relevant that the TOL catabolic pathways are relaxed in substrate specificity and that the regulators can readily undergo alterations in effector specificity (19, 113). This has already led to the use of the TOL catabolic segments to expand the range of growth substrates that different microbes can use (see References 107, 108 for reviews). Given the promiscuity of pWW0 among true *Pseudomonas* spp., and the metabolic versatility of the TOL catabolic pathways, the *xyl* genes of pWW0 remain important modules for the natural and laboratory construction of microbes with new activities.

ACKNOWLEDGMENTS

Work in the authors' laboratories was supported by EC grants BIOTECH CT92-0084 and ENVIRONMENT CT94-0539 and by the CSIC-GBF agreement for the exchange of personnel. Work in Granada was also supported by grant

AMB-1038-CO2-01 from CICYT. We are indebted to Victor de Lorenzo, José Pérez-Martín, Mari-Trini Gallegos, Maxi Manzanera, Asun Delgado, and Rafa Salto for communication of results prior to publication. We also thank Pepe Pérez-Martín, Fernando Rojo, Wouter Duetz, Mari-Trini Gallegos, and Victor de Lorenzo for comments on this review. Appreciation is expressed to K Shashok for suggestions on improving the readability of the text, and to María del Mar Fandila for preparing the manuscript.

> **Visit the *Annual Reviews* home page at**
> **http://www.annurev.org.**

Literature Cited

1. Abril MA, Buck M, Ramos JL. 1991. Activation of the *Pseudomonas* TOL plasmid upper pathway operon. *J. Biol. Chem.* 266:15832–38

2. Abril MA, Michán C, Timmis KN, Ramos JL. 1989. Regulator and enzyme specificities of the TOL plasmid-encoded upper pathway for degradation of aromatic hydrocarbons and expansion of the substrate range of the pathway. *J. Bacteriol.* 171:6782–90

3. Abril MA, Ramos JL. 1993. Physical organization of the upper pathway operon promoter of the *Pseudomonas* TOL plasmid. Sequence and positional requirements for XylR-dependent activation of transcription. *Mol. Gen. Genet.* 239:281–88

4. Altuvia S, Almirón M, Huisman G, Kolter R, Storz G. 1994. The *dps* promoter is activated by OxyR during growth and by IHF and σ^S in stationary phase. *Mol. Microbiol.* 13:265–72

5. Assinder SJ, Williams PA. 1990. The TOL plasmids: determinants of the catabolism of toluene and xylenes. *Adv. Microb. Physiol.* 31:1–69

6. Austin S, Dixon R. 1992. The prokaryotic enhancer binding protein NtrC has an ATPase activity which is phosphorylation and DNA dependent. *EMBO J.* 11:2219–28

7. Bailey SA, Duggleby C, Worsey MJ, Williams PA, Hardy KG, Broda P. 1977. Two modes of loss of the TOL function from *Pseudomonas putida* mt-2. *Mol. Gen. Genet.* 154:203–4

8. Benson S, Shapiro J. 1978. TOL is a broad-host-range plasmid. *J. Bacteriol.* 135:278–80

9. Berger D, Narberhaus F, Kustu S. 1994. The isolated catalytic domain of NifA, a bacterial enhancer-binding protein, activates transcription *in vitro*: Activation is inhibited by NifL. *Proc. Natl. Acad. Sci. USA* 91:103–7

10. Bradley DE, Williams PA. 1982. The TOL plasmid is naturally-derepressed for transfer. *J. Gen. Microbiol.* 128:3019–24

11. Brunelle A, Schleif RF. 1987. Missing contact probing of DNA-protein interactions. *Proc. Natl. Acad. Sci. USA* 84:6673–76

12. Brunelle A, Schleif RF. 1989. Determining residue-base interactions between AraC protein and *araI* DNA. *J. Mol. Biol.* 209:607–22

13. Bustos SA, Schleif RF. 1993. Functional domains of the AraC protein. *Proc. Natl. Acad. Sci. USA* 90:5638–42

14. Calb R, Davidovitch A, Koby S, Giladi H, Goldenberg D, et al. 1996. Structure and function of the *Pseudomonas putida* integration host factor. *J. Bacteriol.* 178:6319–26

15. Cases I, de Lorenzo V, Pérez-Martín J. 1996. Involvement of σ^{54} in exponential silencing of the *Pseudomonas putida* TOL plasmid Pu promoter. *Mol. Microbiol.* 19:7–17

16. Cass LG, Wilcox G. 1986. Mutations in the *araC* regulatory gene of *Escherichia coli* B/r that affect repressor and activator functions of AraC protein. *J. Bacteriol.* 166:892–900

17. Chakrabarty AM. 1972. Genetic basis of the biodegradation of salicylate in *Pseudomonas*. *J. Bacteriol.* 112:815–23

18. Collado-Vides J, Magasanik B, Gralla JD. 1991. Control site location and transcriptional regulation in *Escherichia coli*. *Microbiol. Rev.* 55:371–94

19. Delgado A, Ramos JL. 1994. Genetic evidence for activation of the positive

transcriptional regulator XylR, a member of the NtrC family of regulators, by effector binding. *J. Biol. Chem.* 269:8059–62

20. Delgado A, Salto R, Marqués S, Ramos JL. 1995. Single amino acid changes in the signal receptor domain of XylR resulted in mutants that stimulate transcription in the absence of effectors. *J. Biol. Chem.* 270:5144–50

21. de Lorenzo V, Eltis L, Kessler B, Timmis KN. 1993. Analysis of *Pseudomonas* gene products using *lacI^q/Ptrp-lac* plasmids and transposons that confer conditional phenotypes. *Gene* 123:17–24

22. de Lorenzo V, Herrero M, Timmis KN. 1991. An upstream XylR and IHF induced nucleoprotein complex regulates the σ^{54}-dependent Pu promoter of TOL plasmid. *EMBO J.* 10:1159–67

23. de Lorenzo V, Pérez-Martín J. 1996. Regulatory noise in prokaryotic promoters: how bacteria learn to respond to novel environmental signals. *Mol. Microbiol.* 19:1177–84

24. Dixon R. 1986. The *xylABC* promoter from the *Pseudomonas putida* TOL plasmid is activated by nitrogen regulatory genes in *Escherichia coli. Mol. Gen. Genet.* 206:129–36

25. Downing RG, Broda PA. 1979. A cleavage map of the TOL plasmid of *Pseudomonas putida* mt-2. *Mol. Gen. Genet.* 68:189–91

26. Du Y, Holtel A, Reizer J, Saier MH Jr. 1996. σ^{54}-dependent transcription of the *Pseudomonas putida xylS* operon is influenced by the IISNtr protein of the phosphotransferase system in *Escherichia coli. Res. Microbiol.* 147:129–32

27. Duetz WA. 1996. *Physiology of toluene-degrading* Pseudomonas *strains under various conditions of nutrient limitation in chemostat culture.* PhD dissertation. Univ. Groningen. Groningen, The Netherlands. 68 pp.

28. Duetz WA, Marqués S, de Jong C, Ramos JL, van Andel JG. 1994. Inducibility of the TOL catabolic pathway in *Pseudomonas putida* (pWW0) growing on succinate in continuous culture: evidence of carbon catabolite repression control. *J. Bacteriol.* 176:2354–61

29. Duetz WA, Marqués S, Wind B, Ramos JL, van Andel JG. 1996. Catabolite repression of the toluene degradation pathway in *Pseudomonas putida* harboring pWW0 under various conditions of nutrient limitation in chemostat culture. *Appl. Environ. Microbiol.* 62:601–6

30. Duetz WA, van Andel JG. 1991. Stability of TOL plasmid pWW0 in *Pseudomonas putida* mt-2 under nonselective conditions in continuous culture. *J. Gen. Microbiol.* 137:1369–74

31. Duetz WA, Winson MK, van Andel JG, Williams PA. 1991. Mathematical analysis of catabolic function loss in a population of *Pseudomonas putida* mt-2 during nonlimited growth on benzoate. *J. Gen. Microbiol.* 137:1363–68

32. Dunn NW, Gunsalus IC. 1973. Transmissible plasmid coding of early enzymes of naphthalene oxidation in *Pseudomonas putida. J. Bacteriol.* 114:974–79

33. Duval-Valentin G, Reiss C. 1990. How *Escherichia coli* RNA polymerase can negatively regulate transcription from a constitutive promoter. *Mol. Microbiol.* 4:1465–75

34. Eustance RJ, Bustos SA, Schleif RF. 1994. Locating and lengthening the interdomain linker in AraC protein. *J. Mol. Biol.* 242:330–38

35. Fernández S, de Lorenzo V, Pérez-Martín J. 1995. Activation of the transcription regulator XylR of *Pseudomonas putida* by release of repression between functional domains. *Mol. Microbiol.* 16:205–13

36. Franklin FCH, Bagdasarian MM, Timmis KN. 1981. Molecular and functional analysis of the TOL plasmid pWW0 from *Pseudomonas putida* and cloning of genes for the entire regulated aromatic ring *meta*-cleavage pathway. *Proc. Natl. Acad. Sci. USA* 78:7458–62

37. Franklin FCH, Lehrbach PR, Lurz R, Rueckert B, Bagdasarian M, Timmis KN. 1983. Localization and functional analysis of transposon mutations in regulatory genes of the TOL catabolic pathway. *J. Bacteriol.* 154:676–85

38. Gallegos MT. 1996. *Caracterización del regulador transcripcional* XylS *del plásmido TOL de* Pseudomonas putida. PhD dissertation. Univ. Granada. Granada, Spain. 254 pp.

39. Gallegos MT, Marqués S, Ramos JL. 1996. Expression of the TOL plasmid *xylS* gene in *Pseudomonas putida* occurs from a σ^{70}-dependent promoter or from F^{70}- and σ^{54}-dependent tandem promoters according to the aromatic compound used for growth. *J. Bacteriol.* 178:2356–61

40. Gallegos MT, Marqués S, Ramos JL. 1997. The TACAN$_4$TGCA motif upstream from the -35 region on the σ^{70}/σ^S-dependent Pm promoter of the TOL plasmid is the minimum DNA

segment required for transcription stimulation by XylS regulators. *J. Bacteriol.* 178:6427–34

41. Gallegos MT, Michán C, Ramos JL. 1993. The XylS/AraC family of regulators. *Nucleic Acids Res.* 21:807–10

42. Gomada M, Imaishi H, Miura S, Inouye S, Nakazawa T, Nakazawa A. 1994. Analysis of DNA bend structure of promoter regulatory regions of xylene-metabolizing genes on the *Pseudomonas* TOL plasmid. *J. Biochem.* 116:1096–104

43. Gomada M, Inouye S, Imaishi H, Nakazawa A, Nakazawa T. 1992. Analysis of an upstream regulatory sequence required for activation of the regulatory gene *xylS* in xylene metabolism directed by the TOL plasmid of *Pseudomonas putida*. *Mol. Gen. Genet.* 233:419–26

44. Grinsted J, de la Cruz F, Schmitt R. 1990. The Tn21 subgroup of bacterial transposable elements. *Plasmid* 24:163–89

45. Gu B, Lee JH, Hoover TR, Scholl D, Nixon BT. 1994. *Rhizobium meliloti* DctD, a σ54-dependent transcriptional activator, may be negatively controlled by a subdomain in the C-terminal end of its two-component receiver module. *Mol. Microbiol.* 13:51–66

46. Gussin GN, Olson C, Igarashi K, Ishihama A. 1992. Activation defects caused by mutations in *Escherichia coli* *rpoA* are promoter specific. *J. Bacteriol.* 174:5156–60

47. Harayama S. 1995. Gene bank. DDB5 Acc. no. D63341 (NID g 939832).

48. Harayama S, Lehrbach PR, Timmis KN. 1984. Transposon mutagenesis analysis of the *meta*-cleavage pathway operon genes of the TOL plasmid of *Pseudomonas putida* mt-2. *J. Bacteriol.* 160:251–55

49. Harayama S, Mermod N, Rekik M, Lehrbach PR, Timmis KN. 1987. Role of the divergent branches of the *meta*-cleavage pathway in degradation of benzoate and substituted benzoates. *J. Bacteriol.* 169:558–64

50. Harayama S, Rekik M. 1990. The *meta*-cleavage operon of TOL degradative plasmid pWW0 comprises 13 genes. *Mol. Gen. Genet.* 221:113–20

51. Harayama S, Rekik M, Wubbolts M, Rose K, Leppik RA, Timmis KN. 1989. Characterization of five genes in the upperpathway operon of TOL plasmid pWW0 from *Pseudomonas putida* and identification of the gene products. *J. Bacteriol.* 171:5048–55

52. Harayama S, Timmis KN. 1989. Catabolism of aromatic hydrocarbons by

Pseudomonas. In *Genetics of Bacterial Diversity*, ed. D Hopwood, K Chater, 151–174. New York: Academic

53. Holtel A, Abril MA, Marqués S, Timmis KN, Ramos JL. 1990. Promoter-upstream activator sequences are required for expression of the *xylS* gene and upperpathway operon on the *Pseudomonas* TOL plasmid. *Mol. Microbiol.* 4:1551–56

54. Holtel A, Goldenberg D, Giladi H, Oppenheim AB, Timmis KN. 1995. Involvement of IHF protein in expression of the Ps promoter of the *Pseudomonas putida* TOL plasmid. *J. Bacteriol.* 177:3312–15

55. Holtel A, Marqués S, Möhler I, Jakubzik U, Timmis KN. 1994. Carbon sourcedependent inhibition of *xyl* operon expression of the *Pseudomonas putida* TOL plasmid. *J. Bacteriol.* 176:1773–76

56. Holtel A, Timmis KN, Ramos JL. 1992. Upstream binding sequences of the XylR activator protein and integration host factor in the *xylS* gene promoter region of the *Pseudomonas* TOL plasmid. *Nucleic Acids Res.* 20:1755–62

57. Hopper S, Bock A. 1995. Effector-mediated stimulation of ATPase activity by the σ54-dependent transcriptional activator FhlA from *Escherichia coli*. *J. Bacteriol.* 177:2798–803

58. Hugouvieux-Cotte-Pattat N, Köhler T, Rekik M, Harayama S. 1990. Growthphase-dependent expression of the *Pseudomonas putida* TOL plasmid pWW0 catabolic genes. *J. Bacteriol.* 172:6651–60

59. Inouye S, Ebina Y, Nakazawa A, Nakazawa T. 1984. Nucleotide sequence surrounding transcription initiation site of *xylABC* operon on TOL plasmid of *Pseudomonas putida*. *Proc. Natl. Acad. Sci. USA* 81:1688–91

60. Inouye S, Gomada M, Sangodar UMX, Nakazawa A, Nakazawa T. 1990. Upstream regulatory sequence for transcriptional activator XylR in the first operon of xylene metabolism on the TOL plasmid. *J. Mol. Biol.* 216:251–60

61. Inouye S, Nakazawa A, Nakazawa T. 1981. Molecular cloning of gene *xylS* of the TOL plasmid: evidence for positive regulation of the *xylDEGF* operon by *xylS*. *J. Bacteriol.* 148:413–18

62. Inouye S, Nakazawa A, Nakazawa T. 1983. Molecular cloning of regulatory gene *xylR* and operator-promoter regions of the *xylABC* and *xylDEGF* operons of the TOL plasmid. *J. Bacteriol.* 155:1192–99

63. Inouye S, Nakazawa A, Nakazawa T.

1984. Nucleotide sequence of the promoter region of the *xyl*DEGF operon on TOL plasmid of *Pseudomonas putida*. *Gene* 29:323–30

64. Inouye S, Nakazawa A, Nakazawa T. 1985. Determination of the transcription initiation site and identification of the protein product of the regulatory gene *xyl*R for *xyl* operons on the TOL plasmid. *J. Bacteriol.* 163:863–69

65. Inouye S, Nakazawa A, Nakazawa T. 1986. Nucleotide sequence of the regulatory gene *xyl*S on the *Pseudomonas putida* TOL plasmid and identification of the protein product. *Gene* 44:235–42

66. Inouye S, Nakazawa A, Nakazawa T. 1987. Expression of the regulatory gene *xyl*S on the TOL plasmid is positively controlled by the *xyl*R gene product. *Proc. Natl. Acad. Sci. USA* 84:5182–86

67. Inouye S, Nakazawa A, Nakazawa T. 1987. Overproduction of the *xyl*S gene product and activation of the *xyl*DLEGF operon on the TOL plasmid. *J. Bacteriol.* 169:3587–92

68. Inouye S, Nakazawa A, Nakazawa T. 1988. Nucleotide sequence of the regulatory gene *xyl*R of the TOL plasmid from *Pseudomonas putida*. *Gene* 66:301–6

69. Inouye S, Yamada M, Nakazawa A, Nakazawa T. 1989. Cloning and sequence analysis of the *ntr*A (*rpo*N) gene of *Pseudomonas putida*. *Gene* 85:145–52

70. Ishihama A. 1993. Protein-protein communication within the transcription apparatus. *J. Bacteriol.* 175:2483–89

71. Jacoby GA, Rogers JE, Jacob AE, Hedges RW. 1978. Transposition of *Pseudomonas* toluene-degrading genes and expression in *Escherichia coli*. *Nature* 274:179–80

72. Jeffrey WH, Cuskey SM, Chapman PJ, Resnick S, Olsen RH. 1992. Characterization of *Pseudomonas putida* mutants unable to catabolize benzoate: cloning and characterization of *Pseudomonas* genes involved in benzoate catabolism and isolation of a chromosomal DNA fragment able to substitute for *xyl*S in activation of the TOL lower-pathway promoter. *J. Bacteriol.* 174:4986–94

73. Kaldalu N, Mandel T, Ustav M. 1996. TOL plasmid transcription factor XylS binds specifically to the Pm operator sequence. *Mol. Microbiol.* 20:569–79

74. Keil H, Keil S, Pickup RW, Williams PA. 1985. Evolutionary conservation of genes coding for meta pathways enzymes within TOL plasmids pWW0 and pWW53. *J. Bacteriol.* 164:887–95

75. Kessler B, de Lorenzo V, Timmis KN. 1992. A general system to integrate *lacZ* fusions into the chromosomes of gramnegative eubacteria: regulation of the Pm promoter of the TOL plasmid studied with all controlling elements in monocopy. *Mol. Gen. Genet.* 233:293–303

76. Kessler B, de Lorenzo V, Timmis KN. 1993. Identification of a *cis*-acting sequence within the Pm promoter of the TOL plasmid which confers XylS-mediated responsiveness to substituted benzoates. *J. Mol. Biol.* 230:699–703

77. Kessler B, Marqués S, Köhler T, Ramos JL, Timmis KN, de Lorenzo V. 1994. Cross talk between catabolic pathways in *Pseudomonas putida*: XylS-dependent and -independent activation of the TOL *meta* operon requires the same *cis*-acting sequences within the Pm promoter. *J. Bacteriol.* 176:5578–82

78. Köhler T, Harayama S, Ramos JL, Timmis KN. 1989. Involvement of *Pseudomonas putida rpo*N sigma factor in regulation of various metabolic functions. *J. Bacteriol.* 171:4326–33

79. Kunz DA, Chapman PJ. 1981. Catabolism of pseudocumene and 3-ethyltoluene by *Pseudomonas putida* (*arvilla*) mt-2: evidence for new functions of the TOL (pWW0) plasmid. *J. Bacteriol.* 146:179–91

80. Kustu S, Santero E, Keener J, Popham D, Weiss D. 1989. Expression of sigma −54 (*ntr*A)-dependent genes is probably united by a common mechanism. *Microbiol. Rev.* 53:367–76

81. Lauble H, Georgalis Y, Heinemann U. 1989. Studies on the domain structure of the *Salmonella typhimurium* AraC protein. *Eur. J. Biochem.* 185:319–25

82. Lee HS, Berger DK, Kustu S. 1993. Activity of purified NifA, a transcriptional activator of nitrogen fixation genes. *Proc. Natl. Acad. Sci. USA* 90:2266–70

83. Marqués S, Gallegos MT, Ramos JL. 1995. Role of σ^S in transcription from the positively controlled Pm promoter of the TOL plasmid of *Pseudomonas putida*. *Mol. Microbiol.* 18:851–57

84. Marqués S, Holtel A, Timmis KN, Ramos JL. 1994. Transcriptional induction kinetics from the promoters of the catabolic pathways of TOL plasmid pWW0 of *Pseudomonas putida* for metabolism of aromatics. *J. Bacteriol.* 176:2517–24

85. Marqués S, Ramos JL. 1993. Transcriptional control of the *Pseudomonas putida* TOL plasmid catabolic pathways. *Mol. Microbiol.* 9:923–29

86. Marqués S, Ramos JL, Timmis KN. 1993. Analysis of the mRNA structure of the

Pseudomonas putida meta-fission pathway operon around the transcription initiation point, the *xylTE* and the *xylFJ* regions. *Biochim. Biophys. Acta* 1216:227–37

87. Menon KP, Lee NL. 1990. Activation of ara operons by a truncated AraC protein does not require inducer. *Proc. Natl. Acad. Sci. USA* 87:3708–12

88. Mermod N, Lehrbach PR, Reineke W, Timmis KN. 1984. Transcription of the TOL plasmid toluate catabolic pathway operon of *Pseudomonas putida* is determined by a pair of co-ordinately and positively regulated overlapping promoters. *EMBO J.* 3:2461–66

89. Mermod N, Ramos JL, Bairoch A, Timmis KN. 1987. The XylS gene positive regulator of TOL plasmid pWW0: identification, sequence analysis and overproducing leading to constitutive expression of *meta* cleavage operon. *Mol. Gen. Genet.* 207:349–54

90. Meulien P, Downing RG, Broda P. 1981. Excision of the 40 kb segment of the TOL plasmid from *Pseudomonas putida* mt-2 involves direct repeats. *Mol. Gen. Genet.* 184:97–101

91. Michán C, Kessler B, de Lorenzo V, Timmis KN, Ramos JL. 1992. XylS domain interactions can be deduced from intraallelic dominance in double mutants. *Mol. Gen. Genet.* 235:406–12

92. Michán C, Zhou L, Gallegos MT, Timmis KN, Ramos JL. 1992. Identification of critical amino-terminal regions of XylS. The positive regulator encoded by the TOL plasmid. *J. Biol. Chem.* 267:22897–901

93. Morett E, Segovia L. 1993. The σ^{54} bacterial enhancer-binding protein family: mechanism of action and phylogenetic relationship of their functional domains. *J. Bacteriol.* 175:6067–74

94. Müller C, Petruschka L, Cuypers H, Burchhardt G, Herrmann H. 1996. Carbon catabolite repression of phenol degradation in *Pseudomonas putida* is mediated by the inhibition of the activator protein PhlR. *J. Bacteriol.* 178:2030–36

95. Nakai C, Kagamiyama H, Nozaki M. 1983. Complete nucleotide sequence of the metapyrocatechase gene on the TOL plasmid of *Pseudomonas putida* mt-2. *J. Biol. Chem.* 258:2923–28

96. Nakazawa T. 1978. TOL plasmid in *Pseudomonas aeruginosa* PAO: thermosensitivity of self-maintenance and inhibition of host cell growth. *J. Bacteriol.* 132:1347–58

97. Nakazawa T, Yokota T. 1973. Benzoate metabolism in *Pseudomonas putida* (*arvilla*) mt-2: demonstration of two benzoate pathways. *J. Bacteriol.* 115:262–67

98. Nozaki M, Ono K, Nakazawa T, Kotani S, Hayaishi O. 1968. Metapyrocatechase. II. The role of iron and sulfhydryl groups. *J. Biol. Chem.* 243:2682–90

99. Pérez-Martín J, de Lorenzo V. 1994. Coregulation by bent DNA: functional substitutions of the IHF site at the σ^{54}-dependent promoter Pu of the *upper*-TOL operon by intrinsically curved sequences. *J. Biol. Chem.* 269:22657–62

100. Peréz-Martín J, de Lorenzo V. 1995. The amino-terminal domain of the prokaryotic enhancer-binding protein XylR is a specific intramolecular repressor. *Proc. Natl. Acad. Sci. USA* 92:9392–96

101. Pérez-Martín J, de Lorenzo V. 1995. The σ^{54}-dependent promoter Ps of the TOL plasmid of *Pseudomonas putida* requires HU for transcriptional activation *in vivo* by XylR. *J. Bacteriol.* 177:3758–63

102. Peréz-Martín J, de Lorenzo V. 1996. Identification of the repressor subdomain within the signal reception module of the prokaryotic enhancer-binding protein XylR of *Pseudomonas putida*. *J. Biol. Chem.* 271:7899–902

103. Pérez-Martín J, de Lorenzo V. 1996. Physical and functional analysis of the prokaryotic enhancer of the σ^{54}-promoters of the TOL plasmid of *Pseudomonas putida*. *J. Mol. Biol.* 258:562–74

104. Pérez-Martín J, de Lorenzo V. 1996. *In vitro* activities of an N-terminal truncated form of XylR, a σ^{54}-dependent transcriptional activator of *Pseudomonas putida*. *J. Mol. Biol.* 258:575–87

105. Pérez-Martín J, de Lorenzo V. 1996. ATP binding to the σ^{54}-dependent activator XylR triggers a protein multimerization cycle catalyzed by UAS DNA. *Cell* 86:331–39

106. Polissi A, Harayama S. 1993. *In vivo* reactivation of catechol 2,3-dioxygenase mediated by a chloroplast-type ferredoxin: a bacterial strategy to expand the substrate specificity of aromatic degradative pathways. *EMBO J.* 12:3339–47

107. Ramos JL, Díaz E, Dowling D, de Lorenzo V, Molin S, et al. 1994. The behavior of bacteria designed for biodegradation. *Bio-Technology* 13:35–37

108. Ramos JL, Haïdour A, Delgado A, Duque E, Fandila MD, et al. 1995. Potential of toluene-degrading systems for the construction of hybrid pathways for nitrotoluene metabolism. In *Biodegradation of Nitroaromatic Compounds*, ed. JC Spain, pp. 53–68. New York: Plenum

109. Ramos JL, Mermod N, Timmis KN. 1987. Regulatory circuits controlling transcription of TOL plasmid operon encoding *meta*-cleavage pathway for degradation of alkylbenzoates by *Pseudomonas. Mol. Microbiol.* 1:293–300

110. Ramos JL, Michán C, Rojo F, Dwyer D, Timmis KN. 1990. Signal-regulator interactions: genetic analysis of the effector binding site of *xyl*S, the benzoate-activated positive regulator of *Pseudomonas* TOL plasmid *meta*-cleavage pathway operon. *J. Mol. Biol.* 211:373–82

111. Ramos JL, Rojo F, Zhou L, Timmis KN. 1990. A family of positive regulators related to the *Pseudomonas putida* Plasmid XylS and the *Escherichia coli* AraC activators. *Nucleic Acids Res.* 18:2149–52

112. Ramos JL, Stolz A, Reineke W, Timmis KN. 1986. Altered effector specificities in regulators of gene expression: TOL plasmid *xyl*S mutants and their use to engineer expansion of the range of aromatics degraded by bacteria. *Proc. Natl. Acad. Sci. USA* 83:8467–71

113. Ramos JL, Wasserfallen A, Rose K, Timmis KN. 1987. Redesigning metabolic routes: manipulation of TOL plasmid pathway for catabolism of alkylbenzoates. *Science* 235:593–96

114. Ramos-González MI, Duque E, Ramos JL. 1991. Conjugational transfer of recombinant DNA in cultures and in soils: host range of *Pseudomonas putida* TOL plasmids. *Appl. Environ. Microbiol.* 57:3020–27

115. Shaw JP, Harayama S. 1990. Purification and characterisation of TOL plasmid-encoded benzyl alcohol dehydrogenase and benzaldehyde dehydrogenase of *Pseudomonas putida. Eur. J. Biochem.* 191:705–14

116. Shaw LE, Williams PA. 1988. Physical and functional mapping of two cointegrate plasmids derived from RP4 and TOL plasmid pDK1. *J. Gen. Microbiol.* 134:2463–74

117. Shingler V, Moore T. 1994. Sensing of aromatic compounds by the DmpR transcriptional activator of phenol-catabolizing *Pseudomonas* sp. strain CF600. *J. Bacteriol.* 176:1555–60

118. Shingler V, Pavel H. 1995. Direct regulation of the ATPase activity of the transcriptional activator DmpR by aromatic compounds. *Mol. Microbiol.* 17:505–13

119. Sinclair MI, Holloway BW. 1991. Chromosomal insertion of TOL transposons in *Pseudomonas aeruginosa* PAO. *J. Gen. Microbiol.* 137:1111–20

120. Spooner RA, Bagdasarian M, Franklin FCH. 1987. Activation of the *xyl*DLEGF promoter of the TOL toluene-xylene degradation pathway by overproduction of the *xyl*S regulatory gene product. *J. Bacteriol.* 169:3581–86

121. Spooner RA, Lindsay K, Franklin FCH. 1986. Genetic, functional and sequence analysis of the *xyl*R and *xyl*S regulatory genes of the TOL plasmid pWW0. *J. Gen. Microbiol.* 132:1347–58

122. Suzuki M, Brenner SE. 1995. Classification of multihelical DNA-binding domains and application to predict the DBD structures of sigma factor, LysR, OmpR/PhoB, CENP-B, Rap1, and XylS/Ada/AraC. *FEBS Lett.* 372:215–21

123. Suzuki M, Hayakawa T, Shaw JP, Rekik M, Harayama S. 1991. Primary structure of xylene monooxygenase: similarities to and differences from the alkane hydroxylation system. *J. Bacteriol.* 173:1690–95

124. Sze CC, Moore T, Shingler V. 1996. Growth phase-dependent transcription of the σ^{54}-dependent Po promoter controlling the *Pseudomonas* derived (methyl)phenol *dmp* operon of pVI150. *J. Bacteriol.* 178:3727–35

125. Tanaka K, Takayanagi Y, Fujita N, Ishihama A, Takahashi H. 1993. Heterogeneity of the principal sigma factor in *Escherichia coli*: the *rpo*S gene product, σ^{38}, is a second principal sigma factor of RNA polymerase in stationary-phase *Escherichia coli. Proc. Natl. Acad. Sci. USA* 90:3511–15

126. Tsuda M, Iino T. 1987. Genetic analysis of a transposon carrying toluene-degrading genes on TOL plasmid pWW0. *Mol. Gen. Genet.* 210:270–71

127. Tsuda M, Iino T. 1988. Identification and characterization of Tn*4653*, a transposon covering the toluene transposon Tn*4651* on TOL plasmid pWW0. *Mol. Gen. Genet.* 213:72–77

128. Tsuda M, Minegishi KJ, Iino T. 1989. Toluene transposons Tn*4651* and Tn*4653* are class II transposons. *J. Bacteriol.* 171:1386–93

129. Weiss D, Batut J, Klose K, Kustu S. 1991. The phosphorylated form of the enhancer-binding protein NtrC has an ATPase activity that is essential for activation of transcription. *Cell* 67:155–67

130. Williams PA, Murray K. 1974. Metabolism of benzoate and the methylbenzoates by *Pseudomonas putida* (*arvilla*) mt-2: evidence for the existence of a TOL plasmid. *J. Bacteriol.* 120:416–23

131. Williams PA, Shaw LM, Pitt CW, Vrecl M. 1996. *xylUW*, two genes at the start of the upper pathway operon of TOL plasmid pWW0, appear to play no essential part in determining its catabolic phenotype. *Microbiology* 143:101–07

132. Wooton JC, Drummond MH. 1989. The Q-linker: a class of interdomain sequences found in bacterial multidomain regulatory proteins. *Protein Eng.* 2:535–43

133. Worsey MJ, Franklin FCH, Williams PA. 1978. Regulation of the degradative pathway enzymes coded for by the TOL plasmid (pWW0) from *Pseudomonas putida* mt-2. *J. Bacteriol.* 134:757–64

134. Worsey MJ, Williams PA. 1975. Metabolism of toluene and xylenes by *Pseudomonas putida* (*arvilla*) mt-2: evidence for a new function of the TOL plasmid. *J. Bacteriol.* 124:7–13

135. Zhou L, Timmis KN, Ramos JL. 1990. Mutations leading to constitutive expression from the TOL plasmid *meta*-cleavage pathway operon are located at the C-terminal end of the positive regulator XylS protein. *J. Bacteriol.* 172:3707–10

Annu. Rev. Microbiol. 1997. 51:375–414

SULFUR TUFT AND TURKEY TAIL: Biosynthesis and Biodegradation of Organohalogens by Basidiomycetes

Ed de Jong[1]

Department of Wood Science, The University of British Columbia, #270-2357 Main Mall, Vancouver, British Columbia V6T 1Z4, Canada; e-mail: edejong@unixg.ubc.ca

Jim A. Field

Division of Industrial Microbiology, Department of Food Science, Agricultural University Wageningen, P.O. Box 8129, 6700 EV Wageningen, The Netherlands; e-mail: jim.field@algemeen.im.wau.nl

KEY WORDS: Basidiomycetes, organohalogens, AOX, biosynthesis, biodegradation, *Phanerochaete chrysosporium*, chlorinated compounds, xenobiotics

ABSTRACT

Chlorinated aliphatic and aromatic compounds are generally considered to be undesirable xenobiotic pollutants. However, the higher fungi, Basidiomycetes, have a widespread capacity for organohalogen biosynthesis. Adsorbable organic halogens (AOX) and/or low-molecular-weight halogenated compounds are produced by Basidiomycetes of 68 genera from 20 different families. Most of the 81 halogenated metabolites identified from Basidiomycetes to date are chlorinated, although brominated and iodated metabolites have also been described. Two broad categories of Basidiomycete organohalogen metabolites are the halogenated aromatic compounds and the haloaliphatic compounds. Some of these organohalogen metabolites have demonstrable physiological roles as antibiotics and as metabolites involved in lignin degradation. Basidiomycetes produce large amounts of low-molecular-weight organohalogens or adsorbable organic halogens (AOX) when grown on lignocellulosic substrates. In our view, Basidiomycetes, as decomposers of forest litter, are a major source of natural organohalogens in terrestrial environments.

Basidiomycetes are also potent degraders of a wide range of chlorinated pollutants, such as bleachery effluent from kraft mills and pentachlorophenol,

---

[1] Current address: Department of Agrofibres and Cellulose, Agrotechnical Research Institute (ATO-DLO), P.O. Box 17, 6700 AA Wageningen, The Netherlands

0066-4227/97/1001-0375$08.00

polychlorinated dioxins, and polychlorinated biphenyls. The extracellular, lignin-degrading enzymes of the Basidiomycetes are involved in the oxidative degra-dation of chlorophenols and dioxin and can cause reductive dechlorination of halomethanes.

There is no clear-cut separation between "polluters" and "clean-uppers" within the Basidiomycetes. Several genera, e.g. *Bjerkandera, Hericium, Phlebia,* and *Trametes,* produce significant amounts of chlorinated compounds but are also highly effective in metabolizing or biotransforming chlorinated pollutants.

CONTENTS

INTRODUCTION

The occurrence of organohalogens in nature is generally ascribed to anthro-pogenic activities (6). However, over 2600 different natural halogenated com-pounds have been identified (66, 67), and a large pool of the bulk parameter, adsorbable organic halogen (AOX), has been detected in unpolluted environ-ments (18, 69, 134). Samples of forest soil collected worldwide contain AOX concentrations of between 20 mg and 360 mg/kg dry weight soil. The ex-tensive pool of natural organohalogens in terrestrial environments indicates a ubiquitous source. Recent analysis of Norway spruce litter enclosed in litter bags and of a coniferous soil profile strongly indicates that organically bound halogens are produced in soil (92). Furthermore, organically bound halogens are dynamic since both their production and mineralization occur during forest litter decay. The microorganisms that degrade lignocellulosic material appear to be implicated in AOX production.

Basidiomycetes are the most ecologically significant group of organisms responsible for converting lignocellulose debris such as wood, straw, and litter (174). These higher fungi produce macroscopic fruiting bodies, often called mushrooms, toadstools, or conks. Two common examples found worldwide are sulfur tuft (*Hypholoma* [syn. *Neamatoloma*] *fasciculare*) and turkey tail (*Trametes* (syn. *Coriolus*) *versicolor*). Basidiomycetes constitute a major fraction of the living biomass degrading forest litter (61). An estimated ton dry weight basidiomycetous mycelium is produced annually per hectare in a typical temperate hardwood forest (174). The fungus *Armillaria bulbosa* (syn. *A. gallica*) is thought to be among the largest and oldest living organisms (164). Many Basidiomycetes produce substantial amounts of AOX (187), and several are also reported to produce *de novo* aliphatic and aromatic organohalogen metabolites, indicating that this group of organisms is a potentially important source of naturally occurring organohalogens in terrestrial environments and in the atmosphere (49, 60, 75).

This review postulates that Basidiomycetes are indeed a major source of organohalogens in terrestrial environments. In addition, the present knowledge on the biosynthesis, physiological role, and environmental fate of Basidiomycete organohalogen metabolites is discussed. Not only are Basidiomycetes implicated in organohalogen biosynthesis but many reports indicate that these organisms can degrade anthropogenic organohalogens as well. In many cases, lignin-degrading enzymes are responsible for the first steps in xenobiotic organohalogen degradation.

UBIQUITY OF ORGANOHALOGEN PRODUCTION AMONG BASIDIOMYCETES

Adsorbable Organic Halogen Screening

Adsorbable organic halogen (AOX) and total organic halogen (TOX) levels have been used as bulk parameters to screen Basidiomycetes. In a survey by Verhagen et al (187) to assess the ubiquity of AOX production among Basidiomycetes, 191 fungal strains were monitored for organohalogen production when grown on defined liquid media. Approximately 50% of the strains tested and 55% of the genera tested produced more than 0.1 mg/L AOX. Thirty eight strains (20%) produced moderate to high levels of 1–67 mg AOX/L. This group was dominated by species belonging to the genera *Hypholoma, Mycena, Bjerkandera, Pholiota,* and *Phellinus* and produced specific AOX in the range 1–30 g AOX/kg dry weight of mycelial biomass. Many highly ecologically significant fungal species were identified among the moderate to high producers. These species also produced AOX when cultivated on natural lignocellulosic substrates. In another screening for TOX production, Öberg et al (139) detected a significant increase in the total amount of organohalogens with eight of the nine white-rot fungi tested on lignocellulosic substrates, including the common turkey tail (*T. versicolor*), artist's conk (*Ganoderma applanatum*), and *Phlebia radiata.*

Biodiversity of Organohalogen-Producing Capacity

The biological capacity for AOX and TOX and/or specific organohalogen metabolite production (references in next section) is widespread among the different families of Basidiomycetes. There exists a broad taxonomic distribution across 68 genera from 20 families within the orders of the Agaricales and Aphyllophorales (Table 1). Most families belonging to the two orders have at least one positive representative. Taxonomic hot spots are found in the order of the Agaricales, which includes the Strophariaceae and the Tricholomataceae. The order of the Aphyllophorales shows high organohalogen production in the family of the Hymenochaetaceae, and several genera of the ill-defined family of the Polyporaceae also produce organohalogens. Molecular biological techniques were used to infer phylogenetic relationships of the Polyporaceae (87). Using sequence data from mitochondrial ribosomal DNA, *Bjerkandera,* a very potent producer of organohalogens, and *Phanerochaete,* a species negative in our AOX test (187), were shown to be very closely related. The broad capacity of basidiomycetes to produce organohalogens, together with the absence of correlation in organohalogen production between closely related species, indicates a high biodiversity of the biohalogenating capacity.

TYPES OF LOW-MOLECULAR-WEIGHT ORGANOHALOGENS PRODUCED

A wide spectrum of low molecular weight metabolites containing halogens has been described. Our review of the literature in 1995 found reports of 53 halogenated metabolites as *de novo* products from 34 genera of Basidiomycetes (60). This number has now increased to 81 halogenated compounds identified from 46 different genera of Basidiomycetes. Table 1 lists all the genera that produce low-molecular-weight organohalogens. Several genera produce a wide range of halogenated aromatic compounds. *Bjerkandera* spp. produce 23 halogenated metabolites, *Lepista nuda* 17 metabolites, *Phellinus* spp. 9 metabolites, *Armillaria* spp. 8 metabolites, *Russula subnigricans* 8 metabolites, *Hericium erinaceus* 6 metabolites, *Mycena* spp. 5 metabolites, and *Hypholoma* spp. 5 metabolites.

The low molecular weight organohalogens are subdivided into two general categories, the haloaliphatic compounds and the halogenated aromatic compounds. Within each category similar compounds, e.g. halomethanes or chlorinated anisyl metabolites, are arranged together and discussed according to the number of related compounds produced, the species and genera implicated, and the environmental significance. When available, the biosynthesis route and physiological role are also discussed.

Chlorinated Aliphatic Compounds

The most important aliphatic organohalogens are chloromethane and the chlorinated nonprotein amino acids. Fifteen aliphatic organohalogens have been identified, produced by 16 genera.

HALOMETHANES   The first reports on natural halomethane production by Basidiomycetes appeared in the early 1970s on the production of chloromethane (39, 96). Subsequent work showed chloromethane [1] [1] to be the most abundant halomethane produced by Basidiomycetes (75); its biosynthesis is particularly common among widely distributed poroid-wood degraders (Hymenochaeteae), such as members of the *Phellinus, Inonotus, Fomitopora, Hymenochaete, Onnia,* and *Phaeolus* genera, and also from *Fomitopsis cytisina* (79). Of the 63 species examined in the family of the Hymenochaeteae, 34 (53%) released detectable amounts of chloromethane. Chloromethane production has also been reported in the commercially important champignon *Agaricus bisporus* (182). In addition to chloromethane, Basidiomycetes produce five other halomethane metabolites when the medium is supplemented with other halide ions: bromomethane [2] and iodomethane [3] by *Phellinus pomaceus* (78), and chloroiodomethane [4], diiodomethane [5], and dichloroiodomethane [6] by *Bjerkandera adusta* (168). The production of chloroform [6a] by *Cantharellus* species has also been reported (153). In defined media with glucose as carbon source, chloromethane release exhibited a typical secondary metabolite pattern, with detectable production limited to the period after exponential growth (75). *P. pomaceus* showed a higher than 90% conversion of 50 mM Cl$^-$ into chloromethane when grown on cellulose medium (78). Wood-rotting fungi could therefore potentially make a significant contribution to the atmospheric chloromethane load (75). However, the actual production of chloromethane from the bracket-like perennial fruiting bodies actively growing on wood of o.a. *Phellinus* and *Inonotus* spp. has not yet been determined.

The pathway for chloromethane biosynthesis is still unclear. For algae and plants it has been proposed that halomethane formation proceeds via methylation of an inorganic halide by a S-adenosyl-L-methionine (SAM)-dependent methyl transferase (19, 192). It was suggested that the same system is active in *P. pomaceus* (192). However, recent results with *P. chrysosporium* indicate that methionine is converted either directly or, more likely, via an intermediate to CH_3Cl, whereas SAM was determined not to be an intermediate in converting L-methionine to CH_3Cl (80). The halogenating enzyme in

---

[1]Numbers in brackets are metabolites.

Table 1 The production of low-molecular-weight organohalogens and/or AOX and TOX by Basidiomycetes

Family	Agaricales[a] Genus	Halogenated metabolite[b]	Family	Aphyllophorales[a] Genus	Halogenated metabolite[b]
Agaricaceae	Agaricus	1, 37, 38, AOX	Cantharellaceae	Cantharellus	6a
Amanitaceae	Amanita	7–11, AOX	Clavariaceae	Macrotyphula	AOX
Bolbitiaceae	Agrocybe	AOX	Coniophoraceae	Coniophora	AOX
				Serpula	AOX
Boletaceae	Boletus	70, AOX	Corticiaceae	Phlebia	TOX
	Xerocomus	AOX		Meripilus	AOX
Coprinaceae	Coprinus	37, AOX		Peniophora	16, 19, 37, 38, AOX
	Panaeolus	AOX		Resinicium	14
	Psathyrella	37, AOX			
Cortinariaceae	Cortinarius	69	Ganodermataceae	Ganoderma	TOX
	Dermocybe	68, 69			
	Galerina	AOX	Gomphaceae	Ramaria	16, 19
	Gymnopilus	AOX			
	Hebeloma	AOX	Hericiaceae	Hericium	51–56, AOX
Lepiotaceae	Lepiota	13	Hymenochaetaceae	Fomitopora	1
	Leucoagaricus	73		Hymenochaete	1
				Inonotus	1
Russulaceae	Lactarius	12		Onnia	1
	Russula	41, 44–50, AOX		Phellinus	1–3, 18, 37–40, 71, AOX, TOX
				Phylloporia	18, AOX

Family	Genus	
Strophariaceae	*Hypholoma*	18–20, 31, 38, AOX
	Kuehneromyces	75, AOX
	Pholiota	18, 19, AOX
	Psilocybe	AOX
	Stropharia	18, 19, AOX
Tricholomataceae	*Armillaria*	57–61, 63–65, AOX, TOX
	Baeospora	AOX
	Calocybe	AOX
	Clitocybe	61, 62, AOX
	Collybia	AOX
	Flammulina	AOX
	Laccaria	AOX
	Lentinellus	72, AOX
	Lepista	15–20, 23, 27, 28, 30, 35, 36, 42, 43, 77–79, AOX, TOX
	Macrocystidia	AOX
	Marasmius	32, AOX
	Mycena	16, 19, 38, 66, 76, AOX
	Oudemansiella	18, 19, AOX
	Pleurotus	16, AOX
	Strobilurus	66
	Tricholomopsis	AOX
	Xerula	66, 67, AOX
Polyporaceae	*Bjerkandera*	4–6, 15–26, 29, 30, 37, 38, 42, 43, 53, 80, AOX, TOX
	Daedaleopsis	16
	Fomes	16, AOX
	Fomitopsis	1, AOX
	Gloeophyllum	AOX
	Heterobasidion	74, AOX
	Ischnoderma	16, AOX
	Phaeolus	1
	Poria	33, AOX, TOX
	Schizopora	AOX
	Trametes	16, 34, AOX, TOX
Schizophyllaceae	*Schizophyllum*	37

aOrder, the Aphyllophorales are classified according to Ellis & Ellis (53).
bNumbers refer to the chemical structures given in the text. AOX and TOX indicate that the genus produced adsorbable organic halogens (AOX) (187) or total amount of organic halogens (TOX) (89, 139).

P. pomaceus has a clear preference for bromide and iodide over chloride, because when these halide ions are present in equimolar concentrations, iodomethane and bromomethane are preferentially formed (78).

Chloromethane was originally considered a secondary metabolite of Basidiomycetes. However, chloromethane is used by Basidiomycetes as a methylating agent in the biosynthesis of benzoate methyl ester and phenolic methyl ether metabolites (77, 127). Apart from the Hymenochaeteae, which produce detectable levels of chloromethane later in their growth curve, other white-rot fungi, such as *Phanerochaete chrysosporium*, *Trametes versicolor*, and *Phlebia radiata*, also utilize chloromethane as a methyl source (76). Recently, Harper and coworkers compared the efficiencies of chloromethane (CH_3Cl), L-methionine, and S-adenosyl-L-methionine (SAM) as methyl precursors in the biosynthesis of veratryl alcohol in *P. chrysosporium* (80), an important cofactor for the lignin-degrading enzyme, lignin peroxidase, and a common secondary metabolite of white-rot fungi (47). Veratryl alcohol is also of great importance in the lignin peroxidase–mediated oxidation of anthropogenic organohalogens such as pentachlorophenol (36). Both chloromethane and methionine caused earlier initiation of veratryl alcohol biosynthesis, but SAM retarded the formation of the compound. A high level of incorporation of deuterated label into both the 3- and 4-O-methyl groups of veratryl alcohol occurred when either deuterated methionine or chloromethane was present. No significant labeling was detected when deuterated S-adenosylmethionine was added, which led to the conclusion that CH_3Cl was utilized as a methyl donor in veratryl alcohol biosynthesis (80). Chlorinated compounds are common intermediates for nonchlorinated products in organic synthesis, and this could also be the case in living organisms (137). The results summarized above indicate a tightly coupled methylating system that uses chloromethane as a transient intermediate in many Basidiomycetes.

CHLORINATED NONPROTEIN AMINO ACIDS   The genus *Amanita* contains some of the most toxic mushrooms and has consequently been the subject of many chemical investigations. Several organohalogen metabolites have been isolated from the subgenus *Lepidella*. They are all chlorinated nonprotein amino acids of the unsaturated norleucine-type (81, 83). The first chlorinated nonprotein amino acid was isolated from *Amanita solitaria* by Chilton & Tsou in 1972 (34). This compound, (2S)-2-amino-5-chloro-4-hexenoic acid [7] (Figure 1 displays compounds 7–11), was purified from fresh mushrooms in a yield of 300 mg/kg. Subsequently, another nonprotein amino acid, (2S)-2-amino-4-chloro-4-pentenoic acid [8], was isolated in 50 ppm yield from fruiting bodies of *Amanita pseudoporphyria* Hongo (82). (2S,4Z)-2-Amino-5-chloro-6-hydroxy-4-hexenoic acid [9] and (2S)-2-amino-5-chloro-4-pentenoic acid [10] have been isolated

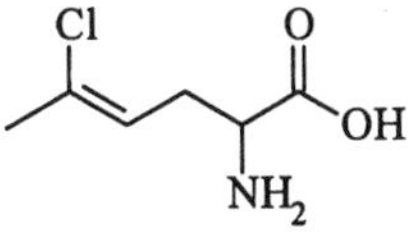

Figure 1   Chlorinated nonprotein amino acids.

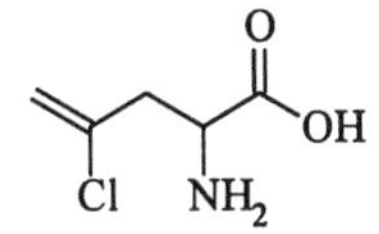

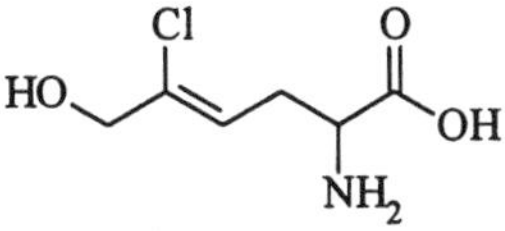

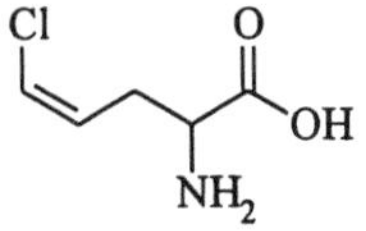

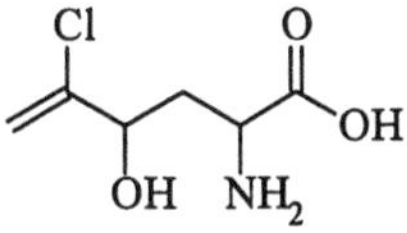

(59 mg/kg) from *Amanita abrupta* and from *Amanita vergineoides,* respectively (140, 141). Recently, another chlorinated amino acid, (2S)-2-amino-5-chloro-4-hydroxy-5-hexenoic acid [11], was isolated from an *Amanita gymnopus* fruiting body (81). The same compound was also isolated from an unidentified *Amanita* sp. of the section *Roanokenses* (Amanitaceae) in a very high yield of the fresh fruiting body (1000 mg/kg) (84).

Structurally, some chlorinated nonprotein amino acids appear to be products of the chlorohydrin reaction. *Amanita* spp. produce several unsaturated nonprotein amino acids, such as 2-amino-4,5-hexadienoic acid and/or 2-amino-4-pentenoic acid, which are likely oxidized by a monooxygenase to an epoxide (83), which then reacts with hydrochloric acid to give the chlorohydrin.

The unsaturated norleucine-type amino acids all have antimicrobial activity. Several are very toxic to guinea pigs and mice and inhibit specific enzymes or cause massive liver cell necrosis (hepatotoxicity) (35, 83, 196). 2-Amino-4-chloro-4-pentenoic acid [8] isolated from *A. pseudoporphyria* has a strong antibacterial activity against several gram-positive and gram-negative bacteria (132). The addition of leucine reversed the growth inhibition of seven bacteria, and it was therefore concluded that this chlorinated amino acid could be regarded as an antimetabolite (132). This amino acid [8] time-dependently and irreversibly inactivates L-methionine gamma-lyase from *Pseudomonas putida* (55).

MISCELLANEOUS CHLORINATED ALIPHATIC COMPOUNDS   Several other chlorinated aliphatic metabolites are produced that show no obvious relationship with the chloromethanes and chlorinated amino acids discussed earlier. A volatile organohalogen, 1-chloro-5-heptadecyne [12] (Figure 2 displays compounds 12–14), was detected in an edible wild milk cap (*Lactarius* spp.) (11). Lepiochlorin [13], an antibacterial lactol, was isolated from liquid cultures of

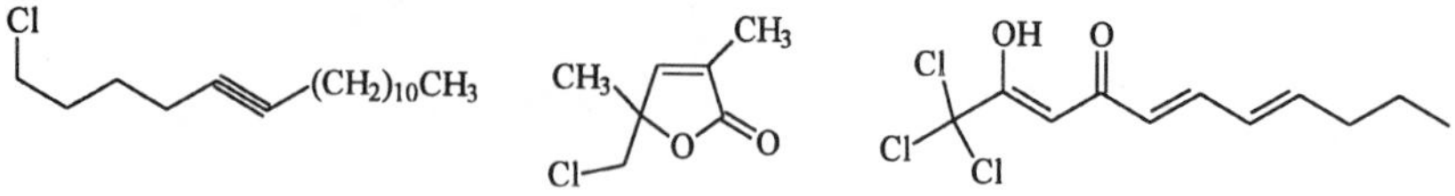

12   **13**   **14**

Figure 2   Miscellaneous chlorinated aliphatic compounds.

a *Lepiota* sp., a fungus cultivated by gardening ants (135). Another aliphatic halogenated compound with an interesting trichloromethyl group was isolated from the mycelium of the fungus *Resinicium pinicola* in 120 mg/kg yield (23). The compound, named pinicoloform [14], possesses antibiotic and cytotoxic activities.

Chlorinated Aromatic Metabolites

Most organohalogens produced by Basidiomycetes have an aromatic structure; important groups include the chlorinated anisyl metabolites (CAM), drosophilins, and other chlorinated hydroquinone methyl ethers (CHME), chlorinated sesquiterpenens (COS), chlorinated anthraquinones, and strobulirins. In total, 66 aromatic organohalogens have been identified, produced by 35 different genera.

CHLORINATED ANISYL METABOLITES   The most commonly encountered chlorinated aromatic compounds produced by Basidiomycetes are the chlorinated anisyl metabolites (CAM). CAM or very closely related compounds [32, 33] (Figure 3 displays compounds 15–34) have now been detected in 18 genera belonging to 4 families from the orders of the Agaricales and the Aphyllophorales (Table 2) (49, 60, 85, 149, 172, 179). Most research has focused on the smokey polypore (*Bjerkandera adusta*), the common wood blewit (*Lepista nuda* (syn. *Clitocybe nuda*)), the sulfur tuft (*Hypholoma fasciculare*), and other *Hypholoma* spp.

Bjerkandera spp. are at present the best-studied Basidiomycetes for the quantity, spectrum, and biosynthesis route of chlorinated metabolites. Two chlorinated anisyl alcohols [15, 18], the corresponding aldehydes [16, 19], and acids [17, 20] are produced by *Bjerkandera adusta* and *Bjerkandera* sp. BOS55 (46, 48, 173). Four other closely related chlorinated benzoic acid derivatives, i.e. methyl 3,5-dichloro-p-anisate [21], 3,5-dichloro-4-hydroxybenzoic acid [24], methyl 3,5-dichloro-4-hydroxybenzoate [25], and 3-chloro-4-hydroxy-benzoic acid [26], were recently isolated from the extracellular fluid of mycelial cultures of *Bjerkandera* spp. (173).

In 1994, Spinnler et al showed that the addition of bromide to the culture medium of *Bjerkandera adusta* led to the production of four brominated

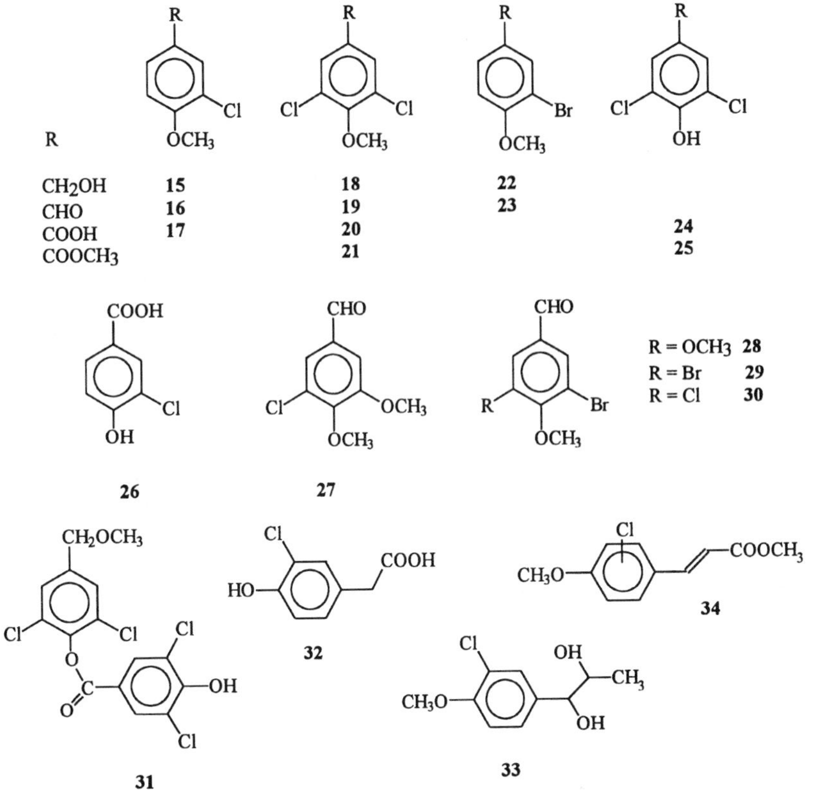

Figure 3 Chlorinated anisyl metabolites and derivatives.

compounds [22, 23, 29, 30] (Table 2) (168). Subsequently, Hjelm and co-workers detected several brominated anisyl metabolites [23, 28, 30] in an unpolluted soil sample located at the arc of a *Lepista nuda* fairy ring (91), the first time that natural brominated compounds had been isolated from terrestrial environments. Many other low-molecular-weight chlorinated compounds have been detected from environmental samples taken in the direct vicinity of fruiting bodies of *L. nuda* [16–20, 27] (49, 91), and from sterilized Norway spruce needle litter incubated with *L. nuda* [16, 19, 27] (89). In addition, dichloro-dimethoxybenzaldehyde [35] was found in soil at the arc of an *L. nuda* fairy ring, and a chlorohydroxybenzaldehyde [36] was observed in the sterilized needle litter (89, 91).

Evidence is accumulating that CAM are biosynthesized via the phenyl-propanoid pathway. The addition of phenylalanine to cultures of *Bjerkandera* sp. increased CAM production tenfold (128). Benzoate is an intermediate in

Table 2 The production of chlorinated anisyl metabolites (CAM) by Basidiomycetes

Strain	Compounds	CAM (mg/liter)[a]	References
Bjerkandera sp. BOS55	15–21, 24–26	17.3	46, 48, 49, 173
Bjerkandera adusta	15–22, 23, 26, 29, 30	50	22, 49, 118, 139, 168, 173, 178
Bjerkandera fumosa	16, 19	2	172
Daedaleopsis confragosa	16	Trace[b]	49
Fomes fomentarius	16	Trace	49
Hypholoma capnoides	18, 19	37.3	49
Hypholoma elongatipes syn. *H. elongatum*	18, 19	108.4	172
Hypholoma fasciculare	18–20, 31	17.0	49, 68, UR[c]
Hypholoma sublateritium	18, 19	12.3	49
Hypholoma subviride	18	NQ[b]	149
Ischnoderma benzonium	16	Trace	49
Lepista deimii	16	NQ	179
Lepista nuda	15–20, 23, 27, 28, 30, 35, 36	NQ	49, 89, 91
Marasmius palmivorus	32	NQ	183
Mycena epipterygia	16, 19	13.9	172
Oudemansiella mucida	18, 19	29.9	49
Peniophora pseudopini	16, 19	13.5	172
Phellinus torulosus	18	2.4	172
Pholiota adiposa	18, 19	5.5	172
Pholiota populnea syn. *P. destruens*	18, 19	11.7	23
Pholiota squarrosa	19	6.8	49
Phylloporia ribis	18	11.1	172
Pleurotus spp.	16	Trace	71
Poria lindbladii syn. *P. cinerescens*	33	NQ	139
Ramaria sp. 158	16, 19	Trace	49
Stropharia aeruginosa	18, 19	NQ	49
Trametes versicolor	16	0.8	49
Trametes sp.	34	NQ	27

[a]When more strains of the same species were tested, the highest values are given.

[b]Trace is less than 0.8 mg of CAM/liter produced. NQ, Not quantified.

[c]UR: FJM Verhagen, BKHL Boekema, FBJ van Assema & JA Field, unpublished results.

the biosynthesis of veratryl alcohol (3,4-dimethoxybenzyl alcohol) from phenylalanine in *P. chrysosporium* (99). Harper and coworkers (80) proposed that the conversion of benzoic acid to veratryl alcohol involves para hydroxylation, methylation of 4-hydroxybenzoic acid, meta hydroxylation of 4-methoxybenzoic acid to form isovanillic acid, and methylation of isovanillic acid to yield veratric acid. Methylation of the hydroxy group can take place with SAM-dependent methyl transferases or with a chloromethane-dependent transferase, as is the case in *P. chrysosporium* (80). Recent work by Mester et al (128) showed that the addition of deuterated benzoate and deuterated 4-hydroxy- and 3-chloro-4-hydroxybenzoates to *Bjerkandera* cultures resulted in an incorporation of the deuterated label into CAM metabolites, as evidenced by GC-MS data. These results combined with the fact that 3-chloro-4-hydroxybenzoate [26] is a *de novo* metabolite from *Bjerkandera* strains suggest that [26] is a likely intermediate in the biosynthesis of CAM. No chlorinating enzyme activity has yet been detected in Basidiomycetes. However, it has been suggested that a chloroperoxidase might be involved because the chlorination takes place ortho of the *p*-hydroxy/methoxy group (60) and chloroperoxidase activity has been detected in other higher fungi such as the organohalogen-producing Ascomycete, *Mollisia caesia* (126). Lignin peroxidase, a common enzyme of white-rot fungi (47), brominates aromatic compounds but is not active with chloride (56, 156). The methylated and chlorinated metabolite can then be reduced by aryl dehydrogenases; these are common enzymes in Basidiomycetes (47) and have been detected in *Bjerkandera* spp. (46). However, chlorination and methylation can also occur at an earlier stage of the phenylpropanoid pathway, as is suggested by the occurrence of 3-chloro-4-hydroxyphenylacetate [32], methyl chloro-4-methoxycinnamate [33], and trametol [34] as metabolites of other Basidiomycetes. Therefore, subsequent work should show whether the biosynthesis routes of veratryl alcohol and CAM are basically the same until chlorination or if chlorination takes place earlier in the biosynthesis route.

Recently, a dimeric product [31] was found in antagonized cultures of *H. fasciculare*. This compound can be formed by an esterase reaction or by oxidative coupling of [18] and/or [19], both very common in *Hypholoma* spp. (49, 85, 172).

Most Basidiomycetes that produce CAM are saprophytes that cause a white rot of wood, and CAM have been shown to possess antibiotic properties (85, 149). However, the CAM alcohols [15] and [18] are also important metabolites in the ligninolytic system of these fungi. Most fungi that synthesize CAMs produce extracellular aryl alcohol oxidases (46, 148). These oxidases have a much higher affinity for the CAM alcohols compared to the structurally similar nonhalogenated secondary metabolites, veratryl alcohol and *p*-anisyl alcohol

(46). They oxidize the alcohols into the corresponding aldehydes [16, 19] to reduce O_2 to H_2O_2, a necessary co-substrate for the extracellular peroxidases, which are key enzymes in lignin degradation (46, 47). Similarly, oxidized AAO reduce quinones and phenoxy radicals and thereby prevent the repolymerization of the phenolic intermediates of lignin degradation (124). The aldehydes that are formed are intracellularly reduced to the corresponding alcohols, generating a physiologically sustainable cycle (46). The chlorinated anisyl alcohols [15, 18] are much better protected against a lignin peroxidase-mediated oxidation than are the nonchlorinated veratryl and anisyl alcohols (46), because the electron-withdrawing character of the chloro group increases the oxidation potential of the methoxy benzyl ring. No specific physiological function has yet been attached to any of the other chlorinated compounds produced by *Bjerkandera* spp. (Table 2). However, the other chlorinated and brominated anisyl alcohol and aldehydes [22, 23, 27–30, 35, 36] are most likely also involved in the H_2O_2-generating cycle, and it has been proposed that the chlorinated acids [17, 20, 24, 26] are intermediates in the biosynthesis route to the physiologically active compounds [15, 18] (173).

CHLORINATED HYDROQUINONE METHYL ETHERS   *Drosophilins* The first chlorinated compound isolated from a Basidiomycete was characterized in 1952 (7, 104). Drosophilin A (p-methoxytetrachlorophenol) [37] (Figure 4 displays compounds 37–50), which strongly resembles the wood preservative pentachlorophenol, was purified in 3 mg/L yield from the culture fluid of the fungus *Drosophila subatrata* now classified as *Psathyrella subatrata*. Drosophilin A was the first natural halogenated compound isolated to contain a halogenated benzene ring (7). Subsequent research has also detected its methylated analog, 1,4-dimethoxy-tetrachlorobenzene [38], in several fungi (Table 3). Drosophilin A [37] and drosophilin A methyl ether [38] have now been detected in 12 species from 9 genera, including *Phellinus*, *Bjerkandera*, and *Coprinus*. Recent research has shown both drosophilins to be present in several species (178). Two other chlorinated hydroquinone methyl ether (CHME) metabolites are produced by *Phellinus* spp. *P. robinea* produces a related and interesting nitro-group containing compound 1,4-dimethoxy-2-nitro-3,5,6-trichlorobenzene [39] (31), whereas *P.* (syn. *Fomes*) *fastuosus* produces 1,2,4-trimethoxy-3,5,6-trichlorobenzene [40] (98). The production of drosophilins seems to be of environmental significance. Several producers are common fungi, producing moderate to high levels of AOX (187), and a compound with a mass-spectrum identical to drosophilin A was detectable in a composite forest litter sample (E de Jong & JA Field, unpublished).

Drosophilin production appears to be inducible. The addition of 3,4-dichloro-aniline induced drosophilin A production in *Schizophyllum commune* (161),

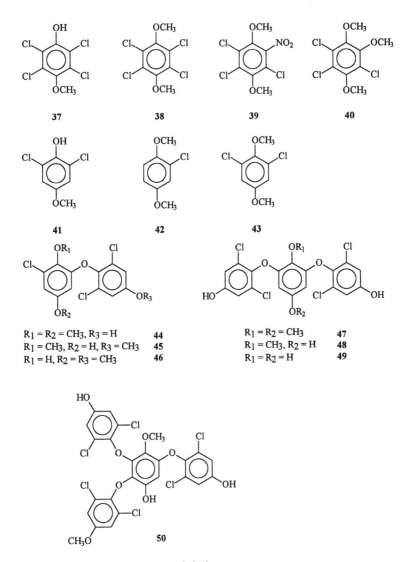

Figure 4   Chlorinated hydroquinone methyl ethers.

Table 3   Drosophilin production by Basidiomycetes

Organism	Drosophilin A [37] (mg/liter)	Drosophilin A methyl ether[a] [38]	References
Agaricus bisporus		NQ[b]	30, 70
Agaricus arvensis	0.5		178
Bjerkandera adusta	0.5	1.1 mg/liter	178
Coprinus plicatilis	20		21
Hypholoma fasciculare		Trace	UR[c]
Mycena megaspora		NQ	186
Peniophora pseudeopini	1.0	1.1 mg/liter	178
Phellinus fastuosus	15.7	11.0 mg/liter	163, 178
		2.4 g/kg	
Phellinus robinea		0.8 g/kg	31
Phellinus yucatensis		0.07 g/kg	94
Psathyrella subatrata	3.2		104
Schizophyllum spp.	NQ		161

[a]Matrices: mg/liter, concentration in a liquid culture; g/kg, amount isolated from the fruiting body of the Basidiomycete indicated.
[b]NQ: Not quantified.
[c]UR: FJM Verhagen, BKHL Boekema, FBJ van Assema & JA Field, unpublished results.

whereas the addition of hydroquinone increased drosophilin concentration in *P. fastuosus* fourfold (178). Hydroquinone might be a substrate for chlorination and methylation reactions and therefore might actually be an intermediate in drosophilin biosynthesis. The antagonizing fungus *Phlebia radiata* increased drosophilin concentration in *P. fastuosus* tenfold (178), indicating a possible role as an antibiotic. Antimicrobial activity of drosophilins has been reported (104).

Russuphelins The toxic mushroom *Russula subnigricans* (Rank Russula) produces a range of chlorinated, biologically active metabolites. Eight metabolites, a monomeric compound, 2,6-dichloro-4-methoxyphenol [41], three diphenyl ethers russuphelin-D [44], russuphelin-E [45] and russuphelin-F [46], three triphenyl ethers russuphelin A [47], russuphelin-B [48], russuphelin-C [49], and a chlorohydroquinone tetramer russuphelol [50] have been isolated from its fruiting bodies (142, 175, 176). Russuphelin A was isolated with a 250 mg yield from 1 kg fresh fruiting bodies, while the other russuphelins were present in much lower concentrations. Also the monomeric compound [41] was isolated in a relatively high yield (125 mg/kg fresh fruiting bodies) (176).

The triphenyl ethers [47, 48, 49] and the tetramer [50] all possess cytotoxic activity against leukemia cells. Of the diphenyl ethers only [44] exhibits cytotoxic activity whereas [45, 46] are inactive (142, 175, 176). The cytotoxic activities of the russuphelins were weakened after dechlorination, and it was suggested

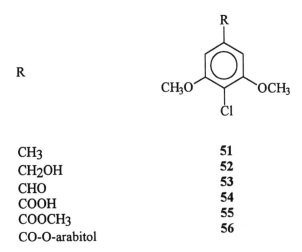

R	
CH3	**51**
CH2OH	**52**
CHO	**53**
COOH	**54**
COOCH3	**55**
CO-O-arabitol	**56**

Figure 5   Chlorinated orcinol methyl ethers.

that the chlorine substitution influences the conformation of the russuphelins and hence their biological activities (176).

2,6-Dichloro-4-methoxyphenol [41] is probably derived from the phenylpropanoid metabolism of phenylalanine or tyrosine, whereas the oligomeric structures are assumed to be synthesized via direct oxidative coupling or, alternatively, via bimolecular nucleophilic substitution (S_N2) reactions of [41] or its demethylated product, 2,6-dichloro-1,4-hydroquinone (176).

Bjerkandera and *Lepista* spp. also produce two chlorinated hydroquinone methyl ethers [42, 43] in trace (91, 168). These compounds are most likely degradation products of the physiologically important chlorinated anisyl alcohols [15, 18] (168).

CHLORINATED ORCINOL METHYL ETHERS   Mycelium of the edible lion's mane fungus, *Hericium erinaceus,* or old man's beard, contains six different kinds of chlorinated orcinol methyl ethers (COME) [51–56] (Figure 5 displays compounds 51–56) (143, 154). The more oxidized alcohol, aldehyde, and acids group [52–54] is probably derived from the metabolism of the orcinol methyl group. Additionally, 4-chloro-3,5-dimethoxybenzaldehyde [53] could be isolated from liquid cultures of *Bjerkandera adusta* (178). Many chlorinated orcinol metabolites showed antibacterial activity (143).

CHLORINATED ORSELLINATE SESQUITERPENES   Over 30 sesquiterpene aryl esters have been isolated from *Armillaria* spp.: of these, eight are esterified with chlorinated orsellinate (Table 4). The honey mushroom, *A. mellea,* produces 70 mg of the chlorinated compound armillaridin [58] (Figure 6 displays

Table 4 Chlorinated orsellinate sesquiterpenes biosynthesized by *Armillaria* spp. and *Clitocybe elegans*

Cl-compound	No.	Species	References
Armellide B	57	*A. novae-zelandiae*	14
Armillaridin	58	*A. mellea*	102
		A. ostoyae	147
Arnamiol	59	*A. mellea*	40, 51
		A. ostoyae	40, 147
		A. tabescens	40
		A. monadelpha	40
		A. gallica	40
		A. cepestipes	40
Melledonal B	60	*A. mellea*	16
		A. novae-zelandiae	14
Melledonal C	61	*A. mellea*	16
		A. ostoyae	147, 167
		A. novae-zelandiae	14
		C. elegans	13
Melledonal D	62	*C. elegans*	13
Melleolide D	63	*A. mellea*	15
		A. ostoyae	167
Melleolide I	64	*A. novae-zelandiae*	14
Melleolide J	65	*A. novae-zelandiae*	14
		A. ostoyae	147

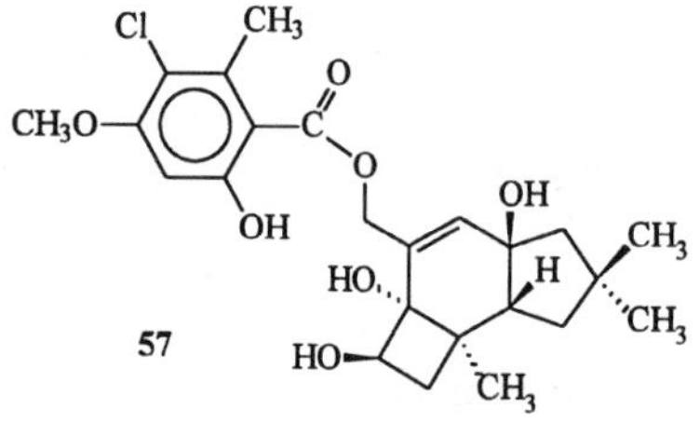

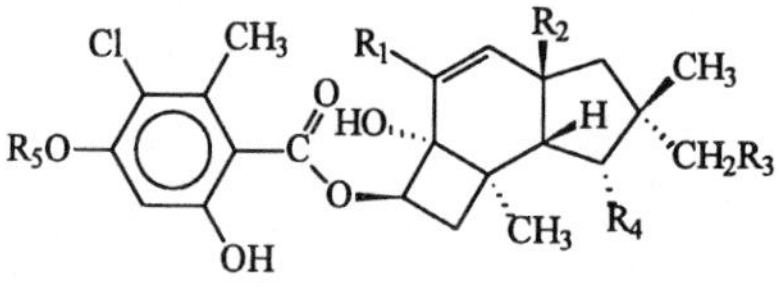

	R₁	R₂	R₃	R₄	R₅	
	CHO	H	H	H	CH3	58
	CHO	OH	H	OH	H	60
	CHO	OH	H	OH	CH3	61
	CHO	OH	OH	OH	CH3	62
	CH2OH	OH	H	OH	CH3	63
	CH2OH	H	H	OH	CH3	64
	CHO	H	H	OH	CH3	65

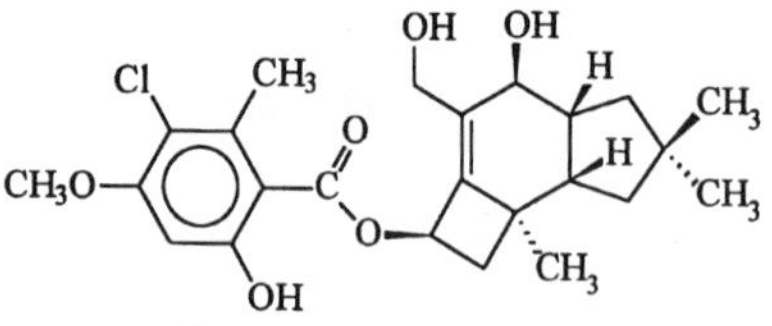

Figure 6 Chlorinated orsellinate sesquiterpenes.

compounds 57–65) per kg dry mycelium (102), while liquid cultures of *A. ostoyae* contain the chlorinated compounds melledonal C (50 mg/L, [61]) and melleolide D (35 mg/L, [63]) (167). *Clitocybe elegans* is the only other genus for which chlorinated sesquiterpene aryl esters have been reported (13), including Melledonal D [62], which has not yet been detected in *Armillaria* spp.

The pathogenic Basidiomycete *Armillaria* causes root disease in both coniferous and deciduous trees. *Armillaria* spp. were the most frequently isolated fungi associated with root rot in living spruce and balsam fir trees in Ontario, Canada (190). It is not known what makes some of the *Armillaria* spp. such virulent parasites, but secondary metabolites are suspected to be the major cause. In a test for fungi- or phytotoxicity of 14 *Armillaria* metabolites, 4 of them chlorinated, the toxicity decreased with increasing hydrophobicity, e.g. methylating the hydroxy group of the orsellinate and/or adding a chlorine atom (147). Sonnenbichler et al found that increased amounts of Melledonal C [61] and other nonchlorinated sesquiterpene aryl esters were produced in cultures of *A. ostoye* growing in the presence of an antagonistic fungus or host plant cells (147, 167). The biosynthesis of the chlorinated metabolite was enhanced up to fivefold upon antagonization (167), indicating that the physiological purpose of the sesquiterpene aryl esters is their antibiotic activity.

CHLORINATED STROBILURIN AND OUDEMANSIN   The strobilurins were first isolated from *Strobilurus tenacellus* (10). The chlorinated strobilurin B [66] (Figure 7 displays compounds 66 and 67) has been isolated from three genera and seven species, e.g. *Mycena alkalina, M. avenacea, M. crocata, M. vitilis, Xerula longipes, X. melanotricha,* and *S. tenacellus.* Chlorinated oudemansin B [67] is produced by *X. melanotricha* (9).

The chlorinated and nonchlorinated strobulirins and the closely related oudemansins are new respiration inhibitors, binding to cytochrome b. Thirteen strobilurins and three oudemansins have been isolated thus far. Their high antifungal activity against phytopathogenic fungi and insects and their low toxicity toward

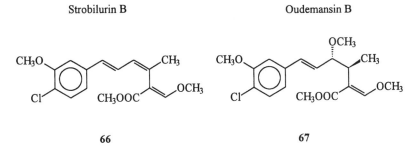

Strobilurin B   Oudemansin B

66   67

Figure 7   Chlorinated strobilurin and oudemansin.

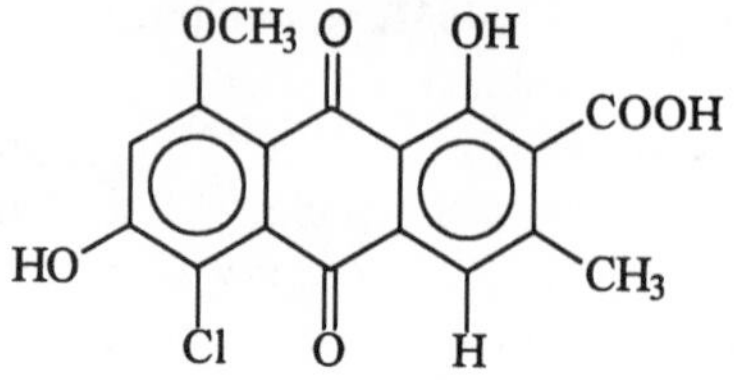

68 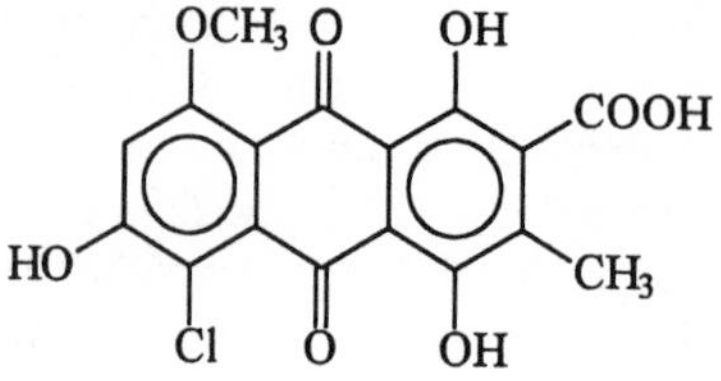   **69**

Figure 8 Chlorinated anthraquinones.

mammals and bacteria make them attractive lead compounds for the synthesis of agricultural fungicides (8).

Most fungi that produce strobilurins and oudemansins grow on wood. *O. mucida* also produces a nonchlorinated strobilurin on sterilized beech wood. Therefore, it appears that the strobilurins play a role in securing nutrient resources for the producers from competing fungi (8).

The aromatic ring and the benzylic carbon atom in the strobilurins and oudemansins are derived from the shikimate pathway, whereas the side chain is built up from acetate units (8). The 3-methoxy-4-chloro position is the same as in one of the *Lepista* metabolites [77] (91, 179).

CHLORINATED ANTHRAQUINONES   The fragile fruiting bodies of the genus *Dermocybe* are blood-red in color and contain many anthraquinones pigments (64, 105, 106). Two chlorinated anthraquinone carboxylic acids, 5-chlorodermolutein [68] (Figure 8 displays compounds 68 and 69) and 5-chlorodermorubin [69], were first isolated from *Dermocybe sanquinea* and *D. semisanquinea* (169). Keller used thin layer chromatography techniques to analyze pigment composition in many *Dermocybe* spp. for systematic purposes. 5-Chlorodermorubin was detected in *D. uliginosa* and *D. sommerfeltii,* and was most likely also present in *D. crocea, D. cinnamomea, D. malicoria, D. phoenica,* and *D. anthracina* (105). 5-Chlorodermorubin [69] was also detected in several *Cortinarius* spp., e.g. *C. aspensis, C. marylandensis,* and *C. pheoniceu* var. *occidentalis* (106). The pigments in anthraquinones are synthesized via the polyketide route (64, 105), with the chlorine molecule possibly originating from a chlorohydrin reaction.

MISCELLANEOUS CHLORINATED AROMATIC COMPOUNDS   Several other chlorinated aromatic metabolites are produced that show no obvious relationship with the metabolites discussed earlier.

Analysis of the flavor compounds of the dried, edible mushroom *Boletus edulis* revealed the production of an unspecified chlorocresol [70] (Figure 9

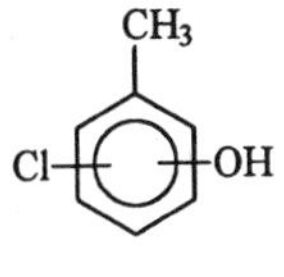

Figure 9   Miscellaneous chlorinated aromatic compounds.

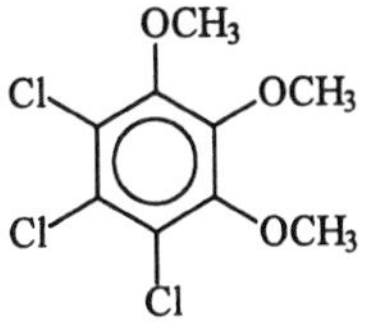

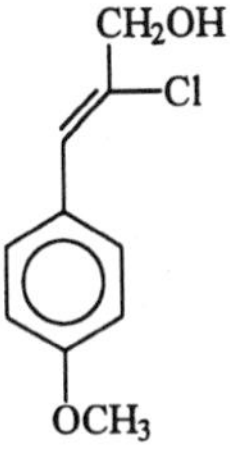

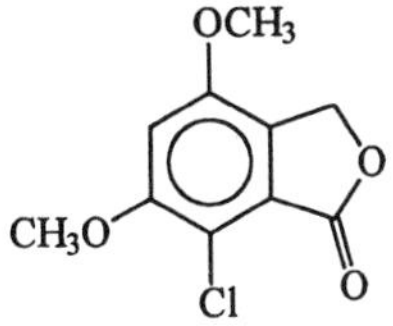

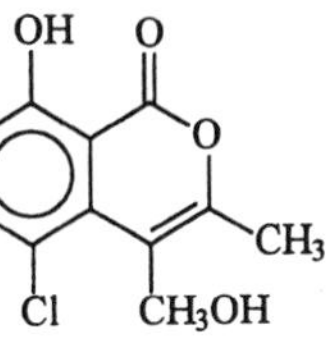

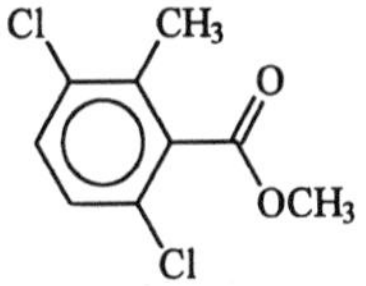

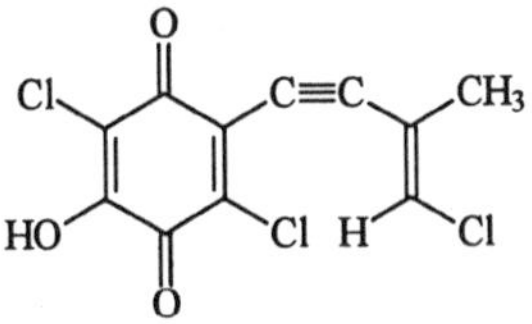

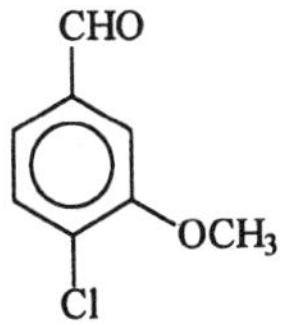

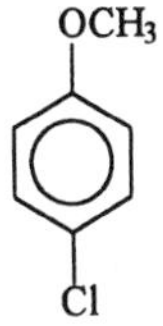

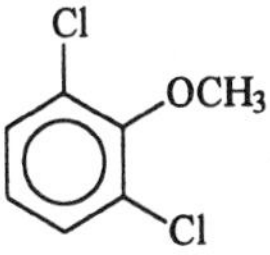

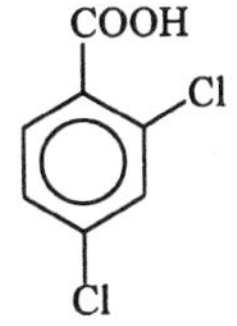

displays compounds 70–80) (180). Another new metabolite, 1,2,3-trimethoxy-4,5,6-trichlorobenzene [71], has recently been isolated from *Phellinus fastuosus* (98); this compound might be related to the drosophilins described earlier. The compound has a selective antibiotic activity against *Botrytis cinerea*. 2-Chloro-3-(4-methoxyphenyl)-2-propen-1-ol [72] was obtained from cultures of *Lentinellus cochleatus* (191). A chlorinated isobenzofuranone [73] was isolated from *Leucoagaricus carneifolia* (95).

The highly pathogenic fungus *Heterobasidion annosum* synthesized a series of specific compounds in the presence of an antagonistic fungus, *Gloeophyllum abietinum,* or host plant cells (166). One compound produced in trace amounts was a chlorinated benzenoid metabolite with a lactone side chain [74]. However, contrary to several related nonchlorinated metabolites, this compound showed no growth-inhibiting effect. The fungus *Kuehneromyces mutabilis* produces methyl 3,6-dichloro-2-methylbenzoate [75] (2). Mycenon [76] is produced by *Mycena* sp. at a concentration of 2.5 mg/L. This is the only metabolite isolated from Basidiomycetes to contain chlorine groups on both the aromatic ring and on the aliphatic side chain. Mycenon shows antibacterial and antifungal activity (86).

3-Methoxy-4-chlorobenzaldehyde [77] was first reported as a metabolite from *Lepista diemii* (179) and has recently also been isolated from *L. nuda* (91). The latter fungus also produces 4-chloroanisole [78] and 2,6-dichloroanisole [79] (91). 2,4-Dichlorobenzoate [80] has also been reported from *Bjerkandera adusta* (139).

ENVIRONMENTAL SIGNIFICANCE OF ORGANOHALOGENS

These results clearly demonstrate that many Basidiomycetes produce organohalogens in the laboratory. Here we try to demonstrate that there is also a substantial production of organohalogens in the natural environment.

Chloromethane is by far the most abundant volatile organohalogen in the atmosphere. Globally, $3–8 \times 10^6$ metric tons/year are derived from natural sources; manmade contributions, estimated at 2.6×10^4 metric tons/year, are negligible in comparison (75). A major part of the chloromethane produced by natural sources is assumed to be of biological origin (75), although the actual contribution of natural terrestrial sources, including the Basidiomycetous fungi, might be relatively low (65).

Recently, it was shown that Basidiomycetes, e.g. *Bjerkandera, Hypholoma,* and *Lepista* spp., produce low-molecular-weight chlorinated anisyl metabolites (CAM) when grown on natural, sterilized lignocellulosic substrates (49, 89). However, analysis of a sum parameter, such as AOX, is much better to screen Basidiomycetes for the production of nonvolatile organohalogens. Verhagen et al showed that many genera produced AOX when cultivated on sterile lignocellulosic material (187); the sulfur tuft (*H. fasciculare*) and *Mycena metata* synthesized up to 132 mg and 193 mg AOX/kg dry weight of forest litter substrate in 6 weeks, respectively (187). The maximum specific rates of organohalogen production for these species on natural substrates is extremely high. Using ergosterol measurements to estimate fungal biomass, values ranging from 630 to 3200 mg/AOX produced per kg mycelium dry weight per day were found with *H. fasciculare* when colonizing wood and forest litter (FJM Verhagen, BKHL Boekema, FBJ van Assema & JA Field, unpublished results). In liquid cultures, the sulfur tuft and other *Hypholoma* spp. produce organically bound halogens up to 3% of their biomass dry weight (187). The maximum rates of organohalogen production occurred during the transition of the fungal culture from primary to secondary metabolism. These results imply that organohalogens may also occur in the environment.

To assess the presence of nonvolatile organohalogens in natural environments, the wood or forest litter in close vicinity of fruiting bodies of Basidiomycetes was tested for chlorinated anisyl metabolites (CAM) (49). In all cases, strains that produced CAM in the laboratory also produced CAM in the environment. The concentrations of the organohalogen metabolites encountered in nature ranged from 15 mg/kg wood produced in the vicinity of *B. adusta,* 39 mg/kg litter by *L. nuda,* 71 mg/kg wood by *P. squarrosa* to 75 mg/kg wood by *Hypholoma* spp. (49). In both laboratory experiments and field samples, the chlorinated anisyl alcohols and aldehydes (16, 18, 19) are encountered in the highest concentrations. These amounts of chloroaromatic metabolites exceed the Dutch and Canadian (British Columbia) norms for analogous chlorophenols in soil, which are set at approximately 1–10 mg/kg for mandatory remediation (60). A large pool of natural AOX has been detected in unpolluted terrestrial environments (18). The origin of these organohalogens is still largely unknown, although evidence is accumulating that Basidiomycetes are responsible for a substantial part of these AOX. When unsterile birch sapwood was incubated with 9 white-rot fungi for 30 weeks, 8 strains showed substantial increases over the blank. The total amount of organically bound halogens was 69 mg/kg dry weight for *B. adusta* and 61 mg/kg dry weight for *P. lindbladii* (139). Species of the genera *Hypholoma, Mycena,* and *Bjerkandera* were among the highest AOX producers in the laboratory (187), and these genera have been reported to be the 1st, 2nd, and 6th most commonly occurring Basidiomycetes in the Netherlands, respectively (136).

Estimating the annual production of AOX is not easy. The specific and total basidiomycetous biomass is not well known. Furthermore, the ratio between high- and low-AOX–producing strains may vary greatly between habitats. Therefore, total AOX production in the specific habitats was estimated by adding up the AOX production of the separate species present, on the basis of the estimated biomass values and specific AOX production (187). Total AOX production in three different forest ecosystems ranged from 100 to 200 g AOX/ha/year based on 13 spp. In a peat bog with a mycelial biomass of *Hypholoma elongatipes* of 80 kg/ha/year, an annual production of 2 kg AOX/ha was calculated (187). However, these calculations are most likely underestimates because they do not take into account all the AOX producing species, because biomass calculations and production rates are uncertain, and because the enhanced production of chlorinated aromatics by antagonistic interactions between competing fungal species or the host plant cells was not taken into account (147, 167). These results, combined with the dominant ecological role of Basidiomycetes as decomposers of forest litter, suggest that they are a major source of natural organohalogens in terrestrial environments. However, the total fungal contribution to the terrestrial AOX pole is probably even greater because other divisions of the fungi, e.g. Ascomycetes and Fungi Imperfecti, have hardly been investigated for their organohalogen production potential. Wood-inhabiting Ascomycetes, e.g. *Xylaria* and *Mollisia* spp., and Fungi Imperfecti, e.g. *Aspergillus* and *Penicillium* spp., are known to produce organohalogens (3, 67, 126, 183).

ENVIRONMENTAL FATE

Mineralization of Chlorinated Aromatics

In the previous section we showed that chlorinated anisyl metabolites (CAM) could be detected in the close vicinity of fruiting bodies of CAM-producing Basidiomycetes (49). However, just outside CAM-producing Basidiomycetous colonies, low-molecular-weight organohalogens were no longer detectable (FJM Verhagen, HJ Swarts & JA Field, unpublished results). This suggests that the compounds are biodegraded or biotransformed. Ligninolytic Basidiomycetes mineralize a wide range of chloroaromatic compounds including chlorophenols (references in next section). However, the chlorinated anisyl alcohols [15, 18] are resistant against oxidative reactions with ligninolytic enzymes (46). Reductive reactions by these ligninolytic enzymes and/or intracellular mineralization may occur. However, many other microorganisms and microbial consortia can completely mineralize a wide range of halogenated aromatic compounds (59, 72, 138, 150). Recently, it was shown that soils from six undisturbed, pristine ecosystems from all over the world had the capacity to mineralize 3-chlorobenzoate and 2,4-dichlorophenoxyacetate (2,4-D) (63). Mineralization of organohalogens must therefore be a global characteristic; it is anticipated that there are microorganisms present in the vicinity of "polluting" basidiomycetes with the potential to degrade the stable, natural organohalogens produced by these fungi. Thus part of the *de novo* biosynthesized organohalogens may be fully mineralized; however, another part may become incorporated into humus according to the following biotransformations.

Biotransformation of Chlorinated Aromatics

How high-molecular-weight organohalogens (chlorohumus) are formed is not known exactly, although two plausible explanations have been advanced. Low-molecular-weight organohalogens biosynthesized by Basidiomycetes and other microorganisms can become incorporated in high-molecular-weight material (49), or high-molecular-weight material is halogenated directly. Basidiomycetes have been shown to produce dimeric and oligomeric organohalogen metabolites. The sulfur tuft produces a dimeric CAM [31] (68), and the polymerization of 2,6-dichloro-4-methoxyphenol [41] in fruiting bodies of *R. subnigricans* resulted in several oligomeric metabolites [44–50] (142, 175, 176). It has also been proposed that a partial metabolism by Basidiomycetes and other microorganisms in the same ecosystem as the methoxy group and alcohol/ aldehyde group of the chlorinated anisyl metabolites (CAM) as well as hydroxylation reactions will yield, among others, 2-chlorophenol and 3-chlorocatechol (49). The latter compounds are readily co-polymerized into humus with phenoloxidizing enzymes, as demonstrated both in vitro (25, 26, 38, 122, 160) and during soil microcosm (33) and composting studies (129). The resulting chlorohumus polymers contain covalently bound organohalogens and are nontoxic (25, 121), and they are poorly biodegraded when incubated in soils (43). Otherwise, extracellular chloroperoxidases can halogenate high molecular weight material directly (93). The presence of chloroperoxidase in soil samples has been indicated (17).

Both mechanisms, the incorporation of low molecular weight compounds and the chloroperoxidase-catalyzed chlorination, are most likely pertinent, and future research should indicate the relative importance of each system. Of significance is the determination of the molecular weight distribution of organohalogens in the litter layer and the different A-horizons. Natural organohalogens increase in molecular weight and in the Cl to carbon ratio with increasing soil depths (90). The chemical structure of high molecular weight organohalogens is basically unknown. However, an oxidative degradation method developed to analyze monomeric building block structures in naturally occurring humus has revealed that the major compounds were derived from 3-chloro-4-hydroxy- and 3,5-dichloro-4-hydroxybenzoate, while in the litter layer 3-methoxy-5-chloro-4-hydroxybenzoate could also be detected (41, 90). However, this method could identify only 2% of all the AOX—basically, only the building block on the leading edge of a branch in the humus polymer could be recovered with the chemical depolymerization procedure employed. Aside from chlorinated compounds, methyl esters of brominated and iodinated 4-hydroxybenzoic acids were recently detected in a sample of unpolluted terrestrial origin (100). Interestingly, bleach kraft mill effluents were most abundant in chlorinated guaiacyl and syringyl units (42). If a chloroperoxidase-catalyzed chlorination reaction is an important mechanism for chlorohumus formation, one would expect a degradation profile more similar to that found in the bleach kraft mill effluents. These data suggest that CAM and other organohalogen metabolites are important precursors to the building blocks for chlorohumus.

The presence of chlorophenolic intermediates suggests that biotoxification reactions could also occur to generate natural hazardous compounds, including polychlorinated dioxins and dibenzofurans (49). It has been shown that a peroxidase-catalyzed oxidation of chlorophenols gives rise to chlorinated dioxins and dibenzofurans (155) and that dioxins are formed in sewage sludge and during composting (155). Interestingly, a recent study in Germany showed significant differences between grassland/plowland and deciduous/coniferous forest soils in rural areas. The amount of all chlorinated dioxins and dibenzofurans found in forest soils totaled around 2.2 μg/kg dry matter, almost 13 times higher than in grasslands (159).

DEGRADATION OF ANTHROPOGENIC ORGANOHALOGENS BY BASIDIOMYCETES

Many Basidiomycetes produce substantial amounts of organohalogens. Interestingly, several Basidiomycetes producing high amounts of low molecular weight organohalogens and AOX, e.g. *Bjerkandera* and *Hypholoma* spp., belong to the white-rot fungi. This group of Basidiomycetes can completely degrade all three major wood components, hemicellulose, lignin, and cellulose. The strategy that these fungi use to degrade lignin is an "enzymatic combustion" (109). Their unique extracellular, oxidative enzyme system causes a rather aspecific degradation of the lignin. This aspecific enzyme system is also effective in the biotransformation of a wide range of chlorinated pollutants by using both oxidative and reductive reactions (20, 29, 58).

Range of Anthropogenic Organohalogens Metabolized by Basidiomycetes

Phanerochaete chrysosporium and the turkey tail, *T. versicolor*, are often used as model organisms to study the biodegradation of anthropogenic organohalogens, although the ability of other white-rot fungi to degrade these xenobiotics has also been investigated. Mineralization of organohalogens in vitro is generally enhanced under culture conditions favorable for the mineralization of lignin, which suggests that the lignin-degrading system may be involved in transforming anthropogenic compounds (29). Fungal extracellular enzymes, including lignin peroxidases, manganese peroxidases, and laccases act via a nonspecific free radical mechanism and are implicated in the degradation of lignin (47, 109); it has been proposed that the same enzymes are involved in the fungal degradation of several organohalogens (58, 74, 101, 146).

DEGRADATION OF ALIPHATIC ORGANOHALOGENS   Surprisingly, several aliphatic organohalogens, including chloroform, trichloroethylene and tetrachloromethane, and some already highly oxidized pollutants are also mineralized by ligninolytic cultures of *P. chrysosporium* under aerobic conditions (108, 162). This suggests a possible role for ligninolytic peroxidases in the mineralization of these chemicals. Recent research has indicated that white-rot fungi can use several different mechanisms to cause reductive reactions (20). However, only with lignin peroxidase has a reductive dehalogenation of aliphatic organohalogens been demonstrated. The system able to generate these reductive reactions contains lignin peroxidase, veratryl alcohol, ethylenedianimetetraacetate (EDTA) or oxalate, and H_2O_2. It has been proposed that the carboxy anion radical is the actual reductive species in this reaction. Since all the components of this reductive system with oxalate as an electron donor are excreted by *P. chrysosporium*, this mechanism may be involved in the reductive dechlorination of these halocarbons by the fungus (108).

CHLOROLIGNINS IN BLEACH KRAFT MILL EFFLUENTS   The traditional bleaching methods of kraft pulps used substantial amounts of elemental chlorine, resulting in the formation of high concentrations of chlorolignin (165). *Trametes versicolor*, in particular, has been used successfully in bioreactors to decolorize and dechlorinate these bleach kraft mill effluents (12). AOX reductions ranging from 52% to 59% were achieved (145). Dechlorination of bleaching effluents has also been observed with other white-rot fungi, e.g. *P. chrysosporium*, *Ganoderma lucidum*, and *Hericium erinaceus* (24, 62, 189). However, recent changes in the bleaching procedures in pulp mills have lowered the priority of this potential application of white-rot fungi (152).

PENTACHLOROPHENOL AND OTHER CHLOROPHENOLS   The biodegradation of the wood preservative pentachlorophenol (PCP) by Basidiomycetes is well documented. A high mineralization (50–70%) of this organohalogen has been described for several white-rot fungi, including *P. chrysosporium, Phanerochaete sordida, Lentinula edodes, Inonotus dryophilus, Irpex lacteus, Bjerkandera adusta, Phellinus badius,* and *Trametes versicolor* (4, 5, 120, 130, 144, 170). The nonwhite rot fungus *Mycena avenacea* can also metabolize PCP (113). In almost all cases a major transformation product was pentachloroanisole. It is a general trend among Basidiomycetes to methylate chlorophenols, most likely as a detoxification mechanism or to prevent polymerization of the phenolic compounds. However, it can also be an aspecific reaction of methyl transferases involved in biosynthetic pathways.

Both lignin peroxidases and laccases have been considered as catalysts in the initial oxidation of chlorophenols and other organohalogens by white-rot fungi (36, 74, 157). These enzymes, as well as other peroxidases, oxidize pentachlorophenol via a phenoxyl radical to the corresponding *para*-quinone (Figure 10) (74, 103, 130, 157) and in other cases, cause the release of chlorine atoms by oxidizing organohalogens. The laccase- or peroxidase-catalyzed dehalogenation of chlorinated phenols can be caused by oxidative coupling (45) or by nucleophilic substitution (74, 97).

Biodegradation of PCP by *P. chrysosporium* has also been examined in sterile and nonsterile soils in the laboratory and in situ (37, 114, 115), and it has been shown that lignin-degrading fungi can be used in the disposal of pentachlorophenol-treated wood (116).

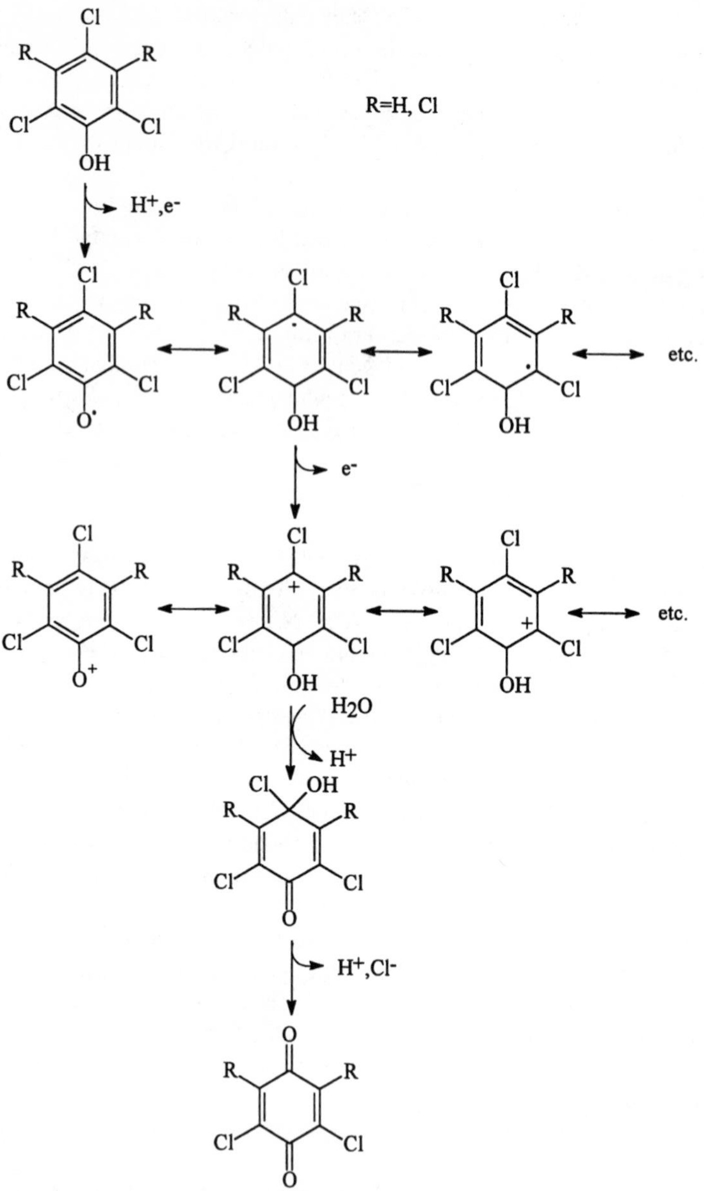

Figure 10 Hypothetical scheme for the mechanism of 2,4,6-trichlorophenol (R=H) and pentachlorophenol (R=Cl) oxidation by peroxidases and laccases. (Adapted from References 74, 97, 157.)

Interestingly, both bacterial spp., e.g. *Mycobacterium* and *Rhodococcus,* and Basidiomycetes, e.g. *Mycena* spp., produce "natural" organohalogen intermediates during the degradation of pentachlorophenol and 2,6-dichlorohydroquinone. The microorganisms caused a methylation of tetrachlorinated hydroquinone degradation intermediates, resulting in the formation of drosophilin A [37] and drosophilin A methyl ether [38]. Likewise, 2,6-dichlorohydroquinone was converted to 3,5-dichloro-4-methoxyphenol, which resembles russuphelin precursor [41] (73, 111, 171).

CHLORINATED DIOXINS   *Phanerochaete chrysosporium* can metabolize di- and tetrachlorinated dioxins (29, 185). Comparable multistep pathways have been proposed for the degradation of 2,7-dichlorodibenzo-dioxin and the phenols 2,4-dichlorophenol and 2,4,5-trichlorophenol (101, 184, 185). In all cases, the involvement of lignin peroxidases, manganese peroxidases, and intracellular enzymes causing the reduction of the formed quinones and subsequent methylation of the hydroquinones with methyl transferases have been postulated. Interestingly, in these pathways all chlorine atoms were removed before ring cleavage occurred.

POLYCHLORINATED BIPHENYLS   Ligninolytic cultures of *P. chrysosporium* degraded some polychlorinated biphenyls (PCB) to CO_2. However, degradation rates were inversely related to the number of chlorine atoms on the biphenyl molecule. There was no correlation between the rate of degradation and the synthesis of lignin peroxidases or of Mn-dependent peroxidases (50, 181). 4,4'-Dichlorobiphenyl was extensively degraded in liquid culture, and 4-chlorobenzoic acid and 4-chlorobenzyl alcohol were identified as the metabolites produced. In contrast, there was negligible mineralization of the tetrachloro- or hexachlorobiphenyl and little evidence for any significant metabolism (50). In other work, *P. chrysosporium* was shown to be capable of a substantial degradation of highly chlorinated PCB mixtures such as aroclor 1260, which has an average of six chlorine molecules on each biphenyl structure. Interestingly, the PCBs were mineralized in both low-nitrogen (ligninolytic) as well as high-nitrogen (nonligninolytic) defined media, with the highest degradation of the congeners in malt extract medium (193), which indicates that lignin-degrading enzymes are not involved in PCB degradation. Other white-rot fungi, e.g. *Pleurotus ostreatus* and *Trametes versicolor,* degraded PCBs in a solid state system (188, 197).

DDT (1,1,1-trichloro-2,2-bis(4-chlorophenyl)ethane) and the so-called "dead-end side product" DDE are extremely persistent environmental pollutants. White-rot fungi are exceptional because they can metabolize both components to CO_2 (28, 29). However, the mineralization is often less than 10%.

The involvement of lignin peroxidase with the degradation of DDT has also been questioned (110). However, only the first reductive dechlorination step in the degradation of DDT is independent of ligninolytic enzymes, but its mineralization only occurs under ligninolytic conditions (20).

ATRAZINE   2-Chloro-4-ethylamino-6-isopropylamino-s-triazine was degraded by both free living and mycorrhizal fungi. The highest incorporation of atrazine carbon in the biomass was obtained with a mycorrhizal fungus (52). Degradation of atrazine in soils was enhanced by inoculation with *P. chrysosporium* or *Pleurotus pulmonarius* (54, 88, 125), and several degradation products have been identified. However, no mineralization of the triazine ring has been found.

OTHER ORGANOHALOGENS DEGRADED BY BASIDIOMYCETES   Several other anthropogenic halogenated compounds can be degraded by Basidiomycetes. These include the amidinohydrazone-type insecticide, hydramethylnon (1), alachlor (57), and chlorobenzenes (195). 3,4-Dichloroaniline was metabolized by several Basidiomycetes including *Schizophyllum, Filoboletus,* and *Auriculariopsis* species (112, 151, 161). The brown-rot fungi *Tyromyces palustris* and *Serpula lacrymans* and the white-rot fungus *T. versicolor* degrade several organoiodine wood preservatives (119). The fluoroquinolone Enrofloxacin was also mineralized by white- and brown-rot fungi (123). The alkyl halides lindane and chlordane were metabolized by the *P. chrysosporium* (107). However, it was concluded that the extracellular peroxidases were not involved in the degradation of these insecticides, but rather the involvement of a P450-monooxygenase was implicated (133). *P. chrysosporium* also had the highest mineralization (50%) of two common phenoxyalkanoic herbicides, 2,4-dichlorophenoxyacetic acid (2,4-D) and 2,4,5-trichlorophenoxyacetic acid (2,4,5-T), in nutrient-rich (nonligninolytic) media (194).

Fate of Anthropogenic Organohalogens

Basidiomycetes caused an almost complete biotransformation of several organohalogens after prolonged incubation times, and a substantial part of the xenobiotic compound was mineralized to carbon dioxide. However, in experiments with both soil spiked with chlorophenols and liquid cultures spiked with chloroanilines, the actual mineralization to carbon dioxide leveled off and remained, in most cases, below 20% after extended incubation with several white-rot fungi (32, 117, 131). In sterile soils spiked with PCP and inoculated with *P. chrysosporium,* a substantial part of the radioactive compounds became soil-bound and thereby unextractable (117). Recently it has been shown that dimers and oligomers were formed as intermediates during the biodegradation of 4-chloroaniline (32) and chlorophenols (101). These oligomers of 4-chloroaniline and chlorophenols were also formed in vivo. Extracellular ligninolytic enzymes, lignin peroxidase and laccase, oxidize these monomers to form several dimers, trimers, and tetramers from chlorophenols and chloroanilines (25, 32, 74, 101, 112, 151). Removal of chlorophenols by ligninolytic enzymes could be enhanced in the presence of reactive co-substrates that had a strong resemblance to lignin building blocks such as guaiacol and 2,6-dimethoxyphenol. Addition of 2,6-dimethoxyphenol enhanced the formation of insoluble polymers from 2,4,5-trichlorophenol by 90% with horseradish peroxidase and by 98% with *T. versicolor* laccase (158). Dechlorination occurs during the oxidative coupling reactions due to the oxidative elimination of *para* chlorine groups, and most of the remaining organochlorine appears to be incorporated into the polymers that are formed (25, 44). Oxidative enzymes can also cause the cross-coupling of several chlorophenols and chloroanilines with high molecular weight material such as humic acid (43, 177). The use of ^{13}C-NMR to detect bound ^{13}C-2,4-dichlorophenol and the release of chloride ions confirmed that covalent bonds are formed between chlorophenols and humic acids (25, 44). The environmental fate of both natural and anthropogenic organohalogens thus seems quite similar.

CONCLUSIONS

There is an ubiquitous capacity among the Basidiomycetes to produce a large variety of organohalogen metabolites. To date, 81 halogenated metabolites have been identified from Basidiomycetes, and this number is likely to rise substantially over the next decade. The chlorinated anisyl metabolites (CAM) are the most important examples because they are detectable in high amounts in the environment and are produced by many different genera of Basidiomycetes. Greater insight into the biosynthesis routes of the different classes of organohalogens is gradually emerging, although the key halogenating enzymes have still not been identified.

Organohalogens are not biological accidents. Some Basidiomycetes produce organically bound halogens up to 3% of their biomass dry weight. Clearly the organohalogens are produced for a purpose. Chloromethane is used by Basidiomycetes as a methylating agent for aromatic carboxyl and phenolic hydroxyl groups. Chlorinated anisyl alcohols have an important physiological role in lignin degradation because they are substrates for extracellular aryl alcohol oxidases generating H_2O_2, a necessary co-substrate for the extracellular peroxidases. Several other organohalogens have a role as a chemical defense substance, protecting colonies from antagonizing organisms. In most cases the chloro group is a vital part of the metabolite because of its physiological role or antibiotic activity. The chlorine group activates methane for transferase reactions in the biosynthesis of important methylated secondary metabolites. Chlorination of the anisyl metabolites makes them better substrates for the

H_2O_2-generating aryl alcohol oxidase and makes them more resistant to the fungus's own ligninolytic enzymes. The chlorine substitution of the russuphelins influences the conformation and hence the biological activities of these compounds.

Basidiomycetes produce significant amounts of low molecular weight organohalogens, e.g. chlorinated anisyl metabolites, or adsorbable organic halogens (AOX) when grown on lignocellulosic substrates. The commonly occurring sulfur tuft (*Hypholoma fasciculare*) and other *Hypholoma* spp. were among the highest organohalogen producers. These halogenated compounds can become incorporated into humus according to known biotransformations. In light of the ecological importance of Basidiomycetes as decomposers of forest litter, we conclude that they are a major source of natural organohalogens in terrestrial environments. Basidiomycetes can also degrade a wide range of anthropogenic organohalogens, and it is proposed that the environmental fate of both natural and many anthropogenic organohalogens is quite similar. Under some conditions the metabolites will be mineralized to carbon dioxide whereas in other cases they will be partially metabolized and become susceptible for incorporation into humus.

Visit the *Annual Reviews home page* at
http://www.annurev.org.

Literature Cited

1. Abernethy GA, Walker JRL. 1993. Degradation of the insecticide hydramethylnon by *Phanerochaete chrysosporium.* *Biodegradation* 4:131–39
2. Abraham BG, Berger RG. 1994. Higher fungi for generating aroma components through novel biotechnologies. *J. Agric. Food Chem.* 42:2344–48
3. Adeboya MO, Edwards RL, Lassoe T, Maitland DJ, Shields L, et al. 1996. Metabolites of the higher fungi. 29. Maldoxin, maldoxone, dihydromaldoxin, isodihydromaldoxin and dechlorodihydromaldoxin-a spirocyclohexadienone, a depsidone and three diphenyl ethers–keys in the depsidone biosynthetic pathway from a member of the fungus genus *Xylaria. J. Chem. Soc. Perkin Trans.* 1:1419–25
4. Alleman BC, Logan BE, Gilbertson RL. 1992. Toxicity of pentachlorophenol to 6 species of white rot fungi as a function of chemical dose. *Appl. Environ. Microbiol.* 58:4048–50
5. Alleman BC, Logan BE, Gilbertson RL.
    1995. Degradation of pentachlorophenol by fixed films of white rot fungi in rotating tube bioreactors. *Water Res.* 29:61–67
6. Amato I. 1993. The crusade against chlorine. *Science* 261:152–54
7. Anchel M. 1952. Identification of drosophilin A as p-methoxytetrachlorophenol. *J. Am. Chem. Soc.* 74:2943
8. Anke T. 1995. The antifungal strobilurins and their possible ecological role. *Can. J. Bot.* 73(Suppl. 1 E-H):S940–45
9. Anke T, Besl H, Mocek U, Steglich W. 1983. Antibiotics from basidiomycetes. XVIII. Strobilurin C and oudemansin B, two new antifungal metabolites from *Xerula* species (Agaricales). *J. Antibiot.* 36:661–66
10. Anke T, Oberwinkler F, Steglich W, Schramm G. 1977. The strobilurins–new antibiotics from the basidiomycete *Strobilurus tenacellus* (Pers. ex Fr.) Sing. *J. Antibiot.* 30:806–10
11. *Determination of volatile organic compounds in mushrooms.* 1993. Short path

thermal desorption application note No. 18. Ringoes, NJ: Sci. Instrum. Serv.

12. Archibald F, Paice MG, Jurasek L. 1990. Decolorization of kraft bleachery effluent chromophores by *Coriolus (Trametes) versicolor. Enzyme Microb. Technol.* 12:846–53

13. Arnone A, Cardillo R, Di Modugno V, Nasini G. 1988. Secondary mould metabolites. XXII. Isolation and structure elucidation of melledonals D and E and melleolides E-H, novel sesquiterpenoid aryl esters from *Clitocybe elegans* and *Armillariamellea. Gaz. Chim. Ital.* 118:517–21

14. Arnone A, Cardillo R, Nasini G. 1988. Secondary mould metabolites. XXIII. Isolation and structure elucidation of melleolides I and J and Armellides A and B, novel sesquiterpenoid aryl esters from *Armillaria nova-zelandiae. Gaz. Chim. Ital.* 118:523–27

15. Arnone A, Cardillo R, Nasini G. 1986. Structures of melleolides B–D, three antibacterial sesquiterpenoids from *Armillariamellea. Phytochemistry* 25:471–74

16. Arnone A, Cardillo R, Nasini G, Valdo Meille S. 1988. Secondary mould metabolites. Part 19. Structure elucidation and absolute configuration of melledonals B and C, novel antibacterial sesquiterpenoids from *Armillariamellea.* X-ray molecular structure of melledonal C. *J. Chem. Soc. Perkin Trans.* 1:503–10

17. Asplund G, Borén H, Carlsson U, Grimvall A. 1991. Soil peroxidase-mediated halogenation of fulvic acids. In *Humic Substances in the Aquatic and Terrestrial Environment,* ed. B Allard, H Borén, A Grimvall, pp. 475–84. Heidelberg: Springer-Verlag

18. Asplund G, Grimvall A. 1991. Organohalogens in nature–more widespread than previously assumed. *Environ. Sci. Technol.* 25:1346–50

19. Attieh JM, Hanson AD, Saini HS. 1995. Purification and characterization of a novel methyltransferase responsible for biosynthesis of halomethanes and methanethiol in *Brassica oleracea. J. Biol. Chem.* 270:9250–57

20. Barr DP, Aust SD. 1994. Mechanisms white rot fungi use to degrade pollutants. *Environ. Sci. Technol.* 28:A78–87

21. Bastian W. 1985. *Vergleichende Untersuchungen zum Sekundärstoffwechsel an Coprophilen und Erd- oder Holzbewohnende Basidiomyceten.* PhD thesis. Univ. Kaiserslautern, Germany (In German)

22. Beck HC, Lauritsen FR, Patrick JS,

Cooks RG. 1996. Metabolism of halogenated compounds in the white rot fungus *Bjerkandera adusta* studied by membrane inlet mass spectrometry and tandem mass spectrometry. *Biotechnol. Bioeng.* 51:23–32

23. Becker U, Anke T, Sterner O. 1994. A novel halogenated compound possessing antibiotic and cytotoxic activities isolated from the fungus *Resinicium pinicola* (J. Erikss.). *Z. Naturforsch. C* 49:772–74

24. Bergbauer M, Eggert C, Kraepelin G. 1991. Degradation of chlorinated lignin compounds in a bleach plant effluent by the white-rot fungus *Trametes versicolor. Appl. Microbiol. Biotechnol.* 34:105–9

25. Bollag J-M, Dec J. 1995. Incorporation of halogenated substances into humic material. In *Naturally-Produced Organohalogens,* ed. A Grimvall, EWB de Leer, pp. 161–69. Dordrecht: Kluwer

26. Bollag J-M, Sjobald RD, Minard RD. 1977. Polymerization of phenolic intermediates of pesticides by a fungal enzyme. *Experienta* 33:1564–66

27. Brambilla U, Nasini G, Depava OV. 1995. Secondary mold metabolites. 49. Isolation, structural elucidation, and biomimetic synthesis of trametol, a new 1-arylpropane-1,2-diol produced by the fungus *Trametes* sp. *J. Nat. Prod.* 58:1251–53

28. Bumpus JA, Powers RH, Sun T. 1993. Biodegradation of DDE (1,1-dichloro-2,2-bis(4-chlorophenyl)ethene) by *Phanerochaete chrysosporium. Mycol. Res.* 97:95–98

29. Bumpus JA, Tien M, Wright D, Aust SD. 1985. Oxidation of persistent environmental pollutants by a white-rot fungus. *Science* 228:1434–36

30. Buss H, Zimmer L. 1974. Natürliche polychlorierte Aromaten in Champignons. *Chemosphere* 3:123–26

31. Butruille D, Dominquez XA. 1972. Un nouveau produit naturel: 1,4-Dimethoxy-2-nitro-3,5,6-trichlorobenzene. *Tetrahedron Lett.* 3:211–12

32. Chang CW, Bumpus JA. 1993. Oligomers of 4-chloroaniline are intermediates formed during its biodegradation by *Phanerochaete chrysosporium. FEMS Microbiol. Lett.* 107:337–42

33. Cheng HH, Haider K, Harper SS. 1983. Catechol and chlorocatechols in soil: degradation and extractability. *Soil Biol. Biochem.* 15:311–17

34. Chilton WS, Tsou G. 1972. A chloro amino acid from *Amanita solitaria. Phytochemistry* 11:2853–57

35. Chilton WS, Tsou G, De Cato L Jr, Malone MH. 1973. The unsaturated norleucines from *Amanita solitaria.* Chemical and pharmacological studies. *Lloydia* 36:169–73

36. Chung NH, Aust SD. 1995. Veratryl alcohol-mediated indirect oxidation of pentachlorophenol by lignin peroxidase. *Arch. Biochem. Biophys.* 322:143–48

37. Chung NM, Aust SD. 1995. Degradation of pentachlorophenol in soil by *Phanerochaete chrysosporium. J. Haz. Mat.* 41:177–83

38. Claus H, Filip Z. 1990. Enzymatic oxidation of some substituted phenols and aromatic amines, and the behaviour of some phenoloxidases in the presence of soil related adsorbents. *Water Sci. Technol.* 22:69–77

39. Cowan MI, Glen AT, Hutchinson SA, Maccartney ME, Mackintosh JM, et al. 1973. Production of volatile metabolites by species of *Fomes. Trans. Br. Mycol. Soc.* 60:347–60

40. Cremin P, Donnelly DMX, Wolfender JL, Hostettmann K. 1995. Liquid chromatographic-thermospray mass spectrometric analysis of sesquiterpenes of *Armillaria* (Eumycota, Basidiomycotina) species. *J. Chrom.* 710:273–85

41. Dahlman O, Morck R, Ljungquist P, Reimann A, Johansson C, et al. 1993. Chlorinated structural elements in high molecular weight organic matter from unpolluted waters and bleached-kraft mill effluents. *Environ. Sci. Technol.* 27:1616–20

42. Dahlman O, Reimann A, Ljungquist P, Mörck R, Johansson C, et al. 1994. Characterization of chlorinated aromatic structures in high molecular weight BKME-materials and in fulvic acids from industrially unpolluted waters. *Water Sci. Technol.* 29:81–91

43. Dec J, Bollag J-M. 1988. Microbial release and degradation of catechol and chlorophenols bound to synthetic humic acid. *Soil Sci. Soc. Am. J.* 52:1366–71

44. Dec J, Bollag J-M. 1994. Dehalogenation of chlorinated phenols during oxidative coupling. *Environ. Sci. Technol.* 28:484–90

45. Dec J, Bollag J-M. 1995. Effect of various factors on dehalogenation of chlorinated phenols and anilines during oxidative coupling. *Environ. Sci. Technol.* 29:657–63

46. De Jong E, Cazemier AE, Field JA, de Bont JAM. 1994. Physiological role of chlorinated aryl alcohols biosynthesized *de novo* by the white rot fungus *Bjerkandera* sp. strain BOS55. *Appl. Environ. Microbiol.* 60:271–77

47. De Jong E, Field JA, De Bont JAM. 1994. Aryl alcohols in the physiology of ligninolytic fungi. *FEMS Microbiol. Rev.* 13:153–88

48. De Jong E, Field JA, Dings JAFM, Wijnberg JBPA, de Bont JAM. 1992. De novo biosynthesis of chlorinated aromatics by the white-rot fungus *Bjerkandera* sp. BOS55. Formation of 3–chloroanisaldehyde from glucose. *FEBS Lett.* 305:220–24

49. De Jong E, Field JA, Spinnler HE, Wijnberg JBPA, de Bont JAM. 1994. Significant biogenesis of chlorinated aromatics by fungi in natural environments. *Appl. Environ. Microbiol.* 60:264–70

50. Dietrich D, Hickey WJ, Lamar R. 1995. Degradation of 4,4′-dichlorobiphenyl, 3, 3′,4,4′-tetrachlorobiphenyl, and 2,2′,4,4′, 5,5′-hexachlorobiphenyl by the white rot fungus *Phanerochaete chrysosporium. Appl. Environ. Microbiol.* 61:3904–9

51. Donnelly DMX, Coveney DJ, Fukuda N, Polonsky J. 1986. New sesquiterpene arylesters from *Armillaria mellea. J. Nat. Prod.* 49:111–18

52. Donnelly PK, Entry JA, Crawford DL. 1993. Degradation of atrazine and 2,4-dichlorophenoxyacetic acid by mycorrhizal fungi at 3 nitrogen concentrations *in vitro. Appl. Environ. Microbiol.* 59:2642–47

53. Ellis MB, Ellis JP. 1990. *Fungi without Gills (Hymenomycetes and Gasteromycetes).* London: Chapman & Hall. 329 pp.

54. Entry JA, Donnelly PK, Emmingham WH. 1996. Mineralization of atrazine and 2,4-d in soils inoculated with *Phanerochaete chrysosporium* and *Trappea darkeri. Appl. Soil Ecol.* 3:85–90

55. Esaki N, Takada H, Moriguchi M, Hatanaka SI, Tanaka H, et al. 1989. Mechanism-based inactivation of L-methionine gamma-lyase by L-2-amino-4-chloro-4-pentenoate. *Biochemistry* 28:2111–16

56. Farhangrazi ZS, Sinclair R, Yamazaki I, Powers LS. 1992. Haloperoxidase activity of *Phanerochaete chrysosporium* lignin peroxidases H2 and H8. *Biochemistry* 31:10763–68

57. Ferrey ML, Koskinen WC, Blanchette RA, Burnes TA. 1994. Mineralization of alachlor by lignin-degrading fungi. *Can. J. Microbiol.* 40:795–98

58. Field JA, de Jong E, Feijoo Costa G, de Bont JAM. 1993. Screening for ligninolytic fungi applicable to the biodegradation of xenobiotics. *Trends Biotechnol.* 11:44–49

59. Field JA, Stams AJM, Kato M, Schraa G. 1995. Enhanced biodegradation of aromatic pollutants in cocultures of anaerobic and aerobic bacterial consortia. *Antonie van Leeuwenhoek* 67:47–77

60. Field JA, Verhagen FJM, de Jong E. 1995. Natural organohalogen production by basidiomycetes. *Trends Biotechnol.* 13:451–56

61. Frankland JC. 1982. Biomass and nutrient cycling by decomposer basidiomycetes. In *Decomposer Basidiomycetes: Their Biology and Ecology*, ed. JC Frankland, JN Hedger, MJ Swift, pp. 241–61. Cambridge: Cambridge Univ. Press

62. Fukui H, Presnell TL, Joyce TW, Chang HM. 1992. Dechlorination and detoxification of bleach plant effluent by *Phanerochaete chrysosporium*. *J. Biotechnol.* 24:267–75

63. Fulthorpe RR, Rhodes AN, Tiedje JM. 1996. Pristine soils mineralize 3–chlorobenzoate and 2,4-dichlorophenoxyacetate via different microbial populations. *Appl. Environ. Microbiol.* 62:1159–66

64. Gill M, Giménez A. 1990. Pigments of fungi. Part 17. (S)-(+)-Dermochrysone, (+)-dermolactone, dermoquinone, and related pigments; new nonoketides from the fungus *Dermocybe sanguinea* (sensu Cleland). *J. Chem. Soc. Perkin Trans.* 1:2585–91

65. Graedel TE, Keene WC. 1996. The budget and cycle of earth's natural chlorine. *Pure Appl. Chem.* 68:1689–97

66. Gribble GW. 1996. The diversity of natural organohalogens in living organisms. *Pure Appl. Chem.* 68:1699–712

67. Gribble GW. 1996. *Naturally Occurring Organohalogen Compounds–A Comprehensive Survey.* New York: Springer Verlag. 498 pp.

68. Griffith GS, Rayner ADM, Wildman HG. 1994. Extracellular metabolites and mycelial morphogenesis of *Hypholoma fasciculare* and *Phlebia radiata* (Hymenomycetes). *Nova Hedwigia* 59:311–29

69. Grimvall A. 1995. Evidence of naturally produced and man-made organohalogens in water and sediments. See Ref. 69a, pp. 3–20

69a. Grimvall A, de Leer EWB, eds. 1995. *Naturally-Produced Organohalogens.* Dordrecht: Kluwer

70. Grove JF. 1981. Volatile compounds from the mycelium of the mushroom *Agaricus bisporus*. *Phytochemistry* 20:2021–22

71. Gutierrez A, Caramelo L, Prieto A, Martinez MJ, Martinez AT. 1994. Anisaldehyde production and aryl-alcohol oxidase and dehydrogenase activities in ligninolytic fungi of the genus *Pleurotus. Appl. Environ. Microbiol.* 60:1783–88

72. Häggblom MM. 1992. Microbial breakdown of halogenated aromatic pesticides and related compounds. *FEMS Microbiol. Rev.* 103:29–72

73. Häggblom MM, Apajalahti JHA, Salkinoja-Salonen MS. 1988. O-Methylation of chlorinated *para*-hydroquinones by *Rhodococcus chlorophenolicus. Appl. Environ. Microbiol.* 54:1818–24

74. Hammel KE, Tardone PJ. 1988. The oxidative 4-dechlorination of polychlorinated phenols is catalyzed by extracellular fungal lignin peroxidases. *Biochemistry* 27:6563–68

75. Harper DB. 1994. Biosynthesis of halogenated methanes. *Biochem. Soc. Trans.* 22:1007–11

76. Harper DB, Buswell JA, Kennedy JT, Hamilton JTG. 1990. Chloromethane, methyl donor in veratryl alcohol biosynthesis in *Phanerochaete chrysosporium* and other lignin-degrading fungi. *Appl. Environ. Microbiol.* 56:3450–57

77. Harper DB, Hamilton JTG, Kennedy JT, McNally KJ. 1989. Chloromethane, a novel methyl donor for biosynthesis of esters and anisoles in *Phellinus pomaceus. Appl. Environ. Microbiol.* 55:1981–89

78. Harper DB, Kennedy JT. 1986. Effect of growth conditions on halomethane production by *Phellinus* species: biological and environmental implications. *J. Gen. Microbiol.* 132:1231–46

79. Harper DB, Kennedy JT, Hamilton JTG. 1988. Chloromethane biosynthesis in poroid fungi. *Phytochemistry* 27:3147–53

80. Harper DB, McRoberts WC, Kennedy JT. 1996. Comparison of the efficacies of chloromethane, methionine, and s-adenosylmethionine as methyl precursors in the biosynthesis of veratryl alcohol and related compounds in *Phanerochaete chrysosporium. Appl. Environ. Microbiol.* 62:3366–70

81. Hatanaka SI, Furukawa J, Aoki T, Akatsuka H, Nagasawa E. 1994. (2S)-2-Amino-5-chloro-4-hydroxy-5-hexenoic acid, a new chloroamino acid, and related compounds from *Amanita gymnopus*. *Mycoscience* 35:391–94

82. Hatanaka SI, Kaneko S, Niimura Y, Kinoshita F, Soma GI. 1974. L-2-Amino-4-chloro-4-pentenoic acid, a new natural amino acid from *Amanita pseudoporphyria* Hongo. *Tetrahedron Lett.* 45:3931–32

83. Hatanaka SI, Niimura Y, Takishima K. 1985. Non-protein amino acids of unsaturated norleucine-type in *Amanita pseudoporphyria*. *Trans. Mycol. Soc. Jpn.* 26:61–68

84. Hatanaka SI, Okada K, Nagasawa E. 1995. Isolation and identification of (2S)-2-Amino-5-chloro-4-hydroxy-5-hexenoic acid from an *Amanita* of the section *Roanokenses* (Amanitaceae). *Mycoscience* 36:395–97

85. Hautzel R, Anke H. 1990. Screening of Basidiomycetes and Ascomycetes for plant growth regulating substances. Introduction of the gibberellic acid induced *de novo* synthesis of hydrolytic enzymes in embryoless seeds of *Triticum aestivum* as test system. *Z. Naturforsch. C* 45:1093–98

86. Hautzel R, Anke H, Sheldrik WS. 1990. Mycenon, a new metabolite from a *Mycena* species TA 87202 (Basidiomycetes) as an inhibitor of isocitrate lyase. *J. Antibiot.* 18:1240–44

87. Hibbett DS, Donoghue MJ. 1995. Progress toward a phylogenetic classification of the polyporaceae through parsimony analysis of mitochondrial ribosomal DNA sequences. *Can. J. Bot.* 73(Suppl. 1 E-H):S853–61

88. Hickey WJ, Fuster DJ, Lamar RT. 1994. Transformation of atrazine in soil by *Phanerochaete chrysosporium*. *Soil Biol. Biochem.* 26:1665–71

89. Hjelm O. 1996. *Organohalogens in coniferous forest soil*. PhD thesis. Linköping Stud. Arts Sci. 139, Linköping, Sweden

90. Hjelm O, Asplund G. 1995. Chemical characterization of organohalogens in a coniferous forest soil. See Ref. 69a, pp. 105–11

91. Hjelm O, Boren H, Öberg G. 1996. Analysis of halogenated organic compounds in coniferous forest soil from a *Lepista nuda* (wood blewitt) fairy ring. *Chemosphere* 32:1719–28

92. Hjelm O, Johansson MB, Oberg-Asplund G. 1995. Organically bound halogens in coniferous forest soil—distribution pattern and evidence of in situ production. *Chemosphere* 30:2353–64

93. Hoekstra EJ, Lassen P, van Leeuwen JGE, de Leer EWB, Carlsen L. 1995. Formation of organic chlorine compounds of low molecular weight in the chloroperoxidase-mediated reaction between chloride and humic material. See Ref. 69a, pp. 149–58

94. Hsu C-S, Suzuji M, Yamada Y. 1971. Chemical constituents of fungi I 1,4-dimethoxy-2,3,5,6-tetrachlorobenzene

(O-Drosophilin A) from *Phellinus yucatensis. Chem. Abstr.* 115864a, Vol. 75

95. Huff T, Kuball HG, Anke T. 1994. 7-Chloro-4,6-dimethoxy-1 (3H)-isobenzofurane and basidalin: Antibiotic secondary metabolites from *Leucoagaricus carneifolia* Gillet (Basidiomycetes). *Z. Naturforsch. C* 49:407–10

96. Hutchinson SA. 1971. Biological activity of volatile fungal metabolites. *Trans. Br. Mycol. Soc.* 57:185–200

97. Iimura Y, Hartikainen P, Tatsumi K. 1996. Dechlorination of tetrachloroguaiacol by laccase of white-rot basidiomycete *Coriolus versicolor. Appl. Microbiol. Biotechnol.* 45:434–39

98. Jain SC, Kumar R, Bharadvaja A, Parmar VS, Errington W, et al. 1996. Novel products of some wood rotting fungi, *Abstr. IUPAC Symp. Chem. Nat. Prod., 20th,* Chicago, Sept. 15–20, p. SE-35

99. Jensen KA, Evans KMC, Kirk TK, Hammel KE. 1994. Biosynthetic pathway for veratryl alcohol in the ligninolytic fungus *Phanerochaete chrysosporium. Appl. Environ. Microbiol.* 60:709–14

100. Johansson C, Borén H, Grimvall A, Dahlman O, Mörck R, et al. 1995. Halogenated structural elements in naturally occurring organic matter. See Ref. 69a, pp. 95–103

101. Joshi DK, Gold MH. 1993. Degradation of 2,4,5-trichlorophenol by the lignin-degrading basidiomycete *Phanerochaete chrysosporium. Appl. Environ. Microbiol.* 59:1779–85

102. JunShan Y, Yuwu C, Xiaozhang F, Dequan Y, Xiaotian L. 1984. Chemical constituents of *Armillariamellea* mycelium I. Isolation and characterization of armillarin and armillaridin. *Planta Med.* 50:288–90

103. Kalyanaraman B. 1995. Radical intermediates during degradation of lignin-model compounds and environmental pollutants–an electron spin resonance study. *Xenobiotica* 25:667–75

104. Kavanagh F, Hervey A, Robbins WJ. 1952. Antibiotic substances from basidiomycetes. IX. *Drosophila subarata* (Batsch:Fr) Quel. *Proc. Natl. Acad. Sci. USA* 38:555–60

105. Keller G. 1982. Pigmentations Untersuchungen bei europäischen Arten aus der Gattung *Dermocybe* (Fr.) Wünsche. *Sydowia* 35:110–26

106. Keller G, Ammirati JF. 1983. Chemotaxonomic significance of anthraquinone derivatives in North America species of *Dermocybe* section *sanguineae. Mycotaxon* 18:357–77

107. Kennedy DW, Aust SD, Bumpus JA. 1990. Comparative biodegradation of alkyl halide insecticides by the white-rot fungus, *Phanerochaete chrysosporium* (BKM-F-1767). *Appl. Environ. Microbiol.* 56:2347–53

108. Khindaria A, Grover TA, Aust SD. 1995. Reductive dehalogenation of aliphatic halocarbons by lignin peroxidase of *Phanerochaete chrysosporium*. *Environ. Sci. Technol.* 29:719–25

109. Kirk TK, Farrell RL. 1987. Enzymatic 'combustion': the microbial degradation of lignin. *Annu. Rev. Microbiol.* 41:465–505

110. Köhler A, Jäger A, Willershausen H, Graf H. 1988. Extracellular ligninase of *Phanerochaete chrysosporium* Burdsall has no role in the degradation of DDT. *Appl. Microbiol. Biotechnol.* 29:618–20

111. Kremer S, Anke H, Sterner O. 1992. *Metabolism of chlorinated phenols by members of the genus Mycena.* Preprints: Soil Decontam. Using Biol. Process., Karlsruhe, FRG, pp. 298–304. 6–9 Dec.

112. Kremer S, Sterner O. 1996. Metabolism of 3,4-dichloroaniline by the basidiomycete *Filoboletus* species TA9054. *J. Agric. Food Chem.* 44:1155–59

113. Kremer S, Sterner O, Anke H. 1992. Degradation of pentachlorophenol by *Mycena avenacea* TA-8480–identification of initial dechlorinated metabolites. *Z. Naturforsch. C* 47:561–66

114. Lamar RT, Davis MW, Dietrich DM, Glaser JA. 1994. Treatment of a pentachlorophenol- and creosote-contaminated soil using the lignin-degrading fungus *Phanerochaete sordida*: a field demonstration. *Soil Biol. Biochem.* 26:1603–11

115. Lamar RT, Dietrich DM. 1990. *In situ* depletion of pentachlorophenol from contaminated soil by *Phanerochaete* spp. *Appl. Environ. Microbiol.* 56:3093–100

116. Lamar RT, Dietrich DM. 1992. Use of lignin-degrading fungi in the disposal of pentachlorophenol-treated wood. *J. Ind. Microbiol.* 9:181–91

117. Lamar RT, Glaser JA, Kirk TK. 1990. Fate of pentachlorophenol (PCP) in sterile soils inoculated with the white-rot basidiomycete *Phanerochaete chrysosporium*: mineralization, volatilization and depletion of PCP. *Soil Biol. Biochem.* 22:433–40

118. Lauritsen FR, Kotiaho T, Lloyd D. 1993. Rapid and direct monitoring of volatile fermentation products in the fungus *Bjerkandera adusta* by membrane inlet tandem mass spectrometry. *Biol. Mass Spectrom.* 22:585–89

119. Lee D, Takahashi M, Tsunoda K. 1992. Fungal detoxification of organoiodine wood perservatives. *Holzforschung* 46:81–86

120. Lestan D, Lamar RT. 1996. Development of fungal inocula for bioaugmentation of contaminated soils. *Appl. Environ. Microbiol.* 62:2045–52

121. Lyr H. 1962. Detoxification of heartwood toxins and chlorophenols by higher fungi. *Nature* 195:289–90

122. Maloney SW, Manem J, Mallevialle J, Flessinger F. 1986. Transformation of trace organic compounds in drinking water by enzymatic oxidative coupling. *Environ. Sci. Technol.* 20:249–53

123. Martens R, Wetzstein H-G, Zadrazil F, Capelari M, Hoffmann P, et al. 1996. Degradation of the fluoroquinolone Enrofloxacin by wood-rotting fungi. *Appl. Environ. Microbiol.* 62:4206–9

124. Marzullo L, Cannio R, Giardina P, Santini MT, Sannia G. 1995. Veratryl alcohol oxidase from *Pleurotus ostreatus* participates in lignin biodegradation and prevents polymerization of laccase-oxidized substrates. *J. Biol. Chem.* 270:3823–27

125. Masaphy S, Levanon D, Henis Y. 1996. Degradation of atrazine by the lignocellulolytic fungus *Pleurotus pulmonarius* during solid-state fermentation. *Biores. Technol.* 56:207–14

126. McInnes AG, Walter JA, Wright JLC, Vining LC, Ranade N, et al. 1990. Biosynthesis of mollisin by *Mollisia caesia*. *Can. J. Chem.* 68:1–4

127. McNally KJ, Harper DB. 1991. Methylation of phenol by chloromethane in the fungus *Phellinus pomaceus*. *J. Gen. Microbiol.* 137:1029–32

128. Mester T, Swarts HJ, Romero i Sole S, de Bont JAM, Field JA. 1997. Stimulation of aryl metabolite production in the basidiomycete *Bjerkandera* sp. strain BOS55 with biosynthetic precursors and lignin degradation products. *Appl. Environ. Microbiol.* 63:1987–94

129. Michel FC Jr, Reddy CA, Forney LJ. 1995. Microbial degradation and humification of the lawn care pesticide 2,4-dichlorophenoxyacetic acid during the composting of yard trimmings. *Appl. Environ. Microbiol.* 61:2566–71

130. Mileski GJ, Bumpus JA, Jurek MA, Aust SD. 1988. Biodegradation of pentachlorophenol by the white-rot fungus *Phanerochaete chrysosporium*. *Appl. Environ. Microbiol.* 54:2885–89

131. Morgan P, Lee SA, Lewis ST, Sheppard AN, Watkinson RJ. 1993. Growth

and biodegradation by white-rot fungi inoculated into soil. *Soil Biol. Biochem.* 25:279–87

132. Moriguchi M, Hara Y, Hatanaka S-I. 1987. Antibacterial activity of L-2-amino-4-chloro-4-pentenoic acid isolated from *Amanita pseudoporphyria* Hongo. *J. Antibiot.* 15:904–6

133. Mougin C, Pericaud C, Malosse C, Laugero C, Asther M. 1996. Biotransformation of the insecticide lindane by the white rot basidiomycete *Phanerochaete chrysosporium. Pestic. Sci.* 47:51–59

134. Muller G, Nkusi G, Scholer HF. 1996. Natural organohalogens in sediments. *J. Prak. Chem.* 338:23–29

135. Nair MSR, Hervey A. 1979. Structure of lepiochlorin, an antibiotic metabolite of a fungus cultivated by ants. *Phytochemistry* 18:326–27

136. Nauta MM, Vellinga EC. 1995. *Atlas van Nederlandse paddestoelen.* Rotterdam: Balkema Uitgevers (In Dutch)

137. Neidleman SL, Geigert J. 1986. *Biohalogenation: Principles, Basic Roles and Applications.* Chichester, UK: Wiley. 203 pp.

138. Neilson AH. 1996. An environmental perspective on the biodegradation of organochlorine xenobiotics. *Int. Biodet. Biodegr.* 37:3–21

139. Öberg G, Brunberg H, Hjelm O. 1996. Production of organically bound chlorine during degradation of Birch wood by common white-rot fungi. *Soil Biol. Biochem.* In press

140. Ohta T, Matsuda M, Takahashi T, Nakajima S, Nozoe S. 1995. (S)-cis-2-amino-5-chloro-4-pentenoic acid from the fungus *Amanita vergineoides. Chem. Pharm. Bull.* 43:899–900

141. Ohta T, Nakajima S, Hatanaka SI, Yamamoto M, Shimmen Y, et al. 1987. A chlorohydrin amino acid from *Amanita abrupta. Phytochemistry* 26:565–66

142. Ohta T, Takahashi A, Matsuda M, Kamo S, Agatsuma T, et al. 1995. Russuphelol, a novel optically active chlorohydroquinone tetramer from the mushroom *Russula subnigricans. Tetrahedron Lett.* 36:5223–26

143. Okamoto K, Shimada A, Shirai R, Sakamoto H, Yoshida S, et al. 1993. Antimicrobial chlorinated orcinol derivatives from mycelia of *Hericium erinaceum. Phytochemistry* 34:1445–46

144. Okeke BC, Smith JE, Paterson A, Watsoncraik IA. 1996. Influence of environmental parameters on pentachlorophenol biotransformation in soil by *Lentinula edo-*

des and *Phanerochaete chrysosporium. Appl. Microbiol. Biotechnol.* 45:263–66

145. Pallerla S, Chambers RP. 1996. New urethane prepolymer immobilized fungal bioreactor for decolorization and dechlorination of kraft bleach effluents. *TAPPI J.* 79(5):155–61

146. Paszczynski A, Crawford RL. 1995. Potential for bioremediation of xenobiotic compounds by the white-rot fungus *Phanerochaete chrysosporium. Biotechnol. Progr.* 11:368–79

147. Peipp H, Sonnenbichler J. 1992. Secondary fungal metabolites and their biological activities. II. Occurrence of antibiotic compounds in cultures of *Armillaria ostoye* growing in the presence of an antagonistic fungus or host plant cells. *Biol. Chem. Hoppe Seyler* 373:675–83

148. Pelaez F, Martinez MJ, Martinez AT. 1995. Screening of 68 species of Basidiomycetes for enzymes involved in lignin degradation. *Mycol. Res.* 99:37–42

149. Pfefferle W, Anke H, Bross M, Steglich W. 1990. Inhibition of solubilized chitin synthase by chlorinated aromatic compounds isolated from mushroom cultures. *Agric. Biol. Chem.* 54:1381–84

150. Pieper DH, Timmis KN, Ramos JL. 1996. Designing bacteria for the degradation of nitro- and chloroaromatic pollutants [Review]. *Naturwissenschaften* 83:201–13

151. Pieper DH, Winkler R, Sandermann H. 1992. Formation of a toxic dimerization product of 3,4-dichloroaniline by lignin peroxidase from *Phanerochaete chrysosporium. Angew. Chem.* 31:68–70

152. Pryke DC, McKenzie DJ. 1996. Substitution of chlorine dioxide for chlorine in Canadian bleached chemical pulp mills. *Pulp Pap. Can.* 97(1):27–29

153. Pyysalo H. 1976. Identification of volatile compounds in seven edible fresh mushrooms. *Acta Chem. Scand.* B 30:235–44

154. Qian FG, Xu GY, Du SJ, Li MH. 1990. Isolation and identification of two new pyrone compounds from the culture of *Hericium erinaceus. Acta Pharm. Sin.* 25:522–25

155. Rappe C. 1996. Sources and environmental concentrations of dioxins and related compounds. *Pure Appl. Chem.* 68:1781–89

156. Renganathan V, Miki K, Gold MH. 1987. Haloperoxidase reactions catalyzed by lignin peroxidase, an extracellular enzyme from the basidiomycete *Phanerochaete chrysosporium. Biochemistry* 26:5127–32

157. Ricotta A, Unz RF, Bollag JM. 1996. Role of a laccase in the degradation of pen-

tachlorophenol. *Bull. Environ. Contam. Toxicol.* 57:560–67

158. Roper JC, Sarkar JM, Dec J, Bollag JM. 1995. Enhanced enzymatic removal of chlorophenols in the presence of co-substrates. *Water Res.* 29:2720–24

159. Rotard W, Christmann W, Knoth W. 1994. Background levels of PCDD/F in soils of Germany. *Chemosphere* 29:2193–200

160. Roy-Arcand L, Archibald FS. 1991. Direct dechlorination of chlorophenolic compounds by laccases from *Trametes (Coriolus) versicolor. Enzyme Microb. Technol.* 13:194–203

161. Schwarz M, Marr J, Kremer S, Sterner O, Anke H. 1992. *Biodegradation of xenobiotic compounds by fungi: metabolism of 3,4-dichloroaniline by* Schizophyllum *species and* Auriculariopsis ampla *and induction of indigo production.* Preprints: Soil Decontam. Using Biol. Proc., 6–9 Dec. Karlsruhe, FRG, pp. 459–64

162. Shah MM, Grover TA, Aust SD. 1993. Reduction of CCl4 to the trichloromethyl radical by lignin peroxidase H2 from *Phanerochaete chrysosporium. Biochem. Biophys. Res. Commun.* 191:887–92

163. Singh P, Rangaswami S. 1966. Occurrence of O-methyl-drosophilin A in *Fomes fastuosus* Lev. *Tetrahedron Lett.* 11:1229–31

164. Smith ML, Bruhn JN, Anderson JB. 1992. The fungus *Armillaria bulbosa* is among the largest and oldest living organisms. *Nature* 356:428–31

165. Solomon KR. 1996. Chlorine in the bleaching of pulp and paper. *Pure Appl. Chem.* 68:1721–30

166. Sonnenbichler J, Bliestle IM, Peipp H, Holdenrieder O. 1989. Secondary fungal metabolites and their biological activities, I. Isolation of antibiotic compounds from cultures of *Heterobasidion annosum* synthesized in the presence of antagonistic fungi or host plant cells. *Biol. Chem. Hoppe Seyler* 370:1295–303

167. Sonnenbichler J, Dietrich J, Peipp H. 1994. Secondary fungal metabolites and their biological activities. 5. Investigations concerning the induction of the biosynthesis of toxic secondary metabolites in Basidiomycetes. *Biol. Chem. Hoppe Seyler* 375:71–79

168. Spinnler HE, de Jong E, Mauvais G, Semon E, Lequere JL. 1994. Production of halogenated compounds by *Bjerkandera adusta. Appl. Microbiol. Biotechnol.* 42:212–21

169. Steglich W, Lösel W, Austel V. 1969. Anthrachinon-pigmente aus *Dermocybe sanguinea* (Wulf. ex Fr.) Wünsche and

D. semisanguinea (Fr.). *Chem. Ber.* 102:4104–18

170. Steiman R, Benoit-Guyod JL, Seigle-Murandi F, Sage L, Toe A. 1994. Biodegradation of pentachlorophenol by micromycetes. 2. Ascomycetes, basidiomycetes, and yeasts. *Environ. Toxicol. Water Qual.* 9:1–6

171. Suzuki T. 1983. Methylation and hydroxylation of pentachlorophenol by *Mycobacterium* sp. isolated from soil. *Pestic. Sci.* 8:419–28

172. Swarts HJ, Teunissen PJM, Verhagen FJM, Field JA, Wijnberg JBPA. 1997. Chlorinated anisyl metabolites produced by basidiomycetes. *Mycol. Res.* 101:372–74

173. Swarts HJ, Verhagen FJM, Field JA, Wijnberg JBPA. 1996. Novel chlorometabolites produced by *Bjerkandera* species. *Phytochemistry* 42:1699–701

174. Swift MJ. 1982. Basidiomycetes as components of forest ecosystems. In *Decomposer Basidiomycetes: Their Biology and Ecology,* ed. JC Frankland, JN Hedger, MJ Swift, pp. 307–37. Cambridge: Cambridge Univ. Press

175. Takahashi A, Agatsuma T, Matsuda M, Ohta T, Nunozawa T, et al. 1992. Russuphelin A, a new cytotoxic substance from the mushroom *Russala subnigricans* Hongo. *Chem. Pharm. Bull.* 40:3185–88

176. Takahashi A, Agatsuma T, Ohta T, Nunozawa T, Endo T, et al. 1993. Russuphelin-B, Russuphelin-C, Russuphelin-D, Russuphelin-E and Russuphelin-F, new cytotoxic substances from the mushroom *Russula subnigricans* Hongo. *Chem. Pharm. Bull.* 41:1726–29

177. Tatsumi K, Freyer A, Minard RD, Bollag J-M. 1994. Enzyme-mediated coupling of 3,4-dichloroaniline and ferulic acid: a model for pollutant binding to humic materials. *Environ. Sci. Technol.* 28:210–15

178. Teunissen PJM, Swarts HJ, Field JA. 1997. Screening of ligninolytic basidiomycetes for the production of Drosophilin A (tetrachloro-4-methoxyphenol) and Drosophilin A methyl ether (tetrachloro-1,4-dimethoxybenzene). *Appl. Microbiol. Biotechnol.* In press

179. Thaller V, Turner JL. 1972. Natural acetylenes Part XXXV. Polyacetylenic acid and benzenoid metabolites from cultures of the fungus *Lepista diemii* Singer. *J. Chem. Soc. Perkin Trans.* 1:2032–34

180. Thomas AF. 1973. An analysis of the flavor of the dried mushroom, *Boletus edulis. J. Agric. Food Chem.* 21:955–58

181. Thomas DR, Carswell KS, Georgiou G. 1992. Mineralization of biphenyl and

PCBs by the white rot fungus *Phanerochaete chrysosporium. Biotechnol. Bioeng.* 40:1395–1402

182. Turner EM, Wright M, Ward T, Osborne DJ, Self RJ. 1975. Production of ethylene and other volatiles and changes in cellulase and laccase activities during the life cycle of the cultivated mushroom *Agaricus bisporus. J. Gen. Microbiol.* 91:167–76

183. Turner WB, Aldridge DC. 1983. *Fungal Metabolites II.* London: Academic. 631 pp.

184. Valli K, Gold MH. 1991. Degradation of 2,4-dichlorophenol by the lignin- degrading fungus *Phanerochaete chrysosporium. J. Bacteriol.* 173:345–52

185. Valli K, Wariishi H, Gold MH. 1992. Degradation of 2,7-dichlorodibenzo-p-dioxin by the lignin-degrading basidiomycete *Phanerochaete chrysosporium. J. Bacteriol.* 174:2131–37

186. van Eijk GW. 1975. Drosophilin A methyl ether from *Mycena megaspora. Phytochemistry* 14:2506

187. Verhagen FJM, Swarts HJ, Kuyper TW, Wijnberg JBPA, Field JA. 1996. The ubiquity of natural adsorbable organic halogen production among basidiomycetes. *Appl. Microbiol. Biotechnol.* 45:710–18

188. Vyas BRM, Sasek V, Matucha M, Bubner M. 1994. Degradation of 3,3′,4,4′-tetrachlorobiphenyl by selected white rot fungi. *Chemosphere* 28:1127–34

189. Wang SH, Ferguson JF, McCarthy JL. 1992. The decolorization and dechlorination of Kraft bleach plant effluent solutes by use of 3 fungi–*Ganoderma lucidum, Coriolus versicolor* and *Hericium erinaceus. Holzforschung* 46:219–23

190. Whitney RD. 1995. Root-rotting fungi in white spruce, black spruce, and balsam fir in Northern Ontario. *Can. J. For. Res.* 25:1209–30

191. Wunder A, Anke T, Klostermeyer D, Steglich W. 1996. Lactarane type sesquiterpenoids as inhibitors of leukotriene biosynthesis and other, new metabolites from submerged cultures of *Lentinellus cochleatus* (Pers. ex Fr.) Karst. *Z. Naturforsch. C* 51:493–99

192. Wuosmaa AM, Hager LP. 1990. Methyl chloride transferase: a carbocation route for biosynthesis of halometabolites. *Science* 249:160–62

193. Yadav JS, Quensen JF, Tiedje JM, Reddy CA. 1995. Degradation of polychlorinated biphenyl mixtures (Aroclors 1242, 1254, and 1260) by the white rot fungus *Phanerochaete chrysosporium* as evidenced by congener-specific analysis. *Appl. Environ. Microbiol.* 61:2560–65

194. Yadav JS, Reddy CA. 1993. Mineralization of 2,4-dichlorophenoxyacetic acid (2,4-D) and mixtures of 2,4-D and 2,4,5-trichlorophenoxyacetic acid by *Phanerochaete chrysosporium. Appl. Environ. Microbiol.* 59:2904–8

195. Yadav JS, Wallace RE, Reddy CA. 1995. Mineralization of mono- and dichlorobenzenes and simultaneous degradation of chloro- and methyl-substituted benzenes by the white rot fungus *Phanerochaete chrysosporium. Appl. Environ. Microbiol.* 61:677–80

196. Yamamura Y, Fukuhara M, Takabatake E, Ito N, Hashimoto T. 1986. Hepatoxic action of a poisonous mushroom, *Amanita abrupta,* in mice and its toxic component. *Toxicology* 38:161–73

197. Zeddel A, Majcherczyk A, Hüttermann A. 1993. Degradation of polychlorinated biphenyls by white-rot fungi *Pleurotus ostreatus* and *Trametes versicolor* in a solid state system. *Toxicol. Environ. Chem.* 40:255–66

Annu. Rev. Microbiol. 1997. 51:415–62

SAFE HAVEN: The Cell Biology of Nonfusogenic Pathogen Vacuoles

Anthony P. Sinai and Keith A. Joiner

Infectious Diseases Section, Department of Internal Medicine, Yale University School of Medicine, 333 Cedar Street, New Haven, Connecticut 06520;
e-mail: anthony.sinai@quickmail.yale.edu

KEY WORDS: *Chlamydia, Toxoplasma, Legionella, Mycobacterium*

ABSTRACT

Our understanding of both membrane traffic in mammalian cells and the cell biology of infection with intracellular pathogens has increased dramatically in recent years. In this review, we discuss the cell biology of the host-microbe interaction for four intracellular pathogens: *Chlamydia* spp., *Legionella pneumophila, Mycobacterium* spp., and the protozoan parasite *Toxoplasma gondii.* All of these organisms reside in vacuoles inside cells that have restricted fusion with host organelles of the endocytic cascade. Despite this restricted fusion, the vacuoles surrounding each pathogen display novel interactions with other host cell organelles. In addition to the effect of infection on host membrane traffic, we focus on these novel interactions and relate them where possible to nutrient acquisition by the intracellular organisms.

CONTENTS

415

0066-4227/97/1001-0415$08.00

INTRODUCTION

All intracellular bacterial, fungal, and protozoan pathogens enter their host cells surrounded by a membrane-bound vacuole (74, 87, 175). Understanding the biogenesis of these vacuoles, and their subsequent interactions with the host cell, are important and increasingly popular areas for investigation, since they are central to successful intracellular parasitism.

The primary nonoxidative host defense of the cells for internalized pathogens is the fusion of lysosomes with the pathogen vacuole, as well as vacuole acidification (137). The paradigm for the last decade has been that pathogens avoid destruction by lysosomes and by acidification in one of three ways: (a) Pathogens such as shigellae (99, 287), Listeria monocytogenes (287), Trypanosoma cruzi (12), and rickettseae (302) rapidly degrade the vacuolar membrane within which they are internalized, allowing free replication in the cytosol; (b) other organisms, including Leishmania spp. and Coxiella burnetti (208), replicate within phagolysosomes; and (c) yet another subset of pathogens, represented by Chlamydia, Legionella, Mycobacteria, and Toxoplasma, block or modulate fusion of the vacuole with lysosomes and also block vacuolar acidification.

It is increasingly clear that these paradigms are overly simplistic. They neither reflect the sophistication of pathogens in circumventing host cell defenses nor take into account much of the new information on membrane traffic in higher eukaryotic cells. For example, although T. cruzi secretes a membranolytic molecule that degrades the vacuolar membrane (13), the initial internalization event is mediated by fusion of lysosomes with the host cell plasma membrane at the site of parasite attachment (284). Hence, the subsequent destruction of the vacuolar membrane by T. cruzi cannot be to avoid lysosomal fusion. As another example, Salmonella typhimurium in epithelial cells resides in a vacuole, which shows limited accessibility to fluid phase endocytic tracers and mature lysosomes, yet rapidly acquires selected lysosomal markers (87), possibly as a consequence of direct delivery from the trans-Golgi network. In contrast, Salmonella within macrophages reside in a compartment indistinguishable from phagolysosomes (186). Survival of the organism in macrophages is dependent on acidification of the vacuole, in a process that may (9) or may not (212) be substantially modulated by the organism. Thus, interactions of the *Salmonella* vacuole with lysosomes defy simple categorization.

This theme is extended by phagosomes containing mycobacteria, long believed to be nonfusogenic (16). Recent evidence indicates that mycobacterial phagosomes interact with and acquire markers of the endosomal and lysosomal pathways (reviewed in 50, 227). Survival and replication of mycobacteria in these vacuoles are apparently achieved by inhibiting phagosomal maturation (52). As a consequence, mycobacteria develop in a compartment that has the characteristics of an endosome rather than a phagolysosome (50, 227).

At the other end of this continuum are the pathogen vacuoles that are the focus of this review. The vacuoles of *Chlamydia* spp., *Legionella pneumophila,* and *Toxoplasma gondii* neither fuse with lysosomes nor acidify. These organisms use different strategies to become essentially invisible to the endocytic pathway. While not truly within a nonfusogenic vacuole, membrane trafficking events surrounding the mycobacterial phagosome are also discussed, since they bear a direct relationship to the situation with nonfusogenic vacuoles.

We briefly review the data on mechanisms of attachment and internalization, then focus on recent literature on membrane traffic between pathogen vacuole and host cell (see Table 1). The interaction of pathogen vacuoles with host cell organelles and nutrient acquisition by intracellular organisms are also covered. Where possible, we attempt to integrate modern concepts in cell biology with the biology of the pathogen vacuoles.

MEMBRANE TRAFFIC—AN OVERVIEW

Membrane traffic within the endocytic cascade is a complex process that is beyond the scope of this article; it forms the basis for several recent reviews (95, 220, 248). What follows is a brief account of recent experiments on phagosome biogenesis and maturation.

Phagosome Interactions with the Endocytic Cascade

From a growing body of work on phagosome biogenesis, the most useful information for subsequent comparison with pathogen vacuoles is on phagosomes containing inert beads or fixed particles (52, 62, 188, 209, 272). Such phagosomes are highly dynamic. Interestingly, however, even the nature of the inert bead itself markedly alters interactions with the endocytic cascade (188). There are more than 200 proteins in mature latex bead phagosomes in U937 cells, including many proteins of cytoskeletal and cytosolic lineage (e.g. annexins, actin-binding proteins) (62).

Table 1 Summary of cell biological events associated with pathogen vacuoles

Organism	Route of entry	Early vacuole	Mature vacuole	NBD-ceramide from Golgi	Vacuolar pH (mature vacuole)	Establishment of organelle association
Chlamydia spp.	Receptor-mediated endocytosis; phagocytosis; pinocytosis (specific kinetics depend on entry mechanism)	Unknown host plasma membrane proteins (0–8 h) (EB-RB conversion)	No known host proteins; parasite proteins (IncA and others); 8-h host lysis (RB growth-EB release)	Yes; 15-min EB release	>6	Mitochondria (*C. psittaci*) (10 h)
Toxoplasma gondii	Active penetration (15–30 s)	No known host proteins; parasite proteins; ROP and GRA proteins (10–60 min)	No known host proteins; parasite proteins; ROP and GRA proteins (hours-days)	Maybe	6.8–7.0	Mitochondria and ER (<15 min)
Legionella pneumophila	Coiling phagocytosis (<3 min); phagocytosis	rab7 (50% of vacuoles); CR3, TfR, MHC class I and II, 5'-nucleotidase (3–60 min)	Parasite proteins? (24 h)	Not known	5.7–6.5	Mitochondria (transient) (1 h) ER (stable) (8 h)
Mycobacterium spp.	Phagocytosis (2–5 min)	MHC class I and II, TfR, HLA-DR, β2-microglobulin; plasma membrane glycoproteins (GM1-gangliosides, galactose glycoconjugates) (3–24 h)	TfR, GM1-gangliosides, galactose glycoconjugates; LAMP, CD63 (less than conventional phagosome); unprocessed; cathepsin D (5–14 days)	Not known	6.3–6.5 (*M. avium*)	None

After internalization, remodeling of the phagosome membrane occurs rapidly (177, 188, 201). In phagocytic cells, plasma membrane proteins (including receptors) are largely removed within the first 3–5 min (188, 202). Fusion with early endosomes also occurs within minutes, resulting in delivery of the monomeric GTPase rab5 to the phagosome. This reflects the capacity of the phagosome to fuse actively with early endosomes, as documented in vivo (272) and in cell free assays (10, 159), in a fashion dependent on the monomeric GTPase rab5 and other components of the 20s vesicle docking complex first described by Sollner et al (266). Even proteins localized predominantly to late endosomes (mannose-6-phosphate receptor, rab7) can be added to steady state within as short a period as 5 min (188, 202, 272) and can be largely removed following an additional 5–10 min of incubation. Fluid phase markers preloaded into lysosomes can reach steady state within the first 10 min. Stepwise acidification of phagosomes presumably results from the delivery of the vacuolar proton ATPase, which can also be added rapidly after phagosome formation (272). These studies all serve to highlight the highly dynamic nature of the phagosome and emphasize that experiments that do not look at early time points may substantially oversimplify the membrane traffic events involved in phagosome maturation.

Recognition and fusion of phagosomes with vesicular organelles of the endocytic cascade are likely to depend either on homologues of the 20s vesicle docking complex (27, 225), on annexins (62, 214), and/or upon the host cell cytoskeleton (G Griffiths, unpublished information) and the myristoylated, alanine-rich C-kinase substrate (8a). Selected members of these protein families are identified on phagosome membranes (96), and their involvement in in vitro phagosome fusion events has been suggested (11). The possibility that rapid protein delivery to phagosomes may reflect intersection of phagosomes with the secretory pathway is also suggested by recent data, mentioned briefly for *Salmonella* above and in more detail for *Chlamydia* below. For example, since lysosome-associated membrane proteins (LAMPs) are trafficked directly from the *trans*-Golgi network to endosomes (125), they could in principle intersect directly with phagosomes.

Macrophages contain an extensive tubuloreticular compartment (136) bearing a high concentration of LAMP (209, 211). Depending on the cell type, these tubular lysosomes may or may not contain late endosomal markers such as rab7 (45), the mannose-6-phosphate receptor, and macrosialin (210). This tubular compartment may be morphologically distinct from the high-density, electron-dense, mannose-6-phosphate receptor–negative terminal lysosomes. Phagosomes containing inert particles in mouse peritoneal macrophages may fuse predominantly with the tubular compartment rather than with terminal lysosomes (63, 270). Furthermore, the delivery of some markers to the phagosome is highly dynamic and may reflect rapid and transient associations with tubular lysosomes (kiss and run) rather than a full-fledged fusion event (63, 270). Therefore, in macrophages the tubular compartment may be the functional lysosome compartment where the bulk of internalized material is degraded but recycling proteins, as well as LAMPs, are spared. Stated another way, older morphologic studies analyzing delivery of luminal components from dense lysosomes to phagosomes may not reflect normal phagosome-lysosome traffic as determined with more modern reagents and approaches.

ATTACHMENT AND INTERNALIZATION

For *Legionella* and *Toxoplasma*, cell entry is a morphologically distinctive process that provides clues to subsequent intracellular events within the vacuole. For *Chlamydia* and mycobacteria, cell entry apparently occurs via more conventional routes. Features of these processes are briefly reviewed (see Table 1).

Chlamydia spp.

The genus *Chlamydia* is made up of four species: *C. trachomatis, C. psittaci, C. pneumoniae,* and the recently recognized *C. percorum* (160). Much of our understanding of the biology of *Chlamydia* comes from the study of several strains of *C. trachomatis* and *C. psittaci.* All chlamydiae display a unique developmental cycle, alternating between the relatively metabolically inert and infectious extracellular form; the elementary body (EB); and the intracellular, metabolically active, noninfectious form, the reticulate body (RB) (297).

Internalization of EBs is preceded by specific attachment to unknown plasma membrane receptors (176, 268). Binding of chlamydiae to cultured cells is inhibitable by trypsin (274, 275, 289), heat (37, 110, 289), and heparin (140, 289, 309). Multiple ligands are implicated (176), including a sulfated oligosaccharide-modified adhesin recognized by heparin sulfate receptors (309), the major outer membrane protein (MOMP) (39, 274, 275), and the EB-specific (109) cysteine-rich outer membrane protein OMP2 (289). Recent evidence suggests that the glycosylation of MOMP (277) may account for its function as an adhesin (141).

Attachment of EBs is followed by internalization into the nascent parasitophorous vacuole, traditionally termed an inclusion. The determinants triggering internalization are intrinsic to the EB outer membrane, as de novo protein synthesis by the pathogen is not required (84, 247, 294), and EB outer membrane ghosts are internalized by the same mechanism as efficiently as intact organisms are (72). All of the endocytic mechanisms of the host cell can be subverted by *Chlamydia* (176). Depending on both the *Chlamydia* strain and host cell and experimental conditions (176, 305), phagocytosis (309, 298), receptor-mediated endocytosis (115, 116, 139, 264, 305), and pinocytosis (217) are reported to mediate internalization.

Since chlamydial internalization per se is neither morphologically distinctive nor known to involve any unique cellular mechanisms, it is not possible at this point to suggest direct links between the entry pathway and subsequent membrane traffic events within the chlamydial inclusion (discussed below).

Toxoplasma gondii

Toxoplasma gondii is a highly cosmopolitan protozoan parasite capable of invading and replicating within nucleated cells of warm-blooded animals (67, 126). It is a member of the family Apicomplexa, a diverse group of protozoans characterized by the presence of a common set of apical structures involved in cell invasion (66).

In the mammalian host, infection usually occurs by the ingestion of either *T. gondii* tissue cysts containing the slow-growing bradyzoite form of the parasite (66, 68) or oocysts, generated as a consequence of sexual reproduction in the ileum of domestic and feral felines (83, 67, 250). Upon entering a new host, differentiation to the rapidly growing tachyzoite form of the parasite occurs (66, 67). Essentially all our knowledge about the biology of the parasitophorous vacuole (PV) is derived from the study of the actively replicating stage of the parasite, the tachyzoite.

The broad host range of *Toxoplasma* suggests that the host cell receptor(s) is fairly common across different cell types or it suggests the presence of multiple potential receptors (260). While no receptor has been identified, the parasite binds laminin (85) and attaches to host cells via β-1 integrin receptors (85, 86). The parasite surface proteins SAG1 (93, 132, 167), SAG2 (94, 263), and SAG3 (293) have been implicated as possible mediators of attachment. Nearly a 90% decrease in attachment is observed in a SAG3 knockout mutant in the additional presence of anti-SAG1 antibodies (293).

Despite the absence of a flagellum, the tachyzoite exhibits a saltatory gliding motility on solid substrates (135, 229, 245). Motility, powered by the actin cytoskeleton (64, 229, 230) of the parasite, plays an essential role in invasion (64, 230, 244). Parasites exhibit gliding motility on the surface of the cell, continually probing it by the extension of a specialized organelle at the tip of the apical end termed the conoid (3, 46, 172).

In response to an as yet unknown signal during the probing process, the parasite reorients itself with its apical end attached to the host cell surface (3, 171, 183, 245), and invasion proceeds. Invasion is a wholly parasite-directed event powered by parasite ATP (133) and the actin cytoskeleton of the parasite (64, 251). The exclusive role of a parasite actin–based motor was elegantly demonstrated using host cells resistant to cytochalasin D, a microfilament depolymerizing agent (64).

Invasion is preceded and accompanied by secretions from two secretory organelles of the parasite, micronemes (1) and rhoptries (182). Among the parasite proteins implicated in invasion are a phospholipase A2 (PLA2) activity (232, 234), the penetration-enhancing factor (149, 150, 185), and the microneme protein MIC2 (55a). At least one component of the penetration-enhancing factor is ROP1, a protein discharged from the rhoptries at the time of invasion (134, 233, 242). A genetically engineered knockout mutant of ROP1 showed no defect in invasion (265).

Invasion is extremely rapid and is completed within 15–30 s (171, 245). The parasite, attached by its conoid, forms an invagination in the plasma membrane, which enlarges as the parasite squeezes through a distinct constriction (3, 46, 166, 183). The site of entry is morphologically identical to the moving junction described for the invasion of erythrocytes by plasmodia (4). The interaction between the invading parasite and the moving junction is extremely tight, resulting in significant deformation of the invading parasite (3, 166, 183) and the sloughing off of antibody bound to parasite surface proteins (70, 263). The nascent vacuole is essentially devoid of host plasma membrane proteins, which are either excluded or rapidly removed (41, 205, 252).

The nascent parasitophorous vacuole membrane (PVM) is rapidly modified by secretions from parasite rhoptries and dense granules, resulting in the replication-competent vacuole (see below). In addition to proteins, rhoptries appear to release membrane whorls into the newly forming vacuole (80, 182). Whether or not parasite lipids contribute to the PVM has been an area of controversy (reviewed in 127). A recent report, based on electrophysiological studies, suggests that the membrane for the PVM is derived completely from the host plasma membrane (276).

Regardless of the precise mechanism involved in invasion and PVM biogenesis, many unique features of this process are already apparent, and they contribute to the equally unique membrane traffic events that occur subsequent to invasion, as discussed below.

Legionella pneumophila

Legionella pneumophila, the etiological agent of Legionnaires disease, is a facultative intracellular pathogen (164). It attaches to and is internalized largely by professional phagocytic cells, including macrophages, monocytes, and neutrophils (65). The organism replicates in alveolar macrophages from many species (123). The organism is also internalized and replicates within various macrophage cell lines, including U937 cells (28) and differentiated HL60 cells (152), as well as in free-living amoebae (reviewed in 65, 79), its natural host.

L. pneumophila attaches to host phagocytes through complement receptors (CR) CR3 and CR1 (192). Attachment is mediated in part through the deposition of the complement components C3b and iC3b on the bacterial surface, which in turn is bound mainly to the major outer membrane protein (MOMP) on the bacterial surface—a 28-kDa porin (26). Attachment and internalization is markedly augmented by opsonization of the organism by either immune or nonimmune serum, reflecting the role of bound C3 on the bacterial surface (26, 192).

Internalization of Legionella pneumophila occurs by a process termed coiling phagocytosis (120, 215). In this process, which is observed with monocytes, macrophages, and neutrophils, bacteria are seen enclosed in a fingerlike projection in one plane and in the center of a coil in the other (120, 215).

Coiling phagocytosis has been variably reported to occur only for the virulent Philadelphia 1 strain of L. pneumophila and for avirulent or environmental isolates (215). Coiling phagocytosis occurs with heat-killed and glutaraldehyde-killed L. pneumophila, but not with immunoglobulin (Ig) G-coated organisms (120) (179). Importantly, IgG-coated organisms multiply in infected cells at the same rate as noncoated organisms, and only 10% of the inoculum is killed by antibody and complement (124, 179).

Coiling phagocytosis is described with a variety of other organisms, including Borrelia burgdorferi, Leishmania donovani, and even inert particles such as quartz crystals (219). With these organisms, the subsequent series of events—described below—resulting in formation of the Legionella replicative phagosome does not occur (219).

Hence, the overall conclusions that can be drawn are that neither coiling phagocytosis nor entry through complement receptors (65, 117) is required for intracellular survival. Whether coiling phagocytosis is necessary for inhibition of phagolysosomal fusion, or whether this process could augment microbial virulence, is not yet known (65, 219).

Mycobacterium spp.

Multiple mechanisms mediate attachment and uptake of Mycobacterium avium and Mycobacterium tuberculosis by professional phagocytes (237). As with most pathogens that reside predominantly in macrophages, complement receptors, particularly CR1 and CR3, play an important role in attachment and uptake (114, 235, 238). Phagocytosis of mycobacteria is enhanced in the presence of nonimmune serum, a source of the complement protein C3 (238). Nonetheless, binding of M. tuberculosis occurs to CR3 in the absence of serum, as is

convincingly demonstrated using Chinese hamster ovary (CHO) cells expressing the CR3 receptor (56).

Mycobacterial attachment and internalization are also mediated by the mannose receptor (MR) (235). Binding by the MR appears to be specific for virulent
strains, potentially because of the differences in the cell surface lipoarabinomannan of fast-growing mycobacteria (207, 236). Simultaneously blocking
the CR and MR pathways suggests hat binding by these pathways may not be
completely independent (237). Internalization is also augmented by the serum
protein, mannose-binding protein, which recognizes lipoarabinomannan and
lipomannan on the bacterial surface (204).

Once attached to the macrophage, internalization occurs by phagocytosis
(238). Barring a single report, where an as yet uncharacterized stretch of
M. tuberculosis DNA enhanced the internalization of noninvasive *Escherichia
coli* into fibroblasts (17), there is no evidence for a *M. tuberculosis* invasion factor. Thus, as is the case for *Chlamydia,* the entry process per se for mycobacteria
cannot be directly linked to subsequent events in phagosome maturation.

ESTABLISHMENT OF THE
REPLICATION-COMPETENT VACUOLE

Chlamydia spp.

Following internalization, chlamydiae reside within a compartment (the inclusion), which neither fuses with lysosomes nor acidifies (see below). That
chlamydial inclusions are distinct from phagolysosomes was first observed in
the elegant experiments of Friis (84). Subcellular fractionation of *Chlamydia
psittaci*–infected L cells demonstrated inclusions banding at a distinct density
from lysosomes (84). When heat-treated EBs were used, the chlamydial signal overlapped with lysosomal markers (84). This result was confirmed by
Zeichner (308) in murine macrophages—in these cells, live EBs formed true
inclusions that did not fuse with lysosomes, while heat-treated EBs were found
in phagolysosomes (308). Examination of the total (308) and surface (307) protein profiles of isolated inclusions (live EB) and phagosomes (heat-killed EB)
cells indicated that although certain proteins were shared, others were unique
to each organelle, supporting a difference in their biogenesis and/or maturation.

Morphological and histochemical studies support the view that the chlamydial inclusion is distinct from phagolysosomes. Chlamydial inclusions regardless of the species, or host cell, do not stain for soluble lysosomal markers (endogenous marker: lysosomal acid phosphatase; exogenous markers: thorotrast,
ferritin) (142, 206, 285, 304). Heat treatment of EBs significantly increased the
proportion of *Chlamydia*-containing vacuoles that fused with thorotrast-loaded

lysosomes (206), supporting the biochemical results of Friis (84) and Zeichner (307, 308). In contrast, neither UV (72) nor antibiotic-treated EB (see below) (84, 294) containing vacuoles fused with lysosomes. In addition, EB outer membrane ghosts remain in a non-lysosome–fused compartment, a property lost following heat treatment (72).

CHLAMYDIAL INCLUSIONS LACK FLUID PHASE AND FIXED MARKERS OF THE ENDOCYTIC PATHWAY   More recent studies have extended these observations, showing the complete segregation of the chlamydial vacuole from the endocytic pathway at all levels and time points. *C. trachomatis* inclusions do not acquire the exogenous fluid-phase endocytic markers lucifer yellow (112, 283), fluorescein dextran (112), or horseradish peroxidase (283) over loading periods of 1–12 h. Similarly, the chlamydial inclusion is not on the recycling pathway for the transferrin receptor, as determined by uptake of Tf-HRP, or immunofluorescent or electron microscopic staining for Tf or the TfR (246, 283). The endogenous soluble lysosomal markers acid phosphatase and cathepsin D and the fixed lysosomal membrane glycoproteins LAMP1 and LAMP2 were not detected in 26- to 40-h vacuoles by immunofluorescence (112, 283). Taken together, these results indicate that the chlamydial inclusion is not on the endocytic pathway. They suggest in general terms either that components necessary for fusion with endosomes are excluded from the inclusion membrane or that secreted chlamydial components block the fusion process. Nonetheless, since vacuoles containing EBs undergo extensive homotypic fusion (158, 297), potentially in an annexin-dependent fashion (153), and since chlamydial inclusions fuse with secretory vesicles from the *trans*-Golgi (see below), the chlamydial vacuole must be categorized as selectively fusogenic rather than nonfusogenic.

ACIDIFICATION   The *C. trachomatis* inclusion has a pH > 6, as determined by measuring emission ratios of the pH-sensitive probe SNAFL conjugated to chlamydiae (239). The pH of vacuoles containing heat-killed organisms was 5.3. This limited acidification of the inclusion is presumably linked to the lack of the vacuolar proton ATPase within the inclusion membrane (112). Limited acidification may also reflect the presence of the Na^+K^+ ATPase within the inclusion (239), although this latter point has not been demonstrated by microscopy.

INHIBITION OF LYSOSOMAL FUSION MAY RESIDE IN DETERMINANTS ON THE EB SURFACE   The factor(s) responsible for the inhibition of lysosomal fusion are linked to determinants on the EB surface (176). This view is based on the observation that internalized EB outer membrane ghosts do not fuse with lysosomes (72). In addition, inhibition of protein synthesis by either UV irradiation (72) or antibiotics (84, 294) did not result in fusion of the inclusion with lysosomes at up to 30 h postinfection. These results are in conflict with a recent report that indicates that blocking early protein synthesis by the EB [although considered relatively metabolically inert, protein synthesis by EBs has been shown to occur very soon after internalization (203)] inhibited both the transport of the inclusion to the peri-nuclear area and entry into the exocytic pathway (both within the first 8 h) and resulted in fusion of nearly all inclusions with lysosomes at 72 h (247). The discrepancy in the results may represent the time point examined, a difference in the cell types, or a difference in the methodology used to assess lysosomal fusion.

ROUTE OF ENTRY AND LYSOSOMAL FUSION   Earlier experiments suggested that opsonization of EBs with polyclonal antibody led to fusion with ferritin-loaded lysosomes in murine macrophages (304) or resulted in cofractionation on renografin gradients with phagolysosomes in infected L cells (84). In contrast, a recent report by Scidmore et al (247) argues that the route of entry does not influence the fate of intracellular *Chlamydia*. In this report, EBs opsonized with monoclonal antibody taken up into Fc receptor expressing HeLa and CHO cells undergo the changes observed with nonopsonized parasites with regard to localization in the perinuclear area and entry into the exocytic pathway (247). The difference between these studies may reflect the different cell types, the use of polyclonal versus monoclonal antibodies, or some other feature.

Inhibition of early chlamydial protein synthesis for up to 72 h is required to observe significant fusion of the inclusion with lysosomes (247). This suggests endocytic membrane trafficking events with regard to the inclusion are retarded (T Hackstadt, personal communication). This is not a general effect on the infected cell but appears to be restricted to the inclusion (71). Defining the molecular basis by which the nascent inclusion selectively blocks or retards membrane traffic may provide insight into the control of membrane traffic in normal cells.

THE CHLAMYDIAL INCLUSION INTERSECTS WITH THE EXOCYTIC ROUTE FOR GLYCOSPHINGOLIPIDS   Recent experiments from the Hackstadt group suggest that the chlamydial inclusion lies on the exocytic pathway for glycosphingolipids. C6-N-(7-nitrobenz-2-oxa-1,3-diazol-tyl) ceramide (C6-NBD-ceramide) is a vital stain for the Golgi apparatus where it concentrates and is converted to sphingomyelin or glucocerebroside before being trafficked to the plasma membrane (147). When *Chlamydia*-infected cells are labeled with C6-NBD-ceramide, approximately 50% of the label is directed to the inclusion and incorporated as sphingomyelin into the chlamydial cell wall (97, 98). Direct confirmation of the fusion of sphingomyelin-containing vesicles with the inclusion was obtained by the photoactivation of the fluorophore in the presence of DAB. Reaction product was detected by electron microscopy in the host Golgi, RB cell walls, inclusion membrane, and vesicular structures adjacent to or fusing with the inclusion (97). The ability to acquire sphingolipids is specific to the chlamydial inclusion as vacuoles of *Coxiella burnetti* fail to acquire the marker (112). Other fluorescently labeled lipid analogues were taken up in the inclusion, but only sphingomyelin localized to the RB cell wall (97).

Multiple lines of evidence suggested that the delivery of sphingomyelin to the inclusion was directly from the Golgi along the exocytic route. C6-NBD sphingomyelin incorporated into the plasma membrane was not targeted to the inclusion. Brefeldin A caused both Golgi dispersal and inhibited sphingolipid accumulation by the inclusion (97). Of note, monensin, which blocks the traffic of newly synthesized sphingomyelin from the Golgi (147), altered the organization of the Golgi but only slightly reduced the fluorescence of the inclusion (97). This host-derived sphingomyelin likely accounts for the presence of this lipid, not normally found in prokaryotes, in chlamydiae (181).

The chlamydial inclusion is, however, separate from the ER and Golgi and is not on the pathway for transport of glycoproteins to the cell surface or to lysosomes. None of the Golgi markers—the engineered ER and ER/Golgi markers (283), the *trans*-Golgi specific marker AP-1 (246), or other Golgi markers including p58, mannosidase II, and β-COP—labeled the inclusion (97). No staining with anti-clathrin heavy-chain antibodies was detected (246) [clathrin-coated vesicles are normally involved in traffic from the Golgi to lysosomes but not the plasma membrane (193)]. Model glycoproteins (vesicular stomatitis virus glycoprotein, transferrin receptor, major histocompatibility complex) were not incorporated into the inclusion or altered in their transport to the cell surface (246, 283), nor was recycling of the transferrin receptor affected (246).

Taken together, these data suggest that chlamydial inclusion may represent an aberrant compartment of the *trans*-Golgi network, selectably capable of fusing with sphingolipid-derived vesicles of Golgi origin (97, 246). They also suggest that vesicular traffic from the Golgi can be fully segregated with respect to glycolipids and glycoproteins, a point of considerable interest (and dispute) in the membrane traffic field (261).

PROTEINS IN THE INCLUSION ARE CHLAMYDIAL IN ORIGIN   The protein composition of the inclusion membrane is largely unknown. Using sera from convalescent guinea pigs infected with *C. psittaci* strain GPIC, Rockey & Rosquist (224) identified a number of proteins specific for *Chlamydia*-infected cells that were not present in EBs. One of these infection-specific proteins, IncA, was localized to the inclusion membrane and exposed to the cytoplasmic (host) side of that membrane (222). IncA is a 39-kDa protein in the RB and exists at two higher molecular weights in infected cells, suggestive of posttranslational modifications (222, 223). In fact, IncA is a ser/thr phosphoprotein that is phos-phorylated by host cell kinases (222). The modification of the IncA protein by host enzymes suggests a role in the chlamydial-host interaction (222). While the function of IncA is unknown, its absence in *C. trachomatis* inclusions suggests that it may be specific to the development of the *C. psittaci* inclusion (221).

To identify chlamydial proteins in the inclusion membrane, Taraska et al (283) raised an antiserum against membrane purified from *C. psittaci*–infected 3T3 Balb/c fibroblasts. The resulting polyclonal antiserum recognizes several *Chlamydia*-specific proteins by immunoblot analysis (29 and 71 kDa) and im-munoprecipitation (39, 42, and 52 kDa). Immunofluorescence studies suggest the antiserum recognizes epitopes on the cytoplasmic aspect of the inclusion membrane. Furthermore, the proteins identified using this antiserum are distinct from IncA.

Taken together, these studies suggest the chlamydial inclusion contains sev-eral chlamydial proteins that likely play a role in the establishment of the replication-competent inclusion (221, 224, 283). Modification of the inclusion and inclusion membrane with chlamydial proteins points to the existence of mechanisms to not only secrete but to also specifically target chlamydial pro-teins to compartments within the inclusion. These modifications presumably establish a unique niche for *Chlamydia* within the infected cell, permitting bacterial growth.

Toxoplasma gondii

The membrane trafficking events that surround the *T. gondii* parasitophorous vacuole are defined by their absence. The nascent PV is relatively devoid of intramembranous particles (205), which are either excluded or rapidly removed (58, 252). The removal of host cell proteins likely excludes recognition signals for fusion with the endocytic pathway (129, 252). Furthermore, shortly after formation, the PVM forms extensive and intimate associations with host cell mitochondria and ER (see below). These interactions could further mask the *T. gondii* PV from the endocytic compartment.

THE *T. GONDII* VACUOLE IS FUSION-INCOMPETENT   The pioneering studies of Jones & Hirsch (130) demonstrated that at least half of the tachyzoites internal-ized by macrophages were in vacuoles that did not fuse with thorotrast-loaded lysosomes. Furthermore, replicating parasites were inevitably found in non-fused vacuoles. In contrast, all heat-killed parasites were found in thorotrast-positive phagolysosomes. Replicating organisms are not essential for inhibi-tion of fusion, however, since UV-irradiated tachyzoites form vacuoles that do not fuse with lysosomes, despite the eventual disappearance of the parasites (73).

Fluid-phase endocytic markers, including lucifer yellow and Texas red dextran, are not incorporated into the *T. gondii* vacuole (129). Similarly, vacuoles do not stain with antibodies to fixed markers, including lysosomal glycoproteins (129), cell surface receptors (58), the transferrin receptor (128), and M6PR (170).

Killing the parasites after entry does not reverse the fusion incompetence of the vacuole, despite the involution of the parasites and their disappearance over 48 h or more (129). Taken as a whole, these results indicate that the *T. gondii* vacuole is fusion incompetent and that this fusion incompetence is established at the time of parasite entry. We presume, given the protein sorting events described above, that the newly formed *T. gondii* PVM lacks protein signals for fusion with endocytic organelles. For example, established *T. gondii* vacuoles in CHO cells stably overexpressing rab4 or rab5 do not stain for either of these markers (K Joiner, I Mellman, unpublished observations), nor do vacuoles stain with antibodies to the V-snare cellubrevin (C Beckers, P de Camilli, K Joiner, unpublished observations). Although the absence of additional components of the 20s vesicle docking complex in the PVM would be expected, reconstitution of vacuole fusion competence by introduction of these components into the PVM will ultimately be required to test the hypothesis that their absence explains vacuole fusion incompetence. Alternatively, generating *T. gondii* mutants conditional for vacuole fusion (and/or organelle association), analogous to the situation in *Legionella* (see below), provides an alternative approach to investigating this central question in *T. gondii* pathogenesis.

Unlike the chlamydial inclusion that fuses with exocytic vesicles, the *T. gondii* PVM is truly nonfusogenic. A recent report demonstrated labeling of intracellular *T. gondii* with C6-NBD-ceramide (61). The kinetics of accumulation of the label in the PVM (4 h) and intravacuolar parasites (5 h) (61) were considerably slower than those seen with the chlamydial inclusion (98). This suggests that rather than being on the exocytic route, the parasites and the PV acquire the label by another mechanism. While isolated from the endocytic and secretory pathways, *T. gondii* establishes interactions with host cell mitochondria and endoplasmic reticulum (see below), which along with its modification of the PV and PVM permit intracellular growth (see below).

THE ROUTE OF ENTRY IS ESSENTIAL FOR FUSION INCOMPETENCE OF THE *T. GONDII* VACUOLE   Upon addition of antibody-opsonized parasites to macrophages, parasites are now found in phagolysosomes (131). This could be the consequence either of altering the route of parasite entry or of killing parasites during the internalization step as a result of triggering the respiratory burst in macrophages. To distinguish between these possibilities, CHO cells transfected with Fc receptors were used. The addition of opsonized parasites to these cells resulted in uptake into a compartment that could be loaded with fluid-phase markers and was positive for both cell surface proteins and lysosomal markers (129). Parasites were killed. These results clearly established that altering the route of parasite entry reversed the fusion incompetence of the vacuole.

Morisaki et al (171) demonstrated incorporation of fixed markers into the macrophage vacuole containing antibody-coated parasites. These studies have been extended to indicate that antibody-opsonized *T. gondii* phagocytosed by macrophages experience a sequence of interactions with the host endocytic pathway essentially indistinguishable from inert particles (170). Of interest, a small proportion of parasites internalized by phagocytosis escaped from the primary vacuole by a mechanism morphologically resembling active invasion (171).

ACIDIFICATION   The *T. gondii* PV does not acidify (259). In contrast, vacuoles containing opsonized or dead parasites are rapidly acidified (259). The lack of PV acidification is likely due to the absence of the H^+-ATPase within the PVM, which cannot be detected by immunofluorescence microscopy (170, 258). Another property of the *T. gondii* PVM, the presence of a nonspecific pore (see below) connecting the vacuolar space to the host cell cytosol, would preclude acidification of the vacuole (241).

THE VACUOLAR SPACE AND PVM ARE MODIFIED BY SECRETION OF PARASITE PROTEINS   While initially devoid of intramembrane particles, the PVM is rapidly modified by secretions from the parasite rhoptries and dense granules (2, 25, 43, 44, 69, 145). In addition, the vacuolar space becomes filled with a tubulovesicular network, elements of which become continuous with the PVM (254, 257). The release of the network may occur as a regulated secretion event from the posterior end of the parasite in response to an as yet undetermined stimulus (257), as well as from release of dense granules. The network is rapidly modified with parasite proteins secreted by dense granules (designated GRA) (42, 257). Although it is likely that dense granule proteins within the network participate in nutrient salvage, evidence for this hypothesis is limited (see below).

Individual dense granules contain a mixture of GRA proteins, which upon secretion into the vacuolar space are differentially targeted (2, 42, 44). Immunolocalization studies indicate differential targeting of GRA proteins following release from the parasite. GRA1, GRA2, GRA4, GRA6, and the nucleoside triphosphate hydrolase (NTPase) localize to the tubulovesicular network by electron microscopy (32, 43, 256, 257), although not necessarily by subcellular fractionation (B Samuel, H Ngo, H Qi, K Joiner, manuscript in preparation). GRA5 associates exclusively with the PVM (144), while GRA3 shows both PVM and network association (144). Short hydrophobic stretches within the

GRA3 sequence may account for its association with membranes (31, 189; B Samuel, H Ngo, H Qi, K Joiner, manuscript in preparation). Sequences with characteristics of membrane spanning domains in GRA4, GRA5 (144a), and GRA6 may explain their membrane localization, although the mechanism remains obscure (42, 143, 144). GRA2 exists as both a soluble and a membrane-associated form; membrane association may be mediated by amphipathic α-helical regions within the protein (165a, 257).

In contrast to the differential localization of the GRA proteins, all rhoptry proteins identified to date localize exclusively to the PVM (25, 69). Of these, ROP2, ROP3, ROP4, and ROP7 are exposed to the host cell cytoplasm (25). These ROP proteins are likely to by processed at both the N and the C termini (JF Dubremetz, personal communication) and putatively possess a transmembrane domain (25).

Modification of the vacuolar space in a differential manner (PVM versus network versus the PV lumen) suggests the presence of either specific targeting signals and/or the secretion of a sorting machinery, such as molecular chaperones within the lumen of the vacuole. At least one potential molecular chaperone, the immunophilin Cyp18, possessing prolyl-peptidyl isomerase activity (113), is secreted into the lumen of the vacuole (JA Silverman, KA Joiner, unpublished results).

Being totally isolated from the endocytic and secretory pathways, parasite-mediated modifications of the PV and PVM presumably play an essential role in the establishment of the replication competent vacuole. These modifications, which include a pore in the PVM and the parasite-directed association of host organelles with the PVM (see below), are likely to provide direct access to nutrients from the host cell.

Legionella pneumophila

SORTING EVENTS DURING LEGIONELLA INTERNALIZATION INTO MACROPHAGES Internalization of *Legionella* through coiling phagocytosis is slower than conventional phagocytosis, suggesting that sorting processes within the phagosome membrane may occur to a greater extent with the coiled than with the conventional phagosome (51). Class I and Class II major histocompatibility complex molecules, as well as alkaline phosphatase, are rapidly excluded from coiled and conventional phagosomes (51). CR3 and the 5'-nucleotidase are included within the coil and conventional phagosome followed by the rapid removal of the 5'-nucleotidase (51).

FORMATION OF THE REPLICATIVE PHAGOSOME   A characteristic set of morphologic events occur following internalization of *Legionella* into monocytes or macrophages (118).

These events, of which association with host cell organelles plays a major role (see below), lead to formation of what has been termed the replicative phagosome (118). The mature replicative phagosome is a ribosome-studded organelle owing to its being surrounded with host endoplasmic reticulum (ER) (118, 278) (see below). Although formalin-killed *Legionella* are internalized by coiling phagocytosis, the replicative phagosome does not form (118, 120). In contrast, with IgG-coated organisms, no coiling phagocytosis occurs, but the replicative phagosome is formed (120).

VIRULENCE FACTORS ASSOCIATED WITH THE FORMATION OF THE REPLICATIVE PHAGOSOME   The use of mutants, altered in their infectivity and intracellular growth, has been vital in defining the biology of the *Legionella*-host cell interaction. The macrophage infectivity potentiator protein (Mip) is a 24-kDa *L. pneumophila* gene product involved in infection of macrophages (48). The protein is an immunophilin, possessing FK506 binding properties and prolyl-peptidyl isomerase activity (48). Mip mutants exhibit attenuated virulence in a guinea pig model. Although the precise mode of Mip action is unclear, the protein appears to be involved in establishing a vacuole permissive for replication.

Mutation analysis indicates that the ability to form a replication-competent vacuole requires multiple gene products. A *L. pneumophila* mutant designated 25D, isolated based on its inability to grow in macrophages, was deficient in the formation of the replicative phagosome (121). This mutant, which entered cells by coiling phagocytosis and did not recruit organelles, was unable to block lysosomal fusion and did not grow intracellularly (121). Additionally, mutant 25D was avirulent in the guinea pig model (33). A clone, capable of complementing these defects (151), was subsequently found to encode an operon [*icm WXYZ* (icm, intracellular multiplication)] (35) as well as the gene for *dotA* (dot, defective in organelle trafficking) (29).

The *dotA* locus had been previously identified in a screen for *L. pneumophila* defective in intracellular replication (29). Mutants of a Thy derivative of *Legionella* (168) were isolated that had augmented survival intracellularly during thymine starvation, as a consequence of defective intracellular replication. Intracellular phenotypes were determined by electron microscopy to assess ribosome association and fusion with ferritin-labeled lysosomes. In macrophages, mutants isolated by this scheme fell into two phenotypic classes: class I mutants were defective both in organelle recruitment and in the inhibition of lysosomal fusion; class II mutants were defective only in intracellular growth. The *Legionella* locus that complements the intracellular growth defects in both mutant classes was designated *dotA*.

Mutants with Tn903dIIlacZ transposon insertions in *icmX, icmY, icmZ,* and *dotA* were unable to kill macrophages, confirming the role of these genes in macrophage killing (231). Mutagenesis of wild-type organisms with the same transposon, followed by selection for mutants unable to kill HL60 cells (Mak, macrophage killing), identified 16 DNA hybridization groups, only one of which was *dotA/icm* (231). Interestingly, all mutants defective in macrophage killing were also resistant to the effects of high salt, although these are linked defects and not the same gene.

Chemical mutagenesis, in combination with the thymineless death enrichment scheme, was used to identify additional phenotypic classes defective in intracellular growth (279). In addition, selection of mutants for salt resistance and chemical mutagenesis has uncovered at least 16 genes that are required for formation of the replicative phagosome (R Isberg, personal communication). In addition to the *icmA* locus (which includes *dotA, icmW, icmX, icmY,* and *icmZ*), three genes (*dotB, dotC,* and *dotD*) are linked to *dotA* but are 10 kb away. Ten genes (including *dotE* through *dotK* and *icmB*) are located in a single contiguous locus that is not linked to *dotA*. Many of these genes have transmembrane domains. Two have lipoprotein signal sequences. Thus, it is possible that these genes are involved in the formation of a secretion system analogous but not identical to the type III secretion apparatus, which is required by *Legionella* for formation of the replicative phagosome.

THE REPLICATIVE PHAGOSOME DOES NOT FUSE WITH THE ENDOCYTIC CASCADE   The first report of the absence of fusion of the *Legionella* vacuole with organelles of the endocytic cascade was by Horwitz, who showed that the majority of phagosomes were not fused with lysosomes preloaded with thorotrast (119). Formalin killing led to nearly complete fusion, whereas coating with antibody or antibody and complement increased fusion substantially (119). Activation of monocytes also increased fusion, but >90% of fused phagosomes still had associated organelles (119). Macrophage activation blocked replication to a far greater extent than was explicable based on fusion (119).

The fusion of *Legionella* vacuoles was examined more recently using additional markers for the endocytic cascade, including LAMP and Texas red ovalbumin loaded lysosomes, in J774 cells (279). Less than 20% of phagosomes containing Thy⁻ *Legionella* contain LAMP, and less than 30% of phagosomes were fused with Texas red ovalbumin–containing compartments by immunofluorescence (279). By electron microscopy, Clemens & Horwitz (52) reported the virtual absence of LAMP or cathepsin D positivity in *Legionella* phagosomes in human macrophages, indicative of the absence of fusion. In contrast, Thy⁻ mutants defective in intracellular growth were in vacuoles in bone marrow–derived macrophages that were LAMP positive (up to 86%) and Texas red ovalbumin positive (up to 68%). It is possible that the higher percentage of LAMP-positive versus Texas red ovalbumin–positive phagosomes reflects the difference noted earlier between lysosomes and LAMP-positive compartments in macrophages.

The kinetics of acquisition of fixed endosomal markers by wild-type and mutant *Legionella* are also known. For wild-type *L. pneumophila*, 20–40% of phagosomes are rab7 positive and LAMP negative at 5 min (C Roy, R Isberg, personal communication). By 15–30 min, essentially all phagosomes are rab7 and LAMP negative. This result implies either that fusion of *Legionella* vacuoles with rab7-positive compartments occurs transiently or that rab7 alone is recruited to the phagosome, independent of a specific fusion event. For the *dotA* mutant, phagosomes are nearly fully LAMP positive at 5 min, and 50% stain with anti rab7. By 15 minutes, phagosomes are routinely rab7 negative but LAMP positive. This result indicates that the defect accorded by *dotA* occurs early in the intersection with the endocytic cascade.

SUMMARY OF REPLICATIVE PHAGOSOME FORMATION   Of the four pathogens covered in this review, *Legionella* is the most amenable to genetic analysis. At least 16 genes needed to establish the replicative phagosome are identified. The genetic link between inhibition of fusion and establishment of organelle association (see below) is most consistent with a role for secreted proteins or other components in both processes. Elucidation of the subcellular localization and function of genes involved in replicative phagosome formation should provide direct evidence to support this hypothesis.

ACIDIFICATION   The pH of *Legionella*-containing phagosomes was determined using FITC-labeled bacteria and found to be 5.7–6.5 (122). Phagosomes containing dead bacteria were, on average, at a pH of about 0.8 lower. Simultaneous measurement, in the same cell, of the pH of *L. pneumophila*–containing and erythrocyte-containing phagosomes showed the pH to be 6.1 and 5.0 or below, respectively (122). Furthermore, the mean pH of *L. pneumophila* vacuoles was the same in activated and nonactivated monocytes (122), indicating that the vacuolar pH is not the basis for the growth inhibition observed with activated monocytes (124).

Mycobacterium spp.

FUSION WITH LYSOSOMES   The paradigm for the last two decades has been that the mycobacterial phagosome does not fuse with lysosomes (15, 16, 81, 82, 100, 101, 253, 306). This was based initially on the pioneering studies of Armstrong & Hart (15), who found that three fourths of the phagosomes containing live (nondamaged by electron microscopy) *M. tuberculosis* (H37Rv) did not fuse with ferritin-loaded or acid phosphatase–containing compartments in resident mouse peritoneal macrophages at day 1 and day 4 after infection. In contrast, nearly all organisms that appeared damaged were in phagosomes that had fused with lysosomes, as had lethally irradiated organisms, whether or not they were intact.

Subsequently, a variety of laboratories demonstrated that lysosomes pre-loaded with electron-dense colloids or detected with acridine orange did not fuse with mycobacterial phagosomes and that the soluble lysosomal marker acid phosphatase was not detected in vacuoles containing other mycobacterial species (81, 82, 100, 101, 253). In stark contrast, McDonough et al (165) suggested that the majority of *M. tuberculosis* vacuoles fused early with lysosomes in J774.16 cells (2 h), but later virulent (H37Rv) *M. tuberculosis* could be identified in secondary nonfused vacuoles and occasionally in the cytosol.

More recent experiments have approached the same question by determining the presence of fixed markers for components of the endocytic cascade within the mycobacterial vacuole. This work has shown that mycobacterial phagosomes have characteristics of an early endosomal compartment that does not further mature into or fuse with either a late endosomal or lysosomal compartment. The mechanism that prevents phagosome maturation is not yet clear, nor are any mycobacterial genes identified that contribute to the delayed maturation.

Sturgill-Koszycki et al (273) demonstrated that vacuole membranes surrounding *M. avium* and *M. tuberculosis* in mouse peritoneal macrophages contained LAMP for up to 14 days of infection. The phagosomes did not contain the vacuolar ATPase and did not acidify (discussed below). This result indicates either selective delivery of LAMP but not the vacuolar ATPase to the phagosome or the selective sorting of V-ATPase but not LAMP from the phagosome; in either event it provides evidence for vesicular traffic into or out of the mycobacterial phagosome.

Subsequently, Clemens & Horwitz (52) showed that the level of LAMP and CD63 staining in the *M. tuberculosis* (Erdman strain) phagosome was intermediate between *Legionella* phagosomes (essentially lacking LAMP) and polystyrene beads. Another lysosomal marker, cathepsin D, was present at low levels in both the mycobacteria and *Legionella* phagosomes. Twenty percent of the *M. tuberculosis* phagosomes lacked any markers whatsoever (analogous to the *Legionella* phagosomes). Additionally, an approximately equal fraction of phagosomes contained the same levels of CD63 as those containing dead organisms, indicating killing of the bacteria.

Compared with *Legionella* and polystyrene beads, the *M. tuberculosis* (Erdman strain) phagosome also stained intensely for major histocompatibility complex class I and II (52). Although HLA-DR was present at the same level at day 1 and day 5, β2-microglobulin was present at 3 h and 1 day but largely gone at day 5, indicating delayed clearance in comparison to polystyrene beads. Of particular interest, the transferrin receptor, the prototypic marker for the recycling (sorting and perinuclear) endosomal compartment, was present at 3 h,

1 day, and 5 days in *M. tuberculosis* but not *Legionella* or polystyrene bead phagosomes.

THE MYCOBACTERIAL VACUOLE IS ON THE ENDOSOMAL RECYCLING PATHWAY
The above studies did not examine the dynamics of intersection of the phago-some with the endocytic or secretory apparatus. In principal, the presence of endocytic markers in the mycobacterial phagosome could reflect direct delivery from the *trans*-Golgi network (TGN) (i.e. the vacuole is present on the secretory pathway rather than the endocytic pathway), retention of selected markers fol-lowing phagocytosis (if vesicular traffic into and out of the vacuole is minimal), or delivery from endosomes if the phagosome is on the endosomal pathway.

The answer is apparent from several studies. Frehel et al (81) demonstrated that 80% of *M. avium* phagosomes in bone marrow macrophages contained horseradish peroxidase (HRP) after 60 min of loading at 37°C. More recently, deChastellier et al (59) showed that fluid-phase HRP and surface-labeled [³H] galactose glycoconjugates were delivered to the *M. avium* phagosome, but only if the organism appeared undigested by electron microscopy. Russell and coworkers (228) used an innovative approach by showing that cholera toxin (biotinylated B subunit) binds to cell surface GM1 gangliosides in macrophages and gains access to mycobacterial vacuoles (both *M. avium* and *M. tuberculo-sis*) within 5 min of addition to cells, and equilibrates within 15 min. Access to IgG phagosomes was considerably slower. Access did not proceed through a brefeldin A–sensitive compartment. The route was different from that of fluid-phase markers, since digoxigenin-labeled dextran was in different vesicles. Access specifically to the recycling endosomal apparatus was also examined by Clemens & Horwitz (53). The *M. tuberculosis* phagosome (but not the polystyrene bead or dead *M. tuberculosis*) acquired exogenously added trans-ferrin in a time-dependent fashion, and transferrin could be chased out of the *M. tuberculosis* phagosomes by incubation in medium lacking transferrin. Finally, Sturgill-Koszycki and colleagues (272) conducted similar experiments quantitating digoxigenin-labeled transferrin entry and exit from isolated *M. avium* phagosomes. There was rapid accumulation of transferrin within the first 5 min, then a rapid decline from 5 to 15 min (which was no different between macrophage lysates and purified vacuoles). Altogether, these results provide convincing evidence that the mycobacterial phagosome is on the recy-cling endosome pathway.

IS THE MYCOBACTERIAL PHAGOSOME ON A DIFFERENT PATHWAY THAN CON-VENTIONAL PHAGOSOMES OR SIMPLY ARRESTED IN DEVELOPMENT AT THE EARLY ENDOSOMAL STAGE?   It seems most likely that the mycobacterial phago-some is arrested at the early endosomal stage, rather than being on a different pathway for phagosome development. Support for this conclusion comes from recent studies on cathepsin D: It is synthesized as a 51- to 55-kDa proenzyme, which is secreted from the TGN and delivered to the endosomal network, where it is sequentially processed in a pH-dependent fashion. Cathepsin D in purified *M. avium* vacuoles is in the partially processed 48-kDa form (even nine days following infection), whereas cathepsin D in IgG-bead phagosomes is in fully processed 31/17-kDa form (272). This result suggests that there is limited proteolytic activity of the *M. avium* phagosomes, lack of exposure to an acidic environment, or both. The fact that acidifying isolated *M. avium* phagosomes to pH 4.5 with acetate buffer resulted in conversion of cathepsin D argues that the major defect is in acidification.

Importantly, with IgG phagosomes, 3–5 min after formation, the 51- to 55-kDA form of procathepsin D is present. The enzyme is fully processed at a later time, but it is not possible to distinguish between conversion acquired during phagosome formation; and fusion with lysosomes/endosomes containing processed cathepsin D. Altogether, these results argue that LAMP and the cathepsin are acquired by phagosomes prior to acquisition of proton-ATPases. Whether they are delivered by the secretory pathway or the endocytic pathway remains to be determined, since the point of addition of proton-ATPases to the endocytic cascade is not known.

SUMMARY OF RESULTS WITH FIXED MARKERS, AND MARKERS OF TRAFFIC THROUGH THE MYCOBACTERIA PHAGOSOME   Altogether, the above results suggest that mycobacterial vacuoles are highly dynamic, fusion-competent vesicles that behave like an extension of the recycling endosomal apparatus. The data suggest the possibility that mycobacterial vacuoles may retain the homotypic endosome-endosome fusion machinery, while not acquiring the machinery for endosome-lysosome fusion.

Of direct relevance to this point, live *Listeria monocytogenes* can recruit and retain the GTPase rab5 on the phagosome membrane, but dead *Listeria* cannot. Early endosomal markers and key fusion factors accumulate on the live but not on the dead Listerial phagosome membrane (10, 11). Although the mechanism is not clear, this suggests a possible strategy for pathogen vacuoles to remain locked in the stage of homotypic early endosomal fusion mediated by rab5 (229a). To date, there are no reports examining rab5 staining of the mycobacterial phagosome.

ROUTE OF ENTRY AND LYSOSOME FUSION   The route of cell entry influences fusion of the *M. tuberculosis* vacuole with lysosomes. Armstrong & Hart showed by electron microscopy that incubation of *M. tuberculosis* H37Rv with immune rabbit serum prior to addition to cells resulted in fusion of vacuoles with ferritin-labeled lysosomes (16). There was no adverse effect on viability or subsequent replicative potential of the organisms, which argues against the hypothesis that inhibition of phagosome-lysosome fusion is a prerequisite for survival of organisms in macrophages. In fact, phagosome-lysosome fusion was reported to stimulate *M. tuberculosis* growth (36). Varying results are reported with *Mycobacterium leprae* (82, 253).

SECRETED MYCOBACTERIAL PRODUCTS AND THE INHIBITION OF FUSION   It was originally reported that sulfatides of *M. tuberculosis,* when preloaded into macrophages, inhibited fusion of yeast phagosomes with lysosomes (92). Subsequently, it became clear that sulfatides and other colloids could render lysosomes generally incapable of fusing, but that the results differed depending on the assay used to monitor the fusion event (169). Gordon et al (91) suggested that ammonia produced by *M. tuberculosis* was sufficient to inhibit fusion, and Hart et al (102) showed that chemical inhibitors of phagosome-lysosome fusion [ammonium chloride, poly α-D-glutamic acid (PGA)] inhibited saltatory movements of lysosomes. Of interest, *M. tuberculosis* synthesizes and secretes in large amounts a glutamine synthetase, which may play a role in modulating ammonia levels in the phagosome (103). Although rapid diffusion of ammonia would preclude the establishment of a substantial gradient in the cell (D Russell, personal communication), perhaps other as yet uncharacterized weak bases produced by mycobacteria accumulate in the mycobacterial vacuole. Since endosome acidification is necessary to recruit endosome coat proteins potentially necessary for downstream fusion events (14), even partial blockade of phagosome acidification by weak bases could contribute to the inhibition of phagosome maturation.

ACIDIFICATION   Vacuoles containing *M. tuberculosis* and *M. avium* are impaired in acidification. Crowle and coworkers (55) used the weak-base DAMP (N-(3-(2,4-dinitrophenyl)amino)propyl)-N-(3-ammopropyl)methyl amine dihydrochloride) (described below) to show that nearly 40% of nascent phagosomes containing live organisms were acidified, although to a lesser extent than neighboring lysosomes. Only at later points were the large majority of phagosomes not acidified. Sturgill-Koszycki et al (273) used carboxyfluorescein-labeled organisms to show that the average pH of nascent (5 min) *M. avium* phagosomes was 6.3–6.5. The proton ATPase was absent from the vacuoles but was demonstrated in a reticular network near the macrophage surface (228). Oh & Straubinger (187) labeled *M. avium* with N-hydroxysuccinimidyl esters of carboxyfluorescein (CF) and rhodamine, then used the ratio of the two to calculate internal pH. These workers showed extremely slow acidification of *Salmonella typhimurium* and zymosan phagosomes—4–6 h were required to lower the pH below 5. During this same period, there was little change in the pH of *M. avium* vacuoles.

The nature of the compartment containing the proton ATPase is generally unclear in all cells, and the density of the proton ATPase in early endosomes is not known. Therefore, it is unclear whether the absence of the proton ATPase in mycobacterial phagosomes reflects simply the absence of phagosome fusion with proton-ATPase–rich compartments or specific removal of the proton ATPase from the phagosome membrane. Recent data support the former possibility (272).

VACUOLAR ASSOCIATION WITH HOST CELL ORGANELLES

The interaction of the PVM with host organelles is a morphologically striking feature of the biology of selected pathogens in nonfusogenic vacuoles. This phenomenon, which we term PVM-organelle association, has been described for *C. psittaci* (158), *T. gondii* (60, 73, 130, 262), and *L. pneumophila* (118) but not for phagosomes containing mycobacteria.

Chlamydia psittaci

The vacuolar membrane of certain *C. psittaci* strains forms tight associations with host mitochondria (158, 191). These morphologically resemble *T. gondii* PVM-mitochondrial associations (see below). Mitochondria associated with the inclusion membrane copurify with inclusions, indicating a stable interaction (156). The minimum distance between the inclusion membrane and associated mitochondria was estimated to be 5.1 nm, although no data regarding the mean separation between the membranes were reported (156, 158). The initiation of mitochondrial association with the *C. psittaci* inclusion is first observed 10–12 h postinfection (158, 271). This coincides with the initiation of the reproductive phase of the chlamydial life cycle (reviewed in 297), suggesting a role for mitochondrial association in the bacterial growth, as a potential mechanism to scavenge ATP (see below).

The molecular mechanism of mitochondrial association with the *C. psittaci* inclusion membrane remains unknown. Association may be mediated by chlamydial proteins localizing to the host cell side of the inclusion membrane. While a protein satisfying these criteria has been identified (222), there is as yet no evidence for its involvement in mitochondrial association. Surface projections, present at the point of contact between RBs (157, 213) and the inclusion membrane, do not mediate mitochondrial association (158). PVM-mitochondrial association, however, is not required for growth by all chlamydiae, as *C. trachomatis* inclusions do exhibit these interactions (158).

Toxoplasma gondii

Associations of the *T. gondii* PVM with host mitochondria and ER were noted in the earliest electron microscope studies (130). Both organelles associate extremely intimately with the PVM, with direct membrane-membrane continuities separated by a mean distance of 12 and 18 nm, respectively, for mitochondria and ER (262). This separation is less than half the diameter of a ribosome (5). Association with host organelles is dependent on active parasite invasion, is established concomitant with or fairly soon after invasion, and is not disrupted if parasites are killed after cell entry (262).

Organelle association was specific to *T. gondii* vacuoles, as the PVs of *Leishmania amazonensis* and *Coxiella burnetti* fail to show these interactions (262). At 4 h postinfection, 18% and 56% of the PVM surface area are associated with mitochondria and ER, respectively (262). By 20 h postinfection, mitochondrial association is essentially unchanged, while that of ER is approximately half that at the 4-h time point. The lack of change in the extent of mitochondrial association most likely reflects the deformation and stretching of PVM-associated mitochondria along the vacuolar membrane as it grows (146, 262). Nocodazole (a microtubule depolymerizing agent) treatment prior to infection reduces by half the extent of mitochondrial but not ER association, suggesting a role for microtubules in establishing the interaction. The role is likely to be passive, as microtubule disruption causes the dispersion of mitochondria to the periphery from the perinuclear area where the *T. gondii* vacuole is established (262). In support of this argument, the maintenance of PVM-associated organelles does not require either parasite or host cell viability nor does it require an intact host microtubule cytoskeleton. Furthermore, PVM markers cofractionate with organelle markers in sucrose density gradients, and cofractionation is not altered by treatment with sodium carbonate (262), indicating a high-affinity interaction.

Taken together, these results indicate that the PVM-organelle association is a specific, intimate, and stable interaction that is established early after parasite invasion. The link to parasite invasion and the early remodeling events in the PV suggest the involvement of a rhoptry function in organelle association (A Sinai, P Webster, K Joiner, unpublished data). Several rhoptry proteins localized to the PVM—including ROP2, ROP3, ROP4, and ROP7 (PO56)—are exposed to the cytoplasm (25), suggesting a potential role in organelle association. Antiserum raised against a rhoptry-enriched fraction of *T. gondii* specifically coimmunoprecipitates two host cell proteins from infected cells but not from uninfected cells or from a mixture of uninfected cells and extracellular parasites (A Sinai, K Joiner, unpublished data).

The extensive interactions with host organelles, particularly early after infection, suggests a potential role in the establishment and/or maintenance of the nonfusogenic state. We believe, however, that the primary role of organelle association in *T. gondii* pathogenesis is the scavenging of nutrients from the host (see below).

Legionella pneumophila

Association with host cell organelles is required for the intracellular growth of *L. pneumophila* (118, 121, 278). Unlike *T. gondii* and *C. psittaci*, the *L. pneumophila* vacuole exhibits transient associations with mitochondria, which give way to stable interactions with the ER (118, 278). The interactions with ER are required to establish the replication-competent *Legionella* phagosome within which bacterial growth occurs (121, 278). As seen with *T. gondii* (130, 262), the face of the ER interacting with the vacuolar membrane is devoid of ribosomes (118). The distance between the vacuolar membrane and ER (ribosomes) was 10 nm (118), which is in the range of proximity recorded for organelle association with the *T. gondii* (262) and *C. psittaci* (156, 158) vacuoles.

Induction of autophagy has been proposed as a mechanism of ER recruitment (278). Autophagy is a mechanism by which cells break down both organelles and cytosolic components, resulting in extensive proteolytic degradation within the autophagosome (125, 249). Autophagy can be triggered in normal cells by starvation for amino acids, where the anabolic needs of the cell are met by the degradation of its proteins (249). *Legionellae* initiate intracellular growth earlier, following infection of cells starved for amino acids (278).

Legionellae exhibit morphological interactions, including coiling phagocytosis and organelle association, upon infecting free-living amoebae, their natural hosts (34, 49, 79). This suggests that the host cell components on the organelle membranes involved in organelle association have been evolutionarily conserved.

The establishment of organelle association is genetically linked to the product of the *dotA* locus (29), as well as to at least 15 other genes (R Isberg, personal communication). Mutants in *dotA* fail to recruit organelles and as a consequence do not establish a replication competent vacuole (29, 30). Vacuoles containing these mutants are fusogenic and acquire both endocytic and lysosomal markers, and the bacteria do not replicate within macrophages (29, 30). Thus, the *dotA* gene product (as well as the additional genes described above) affects organelle association as well as fusion competence, arguing for a central role in the establishment of a replication competent vacuole (29).

Little is known about how DotA mediates organelle association. The protein has been localized to the inner membrane of *L. pneumophila* and has the topological features of ABC transporter proteins (226). Several ABC transporter protein systems have been shown to be involved in export processes (78). A role for DotA in the export apparatus for secreted *Legionella* components modifying the phagosomal membrane is an attractive hypothesis. By serving a transport function, mutations in DotA would be expected to exhibit pleiotropic effects.

To the best of our knowledge, organelle association is observed only with pathogens residing in nonfusogenic vacuoles. Covering the PV with host organelles may hinder access to and fusion with organelles of the endocytic cascade. In addition, organelle association may be an adaptation to increase interaction with potential host nutrient pools. Whether organelle association plays a role in the scavenging of nutrients by these pathogens remains to be determined.

NUTRIENT ACQUISITION BY INTRAVACUOLAR PATHOGENS

While providing protection from host defenses, the vacuolar membrane of nonfusogenic vacuoles restricts access to the nutrient-rich cytoplasm. Growth of the pathogens despite this barrier suggests intuitively that the organisms may modify the PVM, allowing them to scavenge necessary nutrients and metabolites. These activities, none of which has been defined at the molecular level, would facilitate the concentration of nutrients within the PV by both active and passive means.

In this section we explore, in broad terms, the mechanisms by which intravacuolar pathogens in nonfusogenic vacuoles acquire nutrients, in the context of what is known about their nutritional requirements. We broadly group these requirements into the classes of energy metabolism and nucleic acid precursors, amino acids, and lipids. Because of space limitations, we restrict the discussion to chlamydiae, *T. gondii*, and *L. pneumophila*. Readers interested in nutrient acquisition by mycobacteria are referred to a review by Barclay & Wheeler (22).

Energy Metabolism and Nucleic Acid Precursors

CHLAMYDIA SPP.   Of the pathogens described in this review, only the chlamydiae are true energy parasites (173). Infection of cells with chlamydiae results in an increase in both glycolytic and respiratory activity in infected cells (89, 90, 174). Since chlamydiae lack cytochromes and flavoproteins (7, 8), the increase in respiration has to be from the host cell (90). Stimulation of both glycolysis and respiration are host cell responses to infection, as they are maintained despite treatments that block or inhibit the multiplication of *Chlamydia* (90, 174). All the evidence, reviewed by Moulder (176) and McClarty (160), indicates that chlamydial multiplication is dependent on the host cell for all its ATP and other high-energy metabolites (173).

Replicating forms (RBs) of *Chlamydia* possess mechanisms to utilize ATP and other high-energy intermediates from the host cell (160). Chlamydial RBs possess an efficient translocator that takes up ATP while coupling it to the expulsion of ADP (107). Such an arrangement, functionally a reverse mitochondrion, has also been characterized in the intracytoplasmic pathogen *Rickettsia prowazekii* (301). Unlike *Chlamydia, R. prowazekii* does not have the barrier of an inclusion membrane to cross (175).

How, then, is ATP delivered to the intravacuolar RBs? The answer may lie in the observation that replicating RBs within the inclusion lie closely apposed to the luminal wall of the PVM. Morphological studies (157) demonstrated hollow projections (6 nm in diameter) on the RB surface, which penetrate the inclusion membrane and appear to have access to the cytosol. These surface projections are observed on both EBs and RBs (184, 213). The existence of a channel linking the RB surface with the cytosol through the inclusion membrane has led to the proposal that chlamydiae may scavenge ATP and other nutrients using a soup-through-a-straw mechanism (267). Delivery of ATP by such a mechanism would likely be restricted to the organisms in direct contact with the inclusion membrane. This would not account for the ATP demands of *Chlamydia* in the center of the inclusion, which is likely to be required for the reconversion of RBs to EBs. A nonspecific channel in the inclusion membrane, similar to that identified in the *T. gondii* PVM (241), could provide ATP to the center of the inclusion. Recent results, using a microinjection approach that led to the identification of the *T. gondii* pore, show no evidence for such an activity in the chlamydial inclusion membrane (111), suggesting the presence of an alternative mechanism.

A considerable body of work has been done to elucidate the requirements of chlamydiae for nucleic acid precursors, which are obtained from host cell pools. These studies, which have been extensively reviewed by McClarty (160) and Moulder (176), are briefly summarized here. Transport of ribonucleotide triphosphates is restricted to RBs and not observed in EBs (106). EBs, however, do possess limited supplies of NTPs, which are required for the early differentiation events (290). While all four ribonucleotide triphosphates are efficiently scavenged (105, 280), deoxy NTPs are not (106). Chlamydial requirements for DNA synthesis are met entirely by the reduction of ribonucleotide triphosphates (295) using specific ribonucleotide reductases and the thymidylate synthase pathway (75). Despite being able to efficiently scavenge CTP, chlamydiae possess a CTP synthase activity that has recently been cloned (290, 291, 303). The presence of this gene in an operon with CMP-KDO synthase suggests a primary role in lipopolysaccharide biosynthesis (291). In addition to NTPs, chlamydiae efficiently transport purines but not pyrimidines (161, 162), the demand for which are met by scavenging uracil from the host (160, 162).

The above studies on the scavenging of nucleic acid precursors by chlamydiae do not address the question of how the barrier of the inclusion membrane is breached. It is likely that this pathway will be similar to that for the uptake of ATP, as discussed above.

TOXOPLAMSA GONDII   Unlike *Chlamydia* spp., *T. gondii* is able to synthesize its own ATP (243). The parasite contains the full complement of glycolytic enzymes as well as cytochromes (40, 148). Additionally, extracellular parasites generate a mitochondrial membrane potential and accumulate potential sensitive dyes like rhodamine 123 (281, 282). Intracellular parasites fail to accumulate rhodamine 123 (282); however, the observation that they accumulate other potential sensitive dyes (including Mito Tracker, Molecular Probes) (262) and are sensitive to atovaquone, a drug believed to disrupt the electron transport chain, argues for mitochondrial activity intracellularly (196).

While *T. gondii* does not apparently scavenge ATP for energy, the organism may utilize ATP as the source of adenosine. *T. gondii* is a purine auxotroph and preferentially salvages adenosine, which is subsequently incorporated into both parasite adenosine and guanine (138, 194, 243). We have suggested that ATP in the vacuolar space is sequentially dephosphorylated to ADP and AMP by a potent nucleotide triphosphate hydrolase (NTPase) (18, 19, 21, 32), which is secreted into the vacuolar space (32, 256). The enzyme, which accounts for 2–4% of total parasite protein, shows no substrate preferences and is able to degrade all NTPs to NMPs with equal efficiency (18, 19, 21, 32). AMP could then be subsequently dephosphorylated to adenosine by the plasma membrane 5′-nucleotidase and transported into the parasite by a specific transporter (240). Within the parasite, the adenosine is phosphorylated to AMP by a potent adenosine kinase (198, 199, 138). In addition to adenosine, the parasite can utilize inosine, adenine, and hypoxanthine, indicating salvage mechanisms for alternative purine sources (138, 240). *T. gondii*, while not deficient in its synthesis of pyrimidines (20), is able to efficiently accumulate these compounds, suggesting the presence of a specific uptake mechanism (20, 194).

ATP and other small soluble molecules are transported across the PVM by a nonspecific pore (241). Using an approach of microinjecting infected cells with fluorescently labeled dextrans and peptides, Schwab and colleagues (241) determined that the PVM acts as a molecular sieve, allowing the free bidirectional transport of soluble molecules of under 1300 daltons. The presence of this activity in the *T. gondii* PVM allows the free transport of nucleotides, amino acids, and peptides into the vacuolar space (241). As a bidirectional channel, the pore prevents the accumulation of waste metabolites within the PV and may play a role in the elimination of pyruvate, a byproduct of parasite glycolysis (241, 282).

One potential role for PVM-mitochondrial association may be the establishment of a locally high concentration of ATP in the vicinity of the PV (262, 282). This is unlikely to be the sole function of PVM-mitochondrial association, as

ATP generation solely by glycolysis is sufficient to maintain parasite growth (243).

LEGIONELLA SPP.   Little is known about energy metabolism by *Legionella* within macrophages. In vitro studies, however, indicate that these bacteria meet most of their energy needs from amino acids, in particular from serine, threonine, glutamate, and tyrosine (286, 300). Glucose, acetate, and succinate, however, are metabolized slowly and used primarily for biosynthetic functions (286, 300). Amino acids serve as the sole carbon and nitrogen sources in defined chemical media (218, 299). Energy is derived by oxidative metabolism, consistent with legionellae being strict aerobes (286). In addition to amino acids, compounds known to be intermediates in the tricarboxylic acid cycle are able to supply some of the energy needs of legionellae in vitro (286).

The apparent preference for amino acids as the source of energy in vitro may be extended within the infected cells. One mechanism that has been proposed for the association of ER with the *L. pneumophila* vacuole is the induction of autophagy (278). One might speculate that the ability to trigger autophagy is a mechanism of increasing concentrations of free amino acids, which could be scavenged by *L. pneumophila*. Alternative scenarios are active transporters or a passive mechanism, such as a pore within the vacuolar membrane.

While legionellae assimilate exogenous adenine and uridine into nucleic acids, they are able to synthesize all of their purine and pyrimidine needs de novo (286). Unlike chlamydiae and *Toxoplasma*, legionellae do not appear to require host cell nucleosides (see above) but do possess the capacity to accumulate these compounds.

Amino Acid Metabolism

CHLAMYDIA SPP.   The requirements for amino acids by chlamydiae are defined by strain-specific auxotrophies (6, 54, 190) and vary significantly from strain to strain (106, 176). Intracellular chlamydiae compete with the host cells for the amino acid pools (104). Depletion of isoleucine in the media of infected cells inhibits the multiplication of both the host cells and *C. psittaci* (104). Multiplication of either the chlamydiae or the host cells is initiated by the same minimum concentration of the amino acid (104). Interestingly, inhibition of host protein synthesis by cyclohexamide abolished the need for additional isoleucine by chlamydiae, suggesting that in the absence of host protein synthesis amino acids released by normal turnover are sufficient for chlamydial growth (104). The amino acid requirements of other chlamydiae are similarly reduced by cyclohexamide treatment, such that growth in the absence of normally required amino acids (6, 54, 104, 190) is now possible (6).

The depletion of tryptophan induced by the treatment of infected fibroblasts with Interferon gamma (IFNγ) blocks the growth of chlamydiae, indicating an auxotrophy for this amino acid (38). The inhibitory effect of IFNγ on *Chlamydia* is lost in host cells defective in indoleamine 2,3-dioxygenase, in which degradation of tryptophan is not observed (288). In addition, starvation or growth inhibition of intracellular *Chlamydia* can result in the establishment of a persistent state (reviewed in 24).

The effect of cysteine depletion on chlamydial development is particularly interesting. The infectious EB form contains an outer membrane with highly disulfide–cross-linked proteins (23, 180). This constitutes a high demand for cysteine, depletion of which has been shown to block the conversion of RBs to EBs (269).

Little is known about the mechanism of amino acid uptake by intracellular chlamydiae. The function of the major outer membrane protein as a porin could permit the accumulation of amino acids and other small molecules within the periplasmic space (23). This, however, does not provide a mechanism for how amino acids cross the inclusion membrane. One possibility is that the surface projections extending from the surface of inclusion associated RBs into the cytoplasm may facilitate amino acid transfer (157). Alternatively, specific transporters in the RB surface may be employed (160). Two active transport mechanisms, exchanging APT for lysine and methionine, respectively, have been described (107, 108).

While mechanisms for amino acid transport into the RB have been described, nothing is known about how the barrier of the inclusion membrane is crossed (176). In the absence of a pore (111), alternative mechanisms of uptake must be envisioned. These could include specific transporters or even delivery in host exocytic vesicles.

TOXOPLASMA GONDII   Little is known about the requirements of *T. gondii* for amino acids. One well-defined auxotrophy is that for tryptophan (195, 197). Tryptophan degradation can be induced by exposing cells to IFNγ by the induction of indoleamine 2,3-dioxygenase (IDO) (57, 200). As with *Chlamydia*, addition of tryptophan, or IFNγ stimulation of cells deficient in IDO activity, fails to inhibit the growth of *T. gondii* (288). The ability to convert tryptophan auxotrophy to prototrophy by the expression in *T. gondii* of the *E. coli* *trpB* gene, which encodes the β-subunit of tryptophan synthase, identifies the specific activity lacking in wild-type parasites (255).

Auxotrophies for other amino acids have not been identified. The ability to efficiently metabolically label parasites both intracellularly and extracellularly with several labeled amino acids indicates the presence of amino acid uptake mechanisms (260). In addition, the presence of the pore in the PVM allows unrestricted diffusion of amino acids from the host cell into the PV to be scavenged by the parasite (241).

LEGIONELLA SPP.   Amino acids play a vital role in both energy production (see above) and biosynthetic reactions in legionellae (286). Using ^{14}C-labeled amino acids, only 21% of the label was recovered in proteins, 29% and 16%, respectively, were fractionated to lipids and nucleic acids, and the remainder were distributed among low-molecular-weight compounds and polysaccharides (24% and 9%, respectively) (286). While the specific catabolism and synthesis patterns of amino acids of intracellular legionellae have not been determined, amino acids likely play a central role in this environment. One may speculate that the ability of *Legionella* to trigger the autophagous pathway (278) may constitute a mechanism of increasing intracellular amino acid pools. The availability of amino acids to intracellular bacteria may be further enhanced by the ability of the bacteria to reduce host cell competition by inhibiting host protein synthesis (163).

Legionellae are able to synthesize most of their amino acids de novo but are auxotrophic for serine and threonine (218, 299). Interestingly, both serine and threonine can be used efficiently for energy production (286). The formation of the replication competent vacuole (118) defined by its association with the ER and the triggering of autophagy (278) are potentially employed by the pathogen to establish this unique niche.

Lipid Metabolism

CHLAMYDIA SPP.   Bacteria generally do not utilize sphingolipids. Chlamydiae are among the exceptions (181) and efficiently scavenge host sphingolipids by hijacking vesicular traffic from the Golgi to the plasma membrane (97, 98, 246). Sphingomyelin is readily accumulated in the chlamydial inclusion and incorporated into the membranes of replicating RBs (98). The delivery of sphingolipids, and presumably other lipids within the vesicle, is mediated by vesicle fusion events at the inclusion membrane (97, 246). Thus, while nonfusogenic with regard to the endocytic machinery, the inclusion membrane readily fuses with vesicles on the exocytic route (112).

The presence of lipids specific to *Chlamydiae* indicates they possess the ability to synthesize and modify lipids (76, 77, 88, 154, 180). The observation that selective inhibition of host protein synthesis with cyclohexamide inhibits lipid biosynthesis in L cells by over 90% without affecting the growth of *C. trachomatis* strain LGV supports this view (216). Furthermore, in *C. trachomatis*–infected, cyclohexamide-treated L cells, [^{14}C]isoleucine is efficiently incorporated into *Chlamydia*-specific branched-chain fatty acids (216).

Hackstadt et al (97) examined whether fluorescent lipid probes other than labeled ceramide accumulated within intracellular *Chlamydiae*. They report that while NBD-labeled phospholipids readily label cellular membranes, including the inclusion membrane, specific accumulation within replicating chlamydiae was restricted to labeled sphingolipids (97). The reason for the selective accumulation of NBD-labeled ceramide in chlamydiae or for the role of sphingolipids in their biology is not known (97). As exocytic vesicles contain lipids other than sphingolipids, delivery of other lipids to the inclusion can be envisioned by these fusion events (97).

TOXOPLASMA GONDII   Little is known about lipid metabolism in *T. gondii*. Preliminary experiments suggest that extracellular parasites under conditions that permit metabolic labeling of proteins fail to accumulate [^{14}C]acetate into a lipid fraction (A Sinai, K Joiner, unpublished data).

The addition of labeled fatty acids or acetate to infected cells leads to the accumulation of labeled lipids (including those on glycosyl phosphatidylinositol-anchored proteins) in parasites (178, 292). This suggests intracellular *T. gondii* are able to scavenge lipids from the host cell by as yet undetermined pathways. The evidence of lipid-modifying enzymes from the *T. gondii* dbEST sequencing project (http://www.ncbi.nlm.nih.gov/dbEST/) (4a) suggests that scavenged fatty acids and lipids can potentially be modified by the parasite.

We believe the ability of the PVM to intimately associate with organelles involved in lipid biosynthesis (the ER and mitochondria) offers a mechanism by which direct translation can be achieved. While direct evidence cannot be offered, sites of direct membrane-membrane continuity between mitochondria and ER have been implicated in the bulk transfer of lipids between these organelles (reviewed in 296).

A recent report suggests that the addition of fluorescently labeled ceramide to *T. gondii*–infected cells results in the labeling of the PVM and intracellular tachyzoites 4 and 5 h after the addition of the lipid (61). The kinetics of NBD-ceramide delivery to the *T. gondii* PVM (61) are not consistent with those seen with traffic from the Golgi to the chlamydial inclusion (98). Based on the kinetics, delivery of NBD-ceramide to the *T. gondii* vacuole does not appear to occur by vesicular transport. Alternative mechanisms include nonspecific diffusion and direct membrane-to-membrane translocation (155). The latter mechanism, at the sites of PVM-organelle association, have been proposed as a potential mechanism of lipid scavenging by *T. gondii* (262).

LEGIONELLA SPP.   By all indications, legionellae are able to synthesize fatty acids and lipids de novo (286). The carbon for lipid synthesis can be derived from sugars, amino acids, and intermediate metabolites (286). Carbon from pyruvate and acetate is incorporated primarily into a lipid fraction consistent with lipid synthesis by the acetyl coenzyme A–dependent pathway (286, 300). While we know that legionellae are able to synthesize their own lipid, nothing is known about the origins of the lipids for the expanding vacuolar membrane. These are likely to be derived from the host cell and may be delivered by either vesicular traffic, as seen for the chlamydial inclusion 98, or by direct translocation from associated ER as suggested for the *T. gondii* PVM (262).

CONCLUDING REMARKS

While the advances in our understanding of the cell biology of pathogen-host cell interaction have been dramatic, they point to how little we really know. Pathogenic microorganisms, with millennia of evolution behind them, have subverted normal host cell functions to establish their unique niches. Understanding the molecular basis of how pathogens subvert cellular functions offers a new perspective into mammalian cell biology. In addition, defining the cell biological and molecular basis of the pathogen host interaction will permit the development of novel therapeutic strategies.

ACKNOWLEDGMENTS

We gratefully acknowledge the following individuals for providing unpublished data for inclusion in this review: Jean Francois Dubremetz, Gareth Griffiths, Ted Hackstadt, Ralph Isberg, David Russell, Kai Simons, and Phil Stahl. Preparation of this paper was supported by Public Health Service Grant AI RO1 30060 and by the Burroughs Wellcome Fund Scholar Award in Molecular Parasitology to KAJ.

> Visit the *Annual Reviews home page* at
> http://www.annurev.org.

Literature Cited

1. Achbarou A, Mercereau-Puijalon O, Autheman JM, Fortier B, Camus D, Dubremetz JF. 1991. Characterization of microneme proteins of *Toxoplasma gondii*. *Mol. Biochem. Parasitol.* 47:223–33
2. Achbarou A, Mercereau-Puijalon O, Sadak A, Fortier B, Leriche JA, et al. 1991. Differential targeting of dense granule proteins in the parasitophorous vacuole of *Toxoplasma gondii*. *Parasitology* 103:321–29
3. Aikawa M, Komata Y, Asai T, Midorikawa O. 1977. Transmission and scanning electron microscopy of host cell entry by *Toxoplasma*. *Am. J. Pathol.* 87:285–96

4. Aikawa M, Miller LH, Johnson J, Rabbege J. 1978. Erythrocyte entry by malaria parasites. A moving junction between erythrocyte and parasite. *J. Cell Biol.* 77:77–82
4a. Ajioka J, Wan K-L, Hillier L, Maara M, Curruthers V, Sibley D. 1997. *Toxoplasma gondii* expressed sequence tags and gene discovery. Presented at Mol. Cell. Biol. Apicomplexan Protozoa, Park City, UT, Jan. 7–12
5. Alberts B, Bray B, Lewis J, Raff M, Roberts K, Watson JD. 1994. *Molecular Biology of the Cell*. New York: Garland
6. Allan I, Pearce JH. 1983. Amino acid requirements of strains of *Chlamydia trachomatis* and *C. psittaci* growing in

McCoy cells. Relationship with clinical syndrome and host origin. *J. Gen. Microbiol.* 129:2001–7

7. Allen EG, Bovarnick MR. 1957. Association of reduced diphosphopyridine nucleotide cytochrome c reductase activity with meningopneumonitis virus. *J. Exp. Med.* 105:539–47

8. Allen EG, Bovarnick MR. 1962. Enzymatic activity associated with meningopneumonitis virus. *Ann. NY Acad. Sci.* 98:229–33

8a. Allen LH, Aderem A. 1995. A role for MARCKS, the alpha isozyme of protein kinase C and myosin I in zymosan phagocytosis in macrophages. *J. Exp. Med.* 182:829–40

9. Alpuche-Aranda CM, Swanson JA, Loomis WP, Miller SI. 1992. *Salmonella typhimurium* activates virulence gene transcription within acidified macrophage phagosomes. *Proc. Natl. Acad. Sci. USA* 89:10079–83

10. Alvarez-Dominguez C, Barbieri AM, Beron W, Wandinger-Ness A, Stahl PD. 1996. Phagocytosed live *Listeria monocytogenes* influences Rab 5 regulated *in vitro* phagosome-endosome fusion. *J. Biol. Chem.* 271:13834–43

11. Alvarez-Dominguez C, Roberts R, Stahl PD. 1997. Internalized *Listeria monocytogenes* modulates intracellular trafficking and delays maturation of the phagosome. *J. Cell Sci.* 110:73–43

12. Andrews NW, Abrams CK, Slatin SL, Griffiths G. 1990. A *T. cruzi*-secreted protein immunologically related to complement component C9: Evidence for membrane pore forming activity at low pH. *Cell* 61:1277–87

13. Andrews NW, Whitlow MB. 1989. Secretion by *Trypanosoma cruzi* of a hemolysin active at low pH. *Mol. Biochem. Parasitol.* 33:249–56

14. Aniento F, Gu F, Parton RG, Gruenberg J. 1996. An endosomal beta-COP is involved in the pH-dependent formation of transport vesicles destined for late endosomes. *J. Cell Biol.* 133:29–41

15. Armstrong J, Hart P. 1971. Response of cultured macrophages to *Mycobacterium tuberculosis* with observations on fusion of lysosomes with phagosomes. *J. Exp. Med.* 134:713–40

16. Armstrong J, Hart P. 1975. Phagosome-lysosome interactions in cultured macrophages infected with virulent tubercle bacilli. *J. Exp. Med.* 142:1–16

17. Arruda S, Bomfim G, Knights R, Huima-Byron T, Riley LW. 1993. Cloning of a *M. tuberculosis* DNA fragment associ-ated with entry and survival inside cells. *Science* 261:1454–57

18. Asai T, Miura S, Sibley D, Okabayashi H, Tsutomu T. 1995. Biochemical and molecular characterization of nucleoside triphosphate hydrolase isozymes from the parasitic protozoan *Toxoplasma gondii*. *J. Biol. Chem.* 270:11391–97

19. Asai T, O'Sullivan WJ. 1983. A potent nucleoside triphosphate hydrolase from the parasitic protozoan *Toxoplasma gondii*. *J. Biol. Chem.* 258:6816–22

20. Asai T, O'Sullivan WJ, Kobayashi M, Gero AM, Yokogawa M, Tatibana M. 1983. Enzymes of the de novo pyrimidine biosynthetic pathway in *Toxoplasma gondii*. *Mol. Biochem. Parasitol.* 7:89–100

21. Asai T, Suzuki Y. 1990. Remarkable activities of nucleoside triphosphate hydrolase in the tachyzoites of both virulent and avirulent strains of *Toxoplasma gondii*. *FEBS Microbiol. Lett.* 72:89–92

22. Barclay R, Wheeler PR. 1989. Metabolism of mycobacteria in tissues. In *Metabolism of Mycobacteria in Tissues,* ed. C Ratledge, J Stanford, JM Grange, pp. 37–196. London: Academic

23. Bavoil P, Ohlin O, Schacter J. 1984. Role of disulphide bonding in outer membrane structure and permeability of *Chlamydia trachomatis*. *Infect. Immun.* 44:479–85

24. Beatty WL, Morrison RP, Byrne GI. 1994. Persistent *Chlamydiae*: from cell culture to a paradigm for chlamydial pathogenesis. *Microbiol. Rev.* 58:686–99

25. Beckers CJM, Dubremetz JF, Mercereau-Puijalon O, Joiner KA. 1994. The *Toxoplasma gondii* rhoptry protein ROP2 is inserted into the parasitophorous vacuole membrane, surrounding the intracellular parasite, and is exposed to the host cell cytoplasm. *J. Cell Biol.* 127:947–61

26. Bellinger-Kawahara C, Horwitz MA. 1990. Complement component C3 fixes selectively to the major outer membrane protein (MOMP) of *Legionella pneumophila* and mediates phagocytosis of liposome MOMP complexes by human monocytes. *J. Exp. Med.* 172:1201–10

27. Bennett M. 1995. SNAREs and the specificity of transport vesicle targeting. *Curr. Opin. Cell Biol.* 7:581–6

28. Berger K, Isberg RR. 1994. Intracellular survival by *Legionella*. *Methods Cell Biol.* 45:247–59

29. Berger KH, Isberg RR. 1993. Two distinct defects in intracellular growth complemented by a single genetic locus in *Legionella pneumophila*. *Mol. Microbiol.* 7:7–19

30. Berger KH, Merriam JJ, Isberg RR. 1994. Altered intracellular targeting properties associated with mutations in the *Legionella pneumophila dotA* gene. *Mol. Microbiol.* 14:809–22

31. Bermudes D, Dubremetz JF, Joiner KA. 1994. Molecular characterization of the dense granule protein GRA3 from *Toxoplasma gondii*. *Mol. Biochem. Parasitol.* 68:247–57

32. Bermudes D, Peck KR, Afifi-Afifi M, Beckers CJM, Joiner KA. 1994. Tandemly repeated genes encode nucleoside triphosphate hydrolase isoforms secreted into the parasitophorous vacuole of *Toxoplasma gondii*. *J. Biol. Chem.* 269:29252–60

33. Blander SJ, Breiman RF, Horwitz MA. 1989. A live avirulent mutant *Legionella* vaccine induces protective immunity against lethal aerosol challenge. *J. Clin. Invest.* 83:810–15

34. Bozue JA, Johnson W. 1996. Interaction of *Legionella pneumophila* with *Acanthamoeba castellani:* uptake by coiling phagocytosis and inhibition of phagosome-lysosome fusion. *Infect. Immun.* 64: 668–73

35. Brand BC, Sadosky AB, Shuman HA. 1994. The *Legionella pneumophila icm* locus: a set of genes required for intracellular multiplication in human macrophages. *Mol. Microbiol.* 14:797–808

36. Brown CA, Draper P, Hart PD. 1969. Mycobacteria and lysosomes: a paradox. *Nature* 221:658–60

37. Byrne GI. 1976. Requirements for ingestion of *Chlamydia psittaci* elementary bodies by L cells. *Infect. Immun.* 14:645–51

38. Byrne GI, Lehmann LK, Landry GJ. 1986. Induction of tryptophan catabolism is the mechanism for gamma-interferon-mediated inhibition of intracellular *Chlamydia psittaci* replication in T24 cells. *Infect. Immun.* 53:347–51

39. Caldwell HD, Perry LJ. 1982. Neutralization of *Chlamydia trachomatis* infectivity with antibodies to the major outer membrane protein. *Infect. Immun.* 38:745–54

40. Capella JA, Kaufman HE. 1964. Enzyme histochemistry of *Toxoplasma gondii*. *Am. J. Trop. Med. Hyg.* 13:664–68

41. Carvalho LD, Souza WD. 1989. Cyto-chemical localization of plasma membrane enzyme markers during interiorization of tachyzoites of *Toxoplasma gondii* by macrophages. *J. Protozool.* 36:164–70

42. Cesbron-Delauw MF. 1994. Dense-granule organelles of *Toxoplasma gondii:* their role in the host-parasite relationship. *Parasitol. Today* 10:293–96

43. Cesbron-Delauw MF, Guy B, Torpier G, Pierce RJ, Lenzen G, et al. 1989. Molecular characterization of a 23-kilodalton major antigen secreted by *Toxoplasma gondii*. *Proc. Natl. Acad. Sci. USA* 86:7537–41

44. Charif H, Darcy F, Torpier G, Cesbron-Delauw MF, Capron A. 1990. Characterization and localization of antigens secreted from tachyzoites. *Exp. Parasitol.* 71:114–24

45. Chavier P, Parton RG, Hauri HP, Simons K, Zerial M. 1990. Localization of low molecular weight GTP binding proteins to exocytic and endocytic compartments. *Cell* 62:317–29

46. Chiappino ML, Nichols BA, O'Connor GR. 1984. Scanning electron microscopy of *Toxoplasma gondii:* parasite torsion and host-cell responses during invasion. *J. Protozool.* 31:288–92

47. Cianciotto NP, Eisenstein BI, Mody CH, Engelberg NC. 1990. A mutation in the *mip* gene results in an attenuation of *Legionella pneumophila* virulence. *J. Infect. Dis.* 162:121–26

48. Cianciotto NP, Eisenstein BI, Mody CH, Toews GB, Engleberg NC. 1989. A *Legionella pneumophila* gene encoding a species specific surface protein potentiates initiation of intracellular infection. *Infect. Immun.* 57(4):1225–62

49. Cirillo JD, Falkow S, Tompkins LS. 1994. Growth of *Legionella pneumophila* in *Acanthamoeba castellanii* enhances invasion. *Infect. Immun.* 62: 3254–61

50. Clemens DL. 1996. Characterization of the *Mycobacterium tuberculosis* phagosome. *Trends Microbiol.* 4:113–18

51. Clemens DL, Horwitz MA. 1992. Membrane sorting during phagocytosis: selective exclusion of major histocompatibility complex molecules but not complement receptor CR3 during conventional and coiling phagocytosis. *J. Exp. Med.* 175:1317–26

52. Clemens DL, Horwitz MA. 1995. Characterization of the *Mycobacterium tuberculosis* phagosome and evidence that phagosome maturation is inhibited. *J. Exp. Med.* 181:257–70

53. Clemens DL, Horwitz MA. 1996. The *Mycobacterium tuberculosis* phagosome interacts with early endosomes and is accessible to exogenously administered transferrin. *J. Exp. Med.* 184:1349–55

54. Coles AM, Pearce JH. 1987. Regulation of *Chlamydia psittaci* (strain guinea pig inclusion conjunctivities) growth in McCoy cells by amino acid antagonism. *J. Gen. Microbiol.* 133:701–8

55. Crowle AJ, Dahl R, Ross E, May MH. 1991. Evidence that vesicles containing living, virulent *Mycobacterium tuberculosis* or *Mycobacterium avium* in cultured human macrophages are not acidic. *Infect. Immun.* 59:1823–31

55a. Curruthers VB, Sibley LD. 1997. MIC2 is a multispecific adhesin for *Toxoplasma gondii* attachment and invasion of host cells. Presented at Mol. Cell. Biol. Apicomplexan Protozoa, Park City, UT, Jan. 7–12

56. Cywes C, Godenir NL, Hoppe HC, Scholle RR, Styne LM, et al. 1996. Nonopsonic binding of *Mycobacterium tuberculosis* to human complement receptor type 3 expressed in Chinese hamster ovary cells. *Infect. Immun.* 64:5373–83

57. Dai W, Pan H, Kwok O, Dubey JP. 1994. Human idoleamine 2,3-dioxygenase inhibits *Toxoplasma gondii* growth in fibroblast cells. *J. Interf. Res.* 14:313–17

58. De Carvalho L, deSouza W. 1989. Cytochemical localization of plasma membrane enzyme markers during interiorization of tachyzoites of *Toxoplasma gondii* by macrophages. *J. Protozool.* 36:164–70

59. deChastellier C, Lang T, Thilo L. 1995. Phagocytic processing of the macrophage endoparasite *Mycobacterium avium*, in comparison to phagosomes which contain *Bacillus subtilis* or latex beads. *Eur. J. Cell Biol.* 68:167–82

60. DeMelo EJT, Carvalho TU, Souza W. 1992. Penetration of *Toxoplasma gondii* into host cells induces changes in the distribution of mitochondria and the endoplasmic reticulum. *Cell Struct. Funct.* 17:311–17

61. DeMelo EJT, Souza W. 1996. Pathway for C6-NBD-ceramide on the host cell infected with *Toxoplasma gondii*. *Cell Struct. Funct.* 21:47–52

62. Desjardins M, Celis JE, Meer GV, Dieplinger H, Jahraus A, et al. 1994. Molecular characterization of phagosomes. *J. Biol. Chem.* 269:32194–200

63. Desjardins M, Huber L, Parton R, Griffiths G. 1994. Biogenesis of phagolysosomes proceeds through a sequential series of interactions with the endocytic apparatus. *J. Cell Biol.* 124:677–88

64. Doborowski JM, Sibley LD. 1996. *Toxoplasma* invasion of mammalian cells is powered by the actin cytoskeleton. *Cell* 84:933–39

65. Dowling JN, Saha AK, Glew RH. 1992. Virulence factors of the family *Legionellaceae*. *Microbiol. Rev.* 56:32–60

66. Dubey JP. 1977. *Toxoplasma, Hammondia, Besnotia, Sarcocystis* and other tissue cyst forming coccidia of man and animals. In *Toxoplasma, Hammondia, Besnotia, Sarcocystis and Other Tissue Cyst Forming Coccidia of Man and Animals*, ed. JP Kreiger, pp. 101–237. New York: Academic

67. Dubey JP, Beattie CP. 1988. *Toxoplasmosis of Animals and Man*. Boca Raton, FL: CRC

68. Dubey JP, Murrell KD, Fayer R, Schad GA. 1986. Distribution of *Toxoplasma gondii* tissue cysts in commercial cuts of pork. *J. Am. Vet. Med. Assoc.* 188:1035–37

69. Dubremetz JF, Achbarou A, Bermudes D, Joiner KA. 1993. Kinetics of apical organelle exocytosis during *Toxoplasma gondii* host cell interaction. *Parasitol. Res.* 79:402–8

70. Dubremetz JF, Rodriguez C, Ferreira E. 1985. *Toxoplasma gondii*: redistribution of monoclonal antibodies on tachyzoites during host cell invasion. *Exp Parasitol.* 59:24–32

71. Eissenberg LG, Wyrick PB. 1981. Inhibition of phagolysosomal fusion is localized to the *Chlamydia psittaci* laden vacuoles. *Infect. Immun.* 32:880–96

72. Eissenberg LG, Wyrick PB, Davis CH, Rumpp JW. 1983. *Chlamydia psittaci* elementary bodies envelopes: ingestion and inhibition of lysosomal fusion. *Infect. Immun.* 40:741–51

73. Endo T, Pelster B, Peikarski G. 1981. Infection of murine peritoneal macrophages with *Toxoplasma gondii* exposed to ultraviolet light. *Z. Parasitenkd.* 65:121–29

74. Falkow S, Isberg RR, Portnoy DA. 1992. The interaction of bacteria with mammalian cells. *Annu. Rev. Cell Biol.* 8:333–63

75. Fan H, McClarty G, Burman RC. 1991. Biochemical evidence for the existence of thymidylate synthase in the obligate intracellular parasite *Chlamydia trachomatis*. *J. Bacteriol.* 173:6670–77

76. Fan VSC, Jenkins HM. 1974. Lipid

metabolism of monkey kidney cells (LLC-MK2) infected with *Chlamydia trachomatis* strain lymphogranuloma venereum. *Infect. Immun.* 10:464–510

77. Fan VSC, Jenkins HM. 1975. Biosynthesis of phospholipids and neutral lipids of monkey kidney cells (LL-MK–2) infected with *Chlamydia trachomatis* strain lymphogranuloma venereum (38538). *Proc. Soc. Exp. Biol. Med.* 148:351–57

78. Fath MJ, Kolter R. 1992. ABC transporters: bacterial exporters. *Microbiol. Rev.* 57:995–1017

79. Fields BS. 1993. *Legionella* and protozoa: interaction of a pathogen and its natural host. In *Legionella: Current Status and Emerging Perspectives,* ed. JM Barbaree, RF Breiman, AP Doufour, pp. 129–36. Washington, DC: Am. Soc. Microbiol.

80. Foussard F, Leriche MA, Dubremetz JF. 1991. Characterization of the lipid content of *Toxoplasma gondii* rhoptries. *Parasitology* 102:367–70

81. Frehel C, Chastellier C, Lang T, Rastogi N. 1986. Evidence for inhibition of fusion of lysosomal and prelysosome compartments with phagosomes in macrophages infected with pathogenic *Mycobacterium avium. Infect. Immun.* 52:252–62

82. Frehel C, Rastogi N. 1987. *Mycobacterium leprae* surface components intervene in the early phagosome-lysosome fusion inhibition event. *Infect. Immun.* 55:2916–21

83. Frenkel JK, Dubey JP, Miller NL. 1970. *Toxoplasma gondii* in cats: fecal stages identified as coccidian oocysts. *Science* 167:893–96

84. Friis RR. 1972. Interactions of L cells and *Chlamydia psittaci:* entry of the parasite and host responses to its development. *J. Bacteriol.* 110:706–21

85. Furtado GC, Cao J, Joiner KA. 1992. Laminin on tachyzoites of *Toxoplasma gondii* mediates parasite binding to the β1 integrin receptor a6β1 on human foreskin fibroblasts and Chinese hamster ovary cells. *Infect. Immun.* 60:4925–31

86. Furtado GC, Slowik M, Kleinman HK, Joiner KA. 1992. Laminin enhances binding of *Toxoplasma gondii* tachyzoites to J774 murine macrophage cells. *Infect. Immun.* 60:2337–42

87. Garcia-delPortillo F, Finlay BB. 1995. The varied lifestyles of intracellular pathogens within eukaryotic vacuolar compartments. *Trends Microbiol.* 3:373–80

88. Gaugler RW Jr, Neptune EM, Adams GM, Sallee TL, Weiss E, Wilson NN. 1969. Lipid synthesis by isolated *Chlamydia psittaci. J. Bacteriol.* 100:823–26

89. Gill SD, Stewart RB. 1970. Glucose requirements of L cells infected with *Chlamydia psittaci. Can. J. Microbiol.* 15:997–1001

90. Gill SD, Stewart RB. 1970. Respiration of L cells infected with *Chlamydia psittaci. Can. J. Microbiol.* 16:1079–85

91. Gordon AH, Hart PD, Young MR. 1980. Ammonia inhibits phagosome-lysosome fusion in macrophages. *Nature* 286:79–80

92. Goren M, Hart P, Young M, Armstrong J. 1976. Prevention of phagosome-lysosome fusion in cultured macrophages by sulfatides of *Mycobacterium tuberculosis. Proc. Natl. Acad. Sci. USA* 73:2510–14

93. Grimwood J, Smith JE. 1992. *Toxoplasma gondii:* the role of a 30-kDa surface protein in host cell invasion. *Exp. Parasitol.* 74:106–11

94. Grimwood J, Smith JE. 1996. *Toxoplasma gondii:* the role of the parasite surface and secreted proteins in host cell invasion. *Int. J. Parasitol.* 26:169–73

95. Gruenberg J, Maxfield FR. 1995. Membrane transport in the endocytic pathway. *Curr. Opin. Cell Biol.* 7:552–63

96. Hackam DJ, Rotstein OD, Bennett MK, Klip A, Grinstein S, Manolson MF. 1996. Characterization and subcellular localization of target membrane soluble NSF attachment protein receptors (t-SNAREs) in macrophages. Syntaxins 2, 3 and 4 are present on phagosomal membranes. *J. Immunol.* 156:4377–83

97. Hackstadt T, Rockey DD, Heizen RA, Scidmore MA. 1996. *Chlamydia trachomatis* interrupts an exocytic pathway to acquire endogenously synthesized sphingomyelin in transit from the Golgi apparatus to the plasma membrane. *EMBO J.* 15:964–77

98. Hackstadt T, Scidmore M, Rockey D. 1995. Lipid metabolism in *Chlamydia trachomatis*-infected cells:directed trafficking of Golgi-derived sphingolipids to the chlamydial inclusion. *Proc. Natl. Acad. Sci. USA* 92:4877–81

99. Hale TL. 1991. Genetic basis of virulence in *Shigella* species. *Microbiol. Rev.* 55(2):206–24

100. Hart P, Armstrong J, Brown C, Draper P. 1972. Ultrastructural study of the behavior of macrophages toward parasitic

mycobacteria. *Infect. Immun.* 5:803–7

101. Hart P, Young M. 1991. Ammonium chloride, an inhibitor of phagosome-lysosome fusion in macrophages, concurrently induces phagosome-endosome fusion, and opens a novel pathway: studies of a pathogenic *Mycobacterium* and a nonpathogenic yeast. *J. Exp. Med.* 174:881–89

102. Hart P, Young M, Jordan M, Perkins W, Geisow M. 1983. Chemical inhibitors of phagosome-lysosome fusion in cultured macrophages also inhibit saltatory lysosomal movements. *J. Exp. Med.* 158:477–92

103. Harth G, Clemens DL, Horwitz MA. 1994. Glutamine synthetase of *Mycobacterium tuberculosis:* extracellular release and characterization of its enzymatic activity. *Proc. Natl. Acad. Sci. USA* 91:9342–46

104. Hatch T. 1975. Competition between *Chlamydia psittaci* and L cell for host isoleucine pools: a limiting factor in chlamydial multiplication. *Infect. Immun.* 12:211–20

105. Hatch T. 1975. Utilization of L-cell nucleoside triphosphates by *Chlamydia psittaci* for ribonucleic acid synthesis. *J. Bacteriol.* 122:393–400

106. Hatch TP. 1988. Metabolism of *Chlamydia.* In *Metabolism of Chlamydia*, ed. AL Barron, pp. 97–110. Boca Raton, FL: CRC

107. Hatch TP, Al-Houssainy E, Silverman JA. 1982. Adenine nucleotide and lysine transport in *Chlamydia psittaci. J. Bacteriol.* 150:662–70

108. Hatch TP, Micelli M, Silverman JA. 1985. Synthesis of protein in host free reticulate bodies of *Chlamydia psittaci* and *Chlamydia trachomatis. J. Bacteriol.* 162:938–42

109. Hatch TP, Miceli M, Sublett JE. 1986. Synthesis of disulphide bonded outer membrane proteins during the developmental cycle of *Chlamydia psittaci* and *Chlamydia trachomatis. J. Bacteriol.* 165:379–85

110. Hatch TP, Vance DW Jr, Al-Houssainy E. 1981. Attachment of *Chlamydia psittaci* to formaldehyde fixed and unfixed L cells. *J. Gen. Microbiol.* 125:273–83

111. Heizen RA, Hackstadt T. 1996. The *Chlamydia trachomatis* parasitophorous vacuolar membrane is not passively permeable to low molecular weight compounds. *Infect. Immun* 65:1088–94

112. Heizen RA, Scidmore MA, Rockey DD, Hackstadt T. 1996. Differential interactions with the endocytic and exocytic pathways distinguish the parasitophorous vacuoles of *Coxiella burnetti* and *Chlamydia trachomatis. Infect. Immun.* 64:796–809

113. High KP, Joiner KA, Handschumacher RE. 1994. Isolation, cDNA sequences, and biochemical characterization of the major cyclosporine binding proteins of *Toxoplasma gondii. J. Biol. Chem.* 269:9105–12

114. Hirsch SH, Ellner JJ, Russell DG, Rich EA. 1994. Complement receptor-mediated uptake and tumor necrosis factor-a-mediated growth inhibition of *Mycobacterium tuberculosis* by human alveolar macrophages. *J. Immunol.* 152:743–53

115. Hodinka RL, Davis CH, Choong J, Wyrick PB. 1988. Ultrastructural study of endocytosis of *Chlamydia trachomatis* by McCoy cells. *Infect. Immun.* 56:1456–63

116. Hodinka RL, Wyrick PB. 1986. Ultrastructural study of mode of entry of *Chlamydia psittaci* into L-929 cells. *Infect. Immun.* 54:855–63

117. Horwitz M. 1992. Interactions between macrophages and *Legionella pneumophila. Curr. Top. Microbiol.* 181:265–82

118. Horwitz MA. 1983. Formation of a novel phagosome by the Legionnaires disease bacterium (*Legionella pneumophila*) in human monocytes. *J. Exp. Med.* 158:1319–31

119. Horwitz MA. 1983. The Legionnaires disease bacterium (*Legionella pneumophila*) inhibits phagosome-lysosome fusion in human monocytes. *J. Exp. Med.* 158:2108–26

120. Horwitz MA. 1984. Phagocytosis of the Legionnaires disease bacterium (*Legionella pneumophila*) occurs by a novel mechanism: engulfment within a pseudopod coil. *Cell* 36:27–33

121. Horwitz MA. 1987. Characterization of avirulent mutant *Legionella pneumophila* that survive but do not multiply within human monocytes. *J. Exp. Med.* 166:1310–28

122. Horwitz MA, Maxfield FR. 1984. *Legionella pneumophila* inhibits acidification of its phagosome in human monocytes. *J. Cell. Biol.* 99:1963–43

123. Horwitz MA, Silverstein SC. 1980. Legionnaires disease bacterium (*Legionella pneumophila*) multiplies intracellularly in human monocytes. *J. Clin. Invest.* 66:441–50

124. Horwitz MA, Silverstein SC. 1981. Activated human monocytes inhibit the intracellular multiplication of Legionnaires disease bacteria. *J. Exp. Med.* 154:1618–35

125. Hunziker W, Gueze HJ. 1996. Intracellular trafficking of lysososmal membrane proteins. *BioEssays* 18:379–89

126. Jacobs L. 1956. Propagation, morphology and biology of *Toxoplasma. Ann. NY Acad. Sci.* 64:154–79

127. Joiner KA. 1991. Rhoptry lipids and parasitophorous vacuole formation: a slippery issue. *Parasitol. Today* 7:226–27

128. Joiner KA. 1992. The parasitophorous vacuole membrane surrounding *Toxoplasma gondii:* a specialized interface between parasite and cell. In *Toxoplasmosis*, ed. JL Smith, pp. 73–81. Berlin: Springer-Verlag

129. Joiner KA, Fuhrman SA, Mietinnen H, Kasper LL, Mellman I. 1990. *Toxoplasma gondii:* fusion competence of parasitophorous vacuoles in Fc receptor transfected fibroblasts. *Science* 249:641–46

130. Jones TC, Hirsch JG. 1972. The interaction between *Toxoplasma gondii* and mammalian cells. II. The absence of lysosomal fusion with phagocytic vacuoles containing living parasites. *J. Exp. Med.* 136:1173

131. Jones TC, Len L, Hirsch JG. 1975. Assessment *in vitro* of immunity against *Toxoplasma gondii. J. Exp. Med.* 141:466–82

132. Khan IA, Kasper LH. 1993. Role of P30 in host immunity and pathogenesis of *T. gondii* infection. *Res. Immunol.* 144:45–48

133. Kimata I, Tanabe K. 1982. Invasion by *Toxoplasma gondii* of ATP-depleted and ATP-restored chick embryo erytrocytes. *J. Gen. Microbiol.* 128:2499–501

134. Kimata I, Tanabe K. 1987. Secretion of *Toxoplasma gondii* of an antigen that appears to become associated with the parasitophorous vacuole membrane upon invasion of the host cell. *J. Cell. Sci.* 88:231–39

135. King CA. 1988. Cell motility of sporozoan protozoa. *Parasitol. Today* 4:315–19

136. Knapp PE, Swanson JA. 1990. Plasticity of the tubular lysosomal compartment in macrophages. *J. Cell. Sci.* 95:433–39

137. Kornfeld S, Mellman I. 1989. The biogenesis of lysosomes. *Annu. Rev. Cell Biol.* 5:483–525

138. Krug EC, Marr JJ, Berens RL. 1989. Purine metabolism in *Toxoplasma gondii. J. Biol. Chem.* 264:10601–7

139. Kuo CC, Chi EW, Grayston JT. 1988. Ultrastructural study of entry of *Chlamydia* strain TWAR into HeLa cells. *Infect. Immun.* 56:1668–72

140. Kuo CC, Grayston JT. 1976. Interaction of *Chlamydia trachomatis* organisms and HeLa 229 cells. *Infect. Immun.* 13:1103–9

141. Kuo CC, Takahashi N, Swanson A, Ozeki Y, Hakomori S-I. 1996. An N-linked high mannose type oligosaccharide, expressed at the major outer membrane protein of *Chlamydia trachomatis*, mediates attachment and infectivity of the microorganism to HeLa cells. *J. Clin. Invest.* 98:2813–18

142. Lawn AM, Blythe WA, Taverne J. 1973. Interactions of TRIC agents with macrophages and BHK-21 cells observed by electron microscopy. *J. Hyg.* 71:515–28

143. Lecordier L, Meleon-Borodowski I, Dubremetz JF, Tourvielle B, Mercier C, Deslee D, et al. 1995. Characterization of a dense granule antigen of *Toxoplasma gondii* (GRA6) associated to the network of the parasitophorous vacuole. *Mol. Biochem. Parasitol.* 70:85–94

144. Lecordier L, Mercier C, Torpier G, Tourvielle B, Darcy F, et al. 1993. Molecular structure of a *Toxoplasma gondii* dense granule antigen (GRA 5) associated with the parasitophorous vacuole membrane. *Mol. Biochem. Parasitol.* 59:143–54

144a. Lecordier L, Sibley LD, Cesbron-Delauw M-F. 1996. Targeting of the GRA5 protein to the parasitophorous vacuole membrane. Presented at 4th Int. Bienn. Toxoplasma Conf., Drymen, Scotland, July 22–26

145. Leriche MA, Dubremetz JF. 1991. Characterization of the protein contents of rhoptries and dense granules of *Toxoplasma gondii* tachyzoites by subcellular fractionation and monoclonal antibodies. *Mol. Biochem. Parasitol.* 45:249–60

146. Lindsay DS, Mitschler RR, Toivio-Kinnucan MA, Upton SJ, Dubey JP, Blagburn BL. 1993. Association of host cell mitochondria with developing *Toxoplasma gondii* tissue cysts. *Am. J. Vet. Res.* 54:1663–67

147. Lipsky NG, Pagano RE. 1985. Intracellular translocation of fluorescent sphingolipids in cultured fibroblasts: endogenously synthesized sphingomyelin and glucocerebroside analogues pass through the Golgi apparatus en route

to the plasma membrane. *J. Cell. Biol.* 100:27–34

148. Lund E, Hasson HA, Lycke E, Sourander P. 1966. Enzymatic activities of *Toxoplasma gondii. Acta Pathol. Microbiol. Scand.* 68:59–67

149. Lycke E, Carlberg K, Norrby R. 1975. Interactions between *Toxoplasma gondii* and its host cell: function of the penetration host factor to *Toxoplasma. Infect. Immun.* 11:855–61

150. Lycke N, Norrby R. 1966. Demonstration of a factor of *Toxoplasma gondii* enhancing the penetration of *Toxoplasma* parasites into culture host cells. *Br. J. Exp. Pathol.* 47:248–56

151. Maara A, Blander S, Horwitz MA, Shuman HA. 1992. Identification of a *Legionella pneumophila* locus required for intracellular multiplication in human macrophages. *Proc. Natl. Acad. Sci. USA* 89:9607–11

152. Maara A, Horwitz MA, Shuman HA. 1990. The HL-60 model for the interaction of human macrophages with the Legionnaires disease bacterium. *J. Immunol.* 144:2738–44

153. Majeed M, Earnst JD, Magnusson K-E, Kihlstrom E, Stendahl O. 1994. Selective translocation of annexins during intracellular redistribution of *Chlamydia trachomatis* in HeLa and McCoy cells. *Infect. Immun.* 62:126–32

154. Makino S, Jenkins HM, Yu HM, Townsend D. 1970. Lipid composition of *Chlamydia psittaci* grown in monkey kidney cells in a defined medium. *J. Bacteriol.* 103:62–70

155. Martin OC, Pagano RE. 1994. Internalization and sorting of a fluorescent analogue of glycosylceramide to the Golgi apparatus in human skin fibroblasts: utilization of endocytic and non-endocytic transport mechanisms. *J. Cell. Biol.* 125: 769–81

156. Matsumoto A. 1981. Isolation and electron microscopic observations on intracytoplasmic inclusions containing *Chlamydia psittaci. J. Bacteriol.* 145: 605–12

157. Matsumoto A. 1988. Structural characteristics of chlamydial bodies. In *Microbiology of Chlamydia,* ed. AL Barron, pp. 21–45. Boca Raton, FL:CRC

158. Matsumoto A, Bessho I, Uehira K, Suda T. 1991. Morphological studies on the association of mitochondria with chlamydial inclusions. *J. Electron Micros.* 40:356–63

159. Mayorga LS, Francisco B, Stahl PD.

1991. Fusion of newly formed phagosomes with endosomes in intact cells and in a cell-free system. *J. Biol. Chem.* 266:6511–17

160. McClarty G. 1994. *Chlamydiae* and the biochemistry of intracellular parasitism. *Trends Microbiol.* 2:157–64

161. McClarty G, Fan H. 1993. Purine metabolism by intracellular *Chlamydiae. J. Bacteriol.* 175:4662–69

162. McClarty G, Qin B. 1993. Pyrimidine metabolism by intracellular *Chlamydiae. J. Bacteriol.* 175:4652–61

163. McCusker KT, Braaten BA, Cho MW, Low DA. 1991. *Legionella pneumophila* inhibits protein synthesis in Chinese hamster ovary cells. *Infect. Immun.* 59:240–46

164. McDade JE, Shepherd CC, Frase DW, Tsai TR, Redus MA, et al. 1977. Legionnaires disease: isolation of a bacterium and demonstration of its role in other respiratory diseases. *N. Engl. J. Med.* 297:1197–203

165. McDonough K, Kress Y, Bloom B. 1993. Pathogenesis of tuberculosis: interaction of *Mycobacterium tuberculosis* with macrophages. *Infect. Immun.* 61:2763–73

165a. Mercier C, Cesbron-Delauw M-F, Sibley LD. 1996. Specific membrane targeting of the *Toxoplasma gondii* protein GRA2 following secretion into the host cell. Presented at Am. Soc. Cell. Biol. Annu. Meet., San Francisco, Dec. 7–11

166. Michel R, Schupp S, Raether W, Bierther FW. 1980. Formation of a close junction during invasion of eythrocytes by *Toxoplasma gondii in vitro. Int. J. Parasitol.* 10:309–13

167. Mineo JR, McLeod R, Mack D, Khan IA, Ely KH, et al. 1993. Antibodies to *Toxoplasma gondii* major surface protein (SAG-1, P30) inhibit infection of host cells and are produced in murine intestine after per-oral infection. *Immunology* 150:3951–64

168. Mintz CS, Chen J, Shuman HA. 1988. Isolation and characterization of auxotrophic mutants of *Legionella pneumophila* that fail to multiply in human monocytes. *Infect. Immun.* 56:1449–55

169. Mor N, Goren M. 1987. Discrepancy in assessment of phagosome-lysosome fusion with two lysosomal markers in murine macrophages infected with *Candida albicans. Infec. Immun.* 55:1663–67

170. Mordue D, Sibley LD. 1996. Kinetics of phagosome processing and its avoidance following active *Toxoplasma* inva-

sion. Presented at Am. Soc. Cell Biol. Annu. Meet., San Francisco, Dec. 7–11

171. Morisaki JH, Heuser JE, Sibley LD. 1995. Invasion of *Toxoplasma gondii* occurs by active penetration of the host cell. *J. Cell Sci.* 108:2457–64

172. Morrissette NS, Bedian V, Webster P, Roos DS. 1994. Characterization of extreme apical antigens from *Toxoplasma gondii*. *Exp. Parasitol.* 79:445–59

173. Moulder JW. 1962. *The Biochemistry of Intracellular Parasitism,* p.122–24. Chicago: Univ. Chicago Press

174. Moulder JW. 1970. Glucose metabolism of L cells before and after infection with *Chlamydia psittaci. J. Bacteriol.* 104:1189–96

175. Moulder JW. 1985. Comparative biology of intracellular parasitism. *Microb. Rev.* 49:298–337

176. Moulder JW. 1991. Interactions of *Chlamydiae* with host cells *in vitro. Microbiol. Rev.* 55:143–90

177. Muller WA, Steinman RM, Cohn ZA. 1980. The membrane proteins of the vacuolar system II. Bi-directional flow between secondary lysosomes and plasma membrane. *J. Cell Biol.* 86:304–13

178. Nagel SD, Boothroyd JC. 1989. The major surface antigen, P30, of *Toxoplasma gondii* is anchored by a gycolipid. *J. Biol. Chem.* 264:5569–74

179. Nash TW, Libby DM, Horwitz MA. 1984. Interaction between the Legionnaires disease bacterium (*Legionella pneumophila*) and human alveolar macrophages. Influence of antibody, lymphokines and hydrocortisone. *J. Clin. Invest.* 74:771–82

180. Newhall WJ. 1987. Biosynthesis and disulphide cross-linking of outer membrane components during the growth cycle of *Chlamydia trachomatis. Infect. Immun.* 55:162–68

181. Newhall WJ. 1988. Macromolecular and antigenic composition of *Chlamydiae.* In *Microbiology of Chlamydia,* ed. AL Barron, pp. 48–66. Boca Raton FL: CRC

182. Nichols BA, Chiappino ML, O'Connor GR. 1983. Secretion from the rhoptries of *Toxoplasma gondii* during host-cell invasion. *J. Ultrastruct. Res.* 83:85–98

183. Nichols BA, O'Connor GR. 1981. Penetration of mouse peritoneal macrophages by the protozoon *Toxoplasma gondii. Lab. Invest.* 44:324–34

184. Nichols BA, Setzer PY, Pang G, Dawson CR. 1985. New view of the surface projections of *Chlamydia trachomatis. J. Bacteriol.* 164:344–49

185. Norrby R. 1971. Immunological study

on the host cell penetration factor of *Toxoplasma gondii. Infect. Immun.* 3:278–86

186. Oh YK, Alpuche-Aranda C, Berthaiume E, Jinks T, Miller SI, Swanson JR. 1996. Rapid and complete fusion of macrophage lysosomes with phagosomes containing *Salmonella typhimurium. Infect. Immun.* 64:3877–83

187. Oh YK, Straubinger RM. 1996. Intracellular fate of *Mycobacterium avium:* use of dual label spectrophotometry to investigate the influence of bacterial viability and opsonization on phagosomal pH and phagosome-lysosome interaction. *Infect. Immun.* 64:319–25

188. Oh YK, Swanson JA. 1996. Different fates of phagocytosed particles after delivery into macrophage lysosomes. *J. Cell Biol.* 132:585–93

189. Ossorio PN, Dubremetz JF, Joiner KA. 1994. A soluble secretory protein of the intracellular parasite *Toxoplasma gondii* associates with the parasitophorous vacuole membrane through hydrophobic interactions. *J. Biol. Chem.* 269:15350–57

190. Ossowski L, Becker Y, Bernkopf H. 1965. Amino acid requirements of trachoma strains and other agents in the PLT group in cell culture. *Isr. J. Med. Sci.* 1:186–93

191. Paterson EM, de la Maza LM. 1988. *Chlamydia* parasitism: ultrastructural characterization of the interaction between the chlamydial cell envelope and the host cell. *J. Bacteriol.* 170:1389–92

192. Payne NR, Horwitz MA. 1987. Phagocytosis of *Legionella pneumophila* is mediated by human monocyte complement receptors. *J. Exp. Med.* 166:1377–89

193. Pearce BMF, Robinson MS. 1990. Clathrin, adaptors and sorting. *Annu. Rev. Cell Biol.* 6:151–71

194. Perroto J, Kreister DB, Gelderman AH. 1977. Incorporation of precursors into *Toxoplasma* DNA. *J. Protozool.* 18:470–73

195. Pfefferkorn ER. 1984. Interferon gamma blocks the growth of *Toxoplasma gondii* in human fibroblasts by inducing the host cells to degrade tryptophan. *Proc. Natl. Acad. Sci. USA* 81:908–12

196. Pfefferkorn ER, Borotz SE, Nothnagel RF. 1993. Mutants of *Toxoplasma gondii* resistant to atovaquone (566C80) or decoquinate. *J. Parasitol.* 79:559–64

197. Pfefferkorn ER, Eckel M, Rebhun S. 1986. Interferon-γ suppresses the growth of *Toxoplasma gondii* in human fi-

broblasts through starvation for trypto-phan. *Mol. Biochem. Parasitol.* 20:215–24

198. Pfefferkorn ER, Pfefferkorn LC. 1976. Arabinosyl nucleosides inhibit *Toxoplasma gondii* and allow the selection of resistant mutants. *J. Parasitol.* 62:994–99

199. Pfefferkorn ER, Pfefferkorn LC. 1978. The biochemical basis for resistance to adenine in a mutant of *Toxoplasma gondii. J. Parasitol.* 64:486–92

200. Pfefferkorn ER, Rebhun S, Eckel M. 1986. Characterization of an indoleamine 2,3-dioxygenase induced by gamma-interferon in cultured human fibroblasts. *J. Interf. Res.* 6:267–79

201. Pitt A, Mayorga L, Stahl P, Schwartz A. 1992. Alterations in the protein composition of maturing phagosomes. *J. Clin. Invest.* 90:1978–83

202. Pitt A, Mayorga LS, Schwartz AL, Stahl PD. 1992. Transport of phagosomal components to an endosomal compartment. *J. Biol. Chem.* 267:126–32

203. Plaunt MR, Hatch TP. 1988. Protein synthesis early in the developmental cycle of *Chlamydia psittaci. Infect. Immun.* 56:3021–25

204. Polotsky VY, Belisle JT, Mikusova K, Ezekowitz RAB, Joiner KA. 1997. Interaction of human mannose binding protein with *Mycobacterium avium. J. Infect. Dis.* In press

205. Porchet-Hennere E, Torpier G. 1983. Relations entre *Toxoplasma* et sa cellule-hote. *Protistologica* 19:357–70

206. Prain CJ, Pearce JH. 1989. Ultrastructural studies on the intracellular fate of *Chlamydia psittaci* (strain GPIC) and *Chlamydia trachomatis* (strain lymphogranuloma venereum 434): modulation of intracellular events and relationship with the endocytic mechanism. *J. Gen. Microbiol.* 135:2107–23

207. Prinzis S, Chatterjee D, Brennan P. 1993. Structure and antigenicity of lipoarabinomannan from *Mycobacterium bovis* BCG. 139:2649–58

208. Rabinovitch M, Veras PST. 1996. Cohabitation of *Leishmania amazonensis* and *Coxiella burnetti. Trends Microbiol.* 4:158–61

209. Rabinowitz S, Horstmann H, Gordon S, Griffiths G. 1992. Immunocytochemical characterization of the endocytic and phagolysosomal compartments in peritoneal macrophages. *J. Cell Biol.* 116:95–112

210. Rabinowitz SS, Gordon S. 1991. Macrosialin, a macrophage restricted membrane glycoprotein differentially glycosylated in response to inflammatory stimuli. *J. Exp. Med.* 174:827–37

211. Racoosin E, Swanson J. 1993. Macropinosome maturation and fusion with tubular lysosomes in macrophages. *J. Cell. Biol.* 121:1011–20

212. Rathman M, Sjaastad MD, Falkow S. 1996. Acidification of phagosomes containing *Salmonella typhimurium* in murine macrophages. *Infect. Immun.* 64:2765–73

213. Raulston JE. 1995. Chlamydial envelope components and pathogen-host cell interactions. *Mol. Microbiol.* 15:607–16

214. Raynal P, Pollard HB. 1994. Annexins: the problem of assessing the biological role for a gene family of multifunctional calcium- and phospholipid-binding proteins. *Biochim. Biophys. Acta* 1197:63–93

215. Rechitzer C, Blum J. 1989. Engulfment of the Philadelphia strain of *Legionella pneumophila* within pseudopod coils in human phagocytes. Comparison with other *Legionella* strains and species. *Acta Pathol. Microbiol. Immunol. Scand.* 97:105–14

216. Reed SI, Anderson LA, Jenkins HM. 1981. Use of cyclohexamide to study independent lipid metabolism of *Chlamydia trachomatis* cultivated in mouse L cells in serum free medium. *Infect. Immun.* 31:668–73

217. Reynolds DJ, Pearce JB. 1990. Characterization of the cytochalasin-D resistant (pinocytic) mechanism of endocytosis utilized by chlamydiae. *Infect. Immun.* 58:3208–16

218. Ristoph JD, Hedlund KW, Gowda S. 1981. Chemically defined medium for *Legionella pneumophila* growth. *J. Clin. Microbiol.* 13:115–19

219. Rittig MG, Haupl T, Burmester GR. 1994. Coiling phagocytosis: a way for MHC class I presentation of bacterial antigens? *Int. Arch. Allergy Immunol.* 103:4–10

220. Robinson MS, Watts C, Zerial M. 1996. Membrane dynamics in endocytosis. *Cell* 84:13–21

221. Rockey DD, Fischer ER, Hackstadt T. 1996. Temporal analysis of the developing *Chlamydia psittaci* inclusion using fluorescence and electron microscopy. *Infect. Immun.* 64:4269–78

222. Rockey DD, Grosenbach D, Hruby DE, Peacock MG, Heizen RA, Hackstadt T. 1997. *Chlamydia psittaci* IncA is phosphorylated by the host cell and is ex-

posed on the cytoplasmic face of the developing inclusion. *Mol. Microbiol.* In press

223. Rockey DD, Heizen RA, Hackstadt T. 1995. Cloning and characterization of a *Chlamydia psittaci* gene coding for a protein localized in the inclusion membrane of infected cells. *Mol. Microbiol.* 15:617–26

224. Rockey DD, Rosquist JL. 1994. Protein antigens of *Chlamydia psittaci* present in infected cells but not detected in the infectious elementary body. *Infect. Immun.* 62:106–12

225. Rothman JE. 1994. Mechanisms of intracellular protein transport. *Nature* 372:55–63

226. Roy CR, Isberg RR. 1996. The topology of *Legionella pneumophila* DotA: an inner membrane protein required for replication in macrophages. *Infect. Immun.* 65:571–78

227. Russell DG. 1995. *Mycobacterium* and *Leishmania*: stowaways in the endosomal network. *Trends Cell Biol.* 5:125–28

228. Russell DG, Dant J, Sturgill-Koszycki S. 1996. *Mycobacterium avium-* and *Mycobacterium tuberculosis*-containing vacuoles are dynamic, fusion competent vesicles that are accessible to glycosphingolipids from the host cell plasmalemma. *J. Immunol.* 156:4764–73

229. Russell DG, Sinden RE. 1981. The role of the cytoskeleton in the motility of coccidian parasites. *J. Cell Sci.* 50:345–59

229a. Rybin V, Ullrich O, Rubino M, Alexandrov K, Simon I, et al. 1996. GTPase activity of Rab5 acts as a timer for endocytic membrane fusion. *Nature* 383:266–69

230. Ryning FW, Remington JS. 1978. Effect of cytochalasin D on *Toxoplasma gondii* cell entry. *Infect. Immun.* 20:739–43

231. Sadosky AB, Wiater LA, Shuman HA. 1993. Identification of *Legionella pneumophila* genes required for growth within and killing of human macrophages. *Infect. Immun.* 61:5361–73

232. Saffer LD, Long-Krug SA, Schwartzman JD. 1989. The role of phospholipase in host cell penetration by *Toxoplasma gondii*. *Am. J. Trop. Med. Hyg.* 40:145–49

233. Saffer LD, Mercereau-Puijalon O, Dubremetz JF, Schwartzman J. 1992. Localization of a *Toxoplasma gondii* rhoptry protein by immunoelectron microscopy during and after host cell penetration. *J. Protozool.* 39:526–30

234. Saffer LD, Schwartzman JD. 1991. A soluble phospholipase of *Toxoplasma gondii* associated with host cell penetration. *J. Protozool.* 38:454–60

235. Schlesinger L. 1993. Macrophage phagocytosis of virulent but not attenuated strains of *Mycobacterium tuberculosis* is mediated by mannose receptors in addition to complement receptors. *J. Immunol.* 150:2920–30

236. Schlesinger L, Hull S, Kaufman T. 1994. Binding of the terminal mannosyl units of lipoarabinomannan from a virulent strain of *Mycobacterium tuberculosis* to human macrophages. *J. Immunol.* 152:4070–79

237. Schlesinger LS. 1996. Entry of *Mycobacterium tuberculosis* into mononuclear phagocytes. *Curr. Top. Microbiol. Immunol.* 215:71–96

238. Schlesinger LS, Bellinger-Kawahara CG, Payne NR, Horwitz MA. 1990. Phagocytosis of *Mycobacterium tuberculosis* is mediated by human monocyte complement receptors and complement component C3. *J. Immunol.* 144(7):2771–80

239. Schramm N, Bagnell CR, Wyrick PB. 1996. Vesicles containing *Chlamydia trachomatis* serovar L2 remain above pH 6 within HEC-1B cells. *Infect. Immun.* 64:1208–14

240. Schwab JC, Afifi M, Pizzorno G, Handschumacher RE, Joiner KA. 1995. *Toxoplasma gondii* tachyzoites possess an unusual plasma membrane adenosine transporter. *Mol. Biochem. Parasitol.* 70:59–69

241. Schwab JC, Beckers CJM, Joiner KA. 1994. The parasitophorous vacuole membrane surrounding intracellular *Toxoplasma gondii* functions as a molecular sieve. *Proc. Natl. Acad. Sci. USA* 91:509–13

242. Schwartzman JD. 1986. Inhibition of a penetration enhancing factor of *Toxoplasma gondii* by monoclonal antibodies specific for rhoptries. *Infect. Immun.* 51:760–64

243. Schwartzman JD, Pfefferkorn ER. 1982. *Toxoplasma gondii*: purine synthesis and salvage in mutant host cells and parasites. *Exp. Parasitol.* 53:77–86

244. Schwartzman JD, Pfefferkorn ER. 1983. Immunofluorescent localization of myosin at the anterior pole of the coccidian *Toxoplasma gondii*. *J. Protozool.* 30:657–61

245. Schwartzman JD, Saffer LD. 1992. How *Toxoplasma gondii* gets into and out of host cells. *Subcell. Biochem.* 18:333–64

246. Scidmore MA, Fischer ER, Hackstadt T. 1996. Sphingolipids and glycoproteins are differentially trafficked to the *Chlamydia trachomatis* inclusion. *J. Cell Biol.* 134:363–74

247. Scidmore MA, Rockey DD, Fischer ER, Heizen RA, Hackstadt T. 1996. Vesicular interactions of the *Chlamydia trachomatis* inclusion are determined by early protein synthesis rather than route of entry. *Infect. Immun.* 64:5366–72

248. Seaman MN, Byrd CG, Emr S. 1996. Receptor signalling and the regulation of endocytic membrane traffic. *Curr. Opin. Cell Biol.* 8:549–56

249. Seglen PO, Bohley P. 1992. Autophagy and other vacuolar protein degradation mechanisms. *Experimentia* 48:158–72

250. Sheffield HG, Melton ML. 1970. *Toxoplasma gondii:* the oocyst sporozoite and infection of cultured cells. *Science* 167:892–93

251. Sibley LD. 1995. Invasion of vertebrate cells by *Toxoplasma gondii. Trends Cell Biol.* 5:129–32

252. Sibley LD, Dobrowolski J, Morisaki JH, Heuser JE. 1994. Invasion and intracellular survival *by Toxoplasma gondii.* In *Invasion and Intracellular Survival by Toxoplasma gondii,* ed. DG Russell, pp. 245–64. London: Bailleire Tindall

253. Sibley LD, Franzblau SG, Krahenbuhl JL. 1987. Intracellular fate of *Mycobacterium leprae* in normal and activated mouse macrophages. *Infect. Immun.* 55(3):680–85

254. Sibley LD, Krahenbuhl JL, Adams GMW, Weidner E. 1986. *Toxoplasma* modifies macrophage phagosomes by secretion of a vesicular network rich in surface proteins. *J. Cell Biol.* 103:867–74

255. Sibley LD, Messina M, Niesman IR. 1994. Stable DNA transformation in the obligate intracellular parasite *Toxoplasma gondii* by complementation of tryptophan auxotrophy. *Proc. Natl. Acad. Sci. USA* 91:5508–12

256. Sibley LD, Niesman IR, Asai T, Takeuchi T. 1994. *Toxoplasma gondii:* secretion of a potent nucleoside triphosphate hydrolase into the parasitophorous vacuole. *Exp. Parasitol.* 79:301–11

257. Sibley LD, Niesman IR, Parmley SF, Cesbron-Delauw M-F. 1995. Regulated secretion of multi-lamellar vesicles leads to formation of a tubulovesicular network in host-cell vacuoles occupied by *Toxoplasma gondii. J. Cell Sci.* 108:1669–77

258. Sibley LD, Pouletty C, Boothroyd JC. 1993. Formation and modification of the parasitophorous vacuole occupied by *Toxoplasma gondii.* In *Toxoplasmosis,* ed. JE Smith, pp. 63–72. Berlin: Springer-Verlag

259. Sibley LD, Weidner E, Krahenbuhl JL. 1985. Phagosome acidification blocked by intracellular *Toxoplasma gondii. Nature* 315:416–19

260. Silverman JA, Joiner KA. 1996. *Toxoplasma*-host cell interaction. In *Host Response to Intracellular Pathogens,* ed. SHE Kaufman, pp. 313–38. Austin, TX: Landers

261. Simons K, Ikonen E. 1997. Sphingolipid-cholesterol rafts in membrane trafficking, signalling and disease. *Nature.* In press

262. Sinai A, Webster P, Joiner KA. 1997. Association of host cell endoplasmic reticulum and mitochondria with the *Toxoplasma gondii* parasitophorous vacuole membrane—a high affinity interaction. *J. Cell Sci.* In press

263. Smith J. 1995. A ubiquitous intracellular parasite: the cellular biology of *Toxoplasma gondii. Int. J. Parasitol.* 25:1301–9

264. Soderlund G, Kihlstrom E. 1983. Effect of methylamine and monodansylcadaverine on the susceptibility of McCoy cells to *Chlamydia trachomatis* infection. *Infect. Immun.* 40:534–41

265. Soldati D, Kim K, Kampmeier J, Dubremetz JF, Boothroyd JC. 1995. Complementation of a *Toxoplasma gondii* ROP1 knock-out mutant using phleomycin selection. *Mol. Biochem. Parasitol.* 74: 87–97

266. Sollner T, Whiteheart SW, Brunner M, Erdjument-Bromage H, Geromanos S, et al. 1993. SNAP receptors implicated in vesicle targeting and fusion. *Nature* 362: 318–24

267. Stephens RS. 1993. Challenge of *Chlamydia* research. *Infect. Agents Dis.* 1:279–93

268. Stephens RS. 1994. Molecular mimicry and *Chlamydia trachomatis* infection of eukaryotic cells. *Trends Microbiol.* 2:99–101

269. Stirling P, Allan I, Pearce JH. 1983. Interference with transformation of *Chlamydiae* from reproductive to infective body forms by deprivation of cysteine. *FEMS Microbiol. Lett.* 19:133–36

270. Storrie B, Desjardin M. 1996. The biogenesis of lysosomes: is it a kiss and run, continuous fusion and fission process? *BioEssays* 18:895–903

271. Storz J, Spears P. 1978. *Chlamydiales:* properties, cycles of development and effect on eukaryotic host cells. *Curr. Top. Microbiol. Immunol.* 77:168–214

272. Sturgill-Koszycki S, Schaible UE, Russell DG. 1996. *Mycobacterium* containing phagosomes are accessible to early endosomes and reflect a transitional state in normal phagosome biogenesis. *EMBO J.* 15:6960–68

273. Sturgill-Koszycki S, Schlesinger PH, Chakraborty P, Haddix PL, Collins HL, et al. 1994. Lack of acidification in *Mycobacterium* phagosomes produced by exclusion of the vesicular proton ATPase. *Science* 263:678–81

274. Su H, Watkins NG, Zhang Y-X, Caldwell HD. 1990. *Chlamydia trachomatis-* host cell interactions: role of *Chlamydia* major outer membrane protein as an adhesin. *Infect. Immun.* 58:1017–25

275. Su H, Zhang Y-X, Barrera O, Watkins NG, Cladwell HD. 1988. Differential effect of trypsin on infectivity of *Chlamydia trachomatis:* loss of infectivity requires cleavage of major outer membrane protein variable domains II and IV. *Infect. Immun.* 56:2094–100

276. Suss-Toby E, Zimmerberg J, Ward GE. 1996. *Toxoplasma* invasion: the parasitophorous vacuole is formed from host cell plasma membrane and pinches off via a fission pore. *Proc. Natl. Acad. Sci. USA* 93:8413–18

277. Swanson AF, Kuo C-C. 1991. Evidence that the major outer membrane protein of *Chlamydia trachomatis* is glycosylated. *Infect. Immun.* 59:2120–25

278. Swanson MS, Isberg RR. 1995. Association of *Legionella pneumophila* with the macrophage endoplasmic reticulum. *Infect. Immun.* 63:3609–20

279. Swanson MS, Isberg RR. 1996. Identification of *Legionella pneumophila* mutants that have aberrant intracellular fates. *Infect. Immun.* 64:2585–94

280. Tamura A, Imanaga M. 1965. RNA synthesis in cells infected with the meningopneumonitis virus. *J. Mol. Biol.* 11:97–108

281. Tanabe K. 1985. Visualization of the mitochondria of *Toxoplasma gondii-* infected mouse fibroblasts by the cationic permeant fluorescent dye rhodamine 123. *Experientia* 41:101–2

282. Tanabe K, Murakami K. 1984. Reduction in the mitochondrial membrane potential of *Toxoplasma gondii* after invasion of host cells. *J. Cell Sci.* 70:73–81

283. Taraska T, Ward DM, Ajioka RS, Wyrick PB, Davis-Kaplan SR, et al. 1996. The late chlamydial inclusion membrane is not derived from the endocytic pathway and is relatively deficient in host proteins. *Infect. Immun.* 64:3713–27

284. Tardieux I, Webster P, Ravesloot J, Boron W, Lunn JA, et al. 1992. Lysosome recruitment and fusion are early events required for trypanosome invasion of mammalian cells. *Cell* 71:1117–30

285. Tavare J, Blythe WA, Ballard RC. 1974. Interactions of TRIC agents with macrophages: effects on lysosomal enzymes of the cell. *J. Hyg.* 72:297–309

286. Tesh MJ, Morse SA, Miller RD. 1983. Intermediary metabolism in *Legionella pneumophila:* utilization of amino acids and other compounds as energy sources. *J. Bacteriol.* 154:1104–9

287. Theriot JA. 1995. The cell biology of infection by intracellular bacterial pathogens. *Annu. Rev. Cell Dev. Biol.* 11:213–39

288. Thomas SM, Garrity LF, Brandt CR, Schobert CS, Feng GS, et al. 1993. IFN-γ mediated antimicrobial response: indoleamine 2,3-dioxygenase deficient mutant host cells no longer inhibit *Chlamydia* spp. or *Toxoplasma gondii. J. Immunol.* 150:5529–34

289. Ting LM, Hsia RC, Haidaris CG, Bavoil PM. 1995. Interactions of outer envelope proteins of *Chlamydia psittaci* GPIC with the HeLa cell surface. *Infect. Immun.* 63:3600–8

290. Tipples G, McClarty G. 1993. The obligate intracellular bacterium *Chlamydia trachomatis* is auxotrophic for three of the four ribonucleotides. *Mol. Microbiol.* 8:1105–14

291. Tipples G, McClarty G. 1995. Cloning and expression of the *Chlamydia trachomatis* gene for CTP synthase. *J. Biol. Chem.* 270:7908–14

292. Tomavo S, Schwarz R, Dubremetz JF. 1989. Evidence for glycosyl-phosphatidyl inositol anchor of *Toxoplasma gondii* surface antigens. *Mol. Cell. Biol.* 9:4576–80

293. Tomavo S, Wyls V, Toursel C. 1996. Targeted disruption of P63 gene of *Toxoplasma gondii* reduces attachment to host cells and invasion. Presented at VIIth Mol. Parasitol. Meet., Woods Hole, MA

294. Tribby IIE, Friis RR, Moulder JW. 1973. Effect of chloramphenicol, rifampin, and naladixic acid on *Chlamydia psittaci* grown in L cells. *J. Infect. Dis.* 127:155–63

295. Tribby IIE, Moulder JW. 1996. Availability of bases and nucleosides as precursors of nucleic acids in L cells and in the agent of meningopneumonitis. *J. Bacteriol.* 91:2362–67

296. Vance JE, Shiao Y-J. 1996. Intracellular trafficking of phospholipids: import of phosphatidylserine into mitochondria. *Anticancer Res.* 16:1333–40

297. Ward ME. 1988. The chlamydial developmental cycle. In *The Microbiology of Chlamydia,* ed. AL Baron, pp. 71–98. Boca Raton, FL: CRC

298. Ward ME, Murray A. 1984. Control mechanisms governing the infectivity of *Chlamydia trachomatis* for HeLa cells: mechanisms of endocytosis. *J. Gen. Microbiol.* 130:1765–80

299. Warren WJ, Miller RD. 1979. Growth of Legionnaires disease bacterium (*Legionella pneumophila*) in chemically defined medium. *J. Gen. Microbiol.* 10:50–55

300. Weiss E, Peacock MG, Williams JC. 1980. Glucose and glutamate metabolism of *Legionella pneumophila. Curr. Microbiol.* 4:1–6

301. Winkler HH. 1976. Rickettsial permeability. An ADP-ATP transport system. *J. Biol. Chem.* 251:389–96

302. Winkler HH. 1990. Rickettsia species (as organisms). *Annu. Rev. Microbiol.* 44:131–53

303. Wylie JL, Berry JD, McClarty G. 1996. *Chlamydia trachomatis* CTP synthase: molecular characterization and developmental regulation of expression. *Mol. Microbiol.* 22:631–42

304. Wyrick PB, Brownridge EA. 1978. Growth of *Chlamydia psittaci* in macrophages. *Infect. Immun.* 19:1054–60

305. Wyrick PB, Choong J, Davis CH, Knight ST, Royal MO, et al. 1989. Entry of genital *Chlamydia trachomatis* into polarized human epithelial cells. *Infect. Immun.* 57:2378–89

306. Xu S, Cooper A, Sturgill-Koszycki S, vanHeyningen T, Chatterjee D, et al. 1994. Intracellular trafficking in *Mycobacterium tuberculosis* and *Mycobacterium avium* infected macrophages. *J. Immunol.* 153:2568–78

307. Zeichner SL. 1982. Isolation and characterization of phagosomes containing *Chlamydia psittaci* from L cells. *Infect. Immun.* 38:325–42

308. Zeichner SL. 1983. Isolation and characterization of macrophage phagosomes containing infectious and heat-inactivated *Chlamydia psittaci:* two phagosomes with different intracellular behaviors. *Infect. Immun.* 40:956–66

309. Zhang JP, Stephens RS. 1992. Mechanism of attachment of *Chlamydia trachomatis* to eukaryotic cells. *Cell* 69:861–69

Annu. Rev. Microbiol. 1997. 51:463–89

TRANSCRIPTION OF PROTEIN-CODING GENES IN TRYPANOSOMES BY RNA POLYMERASE I

Mary Gwo-Shu Lee

Department of Pathology, New York University, 550 First Avenue, New York, New York 10016

Lex H. T. Van der Ploeg

Merck Research Laboratories (RY80M-213), Department of Genetics and Molecular Biology, PO Box 2000, Rahway, New Jersey, 07065

KEY WORDS: RNA polymerase, trypanosome, transcription, α-amanitin, rRNA, protein-coding gene

ABSTRACT

In eukaryotes, RNA polymerase (pol) II transcribes the protein-coding genes, whereas RNA pol I transcribes the genes that encode the three RNA species of the ribosome [the ribosomal RNAs (rRNAs)] at the nucleolus. Protozoan parasites of the order *Kinetoplastida* may represent an exception, because pol I can mediate the expression of exogenously introduced protein-coding genes in these single-cell organisms. A unique molecular mechanism, which leads to pre-mRNA maturation by trans-splicing, facilitates pol I–mediated protein-coding gene expression in trypanosomes. Trans-splicing adds a capped 39-nucleotide mini-exon, or spliced leader transcript, to the 5' end of the main coding exon posttranscriptionally. In other eukaryotes, the addition of a 5' cap, which is essential for mRNA function, occurs exclusively as a result of RNA pol II–mediated transcription. Given the assumption that cap addition represents the limiting factor, trans-splicing may have uncoupled the requirement for RNA pol II–mediated mRNA production. A comparison of the α-amanitin sensitivity of transcription in naturally occurring trypanosome protein-coding genes reveals that a unique subset of protein-coding genes—the variant surface glycoprotein (VSG) expression sites and the procyclin or the procyclic acidic repetitive protein (PARP) genes—are transcribed by an RNA polymerase that is resistant to the mushroom toxin α-amanitin, a characteristic of transcription by RNA pol I. Promoter analysis and a pharmacological characterization of the RNA polymerase that transcribes these genes have

463

0066-4227/97/1001-0463$08.00

strengthened the proposal that the VSG expression sites and the PARP genes represent naturally occurring protein-coding genes that are transcribed by RNA pol I.

CONTENTS

INTRODUCTION

Trypanosomes are unicellular eukaryotic flagellates. African trypanosomes are the causative agents of sleeping sickness in humans and nagana in cattle, both of which are diseases endemic to large parts of tropical Africa. *Trypanosoma cruzi* is the causative agent of Chagas' disease, which is endemic to South America. Untreated infections with these organisms are often fatal. The medical relevance of the molecular genetics of trypanosomatids has made it a focus for numerous research programs. These evolutionarily divergent organisms (146) exhibit many exciting biochemical novelties not previously encountered in other eukaryotes. These discoveries are covered in several recent reviews on RNA editing (8, 142, 150), the kinetoplast DNA (15, 44, 133), antigenic variation (16–18, 36, 116, 160, 162), trans-splicing (1, 83, 161), and aspects of genome organization and gene expression (27, 29, 30, 45, 112, 166). Here, we restrict our discussion to RNA pol I–mediated protein-coding gene expression (27, 130, 181).

PARASITE LIFE CYCLE

Trypanosoma brucei lives freely in the blood of its mammalian host, where it is subject to continuous immune surveillance. Its escape from immune-attack in the blood is accomplished in an ingenious manner. The bloodstream-form of *T. brucei* is covered with a dense surface coat, made up of identical glycoproteins known as variant surface glycoproteins (VSGs). By periodically switching to the expression of a new VSG gene, and thus changing the antigenic identity of the VSG coat, individual trypanosomes survive immune surveillance. The infection can thus persist in blood, despite the ongoing elimination of those parasites that failed to switch to the expression of a new VSG coat. Aspects of antigenic variation of African trypanosomes have been reviewed and we will not spend much effort elaborating the details of its mechanisms (16, 18, 36, 112, 116, 160, 162). The VSG gene is important, however, to the main theme of this review because current evidence indicates that it represents the first example of a naturally occurring protein-coding gene that may be transcribed by RNA pol I.

Transmission of the African trypanosome *T. brucei* is initiated by the ingestion of an infected blood meal by the tsetse fly (45, 169, 170). When the tsetse fly takes a blood meal from an infected animal, trypanosomes are transferred to the midgut of the fly. The trypanosomes then differentiate and lose their VSG coat, which is replaced by the procyclic acidic repetitive protein (PARP or procyclin) (6, 124, 125). In the tsetse fly, insect-form (procyclic) trypanosomes migrate to the salivary glands, where they differentiate into nondividing metacyclic forms that are pre-adapted to life in the bloodstream of mammalian hosts. These trypanosomes again express a VSG coat. When the fly initiates probing for a blood meal, metacyclic trypanosomes present in the fly's saliva can be injected at the site of skin puncture. From here they can migrate into the bloodstream and initiate infection.

Both the VSG coat and the PARP protein appear essential for the survival of the trypanosome. The VSG accounts for about 10% (10^7 molecules per cell) of the total protein of the bloodstream-form of *T. brucei* (24, 37, 38, 158), whereas there are approximately 6×10^6 (representing ~1% of total protein) PARP protein molecules per cell (32). The application of several molecular genetic tools, including the efficient introduction of foreign DNA into trypanosomes (either transiently or stably by homologous recombination), has facilitated dissecting the process of differential–surface coat expression in trypanosomes and the analysis of the RNA polymerase that transcribes these genes (43, 91, 153).

NUCLEAR GENE EXPRESSION IN TRYPANOSOMES

Discontinuous Transcription

Transcription of protein-coding genes and mRNA maturation in trypanosomes and other kinetoplastids involve mechanisms distinct from those observed in most other eukaryotes. Every mRNA in trypanosomes contains two exons, a 5′ mini-exon and a main coding exon, which are transcribed from two separate

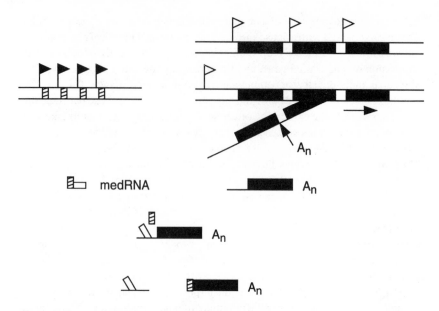

Figure 1   Trans-splicing and polycistronic transcription in trypanosomes. Tandem arrays of mini-exon genes (*left, hatched, and open boxes*), transcribed from promoters (*black flags*) located directly in front of each mini-exon gene, generate 140-nt medRNAs. Polycistronic transcription of protein-coding genes (*black boxes, direction of transcription indicated by the arrow facing right*), from promoters (*open flags*) located directly in front of each gene of the array, or at each of the intergenic regions, is shown schematically in the upper right corner. Precursor RNAs are matured by trans-splicing and cleavage for poly A addition (*arrow with A_n symbol*). The precursor RNAs are joined by trans-splicing (*schematically indicated by the three bottom lines*), thereby generating mature mRNA and a Y-shaped by-product (covalently joined introns) of the trans-splicing reaction.

genes (161). The common 5′ capped 39-nt noncoding mini-exon is derived from a 140-nt mini-exon donor RNA (medRNA), which is encoded by a 200-fold repeated gene (1, 14, 23, 40, 41, 104, 163). The cap structure of the 5′ end of mini-exon sequences contains $m7GpppA^*A^*C^*U^*AA^*CG$ (* indicates base modification) (50, 98). The exon-intron boundaries of the medRNA and the main coding precursor RNA are flanked by the conserved GT-AG dinucleotides of intron boundaries in cis-splicing. Analysis of processing intermediates has shown that trans-splicing of the mini-exon and the main coding exon occurs via a mechanism similar to that of cis-splicing in higher eukaryotes (21, 22, 83, 99, 101, 103, 151, 155, 156, 161) (Figure 1). The exact function of the mini-exon, which is found at the 5′ ends of all mRNAs, has remained unclear. Because of the discontinuous synthesis of nuclear mRNA, the 5′ end of the mRNA does not identify the transcription initiation site of the protein-coding genes, but instead it identifies that of the medRNA genes. This complicates the identification of transcription initiation sites for most RNA pol II–transcribed protein-coding genes. Thus far, only two putative RNA pol II promoters of *T. brucei* have been characterized (7, 86). In these studies, UV inactivation of transcription and promoter analysis by transient transformation demonstrated that: (*a*) the actin promoter mapped to a region located ~4 kb upstream of its coding sequence (7), and (*b*) the intergenic regions of the tandemly arrayed hsp 70 genes could function as RNA pol II promoters (86).

Polycistronic Transcription of Tandemly Arrayed Genes

Most genes in trypanosomes are tightly packed in tandem arrays, encoding copies of the same or similar genes or encoding clusters of different genes. These tandemly linked genes are separated by short intergenic region sequences and are polycistronically transcribed (Figure 1). Consult the following references: as published for the tubulin gene cluster (71); for phosphoglycerate kinase (PGK) genes (53); for calmodulin genes, (157, 177); for hsp 70 gene clusters (54, 92, 90); for VSG gene expression sites (ESs) (2, 58, 76, 82, 115, 183); and for PARP genes (31, 131).

Polycistronic transcription of genes could occur either from promoters located upstream of the tandem arrays of genes, or from individual promoters located in front of each individual gene. Through processing events involving polyadenylation and trans-splicing, intergenic regions are excised from the polycistronic precursor RNA, which results in the generation of multiple, mature, capped mRNAs. The length of the intergenic regions measures, on average, 200–400 nt. Because nonpolyadenylated, branched (Y-shaped), intron-derived RNA of relatively small size is released from the trans-splicing reaction (~180–300 nt; 120), the trypanosome genes may be preferentially encoded into polycistrons. Recent evidence suggests that the length of intergenic regions may be of importance for the efficiency of RNA-processing events (11, 85, 167). The evolutionary forces that molded the *T. brucei* genome into its current structure, and the functional significance of polycistronic transcription of protein-coding genes, remain undefined.

Posttranscriptional Control of mRNA Abundance

As a result of the polycistronic transcription of protein-coding genes, the differential expression of steady state mRNA of individual genes derived from a single polycistronic precursor requires posttranscriptional control of mRNA expression. Only the expression of VSG and PARP genes appears to involve transcriptional control, whereas other differentially expressed protein-coding genes in trypanosomes rely on posttranscriptional control mechanisms (5, 12, 53, 68 74, 87, 152, 154). For instance, the tandemly arrayed PGK A, B, and C genes are transcribed at the same rate in both bloodstream-form and procyclic trypanosomes, whereas the various PGK steady state mRNA levels are differentially regulated in each developmental stage (53). The mechanisms of post-transcriptional control are being elucidated: Poly A addition appears to be partially specified by the location of the downstream 3′ splice acceptor site, indicating that trans-splicing and poly A addition may be coupled (11, 85, 167). The intergenic region polypyrimidine tract, which is required for trans-splicing, may participate in the choice of the upstream polyadenylation site. Mutations, including deletions and substitutions, of the intergenic region polypyrimidine tract can lead to aberrant RNA processing (polyadenylation and trans-splicing), resulting in length-variant untranslated regions and a vastly different mRNA abundance. A plethora of evidence has shown that the 3′ untranslated region (UTR) of *T. brucei* mRNA plays an important role in post-transcriptional regulation of mRNA abundance (66, 68, 70, 117). Similar control mechanisms were also observed in related parasites (4, 107). The mechanisms involved in control of the mRNA abundance that is exerted by the 3′ UTR remain to be determined.

Cap Structure

In other eukaryotes, the 5′ ends of transcripts that are generated by RNA pol I and RNA pol II are structurally distinct, reflecting important functional specialization. RNA Pol I–derived precursor RNAs are characterized by an unmodified triphosphate group at their 5′ ends (147, 148). In the case of RNA pol II–primary transcripts, this triphosphate is rapidly modified by the addition of 7-methyl guanosine triphosphate, joined in a 5′-5′ triphosphate linkage, followed by methylation to form the characteristic cap structure of mRNAs (7mGpppmXpmY...) (134). This process is conducted by an RNA pol II–associated capping activity. This 5′ cap is believed to be essential for pre-mRNA splicing, mRNA transport, mRNA stability, and mRNA translatability. We now believe that the inability of RNA pol I to generate functional mRNA in eukaryotes may be due to the absence of an RNA pol I–associated capping activity (95; 130, 181).

In *T. brucei*, the medRNA and at least four species of small nuclear RNAs (snRNAs U2, U4, U6, and RNA B) contain a 5′ cap [trimethyl-Guanosine (TMG)]. In other eukaryotes, the genes encoding the U2, U6, and RNA B are normally transcribed by RNA pol II, whereas in *T. brucei* they appear to be transcribed by RNA pol III and are capped (48, 63, 94, 97). As expected, the trypanosome RNA pol III–transcribed 5S rRNA is not capped, and it contains a di- or triphosphate at its 5′ end, as described for other eukaryotes (84). The RNA polymerase that controls the transcription of the mini-exon genes is not clearly defined (61, 136). The RNA pol III–specific-inhibitor tagetitoxin failed to affect transcription of the medRNA genes in *Leishmania tarentolae*, which indicates that RNA pol II may be responsible for transcription of medRNA genes in this

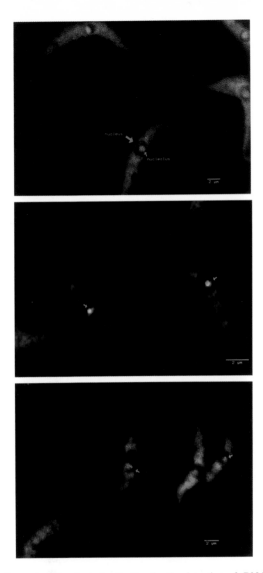

Figure 2 Dual fluorophore in situ hybridization for the detection of rRNA transcripts and *neo* RNA generated from an rRNA promoter that is fused to a PARP 3' splice acceptor site and a neomycin phosphotransferase gene. Pseudocolored confocal-laser scanning images of *neo* RNA, (visualized with rhodamine in *red*) and rRNA (visualized with fluorescein in *green*). *(Top panel)*: Detection of rRNA in wildtype trypanosomes; *(middle panel)*: visualization of the rRNA *(green)* and *neo* RNA *(red)* from a tandem array of PARP promoter *neo* genes integrated at the αß-tubulin locus; *(bottom panel)*: *neo* RNA *(red)* and rRNA *(green)*. *Neo* RNA is transcribed from rRNA promoter *neo* gene tandem arrays that are integrated at an rRNA locus. *(Arrows outline the location of the neo RNA signal at a single discrete site at the edge of the nucleolus.)*

related protozoan (136). Given that all mRNAs in trypanosomatids contain a mini-exon sequence at their 5′ end, it has not yet been feasible to determine the structure of the 5′ end of the RNA pol II–derived pre-mRNA. It is therefore not known whether trypanosomes possess an RNA pol II–associated capping activity.

RNA POLYMERASES IN TRYPANOSOMES

Transcription in eukaryotes, including trypanosomes, is controlled by at least three DNA-dependent RNA polymerases that can be distinguished on the basis of their sensitivity to the mushroom-derived toxin α-amanitin (42, 81). RNA pol I (which is insensitive to α-amanitin up to a concentration of 1 mg/ml) transcribes rRNA genes. RNA pol II (which is inhibited by low concentrations of α-amanitin) transcribes mainly protein-coding genes. RNA pol III (which has an intermediate sensitivity to α-amanitin) transcribes tRNAs, 5S rRNA, and several other small nuclear RNAs. The RNA polymerases are complex multi-subunit enzymes of high molecular weight (generally about 550–700 kDa). An overview of RNA polymerase function, conserved trans-acting protein factors associated with RNA polymerases and mechanisms involved in transcription initiation, elongation, and termination can be found in numerous excellent reviews (33, 51, 52, 96, 111, 122, 126, 147, 148, 168, 178)). Only limited information is available regarding the RNA polymerases in trypanosomes, and only genes that encode the largest subunits of RNA pol -I, -II, and -III have been characterized from T. brucei; other associated protein factors remain to be identified. The following discussion focuses on a structural comparison of the T. brucei RNA polymerase–largest subunit genes, which turned out to be important to our understanding of pol I–mediated protein-coding gene expression.

The deduced amino acid sequence of the largest subunits of RNA pol -I, -II, and -III of T. brucei is conserved when compared to other eukaryotes (34, 35, 47, 75, 77, 78, 144, 145). The largest subunits of each nuclear RNA polymerase are similar among different eukaryotes, and the conserved domains are clustered into 6–8 regions (A–G; from N-terminal to C-terminal). Each of the T. brucei RNA pol I and RNA pol III–largest subunits are encoded by a single gene, whereas the RNA pol II–largest subunit is encoded by two polymorphic genes (46, 47, 145). The deduced amino acid sequences of T. brucei RNA pol -I, -II, and -III–largest subunits are 47%, 61%, and 58% identical to their counterparts in the yeast Saccharomyces cerevisiae, respectively (75, 144). The largest subunit of RNA pol I is distinguished from the largest subunits of RNA pol II and RNA pol III by two domains that are located in nonconserved spacer regions. The first RNA pol I–specific spacer region, which separates the conserved regions A and B, is ~115 amino acids in yeast and ~201 amino acids in *T. brucei*, making it longer than that of other RNA polymerase–largest sub-units. The second putative RNA pol I-specific region is in the hydrophilic acidic spacer region between the conserved regions F and G, and this spacer is 110 and 205 amino acids in yeast and in trypanosome RNA pol I, respectively (again, longer than that found in other largest subunits). Comparison of the sequences of the yeast and trypanosome RNA pol III–largest subunits revealed four short blocks that are potentially RNA pol III specific.

Several unique properties were found in the RNA pol II–largest subunit of *T. brucei*. The *T. brucei* RNA pol II–largest subunits are encoded by two distinct loci (pol II-A and pol II-B, with two alleles each) (46, 47, 145). The proteins encoded at these two loci differ by three amino acid substitutions when compared with different geographic isolates, as characterized in the MiTat serodeme (stock 427-60) and by four amino acid substitutions in the IsTat serodeme (Val to Met, Val to Ile, and Ser to Asn at amino acid positions 130, 390, and 471, respectively, in Mitat 1.1c; Tyr to His, Val to Ile, Ser to Arg, and Met to Leu at positions 317, 390, 1563, and 1589 in IsTat 1.1). Analysis of RNA pol II genes among different antigenic variants of the MiTat serodeme demonstrated that even individual trypanosome variants exhibited allelic heterogeneity at the RNA pol II-A locus (28). In MiTat 118a bloodstream-form and 427-60 procyclic-form trypanosomes, amino acid substitutions were identified at positions 130 and 390, but not at position 471, of their RNA pol II–largest subunit genes. The potential biological consequences of the amino acid substitutions in different RNA pol II–largest subunit genes remain to be elucidated, though they do not appear to generate an α-amanitin–resistant pol II (see next sections).

The presence of two copies of the RNA pol II–largest subunit genes was restricted to salivarian trypanosomes that exhibit antigenic variation, with the possible exception of *T. vivax* (143). In contrast, only one RNA pol II gene was observed in species that do not exhibit antigenic variation, such as *T. cruzi*, *Crithidia fasciculata*, and *Leishmania mexicana mexicana* (47). It is unclear whether the correlation between antigenic variation and this particular RNA Pol II–largest subunit organization will be relevant to the mechanism of antigenic variation, or whether it represents the coevolution of otherwise unrelated parameters. The two sets of RNA pol II–largest subunit genes are not essential for insect-form trypanosomes, because the deletion of both pol II-B alleles did not affect cell viability (28). Finally, trypanosome RNA pol II–largest subunit polypeptides lack the common tandem heptapeptide repeats (Tyr-Ser-Pro-Thr-Pro-Ser) at the carboxyl terminal domain (CTD) of RNA pol II–largest subunits (47, 145). This C-terminal repeat domain is present in RNA pol II–largest subunits of yeast, *Drosophila*, mouse, *Caenorhabditis elegans, Arabidopsis*, and *Plasmodium*, and it is a common feature of eukaryotic RNA pol II. Genetic analysis reveals that the CTD is essential for cell viability. Extensive phosphorylation within the CTD may serve a role in transcription initiation, leading to the release of RNA pol II from the pre-initiation complex (108). The trypanosome RNA pol II C terminus is characterized by a high proportion of proline residues and acidic amino acid residues, and by a high frequency of amino acids Ser and Tyr, which are potential phosphorylation sites. The C-terminal domain of the *T. brucei* RNA pol II–largest subunit may also be associated with release of the elongation complex because it is extensively phosphorylated, despite the absence of a heptapeptide repeat (26).

Protein-Coding Gene Expression by an α-Amanitin-Resistant RNA Polymerase

In 1984, Kooter & Borst reported that transcription of the VSG genes in bloodstream-form trypanosomes is insensitive to α-amanitin, a characteristic of rRNA gene transcription by RNA pol I (81). Subsequently, PARP gene transcription in insect-form trypanosomes was shown to be similarly resistant to α-amanitin (31, 127). A subset of protein-coding genes in trypanosomes is therefore transcribed by an RNA polymerase that differs from the conventional RNA pol II. This raised the possibility that the VSG and the PARP genes might be transcribed by RNA pol I, or by a pol I-like RNA polymerase, rather than by RNA pol II. The finding of two RNA pol II–largest subunit loci in trypanosomes that exhibit antigenic variation led to an alternative hypothesis that invoked a role for the second copy of the RNA pol II–largest subunit genes in α-amanitin-resistant VSG and PARP gene transcription. The latter notion was disproved by the inactivation of both alleles of the pol II-B genes by homologous recombination, which did not affect the α-amanitin sensitivity of gene expression in insect-form trypanosomes. Indirect evidence now supports the notion that RNA pol I, which normally only transcribes the rRNA genes, transcribes VSG and PARP genes (28, see following sections).

RNA POL I–MEDIATED GENE EXPRESSION IN TRYPANOSOMES

Transcription of rRNA Genes in Trypanosoma brucei

In most organisms, rRNA genes are repetitive and they are arranged at multiple loci in clusters of tandem repeats, separated by intergenic spacers. The rRNA repeat in most eukaryotes encodes three mature (18S, 5.8S, and 28S) rRNA species. One unit of the rRNA repeat of *T. brucei* is approximately 18 kb and encodes from 5′ to 3′: the 18S, 5.8S, 28S α, a 200-nt SrRNA, 28S β, a 180-nt SrRNA, a 70-nt SrRNA, a 140-nt SrRNA, and an approximately 8-kb spacer (64, 171). The rRNA loci in *T. brucei* are encoded on at least five different size-classes of larger chromosomes (chromosome bands 19', 16, 15', 10, and 9') and mini-chromosomes of ~60 kb (59, 164, 180).

Transcription of the rRNA repeats starts 1.2 kb in front of the 18S rRNA gene and the entire transcription unit from the promoter to the terminator measures ~10 kb. The primary transcript is first processed to a 3.4-kb precursor RNA encoding 18S rRNA, and a 5.6-kb precursor containing the 5.8S rRNA and the 28S rRNA. The latter is further processed to yield the 5.8S rRNA, and a 5.0-kb precursor that encodes 28S rRNA and five other rRNAs. As in other eukaryotes, the transcribed spacer regions between rRNAs are AT-rich. Most functional domains at the nontranscribed spacer other than the promoter (i.e. terminators, enhancers) remain to be characterized.

Eukaryotic rRNA promoters appear structurally similar, despite the absence of a recognizable sequence homology (111, 122, 147, 148). The rRNA promoter generally contains a core promoter, extending from ~10 bp downstream (+10) to ~40 bp upstream (−40) of the transcription initiation site, and an upstream control element (UCE) from the core promoter to ~−150 nt). The core promoter is required for accurate transcription initiation, and it contains a domain that spans the transcription initiation site and an upstream (or distal) domain that functions in the binding of trans-acting factors. The UCE is not absolutely required for transcription initiation, but it stimulates transcription in vivo, or in vitro, under stringent conditions. The structure of the *T. brucei* rRNA promoter is similar to that of other eukaryotes (see next section).

Transcription of Protein-Coding Genes by RNA Pol I

In most eukaryotes, RNA pol I only directs the transcription of rRNA genes and cannot mediate the expression of functional mRNA, which basically differs from pol I–derived RNA in the presence of an essential 5' cap structure (95). In trypanosomes, the addition of a 5' cap to each mRNA can occur post-transcriptionally by trans-splicing. Thus, in trypanosomes, theoretically any RNA polymerase can be utilized to generate precursor RNAs whose 5' ends resemble capped mRNA. The notion that trypanosomes can efficiently express protein-coding genes by pol I was subsequently tested and proved (130, 181).

Transcription that is mediated by an rRNA promoter was analyzed by fusing this promoter to a PARP gene 3' splice acceptor site and to a reporter gene (72, 130, 181, 182). The rRNA promoter efficiently directed the transcription of a chloramphenicol acetyl transferase (CAT) gene or a luciferase reporter gene in transiently transformed trypanosomes. The production of CAT or luciferase mRNAs was dependent on rRNA promoter control elements, and the expression of CAT or luciferase activity required the presence of a functional 3' splice acceptor site. Mutational analysis demonstrated that sequences from nt −258 to +10, relative to the rRNA transcription initiation site, are essential for full rRNA promoter function. The sequence from nt -75 to $+10$ represents the minimal core promoter sequence (72). Transitions at the transcription start site, at residue ($+1$) (T-C), severely reduced $T.$ $brucei$ rRNA promoter activity (130, 181). Further analysis showed that the rRNA core promoter contains two essential domains, extending from position -42 to -13 and from position -53 to -62 (72). The UCE and other enhancer-like elements of the rRNA promoter have not yet been accurately mapped.

In trypanosome cell lines that were stably transformed with a neomycin phosphotransferase (neo) or hygromycin phosphotransferase (hph) gene under the control of an rRNA promoter and a PARP 3' splice acceptor site, the neo gene transcription was α-amanitin resistant, and the capped 39-nt mini-exon was trans-spliced onto the 5' end of the neo mRNA. Transcription started at the rRNA promoter initiation site and neo RNA could be located to the nucleolus, the site of RNA pol I–mediated rRNA gene expression and ribosomal subunit biogenesis (27, 130, 181) (Figure 2). These results provided conclusive evidence that pol I can mediate protein-coding gene expression in trypanosomes.

Organization of VSG Gene Transcription Units

The VSG and PARP gene transcription units might be controlled by RNA pol I or a pol I–like RNA polymerase, because their transcription is insensitive to the addition of α-amanitin. In this section, we describe the organization of the VSG and the PARP gene transcription units and we summarize the evidence indicating that the ES and the PARP genes may be transcribed by RNA pol I. We do not describe mechanisms involved in antigenic variation of the VSG. A detailed discussion of VSG gene switching can be found in many excellent reviews (16, 18, 36, 112, 116, 160, 162).

The genome of $T.$ $brucei$ encodes potentially as many as 1000 different transcriptionally silent, basic-copy VSG (BC) genes. The BC genes are located in tandem arrays at internal positions of the chromosomes, and at many of the trypanosome telomeres. The transcriptionally active VSG gene is invariably located at a telomeric ES. For a VSG gene to be expressed, a silent BC gene needs to be translocated to an active ES, generating an expression-linked copy (ELC) of the gene. Approximately 20 ESs have been identified in the trypanosome genome. However, only one is active at a time, thus assuring the mutually exclusive transcription of only a single VSG gene. The expression of a new VSG gene can also be brought about by the differential transcriptional control of the ESs, in which the coordinated activation of an ES and the inactivation of the old ES orchestrate an antigenic switch.

Only three ESs of bloodstream-form trypanosomes have been analyzed in detail. These ESs encode polycistronic transcription units that, in addition to the telomerically located VSG gene, encode up to ten expression-site–associated genes (ESAGs) (2, 3, 25, 39, 49, 57, 58, 76, 82, 93, 100, 109, 115, 121, 123, 135, 138, 149, 183). The promoter of the ES is located about 45–55 kb upstream of the VSG gene. A large array of 50-bp repeats, whose function remains to be determined, is located at the 5' end of the VSG gene promoter. From 5' to 3', the ES thus contains a promoter, ~10 ESAGs, a large array of 70-bp repeats, and the VSG gene, flanked by subtelomere and telomere repeats.

Deletional analysis allowed Pham et al and Vanhamme et al to demonstrate that the fully active VSG promoter mapped to a small fragment, located between nucleotide 67 and the transcription initiation site (118, 165). A UCE as described for most eukaryotic RNA pol I promoters could not be identified for the VSG promoter. Linker-scanning mutagenesis revealed that the core VSG promoter contains at least two essential regulatory elements, including sequences within the region −67/−60 and the region −35/−20, upstream of the transcription initiation site. In contrast to the behavior of the rRNA promoter, transitions at the VSG promoter–transcription initiation site did not reduce its activity in transient transformation assays (118). Applying point mutations and double and triple mutations, Vanhamme et al showed that three short nucleotide sequences at positions −61 to −59 (box 1), −38 to −35 (box 2), and −1 to +1 (start site) were essential for the VSG promoter activity (165). Thus, overall, the structure of the VSG core promoter shares the domain organization of RNA pol I promoters. Interestingly, the box 2 element of the VSG promoter is interchangeable with its counterpart derived from the rRNA promoter, again suggesting that the VSG promoter is controlled by RNA pol I. Vanhamme et al identified a specific DNA binding protein of the noncoding strand of box 2. Binding could be competed by box 2–like sequences derived from the metacyclic VSG, rRNA, or PARP promoters (165). Double-stranded DNA binding activity could not be identified.

Transcription of ESs; Exchange of VSG and rRNA Promoters

The mechanism of antigenic variation and the control of ES transcription are intricately linked, and we assume that the understanding of antigenic variation will require the determination of the RNA polymerase that transcribes the VSG genes. The bloodstream-form trypanosome normally activates only one ES, silencing all other ESs. The mechanisms that control the mutually exclusive activation of one ES at a time remain elusive. DNA rearrangement events that might control ES activation or inactivation have not been identified. The only distinguishing feature is the presence of a novel modified nucleotide base "J" (5-β-D-glucopyranosyloxy-methyl-uracil), only found at the inactive ESs (55, 56). Interestingly, when the VSG promoter in a bloodstream-form trypanosome was replaced with an rRNA promoter that is normally constitutively active, the ES rRNA promoter was still efficiently inactivated and reactivated (128), indicating that the type of RNA pol I promoter that drives ES transcription is not crucial to the transcriptional status of the ES.

In insect-form trypanosomes, virtually all ESs are inactivated. It is assumed that the down-regulation of the ESs in insect-form trypanosomes occurs partly at the level of transcription initiation and partly by transcription attenuation (114, 113, 129). As determined by nuclear-run assays, in insect-form trypanosomes, transcription at the VSG promoter appears to terminate in a region ~700 bp downstream from the transcription initiation site. Because VSG promoter sequences are highly conserved and repetitive, it is not simple to determine which promoters are active. DNA nucleotide sequence analysis of transcripts derived from insect-form VSG promoters suggest that many of them are simultaneously active at a low level. This was confirmed by the low level of transcription of an *hph* gene introduced into the region directly downstream of a VSG promoter in insect-form trypanosomes (129). Replacing the VSG promoter with an rRNA promoter restored full activity of the reporter gene at that ES (129) and the PARP promoter, and even the *hsp* 70 intergenic region promoters can direct transcription at the inactivated insect-form ES (86, 88). These results suggest that the repression of the VSG promoter in insect-form trypanosomes has a VSG promoter-sequence–specific character. Additional evidence for VSG promoter-specific repression at the ES in insect-form trypanosomes was provided by Horn & Cross (67). These investigators showed that the PARP, VSG, and rRNA promoters were transcriptionally repressed when integrated into a region between the 70-bp repeats and the 221 VSG gene at the inactive 221 ES of the bloodstream-form variant 118. When these transformed bloodstream-form cell lines differentiated into insect-form trypanosomes, only transcription from the VSG promoter remained repressed, whereas transcription of the rRNA and the PARP promoters at the 221 ES in insect-form trypanosomes was derepressed.

In insect-form trypanosomes, down-regulation of VSG gene promoter activity is also affected by the genomic context in which it resides. First, the VSG promoter is generally repressed at the ES, or when integrated at several RNA pol II–transcribed loci (119, 159, 179); second, the VSG promoter is active when retained on episomes or when tested in transient transformation assays (118, 165, 182); third, the VSG promoter is derepressed when targeted into the rRNA nontranscribed spacer region of insect-form trypanosomes (129); fourth, the VSG gene promoter can be up-regulated in cis by the presence of a fully active PARP promoter (159). Finally, the addition of a fragment derived from the 5′ end of the PARP promoter (nts −743 to −111) that is fused to the 5′ end of the VSG gene promoter resulted in the up-regulation of the VSG promoter, when integrated at chromosomal locations at which the VSG promoter is normally repressed in insect-form trypanosomes (119). As mentioned earlier, in bloodstream-form trypanosomes, the VSG promoter is only active at the transcribed ES or when integrated at an rRNA locus, whereas it is repressed when integrated at many other RNA pol II–transcribed loci (13; MGS Lee, unpublished information). Taken together, these data outline that the genomic environment in which the VSG promoter is located represents a major determining factor of VSG promoter activity. Finally, the VSG, PARP, and rRNA promoters may share control elements.

PARP Gene Transcription Units

The PARP genes, which are also transcribed in an α-amanitin-resistant manner, encode the major surface-coat protein of the insect stage of trypanosomes. These genes are organized in tandem arrays of pairs of genes (PARP α and β), at two distinct loci (PARP A and PARP B) (31, 65, 79, 102, 131). Two alleles of the PARP A locus are identical, whereas the two alleles encoding the PARP B locus are polymorphic (PARP B1 and PARP B2). The protein that is encoded by the PARP Aα gene consists of a domain of tandem pentapeptide repeats (Glycine-proline-glutamic acid-glutamic acid-threonine), and proteins encoded by PARP Aβ and PARP B locus contain dipeptide repeats (Glutamic acid-Proline; 102). Downstream of the PARP genes, one finds several genes that share homology with the ESAGs of the VSG ESs (10, 80, 166).

The PARP loci are also polycistronically transcribed. In contrast to the ES, each PARP polycistronic transcription unit is relatively short (8–10 kb) and controlled by a single promoter located immediately upstream of the PARP gene array (31, 131). Transcription initiates approximately −82 to −86 nts upstream of the 3′ splice acceptor site of the α PARP gene (19, 69, 141). The two ends of each PARP transcription unit are characterized by α-amanitin–sensitive transcription units. Recently, Berberof et al identified a putative transcription termination region that extends approximately 2 kb downstream of the PARP A locus (9). It appears that sequences derived from this putative termination region efficiently inhibit CAT gene expression by the PARP, the rRNA, and the VSG promoter in an orientation-dependent manner. This region did not inhibit RNA pol II transcription.

Structure of the PARP Promoter

The structure of the PARP promoter was analyzed by Brown et al and Sherman et al (19, 141). The minimal PARP promoter extends roughly from position −250 to the transcription initiation site. Linker scanning mutagenesis analysis revealed that the PARP promoter contains at least two critical regions required for full promoter activity. The first region (the core promoter) is located close to the transcription initiation site, spanning nucleotides −69 to +12 with three domains of importance for transcriptional efficiency (nt −69 to −56, −37 to

-11, and -11 to $+12$). The second region locates between nt -140 and -131, and a third locates between nucleotides -228 and -205. Scanning this region with 10-bp replacement analysis, Sherman et al found that mutations at nt -43 to -14 and at nt -73 to -64 of the core promoter region drastically reduced promoter activity and mutations at nts -113 to -94, -139 to -124, and -143 to -164 reduced promoter activity by roughly 50%. In addition, the spacing between the elements of the core promoter significantly affected PARP promoter activity. Mutations at the transcription initiation site affected both promoter activity and the location of the transcription initiation site, effects that are comparable to those observed for the rRNA promoter. Thus, the PARP promoter consists of a core region that has two important domains and a UCE. This structural organization is similar to that of the rRNA promoter. Swapping the UCE and the distal element of the core region of the PARP promoter with corresponding domains of the rRNA promoter also resulted in fully active promoters, which indicates functional similarity in the PARP and rRNA promoters (72). Proteins that recognize the PARP core promoter elements were identified by mobility shift assays. Janz et al identified proteins that bind to double-stranded DNA fragments from the PARP core promoter region (73). However, this binding activity could not be competed by VSG and rRNA gene promoter fragments. Subsequently, a sequence-specific, single-stranded DNA binding complex was identified that interacted with the coding strand of the $-69/-55$ element of the PARP core promoter (20). The presence of specific double-stranded DNA binding elements remains to be resolved (73).

The PARP proteins are exclusively expressed in insect-form *T. brucei*. PARP mRNA is abundant in insect-form trypanosomes, whereas it is found at low levels in bloodstream-form trypanosomes. Nuclear run-on assays demonstrated that the PARP promoter is active in both insect- and bloodstream-form trypanosomes. However, promoter activity is about five- to ten–fold lower in the bloodstream-form than in the insect-form (13, 113, 127). In the insect-form, PARP promoter activity does not seem to be affected by different genomic contexts in which it may be integrated. In contrast, the PARP promoter is generally down-regulated when integrated at different chromosomal loci of bloodstream-form trypanosomes. Unlike the VSG promoter, which is functionally active at the rRNA nontranscribed spacer region of both insect- and bloodstream-form trypanosomes, the PARP promoter remains down-regulated at the nontranscribed spacer region of rRNA genes in bloodstream-form trypanosomes (13). It therefore seems that the PARP promoter cannot be derepressed by placing it within an rRNA repeat array of bloodstream-form trypanosomes. Repression of the PARP promoter appears specific for the bloodstream stage of the parasite.

EVIDENCE FOR VSG AND PARP GENE EXPRESSION BY RNA POL I

Several lines of indirect evidence support the notion that the PARP and VSG genes may be transcribed by RNA pol I. First, the hypothesis that the α-amanitin–resistant transcription of the PARP gene and the VSG promoter sequences in insect-form trypanosomes may be directed by one of the two sets of the RNA pol II–largest subunit genes was excluded by the study of RNA pol II-B–largest subunit knock-out mutant cell lines (28).

Second, the PARP, VSG, and rRNA promoters share structural similarities (19, 72, 118, 141, 165, 182). The RNA pol I promoters analyzed to date are characterized by the presence of a core promoter region that extends from -70 to $+10$ nt and UCE, located at nt -130 to -140. Comparing the location of essential elements of the PARP and rRNA promoters reveals a similar organization when compared to eukaryotic RNA pol I promoters, and these promoters are markedly different from the RNA pol II and RNA pol III promoters. A UCE region was not found in the VSG promoter. Nevertheless, the structure of the VSG core promoter resembles those of the PARP and rRNA promoters, and essential control elements of the PARP and VSG gene promoters are functionally interchangeable with those of the rRNA promoter (72, 165).

Third, the PARP and VSG genes of *T. brucei* do not resemble RNA pol II transcription units in their sensitivity to α-amanitin or to the nonionic detergent N-lauroyl sarcosine (Sarkosyl) in nuclear run-on assays (81, 127, 132). Transcription of the PARP, VSG and rRNA genes is equally resistant to the addition of 1 mg/ml of α-amanitin, whereas transcription of the RNA pol II–transcribed protein-coding genes, mini-exon genes, and the RNA pol III–transcribed 5S rRNA gene is markedly reduced by 90–95%, owing to the addition of 1 mg/ml of α-amanitin. Addition of Sarkosyl (0.6%) in nuclear run-on assays does not affect (or increase) the transcription of the PARP, VSG, and rRNA genes (132). In contrast, the transcription of other protein-coding genes by RNA pol II is completely abolished by the addition of 0.6% of Sarkosyl. These data indicate that rRNA, PARP, and VSG genes might indeed be transcribed by the same RNA polymerase, i.e. RNA pol I. This series of experiments should be interpreted carefully, because by using the differential sensitivity of RNA polymerases to Mn^{2+} and 1,10-phenanthroline to discriminate between RNA pol I and RNA pol II, Grondal et al demonstrated that VSG gene expression more closely resembled RNA pol II transcription (61). These discrepancies may be explained by the different experimental designs, the compounds used, and the controls included. From this comparison it is clear that an unambiguous polymerase identification should not be based solely on pharmacological criteria.

Fourth, the sensitivity of PARP and VSG gene transcription to temperature (or environmental) changes resembles that of the rRNA genes (87). A short heat shock down-regulated the transcription of the α-amanitin-resistantly transcribed rRNA and PARP genes in insect-form trypanosomes and did not affect transcription of RNA pol II– and RNA pol III–transcribed genes. In bloodstream-form trypanosomes, the rRNA and VSG genes were equally sensitive to the transfer of the cells from culture medium at 37°C to 41°C.

Fifth, *neo* RNA derived from rRNA or PARP promoters located to the nucleolus (27, 130). In all eukaryotes analyzed to date, transcription by RNA pol I occurs at the nucleolus, a highly organized subnuclear compartment to which all RNA pol I and nascent rRNA is confined (137, 139, 140). By in situ hybridization, the location of trypanosome nucleoli was defined by the distribution of subnuclear rRNA transcripts (Figure 2). With this method, nucleoli could be identified in 80–90% of interphase trypanosomes and the volume of the nucleolus seemed variable in the unsynchronized trypanosome population, which may relate to the metabolic and/or cell cycle stage of each interphase cell.

Attempts to locate the PARP mRNA or VSG mRNA at the trypanosome nucleolus failed; the PARP and VSG mRNA in insect- or bloodstream-form trypanosome nuclei, respectively, could not be detected at any unique subnuclear compartment (HM Chung, unpublished information). It is possible that the PARP and the VSG mRNA are rapidly processed and transported, and that they therefore do not represent reliable markers for localization of the site of their transcription. By dual-fluorophore in situ hybridization, Chung et al compared the location of *neo* RNA transcripts in transformed insect-form cell lines containing (Figure 2): (*a*) an rRNA promoter, driving a *neo* gene that is integrated at an rRNA locus or a PARP locus (*bottom panel*); (*b*) the PARP promoter, driving a *neo* gene that is integrated at a tubulin or a RNA pol II–largest subunit gene locus (*middle panel*); and (*c*) a *neo* gene that is under the control of an RNA pol II promoter (data not shown) (27, 130). In cell lines generated with constructs containing the rRNA promoter that drives the *neo* gene, nuclear *neo* RNA was located at the nucleolus, with diffuse *neo* staining throughout the cytoplasm. Nucleolar *neo* RNA signals could be detected in ~60% (integrated at the rRNA locus) and 20% (integrated at the PARP locus) of the interphase nuclei. In cell lines within which the construct of the PARP promoter controlling the *neo* gene was integrated at the tubulin locus, the *neo* RNA signal was also specifically located at the nucleolus or the periphery of the nucleolus. Nucleolar *neo* RNA could be detected in ~20% of the nuclei, whereas in other cells, specific nuclear hybridization could not be observed. Similar results were also observed in cell lines in which the PARP-*neo* construct was integrated into an RNA pol II–largest subunit gene locus (Chung et al, unpublished information). It is unclear whether the inability to detect *neo* RNA in some of the cells was the result of more efficient *neo* pre-mRNA processing, of transport, or of nonconstitutive expression of the *neo*-gene in these cells. In contrast, *neo* RNA could never be detected at the nucleolus in cells where the *neo* gene was driven by an RNA pol II promoter (the tubulin gene or the calmodulin gene locus). Consistent results were also obtained using the *hph* gene as a marker. These findings suggest that the rRNA and the PARP promoters direct the synthesis of *neo* RNA at the nucleolus in insect-form trypanosomes.

Different results were obtained for the subnuclear location of ES transcripts in bloodstream-form trypanosomes (JCBM Zomerdijk, I Chaves, R Dirks & P Borst, personal communication). Using a combination of probes representing the majority of sequences of the ES, its transcripts were localized to a region outside the nucleolus. Similarly, transcripts from the active 221 ES with an *hph* and a *neo* gene inserted between ESAG 1 and the 221 VSG gene could not be co-localized with the nucleolus (I Chaves, R Dirks & P Borst, personal communication). In this same series of experiments, an rRNA promoter control construct inserted in a ribosomal array gave significant reporter mRNA nucleolar co-localization. Based on these data, bloodstream-form ES transcription could not be confirmed to occur at the nucleolus. PARP gene transcription in insect-form trypanosomes and VSG gene transcription in bloodstream-form trypanosomes may therefore differ in their nucleolar dependence.

Transcription of the PARP, the VSG, and the rRNA genes is differentially regulated when these promoters are located at different chromosomal contexts, or when their activity is compared in different life-cycle stages (13, 67, 129, 159, 179; Lee, unpublished information). The mechanism involved in this positional control of transcription is not clear. However, PARP and rRNA gene promoters that are integrated at different chromosomal contexts in bloodstream-form trypanosomes can be derepressed in procyclic-form trypanosomes, which indicates that stage-specific factors may be involved in their differential transcriptional control. The VSG gene promoter also appears restricted in its ability to direct the expression of reporter genes when integrated at different chromosomal contexts in each of the two life-cycle stages. This may reflect sensitivity of the VSG promoter to its chromatin environment.

CONCLUDING REMARKS

Even though a significant body of evidence has accumulated in favor of the notion that the PARP, VSG, and rRNA genes share a common RNA pol I–transcriptional machinery, some observations are not in agreement with this hypothesis. Additional proof is required to unambiguously settle the proposed role of RNA pol I in the transcription of PARP and VSG genes. Given the body of evidence that has accumulated, we consider it reasonable to operate under the hypothesis that RNA pol I is responsible for the transcription of the VSG and PARP genes.

In *T. brucei*, the only naturally occurring protein-coding genes that may be transcribed by RNA pol I are derived from the PARP and VSG gene transcription units, whereas all other protein-coding genes appear to be transcribed by RNA pol II. The fact that these surface-coat genes represent the only known transcriptionally controlled genes of these protozoa may be critical to our understanding of the use of RNA pol I for protein-coding gene expression in these organisms. In addition, both the PARP and VSG mRNAs are among the most abundant mRNAs of trypanosomes, and it has been postulated that making use of the efficient RNA pol I machinery may facilitate the rapid generation of large quantities of these mRNAs and their proteins.

Several new approaches may potentiate the investigation of the polymerases that transcribe the PARP and the VSG genes. An in vitro transcription system from cultivated insect-form trypanosomes of *T. brucei* recently developed by Guntz et al (62) directed the transcription of a U_2, G-less cassette by RNA pol III. The efficiency of this system for RNA pol I and RNA pol II transcription remains to be determined and may prove valuable in the identification of the polymerases that transcribe VSG and PARP genes. Also, the establishment of RNA pol I–temperature-sensitive (ts) mutants and cell lines in which RNA pol I expression is placed under the control of an inducible promoter will prove useful (172–174). RNA polymerase ts mutants have been successfully established in yeast by mutagenesis of the cloned genes (175, 176). Because RNA polymerases are conserved, mutagenesis approaches applied in yeast may be applicable to the generation of trypanosome RNA pol I ts mutants. In yeast, the necessity for pol I–mediated rRNA gene expression was circumvented by placing the rRNA gene under the control of a GAL7 promoter. Subsequently, RNA pol I mutants that negatively affected the function of RNA pol I and the synthesis of rRNA could be rescued in the presence of galactose (106). This result suggests that it is possible to generate RNA pol I knock-out cell lines, once the synthesis of rRNA is rescued by transcription of the rRNA genes from a different RNA polymerase. Similar strategies may aid the analysis of RNA pol I–mediated gene expression in trypanosomes.

α-Amanitin–resistant RNA pol I-like transcription of protein-coding genes has thus far only been identified in African trypanosomes. Because RNA pol I–mediated protein-coding gene expression is dependent on trans-splicing, and because trans-splicing has been identified in several other organisms including nematodes (105), the potential role of RNA pol I for protein-coding gene expression in other organisms warrants further analysis. The functional dissection of the nucleolus and its significance for gene expression may thus be explored in protozoa, and possibly in more complex eukaryotes, which will shed light on critical parameters that explain RNA polymerase specialization.

ACKNOWLEDGMENT

We thank Doris Cully for her critical reading of the manuscript.

Visit the *Annual Reviews home page* at
http://www.annurev.org.

Literature Cited

1. Agabian N. 1990. Transsplicing of nuclear pre-mRNAs. *Cell* 61:1157–60
2. Alexandre S, Guyaux M, Murphy NB, Coquelet H, Pays A. 1988. Putative genes of a variant-specific antigen gene transcription unit in *Trypanosoma brucei. Mol. Cell. Biol.* 8:2367–78
3. Alexandre S, Paindavoine P, Tebabi P, Pays A, Halleux S, et al. 1990. Differential expression of a family of putative adenylate/guanylate cyclase genes in *Trypanosoma brucei. Mol. Biochem. Parasitol.* 43:279–88
4. Alyzzz R, Argaman M, Halman S, Shapira M. 1994. A regulatory role for the 5′ and 3′ untranslated regions in differential expression of hsp 83 in *Leishmania. Nucleic Acids Res.* 22:2922–29
5. Argaman M, Aly R, Shapira M. 1994. Expression of heat-shock protein-83 in *Leishmania* is regulated post-transcriptionally. *Mol. Biochem. Parasitol.* 64:95–110
6. Bayne RAL, Kilbride EA, Lainson FA, Tetley L, Barry JD. 1993. A major surface antigen of procyclic stage *Trypanosoma congolense. Mol., Biochem. Parasitol.* 61:295–310
7. Ben Amar MF, Jeffries D, Pays A, Bakalara N, Kendall G, Pays E. 1991. The actin gene promoter of *Trypanosoma brucei. Nucleic Acids Res.* 19:5857–62
8. Benne R. 1990. RNA editing in trypanosomes: Is there a message? *Trends Genet.* 6:177–81
9. Berberof M, Pays A, Lips S, Tebabi P, Pays E. 1996. Characterization of a transcription terminator of the procyclin PARP A unit of *Trypanosoma brucei. Mol. Cell. Biol.* 16:914–24
10. Berberof M, Pays A, Pars E. 1991. A similar gene is shared by both the VSG and procyclin gene transcription units of *Trypanosoma brucei. Mol. Cell. Biol.* 11:1473–79

11. Berberof M, Vanhamme L, Pars E. 1995. *Trypanosoma brucei*: a preferential splicing at the inverted polyadenylation site of the VSG mRNA providing further evidence for coupling between trans-splicing and polyadenylation. *Exp. Parasitol.* PO(3), 565–67
12. Berberof M, Vanhamme L, Tebabi P, Pays A, Jefferies D, et al. 1995. The 3′ terminal region of the mRNAs for VSG and procyclin can confer stage specificity to gene expression in *Trypanosoma brucei. EMBO J.* 14:2925–34
13. Biebinger S, Rettenmaier S, Flaspohler J, Hartmann C, Pena-Diaz J, et al. 1996. The PARP promoter of *Trypanosoma brucei* is developmentally regulated in a chromosomal context. *Nucleic Acids Res.* 24 (7), 1202–11
14. Boothroyd JC, Cross GAM. 1982. Transcripts coding for different variant surface glycoproteins of *Trypanosoma brucei* have a short, identical exon at their 5′ end. *Gene* 20:281–89
15. Borst P. 1991. Why kinetoplast DNA networks? *Trends Genet.* 7: 139–41
16. Borst P. 1991. Molecular genetics of antigenic variation. *Immunoparasitol. Today* 3:A29–A33
17. Borst P. 1991. Transferrin receptor, antigenic variation and the prospect of a trypanosome vaccine. *Trends Genet.* 7:307–9
18. Borst P. 1986. Discontinuous transcription and antigenic variation in trypanosomes. *Annu. Rev. Biochem.* 55:701–32
19. Brown SD, Huang J, Van der Ploeg LHT. 1992. The promoter for the procyclic acidic repetitive protein (PARP) genes of *Trypanosoma brucei* shares features with RNA polymerase I promoters. *Mol. Cell. Biol.* 12:2644–52
20. Brown SD, Van der Ploeg LHT. 1994. Single-stranded DNA-protein binding in

the procyclic acidic repetitive protein (PARP) promoter of *Trypanosoma brucei. Mol. Biochem. Parasitol.* 56:109–22

21. Bruzik JP, Maniatis T. 1992. Spliced leader RNAs from lower eukaryotes are trans-spliced in mammalian cells. *Nature* 360:692–95

22. Bruzik JP, Van Doren K, Hirsh D, Steitz JA. 1988. Trans-splicing involves a novel form of small nuclear ribonucleoprotein particles. *Nature* 335:559–62

23. Campell DA, Thornton DA, Boothroyd JC. 1984. Apparent discontinuous transcription of *Trypanosoma brucei* surface antigen genes. *Nature* 311:350–55

24. Carrington M, Miller N, Blum M, Roditi I, Wiley D, Turner M. 1991. Variant specific glycoprotein of *Trypanosoma brucei* consists of two domains each having an independently conserved pattern of cysteine residues. *J. Mol. Biol.* 221:823–35

25. Carruthers VB, Navarro M, Cross GAM. 1996. Targeted disruption of expression site associated gene–1 in bloodstream-form *Trypanosoma brucei. Mol. Biochem. Parasitol.* 81:65

26. Chapman AB, Agabian N. 1994. *Trypanosoma brucei* RNA polymerase II is phosphorylated in the absence of carboxyl-terminal domain heptapeptide repeats. *J. Biol. Chem.* 269:4754–60

27. Chung HM, Lee MGS, Van der Ploeg LHT. 1993. RNA polymerase I-mediated protein-coding gene expression in *Trypanosoma brucei. Parasitol. Today* 8:414

28. Chung HMM, Lee MGS, Dietrich P, Huang J, Van der Ploeg. LHT. 1993. Disruption of large subunit RNA polymerase II genes in *Trypanosoma brucei. Mol. Cell. Biol.* 13:3734–43

29. Clayton C. 1992. Developmental regulation of gene expression in Trypanosoma brucei. *Prog. Nucleic Acid Res. Mol. Biol.* 43:37–66

30. Clayton CE. 1988. The molecular biology of kinetoplastidae. *Genet. Eng. News* 7:1

31. Clayton CE, Fueri JP, Itzakhi JE, Bellofatto V, Sherman DR, et al. 1990. Transcription of the procyclic acidic repetitive protein genes of *Trypanosoma brucei. Mol. Cell. Biol.* 10:3036–47

32. Clayton CE, Mowatt M. 1989. The procyclic acidic repetitive proteins of *Trypanosoma brucei. J. Biol. Chem.* 264:15088–93

33. Conaway RC, Conaway JW. 1993. General initiation factors for RNA polymerase II. *Annu. Rev. Biochem.* 62:161–90

34. Cornelissen AWCA, Evers R, Grondal E, Hammer A, Jess W, Kock J. 1988. Transcription and RNA polymerases in *Trypanosoma brucei. Nova Acta Leopold.* 279:215–16

35. Cornelissen AWCA, Evers R, Kock J. 1988. Structure and sequences encoding subunits of eukaryotic RNA polymerases. *Oxford Surv. Eukaryot. Genet.* 5:91

36. Cross GAM. 1975. Identification, purification and properties of clone specific glycoprotein antigens constituting the surface coat of *Trypanosoma brucei. Parasitology* 71:393–417

37. Cross GAM. 1984. Structure of the variant glycoprotein and surface coat of *Trypanosoma brucei. Philos. Trans. R. Soc. London Ser. B* 370:3–12

38. Cross GAM. 1990. Cellular and genetic aspects of antigenic variation in trypanosomes. *Annu. Rev. Immunol.* 8:83–110

39. Cully DF, Ip HS, Cross GAM. 1985. Coordinate transcription of variant surface glycoprotein genes and an expression site associated gene family in *Trypanosoma brucei. Cell* 42:173–82

40. De Lange T, Liu AYC, Van der Ploeg LHT, Borst P, Tromp MC, Van Boom JH. 1983. Tandem repetition of the 5′ mini-exon of variant surface glycoprotein genes: a multiple promoter for VSG gene transcription. *Cell* 34:891–900

41. Dorfman D, Donelson J. 1984. Characterization of the 1.35 kilobase DNA repeat unit containing the conserved 35 nucleotides at the 5′-termini of variable surface glycoprotein mRNAs in *Trypanosoma brucei. Nucleic Acids Res.* 12:4907–20

42. Earnshaw DL, Beebee TJC, Gutteridge WE. 1987. Determination of RNA multiplicity in *Trypanosoma brucei.* Characterization and purification of alpha-amanitin-resistant and sensitive enzymes. *Biochem. J.* 241:649–55

43. Eid J, Sollner-Webb B. 1991. Stable integrative transformation of *Trypanosoma brucei* that occurs exclusively by homologous recombination. *Proc. Natl. Acad. Sci. USA* 88:2118–21

44. Englund PT, Ferguson M, Guilbride DL, Johnson CE, Li C, et al. 1994. The replication of kinetoplast DNA. In *Molecular Approaches to Modern Parasitology*, ed. JC Boothroyd, R Komuniecki, p. 147. New York: Wiley-Liss

45. Englund PT, Hajduk SL, Marini JC. 1982. The molecular biology of trypanosomes. *Annu. Rev. Biochem.* 51:695–726

46. Evers R, Cornelissen AWCA. 1990. The *Trypanosoma brucei* protein phosphatase gene: polycistronic transcription with the

RNA polymerase II largest subunit gene. *Nucleic Acids Res.* 18:5089–95

47. Evers R, Hammer A, Kock J, Jess W, Borst P, et al. 1989. *Trypanosoma brucei* contains two RNA polymerase II largest subunit genes with an altered C-terminal domain. *Cell* 56:585–97

48. Fantoni A, Dare A, Tschudi C. 1994. RNA polymerase III-mediated transcription of the trypanosome U_2 small nuclear RNA gene is controlled by both intergenic and extragenic regulatory elements. *Mol. Cell. Biol.* 14:2021–28

49. Florent I, Raibaud A, Eisen H. 1991. A family of genes related to a new expression site-associated gene in *Trypanosoma equiperdum. Mol. Cell. Biol.* 11:2180–88

50. Freistadt MS, Cross GAM, Branch AD, Robertson HD. 1987. Direct analysis of the mini-exon donor RNA of *Trypanosoma brucei*: detection of a novel cap structure also present in messenger RNA. *Nucleic Acids Res.* 15:9861–79

51. Geiduschek EP. 1988. Transcription by RNA polymerase III. *Annu. Rev. Biochem.* 57:873–914

52. Geiduschek EP, Kassavetis GA. 1995. Comparing transcriptional initiation by RNA polymerases I and III. *Curr. Opin. Cell Biol.* 7:344–51

53. Gibson WC, Swinkels BW, Borst P. 1988. Post-transcriptional control of the differential expression of phosphoglycerate kinase genes in *Trypanosoma brucei. J. Mol. Biol.* 201:315

54. Glass DJ, Polvere RI, Van der Ploeg LHT. 1986. Conserved sequences and transcription of the hsp 70 gene family in *Trypanosoma brucei. Mol. Cell. Biol.* 6:4657–66

55. Gommers-Ampt JH, Teixeira AJR, van der Werken G, van Dijk WJ, Borst P. 1993. The identification of hydroxymethyluracil in DNA of *Trypanosoma brucei. Nucleic Acids Res.* 21:2039–43

56. Gommers-Ampt JH, Van Leeuwen F, deBeer ALJ, Vliegent-Hart JFG, Dizdaroglu M, et al. 1993. β-D-Glucosylhydroxymethyluracil: a novel modified base present in the DNA of the parasitic protozoan *Trypanosoma brucei. Cell* 75:1129–36

57. Gottesdiener K. 1994. A new VSG expression site-associated gene (ESAG) in the promoter region of *Trypanosoma brucei* encodes a protein with 10 potential transmembrane domains. *Mol. Biochem. Parasitol.* 63:143–51

58. Gottesdiener K, Chung H-M, Brown SD, Lee MGS, Van der Ploeg LHT. 1991. Characterization of VSG gene expression site promoters and promoter-associated DNA rearrangement events. *Mol. Cell. Biol.* 11:2467–80

59. Gottesdiener K, Garcia-Anoveros J, Lee MG, Van der Ploeg LHT. 1990. Chromosome organization of the protozoan *Trypanosoma brucei. Mol. Cell. Biol.* 10:6079–83

60. Gottesdiener K, Goriparthi L, Masucci J, Van der Ploeg LHT. 1992. A proposed mechanism for promoter-associated DNA rearrangement events at a variant surface glycoprotein gene expression site. *Mol. Cell. Biol.* 12:4784–94

61. Grondal EJM, Evers R, Kosubek K, Cornelissen AWCA. 1989. Characterization of the RNA polymerases of *Trypanosoma brucei*: trypanosomal mRNAs are composed of transcripts derived from both RNA polymerase II and III. *EMBO J.* 8:3383–89

62. Gunzl A, Tschudi C, Nakaar V, Ullu E. 1995. Accurate transcription of the trypanosome U_2 small nuclear RNA gene in a homologous extract *J. Biol. Chem.* 270:17287–91

63. Hartshorne T, Agabian N. 1993. RNA-B is the major nucleolar trimethylguanosine-capped small nuclear RNA associated with fibrillarin and pre-rRNAs in *Trypanosoma brucei. Mol. Cell. Biol.* 13:144–54

64. Hasan G, Turner MJ, Cordingley JS. 1984. Ribosomal RNA genes of *Trypanosoma brucei*: mapping the regions specifying the six small ribosomal RNAs. *Gene* 27:75–86

65. Hehl A, Roditi I. 1994. The regulation of procyclin expression in *Trypanosoma brucei*: making or breaking the rules? *Parasitol. Today* 10:442–45

66. Hehl A, Vassella E, Braun R, Roditi I. 1994. A conserved stem-loop structure in the 3′ untranslated region of procyclin mRNAs regulates expression in *Trypanosoma brucei. Proc. Natl. Acad. Sci. USA* 91:370–74

67. Horn D, Cross GAM. 1995. A developmentally regulated position effect at a telomeric locus in *Trypanosoma brucei. Cell* 83:555–61

68. Hotz HR, Lorenz P, Fischer R, Krieger S, Clayton C. 1995. Role of 3′ untranslated region in the regulation of hexose transporter mRNAs in *Trypanosoma brucei. Mol. Biochem. Parasitol.* 75:1–14

69. Huang J, Van der Ploeg LHT. 1991. Requirement of a polypyrimidine tract for trans-splicing in trypanosomes: discriminating the PARP promoter from the

immediately adjacent 3′ splice acceptor site. *EMBO J.* 10:3877–85

70. Hug M, Carruthers VB, Hartmann C, Sherman DS, Cross GAM, Clayton C. 1993. A possible role for the 3′ untranslated region in developmental regulation in *Trypanosoma brucei. Mol. Biochem. Parasitol.* 61:87–96

71. Imboden MA, Laird PW, Affolter M, Seebeck T. 1987. Transcription of the intergenic regions of the tubulin gene cluster of *Trypanosoma brucei:* evidence for a polycistronic transcription unit in a eukaryote. *Nucleic Acids Res.* 15:7357–68

72. Janz L, Clayton C. 1994. The PARP and rRNA promoters of *Trypanosoma brucei* are composed of dissimiliar sequence elements that are functionally interchangeable. *Mol. Cell. Biol.* 14:5804–11

73. Janz L, Hug M, Clayton C. 1994. Factors that bind to RNA polymerase I promoter sequences of *Trypanosoma brucei. Mol. Biochem. Parasitol.* 65:99–108

74. Jefferies D, Tebabi P, Pays E. 1991. Transient activity assays of the *Trypanosoma brucei* VSG gene promoter: control of gene expression at the posttranscriptional level. *Mol. Cell. Biol.* 11:338–43

75. Jess W, Hammer A, Cornelissen AWCA. 1989. Complete sequence of the gene encoding the largest subunit of RNA polymerase I of *Trypanosoma brucei. FEBS Lett.* 249:123–28

76. Johnson P, Kooter J, M. Borst P. 1987. Inactivation of transcription by UV-irradiation of *T. brucei* provides evidence for a multicistronic transcription unit including a VSG gene. *Cell* 51:273–81

77. Kitchin PA, Ryley JF, Gutteridge W. 1984. The presence of unique DNA-dependent RNA polymerase in trypanosomes. *Comp. Biochem. Physiol. B* 77:223–31

78. Kock J, Evers R, Cornelissen AWCA. 1988. Structure and sequence of the gene for the largest subunit of trypanosomal RNA polymerase III. *Nucleic Acids Res.* 16:8753–72

79. Koenig E, Delius H, Carrington M, Williams RO, Roditi I. 1989. Duplication and transcription of procyclin genes in *Trypanosoma brucei. Nucleic Acids Res.* 17:8727–39

80. Koenig-Martin E, Yamage M, Roditi I. 1992. *Trypanosoma brucei:* a procyclin-associated gene encodes a polypeptide related to ESAG 6 and 7 proteins. *Mol. Biochem. Parasitol.* 55:135–45

81. Kooter JM, Borst P. 1984. Alpha-amanitin-insensitive transcription of variant surface glycoprotein genes provides further evidence for discontinuous transcription in trypanosomes. *Nucleic Acids Res.* 12:9457–72

82. Kooter JM, Van der Spek HJ, Wagter R, d'Oliveira CE, Van der Hoeven F, et al. 1987. The anatomy and transcription of a telomeric expression site for variant-specific surface antigens in *Trypanosoma brucei. Cell* 51:261–72

83. Laird PW. 1989. Trans-splicing in trypanosomes—archaism or adaption? *Trends Genet.* 5:204–8

84. Laird PW, Kooter JM, Loosbroek N, Borst P. 1985. Mature mRNAs of *Trypanosoma brucei* possess a 5′ cap acquired by discontinuous RNA synthesis. *Nucleic Acids Res.* 13:4253–66

85. Lebowitz JH, Smith HQ, Rusche L, Beverley SM. 1993. Coupling of poly(A) site selection and trans-splicing in *Leishmania. Genes Dev.* 7:996–1007

86. Lee MGS. 1996. An RNA polymerase II promoter in the hsp 70 locus of *Trypanosoma brucei. Mol. Cell. Biol.* 16:1220–30

87. Lee MGS. 1995. Heat shocks does not increase the transcriptional efficiency of the hsp 70 genes of *Trypanosoma brucei. Exp. Parasitol.* 81:608–13

88. Lee MGS. 1995. A foreign transcription unit in the inactivated VSG gene expression site of the procyclic form of *Trypanosoma brucei* and formation of large episomes in stably transformed trypanosomes. *Mol. Biochem. Parasitol.* 69:223–38

89. Lee MGS, Axelrod N. 1995. Construction of trypanosome artificial mini-chromosomes. *Nucleic Acids Res.* 23:4893–99

90. Lee MGS, Polvere RI, Van der Ploeg LHT. 1990. Evidence for segmental gene conversion between a cognate hsp 70 gene and the temperature-sensitively transcribed hsp 70 genes of *Trypanosoma brucei. Mol. Biochem. Parasitol.* 41:213–20

91. Lee MGS, Van der Ploeg LHT. 1990. Homologous recombination and stable transfection in the parasitic protozoan *Trypanosoma brucei. Science* 250:1583–87

92. Lee MGS, Van der Ploeg LHT. 1990. Transcription of the heat shock 70 locus in *Trypanosoma brucei. Mol. Biochem. Parasitol.* 41:221–31

93. Ligtenberg MJL, Bitter W, Kieft R, Steverding D, Janssen H, et al. 1994. Reconstitute of a surface transferrin binding complex in insect-form *Trypanosoma brucei. EMBO J.* 13:2565–73

94. Lobo SL, Hernandez NT. 1994. Transcription of snRNA genes by RNA polymerases II and III. In *Transcription: Mechanisms and Regulation*, ed. RC Conway, JW Conway, pp. 127–59. New York: Raven

95. Lopata M, Cleveland DW, Sollner-Webb B. 1986. RNA polymerase specificity of mRNA production and enhancer action. *Proc. Acad. Natl. Sci. USA* 83:6377–81

96. Maldonado E, Reinberg D. 1995. News on initiation and elongation of transcription by RNA polymerase II. *Curr. Opin. Cell Biol.* 7:352–61

97. Marchetti M, Silva E, Tschudi C, Lee MGS, Ullu E. 1995. RNA polymerase II and RNA pol III transcription units are closely interspersed in the trypanosome genome. *Woods Hole Mol. Parasitol. Meet.* p. 26 (Abstr.)

98. McNally KP, Agabian N. 1992. *Trypanosoma brucei* spliced-leader RNA methylations are required for transsplicing in vivo. *Mol. Cell. Biol.* 12:4844–51

99. Miller AI, Wirth DF. 1988. Trans splicing in *Leishmania enriettii* and identification of ribonucleoprotein complexes containing the spliced leader and U_2 equivalent RNAs. *Mol. Cell. Biol.* 8:2597–2603

100. Moore JB, Beverley SM. 1996. Pteridine-receptor and recurrent amplification of extrachromosomal DNAs in *Leishmania*. *Woods Hole Mol. Parasitol. Meet.* p. 107 (Abstr.)

101. Mottram J, Perry KL, Lizardi PM, Luhrmann R, Agabian N. 1989. Isolation and sequence of four small nuclear U RNA genes of *Trypanosoma brucei* sudsp. *brucei*: identification of the U_2, U_4, and U_6 RNA analog. *Mol. Cell. Biol.* 9:1212–23

102. Mowatt MR, Wisdom GS, Clayton CE. 1989. Variation of tandem repeats in the developmentally regulated procyclic acidic repetitive proteins of *Trypanosoma brucei*. *Mol. Cell. Biol.* 9:1332–35

103. Murphy WJ, Watkins KP, Agabian N. 1986. Identification of a novel Y branch structure as an intermediate in trypanosome mRNA processing: evidence for trans splicing. *Cell* 47:517–25

104. Nelson RG, Parsons M, Barr PJ, Stuart K, Selkirk M, Agabian N. 1983. Sequences homologous to variant antigen mRNA spliced leader are located in tandem repeats and variable orphons in *Trypanosoma brucei*. *Cell* 34:901–10

105. Nilsen TW. 1993. Trans-splicing of nematode premessenger RNA. *Annu. Rev. Microbiol.* 47:413–40

106. Nogi Y, Vu L, Nomura M. 1991. An approach for isolation of mutants defective in 35S ribosomal RNA synthesis in *Saccharomyces cerevisiae*. *Proc. Natl. Acad. Sci. USA* 88:7026–30

107. Nozaki T, Cross GAM. 1995. Effects of $3'$ untranslated and intergenic regions on gene expression in *Trypanosoma cruzi*. *Mol. Biochem. Parasitol.* 75:55

108. O'Brien T, Hardin S, Greenleaf A, Lis JT. 1994. Phosphorylation of RNA polymerase II C-terminal domain and transcriptional elongation. *Nature* 370:75–77

109. Paindavoine P, Rolin S, Van Assel S, Geuskens M, Jauniaux JC, et al. 1992. A gene from the VSG expression site encodes one of several transmembrane adenylate cyclases located on the flagellum of *Trypanosoma brucei*. *Mol. Cell. Biol.* 12:1218–25

110. Parsons M, Nelson RG, Watkins KP, Agabian N. 1984. Trypanosomes mRNA share a common $5'$ spliced leader sequence. *Cell* 38:309–16

111. Paule MR. 1994. Transcription of ribosomal RNA by eukaryotic RNA pol I. In *Transcription: Mechanism and Regulation*, ed. RC Conway, JW Conway, p. 83. New York: Raven

112. Pays E. 1992. Genome organization and control of gene expression in trypanosomatids. In *The Eukaryotic Genome Organization and Regulation*, ed. PMA Broda, SG Oliver, PFG Sims. *Symp. Soc. Gen. Microbiol* 50:127–60

113. Pays E, Coquelet H, Pays A, Tebabi P, Steinert M. 1989. *Trypanosoma brucei*: post-transcriptional control of the variable surface glycoprotein gene expression site. *J. Mol. Cell. Biol.* 9:4018–21

114. Pays E, Coquelet H, Tebabi P, Pays A, Jefferies D, et al. 1990. *Trypanosoma brucei*: constitutive activity of the VSG and procyclin gene promoters. *EMBO J.* 9:3145–51

115. Pays E, Tebabi P, Pays A, Coquelet H, Revelard P, Salmon D, Steinert M. 1989. The genes and transcripts of an antigen gene expression site from *T. brucei*. *Cell* 57:835–45

116. Pays E, Vanhamme L, Berberof M. 1994. Genetic controls for the expression of surface antigens in African trypanosomes. *Annu. Rev. Microbiol.* 48:25–52

117. Pelle R, Murphy NB. 1993. Stage-specific differential polyadenylation of mini-exon derived RNA in African trypanosomes. *Mol. Biochem. Parasitol.* 59:277–86

118. Pham VP, Qi CC, Gottesdiener KM. 1996. A detailed mutational analysis of the VSG gene expression site promoter. *Mol. Biochem. Parasitol.* 75:241

119. Qi CC, Urmenyi T, Gottesdiener KM. 1996. Analysis of a hybrid PARP/VSG ES promoter in procyclic trypanosomes. *Mol. Biochem. Parasitol.* 77:147

120. Ralph D, Huang J, Van der Ploeg LHT. 1988. Physical identification of branched intron side-products of splicing in *Trypanosoma brucei. EMBO J.* 7:2539–45

121. Redpath M, Windle H, Nolan D, Pays E, Voorheis H, Carrington M. 1996. ESAGs 11, a gene from an expression site of *Trypanosome brucei brucei* Antat 1. 1a. *Woods Hole Mol. Parasitol. Meet.* p. 83 (Abstr.)

122. Reeder RH. 1992. Regulation of transcription by RNA polymerase I. In *Transcriptional Regulation*, ed. SL McKnight, KR Yamamoto, p. 315. Cold Spring Harbor, NY: Cold Spring Harbor Lab. Press

123. Revelard P, Lips S, Pays E. 1990. A gene from the VSG gene expression site of *Trypanosoma brucei* encodes a protein with both leucine-rich repeats and a putative zinc finger. *Nucleic Acids Res.* 18:7299–303

124. Roditi I, Pearson TW. 1990. The procyclin coat of African trypanosomes. *Parasitol. Today* 6:79–82

125. Roditi I, Schwarz H, Pearson TW, Beecroft RP, Liu MK, et al. 1989. Procyclin gene expression and loss of the variant surface glycoprotein during differentiation of *Trypanosoma brucei. J. Cell. Biol.* 108:737–46

126. Roeder R. 1996. The role of general initiation factors in transcription by RNA polymerase II. *Trends Biochem. Sci.* 21:327

127. Rudenko G, Bishop D, Gottesdiener K, Van der Ploeg LHT. 1989. Alpha-amanitin resistant transcription of protein-coding genes in insect and bloodstream form *Trypanosoma brucei. EMBO J.* 13:4259–63

128. Rudenko G, Blundell PA, Dirks-Mulder A, Kieft R, Borst P. 1995. A ribosomal DNA promoter replacing the promoter of a telomeric VSG gene expression site can be efficiently switched on and off in *T. brucei. Cell* 83:547

129. Rudenko G, Blundell PA, Taylor MC, Kieft R, Borst P. 1994. VSG gene expression site control in insect-form *Trypanosoma brucei. EMBO J.* 13:5470–82

130. Rudenko GH, Chung HM, Pham VP, Van der Ploeg LHT. 1991. RNA polymerase I can mediate expression of CAT and *neo* protein-coding genes in the parasitic protozoan *Trypanosoma brucei. EMBO J.* 10:3387–97

131. Rudenko G, Le Blancq S, Smith J, Lee MGS, Rattray A, Van der Ploeg LHT.

1990. Procyclic acidic repetitive protein (PARP) genes located in an unusually small α-amanitin-resistant transcription unit: PARP promoter activity assayed by transient DNA transfection of *Trypanosoma brucei. Mol. Cell. Biol.* 10:3492–504

132. Rudenko G, Lee MGS, Van der Ploeg LHT. 1992. The PARP and VSG genes of *Trypanosoma brucei* do not resemble RNA polymerase II transcription units in sensitivity to Sarkosyl in nuclear run-on assays. *Nucleic Acids Res.* 20:303–6

133. Ryan KA, Sapiro TA, Rauch CA, Englund PT. 1988. Replication of kinetoplast DNA in trypanosomes. *Annu. Rev. Microbiol.* 42:339–58

134. Salditt-Georgieff M, Harpold M, Chen-Kiang S, Darnell JE Jr. 1980. The addition of 5' cap structures occurs early in hnRNA synthesis and prematurely terminated molecules are capped. *Cell* 19:69

135. Salmon D, Geuskens M, Hanocq F, Hanocq-Quertier J, Nolan D, et al. 1994. A novel heterodimeric transferrin receptor encoded by a pair of VSG expression site-associated genes in *T. brucei. Cell* 78:75–86

136. Satio RM, Elgort MG, Campbell DA. 1994. A conserved upstream element is essential for transcription of the *Leishmania tarentolae* mini-exon gene. *EMBO J.* 13:5460–69

137. Scheer U, Benavente R. 1990. Functional and dynamic aspects of the mammalian nucleolus. *BioEssays* 12:14

138. Schell D, Ever R, Pries D, Ziegelbauer K, Kiefer H, et al. 1991. A transferrin-binding protein of *Trypanosoma brucei* is encoded by one of the genes in the variant surface glycoprotein gene expression site. *EMBO J.* 10:1061–66

139. Shaw PJ, Jordan EG. 1995. The nucleolus. *Annu. Rev. Cell Dev. Biol.* 11:93–121

140. Sheer U, Weisenberger D. 1994. The nucleolus. *Curr. Opin. Cell Biol.* 6:354

141. Sherman DR, Janz L, Hug M, Clayton CE. 1991. Anatomy of the PARP gene promoter of *Trypanosoma brucei. EMBO J.* 10:3379–86

142. Simpson L. 1990. RNA editing—a novel phenomenon? *Science* 250:512

143. Smith JL, Chapman AB, Agabian N. 1993. *Trypanosoma vivax:* evidence for only one RNA polymerase II largest subunit gene in a trypanosome which undergoes antigenic variation. *Exp. Parasitol.* 76:242

144. Smith JL, Levin JR, Agabian N. 1989. Molecular characterization of the *Trypanosoma brucei* RNA polymerase I and

III largest subunit genes. *J. Biol. Chem.* 264:18091–99

145. Smith JL, Levin JR, Ingles CJ, Agabian N. 1989. In trypanosomes the homologue of the largest subunit of RNA polymerase II is encoded by two genes and has a highly unusual C-terminal domain structure. *Cell* 56:815

146. Sogui ML, Elwood HJ, Gercerson JH. 1986. Evolutionary diversity of eukaryotic small-subunit rRNA genes. *Proc. Natl. Acad. Sci. USA* 83:1383–86

147. Sollner-Webb B, Mougey EB. 1991. News from the nucleolus: rRNA gene expression. *Trends Biochem. Sci.* 16:58

148. Sollner-Webb, B, Tower J. 1986. Transcription of cloned eukaryotic ribosomal RNA genes. *Annu. Rev. Biochem.* 55:801–30

149. Steverding D, Chaudhri M, Stierhof YD, Overath P, Ligtenberg, M, Borst P. 1994. ESAG 6 and 7 products of *Trypanosoma brucei* form a transferrin binding protein complex. *J. Cell Biol.* 64:78

150. Stuart K. 1991. RNA editing in mitochondrial mRNA of trypanosomatids. *Trends Biochem. Sci.* 16:68

151. Sutton RE, Boothroyd JC. 1986. Evidence for trans splicing in trypanosomes. *Cell* 47:527

152. Teixeira SMR, Kirchhoff LV, Donelson JE. 1995. Post-transcriptional elements regulating expression of mRNAs from the amastin/Tuzin gene cluster of *Trypanosoma cruzi.* *J. Biol. Chem.* 270:22586–94

153. ten Asbroek ALMA, Ouellette M, Borst P. 1990. Targeted insertion of the neomycin phosphotransferase gene into the tubulin gene cluster of *Trypanosoma brucei. Nature* 348:174

154. Torri AF, Hajduk SL. 1988. Posttranscriptional regulation of cytochrome C expression during the developmental cycle of *Trypanosoma brucei. Mol. Cell. Biol.* 8:4625–33

155. Tschudi C, Richards FF, Ullu E. 1986. The U2 RNA of *Trypanosoma brucei gambiense*: implication for a splicing mechanism in trypanosomes. *Nucleic Acids Res.* 14:8893–903

156. Tschudi C, Ullu E. 1990. Destruction of U2, U4, U6 small nuclear RNA blocks trans splicing in trypanosome cells. *Cell* 61:459

157. Tschudi C, Ullu E. 1988. Polygene transcripts are precursors to calmodulin mRNAs in trypanosomes. *EMBO J.* 7:455

158. Turner MJ. 1982. Biochemistry of the variant surface glycoproteins of salivarian trypanosomes. *Adv. Parasitol.* 21:69

159. Urmenyi TP, Van der Ploeg LHT. 1995. PARP promoter-mediated activation of a VSG expression site in insect-form *Trypanosoma brucei. Nucleic Acids Res.* 23:1010–18

160. Van der Ploeg LHT. 1986. Discontinuous transcription and splicing in trypanosomes. *Cell* 47:479

161. Van der Ploeg LHT. 1990. Antigenic variation in African trypanosomes: genetic recombination and transcriptional control of VSG genes. In *Frontiers in Molecular Biology: Genome Rearrangements and Amplification*, ed. BD Hames, DM Glover, p. 51. New York: IRL Press

162. Van der Ploeg LHT, Gottesdiener KM, Lee GSM. 1992. Antigenic variation in African trypanosomes. *Trends Genet.* 8:452

163. Van der Ploeg LHT, Liu AYC, Michel PAM, De Lange T, Borst P, et al. 1982. RNA splicing is required to make the messenger RNA for a variant surface antigen in trypanosomes. *Nucleic Acids Res.* 10:3591–604

164. Van der Ploeg LHT, Smith CL, Polvere RI, Gottesdiener KM. 1989. Improved separation of chromosome-sized DNA from *Trypanosoma brucei* stock 427. *Nucleic Acids Res.* 17:3217–27

165. Vanhamme L, Pays A, Tebabi P, Alexandre S, Pays E. 1995. Specific binding of proteins to the noncoding strand of a crucial element of the variant surface glycoprotein, procyclin, and ribosomal promoters of *Trypanosoma brucei. Mol. Cell. Biol.* 15:5598–606

166. Vanhamme L, Pays E. 1995. Control of gene expression in trypanosomes. *Microbiol. Rev.* 59:223

167. Vassella E, Braun R, Roditi I. 1994. Control of polyadenylation and alternative splicing of transcripts from adjacent genes in a procyclin expression site: a dual role for polypyrimidine tracts in trypanosomes? *Nucleic Acids Res.* 22:1359–64

168. Verrijzer CP, Tijian R. 1996. TAFs mediate transcriptonal activation and promoter selectivity. *Trends Biochem. Sci.* 21:338

169. Vickerman K. 1985. Developmental cycles and biology of pathogenic trypanosomes. *Med. Bull.* 41:105

170. Vickerman K, Tetley L, Hendry KAK, Turner CMR. 1988. Biology of African trypanosomes in the tse-tse fly. *Biol. Cell* 64:109

171. White TCG, Rudenko G, Borst P. 1986. Three small RNAs within the 10 kb trypanosome rRNA transcription unit are

analogous to Domain VII of other eukaryotic 28S rRNAs. *Nucleic Acids Res.* 14:9471–89

172. Wirtz E, Clayton C. 1995. Inducible gene expression in trypanosomes mediated by a prokaryotic repressor. *Science* 268:1179–82

173. Wirtz E, Gottesdiener KM. 1995. A reverse genetic approach to the identitiy of the polymerase mediating PARP and VSG transcription in trypanosomes *Woods Hole Mol. Parasitol. Meet.* p. 23 (Abstr.)

174. Wirtz E, Hartman C, Clayton C. 1994. Use of T3 and T7 RNA polymerase to drive gene expression in transgenic trypanosomes. *Nucleic Acids Res.* 22:3387–94

175. Wittekind M, Dodd J, Vu L, Kolb JM, Sentenac A, Nomura M. 1988. Isolation and characterization of temperature sensitive mutations in RPA190, the gene encoding the largest subunit of RNA polymerase I from *Saccharomyces cerevisiae*. *Mol. Cell. Biol.* 8:3997–4008

176. Wittekind M, Kolb JM, Dodd J, Yamagishi M, Memet S, et al. 1990. Conditional expression of RPA190, the gene encoding the largest subunit of yeast RNA polymerase I: effects of decreased rRNA synthesis on ribosomal protein synthesis. *Mol. Cell. Biol.* 10:2049–59

177. Wong S, Morales TH, Neigel JE, Campbell DA. 1993. Genomic and transcriptional linkage of the genes for camodulin, EF-hand 5 protein, and ubiquitin extension protein 52 in *Trypanosoma brucei*. *Mol. Cell. Biol.* 13:207–16

178. Young RA. 1991. RNA polymerase II. *Annu. Rev. Biochem.* 60:689–715

179. Zomerdijk JCBM, Kieft R, Borst P. 1993. Insertion of the promoter for a variant surface glycoprotein gene expression site in an RNA pol II transcription unit of procyclic *Trypanosoma brucei*. *Mol. Biochem. Parasitol.* 57:295–304

180. Zomerdijk JCBM, Kieft R, Borst P. 1992. A ribosomal RNA gene promoter at the telomere of a mini-chromosome in *Trypanosoma brucei*. *Nucleic Acids Res.* 20:2725–34

181. Zomerdijk JCBM, Kieft R, Borst P. 1991. Efficient production of functional mRNA mediated by RNA polymerase I in *Trypanosoma brucei*. *Nature* 353:772–75

182. Zomerdijk JCBM, Kieft R, Shiels PG, Borst P. 1991. Alpha-amanitin resistant transcription units in trypanosomes: a comparison of promoter sequences for a VSG gene expression site and for the ribosomal RNA genes. *Nucleic Acids Res.* 19:5153–58

183. Zomerdijk JCBM, Ouellette M, Ten Asbroek ALMA, Kieft R, Bommer AMM, Clayton CE, Borst P. 1990. The promoter for a variant surface glycoprotein gene expression site in *Trypanosoma brucei*. *EMBO J.* 9:1791–1802

Annu. Rev. Microbiol. 1997. 51:491–525

SYNTHESIS OF ENANTIOPURE EPOXIDES THROUGH BIOCATALYTIC APPROACHES

A. Archelas and R. Furstoss

Groupe Biocatalyse et Chimie Fine, ERS 157 associée au CNRS, Faculté des Sciences de Luminy, Case 901,163 Avenue de Luminy, 13288 Marseille Cedex 9, France

KEY WORDS: chiral epoxides, biotransformations, epoxidation, epoxide hydrolysis

ABSTRACT

Enantiopure epoxides, as well as their corresponding vicinal diols, are valuable intermediates in fine organic synthesis, in particular for the preparation of biologically active compounds. The necessity of preparing such target molecules in an optically pure form has triggered much research, leading to the emergence of various new methods based on either conventional chemistry or enzymatically catalyzed reactions. In this review, we focus on the biocatalytic approaches, which include direct epoxidation of olefinic double bonds as well as indirect biocatalytic methods, and which allow for the synthesis of these important chiral building blocks in enantiomerically enriched or even enantiopure form.

CONTENTS

491

0066-4227/97/1001-0491$08.00

INTRODUCTION

Enantiopure epoxides (often incorrectly called "chiral" epoxides) are valuable intermediates in organic synthesis. Their corresponding vicinal diols, which can be either transformed into the epoxide itself or used as highly reactive cyclic sulfates or sulfites, are similarly valuable epoxide-like building blocks (91). This, essentially, is due to the versatility of the oxirane function, which can be chemically transformed into numerous—more elaborated—enantiopure intermediates en route to biologically active targets (Figure 1).

In recent years, considerable effort (76, 143) has been devoted to the synthesis of these intermediates in enantiomerically pure form, because chiral molecules often show different biological activity for each enantiomer (41, 147). Both purely chemical and enzyme-catalyzed methodologies have been developed, and several reviews and monographs have been devoted to this topic (18, 75, 82, 130). The Katsuki-Sharpless method was the first conventional chemical approach. It was a breakthrough, which allowed the asymmetric epoxidation of allylic alcohols (78, 134). More recently, catalysts allowing for the epoxidation of nonfunctionalized olefins have been developed, leading to high stereoselectivities when applied to the epoxidation of *cis*-alkenes. For *trans* and terminal olefins, however, the selectivity achieved with these methods is less satisfactory; the Jacobsen-Katsuki catalysts are more successful with those (75). To these

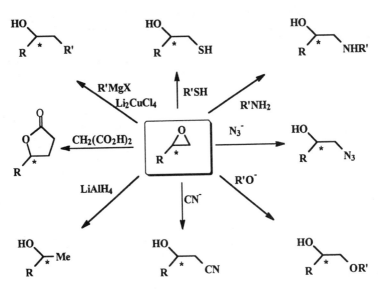

Figure 1   Reaction of epoxides with nucleophiles. (Adapted from 40.)

chemical approaches a determining achievement has been added: The Sharpless dihydroxylation process allows direct dihydroxylation of olefins (82). This is particularly satisfactory—for yield and optical purity—when applied to *trans*-substituted olefins, thus affording an alternative method for synthesizing the corresponding *trans*-epoxide (81). However, both approaches suffer from limitations: they require the use of heavy metal–based catalysts, possible sources of industrial pollution; and all these procedures have patents pending and, thus, are cost-generating processes.

As alternatives to these chemical approaches, a number of enzyme-catalyzed methods studied during the last decade now provide various new and efficient tools. Here again, several reviews and monographs have focused on the possible use of enzymes for fine organic synthesis, including ways to synthesize optically enriched epoxide (17, 32, 130). This review does not aim to achieve an exhaustive catalog of the possible epoxidation reactions but rather to discuss more recent information about these biotechnological techniques, which—by being environmentally gentle—avoid some of the drawbacks of the chemical methods.

DIRECT ENZYMATIC EPOXIDATION OF DOUBLE BONDS

Numerous examples of metabolic pathways, describing the catabolism of molecules that show very diverse structures, have been described (61). Many imply the direct incorporation of an oxygen atom, which can originate from either water or molecular oxygen. Few of the enzymes in these transformations have been identified, however, so many are not yet classified. Some of these are described below. However, some studies have indicated that at least three types of enzymes can be involved in these bioepoxidations. These are heme monooxygenases (cytochrome P-450), ω-monooxygenases, and methane monooxygenases.

Heme-Dependent Monooxygenases

One of the most important classes of oxygenating enzymes is the cytochrome P-450 family. These hemoproteins have been detected in all types of living cells, including insects, plants, yeast, bacteria, and fungi, but because of their involvement in detoxication processes, the mammalian enzymes have been by far the most thoroughly studied (54). These enzymes oxygenate lipophilic xenobiotics, the first and essential step in the detoxication process; the second step is conjugation with various natural counterparts, such as glutathione, glycosides, and sulfates. This leads to water-soluble—and thus excretable—metabolites. The prosthetic moiety implied in these cytochrome P-450 enzymes is a ferroporphyrin entity called heme, which can activate molecular oxygen and incorporate one of its two oxygen atoms into an organic molecule (135, 146). Depending on the molecular characteristics of the substrates, these monooxygenases will allow, for instance, hydroxylation of nonactivated carbon atoms or epoxidation of olefinic double bonds (63, 96, 122, 136).

Epoxidation of various olefins by cytochrome P-450 enzymes has been studied using rat liver microsomes (92, 141). Microbial epoxidation implying these enzymes has also been described. Ruettinger & Fulco (139) reported the epoxidation of fatty acids such as palmitoleic acid by a cytochrome P-450 from *Bacillus megaterium*. Their results indicate that both the epoxidation and the hydroxylation processes are catalyzed by the same NADPH-dependent monooxygenase. More recently, other researchers demonstrated that the cytochrome P-450$_{cam}$ from *Pseudomonas putida*, which is known to hydroxylate camphor at a nonactivated carbon atom, is also responsible for stereoselective epoxidation of *cis-β*-methylstyrene (124). The (1S,2R) enantiomer epoxide obtained showed an enantiomeric purity (ee) of 78%. This result fits the predictions based on a theoretical approach (Figure 2).

During recent years, several cytochrome P-450 enzymes have been cloned and overexpressed in various hosts (173). However, to the best of our knowledge, no synthetic application of such overexpressed enzymes has been described. Interestingly, however, an elegant result has been obtained concerning another heme-dependent protein, human myoglobin (88). Thus, the phenylalanine-43 in the heme pocket of this protein was replaced by tyrosine using site-directed mutagenesis: The tyrosine-43 mutant was shown to be approximately 25 times more active than the wild-type protein in mediating the oxidation of styrene by hydrogen peroxide. Moreover, it was shown to afford (R)-styrene oxide (60% yield), which had an ee as high as 98%, whereas the wild-type protein produced the racemic product. This undoubtedly is an early illustration of the potential that molecular biology approaches have for direct epoxidation of olefinic double bonds. It is not clear whether this particular example could be scaled up to multigram—or industrial—quantities.

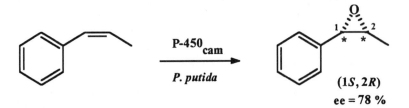

Figure 2   Stereoselective epoxidation of *cis-β*-methylstyrene using cytochrome P-450$_{cam}$ from *Pseudomonas putida* (124).

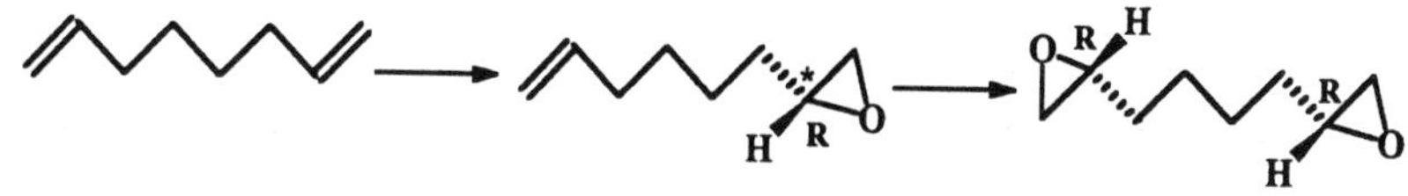

Figure 3   Bioepoxidation of 1,7-octadiene to 7,8-epoxy-1-octene and 1,2-7,8-diepoxyoctane (99).

ω-Hydroxylases

The ability to activate and transfer molecular oxygen into organic molecules is not restricted to the cytochrome P-450 family, and some non-heme monooxygenases have been described as involved in such processes as well. This is the case for ω-hydrolases. These enzymes have been detected in several microorganisms (97). As early as 1973, Abbott et al (1, 97) unequivocally established that the *Pseudomonas oleovorans* strain was able to perform ω-hydroxylation of aliphatic molecules as well as stereoselective epoxidation of a terminal double bond in the corresponding olefins. These authors observed that, for instance, 1,7-octadiene led exclusively to 7,8-epoxy-1-octene, which further on could be reepoxidized to the corresponding diepoxide (99) (Figure 3). It has been established that the first epoxidation was stereoselective—leading to the (*R*)-(+)-7-epoxide in 80% ee. Further epoxidation of this primary metabolite again led to an (*R*)-configuration of the second epoxide. Interestingly, starting from racemic, 7,8-epoxy-1-octene led to a preferred formation of the (*R*)-diepoxide (ee 20%), a surprising fact given that the epoxide function and the olefinic double bond are far away from each other and afford very slight chiral induction for the second step. The enzymatic mechanism of these reactions has been studied by Coon et al (140, 151), who isolated the so-called ω-hydroxylase enzyme. They showed that this enzyme was capable of performing either hydroxylation or epoxidation. Furthermore, they observed that both types of reactions can be in competition on an olefinic substrate, depending upon the structure of the substrate. However, cyclic olefins were neither hydroxylated nor epoxidized by this enzyme (98).

These results have led to an interesting industrial application for the synthesis of the β-blockers Metoprolol® and Atenolol®. Thus, epoxidation of the prochiral allyl ethers by several bacteria, including the *P. oleovorans* strain mentioned above, led to the corresponding (*S*)-epoxides, which showed excellent enantiomeric purities (Figure 4). Further on, these chirons (i.e. chiral building blocks) were transformed into the corresponding (*S*)-enantiomers of the drugs developed by the Shell & Gist-Brocades Companies (see 71). Refinement of this approach has led to the development of interesting molecular biology strategies, which led to the construction of a *P. putida* strain equipped with an *alk* gene from *P. oleovorans*. Thus, this recombinant biocatalyst can accumulate

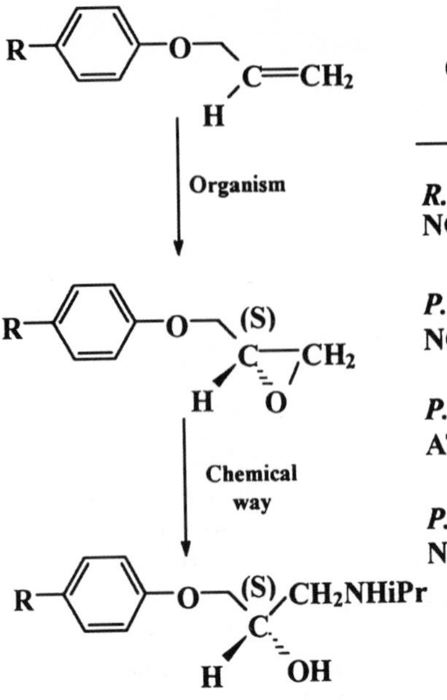

Organism	Metoprolol o.p. (%)
R. equi NCIB 12035	95.4
P. putida NCIB 9571	98
P. oleovorans ATCC 29347	98.4
P. aeruginosa NCIB 8704	98.8

R = CH₃OCH₂CH₂ : Metoprolol

Figure 4   Synthetic route to Metoprolol® incorporating stereospecific microbial epoxidation (71). o.p., optical purity.

the products that are not oxidized further. This *P. oleovorans* monooxygenase has been shown to epoxidize various types of olefins, including allylphenyl ethers and allylbenzene with high stereoselectivity (170).

Other bacteria have also achieved asymmetric epoxidation of substrates such as straight chain aliphatic terminal or subterminal alkenes or aromatic olefins (118, 121, 138). In this context, one of the most thoroughly studied enzymatic systems implied in such epoxidation is xylene oxygenase (XO), which originates from a *P. putida* strain capable of utilizing toluene and xylenes as carbon sources for growth. This XO system essentially acts in hydroxylation reactions of aromatic substrates (170), and it has been introduced by genetic engineering into an *Escherichia coli* host. Interestingly, it has been observed that whereas this enzyme generally does not epoxidize olefinic double bonds, it operates nicely on styrene to produce (*S*)-styrene oxide, which is obtained in 93% enantiomeric purity. Enantiopure styrene oxide is a useful chiral building block for the synthesis of optically active α-substituted benzylalcohols (169) and has been used to synthesize enantiopure pharmaceuticals (23, 31). Similarly, *m*-chlorostyrene was epoxidized with high stereoselectivity (ee > 95%). However *p*-chlorostyrene only led to a 37% ee whereas *p*-methylstyrene was preferentially oxidized at the methyl substituent. Biotechnological improvements of these monooxygenase-based transformations, using two-liquid phase fermentations, have been elaborated to enhance the practicability and yields of these biooxidations (168). However, although elegant, the real applicability of these techniques to large-scale production is still to be demonstrated.

Methane Monooxygenases

Other interesting non-heme oxygenases are the methane-monooxygenases (MMOs) (156). These enzymes can transform methane into methanol, a reaction extremely difficult to carry out using conventional synthesis. These MMO enzymes have been isolated from certain methanotrophic microorganisms, such as *Methylosinus trichosporium, Methylococcus capsulatus* or *Methylococcus organophilum.* They are able to introduce oxygen into a large number of hydrocarbons, such as alkanes and alkenes, as well as alycyclic and aromatic substrates (45, 87). Hou et al (65) reported that short-chain alkenes (C2-C4) can thus be transformed into their corresponding epoxides. The MMOs from *M. capsulatus* and *M. trichosporium* have been purified and their enzymatic properties have been studied. It has been shown that the same enzymatic system was responsible for the hydroxylation and the epoxidation process. Ohno & Okura (119) also observed epoxidation of short-chain alkenes, but the ees they obtained were low (14 < ee < 28%). More recently, Seki et al studied the epoxidation of halogenated allyl derivatives by the bacteria *M. trichosporium.* Interesting results—i.e. inversion of stereoselectivities—were observed, depending on the substrate substitution. However, the observed ee of the products were again low (144).

Miscellaneous Microbial Epoxidations

Numerous epoxidation reactions have been described throughout the literature. However, in many cases, the exact nature of the enzyme operating in these reactions has not been established. For instance, de Bont et al (56, 159) isolated from soil several bacterial strains from, in particular, the *Mycobacterium, Nocardia,* and *Xanthobacter* genuses, which are able to grow on ethylene, propene, 1-butene (or even butane) as sole carbon source. These were shown to achieve the stereoselective epoxidation of these substrates. Detailed mechanistic studies using a *Mycobacterium* strain and $^{18}O_2$ have revealed that the oxygen incorporated indeed originated from molecular oxygen, and that the enzymes implied for hydroxylation were different from those implied for epoxidation (33). Furthermore, interesting ees were obtained in some cases (ee ≈ 80%). In this context, it was shown by Archelas et al (8) that some of these strains are also able to epoxidize substituted alkenes, again leading to variations in the observed stereoselectivities. The *Xanthobacter* strains also allowed accumulation of 2,3-epoxybutane, from the *cis* or *trans* olefine, although with low to moderate yields (152). However, in spite of interesting technological improvements, all these biooxygenation reactions are of low preparative value because they show both low turnover values and severe inhibition of the monooxygenase as a result of product concentration (55).

Similar results were described by Mahmoudian & Michael, who isolated 18 bacterial strains able to produce optically enriched epoxides showing excellent ees (up to 98%) (94, 95). However, in the case of *trans*-(2R,3R)-epoxybutane, it was shown that the enantiomeric enrichment is in fact due to a second-step enantioselective hydrolysis of the epoxide, which is first produced in racemic form. This, interestingly, is an unexpected example of the possible use of microbial epoxide hydrolases for the synthesis of enantiopure epoxides (see below). This is not the case for a *Rhodococcus rhodochrous* strain that affords 1,2-epoxyalcanes from short-chain terminal olefins. Indeed, in this case, no product inhibition has been observed (166). However, the obtained ees were not determined, and it seems probable that they were quite low. On the other hand, a *Nocardia corallina* strain was described that afforded the corresponding 1,2(R)-epoxy-2-methylalkane with ees as high as 90%, depending on the chain length (149) (Figure 5). These epoxides were used as chiral building block to prepare prostaglandin ω-chains.

Numerous other examples indicating the ability of various different microbial strains have been described (120, 161), and several patents have been filed in this context (46, 64). However, although interesting, it is not clear whether some of these processes are used for the production of enantiopure epoxides. (See Table 1 for a summary of bacterial epoxidations.)

One of the drawbacks of using bacterial strains is further degradation of the epoxide produced. More promising results have been described recently in

n = 3	Yld 32 % ee 76 %
n = 4	Yld 55 % ee 90 %
n = 5	Yld 56 % ee 88 %

Figure 5   Stereoselective microbial oxidation of 2-methyl-1-alkene (149). n, Number; Yld, yield.

Table 1   Miscellaneous microbial epoxidation

Olefines	Bacteria strains	Epoxides		Reference
		R	S	
3-Chloro-propene	B3	4	96	56
3-Chloro-propene	B4	3	97	56
3-Chloro-propene	2W	1	99	56
3-Chloro-propene	Py2	20	80	56
1,7-Octene	Pseudomonas oleovorans	83	17	99
1,2-Hexadecene	Corynebacterium equi IFO 3730	100	0	120
2-Methyl-1-hexene	Nocardia corallina	86	14	149
Propene	Methylosinus trichosporium OB3b	57	43	119
1,3-Butadiene	Methylosinus trichosporium	36	64	119
4-Bromo-1-butene	Mycobacterium L1	82	18	8
4-Bromo-1-butene	Mycobacterium E3	94	6	8
4-Bromo-1-butene	Nocardia Ip1	90	10	8
3-Butene-1-ol	Mycobacterium L1	70	30	8
" " " "	Mycobacterium E3	79	21	8
" " " "	Nocardia Ip1	92	8	8

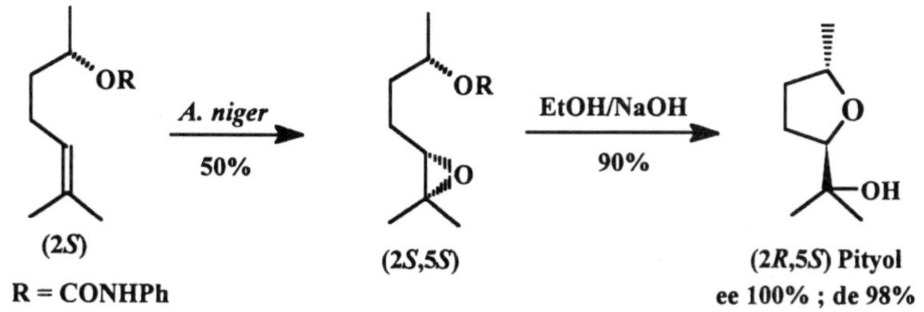

Figure 6   Epoxidation of a (S)-sulcatol derivative by the fungus *Aspergillus niger*. Synthesis of the biologically active pheromone pityol (7). ee, Enantiomeric purity; de, diastereomeric purity.

this context using fungi instead of bacteria. Archelas & Furstoss (7) described the epoxidation of an (S)-sulcatol derivative by the fungus *Aspergillus niger*, which affords the corresponding (2S,5S) enantiomer in 100% ee and 50% yield. This can be further transformed into the natural enantiomer of the biologically active pheromone pityol (Figure 6). However, it is to be emphasized that such an accumulation of epoxide in the presence of a whole cell fungi appears to be an exception. Indeed, much more common is the observation that the epoxide thus formed is metabolized further on into the corresponding vicinal diol. Depending on reaction conditions (medium pH), this can be achieved either spontaneously (acidic medium) or enzymatically (neutral pH). An interesting application of this approach is allowance of the "tuned" preparation of either enantiomer of

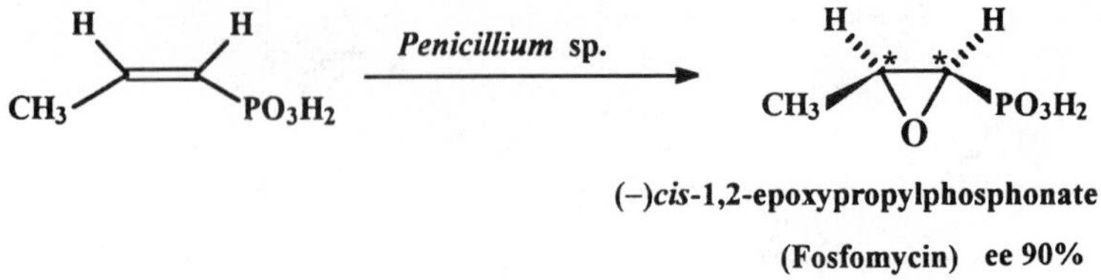

(−)*cis*-1,2-epoxypropylphosphonate

(Fosfomycin) ee 90%

Figure 7 Microbial epoxidation of *cis*-propenylphosphonic acid to fosfomycin (163). ee, Enantiomeric purity.

this diol (175). As was previously emphasized, these diols can themselves either lead to the corresponding epoxide without loss of enantiomeric integrity or be used as their cyclic sulfates or sulfites (91) (see below).

Other fungal strains, like *Cunninghamella elegans* and *Syncephalastrum racemosum*, also epoxidize aromatic substrates, but again the intermediate epoxides are directly processed into the less-reactive (and less toxic) corresponding *trans*-diols. The enzymes implied in these oxidations were reported to be cytochrome P-450 type enzymes, which possessed stereo and regioselectivities different from the mammalian enzymes (27, 101).

An interesting application of such a fungal epoxidation has been described with 20 fungi that carry out the epoxidation of the olefinic precursor of fosfomycin, a large-spectrum antibiotic (163). This can be obtained (although in low concentration, i.e. 0.5 g/liter) in 90% enantiomeric purity using this very simple procedure (162) (Figure 7).

As mentioned above, direct epoxidation of olefinic double bonds is possible by using various monooxygenases from different types and origins, either as purified enzymes or as wild or genetically engineered strains. Furthermore, these can in certain cases lead to interesting, although not sufficient, enantiomeric purities. However, it is clear that severe practical difficulties are linked to the fact that these enzymes are essentially cofactor dependent and that one often encounters product inhibition so only low concentrations of epoxides can be accumulated in the medium. In spite of some biotechnological improvements based on the use of biphasic systems implying organic solvents (22, 47, 168), no practical breakthrough process seems to have been achieved, thus hampering the use of this direct epoxidation approach for large-scale laboratory and, worse, industrial production.

INDIRECT WAYS TO ENANTIOPURE EPOXIDES

Because of the problems encountered in the production of enantiopure epoxides using direct epoxidation of olefinic double bonds, several indirect approaches

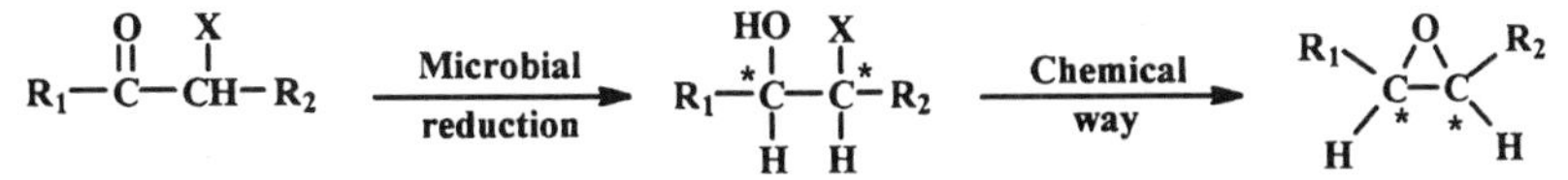

$$R_1-\overset{\overset{O}{\|}}{C}-\overset{\overset{X}{|}}{C}H-R_2 \quad \xrightarrow[\text{reduction}]{\text{Microbial}} \quad R_1-\overset{\overset{HO}{|}_*}{\underset{\overset{|}{H}}{C}}-\overset{\overset{X}{|}_*}{\underset{\overset{|}{H}}{C}}-R_2 \quad \xrightarrow[\text{way}]{\text{Chemical}} \quad \overset{R_1}{\underset{H}{}}\diagdown\overset{O}{\underset{*}{C}}-\overset{}{\underset{*}{C}}\diagup\overset{R_2}{\underset{H}{}}$$

Figure 8   Chemoenzymatic synthesis of chiral epoxides.

have been explored in recent years. These include microbial reduction of α-halo-ketones as well as the use of enzymes such as haloperoxidases, halohydrine epoxydases, and epoxide hydrolases. Various other approaches, based on the use of miscellaneous enzymatic reactions, have also been explored. As is seen, some of these enzymes led to interesting results, and it can be predicted that future industrial applications will be based on the use of one of these "easy-to-use" catalysts.

Microbial Reduction of α-Haloketones

Because of the enormous amount of work devoted to the study of alcohol dehydrogenase–mediated reduction of various substituted carbonyl compounds, it appears a priori that a strategy that would imply stereoselective reduction of an α-halogenated ketone and further cyclization of the halohydrine obtained into the corresponding epoxide would be a well-explored way for the synthesis of enantiopure epoxides (Figure 8). Surprisingly enough, scant information is available about this. One possible explanation may be the intrinsic reactivity of α-haloketones, which act more like enzyme inhibitors than as "normal" substrates.

Nevertheless, examples have been described that illustrate this approach. Imuta et al achieved the reduction of α-bromo and α-chloroacetophenone using the bacteria *Cryptococcus macerans*, which leads to the correspond-ing (R)-2-halohydrines as intermediates, further cyclized into enantiomerically pure styrene oxide (68). More recently, Weijers et al studied the reduction of α-chloroacetone into 1-chloro-2-propanol, with the aim to prepare enantiop-ure 1,2-epoxypropane (160). This could be obtained using several bacterial or yeast cells. It has been shown, however, that racemic 3-chloro-2-butanone led to a 1/1 mixture of the $(2S,3R)$ and $(2S,3S)$ diastereoisomers. Thus, the reduction of this substrate occurred in a non-enantioselective manner (i.e. each of the two enantiomers was reduced), but the reduction itself was stereose-lective, leading to the two diastereoisomeric-corresponding halohydrines, thus obtained in high enantiomeric purity (ee > 97%). As a consequence, separa-tion of these two diastereoisomers could afford a route to both the *cis* or the *trans* epoxide in high optical purity. A similar approach has been developed by Besse & Veschambre, who screened numerous microorganisms allowing the reduction of α-haloketones, e.g. 3-bromo-2-octanone. Here again, a mixture

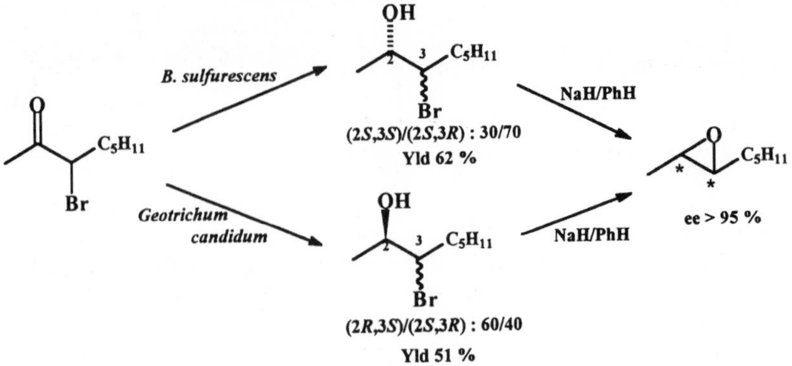

Figure 9   Microbial reduction of α-bromoketone. Synthesis of chiral 1-phenyl-1,2-epoxypropane (18). ee, Enantiomeric purity; Yld, yield.

of diastereoisomers were obtained systematically (18). Interestingly, these authors observed that it was possible to obtain a mixture of either the (2S,3SR) or the (2R,3SR) diastereoisomers, depending on the microbial strain used. Also interestingly, these diastereoisomers could be further cyclized into the corresponding epoxides using conventional synthetic methods (i.e. NaH/Benzene) without noticeable loss of stereochemical integrity (Figure 9).

In certain cases one of these diastereoisomers could be obtained selectively. Reduction of 4-phenyl-3-bromo-2-butanone achieved using the fungus *A. niger* led to a 48% yield of the single (2R,3R) stereoisomer in excellent enantiomeric purity (ee > 98%). In this case, the remaining (unreacted) α-haloketone substrate was also optically pure. Thus, in this particular case, the reduction was both enantio- and stereoselective. Further development of these studies led these authors to raise this method to a more general level, showing that starting from various α-haloketones, it was possible to prepare—although through an intermediate chromatographic separation—the four stereoisomers of a given substrate, such as 4-phenyl-2,3-epoxybutane or *trans-β*-methylstyrene oxide. The yields were moderate, but the enantiomeric purities were excellent (ee > 95%) (16).

Such a strategy has been also developed (24) starting from 2-chloro-3-oxo-esters, which allow the synthesis of valuable glycidic ester derivatives. For example, using a reduction by *Mucor plumbeus*, the *anti* (2R,3R)-ethyl-2-chloro-3-hydroxyhexanoate enantiomer was obtained in 50% yield and 92% ee from the corresponding chlorooxoester. On the other hand, both *syn* (2S,3R) and *anti* (2R,3R) enantiomers could also be formed from ethyl-2-chloro-3-hydroxy-3-phenylproprionates and were obtained in 45–50% yield (ee > 95%) using *Mucor racemosus* or *Rhodotorula glutinis* strains, respectively. These

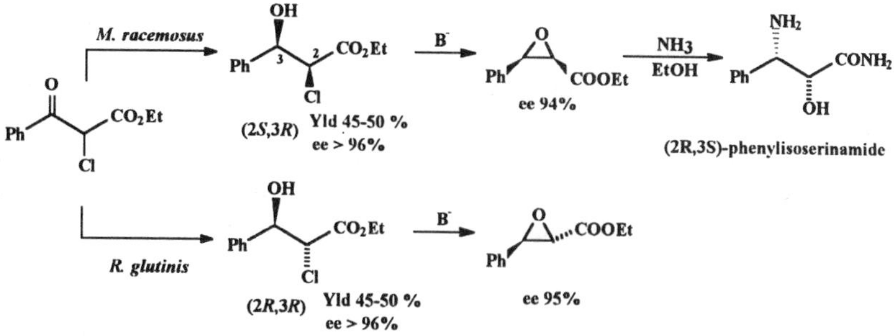

Figure 10 Stereoselective microbial reduction of 2-chloro-3-oxoester. Synthesis of (2R,3S)-3-phenylisoserine (25). ee, Enantiomeric purity; Yld, yield.

intermediates could be efficiently cyclized into the corresponding *cis* or *trans* epoxyesters depending on the reaction conditions without loss of enantiomeric purity (25). This allowed the synthesis of enantiopure N-benzoyl and N-*tert*-butoxycarbonyl(2R,3S)-3-phenylisoserine, the side chain of taxol and taxotere® (Figure 10).

Reaction Mediated by Haloperoxidases and Halohydrine Epoxydases

Another way to synthesize halohydrines is the enzymatic addition of hypohalous acid (XOH) on an olefinic double bond. A class of enzymes, haloperoxydases, are known to catalyze the formation of α-halohydrins from alkenes in the presence of a halogen and a hydroperoxide (50). These enzymes are available from a variety of sources (44), and according to the halide ion they can utilize, they can be separated into three groups: chloroperoxydases, bromoperoxydases (algae and bacteria), and iodoperoxydases (algae). Iodoperoxydases catalyze the formation of carbon-iodine bonds, whereas bromoperoxidases catalyze iodination and bromination reactions, chloroperoxidases being able to handle either chlorine, bromine, or iodine ions. These enzymes can be either heme-containing enzymes (44) or non-heme enzymes (83, 154, 164). They can react with various types of organic molecules, and a recent review has been devoted to these different aspects (153). However, the most commonly used haloperoxidase is the chloroperoxidase from *Caldariomyces fumago*, available as a commercial preparation. This hemoprotein has been crystallized and shown to exist in solution at neutral pH as a heavily glycosylated ($\approx$25–30% by weight) monomer (58, 107), and its exact structure has been studied by resonance Raman studies and by extended X-ray fine structure spectroscopy (10). In the presence of hydrogen peroxide, this enzyme converts a large variety of olefins into the corresponding halohydrin (89, 108). Several types of olefins can thus be transformed, including alkenes (50), dienes (137), and allenes (50), as well as more complex substrates (115). Interestingly, the products formed from these reactions are those that could be predicted for chemical attack of hypohalous acid (XOH) on the olefine substrate. Moreover, it happens that the obtained products are racemic, giving a clue to the mechanism implied. In fact, it appears that these enzymes, in the presence of halide ions, act as an enzymatic source of hypohalous acid, which then adds non-enzymatically to the substrate (44, 53). The synthetic utility of this approach for the synthesis of enantiopure epoxides is linked to the further use of these intermediates. In particular, cyclization into epoxides possibly are achieved by using halohydrine epoxidase enzymes. This type of enzymes has been detected in several bacteria, such as *Flavobacterium*, *Pseudomonas*, and *Arthrobacter* species (35). This last one has been purified and characterized (26). It allows the cyclization of 3-chloro-1,2-propanediol and of 1,3-dichloro-2-propanol, thereby affording either glycidol or epichlorhydrine, respectively. However, the enantioselectivity of these reactions is not indicated, presumably because the obtained products were racemic.

It was recently described that such a two step process was occurring in the course of the microbial oxidation of the olefinic precursor of fosfomycin, a clinically important, broad spectrum antibiotic. Fifteen strains of aerobic bacteria as well as two fungal strains were shown to produce optically active fosfomycin (69). Despite the lack of enantioselectivity, such reactions may still be interesting for an individual process. E.g. Figure 11 displays an elegant and efficient industrial application for the proposed production of—racemic—propylene oxide (Cetus procedure) by Neidelman (114). However, to the best

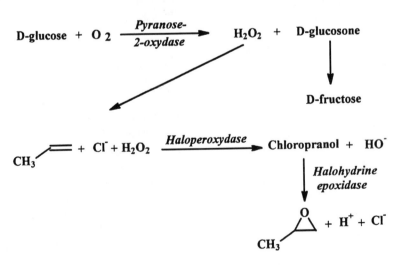

Figure 11   Production of propylene oxide through the Cetus procedure (114).

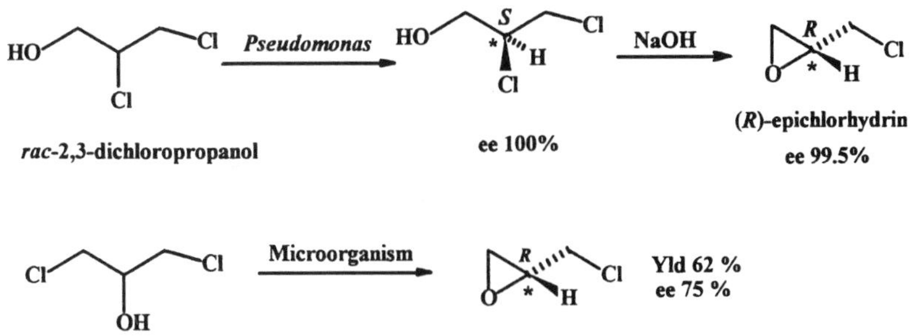

Figure 12   Enantioselective degradation of halohydrines. Synthesis of (R)-epichlorohydrine (74). ee, Enantiomeric purity; Yld, yield.

of our knowledge, severe competition with conventional chemical processes occurred, and this procedure has not been used industrially.

Enantioselective degradation of such halohydrines is one way to prepare optically enriched epoxides. Kasai et al (74) used an immobilized *Pseudomonas* strain to degrade racemic 2,3-dichloro-1-propanol. The residual (S)-substrate, which showed an ee as high as 99%, was then further transformed chemically into the corresponding (R)-epichlorohydrine (Figure 12).

Obviously, in such transformations, yield is limited to 50%. On the other hand, enzymatic conversion of prochiral substrates can allow a theoretical 100% yield. Such an approach was illustrated by Nakamura et al (110, 113), starting from the prochiral 1,3-dichloropropan-2-ol, which was converted into a 62% yield of (R)-epichlorohydrine (75% ee). Although this is not sufficient for further synthesis of biologically active compounds, it indicates that such transformations, leading to enantiomerically enriched epoxides, may be useful (Figure 12).

Interestingly, it has also been shown that the *C. fumago* chloroperoxidase is able to yield epoxides directly from olefins in the absence of halide ions. For instance, styrene, as well as some of its analogs, is epoxidized in a stereoselective way in the presence of either *t*-butyl peroxide (30) or peroxide (3, 49, 100, 123), leading to their corresponding epoxides. Extensive studies developed recently by Allain & Hager (3) and by Zaks & Dodds (172) indicate that the best substrates are aliphatic *cis*-disubstituted olefins. In these cases, ees as high as 95% can be reached, but yields are generally moderate. Even lower yields and ees were obtained from differently substituted olefins (30). Interestingly, 2-methylalkenes appeared to lead in certain cases to interesting enantiomeric purities (37) (Table 2).

A first multistep synthesis featuring the use of chloroperoxidase for the synthesis of a biologically important target was described recently by Lakner & Hager (86). (R)-(−)-mevalonolactone was synthesized through stereoselective

Table 2   Asymmetric epoxidation reactions catalyzed by CPO

Substrates	ee Epoxides (%)	Absolute configuration (epoxide)	Epoxide yield (%)	Reference
cis-2-Heptene	96	2R,3S	78	3
cis-2-Octene	92	2R,3S	82	3
cis-2-Methylstyrene	96	1S,2R	67	3
1,2-Dihydronaphtalene	97	1R,2R	85	3
Styrene	49	R	23	30
p-Chlorostyrene	66	R	35	30
p-Bromostyrene	68	R	30	30
p-Nitrostyrene	28	R	5	30
α-Methylstyrene	89	R	55	37
$PhOCH_2C(CH_3)=CH_2$	89	R	22	37
2-Methyl-heptene	95	R	23	37
$EtCO_2CH_2C(CH_3)=CH_2$	94	R	34	37

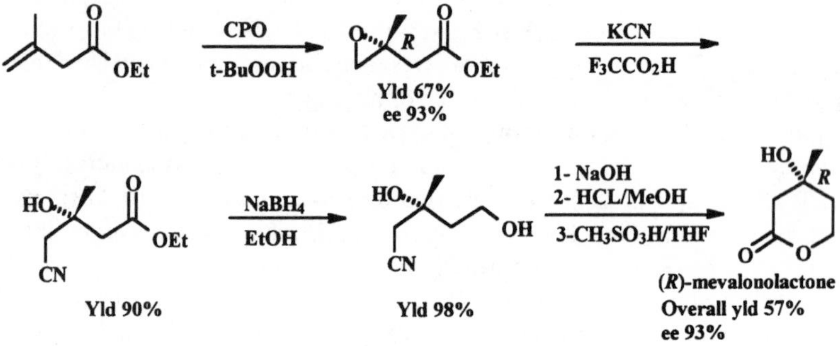

Figure 13   Chloroperoxidase. Synthesis of (R)-mevalonolactone through stereoselective bioepoxidation of methallylpropionate (86). ee, Enantiomeric purity; Yld, yield.

epoxidation of methallylpropionate followed by chemical transformation. The lactone was obtained in 57% overall yield and 93% ee (Figure 13).

The use of such enzymes seems to be hampered by enzyme deactivation through, for instance, reaction with a number of aliphatic terminal olefins to give an inactive N-alkyl derivative (36), as well as by the sensitivity of the enzyme toward the hydrogen peroxide necessarily used. However, interesting biotechnological approaches—i.e. continuous flow bioreactors—have been developed (19). Also, industrial-scale production has recently been claimed (57). Chloroperoxidase from *C. fumago*, modified by genetic engineering of the cloned gene, was used in the presence of hydrogen peroxide and an organic solvent to epoxidize subterminal olefins. This, however, only affords low to fair yields of epoxides showing variable ees (74 < ee < 97).

Epoxide Hydrolases

The use of epoxide hydrolases—enzymes able to enantioselectively hydrolyze several types of epoxides—is a new, very actively emerging strategy for the access to enantiopure epoxides. In fact, these enzymes have been known for several decades, and it now appears that they are ubiquitous in nature, having been found in various living cells, such as plants (20), insects (90), bacteria (104), filamentous fungi (106), and mammalian cells (165). Intensive and interesting studies were conducted by Hammock et al (59), as well as by Knehr et al (79), on several of these enzymes, from which important knowledge of their biological properties was gained. Several of these biocatalysts were recently purified as well as cloned and overexpressed, and elegant studies have been devoted to the determination of the intimate mechanism involved in these reactions (5, 21, 84, 150). However, although the ability of these enzymes to hydrolyze oxirane rings has been explored, it was not until recently that the results were applied to organic synthesis, presumably because the studies were conducted by biochemist or biologist teams not directly concerned with possible practical application to fine organic synthesis. The only noticeable exception to this fact is the use of such bacterial epoxide hydrolases for the industrial synthesis of L- (142) and *meso*-tartaric acid (4) from the precursor epoxide. Such an enzyme, isolated from a *P. putida* strain, was isolated and crystallized as early as 1969 by Allen & Jakoby (4), but apparently no further application of this biocatalyst has been studied. This is not the case for mammalian epoxide hydrolases, intensively explored by Berti (15). This pioneering work highlighted the fact that several substrates, including racemic or prochiral aromatic or aliphatic compounds, can be efficiently processed by epoxide hydrolases, which often leads to enantiomerically enriched—or enantiomerically pure—epoxides (the unreacted enantiomer) and/or to the corresponding vicinal diols (13, 14) (Figure 14). Moreover, in the cases of β-alkyl–substituted styrene oxides (with the exception of the methyl substituted derivatives), each of the two enantiomers was attacked by the microsomal epoxide hydrolase with an opposite regioselectivity [i.e. always on the (S)-carbon atom], leading to an almost quantitative yield of the corresponding (1R,2R) diol (12). In spite of these very interesting results, use of these mammalian epoxide hydrolases is still severely hampered—or even impossible—for large-scale industrial production. Even overexpressed enzymes are not currently available, at a reasonable price, in large quantities.

   The biohydrolysis of epoxides by microbial epoxide hydrolases have been known for a long time. Niehaus & Schroepfer (117) described the enantioselective hydrolysis of *cis*- and *trans*-9,10-epoxystearic acids by a *Pseudomonas*

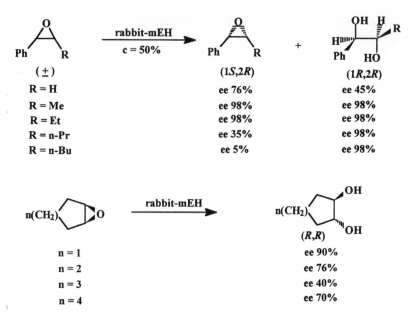

Figure 14 Mammalian epoxide hydrolase. Enantioselective hydrolysis of racemic epoxide and stereoselective hydrolysis of *meso*-epoxide (13, 14). c, Cytosolic; mEH, microsomal EH; ee, Enantiomeric purity; n, number.

sp., whereas Michaels et al (102) observed the hydrolysis of epoxypalmitate by *Bacillus megaterium.* Since then, De Bont et al (34), Escoffier & Prome (39), Jacobs et al (70), and Nagasawa et al (111, 112) have described similarly the enzyme-catalyzed hydrolysis of either short-chain epoxides or an epoxysteroid. All these epoxide hydrolases were originated from bacterial strains, i.e. *Nocardia* sp., *Mycobacterium aureum, Pseudomonas* sp., or *Corynebacterium* sp., respectively. However, these enzymes were either not explored as far as their enantioselectivity was concerned or showed only very low enantioselectivity. Similar results have also been described using various fungi shown to be equipped with epoxide hydrolases (60, 67, 80, 148, 155), which interestingly exhibited promising enantioselectivities.

However, the real breakthrough concerning use of such epoxide hydrolase applied to fine organic chemistry was only recently published, by two groups independently, Faber et al (40, 62) and our group (29, 128). It has been found that both bacteria and filamentous fungi are excellent biocatalysts for achieving the resolution of several racemic epoxides. The determining advantage of these applications is the fact that these microbial enzymes are obtained easily in almost unlimited amounts.

In their preliminary studies, Faber et al (62) observed that the immobilized enzyme preparation derived from *Rhodococcus* sp. (sold by Novo Industry, Denmark), designed for the enzymatic hydrolysis of nitriles, also contained an epoxide hydrolase activity. This *Rhodococcus* enzyme was further purified and characterized to be a cofactor-independent, soluble protein (104). Straight-chain terminal epoxides as well as glycidyl derivatives were well accepted as substrates, but the enantioselectivities were low. The best results were obtained with 2-methyl-2-pentyl oxirane, which was obtained in 72% ee (105). Further work, using such 2-alkyl–substituted straight-chain oxiranes of different lengths, indicates that chiral recognition by the enzyme depends on the size difference between the two alkyl groups (157). Further screening of alkene-utilizing bacteria led to the discovery of other bacteria equipped similarly with interesting epoxide hydrolase activities (125). One of these, *Nocardia* EH1, showed almost absolute enantioselectivity on 2-methyl-1,2-epoxy heptane, leading to both the enantiopure (*R*)-diol and epoxide.

Unexpectedly, as compared with microsomal mammalian epoxide hydrolases, all the *meso*-epoxides tested proved to be nonsubstrates. When sodium azide was added to the medium, enantiomerically enriched (*R*)-azido alcohol generated from the unreacted (*R*)-epoxide was also obtained (103). Although the authors claim that this should be due to an enzyme-catalyzed reaction, it is more reasonable to consider this reaction a pure chemical addition of the azide on the substrate in the course of the reaction (Figure 15). This interpretation is in accordance with the recently published mechanism described for either cytosolic or microsomal rat liver epoxide hydrolases (21, 84). Similar results have also been observed for aminolysis of aryl glycidyl ethers by hepatic microsomes from rat (73) as well as from lipase-catalyzed epoxide aminolysis (72). It, therefore, might well be that these reactions are catalyzed by the chiral protein surface instead of being real enzymatic reactions.

An elegant application of this approach was the synthesis of the biologically active enantiomer of disparlure, the sex pheromone of the moth *Lymantria dispar*, which causes severe damage to trees in parts of the world (126). This was achieved via enantioselective hydrolysis, by a *Pseudomonas* strain, of 9,10-epoxy-15-methyl hexadecanoic acid, a precursor of the pheromone. However, the optical purity of the product, calculated on the basis of the measured optical rotations, is quite low ($\approx$25%) (Figure 16).

Interestingly, the *Nocardia* EH1 enzyme allowed enantioconvergent preparation of the (1*R*,2*R*)-1-phenylpropane-1,2-diol thus obtained in 85% yield and 98% ee (125). This is a valuable result, one that has been observed on other substrates as well (11, 20, 132), since it allows a theoretical 100% yield preparation of a single (enantiopure) antipode of the formed diol starting from a racemic epoxide.

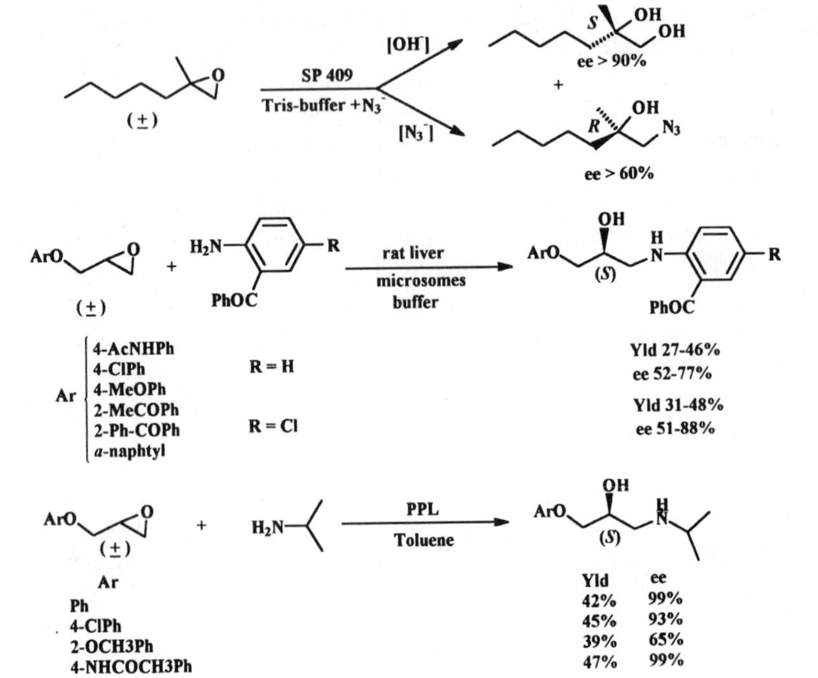

Figure 15 Enantioselective enzyme-catalyzed azidolysis and aminolysis of epoxides (72, 73, 103). ee, Enantiomeric purity; Yld, yield.

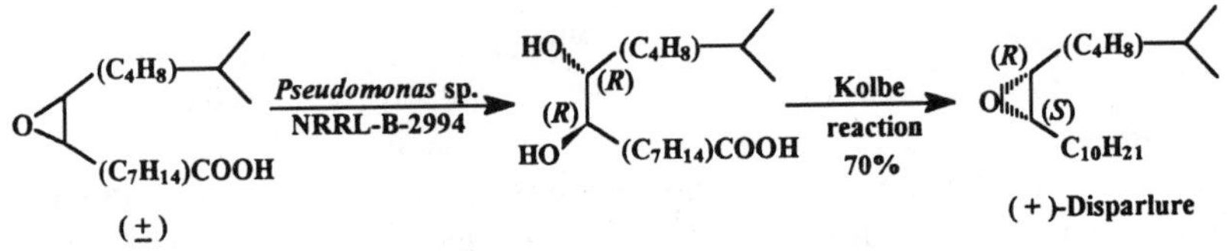

Figure 16 Bacterial epoxide hydrolase. Application to the synthesis of (+)-disparlure (126).

Interestingly, most of the results with bacterial epoxide hydrolases indicate that whereas they often accept aliphatic epoxides as substrates, their aromatic counterparts are badly—or not at all—hydrolyzed by these enzymes. This is not the case for the epoxide hydrolases from fungal origin. We have obtained very interesting results in this context. Racemic geraniol N-phenylcarbamate was efficiently hydrolyzed by a culture of the fungus A. *niger*, leading to a 42% isolated yield of remaining (6S)-epoxide, which showed a 94% ee. This could easily be conducted on 5-g substrate using a 7-liter fermentor (29). We

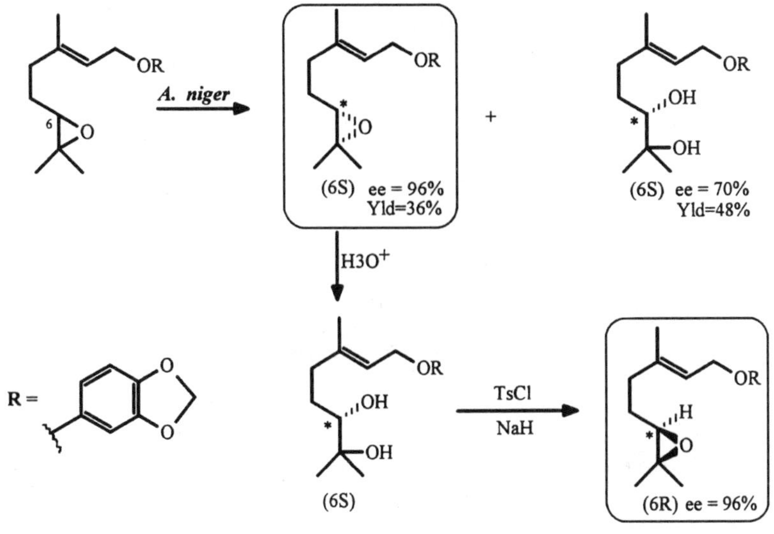

Figure 17 Fungal epoxide hydrolase. Application to the synthesis of (*R*)- and (*S*)-Bower compounds (6). ee, Enantiomeric purity; Yld, yield.

have used this methodology to achieve the preparation of both enantiopure (ee > 96%) enantiomers of the biologically active Bower's compound, a potent analog of insect juvenile hormone (6) (Figure 17). Biological tests using both enantiomers showed that the 6(*R*) antipode was about 10 times more active than the 6(*S*) antipode against the yellow meal worm *Tenebrio molitor*. Moreover, we have shown that, similarly, a diastereoselective hydrolysis of the exocyclic limonene epoxides can be achieved using this same fungi, thus opening the way to the synthesis of either enantiopure bisabolol stereoisomer (29). One of these enantiomers, (4*S*,8*S*)-α-bisabolol, is used on an industrial scale for the preparation of various skin-care creams, lotions, and ointments.

Similar results were obtained with styrene oxide, which was efficiently hydrolyzed by *A. niger*, affording the (*S*)-enantiomer in 96% ee within a few hours (128). Moreover, another fungus, *Beauveria sulfurescens* (presently *Beauveria bassiana*) showed an opposite enantioselectivity, leading to the (*R*)-enantiomer in 98% ee. Both hydrolyses leading to the corresponding *R*-diol, an enantioconvergent process, obtained using a mixture of the two fungi, led to an overall 92% yield of the (*R*)-diol in 89% ee (Figure 18). Further work elicited interesting information concerning the mechanism implied in these transformations (129) and the scope of the substrates admitted. Thus, it has been shown that substituted styrene derivatives—i.e. *para*-substituted styrene oxide (133)

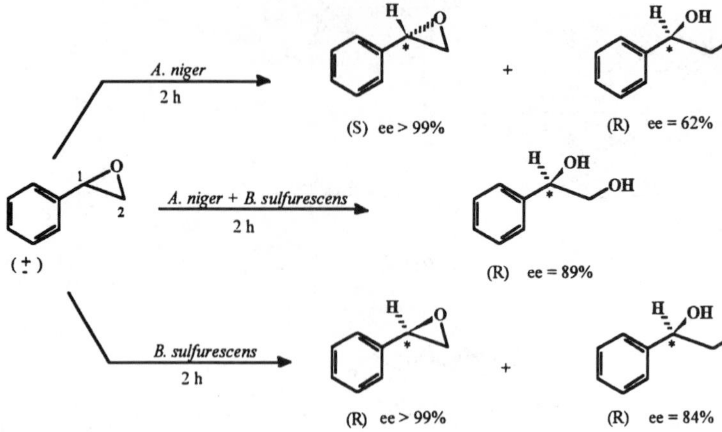

Figure 18   Resolution of styrene oxide by *Aspergillus niger* and *Beauveria sulfurescens* epoxide hydrolases. Deracemization of styrene oxide by using a mixing of these two fungi (128). ee, Enantiomeric purity.

as well as β-substituted derivatives (132)—could be accomodated by one or both of these fungi. In this last case, an interesting enantioconvergent hydrolysis was observed again, affording an 85% preparation yield of enantiopure (1R,2R)-diol. In order to set up an efficient and easy-to-use biotechnological process, a crude lyophilized extract of *A. niger* was prepared and shown to be stable for weeks (or even months) upon storage in the refrigerator. This, then, could be conveniently used in a batch reactor. A 330 mM (54 g/liter) concentration of *p*-nitrostyrene could be hydrolyzed within 6 h to a (analytical) yield of 49% and an ee of 99% of the remaining (S)-epoxide (E ≈ 40) (106, 116). Moreover, controlled acid hydrolysis of the reaction mixture, which contained unreacted (S)-epoxide and the (R)-diol resulting from the enzymatic hydrolysis, led to an overall yield of 94% of (R)-diol (ee 80%), as a result of steric inversion upon acid hydrolysis of the (S)-epoxide. After recrystallization, this (R)-diol (ee 99%) could be recycled and transformed into the biologically active enantiomer of Nifenalol®, known as having β-blocker activity (109) (Figure 19).

Another application of these fungal enantioselective hydrolyses was the synthesis of indene oxide and of its corresponding diol, these intermediates being of crucial importance for the synthesis of the orally active HIV protease inhibitor Indinavir. When submitted to a culture of *B. sulfurescens*, racemic epoxyindene was rapidly hydrolyzed, leading to a 20% yield of recovered enantiomerically pure (ee 98%) (1R,2S) epoxide, and to a 48% yield of the corresponding (1R,2R) *trans*-diol showing a 69% ee (131). Since this diol has the absolute configuration

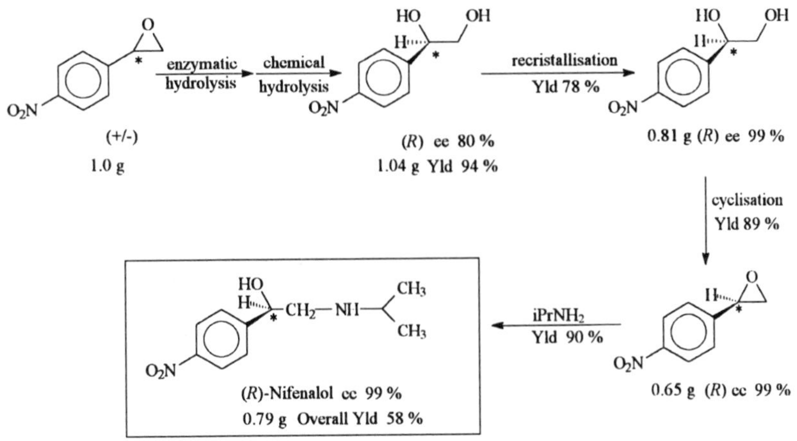

Figure 19   Deracemization of *p*-nitrostyrene oxide by a chemoenzymatic process. Application to the synthesis of (*R*)-Nifenalol®. ee, Enantiomeric purity; Yld, yield;

desired for synthesizing the biologically active HIV protease inhibitor, it might be good to prepare this key-chiral building block intermediate by optimizing this biotransformation. A more extensive study of this biotransformation has been performed by the Merck Company (28, 174). Eighty fungal strains were evaluated for their ability to enantioselectively hydrolyze racemic epoxyindene. Epoxydihydronaphtalene was hydrolyzed similarly to the (1*R*,2*R*)-diol in excellent enantiomeric purity (131).

Miscellaneous Transformations of Epoxide-Bearing Substrates

In addition to the direct or indirect synthesis or transformation of the epoxide moiety itself, several other approaches have been explored in order to obtain enantiopure epoxide-bearing substrates. Some of these approaches seem relatively promising or have already been applied on an industrial scale.

The first, and most efficient, synthetic process is based on the use of lipase—i.e. hydrolytic enzymes, which have been extensively studied, in particular because of their commercial availability, surprising stability (use of organic solvents), and ease of use, being cofactor-independent enzymes. For instance, lipase-catalyzed hydrolysis of various glycidic esters has been studied, because the enantiomerically pure products thus obtainable are valuable chiral synthons for the pharmaceutical industry (9, 48, 77). Porcine pancreatic lipase was shown to enantioselectively hydrolyze several glycidyl esters, with the butyrate derivative leading to the best results (85). Further improvement of this method, using a multiphase membrane enzyme reactor, has been described recently (167). This approach has been patented by a Japanese industry, using whole cell cultures of *Rhodotorula, Pseudomonas*, or *Rhizopus* sp., which may be a practical and much simpler way to achieve these resolutions (2). Similar approaches have also been applied to the synthesis of other glycidyl derivatives, these being either racemic or prochiral substrates. In the case of racemic 2-(methylenhydroxy)-1,2-epoxybutan-4-ol, a transesterification with vinyl acetate was performed by the *Pseudomonas cepacia* lipase in the presence of organic solvents. This allowed the recovery of unreacted (*S*)-diol with an ee of 86% for a 60% conversion ratio, which implies a low yield of the desired product (43). Similar results were obtained using the *Pseudomonas fluorescens* lipase–catalyzed reaction (ee 90%) (42). Better results could be obtained starting from a prochiral glycidyl derivative. Thus, the optically active epoxy alcohol (*R*)-2-butyryloxymethylglycidol was obtained in high enantiomeric purity (ee > 98%) by lipase-catalyzed hydrolysis using a phosphate buffer and an organic cosolvent system. The best results were obtained with *Pseudomonas* sp. lipase (95% yield; ee > 98%) (145) (Figure 20).

The use of lipase-catalyzed ester hydrolyses or transesterification reactions has led, interestingly, to an efficient method for synthesizing the N-benzoyl-(2*R*,3*S*)-3-phenyl isoserine moiety, the C-13 side chain of taxol. The *Mucor miehei* lipase showed itself to be uniquely suited for the stereospecific transesterification of racemic methyl trans-β-phenyl glycidate. Further improvement

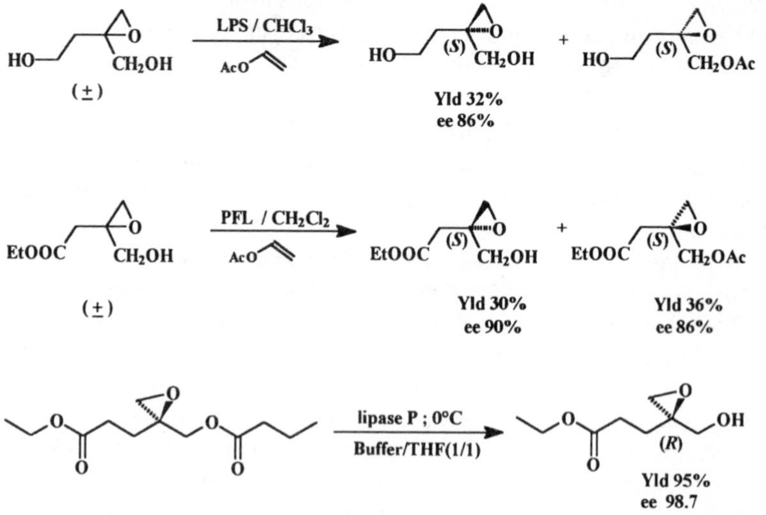

Figure 20 Lipase. Resolution and deracemization of glycidyl derivatives (42, 43, 145). ee, Enantiomeric purity; Yld, yield.

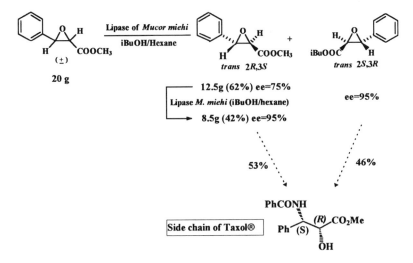

Figure 21   Lipase. Sterespecific transesterification of methyl *trans*-β-phenyl glycidate, precursor of the side chain of Taxol® (52). ee, Enantiomeric purity.

of this reaction led to the choice of an isobutyl alcohol-hexane (1:1; vol/vol) mixture as the reaction medium. Thus, this enzymatic process, coupled with substrate recycling, afforded both the recovered levorotatory-starting substrate and the formed isobutyl ester in 42 and 43% yield, respectively, with both compounds showing ees higher than 95%. This experiment could be conducted without problem on a 20-g lab scale. Interestingly, the product and substrate could be separated by fractional distillation, and the enzyme could be recycled without too much loss of activity (85% recovered activity) (52) (Figure 21).

A similar methodology has been also applied to the synthesis of (2*R*,3*S*) *p*-methoxyphenyl glycidic acid esters, the key chiral building block of Diltiazem®, a drug widely used as an antihypertensive agent because of its calcium antagonist activity (38). Although this optically active intermediate is currently available through conventional synthesis—i.e. resolution (171) or asymmetric synthesis (127)—an enzymatic approach, which in fact is used on the industrial scale, has been devised. The racemic *p*-methoxyphenyl glycidic methyl ester (51, 66) was resolved using lipase mediated hydrolysis, affording the unreacted (and desired) *trans*-(2*R*,3*S*) enantiomer to the exclusion of other stereoisomers. Although the corresponding acid formed during this hydrolysis decomposes in the reaction conditions, and thus cannot be recycled by racemization, this synthetic scheme appears to be the least expensive and most efficient. The cost resulting from the loss of half the racemic substrate, which via conventional

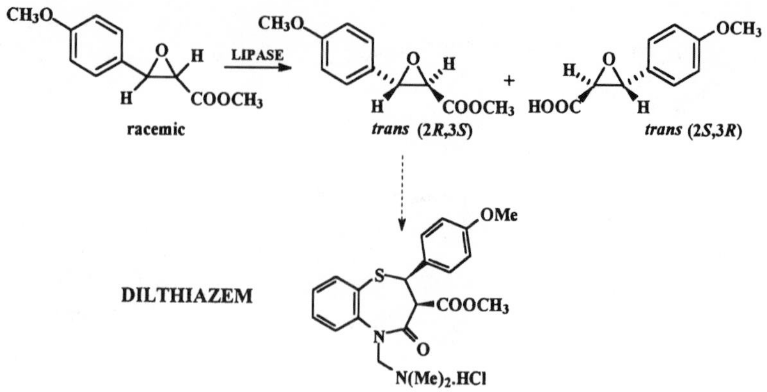

Figure 22 Lipase. Enzymatic preparation of (2*R*,3*S*)-phenyl glycidic esters, the key building-block of Diltiazem® (51).

chemistry is in fact relatively inexpensive to prepare, is balanced by the fact that this resolution is achieved at an early stage of the process, avoiding the so-called enantiomeric ballast over the further synthetic steps (Figure 22).

The kinetic resolution of epoxide-bearing substrates has also been achieved via biotransformation different from hydrolysis. Weijer et al (158) described the enantioselective degradation of short-chain *cis* and *trans* 2,3-epoxyalkanes by a *Xanthobacter* sp. bacteria able to grow on propene. Only the 2(*S*) enantiomers of these substrates were degraded, leading to accumulation of (2*R*)-2,3-epoxyalkanes (ee > 98%) in the reaction medium. On the other hand, the 1,2-epoxyalkanes isomers were degraded without enantioselectivity. Other bacterial strains—i.e. some *Nocardia* species—did, however, allow for the preparation of the remaining epoxide with excellent enantiomeric purity. However, this high ee value was paid for by the low yields obtained. Here again, *trans*-2,3-epoxybutane gave much better results (Figure 23).

A similar approach has been developed recently by a Dutch team for the preparation of glycidyl derivatives (Figure 24) (93). The strain *Acetobacter pasteurianus* achieves the kinetic resolution of racemic glycidol, affording the unreacted (*R*)-glycidol enantiomer. From the calculated *E* value ($E = 16$), it can be predicted that a 99.5% ee value for this product should be attained at 64% conversion. The reactive (*S*)-antipode was in fact oxidized via the corresponding aldehyde, into the glycidic acid. The preliminary studies conducted in the course of this work indicate that this method could be attractive (although the yield of product will be quite low) as compared with chemical approaches. However, more work must be accomplished to judge the economic feasibility of this process.

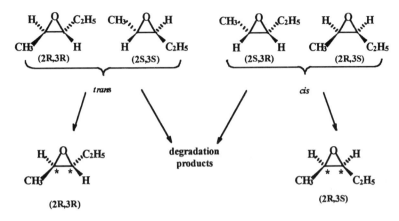

Figure 23   Enantioselective degradation of racemic *cis*- and *trans*-2-ethyl-3-methyloxirane (158).

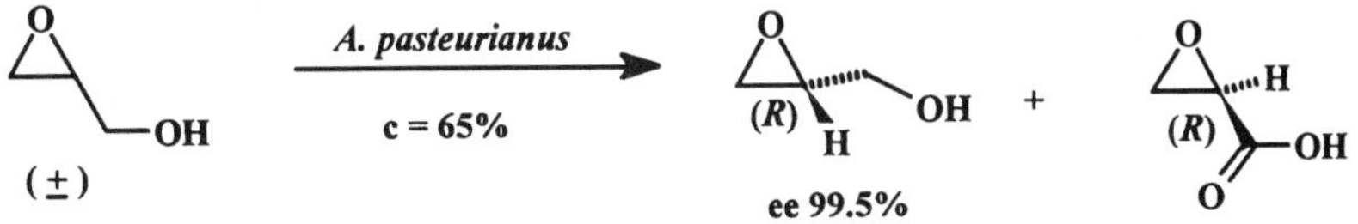

Figure 24   Production of (R)-glycidol by *Acetobacter pasteurianus* (93). ee, Enantiomeric purity; c, conversion ratio,

SUMMARY AND OUTLOOK

As emphasized at the beginning of this review, our intention was not to establish a new, encyclopedic coverage of current literature related to the preparation of enantiopure epoxides. We rather intended to draw an overall picture of the most developed strategies allowing for such syntheses.

It can be drawn from the various results described that if several fundamentally interesting and elegant approaches have been devised, many of them suffer from severe limitations as far as large-scale (industrial) applications are concerned. This is the case for the direct epoxidation of alkenes, which necessitates either cytochrome P-450 enzymes, implying a multienzymatic cascade, or for ω-hydroxylases, which require highly sophisticated processes to achieve reasonable productivities.

Some now-available indirect ways are, on the other hand, more attractive from this point of view. This is the case for lipase-catalyzed resolutions (or asymmetric synthesis starting from prochiral substrates), which have been extensively studied on all kinds of substrates. The advantages of this approach are as follows: (*a*) Several of these enzymes are currently commercially available;

(b) they are cheap; (c) they are stable under various experimental conditions; (d) they do not involve cofactors; and (e) they can be used (sometimes with net improvement of the yields or ees) in the presence of organic solvents. However, the obvious drawback of this approach is the fact that the substrates used have to be epoxide-bearing substrates with another function—ester or alcohol—near the oxirane ring.

In this respect, the most promising method seems now to be the use of microbial epoxide hydrolases. Indeed, these enzymes, which act directly on the epoxide ring, independently of any other functionality, seem to offer the same advantages as lipases: (a) They have been shown recently to be ubiquitous in nature; (b) they are cofactor-independent enzymes; (c) they can be produced easily from various microorganisms; (d) they can be partly purified and used as an enzymatic powder without noticeable loss of enzymatic activity upon storage: (e) they can act in the presence of organic solvents, allowing use of water insoluble substrates; and (f) they lead often to excellent ees of the remaining epoxide, but also in certain cases of the formed diol, which can itself be used as such or can be either cyclized back to the enantiopure epoxide or derivatized into reactive epoxide-like chiral synthons (cyclic sulfite or sulfates).

Any of the methods described in this review can be "the best," depending on the circumstances, but in the near future lipases or, better, epoxide hydrolases should prove to be "the" method of choice, in particular as far as industrial applications are concerned. Research is ongoing in diverse laboratories to explore the scope and limitations of these very promising enzymes.

Visit the *Annual Reviews home page* at
http://www.annurev.org.

Literature Cited

1. Abbott BJ, Hou CT. 1973. Oxidation of 1-alkenes to 1,2-epoxyalkanes by *Pseudomonas oleovorans*. *Appl. Microbiol.* 26:86–91
2. Agase Agrochemicals. 1995. *Jpn. Patent No. Jp 07099-993*
3. Allain EJ, Hager LP. 1993. Highly enantioselective epoxidation of disubstituted alkenes with hydrogen peroxide catalysed by chloroperoxidase. *J. Am. Chem. Soc.* 115:4415–16
4. Allen RH, Jakoby WB. 1969. Tartaric acid metabolism. IX. Synthesis with tartrate epoxidase. *J. Biol. Chem.* 244:2078–84
5. Arand M, Wagner H, Oesch F. 1996. Asp333, Asp495, and His523 form the catalytic triad of rat soluble epoxide hydrolase. *J. Biol. Chem.* 271:4223–29

6. Archelas A, Delbecque JP, Furstoss R. 1993. Microbiological transformations. 30. Enantioselective hydrolysis of racemic epoxides: the synthesis of enantiopure insect juvenile hormone analogs (Bower's compound). *Tetrahedron Asymmetry* 4:2445–46
7. Archelas A, Furstoss R. 1992. Synthesis of optically pure pityol—a pheromone of the bark beetle *Pityophthorus pityographus*—using a chemoenzymatic route. *Tetrahedron Lett.* 33:5241–42
8. Archelas A, Hartmans S, Tramper J. 1988. Stereoselective epoxydation of 4-bromo-1-butene and 3-butene-1-ol with three alkene-utilizing bacteria. *Biocatalysis* 1:283–92
9. Avignon-Tropis M, Treilhou M, Pougny

JR, Frechard-Ortuno I, Linstrumelle G. 1991. Total synthesis of (+)-leukotriene B4 methyl ester and its 5-epimer from (*R*)-glycidol. *Tetrahedron* 47:7279–86

10. Bangcharoenpaurpong O, Champion PM, Hall KS, Hager LP. 1986. Resonance raman studies of isotopically labeled chloroperoxidase. *Biochemistry* 25:2374–78

11. Bellucci G, Berti G, Catelani G, Mastrorilli E. 1981. Unusual steric course of the epoxide hydrolase catalyzed hydrolysis of ± -3,4-epoxytetrahydropyran. A case of complete stereo-convergence. *J. Org. Chem.* 46:5148–50

12. Bellucci G, Chiappe C, Cordoni A. 1996. Enantioconvergent transformation of racemic *cis*-β-alkyl substituted styrene oxides to (*R,R*) *threo* diol by microsomal epoxide hydrolase catalysed hydrolysis. *Tetrahedron Asymmetry* 7:197–202

13. Bellucci G, Chiappe C, Cordoni A, Marioni F. 1994. Different enantioselectivity and regioselectivity of the cytosolic and microsomal epoxide hydrolase catalyzed hydrolysis of simple phenyl substituted epoxides. *Tetrahedron Lett.* 35:4219–22

14. Bellucci G, Chiappe C, Ingrosso G, Rosini C. 1995. Kinetic resolution by epoxide hydrolase catalyzed hydrolysis of racemic methyl substituted methylenecyclohexene oxides. *Tetrahedron Asymmetry* 6:1911–18

15. Berti G. 1986. Enantio- and diastereoselectiviy of microsomal epoxide hydrolase: potential applications to the preparation of non-racemic epoxides and diols. In *Enzymes as Catalysts in Organic Synthesis,* ed. MP Schneider, 178:349–54. Dordrecht, Holland: D. Reidel

16. Besse P, Renard MF, Veschambre H. 1994. Chemoenzymatic synthesis of chiral epoxides. Preparation of 4-phenyl-2,3-epoxybutane and 1-phenyl-1,2-epoxypropane. *Tetrahedron Asymmetry* 5:1249–68

17. Besse P, Veschambre H. 1993. Chemoenzymatic synthesis of "α-bichiral" synthons. Application to the preparation of chiral epoxides. *Tetrahedron Asymmetry* 4:1271–85

18. Besse P, Veschambre H. 1994. Chemical and biological synthesis of chiral epoxides. *Tetrahedron* 50:8885–927

19. Blanke SR, Yi S, Hager LP. 1989. Development of semi-continuous and continuous-flow bioreactors for the high-level production of chloroperoxidase. *Biotechnol. Lett.* 11:769–74

20. Blee E, Schuber F. 1995. Stereocontrolled hydrolysis of the linoleic acid monoepoxide regioisomers catalyzed by soybean epoxide hydrolase. *Eur. J. Biochem.* 230:229–34

21. Borhan B, Jones DA, Pinot F, Grant DF, Kurth MJ, Hammock BD. 1995. Mechanism of soluble epoxide hydrolase. Formation of an α-hydroxy ester-enzyme intermediate through Asp333. *J. Biol. Chem.* 270:26923–30

22. Brink LES, Tramper J. 1985. Optimization of organic solvent in multiphase biocatalysis. *Biotechnol. Bioeng.* 27:1258–69

23. Brown HC, Pai GG. 1983. Asymmetric reduction of prochiral α-halo ketones with B-3-pinanyl-9-borabicyclo[3.3.1]-nonane. *J. Org. Chem.* 48:1784–86

24. Cabon O, Buisson D, Larcheveque M, Azerad R. 1995. The microbial reduction of 2-chloro-3-oxoesters. *Tetrahedron Asymmetry* 6:2199–210

25. Cabon O, Buisson D, Larcheveque M, Azerad R. 1995. Stereospecific preparation of glycidic esters from 2-chloro-3-hydroxyesters. Application to the synthesis of (2*R*,3*S*)-3-phenylisoserine. *Tetrahedron Asymmetry* 6:2211–18

26. Castro CE, Bartnicki EW. 1968. Biohalogenation. Epoxidation of halohydrins, epoxide opening, and transhalogenation. *Biochemistry* 7:3213–18

27. Cerniglia CE, Yang SK. 1984. Stereoselective metabolism of anthracene and phenanthrene by the fungus *Cunninghamella elegans. Appl. Environ. Microbiol.* 47:119–24

28. Chartrain M, Senanayake CH, Rosazza JPN, Zhang J. 1996. *Int. Patent No. WO96/12818*

29. Chen X-J, Archelas A, Furstoss R. 1993. Microbial transformations. 27. The first examples for preparative-scale enantioselective or diastereoselective epoxides hydrolyses using microorganisms. An unequivocal access to all four bisaboloal stereoisomers. *J. Org. Chem.* 58:5528–32

30. Colonna S, Gaggero N, Casella L, Carrea G, Pasta P. 1993. Enantioselective epoxidation of styrene derivatives by chloroperoxidase catalysis. *Tetrahedron Asymmetry* 4:1325–30

31. Coote SJ, Davies SG, Middlemiss D, Naylor A. 1989. Enantiospecific synthesis of (+)-(*R*)-6,7-dimethoxy-2-methyl-4-phenyl-1,2,3,4-tetrahydroisoquinoline from (+)-(*S*)-2-methylamino-1-phenylethanol (halostachine). *J. Chem. Soc. Perkin Trans.* 1:2223–28

32. De Bont JAM. 1993. Bioformation of

optically pure epoxides. *Tetrahedron Asymmetry* 4:1331–40

33. De Bont JAM, Attwood MM, Primrose SB, Harder W. 1979. Epoxidation of short chain alkenes in *Mycobacterium* E20: the involvement of a specific monooxygenase. *Fems Microbiol. Lett.* 6:183–88

34. De Bont JAM, van Dijken JP, van Ginkel KG. 1982. The metabolism of 1,2-propanediol by the propylene oxide utilizing bacterium Nocardia A60. *Biochim. Biophys. Acta* 714:465–70

35. Den Wijngaard AJ, Reuvekamp PTW, Janssen DB. 1991. Purification and characterisation of haloalcohol dehalogenase from *Arthrobacter* sp. strain AD2. *J. Bacteriol.* 173:124–29

36. Dexter AF, Hager LP. 1995. Transient heme N-alkylation of chloroperoxidase by terminal alkenes and alkynes. *J. Am. Chem. Soc.* 117:817–18

37. Dexter AF, Lakner FJ, Campbell RA, Hager LP. 1995. Highly enantioselective epoxidation of 1,1-disubstituted alkenes catalyzed by chloroperoxidase. *J. Am. Chem. Soc.* 117:6412–13

38. Elks J, Ganellin CR. 1990. *Diltiazem in Dictionary of Drugs*, p. 426. London: Chapmann & Hall

39. Escoffier B, Prome JC. 1989. Conversion of steroids and triterpenes by Mycobacteria—stereospecific hydrolysis of steroidal spiro-3-ξ-oxiranes by *Mycobacterium aurum*. *Bioorg. Chem.* 17:53–63

40. Faber K, Mischitz M, Kroutil W. 1996. Microbial epoxide hydrolases. *Acta Chem. Scand.* 50:249–58

41. Fedresel HF. 1993. *Drug Chirality-Scale-Up, Manufacturing, and Control. A Comprehensive Review of All the Things You Have to Consider When Making the Drug You Want*, ed. I Ojima, 23:24–33. New York: Verlag Chemie

42. Ferraboschi P, Casati S, Grisenti P, Santaniello E. 1994. Selective enzymatic transformations of itaconic acid derivatives: an access to potentially useful building blocks. *Tetrahedron* 50:3251–58

43. Ferraboschi P, Grisenti P, Manzocchi A, Santaniello E. 1994. Regio- and enantioselectivity of *Pseudomonas cepacia* lipase in the transesterification of 2-substituted-1,4-butanediols. *Tetrahedron Asymmetry* 5:691–98

44. Fetzner S, Lingens F. 1994. Bacterial dehalogenases: biochemistry, genetics and biotechnological applications. *Microbiol. Rev.* 58:641–85

45. Fox BG, Borneman JG, Wackett LP,

Lipscomb JD. 1990. Haloalkene oxidation by the soluble methane monooxygenase from *Methylosinus trichosporium* OB3b: mechanistic and environmental implications. *Biochemistry* 29:6419–27

46. Furuhashi K. 1989. *Jpn. Patent No. 01075479A*

47. Furuhashi K, Shintani M, Takagi M. 1986. Effects of solvents on the production of epoxides by *Nocardia corallina* B-276. *Appl. Microbiol. Biotechnol.* 23:218–23

48. Geerlof A, Jongejan JA, van Dooren TJGM, Raemakers-Franken PC, den Tweel WJJ, Duine JA. 1994. Factors relevant to the production of (*R*)-(+)-glycidol (2,3-epoxy-1-propanol) from racemic glycidol by enantioselective oxidation with *Acetobacter pasteurianus* atcc 12874. *Enzyme Microb. Technol.* 16:1059–66

49. Geigert J, Lee TD, Dalietos DJ, Hirano DS, Neidleman SL. 1986. Epoxidation of alkenes by chloroperoxidase catalysis. *Biochem. Biophys. Res. Commun.* 136:778–82

50. Geigert J, Neidleman SL, Dalietos DJ, DeWitt SK. 1983. Haloperoxidases: enzymatic synthesis of α,β-halohydrins from gaseous alkenes. *Appl. Environ. Microbiol.* 45:366–74

51. Gentile A, Giordano C, Fuganti C, Ghirotto L, Servi S. 1992. The enzymatic preparation of (2*R*,3*S*) phenyl glycidic acid esters. *J. Org. Chem.* 57:6635–37

52. Gou DM, Liu YC, Chen CS. 1993. A practical chemoenzymatic synthesis of the taxol C-13 side chain N-benzoyl-(2*R*,3*S*)-3-phenylisoserine. *J. Org. Chem.* 58:1287–89

53. Griffin BW. 1983. Mechanism of halide-stimulated activity of chloroperoxidase evidence for enzymatic formation of free hypohalous acid. *Biochem. Biophys. Res. Commun.* 116:873–79

54. Guengerich FP. 1991. Reactions and significance of cytochrome P-450 enzymes. *J. Biol. Chem.* 266:10019–22

55. Habets-Crutzen AQH, Brink LES, Van Ginkel CG, de Bont JAM, Tramper J. 1984. Production of epoxides from gaseous alkenes by resting-cell suspensions and immobilized cells of alkene-utilizing bacteria. *Appl. Microbiol. Biotechnol.* 20:245–50

56. Habets-Crutzen AQH, Carlier SJN, de Bont JAM, Wistuba D, Schurig V, et al. 1985. Stereospecific formation of 1,2-epoxypropane, 1,2-epoxybutane and 1-chloro-2,3-epoxypropane by alkene-

utilizing bacteria. *Enzyme Microb. Technol.* 7:17–21

57. Hager LP et al. 1994. *US Patent No. 5358860*

58. Hager LP, Morris DR, Brown FS, Eberwein H. 1966. Chloroperoxidase. Utilisation of halogen anions. *J. Biol. Chem.* 241:1769–77

59. Hammock BD, Grant DF, Storms DH. 1996. Epoxide hydrolases. In *Comprehensive Toxicology*, ed. I Sipes, C McQueen, A Gandolfi, Ch 18.3. Oxford: Pergamon

60. Hartmann GR, Frear DS. 1963. Enzymatic hydration of *cis*-9,10-epoxyoctadecanoic acid by cell-free extracts of germinating flax rust uredospores. *Biochem. Biophys. Res. Commun.* 10:366–72

61. Hawkins DR, ed. 1994. *A Survey of the Biotransformations of Drugs and Chemicals in Animals*, Vol. 1–5. Cambridge: Royal Soc. Chem.

62. Hechtberger P, Wirnsberger G, Mischitz M, Klempier N, Faber K. 1993. Asymmetric hydrolysis of epoxides using an immobilized enzyme preparation from Rhodococcus sp. *Tetrahedron Asymmetry* 4:1161–64

63. Holland HL. 1992. *Organic Synthesis with Oxidative Enzymes*, ed. HL Holland, pp. 55–199. New York: VCH

64. Hou CT, Patel RN, Laskin AI. 1980. *US Patent No. 4368267*

65. Hou CT, Patel RN, Laskin AI, Barnabe N. 1979. Microbial oxidation of gaseous hydrocarbons: epoxidation of C2 to C4 n-alkenes by methylotrophic bacteria. *Appl. Environ. Microbiol.* 38:127–34

66. Huylshof LA, Hendrik J. 1989. *Eur. Patent No. 0343714A1*

67. Imai K, Marumo S, Mori K. 1974. Derivation of (+)- and (−)-C17-juvenile hormone from its racemic alcohol derivative via fungal metabolism. *J. Am. Chem. Soc.* 96:5925–27

68. Imuta M, Kawai KI, Ziffer H. 1980. Product stereospecificity in the microbial reduction of α-haloaryl ketones. *J. Org. Chem.* 45:3352–55

69. Itoh M, Kusaka M, Hirota T, Nomura A. 1995. Microbial production of antibiotic fosfomycin by a stereoselective epoxidation and its formation mechanism. *Appl. Microbial. Biotechnol.* 43:394–401

70. Jacobs MH, van den Wijngaard AJ, Pentenga M, Janssen DB. 1991. Characterization of the epoxide hydrolase from an epichlorohydrin-degrading *Pseudomonas* sp. *Eur. J. Biochem.* 202:1217–22

71. Johnstone SL, Phillips GT, Robertson

BW, Watts PD, Bertola MA, et al. 1987. Stereoselective synthesis of S-(−)-β-blockers via microbially produced epoxide intermediates. In *Biocatalysis in Organic Media*, ed. C Laane, J Tramper, MD Lilly, pp. 387-92. Elsevier: Amsterdam

72. Kamal A, Damayanthi Y, Rao MV. 1992. Stereoselective synthesis of (S)-propanol amines: lipase catalyzed opening of epoxides with 2-propylamine. *Tetrahedron Asymmetry* 3:1361–64

73. Kamal A, Rao AB, Rao MV. 1992. Stereoselective synthesis of (S)-propanol amines: liver microsomes mediated opening of epoxides with arylamines. *Tetrahedron Lett.* 33:4077–80

74. Kasai N, Tsujimura K, Unoura K, Suzuki T. 1992. Preparation of (S)-2,3-dichloro-1-propanol by *Pseudomonas sp.* and its use in the synthesis of (R)-epichlorhydrin. *J. Indust. Microbiol.* 9:97–101

75. Katsuki T. 1995. Catalytic asymmetric oxidations using optically active (salen)manganese(III) complexes as catalysts. *Coord. Chem. Rev.* 140:189–214

76. Katsuki T. 1995. Mn-Salen catalyzed asymmetric oxidation of simple olefins and sulfides. *J. Synth. Org. Chem. Jpn.* 53:940–51

77. Kloosterman M, Elferink VHM, van Iersel J, Roskam JH, Meijer EM, et al. 1988. Lipase in the preparation of β-blockers. *Trends Biotechnol.* 6:251–56

78. Klunder JM, Ko SY, Sharpless KB. 1986. Asymmetric epoxidation of allyl alcohol: efficient routes to homochiral β-adrenergic blocking agents. *J. Org. Chem.* 51:3710–12

79. Knehr M, Thomas H, Arand M, Gebel T, Zeller HD, Oesch F. 1993. Isolation and characterization of a cDNA encoding rat liver cytosolic epoxide hydrolase and its functional expression in *Escherichia coli*. *J. Biol. Chem.* 268:17623–27

80. Kolattukudy PE, Brown L. 1975. Fate of naturally occurring epoxy acids: a soluble epoxide hydrase, which catalyzes cis hydration, from *Fusarium solani pisi*. *Arch. Biochem. Biophys.* 166:599–607

81. Kolb HC, Sharpless KB. 1992. A simplified procedure for the stereospecific transformation of 2-diols into epoxides. *Tetrahedron* 48:10515–30

82. Kolb HC, van Nieuwenhze MS, Sharpless KB. 1994. Catalytic asymmetric dihydroxylation. *Chem. Rev.* 94:2483–547

83. Krenn BE, Plat H, Wever R. 1988. Purification and some characteristics of a non-haem bromoperoxidase from *Strep-*

tomyces aureofaciens. Biochim. Biophys. Acta 952:255–60

84. Lacourciere GM, Armstrong RN. 1993. The catalytic mechanism of microsomal epoxide hydrolase involves an ester intermediate. *J. Am. Chem. Soc.* 115:10466–67

85. Ladner WE, Whitesides GM. 1984. Lipase-catalyzed hydrolysis as a route to esters of chiral epoxy alcohols. *J. Am. Chem. Soc.* 106:7250–51

86. Lakner FJ, Hager LP. 1996. Chloroperoxidase as enantioselective epoxidation catalyst: an efficient synthesis of (*R*)-(−)-mevanolactone. *J. Org. Chem.* 61:3923–25

87. Leak DJ, Dalton H. 1987. Studies on the regioselectivity and stereoselectivity of the soluble methane monooxygenase from *Methylococcus capsulatus* (bath). *Biocatalysis* 1:23–36

88. Levinger DC, Stevenson JA, Wong LL. 1995. The catalytic activity of human myoglobin is enhanced by a single active site mutation: F42Y. *J. Chem. Soc., Chem. Commun.* 22:2305–6

89. Libby RD, Thomas JA, Kaiser LW, Hager LP. 1982. Chloroperoxidase halogenation reactions. Chemical versus enzymic halogenating intermediates. *J. Biol. Chem.* 257:5030–37

90. Linderman RJ, Walker EA, Haney C, Roe RM. 1995. Determination of the regiochemistry of insect epoxide hydrolase catalyzed epoxide hydration of juvenile hormone by ^{18}O-labeling studies. *Tetrahedron* 51:10845–56

91. Lohray BB. 1992. Cyclic sulfites and cyclic sulfates: epoxide like synthons. *Synthesis* 11:1035–52

92. Lu AYH, West SB. 1980. Multiplicity of mammalian microsomal cytochromes P-450. *Pharmacol. Rev.* 31:277–95

93. Machado SS, Wandel U, Straathof AJJ, Jongejan JA, Duine JA. 1996. Production of (*R*)-glycidol by *Acetobacter pasteurianus*. Presented at Int. Conf. Biotech. Ind. Prod. Fine Chem., Zermatt

94. Mahmoudian M, Michael A. 1992. Biocatalysts for production of chiral epoxides. *Appl. Microbiol. Biotechnol.* 37:23–27

95. Mahmoudian M, Michael A. 1992. Stereoselective epoxidation of phenyl allyl ether by alkene-utilizing bacteria. *Appl. Microbiol. Biotechnol.* 37:28–31

96. Mansuy D, Battioni P, Battioni JP. 1989. Chemical model systems for drug-metabolizing cytochrome P-450-dependent monooxygenases. *Eur. J. Biochem.* 184:267–85

97. May SW, Abbott BJ. 1973. Enzymatic epoxidation. II. Comparison between the epoxidation and hydroxylation reactions catalyzed by the ω-hydroxylation system of *Pseudomonas oleovorans. J. Biol. Chem.* 248:1725–30

98. May SW, Schwartz RD, Abbott BJ, Zaborsky OR. 1975. Structural effects on the reactivity of substrates and inhibitors in the epoxidation system of *Pseudomonas oleovorans. Biochim. Biophys. Acta* 403:245–55

99. May SW, Steltenkamp MS, Schwartz RD, McCoy CJ. 1976. Stereoselective formation of diepoxides by an enzyme system of *Pseudomonas oleovorans. J. Am. Chem. Soc.* 98:7856–58

100. McCarthy MB, White RE. 1983. Functional differences between peroxidase compound I and the cytochrome P-450 reactive oxygen intermediate. *J. Biol. Chem.* 258:9153–58

101. McMillan DC, Cerniglia CE, Fu PP. 1987. Stereoselective fungal metabolism of 7,12-dimethylbenz(α)anthracene—identification and enantiomeric resolution of a K-region dihydrodiol. *Appl. Environ. Microbiol.* 53:2560–66

102. Michaels BC, Fulco AJ, Ruettinger RT. 1980. Hydration of 9,10-epoxypalmitic acid by a soluble enzyme from *Bacillus megaterium. Biochem. Biophys. Res. Commun.* 92:1189–95

103. Mischitz M, Faber K. 1994. Asymmetric opening of an epoxide by azide catalyzed by an immobilized enzyme preparation from *Rhodococcus* sp. *Tetrahedron Lett.* 35:81–84

104. Mischitz M, Faber K, Willetts A. 1995. Isolation of a highly enantioselective epoxide hydrolase from *Rhodococcus* sp. MAJ 11216. *Tetrahedron Asymmetry* 17:893–98

105. Mischitz M, Kroutil W, Wandel U, Faber K. 1995. Asymmetric microbial hydrolysis of epoxides. *Tetrahedron Asymmetry* 6:1261–72

106. Morisseau C, Nellaiah H, Archelas A, Furstoss R, Baratti JC. 1997. Asymmetric hydrolysis of racemic *para*-nitrostyrene oxide using an epoxide hydrolase preparation from *Aspergillus niger. Enzyme Microb. Technol.* 20:446–52

107. Morris DR, Hager LP. 1966. Chloroperoxidase. I. Isolation and properties of the crystalline glycoprotein. *J. Biol. Chem.* 241:1763–68

108. Morrison M, Schonbaum GR. 1976. Peroxidase-catalyzed halogenation. *Annu. Rev. Biochem.* 45:861–88

109. Murmann W, Rumore G, Gamba A.

1967. Pharmacological properties of 1-(4'-nitrophenyl)-2-isopropylamino-ethanol (INPEA). A new β-adrenergic receptor antagonist. *Bull. Chim. Farm.* 106:251–68

110. Nagasawa T, Nakamura T, Yu F, Watanabe I, Yamada H. 1992. Purification and characterization of halohydrin hydrogenhalide lyase from a recombinant *Escherichia coli* containing the gene from a *Corynebacterium* sp. *Appl. Microbiol. Biotechnol.* 36:478–82

111. Nakamura T, Nagasawa T, Yu F, Watanabe I, Yamada H. 1992. Resolution and some properties of enzymes involved in enantioselective transformation of 1,3-dichloro-2-propanol to (*R*)-3-chloro-1,2-propanediol by *Corynebacterium* sp. strain N-1074. *J. Bacteriol.* 174:7613–19

112. Nakamura T, Nagasawa T, Yu F, Watanabe I, Yamada H. 1994. Purification and characterization of two epoxide hydrolases from *Corynebacterium* sp. strain N-1074. *Appl. Environ. Microbiol.* 60:4630–33

113. Nakamura T, Yu F, Mizunashi W, Watanabe I. 1991. Microbial transformation of prochiral 1,3-dichloro-2-propanol into optically active 3-chloro-1,2-propanediol. *Agric. Biol. Chem.* 55:1931–33

114. Neidleman SL. 1980. Use of enzymes as catalysts for alkene oxide production. *Hydrocarb. Process.* pp. 135–38

115. Neidleman SL, Levine SD. 1968. Enzymatic bromohydrin formation. *Tetrahedron Lett.* 37:4057–59

116. Nellaiah H, Morisseau C, Archelas A, Furstoss R, Baratti JC. 1996. Enantioselective hydrolysis of *p*-nitrostyrene oxide by an epoxide hydrolase preparation from *Aspergillus niger. Biotechnol. Bioeng.* 49:70–77

117. Niehaus WG, Schroepfer GJ. 1967. Enzymatic stereospecificity in the hydration of epoxy fatty acids. Stereospecific incorporation of the oxygen of water. *J. Biol. Chem.* 89:4227–28

118. Nöthe C, Hartmans S. 1994. Formation and degradation of styrene oxide stereoisomers by different microorganisms. *Biocatalysis* 10:219–25

119. Ohno M, Okura I. 1990. On the reaction mechanism of alkene epoxidation with *Methylosinus trichosporium* (OB3b). *J. Mol. Catal.* 61:113–22

120. Ohta H, Tetsukawa H. 1979. Asymmetric epoxidation of long chain terminal olefins by *Corynebacterium equi* IFO 3730. *Agric. Biol. Chem.* 43:2099–104

121. Onumonu AN, Colocoussi A, Matthews C, Woodland MP, Leak DJ. 1994. Microbial alkene epoxidation—merits and limitations. *Biocatalysis* 10:211–18

122. Ortiz de Montellano PR. 1987. *Cytochromes P-450.* New York: Plenum. 556 pp.

123. Ortiz de Montellano PR, Choe YS, Depillis G, Catalano CE. 1987. Structure-mechanism relationships in hemoproteins. Oxygenations catalyzed by chloroperoxidase and horseradish peroxidase. *J. Biol. Chem.* 262:11641–46

124. Ortiz de Montellano PR, Fruetel JA, Collins JR, Camper DL, Loew GH. 1991. Theoretical and experimental analysis of the absolute stereochemistry of *cis*-β-methylstyrene epoxidation by cytochrome P-450cam. *J. Am. Chem. Soc.* 113:3195–96

125. Osprian I, Kroutil W, Mischitz M, Faber K. 1997. Biocatalytic resolution of 2-methyl-2-(aryl)alkyloxiranes using novel bacterial epoxide hydrolases. *Tetrahedron Asymmetry.* 18:65–71

126. Otto PPJHL, Stein F, van der Willigen CA. 1988. A bio-organic synthesis of (+)-disparlure, sex pheromone of *Lymantria dispar* (gypsy moth). *Agric. Ecosys. Environ.* 21:121–23

127. Palmer JT. 1985. *US Patent No. 4552695*

128. Pedragosa-Moreau S, Archelas A, Furstoss R. 1993. Microbiological transformations. 28. Enantiocomplementary epoxide hydrolyses as a preparative access to both enantiomers of styrene oxide. *J. Org. Chem.* 58:5533–36

129. Pedragosa-Moreau S, Archelas A, Furstoss R. 1994. Microbiological transformations. 29. Enantioselective hydrolysis of epoxides using microorganisms: a mechanistic study. *Bioorg. Med. Chem.* 2:609–16

130. Pedragosa-Moreau S, Archelas A, Furstoss R. 1995. Epoxydes enantiopurs: obtention par voie chimique ou par voie enzymatique. *Bull. Soc. Chim. Fr.* 132:769–800

131. Pedragosa-Moreau S, Archelas A, Furstoss R. 1996. Microbial transformations. 31. Synthesis of enantiopure epoxides and vicinal diols using fungal epoxide hydrolase mediated hydrolysis. *Tetrahedron Lett.* 37:3319–22

132. Pedragosa-Moreau S, Archelas A, Furstoss R. 1996. Microbial transformations. 32. Use of epoxide hydrolase mediated biohydrolysis as a way to enantiopure epoxides and vicinal diols: application to substituted styrene oxide derivatives. *Tetrahedron* 52:4593–606

133. Pedragosa-Moreau S, Morisseau C, Zylber J, Archelas A, Baratti JC, Furstoss R. 1996. Microbiological transformations. 33. Fungal epoxide hydrolases applied to the synthesis of enantiopure para-substituted styrene oxides. A mechanistic approach. *J. Org. Chem.* 61:7402–7

134. Pfenninger A. 1986. Asymmetric epoxidation of allylic alcohols: the Sharpless epoxidation. *Synthesis* 2:89–116

135. Poulos TL, Raag R. 1992. Cytochrome P-450cam: crystallography, oxygen activation, and electron transfer. *Faseb J.* 6:674–79

136. Raag R, Poulos TL. 1991. Crystal structures of cytochrome P-450cam complexed with camphane, thiocamphor, and adamantane: factors controlling P-450 substrate hydroxylation. *Biochemistry* 30:2674–84

137. Ramakrishnan K, Oppenhuizen ME, Saunders S, Fisher J. 1983. Stereoselectivity of chloroperoxidase-dependent halogenation. *Biochemistry* 22:3271–77

138. Rigby SR, Matthews CS, Leak DJ. 1994. Epoxidation of styrene and substituted styrenes by whole cells of *Mycobacterium* sp. M156. *Bioorg. Med. Chem.* 2:553–56

139. Ruettinger RT, Fulco AJ. 1981. Epoxidation of unsaturated fatty acids by a soluble cytochrome P-450-dependent system from Bacillus megaterium. *J. Biol. Chem.* 256:5728–34

140. Ruettinger RT, Griffith GR, Coon MJ. 1977. Characterisation of the ω-hydroxylase of *Pseudomonas oleovorans* as a non heme iron protein. *Arch. Biochem. Biophys.* 183:528–37

141. Ryan DE, Levin W, Reik LM, Thomas PE. 1982. Enzyme. *Xenobiotica* 12:727–44

142. Sato H. 1975. *Jpn. Patent No. 75140684*

143. Schurig V, Betschinger F. 1992. Metal-mediated enantioselective access to unfunctionalized aliphatic oxiranes—prochiral and chiral recognition. *Chem. Rev.* 92:873–88

144. Seki Y, Shimoda M, Sugimori D, Okura I. 1994. Epoxidation of allyl compounds with *Methylosinus trichosporium* (OB3b): stereoselectivity of methane monooxygenase. *J. Mol. Catal.* 87L:17–19

145. Seu YB, Lim TK, Kim CJ, Kang SC. 1995. Preparation of enantiomerically pure (R)-2-butyryloxymethylglycidol by lipase-catalyzed asymmetric hydrolysis. *Tetrahedron Asymmetry* 6:3009–14

146. Sono M, Roach MP, Coulter ED, Dawson JH. 1996. Heme-containing oxygenases. *Chem. Rev.* 96:2841–87

147. Stinson SC. 1993. Chiral drugs. Wave of new enantiomer products set to flood market. *Chem. Eng. News* 27:38–63

148. Suzuki Y, Marumo S. 1972. Fungal metabolism of (+/−)-epoxyfarnesol and its absolute stereochemistry. *Tetrahedron Lett.* 19:1887–90

149. Takahashi O, Umezawa J, Furuhashi K, Takagi M. 1989. Stereocontrol of a tertiary hydroxyl group via microbial epoxidation. A facile synthesis of prostaglandine ω-chains. *Tetrahedron Lett.* 30:1583–84

150. Tseng HF, Laughlin LT, Lin S, Armstrong RN. 1996. The catalytic mechanism of microsomal epoxide hydrolases involve reversible formation and rate-limiting hydrolysis of the alkyl-enzyme intermediate. *J. Am. Chem. Soc.* 118:9436–37

151. Ueda T, Coon JM. 1972. Enzymatic ω oxidation. VII. Reduced diphosphopyridine nucleotide-rubredoxin reductase: properties and function as an electron carrier in ω hydroxylation. *J. Biol. Chem.* 247:5010–16

152. Van Ginkel CG, Welten HGJ, de Bont JAM. 1986. Epoxidation of alkenes by alkene-grown *Xanthobacter* sp. *Appl. Microbiol. Biotechnol.* 24:334–37

153. Van Pee KH. 1995. B.C.G. halogenation. In *Enzyme Catalysis in Organic Synthesis. A Comprehensive Handbook*, ed. K Drauz, H Waldmann, 2:783–807. Weinheim: VCH

154. Vilter H. 1983. Peroxidases from phaeophyceae. 4. Fractionation and location of peroxidase isoenzymes in *Ascophyllum nodosum*. *Bot. Mar.* 26:451–55

155. Wackett LP, Gibson DT. 1982. Metabolism of xenobiotic compounds by enzymes in cell-extracts of the fungus *Cunninghamella elegans*. *Biochem. J.* 205:117–22

156. Wallar BJ, Lipscomb JD. 1996. Dioxygen activation by enzymes containing binuclear non-heme iron clusters. *Chem. Rev.* 96:2625–57

157. Wandel U, Mischitz M, Kroutil W, Faber K. 1995. Highly selective asymmetric hydrolysis of 2,2-disubstituted epoxides using lyophilized cells of *Rhodococcus* sp. ncimb 11216. *J. Chem. Soc. Perkin Trans.* 1:735–36

158. Weijers CAGM, de Haan A, de Bont JAM. 1988. Chiral resolution of 2,3-epoxyalkanes by *Xanthobacter* Py2. *Appl. Microbiol. Biotechnol.* 27:337–40

159. Weijers CAGM, de Haan A, de Bont JAM. 1988. Microbial production and metabolism of epoxides. *Microbiol. Sci.* 5:156–59

160. Weijers CAGM, Litjens MJJ, de Bont JAM. 1992. Synthesis of optically pure 1,2-epoxypropane by microbial asymmetric reduction of chloroacetone. *Appl. Microbiol. Biotechnol.* 38:297–300

161. Weijers CAGM, Van Ginkel CG, de Bont JAM. 1988. Enantiomeric composition of lower epoxyalkanes produced by methane-, alkane-, and alkene-utilizing bacteria. *Enzyme Microb. Technol.* 10:214–18

162. White RF. 1980. *US Patent No. 2054310*

163. White RF, Birnbaum J, Meyer RT, Broeke JT, Chemerda JM, Demain AL. 1971. Microbial epoxidation of *cis*-propenylphosphonic to (−)-*cis*-1,2-epoxypropylphosphonic acid. *Appl. Microbiol.* 22:55–60

164. Wiesner W, Van Pee KH, Lingens F. 1988. Purification and characterisation of a novel bacterial non-heme chloroperoxidase from *Pseudomonas pyrrocinia*. *J. Biol. Chem.* 263:13725–32

165. Wixtrom RN, Hammock BD. 1985. Membrane-bound and soluble fraction epoxide hydrolases: methodological aspects. In *Biochemical Pharmacology and Toxicology*, ed. D Zakim, DA Vessey, 1:1–93. New York: Wiley

166. Woods NR, Murrell JC. 1991. Epoxidation of gaseous alkenes by a *Rhodococcus* sp. *Biotechnol. Lett.* 12:409–14

167. Wu DR, Cramer SM, Belfort G. 1993. Kinetic resolution of racemic glycidyl butyrate using a multiphase membrane enzyme reactor. Experiments and model verification. *Biotechnol. Bioeng.* 41:979–90

168. Wubbolts MG, Hoven J, Melgert B, Witholt B. 1994. Efficient production of optically active styrene epoxides in two-liquid phase cultures. *Enzyme Microb. Technol.* 16:887–94

169. Wubbolts MG, Noordman R, Van Beilen JB, Witholt B. 1995. Enantioselective oxidation by non-heme iron monooxygenases from *Pseudomonas*. *Recueil des Travaux Chimiques des Pays-Bas* 114:139–44

170. Wubbolts MG, Panke S, Beilen JB, Witholt B. 1996. Enantioselective oxidation by non-heme iron monooxygenases from *Pseudomonas*. *Chimia* 50:436–37

171. Wynberg H, ten Hoeve W. 1989. *Eur. Patent Appl. 0342903A1*

172. Zaks A, Dodds DR. 1995. Chloroperoxidase-catalyzed asymmetric oxidations: substrate specificity and mechanistic study. *J. Am. Chem. Soc.* 117:10419–24

173. Zeldin DC, Dubois RN, Falck JR, Capdevila JH. 1995. Molecular cloning, expression and characterization of an endogenous human cytochrome P-450 arachidonic acid epoxygenase isoform. *Arch. Biochem. Biophys.* 322:76–86

174. Zhang J, Reddy J, Roberge C, Senanayake C, Greasham R, Chartrain M. 1995. Chiral bio-resolution of racemic indene oxide by fungal epoxide hydrolases. *J. Ferment. Bioeng.* 80:244–46

175. Zhang XM, Archelas A, Furstoss R. 1991. Microbiological transformations. 19. Asymmetric dihydroxylation of the remote double bond of geraniol: a unique stereochemical control allowing easy access to both enantiomers of geraniol-6,7-diol. *J. Org. Chem.* 56:3814–17

Annu. Rev. Microbiol. 1997. 51:527–64

CELL-CELL COMMUNICATION IN GRAM-POSITIVE BACTERIA

Gary M. Dunny

Department of Microbiology, University of Minnesota Medical School, Minneapolis, Minnesota 55455; email: gary-d@biosci.cbs.umn.edu

Bettina A. B. Leonard

Institut für Mikrobiologie und Immunologie, Universitatsklinikum Ulm, D-89081 Ulm, Germany

KEY WORDS: pheromone, quorum sensing, 2-component system, oligopeptide permease, gene regulation

ABSTRACT

In gram-positive bacteria, many important processes are controlled by cell-to-cell communication, which is mediated by extracellular signal molecules produced by the bacteria. Most of these signaling molecules are peptides or modified peptides. Signal processing, in most cases, involves either transduction across the cytoplasmic membrane or import of the signal and subsequent interaction with intracellular effectors. Concentrations of signal in the nanomolar range or below are frequently sufficient for biological activity. The microbial processes controlled by extracellular signaling include the expression of virulence factors, the expression of gene transfer functions, and the production of antibiotics.

CONTENTS

0066-4227/97/1001-0527$08.00

INTRODUCTION

More than 50 years ago, Avery, McLeod & McCarty (5) obtained the first set
of data demonstrating that DNA was the genetic material. The experimental
system they used was the transformation of the smooth-colony morphology
trait that is associated with capsular polysaccharide production in *Streptococ-*
cus pneumoniae, which initially had been observed in experimental animals
by Griffith in the 1920s (38a). Further analysis of this system by Hotchkiss &
Tomasz (120, 121) led to the striking notion that these single-celled prokaryotic
organisms engaged in multicellular behavior. It was inferred that the regulation
of competence for transformation in these organisms was mediated by extracellular chemical signals of pneumococcal origin, which appeared to have a
hormone- or pheromone-like function. This enabled individual cells in a population of pneumococci to monitor the density of the population in a specific
ecological niche. The molecular identity of these signaling molecules (originally termed Competence Factors) has only recently been determined (42; see
below); however, the early work on pneumococcal competence pioneered an
area of research that was focused on microbial processes, which appear to be
affected by cell density as communicated by extracellular signaling molecules
that are generated by the bacteria.

Recently, the term quorum sensing has been used to describe the collection of
molecular mechanisms that is employed by the bacteria to monitor density (35).
Figure 1A shows the conventional view of quorum sensing, in which a population of cells of a bacterial strain occupies an ecological niche that is closed (or
in which flow-through is restricted), but that is large in volume, relative to the
cell size. Constitutive low-level synthesis of the signal causes its concentration

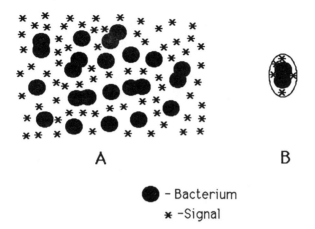

A   B

● – Bacterium

＊ –Signal

Figure 1 Situations in which extracellular indicators of cell density affect the behavior of large or small bacterial populations. In both of these illustrations, the organisms are growing in an environment that either is closed or has restricted flow-through. In *A*, there is a large population of organisms growing in a volume that is very large in comparison with the size of a single cell. Low level, constitutive production of the signal by all cells in the population results in an increase in concentration that parallels the population growth of the culture. The concentration of the signal reaches a level that is sufficient for biological activity when the density of the culture is optimal for the process that is induced by the signal. In *B*, a single cell is growing in a very small volume, i.e. a phagocytic vesicle. In this case, a relatively small number of signal molecules is sufficient to achieve the inducing concentration, and the only bacterial cell with altered behavior is the cell that produces the signal.

to rise in synchrony with the increase in cellular population. The systems have apparently evolved so that the signal reaches a concentration sufficient for biological activity at a cell density that is appropriate for the induced activity to occur efficiently. Figure 1*B* depicts a scenario in which a single diplococcal cell has been trapped in a very small enclosed space, e.g. in a phagocytic vesicle. Because of the small size of the niche, the inducing concentration of signal that is required is achieved very quickly by the biosynthetic activities of the single cell. This mechanism could be used by the bacterium in order to sense that it had become intracellular, and that it should switch on the genes required for such functions as resistance to killing, intracellular growth, and escape into the cytoplasm, etc. Both of these situations represent density-dependent phenomena, but the former is essentially a population behavior, whereas the latter is primarily an environmental monitoring system for single cells. Other forms of signaling include communication between donor and recipient cells, such as that which occurs in conjugation, or in situations in which a signal molecule with demonstrable biological activity may be found in culture medium, but in which the

Table 1 Examples of chemical communication systems in gram-positive bacteria

Organism(s)	Process regulated	signal(s)[a]	Mechanism target[b]
Several lactic acid bacterial spp.	Peptide bacteriocin production	Peptide	2-component system— biosynthetic promoters
Streptococcus pneumoniae and *S. sanguis*	Competence	Peptide	2-component signal— transduction- *com* promoters
Lactococcus lactis	Lantibiotic bac- teriocin production (nisin)	Peptide[c] (nisin)	2-component system— biosynthetic promoters
Bacillus subtilis	Competence (sporulation)	Peptide[c] and peptides	2-component system and signal import- *com* promoters and rap phosphatases
Staphylococcus aureus	Virulence (agr)	Peptide[c]	2-component system— regulatory RNA promoter
Enterococcus faecalis	Conjugation	Peptide	Internalization—trans- cription or translation machinery
Streptomyces griseus	Secondary metabolism	γ-Butyrolactone	Internalization—pro- moters for antibiotic biosynthetic genes

[a]Refers to the functional form of the extracellular signaling molecule.

[b]The known signaling molecules act either by binding to the cell surface and transducing a signal via a 2-component system or by internalization followed by interaction with intracellular effectors. All of the known signaling pathways with the possible exception of the *E. faecalis* mating systems appear to result ultimately in activation of a transcription factor to increase the activity of target promoters.

[c]Asterisk indicates a modified peptide. A mixture of peptides regulate competence in *B. subtilis*.

molecule has an alternative, or even primary function in the same cell in which it is synthesized (see sections on *Bacillus subtilis* and *Enterococcus faecalis*, below). In gram-negative bacteria, acyl-homoserine lactone molecules often serve as the signals, and there is a well conserved mechanism by which the signals are recognized and converted to a functional response by the organisms; there is also evidence for lipid, peptide, and amino acid–based signaling (61a, 63). In this review, we will describe several cell-cell signaling systems in gram-positive bacteria. In these organisms, peptides or modified peptides are often employed as signals, and they appear to function by several different mechanisms.

Table 2   Molecular structures of some extracellular peptide and modified peptide signals of gram-positive bacteria

Organism	Name of signal	Sequence of mature molecule	Length of structural gene (AA)
Lactococcus lactis	Nisin A	ITSISLCTPGCKTGALMGCN-MKTATCHCSIHVSK[a]	57
Lactobacillus plantarum	IF (PlnA)	KSSAYSLQMGATAIKQVK-KLFKKWGW	48
Streptococcus pneumoniae	CSP (ComC)	EMRLSKFFRDFILQRKK	41
Streptococcus gordonii	CSP (ComC1)	DVRSNKIRLWWENIFFNKK	50
Bacillus subtilis	ComX pheromone	ADPITRQW[b]GD	55
	CSF (PhrC)	ERGMT	40
Staphylococcus aureus	AgrD	YSTCDFIM[a]	44
Enterococcus faecalis	cCF10	LVTLVFV	??
	iCF10	AITLIFI	23
	cAD1	LFSLVLAG	??
	iAD1	LFVVTLVG	23

[a]Multiple posttranslational modifications. Nisin has extensive posttranslational modifications of several amino acid residues. AgrA appears to be posttranslationally altered such that it contains a cyclic anhydride structure.

[b]ComX pheromone has a postranslational modification of a tyrosine.

In Table 1, several systems are listed for which there is considerable information about the signal, the response pathway, and the biological process regulated by signaling. This list illustrates the importance of multicellular communication in a variety of processes, including genetic transfer, antibiotic production, growth, and pathogenesis. A great deal of important information about these systems has been obtained very recently, making this one of the most active and exciting research areas in microbiology. In the following sections, we will summarize what is known about the signaling molecule, the signaling process, and the biological significance of each of the systems that is shown in Table 1. The review will conclude with some mention of emerging systems in which similar signaling processes appear to occur, and with comments about directions and opportunities for future work in this area. Although we have tried to include much of the relevant work in this area, we do not claim to provide a comprehensive treatise, and we apologize to any colleagues whose work may not have been included. We have cited more specialized reviews of most of the systems discussed, which may be consulted for further information.

As indicated in Table 1, many of these systems use a 2-component–signal transduction mechanism. Also, the majority of the signaling molecules require a specialized export mechanism. All peptide or peptide-derived signaling molecules analyzed thus far are derived by posttranslational processing of a larger precursor peptide (Table 2). In the cases in which modified peptides serve as the signals, the structural gene is linked in an operon to one or more genes that encode export and posttranslational modification functions. However, the structural genes for the unmodified signals are most often linked to the genes that encode the 2-component signal transducers. We have used these apparent evolutionary relationships for categorizing topics in the discussion of the various systems. For both the modified and the unmodified peptide systems, we first discuss control of bacteriocin production, because this has been the subject of some of the most complete studies.

EXTRACELLULAR SIGNALS AND REGULATION OF EXPRESSION OF UNMODIFIED PEPTIDE BACTERIOCINS

Different strains of lactic acid bacteria produce a variety of peptide bacteriocins and, in a number of cases, the active molecular species is an unmodified peptide or a combination of two such peptides. This class of bacteriocins has been termed Group II, and was the subject of a recent review (88). Several different levels of complexity have evolved in different Group II bacteriocin systems, both in the way that production of the antibiotic is encoded genetically, and in the way that extracellular regulation is achieved.

In many of these systems (Figure 2) there is a gene that encodes a "bacteriocin-like" peptide and is cotranscribed with genes that encode histidine kinase (HK) and response-regulating (RR) proteins of the 2-component signal transduction family (51). This situation is analogous to the pneumococcal competence system that will be described below. A proteolytic fragment of the peptide is exported into the growth medium, in which it serves as an induction factor (IF), i.e. an extracellular signal that can activate expression of bacteriocin production once concentrations in the nanogram/ml range are reached. The IF peptide is generally distinct from the bacteriocin, although the Plantaricin A IF was originally believed to be the bacteriocin, rather than the inducer (89). In

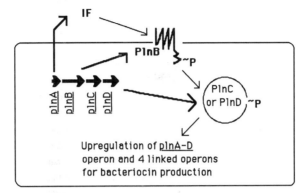

Lactobacillus plantarum

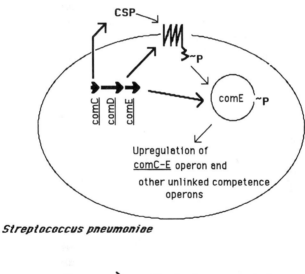

Streptococcus pneumoniae

$\longrightarrow$ synthesis of a gene product

$\longrightarrow$ Stimulation

Figure 2   Comparison of gene structures and signaling pathways for peptide-induced bacteri-ocin production in *Lactobacillus plantarum*, and for peptide-induced competence in *Streptococcus pneumoniae*. In both cases, the gene for the peptide signal is linked to the signal transduction genes, and other operons are upregulated via a 2-component phosphorylation cascade in response to accumulation of the peptide in the culture medium (88).

addition to the IF-RR-HK cassettes, one or more operons (often linked) that encode bacteriocin biosynthesis and immunity are coregulated with IF production, i.e. they are transcriptionally activated in response to the IF-mediated signal transduction pathway. Relatively simple forms of genetic organization for Group II bacteriocin production include the gene clusters for Sakacin A and Sakacin P production—that both seem to contain one operon for IF production—and an adjacent operon that encodes bacteriocin biosynthetic genes (25, 31, 54). In the case of the plantaricins of *Lactobacillus plantarum,* a cluster of over 20 linked *pln* genes, including at least 5 separate operons, participate in the biosynthesis of an IF called Plantaricin A (Figure 2); these genes also participate in the biosynthesis of at least 3 bacteriocins, including single peptide and 2-peptide varieties (61, 88, 89). An even more complex situation exists for bacteriocins BM1 and B2 of *C. pisicola,* in which one set of biosynthetic genes is located on the chromosome, whereas a second biosynthetic cluster—as well as genes for IF production—are plasmid-encoded (106, 107).

The widespread occurrence of 3-gene operons that encode IF, RR, and HK molecules in the biosynthetic gene clusters of the Group II bacteriocins has prompted the use of the term 3-component regulation to describe these extracellular signaling processes (88). Because the signaling molecules have never been identified in many of the previously studied 2-component regulatory systems (51), it is worthwhile to consider the possibility that peptides of bacterial origin are the functional signals in a much larger number of systems than those of which we are currently aware. As in the case of other regulatory systems using the 2-component–signal transduction mechanism described in this review, there is not yet any direct evidence for the binding of a phosphorylated RR molecule to a 5′ target site. However, analysis of sequences that flank functional promoters in the *pln* gene cluster has identified repeat sequences that could serve as potential binding sites for dimeric forms of positive regulatory proteins (25).

INDUCTION OF COMPETENCE
IN *STREPTOCOCCUS PNEUMONIAE*

As noted in the introduction, competent cell transformation in certain strains of pneumococci occurs in a cell density–dependent fashion in batch cultures (and in vivo, as well). Competent cells express a unique set of genes whose products endow the cells to bind DNA, to degrade one strand and import fragments of the complementary strand, and to recombine internalized DNA into homologous regions of the genome at high efficiency (76a, 83). Based on a series of physiological analyses, the existence of an extracellular signal (Competence Factor) of bacterial origin in the regulation of this process was proposed (120, 121). It was suggested that the signal might be perceived by a membrane-bound bacterial receptor (120), and that Competence Factor could have an autocatalytic mode of action. Once a critical level in the growth medium was achieved by constitutive synthesis, a burst of increased Competence Factor synthesis could occur that would be concomitant with, or preceding, competence development. Remarkably, all of these predictions have been verified by experimental results reported during the past two years.

The Pneumococcal Competence-Stimulating Peptide (CSP) and Its Production

Initial biochemical analysis suggests that the signaling molecule might be a small protein (120), but its molecular elucidation proved refractory for many years. A crucial step in this process was the characterization (55) of the *comA* locus, whose insertional inactivation abolished the normal development of competence. Sequence analysis of *comA* indicated that it was an ABC transporter (55), and that it was highly related to a family of proteins that is implicated in the simultaneous processing and export of peptide bacteriocins from several gram-positive species (43). Phenotypic analysis of *comA* mutants supported the notion that the product of this gene was involved in production of the extracellular signal. Havarstein et al (42) therefore used a biochemical approach that had been successful in the isolation of peptide bacteriocins to obtain a pure heptadecapeptide, called CSP (Competence Stimulating Peptide) (see Table 2), from competent cell culture fluids. Both the natural and the synthetic form of CSP stimulated the development of competence when added to appropriate pneumococcal cultures. By using reverse genetics and PCR, a pneumococcal gene *comC*—which encodes the precursor of the mature CSP—was found. Inactivation of this gene abolished the production of the factor and the development of competence. Subsequent analysis has shown that 10 of 17 residues in CSP could be replaced without loss of activity, and of the remaining seven, two terminal-charged residues and five internal-hydrophobic residues were essential for activity. A screen of pneumococcal isolates suggests that they fall into two families with respect to CSP: those that contain a gene identical to that found in the original strain, or those that contain a gene that encodes production of a slightly different peptide of the same size and similar sequence (105). Competent *Streptococcus gordonii* strains also contain a homologous set of competence genes, and they encode one of two additional peptides of the same class (44).

The Signal Transduction Pathway for Pneumococcal Competence

Interestingly, the *comC* gene is not closely linked to *comA*, but rather it is linked to genes that are required for the transduction of the CSP signal. This linkage, along with the available knowledge of mechanisms of regulation of expression of bacteriocins similar to CSP, suggests a molecular mechanism by which the CSP signal acts to induce the expression of competence. Sequence analysis of the DNA that flanks *comC* reveals the presence of two downstream genes, *comD* and *comE*, which likely form an operon with *comC*. These genes encode homologues of the HK sensor and RR components of the 2-component super family of regulatory genes (102). By using comparative sequence analysis of these genes from the strains of *S. pneumoniae* and *S. gordonii* that represent the 4 CSP types, and by carrying out gene switching experiments, Havarstein et al recently obtained compelling genetic evidence that the *comD* (HK) gene in each of the four families encodes the receptor for the cognate CSP (44).

Based on the above results, one would predict that the phosphorylated form of the *comE* (RR) gene would act at one or more promoter sites for genes whose expression is upregulated during the development of competence. Several such genes have been found. A competence-induced operon containing *recA, dinF,* and *cinA* (function unknown) has been independently identified by two groups (76, 94), and high-level expression of *recA* is required for efficient transformation (94, 95). Campbell & Masure (13) have identified a second inducible locus, TYG. They have found common sequence motifs in the promoter regions of these loci, and additional induced loci. Furthermore, Pestova et al have obtained physiological data indicating that the induction of a burst CSP expression immediately precedes the expression of competence (102), which suggests that induction of this locus may be the initial intracellular event. At the time of writing of this review, there was no published evidence for molecular interaction between the phosphorylated ComE protein and the 5′ region of any of these operons. However, the tools are in hand for such studies, and the DNA sequences that serve as the initial molecular target for this signal transduction pathway will likely be identified soon. Interestingly, the window of competence is quite narrow, and the Morrison group has found evidence for an adaptation, or shut down, of competence within minutes of its induction. This process is independent of cell density and appears to represent an intracellular event.

The *ami* locus (originally identified based on a pleiotropic phenotype that included resistance to aminopterin) represents the first oligopeptide permease (Opp) that has been identified in gram-positive bacteria (3); and it has been linked to the development of competence (95). In many microorganisms, this system is encoded by five contiguous genes (48, 49). Opp(Ami)A is an extracellular peptide-binding protein, anchored to the membrane via an amino-terminal lipid in gram-positive bacteria. OppB and OppC represent membrane proteins that are believed to form a channel to allow for import of the ligand. OppD and OppF represent membrane-associated ATPases that drive the active transport of the substrate. In pneumococci, two unlinked *amiA* homologues, *aliA*, and *aliB*, exist, and there is genetic evidence that any of these three binding proteins may interact with the other permease components to form a functional transport system; although *ami* can mediate transport of oligopeptides, the complete list of potential or preferred substrates is not known.

When an insertional mutation in the *amiA* locus was identified that reduced the level of competence (95), there was an inference that this peptide permease system could be involved in the sensing of an extracellular signal. This question was recently examined in detail by using a series of well-characterized polar and nonpolar mutations in various *ami*, and *ali* genes. It was found that a functional *ami* is not required for the sensing of the CSP signal, and that the effects of certain *ami* mutations on competence were indirect, possibly because of strain differences or because of a global regulatory network that is responsive to amino acid or peptide levels (2). Opp systems are directly involved in other forms of bacterial communication, as described below, and it is still formally possible that *ami* could affect the expression of pneumococcal competence by interaction with an as yet unidentified signal distinct from CSP.

CELL-CELL SIGNALING AND REGULATION OF NISIN BIOSYNTHESIS

Nisin is a modified peptide antibiotic that is produced by certain *Lactococcus lactis* strains. It is the archetypal member of the antibiotic family called Lantibiotics (111). As a natural antibacterial agent whose safety and efficacy is well established, it has become heavily used as a food preservative, and it is also used in a variety of other applications (123). The complete sequence of the genes required for the production of, and the immunity to, nisin is available. A great deal of work has been done on the biosynthesis and the structure/function relationships of nisin and related compounds, and on nisin's mode of action on susceptible cells. Thus, nisin serves as an ideal model for the study of postranslationally-modified polypeptides. Another noteworthy feature that has recently been described for this system is that the function of the nisin molecule is not only to act as a bacteriocin against other organisms, but it is also to serve as an apparent quorum-sensing signal for the positive regulation of expression of biosynthetic and immunity genes in the nisin-producing host cell (22, 67).

The molecular organization of the genes required for nisin biosynthesis and immunity are shown in Figure 3. The *nisA* gene encodes the prepeptide; whereas

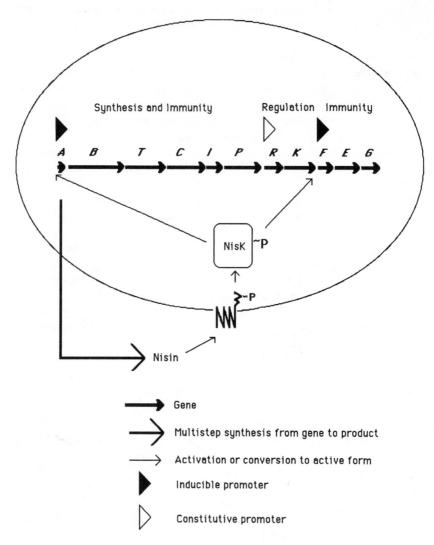

Figure 3   Gene organization and signaling pathway for autoregulatory control of nisin biosynthesis in *Lactococcus lactis*. In this system, the structural gene is part of an operon (*nisA-nisP*) that encodes the postranslational steps of biosynthesis, as well as immunity. A linked, but independently transcribed operon (*nisR-nisK*) encodes the sensing machinery. The signal transduction process activates transcription of the biosynthetic operon and an auxiliary immunity system (*nis F-nisG*), but the operon that encodes the 2-component system is transcribed constitutively, with no induction by nisin (24, 67).

nisB, nisT, and *nisP* genes encode enzymes that are involved in the series of postranslational modifications, export, and proteolytic processing steps that are involved in producing the mature lantibiotic. The molecular biology of this process has been reviewed in detail by de Vos et al (24). The full level of immunity is dependent on the presence of a membrane protein, *nisI,* as well as on a putative ABC transporter that is encoded by *nisFEG* (112). It has been suggested that the latter system could enhance immunity by exporting any nisin that escaped the immune mechanism conferred by NisI (24). The *nisR* and *nisK* genes serve a regulatory function, and they are RR and HK 2-component homologues, respectively.

Based on the results of gene fusion studies and on the molecular characterization of mRNAs that are encoded by the *nis* genes, a molecular model for induction of nisin biosynthesis has been proposed (67). The product of the pathway in this model serves as an extracellular signal for further transcriptional activation of the biosynthetic and immunity genes (Figure 3). The data suggests the presence of three functional promoters in this gene cluster. Promoters 1 and 3 are responsive to the presence of nisin or certain nisin derivatives at concentrations as low as five molecules/cell, i.e. far below the inhibitory concentration of the antibiotic. These promoters show extensive homology in both the -10 and -35 regions and in the intervening segment (22). Promoter 2 is structurally distinct, and it appears to be responsible for a constant level of production of the regulatory proteins. As nisin builds up in the medium, it presumably binds to the HK protein that is encoded by *nisK.* This results in a phosphorylation-dependent signal transduction process, which leads to increased transcription of the operons downstream from promoter 1 (generating more nisin biosynthetic machinery and NNisI) and from promoter 3 (generating increased levels of the accessory immunity proteins). Thus far, there is no evidence for direct molecular interaction between nisin and NisK, or between the phosphorylated NisR protein and the DNA sequences in the promoter 1 and promoter 3 regions. Recently, a nisin-inducible expression vector system that may have broad utility in gram-positive bacteria has been developed (23) using this system.

It is interesting to consider the physiology of nisin production in the context of a normally growing culture of nisin-producing lactococci. In contrast to the pneumococcal system, the available data indicate that there may not be a pronounced shut-down or adaptation step associated with nisin mediated–self-induction. The results of gene fusion studies indicate that, with the addition of increased amounts of nisin to cultures, there is a continuous increase in expression up to the point at which the added nisin becomes inhibitory to normal cell growth (67). However, deletion of the *nisFEG* immunity genes increases the sensitivity of responder cells to added nisin. Thus, the *nisFEG* products could serve to limit the self-induction process in producing strains growing in the absence of exogenously added nisin.

INDUCTION OF COMPETENCE AND SPORULATION IN *BACILLUS SUBTILIS*: CONTROL BY MULTIPLE SIGNALING SYSTEMS.

The following summary of mechanisms that control two important postexponential phase phenomena in *Bacillus subtilis*, sporulation and competence development, illustrates how peptide signals can function in complex regulatory networks. The signaling pathways that lead to the development of competence and sporulation have been used as model systems for understanding how bacteria can integrate a number of regulatory pathways to produce an appropriate response to environmental conditions. The intracellular concentrations, and often the phosphorylation states, of key regulatory molecules, such as Spo0A and ComK, determine whether the cells sporulate or develop competence, respectively (12, 125). Figure 4 shows some of these regulatory pathways in simplified form. Although much is known about the intracellular intermediates that modulate the activity of these molecules [for recent reviews see (26, 51, 101, 113)], until recently the identity of the extracellular signals produced by the bacteria has been elusive.

Initial Evidence for Peptide Signaling

Originally, sporulation of *Bacillus subtilis* was believed to be an individual cellular response to carbon, nitrogen, or phosphate starvation, and to growth stage specific and cell type specific developmental signals. Two observations led to the idea that quorum sensing and peptide signaling may be involved in the initiation of both sporulation and competence. The first indication for a role of quorum sensing was provided by Grossman & Losick (39). They showed that the initiation of sporulation (stage 0) is inefficient at low cell densities in the presence of decoyinine—an inhibitor of GMP synthetase that stimulates sporulation but not competence. They found that the induction of high levels of sporulation depended not only on starvation, but also on a secreted bacterial factor. Medium or sterile filtrates (<1000 mw) from high density cultures could stimulate sporulation at low cell density. The second observation was that cells with delayed sporulation and competence phenotypes contained mutations in the oligopeptide permease operon, *spoOK* (99, 108). In *Bacillus subtilis* strain 168, there is a second oligopeptide permease operon called *app*, which normally contains a mutation that genetically silences the locus. Second site suppressers of *spoOK* mutations have been identified as reversion mutations that reactivate the *app* locus (64).

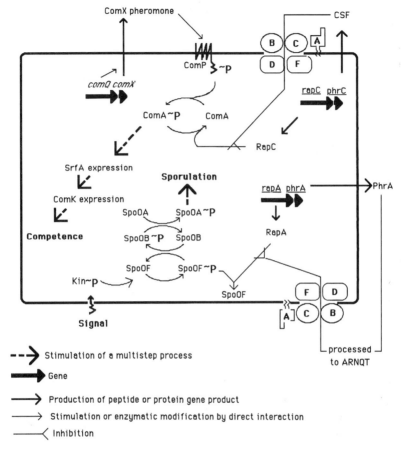

Figure 4 Proposed roles of peptides in the regulatory networks that affect sporulation and competence in *Bacillus subtilis*. In this illustration, the closed shapes containing the letters A–F represent the components of the Opp system (encoded by the *spoOK* operon in strain 168). The zigzag structure is the histidine kinase–sensor protein ComP, which, because of interaction with the ComX pheromone is autophosphorylated (∼**P**), and can subsequently phosphorylate the ComA response regulator. Other gene products are shown in plain font, and the designations for the various-sized arrows are indicated. PhrA is apparently re-internalized via the Opp system, where it can inhibit the RapA phosphatase, which results in more phosphorylated members of the phosphorelay leading to spoOA ∼P and which induces sporulation. The *phrC* gene encodes the competence stimulating peptide (CSF). Internalization of CSF is associated with an elevated phosphorylation state in the relay (possibly via inhibition of RapC), which leads to comK ∼P and competence (114). 2-Component signaling through ComP and mediated by ComX pheromone also stimulates this cascade and up-regulates competence. Interestingly, the stimulatory effect of CSF on competence at low concentrations, as shown in this figure, becomes an inhibitory effect at high CSF concentrations. This indicates a possible role of CSF in the shut-down process, once competence has been achieved (26, 101, 113).

The dependence of competence and sporulation on both an extracellular signal and an oligopeptide permease operon suggest that a peptide signal(s) might accumulate in cell cultures in which they could be internalized through the oligopeptide permease and act intracellularly in the target cell. This type of signal could be a true quorum sensor, in which individual members of a *Bacillus* population gain information about how many other cells are present that must share the limited nutrients, or that could supply DNA to be taken up during competence. Alternatively, a cell wall peptide recycled through the oligopeptide permease could provide a means for the cell to monitor growth rate, which would decrease at the onset of sporulation (99).

Production of the Peptide Signals

Identification of several peptide-derived molecules that can function as extracellular signals has been achieved by amino acid sequencing and mass spectrometry of HPLC-purified molecules. The genes that encode one sporulation-inducing peptide and two competence-inducing peptides were then identified.

A peptide pheromone that enhances competence (ComX pheromone) is a 9 or 10 amino acid peptide, which contains a modification that increases the mass of the molecule by approximately 200 kDa, possibly on a tryptophan residue (74). Competence pheromone is encoded by the *comX-comQ* operon. The *comX* gene encodes a 55 amino acid precursor that contains the competence pheromone sequence as the last 10 amino acids. The 5' end of *comX* overlaps with the 3' end of *comQ*, which is required for the production, maturation, and/or modification of the ComX pheromone.

Competence stimulating factor (CSF) is also produced as a carboxy-terminal cleavage product of a larger protein (114, 115). In the case of CSF, the 5-amino acid peptide (ERGMT) is cleaved from the 40-amino acid precursor, which is encoded by *phrC* (97b). In the case of both CSF and the ComX pheromone, the pathway for secretion and cleavage of the active peptides has not yet been further characterized.

The gene for a third peptide signal, PhrA (100), is transcribed in an operon with the *rapA* phosphatase gene. It is produced as a 44-amino acid precursor that is cleaved by signal peptidase I and secreted as a 19-amino acid signal molecule. Recent evidence suggests that the active product of the *phrA* gene is the pentapeptide ARNQT, which can directly inhibit RapA phosphatase (97a).

Mechanism of Signaling and Integration of the Signals into the Sporulation and Competence Regulatory Networks

The mechanisms by which these peptides signal and the points at which the competence and sporulation peptides enter the signaling cascades are not entirely understood. However, the available data can be incorporated into a schematic model in which the various regulatory pathways that are affected by extracellular signaling are integrated (Figure 4). Due to the dependence of both sporulation and competence on the Opp, the inducing peptides were originally believed to be imported, where they served as intracellular signals. Genetic analysis of the *Bacillus subtilis opp* operon *spoOK* indicated that *spoOKF* was required for competence development, but not for sporulation (108). One possible conclusion from these date is that *spoKF*-encoded ATPase activity is required for import of ComX. However, as discussed below, more recent results suggest that alternative mechanisms of signaling are operative.

Perego et al have shown that sporulation induction by the peptide PhrA requires an active oligopeptide permease (99, 100). If mutations are made in *opp* or *phrA,* then RapA activity in the cell increases and the level of the phosphorylated form of SpoOF decreases. Mutations in *opp* can be suppressed by mutations in the Rap phosphatases RapA and RapB. Their data (97a, 100) suggest that the pentapeptide product of *phrA*, encoded in an operon with the RapA phosphatase, is synthesized, secreted, processed, and re-internalized, where it suppresses the activity of the RapA phosphatase. PhrA-mediated regulation appears to be important for only one of the two central sporulation phosphatases (RapA and RapB). Peptide that is encoded downstream of RapB appears not be transcribed and, therefore, could not function to control RapB activity (100). In any case, the suppression of RapA phosphatase could lead to an accumulation of phosphorylated SpoOF, which in turn would result in increased phosphorylation of SpoOA. The increased phosphorylation of SpoOA serves to initiate sporulation (12, 98).

Activity of the competence pheromone ComX depends on the presence of the 2-component sensor/regulator proteins, ComP (HK) and ComA (RR), which are encoded by genes immediately downstream of the ComX structural gene (115). The sensor HK molecule, ComP, has eight potential transmembrane regions, which suggests that it is localized in the cell membrane. The exact mechanism for signaling is still unknown. It was proposed originally that the competence pheromone (ComX) might be imported through SpoOK, where it could interact with a binding domain on the cytoplasmic side of the ComP protein (74). This would lead to activation of the HK activity and phosphorylation of the RR ComA. More recently, a mechanism involving binding to the cell surface and 2-component signal transduction has been suggested for ComX pheromone signaling (115) (Figure 4). In this model, the dependence on Opp is related to a separate signaling event, such as the internalization of the CSF (PhrC) signal, which in turn affects responsiveness to the competence pheromones. Further molecular studies will need to be completed in order to distinguish between these models.

Signaling by CSF is proposed to be Opp dependent (114, 115). CSF can be transported by Opp and subsequently used as a source of amino acids in amino acid auxotrophs, as well as inducing competence (114, 155). Also CSF addition has been shown to affect the intracellular phosphatase RapC, as well as the transcription of *srfA*, in a RapC independent manner (114). The mechanism by which CSF causes these effects is unknown, because no direct interaction of intracellular effector molecules with CSF has yet been demonstrated.

Together, CSF and the ComX pheromone affect the levels of phosphorylated ComA present in the cell (74, 114, 115). Increased phosphorylation of ComA acts to up-regulate the expression of *srfA*, which ultimately results in a higher expression of the competence transcription factor, ComK, and in competence induction (40, 114). ComX pheromone causes the phosphorylation of the 2-component HK ComP, which leads to the phosphorylation of the RR ComA. Genetic data indicate that the phosphorylation state level of ComA is decreased by the RapC phosphatase (114). CSF addition may decrease RapC activity, in turn also increasing the level of phosphorylated ComA. Low concentrations of CSF also may lead to a ComA independent increase in *srfA* transcription (114) (Figure 4). However, high concentrations of CSF decrease *srfA* expression, which suggests that the peptide may act to down-regulate its activity after competence initiation.

The initiation of competence and of sporulation are both carefully regulated processes that involve the integration of a number of signals and protein phosphorylation cascades. ComX pheromone accumulates in medium and acts in a density-dependent manner, thereby appearing to act as a true pheromone/quorum sensor (74). In contrast, the PhrA peptide does not appear to accumulate in the medium (100). However, mutations in *phrA* can be complemented in mixed cultures of wild-type and *phrA* mutant strains. This suggests that perhaps the PhrA peptide signals only in the case of direct cell-cell contact (very high cell density), or that it is primarily an autocrine peptide signal, affecting the cell that produces it. It is interesting to note that these peptide signals may also permit cross-talk between the sporulation and the competence pathways. The sporulation-inducing peptide, PhrA, also inhibits AbrB, causing an increase in the production of the competence peptide signal CSF (113). Furthermore, Solomon et al have suggested that the signaling peptide CSF may also inhibit the activity of the competence phosphatases, RapA and RapB, which would both enhance the induction of sporulation under certain conditions and activate competence (114).

DENSITY-DEPENDENT REGULATION OF VIRULENCE IN *STAPHYLOCOCCUS AUREUS*: THE ACCESSORY GENE REGULATOR SYSTEM

Quorum Sensing as a Regulator of Virulence Factors

Staphylococcus aureus is a human pathogen that causes a number of severe diseases, including endocarditis, septic arthritis, and toxic shock syndrome.

The accessory gene regulator (*agr*) is a global regulatory locus in *S. aureus* that regulates the expression of a number of virulence factors (65, 77, 78, 97). This includes both secreted factors, such as toxic shock syndrome, toxin-1, alpha-toxin, beta-hemolysin, enterotoxin, lipases and proteases, and cell-associated virulence factors such as protein A, fibronectin binding-protein, and coagulase (1, 57, 65). In late exponential–phase cultures, in a cell density–dependent manner, *agr* causes the down-regulation of cell surface components and the up-regulation of secreted factors (1, 10, 57, 65, 68, 124). The scenarios illustrated in the introduction (Figure 1) likely come into play utilizing the agr system. The in vivo use of a cell density–dependent signal by a pathogen can be envisioned. For example, a quorum sensing mechanism could enable staphylococcal cells to stop producing factors (i.e. fibronectin binding protein) that facilitate binding to a particular surface when the population becomes large, and instead produce secreted factors that might permit the cells to penetrate new sites in which there are more nutrients and less competition. Alternatively, a situation in which even a small number of the bacterial cells occupy a limited space (such as in eukaryotic intracellular vacuoles) may require the bacteria to switch expression from factors that allow for binding and sequestration to factors that will either destroy the tissue or kill the engulfing eukaryotic cell.

Production of the AGR Signal

The cell density–dependent peptide signal in *S. aureus* is encoded by the *agr* locus. The gene cluster contains two divergently transcribed operons (Figure 5) (91). One of these operons, RNAII, encodes the genes that are necessary for the production and detection of the cell density–dependent factor (60). The peptide signal, AgrD (YSTCDFIM), is contained in the middle of a larger 45 amino acid protein (60). AgrD is translationally coupled to a second molecule, AgrB, which is required for its synthesis, in a manner analogous to the ComX/ComQ system of *Bacillus subtilis* (see Figure 4). The exact mechanism for secretion and maturation of the AgrD peptide pheromone is unknown. However, it is thought to contain a cyclic anhydride that is critical for its activity (60).

Mechanism of Signaling

The intracellular effector molecule for the *agr* response is an RNA molecule, RNAIII, that is transcribed divergently from the *agrD* pheromone production and response operon (Figure 5) (56, 91). Expression of this RNA is under positive transcriptional regulation (79). AgrD pheromone is one of three signals that lead to increased transcription of RNAIII. The level of transcription of RNAIII is also controlled by two other loci, *sar* and *xpr* (14, 15, 41). For up-regulation of RNAIII, AgrD pheromone signals by binding to a sensor histidine kinase (HK) that is encoded by *agrC* (60). In the case of the AgrD pheromone, the increased binding of the peptide to cells that express the HK AgrC has been

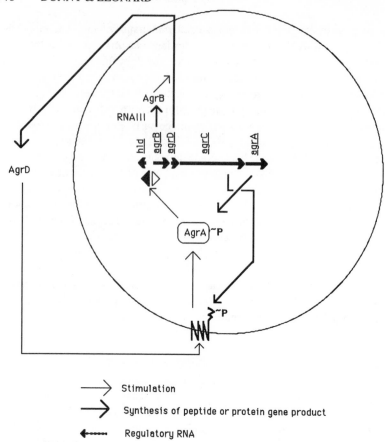

Figure 5 Gene organization and regulatory pathway for the *Staphylococcus aureus agr* system. The genes for synthesis and sensing of the signal peptide (AgrD) are encoded on a single operon (60). The signaling pathway operates through a 2-component system (AgrC-AgrA) and results in increased synthesis of an intracellular effector, RNA III, which increases the expression of several secreted-virulence factors and decreases the expression of certain surface proteins (78).

demonstrated (60). The activation of the HK leads to phosphorylation of the RR protein, AgrA. The mechanism for up-regulation of RNAIII transcription by the phosphorylated form of AgrA is currently unknown, because AgrA lacks a clear DNA binding site, and because direct binding to DNA has not yet been demonstrated (79). However, a second RNAIII regulatory protein, SarA, has been shown to directly bind the RNAIII promoter, and the interaction of AgrA with SarA may facilitate transcriptional activation (79).

Once it is transcribed, RNAIII can affect either the transcription or the translation of target genes (91). The mechanism for the initiation or the repression of transcription by RNAIII is not currently known. In the case of the up-regulation of alpha-toxin translation, the RNAIII molecule binds to the alpha-toxin mRNA (*hla*), preventing the formation of a stem loop that normally interferes with the availability of the ribosome binding site (78).

PHEROMONE-INDUCIBLE CONJUGATION IN *ENTEROCOCCUS FAECALIS*: PEPTIDE SIGNALS WITH INTRACELLULAR TARGETS

The enterococci are well known for their propensity to acquire and disseminate antibiotic resistance determinants (18). Conjugative plasmid transfer in *E. faecalis* has been studied for more than 20 years (20, 27, 30), and the regulation of expression of transfer functions of some of these plasmids by peptide sex pheromones represents the most extensively characterized quorum sensing system in the gram-positive bacterial world. We focus here on the molecular details of the signaling process, and we describe in some detail the recent work that has clarified several important aspects of the system. (For reviews of other aspects of the pheromone plaasmids, see References 20, 30, 128).

According to the model for pheromone-inducible conjugation in *E. faecalis* (28), a donor cell, carrying a pheromone-inducible plasmid, senses the proximity of potential recipient cells by detecting the recipient-produced pheromone. The binding of the pheromone to the donor cell initiates a signaling process that results in the activation of expression of conjugation functions, most notably in the appearance of an inducible surface adhesin called Aggregation Substance (AS). Surface expression of AS on the donor cell facilitates the attachment of the induced donor cell to a recipient-encoded enterococcal binding substance (EBS). The pairs (or aggregates, in high-density cultures) formed by AS-EBS binding can efficiently generate a conjugal mating channel that allows for plasmid transfer to the recipient cell. This is analogous to the situation observed with sex factors of gram-negative bacteria, such as F (19), but apparently without the involvement of sex pili. Once the recipient cell acquires a copy of the plasmid, it assumes a conjugative phenotype identical to that of the original donor, i.e. it expresses transfer functions only in the presence of an exogenous source of pheromone, and excretes no net pheromone activity into its growth medium.

Pheromones and Their Cognate Plasmids: pAD1 and pCF10 as Models

Based on genetic analyses of pheromone production and response in a variety of clinical isolates, it was suggested that it is the *E. faecalis* chromosome that encodes the ability to produce a variety of different molecules that could function as pheromones for cells that carry specific plasmids (28). Because of extensive

biochemical purification and characterization work by Suzuki and coworkers at the University of Tokyo, the identification of a number of pheromones
as hydrophobic peptides, 7–8 amino acids in length, has been accomplished
(81, 82, 117). Further molecular and genetic analysis of many enteroccal plasmids has indicated the existence of several families of plasmids and their cognate
pheromones (129). The acquisition of a plasmid of one family effectively blocks
production of the cognate pheromone (at least in a biologically active form) by
the new donor cell, but the same cell continues to excrete other pheromones
which can, in turn, stimulate the acquisition of plasmids belonging to the other
families. The structural genes required for pheromone production have not been
identified, despite extensive efforts (D Clewell, personal communication). Because the plasmid-encoded–pheromone inhibitor peptides (see below) resemble
pheromones and are synthesized by the processing of a 22 amino acid–signal
peptide-like precursor, it is possible that the pheromones would be encoded similarly. However, it was recently found that membrane fractions of E. faecalis
contain a very large amount of an apparent pro pheromone molecule (~ 3 kDa in
size), as well as a relatively unstable processing activity required for production
of the mature cCF10 pheromone (69). These data suggest that a specialized pro-
cessing/secretion system may be required for the release of mature pheromones
into the culture medium. This could be analogous to the biosynthetic machinery for signaling molecules, such as the pneumococcal CSP and the bacteriocin
IFs described above.

Most of the genetic and molecular analyses of this phenomenon have employed either the hemolysin plasmid pAD1 (122), which determines a response
to the pheromone cAD1, or the tetracycline resistance plasmid, pCF10 (29),
which encodes a response to cCF10 (see Table 2 for amino acid sequences).
Considerable analysis of a third plasmid, pPD1, and its cognate pheromone,
cPD1, has also been done (85, 87, 130). It is worthwhile to consider briefly
the other traits encoded by these plasmids. Both pPD1 and pAD1 encode
cytolysin/bacteriocins, and the pAD1 system has been studied in some detail
in this regard. Interestingly, the cytolysin/bacteriocin activity is conferred by
modified peptides of the lantibiotic family that are similar to nisin (38). Expression of this activity increases the virulence of the host bacteria (59, 116), and it
will be of considerable interest to learn whether the expression of this activity is
regulated by a quorum sensing mechanism. In the case of pCF10, the plasmid
was initially identified in a screen for pheromone-inducible plasmids that carry
antibiotic resistance genes (29). Recently, a recombinant plasmid that confers
high-level vancomycin resistance was identified in conjugation experiments
that employ a clinical isolate of E. faecium (45). The conjugation genes of this
plasmid were highly related to those of pCF10, which was isolated nearly 15
years earlier from the same hospital; this suggests that the same conjugative

plasmid has circulated within this environment for years, picking up different resistance genes over time. Even from this brief discussion, it is obvious that the pheromone plasmids play a major role in the medical aspects of enterococcal biology.

Induction of Conjugation via the Pheromone Signaling Pathway

In contrast to many of the systems described above, the pheromone induction process in enterococcal conjugation does not involve the transduction of a transmembrane signal via an HK-like sensor protein. Rather, the signal molecule itself is imported into the responder cell, and it interacts with intracellular effectors. In *E. faecalis* cells that carry pCF10, the induction of mating at pheromone concentrations ($\sim 5 \times 10^{-12}$ M) that are typically generated by recipient cells (81) likely involves initial binding to PrgZ, which is a homologue of the OppA family of peptide-binding proteins. The pheromone is then internalized via the transmembrane channel that is generated by the OppB and OppC proteins, with the energy for active transport presumably generated by ATP hydrolysis by the OppD and OppF ATPases (71). In Opp$^+$, PrgZ$^-$ cells, signaling can still occur, presumably via pheromone binding to the chromosomally-determined OppA protein. However, this process requires pheromone concentrations exceeding the levels produced naturally in mating mixtures, and it is subject to competitive inhibition by nonrelated oligopeptides. In Opp$^-$ cells, the pheromone response is essentially eliminated, except at very high concentrations ($\sim 10^{-8}$ M), approaching the solubility limits for cCF10 in aqueous solution; at these high concentrations it is possible that internalization might occur by diffusion or via another transport system with low affinity for cCF10.

In both the pCF10 and the pAD1 systems, affinity chromatography studies have demonstrated pheromone binding to putative intracellular regulatory molecules. Several potential intracellular effectors have been identified, depending on the experimental conditions, especially the method used for immobilization of the peptides on the affinity matrix. The data from these experiments, along with the results of extensive additional genetic and molecular analyses, have been incorporated into models that are quite distinct from one another (Figure 6). In the pAD1 system, it is believed that the internal pheromone target is the protein product of the *traA* gene, which is predicted to be a major negative regulator in the system (104). TraA is the only molecule in the pAD1 system that has been shown by biochemical methods to bind cAD1 directly (34). It is believed that, in uninduced cells, this protein interacts with pAD1 DNA sequences at the 5' end of *iad* gene, and that this interaction results in the termination of the transcript at one or both of two stem-loop structures between the *iad* and *traE1* coding sequences. Pheromone, presumably internalized via the concerted activity of the Opp system and the plasmid-encoded TraC protein

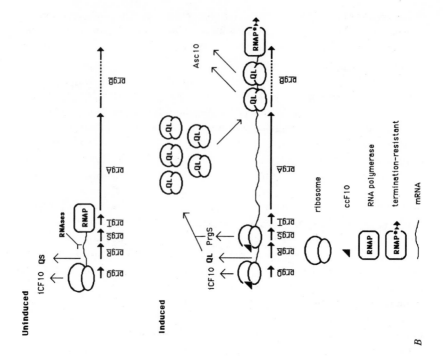

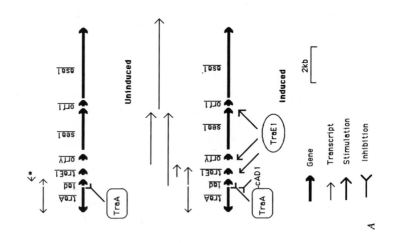

(118), enters the cell, binds to TraA, and thus removes the latter protein from the DNA binding site. Under these conditions, transcription reads through the terminators, resulting in the synthesis of TraE1 (34, 119). This positive regulator is believed to bind to multiple promoters, resulting in its own up-regulation, as well as in the transcriptional activation of conjugation genes, including the *asa1* gene that encodes the Aggregation substance of pAD1 (37, 119). Recently, an additional putative regulatory locus, designated traD, whose message would comprise an antisense RNA for the 3′ segment of the abundant transcript, initiating at *iad* and terminating at TTS1 or TTS2 in the absence of pheromone, has been identified (D Clewell, personal communication). This transcript is detected only in uninduced cells, and it could influence the stability or efficiency of translation of the mRNA that originates from the *iad* promoter.

In the case of pCF10 (Figure 6B), the internalized pheromone is believed to interact with a ribonucleoprotein complex, probably ribosomes or their subunits (6, 71). The initial result of this interaction seems to be the increased translation of Orfs downstream from the Orf (*prgQ*) that encodes the inhibitor peptide iCF10 (86). Initially, this results in the synthesis of the PrgS protein, which appears to promote the stable association of a positive regulatory RNA molecule, Q_L (17), with ribosomes (6, 7). The Q_L-modified ribosomes are able to efficiently translate the message for the target gene *prgB*, which results in the production of Aggregation Substance. Pheromone induction in the pCF10 system is also believed to involve antitermination; but, in this case, it is believed to occur further downstream, with an RNA that is encoded by the intergenic

←——————————————————————————————————————

Figure 6   Comparison of the proposed intracellular phases of induction of conjugation by peptide pheromones in the pAD1 (*A*), and pCF10 (*B*) systems. In both cases, the pheromone is probably internalized by the concerted action of a plasmid-encoded binding protein and the chromosomal Opp system. (*A*) Pheromone cAD1 is proposed to bind to the TraA negative regulator, displacing TraA from a target site on pAD1 and resulting in read-through of transcription into the *TraE1* gene. The TraE1 protein then activates its own transcription and also several additional promoters that encode conjugation products, such as the Asa1 aggregation protein. Similar models have been generated both by the Clewell and the Wirth groups (37, 119), but there is some disagreement about whether there is a separate, inducible promoter for TraE1 at the 5′ end of the structural gene. In uninduced cells carrying pAD1, the binding of TraA to the *iad* promoter region is depicted to cause the termination of transcription upstream of the *TraE1* gene, which precludes the expression of transfer functions. (*B*) Pheromone cCF10 is believed to interact with ribosomes, allowing for translation genes downstream of *prgQ*, especially *prgS*. Synthesis of PrgS is associated with further ribosome modifications, including the stable association of a regulatory RNA, QL, with the ribosomes (6, 7). The stabilization of mRNA and the read-through of transcription to the *prgB* target gene that encodes Aggregation Substance (Asc10) are also enhanced. The modified ribosomes are then able to translate the *prgB* message.

region between *prgS* and *prgT*. This region has been implicated in the antitermination process (7). There is evidence for ribosomal involvement in this process as well, and it is not clear whether antitermination results directly from the induced ribosomal changes described above, or whether there is a direct effect of the pheromone on termination that is distinct from the translational activation. One of the key features of this model is that all the transcripts that encode *prgB* initiate from the single constitutive *prgQ* promoter. This observation is based on the results of molecular analysis of the transcription of this region, using a method for 5' end mapping that distinguishes 5' ends that result from processing from those that are generated by transcription initiation (8). The unusual mode of positive control, involving ribosome modification by both the peptide and a regulatory RNA, is supported by genetic and molecular evidence for the association of the pheromone and of Q_L RNA with ribosomal components (6–8).

Given the similarity of the molecular organization of the regulatory genes of pCF10 and pAD1, and the high degree of homology in some regulatory regions, it may seem surprising that such different models have emerged for the two systems. Some of the differences may have resulted from the different experimental approaches taken. For example, the mRNA species encoded by the positive control regions that have been identified by Northern blotting are fairly similar in the two systems, but the determination of 5' ends has been done by two different methods. Also, it can be very difficult to distinguish the 3' termini of mRNAs derived from processing from those that result from termination. Thus, it is possible that the noncoding mRNA species in the pAD1 system could function similarly to those of pCF10. Moreover, it would not be surprising to find that the pCF10-encoded–negative regulator PrgX (46) (which is likely to have a similar function to that of TraA of pAD1) might bind the pheromone and thus play a role in its interaction with the positive control machinery; in affinity chromatography studies utilizing *E. faecalis* extracts, abundant host proteins in the 40 kDa range were retained that could have obscured PrgX protein, which is apparently expressed at low levels (17). However, there are clearly some important differences between the two systems, including the fact that there is no detectable cross-induction in cells that harbor both plasmids and that are induced with a single pheromone. Additionally, pAD1 encodes a *trans*-acting positive control protein, TraE1 (119), for which no homologue exists in pCF10. Finally, the promoter for the *asa1* gene that encodes the pAD1 Aggregation Substance is located at the 5' end of a short open-reading frame *orf 1*, just upstream from *asa1* (37). A gene corresponding to *orf1* does not exist in pCF10. Based on a comparative analysis of all the published regulatory studies of these plasmids, we suggest that the translational activation mechanism may be similar in all of the pheromone plasmids, but that pAD1 has probably acquired some additional genes, including an additional promoter and a cognate transcriptional activator (TraE1). A recent molecular comparison of various pheromone plasmids by Hirt et al (50) supports the idea of the evolution of these plasmids by acquisition of additional gene modules by an IS-sequence mediated process.

Autocrine Cycles Involving Sex Pheromones and Control of Self Induction

Because pheromone production is encoded by the chromosome, but the expression of conjugation is induced only by exogenously supplied pheromone, the plasmids must encode gene products that block self-induction by endogenous pheromone in donor cells. Two such genes have been found in each of the plasmids analyzed, and this regulatory circuit in the pCF10 system is diagrammed in Figure 7. One class of negative control genes (*iad* of pAD1, and *prgQ* of pCF10) encodes a peptide inhibitor that is excreted into the medium (21, 86). The iad1 inhibitor is present in donor cultures with no cAD1 detected (21, 80), whereas pCF10-containing cells seem to excrete a mixture of cCF10 and iCF10 in a molar ratio that results in no net pheromone or inhibitor activity in the medium (86). The peptide inhibitors of both systems may be viewed as functioning to neutralize any pheromone that might be secreted by the donor cell, while enabling the same cell to sense exogenous pheromone at concentrations of 5 molecules/cell or less (81).

A second type of negative regulator of self induction (PrgY of pCF10 and TraB of pAD1) is predicted to reside in the membrane, and is probably cotranscribed with the extracellular pheromone binding protein (PrgZ and TraC) of each plasmid. The TraB protein of pAD1 has been termed a shutdown protein, because cells expressing this protein excrete no detectable cAD1 (4). In the case of pCF10, the corresponding protein is PrgY. In the presence of a functional *prgY* gene, *E. faecalis* cells produce the same amount of extracellular pheromone as plasmid-free cells, but the membrane-associated pheromone activity described above is not detected (69). The data suggest that PrgY might degrade the putative pro pheromone, prevent its maturation, or interact directly with the PrgZ-Opp complex in such a way that it would inhibit an autocrine cycle that involves simultaneous release of mature cCF10 and delivery of the peptide to the intracellular machinery that activates the mating response.

Although PrgY effectively blocks self induction of conjugation, recent data suggest that donor cells do re-internalize some cCF10 that they produce, and that this process is required for stable maintenance of the plasmid. Several lines of molecular and genetic data suggest a link between pheromone production and plasmid replication, including the fact that the functional replicons for pAD1 and pCF10 overlap the regions that encode negative control and signaling

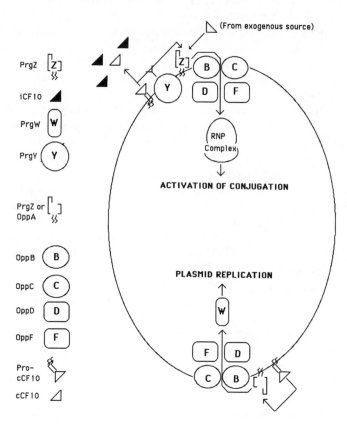

Figure 7 Autocrine and pheromone activities of the cCF10 peptide. Two regulatory circuits utilizing cCF10 are illustrated. A signaling pathway, mediated by exogenous pheromone and leading to the expression of conjugation functions, involves internalization by the concerted action of PrgZ and the chromosomal Opp system. The small amount of cCF10 that is actually released from donor cells is effectively neutralized by the iCF10 inhibitor peptide. The pCF10-encoded–negative regulatory protein PrgY appears to prevent direct transfer of pheromone from the putative membrane-associated pro pheromone to the Opp. Experimental evidence summarized in the text (BAB Leonard & GM Dunny, unpublished observations) suggests that *E. faecalis* cells that carry pCF10 express a second type of signal-processing machinery that does allow some direct re-entry of pheromone back into the cell. PrgY is apparently nonfunctional in this molecular complex. The intracellular target of this pathway is the replication-initiator protein PrgW, and the pheromone appears to be required for stable maintenance of the plasmid.

functions (46, 127). In the case of pCF10, the putative replication initiator protein PrgW also binds pheromone, and the insertion of an engineered gene (derived by mutagenesis of the inhibitor gene *prgQ*) for cCF10 production into *Lactococcus lactis* extends the host range of the plasmid (normally limited to enterococci) to the lactococci (BAB Leonard & GM Dunny, unpublished observations). In addition to the requirement for pheromone synthesis, pCF10 replication requires a functional Opp. These data led to the suggestion that there are two types of cCF10 internalization machinery in *E. faecalis* which are compartmentalized such that, in a pure culture of donor cells, a small amount of endogenous pheromone is recycled into a replication control circuit without inducing a mating response. Interestingly, the same peptide, introduced exogenously (from recipient cells or by the addition of synthetic peptide in the laboratory), can overcome the iCF10 present in the growth medium, can be internalized via the concerted action of PrgZ and Opp, and can induce expression of transfer functions. Spatial compartmentalization (Figure 7), temporal (cell-cycle dependent) compartmentalization, or a combination, can be envisioned.

The control circuits that regulate self induction may also be of considerable relevance to the virulence of plasmid-containing enterococci, which are now major opportunistic pathogens (72, 110). Data from several groups (16, 66, 92) suggest that the expression of Aggregation Substance on the enterococcal cell surface increases the virulence of the organisms in experimental animal models and in tissue culture models. Induction of Aggregation Substance expression by serum has also been noted (66). In the case of pCF10, the serum factor may not act directly as a pheromone, but instead it may act to disrupt the control of self induction. This phenomenon may involve selective binding of iCF10, resulting in an excess of cCF10 when the bacteria are growing in vivo (H Hirt & G Dunny, unpublished results).

CONTROL OF ANTIBIOTIC BIOSYNTHESIS BY A γ-BUTYROLACTONE SIGNAL IN *STREPTOMYCES GRISEUS*

The Actimomycetes are of considerable interest because of their complex differentiation cycles and their production of useful natural products, such as antibiotics. They are phylogenetically distant from the other bacteria discussed in this review, and we will not attempt a comprehensive review of signaling processes in these organisms. The following section represents an example of one such system that has been investigated in some detail and thus represents an interesting contrast to the peptide-based systems described thus far.

In *Streptomyces griseus,* the differentiation process leading to the production of secondary metabolites such as streptomycin, to the formation of aerial mycelia and, ultimately, to sporulation, is enhanced by an extracellular product of the bacteria called A Factor (62). The A factor contains a γ-butyrolactone moiety, and it is the only well-characterized signaling molecule from gram-positive bacteria that closely resembles the homoserine lactone–quorum sensing molecules that are common in gram-negative organisms. The chemistry and the biology of the A factor system have been reviewed, both specifically (53) and in the context of other communication systems in actinomycetes (52). The molecule seems to be produced at a very low constitutive rate during vegetative growth, and when concentrations in the nanomolar range are reached, the molecule can act as an inducer of differentiation. The mechanism of action involves an interaction of A factor with an intracellular receptor protein called ArpA (93). In the absence of the A factor, ArpA functions as a repressor by binding to a specific target DNA sequence in the promoter region of one or more of the operons that are required for antibiotic production and differentiation. The DNA binding activity of the protein is abolished by the A factor binding, allowing for the transcriptional activation of these operons. Interestingly, there is recent evidence that *Streptomyces coelicolor*, which produces the antibiotic actinorhodin, secretes a similar factor, which is capable of complementing *S. griseus* strains with A factor mutations (9). This result, and the work of a number of other groups, suggests that the A factor–like signaling may be widespread.

AN EXPANDING UNIVERSE OF BACTERIAL COMMUNICATION

In the previous sections, a number of relatively well-characterized control systems involving extracellular signals that are produced by gram-negative bacteria have been summarized. There are a variety of reports in the literature that suggest the possibility that we may be just scratching the surface, according to our understanding of the extent to which these mechanisms are used in the gram-positive bacterial world. Many of these reports are recent, and they represent preliminary characterizations of the regulatory mechanisms, or of the potential signaling molecules that could be part of a signaling pathway similar to one of the previously characterized systems, or of a completely novel circuit.

Analysis of several cell-surface proteins that are identified as adhesins, or as potential virulence factors of oral streptococci (32, 58, 75) and enterococci (73), suggests that they share a significant degree of amino acid similarity with OppA lipoproteins. It is conceivable that, in addition to their putative adhesive function, these proteins might be involved in enabling the organism to sense the fact that it has colonized a surface. It has also been recently reported that the initiation of growth on the tooth surface by oral streptococci is a density-dependent process (11), which suggests that signaling between cells could be important in this process, as well.

In pathogenic Group A streptococci, it has been determined that genetic disruption of the chromosomal *opp* operon greatly impairs the ability of the organism to produce the pyrogenic exotoxin SpeB (103). An obvious inference is that toxin production might be regulated by a peptide-based signaling system, although there is as yet no direct evidence for a signal. Another example of the potential role of Opps in signaling was recently reported in *Streptomyces coelicolor* (90). The *opp* cassette was identified as a genetic locus in which mutations resulted in a "bald" (differentiation defective) phenotype. Based on the results of extracellular complementation assays, it has been suggested that an as yet unidentified peptide signal, imported by the Opp system and is part of a regulatory cascade that leads ultimately to sporulation (90). There is also an example of a potential signal whose target, if existent, is not yet known. A gene that encodes a peptide with enterococcal sex pheromone cAD1 activity has been found in a staphylococcal conjugative plasmid (33). However, there is no evidence that the presence of this peptide in culture medium has any effect on the transfer of the plasmid that carries the gene. It is possible that it could be involved in either intracellular regulatory processes in the producing organism, or that it could be involved in control mechanisms requiring direct contact between the communicating cells. As noted previously in this review, some of the peptides that can function as extracellular signals, or their precursors, e.g. the pheromones of the enterococci and the bacilli, may also function within cells or at their surface. This idea gains credibility when one considers that the intracellular pools of some of these molecules may be much higher than the amounts found in the culture medium. Also, many of these bacteria, or the niches they occupy, are very proteolytic, which suggests that the signals might not persist under natural conditions.

Finally, it is worthwhile to consider the possibility that many antibiotics in addition to nisin could function as density signals for the producer cells at subinhibitory concentrations. In the case of daunorubicin synthesis by *Streptomyces peucetius,* a suggestion, based on analysis of the DnrN protein, was made that the product of the antibiotic biosynthetic pathway serves as a negative regulatory factor (36) This protein activates the transcription of a gene that encodes DnrI which, in turn, activates transcription of the biosynthetic genes for daunorubicin production. The antibiotic binds to DnrN, which reduces the DNA binding of this protein. This would be expected to reduce the biosynthetic capability of the cell. This effect of the antibiotic is essentially the opposite from that of nisin, which acts in a postive manner. However, daunorubicin also appears to increase the intracellular level of DnrN, and therefore the regulatory circuit could be complex.

CONCLUSIONS AND FUTURE DIRECTIONS

The past decade has seen a remarkable increase in research on prokaryotic cell-cell communication systems, and these systems have been shown to play critical roles in controlling the expression of some of the most important biological functions of these organisms. It is not unreasonable to assume that as yet undiscovered signaling systems will be shown to affect many more, perhaps even most, of the microbial processes that have a great impact on humans. Some microbial signaling systems may represent useful models for regulatory pathways that were previously believed to be limited to higher organisms. Also, an increased understanding of these systems will be crucial for future advancements in areas such as ecology, physiology, and pathogenesis. Finally, an increased understanding of extracellular signaling mechanisms may lead to many useful applications, including an improved ability to control the expression of useful metabolic functions for biotechnological applications, as well as the identification of novel targets for the development of new drugs or vaccines against microbial pathogens.

ACKNOWLEDGMENTS

Much of this review was written while G Dunny was visiting the laboratory of P Cocconcelli at the Department of Microbiology, Catholic University, Piacenza, Italy. The wonderful hospitality and intellectual stimulation provided in Piacenza contributed greatly to the preparation of this review. The authors wish to thank B Bensing for help with the figures, and all members of the Dunny lab for providing recent data and useful discussions. A number of colleagues, including S Horinouchi, W deVos, I Nes, D Morrison, H Jenkinson, D Clewell, J Nakayama, R Wirth, J Hoch, M Perego, A Grossman, D Dubnau, R Skurray, and R Novick provided reprints or communicated unpublished results. Our research has been supported by NIH grants GM49530 and HL51987 awarded to G Dunny, and by NIH grant AI08742 awarded to BA Leonard, who is currently an Alexander von Humboldt fellow.

> Visit the *Annual Reviews home page* at
> http://www.annurev.org.

Literature Cited

1. Abdelnour A, Arvidson S, Bremell T, Ryden C, Tarkowski A. 1993. The accessory gene regulator (*agr*) controls *Staphylococcus aureus* virulence in a murine arthritis model. *Infect. Immun.* 61:3879–85

2. Alloing G, Granadel C, Morrison DA, Claverys J-P. 1996. Competence pheromone, oligopeptide permease, and induction of competence in *Streptococcus pneumoniae. Mol. Microbiol.* 21:471–78

3. Alloing G, Trombe M-C, Claverys J-P. 1989. Cloning of the *amiA* locus of *Streptococcus pneumoniae* and identification of its functional limits. *Gene* 76:363–68

4. An FY, Clewell DB. 1994. Characterization of the determinant (*traB*) encoding sex pheromone shutdown by the hemolysin/bacteriocin plasmid, pAD1 in *Enterococcus faecalis*. *Plasmid* 31:215–21

5. Avery OT, Macleod CM, McCarty M. 1944. Studies on the chemical nature of the substance inducing transformation of pneumococcal phenotypes. Induction of transformation by a deoxyribonucleic acid fraction isolated from pneumococcus type III. *J. Exp. Med.* 79:137–58

6. Bensing BA, Dunny GM. 1997. Pheromone-inducible expression of an aggregation protein in *Enterococcus faecalis* requires interaction of a plasmid-encoded RNA with components of the ribosome. *Mol. Microbiol.* 24:295–308

7. Bensing BA, Manias DA, Dunny GM. 1997. Pheromone cCF10 and plasmid pCF10-encoded regulatory molecules act post-transcriptionally to activate expression of downstream conjugation functions. *Mol. Microbiol.* 24:285–294

8. Bensing BA, Meyer BJ, Dunny GM. 1996. Sensitive detection of bacterial transcription initiation sites and differentiation from RNA processing sites in the pheromone-induced plasmid transfer system of *Enterococcus faecalis*. *Proc. Natl. Acad. Sci. USA* 93:7794–99

9. Bibb M. 1996. The regulation of antibiotic production in *Streptomyces coelicolor* A3(2). *Microbiology* 142:1335–44

10. Bjorklin A, Arvidson S. 1996. Mutants of *Staphylococcus aureus* affected in the regulation of exoprotein synthesis. *FEMS Microbiol. Lett.* 7:203–6

11. Bloomquist CG, Reilly BE, Liljemark WF. 1996. Adherence, accumulation, and cell division of a natural adherent bacterial population. *J. Bacteriol.* 178:1172–77

12. Burbulys D, Trach KA, Hoch JA. 1991. Initiation of sporulation in B. subtilis is controlled by a multicomponent phosphorelay. *Cell* 64:545–52

13. Campbell EA, Masure HR. 1996. Identification and characterization of two competence-induced promoters in *Streptococcus pneumoniae*. *12th Eur. Meet. Bact. Gene Transf. Expr.* 1:14 (Abstr.)

14. Cheung AL, Koomey JM, Butler CA, Projan SJ, Fischetti VA. 1992. Regulation of exoprotein expression in *Staphylococcus aureus* by a locus (*sar*) distinct from *agr*.

Proc. Natl. Acad. Sci. USA 89:6462–66

15. Cheung AL, Projan SJ. 1996. Cloning and sequencing of *sarA* of *Staphylococcus aureus*, a gene required for the expression of *agr*. *J. Bacteriol.* 176:4168–72

16. Chow JW, Thal LA, Perri MB, Vazquez JA, Donabedian SM, et al. 1993. Plasmid-associated hemolysin and aggregation substance production contribute to virulence in experimental enterococcal endocarditis. *Antimicrob. Agents Chemother.* 37:2474–77

17. Chung JW, Dunny GM. 1995. Transcriptional analysis of a region of the *Enterococcus faecalis* plasmid pCF10 involved in positive regulation of conjugative transfer functions. *J. Bacteriol.* 177:2118–24

18. Clewell DB. 1990. Movable genetic elements and antibiotic resistance in enterococci. *Eur. J. Clin. Microbiol. Infect. Dis.* 9:90–102

19. Clewell DB. 1991. *Bacterial Conjugation.* New York: Plenum

20. Clewell DB. 1993. Bacterial sex pheromone-induced plasmid transfer. *Cell* 73:9–12

21. Clewell DB, Pontius LT, An FY, Ike Y, Suzuki A, Nakayama J. 1990. Nucleotide sequence of the sex pheromone inhibitor (iAD1) determinant of *Enterococcus faecalis* conjugative plasmid pAD1. *Plasmid* 24:156–61

22. De Ruyter PGGA, Kuipers OP, Beerthuyzen MM, Van Alen-Boerrigter I, De Vos WM. 1996. Functional analysis of promoters in the nisin gene cluster of *Lactococcus lactis*. *J. Bacteriol.* 178:3434–39

23. De Ruyter PGGA, Kuipers OP, De Vos WM. 1996. Controlled gene expression systems for *Lactococcus lactis* with the food-grade inducer nisin. *Appl. Environ. Microbiol.* 62:3662–67

24. De Vos WM, Kuipers OP, van der Meer JR, Siezen RJ. 1995. Maturation pathway of nisin and other lantibiotics: post-translationally modified antimicrobial peptides exported by gram-positive bacteria. *Mol. Microbiol.* 17:427–37

25. Diep DB, Havarstein LS, Nes IF. 1996. Characterization of the locus responsible for the bacteriocin production in *Lactobacillus plantarum* C11. *J. Bacteriol.* 178:4472–83

26. Dubnau D. 1991. Genetic competence in *Bacillus subtilis*. *Microbiol. Rev.* 55:395–424

27. Dunny GM, Clewell DB. 1975. Transmissible toxin (hemolysin) plasmid in *Streptococcus faecalis* and its mobilization of

a noninfectious drug resistance plasmid. *J. Bacteriol.* 124:784–90

28. Dunny GM, Craig RA, Carron RL, Clewell DB. 1979. Plasmid transfer in *Streptococcus faecalis*: Production of multiple pheromones by recipients. *Plasmid* 2:454–65

29. Dunny GM, Funk C, Adsit J. 1981. Direct stimulation of the transfer of antibiotic resistance by sex pheromones in *Streptococcus faecalis*. *Plasmid* 6:270–78

30. Dunny GM, Leonard BAB, Hedberg PJ. 1995. Pheromone-inducible conjugation in *Enterococcus faecalis*: interbacterial and host-parasite chemical communication. *J. Bacteriol.* 177:1–2

31. Eisjink VGH, Brurberg MB, Middelhoven PH, Nes IF. 1996. Induction of bacteriocin production in *Lactobacillus sake* by a secreted peptide. *J. Bacteriol.* 178:2232–37

32. Fenno JC, Shaikh A, Spatafora D, Fives-Taylor P. 1995. The *fimA* locus of *Streptococcus parasanguis* encodes an ATP-binding membrane transport system. *Mol. Microbiol.* 15:849–63

33. Firth N, Fink PD, Johnson L, Skurray RA. 1994. A lipoprotein signal peptide encoded by the staphylococcal conjugative plasmid pSK41 exhibits an activity resembling *Enterococcus faecalis* pheromone cAD1. *J. Bacteriol.* 176:5871–73

34. Fujimoto S, Bastos M, Tanimoto K, An F, Wu K, Clewell DB. 1997. The pAD1 sex pheromone response in *Enterococcus faecalis*. In *Streptococci and the Host*, ed. T Horaud, M Sicard, A Bouve, H de Montelos. New York: Plenum. In press

35. Fuqua WC, Winans SC, Greenberg EP. 1994. Quorum sensing in bacteria: the LuxR-LuxI family of cell density-responsive transcriptional regulators. *J. Bacteriol.* 176:269–75

36. Furuya K, Hutchinson CR. 1996. The DnrN protein of *Streptomyces peucetius*, a pseudo-response regulator, is a DNA binding protein involved in the regulation of daunorubicin biosynthesis. *J. Bacteriol.* 178:6310–18

37. Galli D, Friesnegger A, Wirth R. 1992. Transcriptional control of sex-pheromone-inducible genes on plasmid pAD1 of *Enterococcus faecalis* and sequence of a third structural gene for (pPD1-encoded) aggregation substance. *Mol. Microbiol.* 6:1297–1308

38. Gilmore MS, Segarra RA, Booth MC, Bogie CP, Hall LR, Clewell DB. 1994. Genetic structure of the *Enterococcus faecalis* plasmid pAD1-encoded cytolytic toxin system and its relationship to lantibiotic determinants. *J. Bacteriol.* 176:7335–44

38a. Griffith F. 1928. The significance of pneumococcal types. *J. Hyg.* 27:113–159

39. Grossman AD, Losick R. 1988. Extracellular control of spore formation in *Bacillus subtilis*. *Proc. Natl. Acad. Sci. USA* 85:4369–73

40. Hahn J, Luttinger A, Dubnau D. 1996. Regulatory inputs for the synthesis of ComK, the competence transcription factor of *Bacillus subtilis*. *Mol. Microbiol.* 21(4):763–75

41. Hart ME, Smeltzer MS, Iandolo JJ. 1993. The extracellular protein regulator (*xpr*) affects exoprotein and *agr* levels in *Staphylococcus aureus*. *J. Bacteriol.* 175:7875–79

42. Havarstein LS, Coomaraswamy G, Morrison DA. 1995. An unmodified heptadecapeptide pheromone induces competence for genetic transformation in *Streptococcus pneumoniae*. *Proc. Natl. Acad. Sci. USA* 92:11140–44

43. Havarstein LS, Diep DB, Nes IF. 1995. A family of ABC transporters carry out proteolytic processing af their substrates concomitant with export. *Mol. Microbiol.* 16:229–40

44. Havarstein LS, Gaustad P, Nes IF, Morrison DA. 1996. Identification of the streptococcal competence pheromone receptor. *Mol. Microbiol.* 21:863–69

45. Heaton MP, Discotto LF, Pucci MJ, Handwerger S. 1996. Mobilization of vancomycin resistance by transposon-mediated fusion of a VanA plasmid with an *Enterococcus faecium* sex pheromone-response plasmid. *Gene* 171:9–17

46. Hedberg PJ, Leonard BAB, Dunny GM. 1995. Identification and characterzation of the genes of *Enterococcus faecalis* plasmid pCF10 involved in replication and in negative control of pheromone-inducible conjugation. *Plasmid* 34:1–2

47. Deleted in proof

48. Higgins CF, Hiles ID, Salmond GPC, Gill DR, Downie JA, et al. 1986. A family of related ATP-binding subunits coupled to many distinct biological processes in bacteria. *Nature* 323:448–50

49. Hiles ID, Gallagher MP, Jamieson DJ, Higgins CF. 1987. Molecular characterization of the oligopeptide permease of *Salmonella typhimurium*. *J. Mol. Biol.* 195:125–42

50. Hirt H, Wirth R, Muscholl A. 1996. Comparative analysis of 18 sex pheromone plasmids from *Enterococcus faecalis*: detection of a new insertion element on

pPD1 and hypotheses on the evolution of this plasmid family. *Mol. Gen. Genet.* 252:640–47

51. Hoch JA, Silhavy TJ. 1995. *Two-Component Signal Transduction.* Washington, DC: Am. Soc. Microbiol.

52. Horinouchi S, Beppu T. 1992. Autoregulatory factors and communication in actinomycetes. *Annu. Rev. Microbiol.* 46:377–98

53. Horinouchi S, Beppu T. 1994. A-factor as a microbial hormone that controls cellular differentiation and secondary metabolism in *Streptomyces griseus. Mol. Microbiol.* 12:859–64

54. Huehne K, Holck A, Axelson L, Kroeckel L. 1996. Analysis of the sakacin P gene cluster from *Lactobacillus sake* LB674 and its expression in sakacin P negative *L. sake* strains. *Microbiology* 142:1437–48

55. Hui FM, Zhou L, Morrison DA. 1995. Competence for genetic transformation in *Streptococcus pneumoniae*: organization of a regulatory locus with homology to two lactococcin A secretion genes. *Gene* 153:25–31

56. Janzon L, Arvidson S. 1996. The role of the d-lysin gene (*hld*) in the regulation of virulence genes by the accessory gene regulator (*agr*) in *Staphylococcus aureus. EMBO J.* 9(5):1391–99

57. Janzon L, Lofdahl S, Arvidson S. 1986. Evidence for the coordinate control of alpha-toxin and protein A in *Staphylococcus aureus. FEMS Microbiol. Lett.* 33:193–98

58. Jenkinson HF. 1992. Adherence, coaggregation, and hydrophobicity of *Streptococcus gordonii* associated with expression of cell surface lipoproteins. *Infect. Immun.* 60:1225–28

59. Jett BD, Jensen HG, Nordquist RE, Gilmore MS. 1992. Contribution of the pAD1-encoded cytolysin to the severity of experimental *Enterococcus faecalis* endophthalmitis. *Infect. Immun.* 60:2445–52

60. Ji GY, Beavis RC, Novick RP. 1995. Cell density control of staphylococcal virulence mediated by an octapeptide pheromone. *Proc. Natl. Acad. Sci. USA* 92:12055–59

61. Jiménez-Díaz R, Ruiz-Barba JL, Cathcart DP, Holo H, Nes IF, et al. 1995. Purification and partial amino acid sequence of plantaricin *S*, a bacteriocin produced by *Lactobacillus plantarum* LPCO10, the activity of which depends on the complementary action of two peptides. *Appl. Environ. Microbiol.* 61:4459–63

61a. Kaplan HB, Plamann L. 1996. A *Myxococooccus xanthus* cell-density system required for multicellular development. *FEMS Micro. Lett.* 139:89–95

62. Khokhlov AS. 1988. Results and perspectives of actinomycete autoregulator studies. In *Biology of Actinomycetes '88*, ed. Y Okami, T Beppu, H Ogawara, pp. 338–45. Tokyo: Jpn. Sci. Soc. Press

63. Kim SK, Kaiser D, Kuspa A. 1992. Control of cell density and pattern by intercellular signaling in *Myxococcus* development. *Annu. Rev. Microbiol.* 46:117–39

64. Koide A, Hoch JA. 1994. Identification of a second oligopeptide transport system in *Bacillus subtilis* and determination of its role in sporulation. *Mol. Microbiol.* 13:417–26

65. Kornblum J, Kreiswirth B, Projan S, Ross H, Novick R. 1990. *agr*: a polycistronic locus regulating exoprotein synthesis in *Staphylococcus aureus.* In *Molecular Biology of Staphylococci*, ed. RP Novick, pp. 373–402. New York: VCH

66. Kreft B, Marre R, Schramm U, Wirth R. 1992. Aggregation substance of *Enterococcus faecalis* mediates adhesion to cultured renal tubular cells. *Infect. Immun.* 60:25–30

67. Kuipers OP, Beerthuyzen MM, De Ruyter PGGA, Luesink EJ, De Vos WM. 1995. Autoregulation of nisin biosynthesis by signal transduction. *J. Biol. Chem.* 270:27299–304

68. Lebeau C, Vandenesch F, Greenland T, Novick RP, Jerome E. 1994. Coagulase expression in *Staphylococcus aureus* is positively and negatively modulated by an *agr*-dependent mechanism. *J. Bacteriol.* 176:5534–36

69. Leonard BAB, Bensing BA, Hedberg PJ, Dunny GM. 1995. Pheromone-inducible gene regulation and signalling in the control of aggregation substance expression in the conjugative plasmid, pCF10. In *Genetics of Streptococci, Enterococci, and Lactococci, Biol. Stand.*, ed. J Ferretti, M Gilmore, T Klaenhammer, 85:27–34. Washington, DC: Int. Bur. Biol. Stand.

70. Leonard BAB, Colwell A, Manias DA, Dunny GM. 1997. Dual role of a bacterial peptide as a sex pheromone and as an autocrine regulator of plasmid replication. *Mol. Microbiol.* 22:1–2

71. Leonard BAB, Podbielski A, Hedberg PJ, Dunny GM. 1996. *Enterococcus faecalis* pheromone binding protein, PrgZ, recruits a chromosomal oligopeptide permease system to import sex pheromone cCF10 for induction of conjugation. *Proc. Natl. Acad. Sci. USA* 93:260–64

72. Lewis CM, Zervos MJ. 1990. Clinical manifestations of enterococcal infection. *Eur. J. Clin. Microbiol. Infect. Dis.* 9:73–74

73. Lowe AM, Lambert PA, Smith AW. 1994. Cloning of an *Enterococcus faecalis* endocarditis antigen: homology with adhesins from some oral streptococci. *Infect. Immun.* 63:703–6

74. Magnuson R, Solomon J, Grossman AD. 1994. Biochemical and genetic characterization of a competence pheromone from *B. subtilis. Cell* 77:207–16

75. Margolis PS, Driks A, Losick R. 1993. Sporulation gene *spoIIB* from *Bacillus subtilis. J. Bacteriol.* 175:528–40

76. Martin B, Ruellan J-M, Angulo JF, Devoret R, Claverys J-P. 1992. Identification of the *recA* gene of *Streptococcus pneumoniae. Nucleic Acids Res.* 20:6412

76a. Méjean V, Claverys J-P. 1993. DNA processing during entry in transformation of *Streptococcus pneumoniae. J. Biol. Chem.* 268:5594–99

77. Morfeldt E, Janzon L, Arvidson S, Lofdahl S. 1988. Cloning of a chromosomal locus (*exp*) which regulates the expression of several exoprotein genes in *Staphylococcus aureus. Mol. Gen. Genet.* 211:435–40

78. Morfeldt E, Taylor D, Gabain AV, Arvidson S. 1995. Activation of alphatoxin translation in *Staphylococcus aureus* by the *trans*-encoded antisense RNA, RNAIII. *EMBO J.* 14(18):4569–77

79. Morfeldt E, Tegmark K, Arvidson S. 1996. Transcriptional control of the agr-dependent virulence gene regulator, RNAIII, in *Staphylococcus aureus. Mol. Microbiol.* 21(6):1227–37

80. Mori M, Isogai A, Sakagami Y, Fujino M, Kitada C, et al. 1986. Isolation and structure of the *Streptococcus faecalis* sex pheromone inhibitor, iAD1, that is excreted by the donor strain harboring plasmid pAD1. *Agric. Biol. Chem.* 50:539–41

81. Mori M, Sakagami Y, Ishii Y, Isogai A, Kitada C, et al. 1988. Structure of cCF10, a peptide sex pheromone which induces conjugative transfer of the *Streptococcus faecalis* tetracycline resistance plasmid, pCF10. *J. Biol. Chem.* 263:14574–78

82. Mori M, Sakagami Y, Narita M, Isogai A, Fujino M, et al. 1984. Isolation and structure of the bacterial sex pheromone, cAD1, that induces plasmid transfer in *Streptococcus faecalis. FEBS Lett.* 178:97–100

83. Morrison DA, Lacks SA, Guild WR, Hageman JM. 1983. Isolation and

characterization of three new classes of transformation-deficient mutants of *Streptococcus pneumoniae* that are defective in DNA transport and genetic recombination. *J. Bacteriol.* 156:281–90

84. Deleted in proof

85. Deleted in proof

86. Nakayama J, Ruhfel RE, Dunny GM, Isogai A, Suzuki A. 1994. The *prgQ* gene of the *Enterococcus faecalis* tetracycline resistance plasmid, pCF10, encodes a peptide inhibitor, iCF10. *J. Bacteriol.* 176:2003–4

87. Nakayama J, Yoshida K, Kobayashi H, Isogai A, Clewell DB, Suzuki A. 1995. Cloning and characterization of a region of *Enterococcus faecalis* plasmid pPD1 encoding pheromone inhibitor (*ipd*), pheromone sensitivity (*traC*), and pheromone shutdown (*traB*) genes. *J. Bacteriol.* 177:5567–73

88. Nes IF, Diep DB, Havarstein LS, Brurberg MB, Eijsink V, Holo H. 1996. Biosynthesis of bacteriocins in lactic acid bacteria. *Antonie van Leeuwenhoek J. Microbiol. Serol.* 70:113–28

89. Nissen-Meyer J, Larsen AG, Sletten K, Daeschel M, Nes IF. 1993. Purification and characterization of plantaricin A, a *Lactobacillus plantarum* bacteriocin whose activity depends on the action of two peptides. *J. Gen. Microbiol.* 139:1973–78

90. Nodwell JR, McGovern K, Losick R. 1996. An oligopeptide permease responsible for the import of an extracellular signal governing aerial mycelium formation in *Streptomyces coelicolor. Mol. Microbiol.* 22:881–93

91. Novick RP, Ross HF, Projan SJ, Kornblum J, Kreiswirth B, Moghazeh S. 1993. Synthesis of staphylococcal virulence factors is controlled by a regulatory RNA molecule. *EMBO J.* 12:3967–75

92. Olmsted SB, Dunny GM, Erlandsen S, Wells CL. 1994. A plasmid-encoded surface protein on *Enterococcus faecalis* augments its internalization by cultured intestinal epithelial cells. *J. Infect. Dis.* 170:1549–56

93. Onaka H, Ando N, Nihira T, Yamada Y, Beppu T, Horinouchi S. 1995. Cloning and characterization of the A-factor receptor gene from *Streptomyces griseus. J. Bacteriol.* 177:6083–92

94. Pearce BJ, Naughton AM, Campbell EA, Masure HR. 1995. The *rec* locus, a competence-induced operon in *Streptococcus pneumoniae. J. Bacteriol.* 177:86–93

95. Pearce BJ, Naughton AM, Masure HR.

1994. Peptide permeases modulate transformation in *Streptococcus pneumoniae*. *Mol. Microbiol.* 12:881–92

96. Deleted in proof

97. Peng H-L, Novick RP, Kreiswirth B, Kornblum J, Schlievert P. 1988. Cloning characterization, and sequencing of an accessory gene regulator (*agr*) in *Staphylococcus aureus*. *J. Bacteriol.* 170(9):4365–72

97a. Perego M. 1997. Proc. Natl. Acad. Sci. In press

97b. Perego M, Glaser P, Hoch J. 1996. Aspartyl-phosphate phosphatases deactivate the response regulator components of the sporulation signal transduction system in *Bacillus subtilis*. *Mol. Microbiol.* 19:1151–57

98. Perego M, Hanstein C, Welsh KM, Djavakhishvill T, Glaser P, Hoch JA. 1994. Multiple protein-aspartate phosphatases provide a mechanism for the integration of diverse signals in the control of development in B. subtilis. *Cell* 79:1047–55

99. Perego M, Higgins CF, Pearce SR, Gallagher MP, Hoch JA. 1991. The oligopeptide transport system of *Bacillus subtilis* plays a role in the initiation of sporulation. *Mol. Microbiol.* 5:173–85

100. Perego M, Hoch JA. 1996. Cell-cell communication regulates the effects of protein aspartate phosphatases on the phosphorelay controlling development in *Bacillus subtilis. Proc. Natl. Acad. Sci. USA* 93:1549–53

101. Perego M, Hoch JA. 1996. Protein aspartate phosphatases control the output of two-component signal transduction systems. *Trends Genet.* 12(3):97–101

102. Pestova EV, Havarstein LS, Morrison DA. 1996. Regulation of competence for genetic transformation in *Streptococcus pneumoniae* by an auto-induced peptide and a two-component regulatory system. *Mol. Microbiol.* 21:853–62

103. Podbielski A, Pohl B, Woischnik M, Koerner C, Schmidt K-H, et al. 1996. Molecular characterization of group A streptococcal (GAS) oligopeptide permease (Opp) and its effect on cysteine protease production. *Mol. Microbiol.* 21:1087–99

104. Pontius LT, Clewell DB. 1992. Regulation of the pAD1-encoded sex pheromone response in *Enterococcus faecalis*: Nucleotide sequence analysis of *traA. J. Bacteriol.* 174:1821–27

105. Pozzi G, Masala L, Ianelli F, Manganelli R, Havarstein LS, et al. 1996. Competence for genetic transformation in encapsulated strains of pneumococcus: two allelic variants of the peptide pheromone. *12th Eur. Meet. Bact. Gene Transf. Expr.* 1:25 (Abstr.)

106. Quadri LEN, Roy KL, Vederas JC, Stiles ME. 1995. Characterization of the protein conferring immunity to the antibacterial peptide carnobacteriocin B2 and expression of the carnobacteriocins B2 and BM1. *J. Bacteriol.* 177:1144–51

107. Quadri LEN, Sailer M, Roy KL, Vederas JC, Stiles ME. 1994. Chemical and genetic characterization of bacteriocins produced by *Carnobacterium piscicola* LV 17B. *J. Biol. Chem.* 269:12204–21

108. Rudner DZ, LeDeaux JR, Ireton K, Grossman AD. 1991. The *spoOK* locus of *Bacillus subtilis* is homologous to the oligopeptide permease locus and is required for sporulation and competence. *J. Bacteriol.* 173:1388–98

109. Ruhfel RE, Manias DA, Dunny GM. 1993. Cloning and characterization of a region of the *Enterococcus faecalis* conjugative plasmid, pCF10, encoding a sex pheromone binding function. *J. Bacteriol.* 175:5253–59

110. Schaberg DR, Culver DH, Gaynes RP. 1991. Major trends in the microbial ecology of nosocomial infection. *Am. J. Med.* 91(Suppl. 3B):S72–S75

111. Schnell N, Entian K-D, Schneider U, Gotz F, Zahner H, et al. 1988. Prepeptide sequence of epidermin, a ribosomally synthesized antibiotic with four sulphide rings. *Nature* 333:276–78

112. Siegers K, Entian K-D. 1995. Genes involved in immunity to the lantibiotic nisin produced by *Lactococcus lactis* 6F3. *Appl. Environ. Microbiol.* 61:1082–89

113. Soloman JM, Grossman AD. 1996. Who's competent and when: regulation of natural genetic competence in bacteria. *Trends Genet.* 12(4):150–55

114. Soloman JM, Lazazzera BA, Grossman AD. 1996. Purification and characterization of an extracellular peptide factor that affects two different developmental pathways in *Bacillus subtilis. Gene Dev.* 10:2014–24

115. Soloman JM, Magnuson R, Srivastava A, Grossman AD. 1995. Convergent sensing pathways mediate response to two extracellular competence factors in *Bacillus subtilis. Gene Dev.* 9:547–58

116. Stevens SX, Jensen HG, Jett BD, Gilmore MS. 1992. A hemolysin-encoding plasmid contributes to bacterial virulence in experimental *Enterococcus faecalis* endophthalmitis. *Invest. Ophthalmol. Vis. Sci.* 33:1650–56

117. Suzuki A, Mori M, Sakagami Y, Isogai A, Fujino M, et al. 1984. Isolation and structure of bacterial sex pheromone, cPD1. *Science* 226:849–50

118. Tanimoto K, An FY, Clewell DB. 1993. Characterization of the traC determinant of the *Enterococcus faecalis* hemolysin-bacteriocin plasmid pAD1:binding of sex pheromone. *J. Bacteriol.* 175:5260–64

119. Tanimoto K, Clewell DB. 1993. Regulation of the pAD1-encoded sex pheromone response in *enterococcus faecalis*: expression of the positive regulator TraE1. *J. Bacteriol.* 175:1008–18

120. Tomasz A. 1965. Control of the competent state in *Pneumococcus* by a hormone-like cell product: an example for a new type of regulatory mechanism in bacteria. *Nature* 208:155–59

121. Tomasz A, Hotchkiss RD. 1964. Regulation of transformability of pneumococcal cultures by macromolecular cell products. *Proc. Natl. Acad. Sci. USA* 51:480–87

122. Tomich PK, An FY, Damle SP, Clewell DB. 1979. Plasmid related transmissibility and multiple drug resistance in *Streptococcus faecalis* subspecies zymogenes strain DS16. *Antimicrob. Agents Chemother.* 15:828–30

123. Vandenbergh PA. 1993. Lactic acid bacteria, their metabolic products and interference with microbial growth. *FEMS Microbiol. Rev.* 12:221–38

124. Vandenesch F, Kornblum J, Novick R. 1991. A temporal signal, independent of *agr*, is required for *hla* but not *spa* transcription in *Staphylococcus aureus*. *J. Bacteriol.* 173:6313–20

125. van Sinderen D, Luttinger A, Kong L, Dubnau D, Venema G, Hamoen L. 1995. *comK* encodes the compentence transcription factor (CTF), the key regulatory protein for competence development in *Bacillus subtilis*. *Mol. Microbiol.* 15:455–62

126. Deleted in proof

127. Weaver KE, Clewell DB, An F. 1993. Identification, characterization, and nucleotide sequence of a region of *Enterococcus faecalis* pheromone-responsive plasmid pAD1 capable of autonomous replication. *J. Bacteriol.* 175:1900–9

128. Wirth R. 1994. The sex pheromone system of *Enterococcus faecalis*: more than just a plasmid-collection mechanism? *Eur. J. Biochem.* 222:235–46

129. Wirth R, Friesenegger A, Horaud T. 1992. Identification of new sex pheromone plasmids in *Enterococcus faecalis*. *Mol. Gen. Genet.* 233:157–60

130. Yagi Y, Kessler RE, Shaw JH, Lopatin DE, An F, Clewell DB. 1983. Plasmid content of *Streptococcus faecalis* strain 39–5 and identification of a pheromone (cPD1)-induced surface antigen. *J. Gen. Microbiol.* 129:1207–15

Annu. Rev. Microbiol. 1997. 51:565–92

REGULATORS OF APOPTOSIS ON THE ROAD TO PERSISTENT ALPHAVIRUS INFECTION

Diane E. Griffin and J. Marie Hardwick
Department of Molecular Microbiology and Immunology, Johns Hopkins University, School of Hygiene and Public Health, Baltimore, Maryland 21205

KEY WORDS: virus, apoptosis, virus persistence, encephalitis

ABSTRACT

Alphavirus infection can trigger the host cell to activate its genetically programmed cell death pathway, leading to the morphological features of apoptosis. The ability to activate this death pathway is dependent on both viral and cellular determinants. The more virulent strains of alphavirus induce apoptosis with increased efficiency both in animal models and in some cultured cells. Although the immune system clearly plays a central role in clearing virus, the importance of other cellular factors in determining the outcome of virus infections are evident from the observation that mature neurons are better able to resist alphavirus-induced apoptosis than immature neurons are, both in culture and in mouse brains. These findings are consistent with the age-dependent susceptibility to disease seen in animals. Cellular genes that are known to regulate the cell death pathway can modulate the outcome of alphavirus infection in cultured cells and perhaps in animals. The cellular *bax* and *bak* genes, which are known to accelerate cell death, also accelerate virus-induced apoptosis. In contrast, inhibitors of apoptotic cell death such as *bcl-2* suppress virus-induced apoptosis, which can facilitate a persistent virus infection. Thus, the balance of cellular factors that regulate cell death may be critical in virus infections. Additional viral factors also contribute to this balance. The more virulent strains of alphavirus have acquired the ability to induce apoptosis in mature neurons, while mature neurons are resistant to cell death upon infection with less virulent strains. Here we discuss a variety of cellular and viral factors that modulate the outcome of virus infection.

CONTENTS

INTRODUCTION

Recent advances in cellular and molecular biology have increased our under-standing of the relationship between a virus and the host in which the virus replicates. Biological interactions at both the organismal and cellular levels work together to determine the outcome of infection. At the level of the organism, both the ability of the virus to spread beyond the site of initial virus replication and the host immune response determine whether the infecting virus causes pneumonitis, encephalitis, hepatitis, enteritis, or other mild-to-severe manifestations of infection. Interactions of the virus with individual cells are also determined by whether the infection is acute and self-limited, or persistent and progressive. Elimination of a virus from the host requires not only the elimination of circulating cell-free virus but also the elimination of virus-infected cells from tissue. Antibodies can often control intracellular virus replication as well as neutralize cell-free virus, but a genuine cure requires elimination of the infected cell. This can occur either through virus-induced cell death or through immune-mediated death of the virus-infected cell. The corollary to this requirement—for virus-infected cells to be eliminated in order for the host to be cured—is that in order for viruses to cause persistent infection, the virus must not kill the infected cell and it must prevent elimination of these cells by the immune response. We have been exploring, and review, the virus-host interactions that determine the outcome of alphavirus-induced encephalitis.

ALPHAVIRUS OVERVIEW

Alphaviruses are single-strand message-sense RNA viruses that cause mos-quito-borne encephalitis in the Americas (e.g. Eastern equine encephalitis and

Western equine encephalitis) and syndromes of rash and arthritis in much of Europe, Asia, and Africa (e.g. Ockelbo, Ross River, Chikungunya, and O'nyong-nyong viruses). Sindbis virus (SV), the prototype alphavirus, is an Old World virus, but is related to Western equine encephalitis virus (60). It causes rash and arthritis in humans and encephalitis in mice. SV-induced encephalitis in mice provides a convenient animal model for investigating the viral and host determinants of the outcome of infection.

Virus Replication

SV has an icosahedral enveloped virion composed of three structural proteins. The envelope glycoproteins E1 and E2 form heterodimers that trimerize to form flower-like spikes on the virion surface (158). The capsid protein surrounds the 11.7 Kb genome to form a nucleocapsid, and it interacts with the tail of the E2 glycoprotein at the plasma membrane during virion assembly (119). SV replicates in many types of cells in tissue culture but only in muscle, brown fat, and the central nervous system in mice. In the nervous system, neurons are the primary target cells that become infected (82). Replication occurs entirely in the cytoplasm of the infected cell. After attaching to as-yet poorly defined cellular receptors, the virus enters the target cell by receptor-mediated endocytosis. The viral envelope proteins undergo a conformational change that is induced by the acidic pH present in the endosome, resulting in fusion of the viral and endosomal membranes and delivery of the genome-containing nucleocapsid into the cytoplasm (188).

Ribosomes participate in uncoating of the RNA and the genomic RNA is translated to produce the four nonstructural proteins, which form the viral replicase and transcriptase (181, 226). Replication complexes form in association with intracellular smooth membranes. Synthesis of full-length minus strand RNA continues for only the first few hours of infection, whereas synthesis of the full-length genomic and subgenomic RNA continues throughout the replication cycle (169). Structural proteins are translated as a polyprotein from the abundant subgenomic RNA. The SV capsid protein is encoded in the 5' end of the subgenomic RNA and is autocatalytically cleaved from the nascent chain. The N-terminal portion of the precursor of E2 (pE2) serves as a signal sequence for translocation of pE2 into the endoplasmic reticulum (ER). The C-terminal portion of pE2 contains the translocation sequence for 6K, which in turn contains the translocation sequence for E1. Both E1 and pE2 are N-glycosylated, heterodimerize in the ER, and are then processed through the Golgi. pE2, which protects the E1-E2 heterodimer from undergoing premature acid-induced conformational change, is cleaved by cellular proteases into E2 and E3 at a late stage in the transGolgi. Interaction of the viral glycoproteins and assembled nucleocapsids occurs at the plasma membrane, with subsequent budding of mature virions from the cell surface (188).

SV replication profoundly affects the metabolic function of the host cell. Host protein and RNA synthesis ceases within a few hours after infection (188), followed by a decline in phospholipid synthesis (220). Na^+K^+ ATPase function is compromised, resulting in decreased K^+ flux, altered intracellular cation concentrations, and decreased plasma membrane potential (56, 140). Intracellular pH declines within 1–2 h (130).

Sindbis Virus–Induced Disease

The mortality of mice with SV-induced encephalomyelitis is dependent on their age at the time of infection and on the virulence of the strain of SV used for infection (89, 121, 122, 167, 172, 175, 205). Neonatal mice infected with most strains of SV develop fatal infection, whereas increasingly virulent strains are required to induce fatal infection in progressively older mice. In four-week and older mice, induction of fatal disease is dependent on the strain of infected mice as well as on the virulence of the virus strain (204). Outcome in all these situations is correlated with the fate of SV-infected neurons, particularly motor neurons (83, 112). If most infected neurons survive, then the mouse survives and recovers from infection through immunologic control of virus replication. If infected neurons do not survive, the mouse dies; or if it survives, it does not recover neurologic function and remains permanently paralyzed. We have shown that SV kills infected cells, including neurons in the brain and spinal cord, by inducing apoptosis or programmed cell death. Therefore, outcome is ultimately determined by the viral and cellular determinants of programmed cell death in neurons (110). Persistent infection requires survival of the infected cells. This review examines, in depth, these virus-cell interactions during SV infection.

OVERVIEW OF THE CELL BIOLOGY OF APOPTOSIS AND ITS REGULATION

Programmed cell death is recognized as an essential mechanism for the elimination of superfluous or potentially harmful cells in multicellular organisms. Cells between digits must die during development for the formation of fingers (61), approximately 50% of neurons die during maturation of the nervous system (36, 155), and autoreactive lymphocytes are "negatively selected" (186, 187).

Upon activation of the programmed cell death pathway, many cells undergo characteristic morphologic changes, such as chromatin and cytoplasmic condensation, blebbing of cytoplasmic membranes, and fragmentation of the cell into membrane-bound bodies, which is referred to as apoptosis (93). Condensation of chromatin that occurs around the periphery of the nucleus is detectable with DNA stains and by electron microscopy. A cellular endonuclease is activated, cleaving chromosomal DNA into nucleosome-length fragments that are detectable as 180–200 base-pair (bp) ladders on agarose gels (6). The membrane of an apoptotic cell actively blebs but remains intact, ultimately blebbing the cell apart into membrane-bound apoptotic bodies that contain cytoplasmic and/or nuclear material. In tissues, these apoptotic bodies are engulfed by adjacent cells that may not normally be phagocytic (168). In contrast to apoptotic death, necrotic cell death is typified by early loss of membrane integrity, allowing cytosolic spillage into surrounding tissue and random degradation of DNA (47).

A number of cellular genes have been identified that regulate and modulate the cell death pathway. Many of these genes were first identified through the study of viruses. The products of these genes determine the susceptibility of the cell to induction of apoptosis by virus infection, as well as by withdrawal of growth factors and other cell death–inducing stimuli (147, 184).

The Cellular Bcl-2 Family

The Bcl-2 oncogene was first identified at t(14;18) translocations that occur in the majority of follicular B-cell lymphomas (24, 31, 202). This translocation event results in overexpression of Bcl-2, which allows B cells to survive when they would normally die by apoptosis (72, 144, 215). Bcl-2 is an integral membrane protein (24, 25, 102) that protects a wide variety of cell types, both in vivo and in vitro, from many death-inducing stimuli, including serum and growth factor withdrawal, treatment with calcium ionophores, glucose withdrawal, membrane peroxidation, glucocorticoid treatment, chemotherapeutic agents, and virus infection (7, 12, 27, 41, 42, 70, 74, 80, 156, 176, 178, 217). Bcl-2 has homology with the BHRF-1 protein of Epstein Barr virus (31) and CED-9 of the nematode *Caenorhabditis elegans* (68, 69). Transgenic mice that overexpress Bcl-2 in the B-cell lineage exhibit prolonged survival of responsive B cells and they develop an autoimmune disease resembling systemic lupus erythematosus (143, 187). In Bcl-2–deficient mice, the immune system initially appears to develop normally, but with time there is massive apoptosis in the spleen and thymus along with development of polycystic kidneys and hypopigmentation (90, 136, 137, 218). Although Bcl-2 is also abundantly expressed in the nervous system during development, no neurologic abnormalities have been detected in Bcl-2–deficient mice. Bcl-2 is a member of a growing family of related genes, including *bcl-x*, *bcl-w*, and *mcl*-1 (13, 57, 99, 101, 142), which may substitute or compensate for the lack of Bcl-2. *bcl-w* is predominantly expressed in the brain. Mice that are Bcl-x–deficient die during embryogenesis with severe apoptosis of neurons and hematopoietic cells (132).

Although some members of the Bcl-2 family protect cells from cell death, other Bcl-2–related proteins, such as Bax and Bak, promote cell death, either by antagonizing the function of Bcl-2 and Bcl-x$_L$ through heterodimerization (30, 95, 146, 238) or by an independent mechanism. Neurons of Bax knock-out mice are resistant to apoptosis following growth factor withdrawal in vitro or axotomy in vivo, and developmental death of neurons is reduced (39). Thus, resistance to cell death can be produced either by a gain of anti-death activity (e.g. overexpression of Bcl-2) or by downregulation of death-inducers (e.g. decreased expression of Bax). However, the boundary between pro-survival and pro-death members of this family has become blurred, as both Bax and Bak have been reported to have protective effects in some types of cells, including neurons (94, 127).

The domains of Bcl-2 and Bcl-x$_L$ that mediate heterodimerization with Bax and Bak are short 15– to 20–amino acid regions that are highly conserved among Bcl-2 family members, designated BH1 and BH2 (Bcl-2 homology domains) (235). BH1 and BH2 consist of helix-loop-helix motifs that are in close proximity in the three-dimensional structure of the molecule (133). The domain of Bax that is required for heterodimerization with Bcl-2 and Bcl-x$_L$ is a third, more N-terminal homology domain, BH3, which is also in close proximity to BH1 (16, 29, 133). A 46–amino acid peptide containing the BH3 homology domain of Bak is not only sufficient to bind Bcl-x$_L$ but also is sufficient to induce cell death. Others have reported that the BH3 domain may not be required for the death-inducing activity of Bax, suggesting that Bax may also kill cells by an alternate mechanism (180).

The mechanism by which Bcl-2 and other survival-promoting members of the family block apoptosis is not known. Because these proteins form heterodimers with the death-promoting members of the Bcl-2 family such as Bax and Bak, Bcl-2 may sequester these proteins, thereby blocking their death-promoting activity (62, 146, 234, 238). Support for this hypothesis comes from the observation that Bax alone induces cell death (62, 146). In this scenario, the death-promoting members trigger the downstream events that lead to cell death, and the only role of the protective members of the Bcl-2 family is to prevent activation of the death pathway.

An alternative hypothesis suggests that protective Bcl-2–related proteins are capable of blocking cell death by a mechanism other than inhibiting Bax and Bak. This hypothesis suggests that Bcl-2 and Bcl-x$_L$ mediate downstream events to promote cell survival. Support for this hypothesis comes from the observation that mutations in the anti-apoptotic proteins Bcl-x$_L$ and Bcl-2 that render them unable to bind to Bax and Bak do not significantly impair their ability to block apoptosis, which suggests that Bcl-2 and Bcl-x$_L$ can protect cells by a mechanism other than the inhibition of Bax and Bak (23). Thus, a burgeoning conglomerate hypothesis is that Bcl-2 and Bax each have independent mechanisms of blocking or inducing cell death, but that each can regulate the other's activity through heterodimerization. Other proteins are likely to be involved in regulating the function of Bcl-2. Dimer formation between Bcl-2 and Bad, a more distant member of the family, also inhibits the anti-death activity of Bcl-2 and is regulated by the phosphorylation state of Bad (55, 234, 239). Bcl-2 also can recruit Raf-1 kinase to the mitochondrial membrane, but the relevant targets of Raf-1 kinase activity are not known (222).

The molecular mechanisms explaining the pro- and anti-apoptotic effects of Bcl-2 and Bax remain elusive. The localization of Bcl-2 primarily to the outer mitochondrial membrane has focused much attention on this organelle (102, 139). Hence, Bcl-2 was proposed to function in an antioxidant pathway to block apoptosis (73). However, cells grown under anaerobic conditions, where production of reactive oxygen species is unlikely to occur, still die by apoptosis that can be blocked by Bcl-2 (86, 178). In addition, cells that lack mitochondrial DNA and, therefore, lack a functional respiratory chain, still die by apoptosis that can be inhibited by Bcl-2 (85).

Bcl-2 has also been postulated to regulate intracellular calcium. This is an attractive hypothesis, because a Ca^{2+}-dependent endonuclease is often involved in DNA laddering, and calcium ionophores can induce apoptosis (46, 123, 125, 126, 149, 150). Death stimuli, such as interleukin (IL)-3 withdrawal and glucocorticoids, have been reported to deplete calcium stores in the ER, which is blocked by Bcl-2 (42, 104). However, cell death can occur without mobilization of calcium.

Recently, attention has been refocused on the mitochondria to provide clues about the function of Bcl-2. Bcl-2 has been shown to preserve mitochondrial transmembrane potential ($\Delta\psi$) that is lost early in the death process (116, 177). It has been speculated that Bcl-2 may accomplish this task, or that Bax may facilitate loss of mitochondrial membrane potential, by forming pores in the membrane similar to those formed by bacterial colicins. The structure of Bcl-x_L was recently reported and was found to have striking similarity to the B fragment of diphtheria toxin that has ion channel activity and facilitates translocation of the toxin A fragment across membranes. The colicins and the δ-endotoxins of *Bacillus thuringiensis* are believed to exert their cytocidal activities by destroying membrane potential through pore formation. Like these toxins, Bcl-x_L is composed of two central alpha helices (perhaps spanning the membrane), surrounded by five amphipathic alpha helices that potentially shield the hydrophobic core until it is loaded into the membrane (118, 133, 199).

ICE-Family Proteases (Caspases)

IL-1β converting enzyme (ICE) was first identified as the protease responsible for processing the precursor form of IL-1β to its active secreted form (21, 22, 196). A role in apoptosis was suggested when it was recognized that ICE shares significant amino acid sequence homology to CED-3 that is encoded

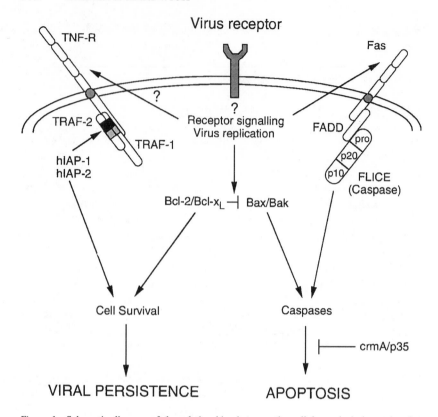

Figure 1 Schematic diagram of the relationships between the cellular and viral proteins that modulate apoptosis in virus infection.

by the nematode *C. elegans* (129, 236). In *C. elegans*, CED-3 has pro-apoptotic activity that is inhibited by CED-9, a homolog of mammalian Bcl-2 (184, 237). Mutations in CED-3 yield adult worms with an extra 131 somatic cells that normally die by programmed cell death during development (43). ICE is a member of a growing family of mammalian ICE-like proteases, which include ICH-1, ICH-2, CPP32, and FLICE, as well as CED-3 (14, 44, 51, 103, 223). These proteins are now known as caspases 1–10 (4). A role for caspases in apoptosis is supported by the ability of the cowpox virus ICE inhibitor, crmA, to block apoptosis that is induced by a variety of death stimuli in cultured cells (54, 128, 157, 195). The baculovirus protein p35 that prevents apoptosis in insect cells is also an inhibitor of caspases, but it is unrelated to crmA (9, 19, 32) (Figure 1). The role of ICE itself in cell death is controversial and other caspases, such as CPP32 (caspase-3) or FLICE/MACH (caspase-8), are more closely linked to apoptosis (14, 38, 134, 140, 232).

iap Proteins

Baculoviruses that do not encode a p35 protein to inhibit caspases have another gene with anti-apoptotic activity, that is, iap (inhibitors of apoptosis) proteins (11, 37). The baculovirus iap proteins all share common amino acid sequence motifs, including a RING finger and two copies of a 60– to 70–amino acid motif called baculovirus iap repeat (BIR). The RING finger ($Cys_3HisCys_4$) motif found near the C terminus in iap is also found in many other proteins and it binds two atoms of zinc, but its function is unknown for any protein (15, 34). The BIRs are located in the N-terminal half of the iap proteins, separated from the RING finger by a short spacer. Each BIR repeat contains a predicted zinc finger ($CysX_2CysX_{16}HisX_6Cys$), as well as other highly conserved residues. The BIR motif is rare, but has been identified in two other viral proteins: the 27kD pA224L protein of African swine fever virus, which also encodes a *bcl-2* homolog (3, 18, 138), and a protein of Chilo iridescent virus (171).

Several mammalian homologs of the baculovirus *iap* genes, designated hIAP-1, hIAP-2, hILP, and two Drosophila homologs (dIAP1, dIAP2), have been identified by independent approaches (45, 65, 115, 212). Although there is still some controversy about whether hIAP-2 possesses anti-apoptotic activity, there is general agreement that hIAP-1, hILP, and both Drosophila homologs are capable of blocking cell death (45, 65, 115). Baculovirus iap can inhibit caspase-induced apoptosis in mammalian cells, which indicates phylogenetic conservation of function (64).

A more distantly related iap-like protein, NAIP, was identified as one of two genes associated with spinal muscular atrophy, a motor neuron disease that results from the loss of spinal cord neurons, presumably due to excessive programmed cell death (166). A deletion of NAIP exon 5 was also recently identified in a case of sporadic amyotrophic lateral sclerosis (ALS), a progressive neurodegenerative disorder affecting motor neurons (84). NAIP contains three N-terminal BIR repeats, but lacks a RING finger and is approximately twice the size of other iap family members. The region of NAIP that shares extensive homology (30% identity) with baculovirus iap includes the region that is deleted in spinal muscular atrophy and sporadic ALS. NAIP appears to be expressed in motor but not in sensory neurons (115). The high degree of sequence conservation between the evolutionarily divergent insect virus *iaps* and the cellular iap-related proteins indicates that they are part of a conserved anti-apoptotic pathway.

Goeddel and associates biochemically purified hIAP-1 and hIAP-2 by virtue of their interaction with a complex of proteins bound to the cytoplasmic tail of tumor necrosis factor receptor-2 (TNF-R2) (162). Fas, TNF-R1, and TNF-R2 are members of a family of receptor molecules that includes the low affinity nerve growth factor receptor (p75 NGF-R), CD30, and CD40 (81, 182). The iap homolog in Drosophila was identified because a deletion of the gene enhanced the death-inducing activity of Reaper, a pro-apoptic Drosophila protein (65). Therefore, it appears that *iap* may function at a step between a cell surface receptor and activation of the caspases, and it may be directly involved in regulating activation of these proteases (Figure 1).

The TNF Receptor Family

TNF is a cytokine produced by activated macrophages that elicits many different biological responses. Cellular responses to TNF are mediated by two different receptors, 55–60 kDa TNF-R1 and 75–80 kDa TNF-R2 (66, 113, 117, 170, 192, 214). The TNF superfamily of receptors shares repeating cysteine-rich repeats in their extracellular domains (182). When engaged by their respective ligands, Fas mediates cell death, whereas the TNF receptors can mediate either cell proliferation or cell death (10,192–194, 200). The cytoplasmic domains of TNF-R1 and Fas (193, 194) contain a sequence termed the "death domain," because overexpression of this domain alone is sufficient to kill cells (230). This death domain appears to function as a multimerization/protein-protein interaction domain. Oligomerization of TNF-R1 via its death domain occurs upon binding to trimeric TNF. Following TNF engagement, the death domain on TNF-R1 associates with the death domain on an intracellular protein called TNF-R–associated death domain (TRADD) (76). TRADD can associate with TNF-R–associated factor (TRAF) 2 or Fas-associating protein with death domain (FADD). Depending on which of these factors is bound to the complex, TNF-R1 can stimulate either NF-kB activation (TRAF2) or it can induce cell death (FADD) (28, 75). Thus, the effects of TNF appear to be modulated by the composition of a protein complex bound to the cytoplasmic tail of the TNF receptors. TRAF2 also mediates activation of NFκB by TNF-R2 and CD40 (164, 189), and this interaction can be inhibited by A20, a zinc finger protein induced by TNF (183).

The death domain of Fas also binds FADD to activate the death pathway (26, 28). In turn, the "death effector domain" of FADD was recently demonstrated to bind to the prodomain of the caspase FLICE-1 (Figure 1) (14, 134). This prodomain targets FLICE-1 to the cytoplasmic domain of Fas via FADD (14, 134). The N-terminal prodomain must be cleaved from the protease domain prior to activation of the protease. The rapid induction of apoptosis following Fas receptor ligation may in part be due to the very short pathway between ligand binding and caspase activation (48).

TRAF1 and TRAF2 were originally identified because they associate with the cytoplasmic tail of TNF-R2 (163). Two additional proteins present in the TNF-R2 complexes with TRAF1 and TRAF2 were identified as hIAP-1 and hIAP-2, which appear to be tethered to TNF-R2 via TRAF1 and TRAF2 (162, 179) (Figure 1). The three iap BIR repeats are sufficient for interaction with TRAF1 and TRAF2, but the RING finger domain is unable to bind to the TRAFs (162). Deletion of the C-terminal RING finger enhances the anti-death function of dIAP-1 when expressed in the Drosophila eye (65). The BIR domain may be responsible for both TRAF interactions and for anti-death activity, whereas the RING finger may modulate the function of the BIRs or it may have an independent pro-death function. Thus, it appears that the cellular iap proteins may be able to modulate a receptor-mediated signal through their interaction with a complex of proteins that are bound to the cytoplasmic domains of cytokine receptors.

INDUCTION OF PROGRAMMED CELL DEATH BY VIRUSES

Recently, it has been recognized that many viruses kill infected cells by inducing apoptosis. For instance, representatives of the herpesvirus (79, 92, 201), adenovirus (156), poxvirus (17), baculovirus (32, 153), parvovirus (20, 131), retrovirus (1, 105, 233), rhabdovirus (98), paramyxovirus (49, 201), orthomyxovirus (70, 191), reovirus (207), bunyavirus (151), picornavirus (87, 198), and alphavirus (110) families are capable of inducing apoptosis in one cultured cell type or another. A wide variety of genes that either induce or prevent cell death are encoded by viruses for the purpose of modulating cellular functions. Necrotic morphologies are characteristic of some virus-infected cells that lyse to release newly assembled virus particles (197). Viruses of this type often are capable of triggering apoptosis, but the apoptotic pathway is blocked by virus-encoded genes that prolong cell survival (32, 67, 156, 228, 229).

It has been suggested that apoptosis is a protective host response for eliminating virus-infected cells (216). That is, prior to the development of a cellular or a humoral immune response, an infected cell commits suicide in response to virus infection, in order to prevent the virus from completing its replication cycle and producing progeny virus that will spread the infection to adjacent cells. This hypothesis is supported by the finding that many viruses encode genes that inhibit the apoptotic pathway. Adenoviruses and baculoviruses with defective anti-apoptotic genes are severely impaired for progeny virus production, apparently because the cell dies prematurely from apoptosis (33). Thus, the ability to inhibit programmed cell death allows the virus to complete its replication cycle before the cell expires. However, other viruses appear to thrive in apoptotic cells (110). Virus-induced apoptosis also can be harmful to the host when the virus infects a population of nonrenewable cells.

There are several potential mechanisms by which viruses activate the cell death pathway. Some viruses do so through the direct action of a specific viral protein. For example, adenovirus E1A alone is sufficient to induce apoptotic cell death (156). E1A stabilizes p53, a transcriptional activator important for cell cycle arrest, which can upregulate the expression of the pro-apoptotic proteins Fas and Bax (120). The VP3/apoptin protein of chicken anemia virus localizes to the nucleus and can induce apoptosis independent of other viral proteins (141). The IE-1 transcription factor encoded by baculoviruses is sufficient to induce apoptosis in insect cells (153).

The $\sigma 1$ protein is responsible for reovirus binding to the cell surface. For the T3D strain, binding appears to be sufficient to cause the rapid induction of apoptotic cell death in murine L-929 cells (207). Retrovirus-encoded proteins, including Tat and gp120/41 of HIV (2, 105) and the HTLV-1 Tax protein (233), induce apoptotic cell death. These effects may be mediated in part by upregulation of Fas ligand (FasL) and/or sensitization of infected cells to TNF-mediated apoptosis (96, 227).

Fas (CD95) is abundantly expressed on activated mature lymphocytes and on lymphocytes infected with HTLV-1, HIV, and Epstein-Barr virus (EBV) (97, 135). Fas is a member of the TNF receptor family that activates apoptotic cell death following interaction with its ligand (FasL) or with anti-Fas antibodies (81, 145, 148, 173). Like TNF, Fas appears to transmit a death signal through the FADD protein (26, 28) that binds the prodomain of a newly identified caspase called FLICE-1 (caspase-8).

Influenza virus infection induces the expression of Fas, and Fas-mediated cell death has been suggested to be an important mechanism of influenza virus killing (190, 191, 219) (Figure 1). HIV-1 Tat protein upregulates FasL expression in T cells and sensitizes T cells to apoptosis that is triggered by T-cell receptor engagement or binding of HIV-1 gp120 to CD4 (227). Peripheral-blood T cells from HIV-infected individuals are very sensitive to Fas-mediated apoptosis (91). Fas-mediated cell death has also been implicated in viral hepatitis (71).

Viruses may also induce apoptosis indirectly through their generalized effects on cellular processes. Baculovirus-mediated cessation of cellular RNA transcription may be a stimulus for apoptosis in baculovirus-infected SF-21 insect cells (34). The shutoff of host protein synthesis has been postulated as a mechanism by which polioviruses activate the death pathway in HeLa human cervical carcinoma cells (198). The observation that uninfected SF-21 and HeLa cells undergo apoptosis following treatment with metabolic inhibitors, such as actinomycin D, supports the hypothesis that viruses could trigger apoptosis in this manner (34, 198). The implication from these studies is that these cells require a labile protective protein to avoid activating the death pathway. In contrast to SF-21 and HeLa cells, neurons, baby hamster kidney (BHK), and a number of other cells are protected from induction of apoptosis by metabolic inhibitors, which suggests that these cells require the expression of new genes to facilitate the death pathway (124). Genes that are activated during neuronal cell death are only now beginning to be identified (50, 52, 114, 224).

SINDBIS VIRUS–INDUCED APOPTOSIS AND ITS ROLE IN FATAL ENCEPHALITIS

SV-Induced Apoptosis

The alphavirus Sindbis is generally propagated and assayed in cultured baby hamster kidney (BHK) cells. SV plaques readily in these cells, replication is rapid, and beginning 4 h after infection, large amounts of virus are released into the supernatant fluid. Beginning 20–24 h after infection, infected cells fail to exclude trypan blue; and essentially all cells are dead, by this criterion, at 48 h after infection (Figure 2). Observation of these cells by videomicroscopy shows evidence of active blebbing of the cytoplasm of infected cells within 24 h after infection, which suggests that SV causes cell death by induction of apoptosis rather than by direct damage to the plasma membrane (110). Evidence for apoptosis being the mechanism of SV-induced cell death includes the condensation of nuclear chromatin and endonucleolytic cleavage of chromosomal DNA into 180 bp repeats, as well as caspase activation (110, 209).

Changes characteristic of apoptosis are also seen after in vitro SV infection of AT-3 rat prostatic adenocarcinoma cells (110), PC12 rat pheochromocytoma cells (88), N18 mouse neuroblastoma cells (110), and HeLa cells (161). Thus, apoptosis is likely to be the mechanism by which SV causes cell death in most mammalian cells. Studies of N18 cells, in which virus production begins 5 h after infection, have shown that flow cytometry can detect changes in nuclear

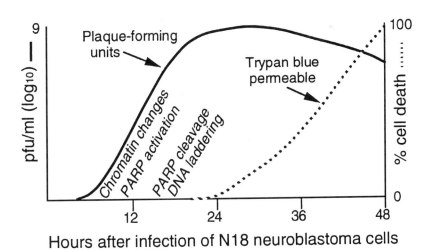

Figure 2 Schematic diagram of the events related to apoptosis that occur in Sindbis virus–infected N18 mouse neuroblastoma cells. PARP = Poly (ADP ribose) polymerase.

chromatin by 6.5 h (209). Activation of the DNA repair enzyme poly (ADP ribose) polymerase (PARP) can be detected by increased consumption of NAD by 10 h. PARP is also a substrate for caspase-3 (CPP32)-mediated proteolytic cleavage. Cleavage of PARP is evident by 16 h, i.e at approximately the same time that DNA fragmentation becomes apparent. At this time early CPE is evident, but the cells are still able to exclude trypan blue (209).

The small surface blebs of SV-infected HeLa cells contain viral structural proteins, ribosomes, fragmented ER, and the RNA-binding protein and autoantigen Ro. Cytopathic vacuoles and budding virus can be seen in association with these blebs (161). It has been proposed that the association of cellular and viral proteins in apoptotic cell fragments may result in autoimmune responses to specific host proteins that are presented in the context of foreign viral proteins during apoptosis (161).

In cycling N18 neuroblastoma cells, SV-induced apoptosis results in a loss of cells primarily from the Go/G_1 population, which suggests that the cells may be particularly susceptible to SV-induced apoptosis during the G1 phase of the cell cycle (209). This is further suggested by studies showing that PC12 cells are less susceptible to SV-induced apoptosis when cell proliferation is inhibited by the expression of a dominant inhibitory form of Ras, a GTP-binding protein that is important for regulation of cell proliferation and neuronal differentiation (88). Because neurons do not proliferate, it is possible that altered expression of cell cycle regulatory proteins may participate in SV-induced apoptosis in these terminally differentiated cells.

A few studies have addressed the role of early cell signalling events. Ras-dependent signalling appears to be important for induction of apoptosis by SV in PC12 cells (88). Although sustained increases in intracellular Ca^{2+} induce apoptosis in some cells (123), SV infection of N18 cells does not lead to an increase in intracellular Ca^{++}, and infection of these cells in the absence of extracellular Ca^{++} and Mg^{++} results in a more rapid progression of the apoptotic process (209). Thus, signalling pathways that rely on increased intracellular Ca^{++} do not appear to play a role in SV-induced apoptosis of N18 cells.

However, the mechanism by which SV induces apoptosis is likely to differ from one mammalian cell type to another. For instance, treatment with the thiol antioxidant, N-acetylcysteine, prevents SV-induced apoptosis in both AT-3 and N18 cells. However, this effect is associated with inhibition of the activation of transcription factor NFκB in AT-3 cells, but not in N18 cells (114). Treatment of cells with oligonucleotide decoys that bind NFκB and prevent activation of downstream genes protects AT-3 cells from SV-induced apoptosis, which indicates that NFκB plays a role in the death pathway activated by SV in this cell type. Because SV effectively shuts off host protein synthesis (188), the requirement for new gene expression to activate the death pathway through the action of NFκB seems somewhat contradictory, but this induction may occur early in the replication cycle. Alternatively, some cellular transcripts may escape the inhibitory effects of the virus. Although NFκB activation has been generally considered to be a cell proliferation signal rather than a cell death signal, the induction of cell proliferation by a variety of proteins, including adenovirus E1A and the cellular transcription factor E2F, can also lead to cell death (154, 156). The identification of NFκB sites in the ICE protease promoter, and the demonstration that the p53 and TNF promoters are activated by NFκB are consistent with a role for NFκB in transcription-dependent induction of cell death (231). However, in other cell types and other death paradigms, NFκB has anti-death activity (8, 213, 221).

Few studies have addressed the question of whether expression of a single SV protein is sufficient to induce apoptosis. Cell death occurs late after infection if the nonstructural proteins alone are synthesized, even though the shutoff of host protein synthesis is efficient (53). It is not clear if the cell death observed here is apoptotic. Characteristic rapid cell death requires synthesis and transport of the viral envelope glycoproteins to the cell surface (53, 211), which suggests that expression of one or more of these proteins is required for induction of apoptosis.

One amino acid change of particular importance for virulence is the change from glutamine to histidine at position 55 of the E2 glycoprotein. Recombinant viruses differing only at this single position have been used for in vitro and in vivo infections, and they have demonstrated the importance of His at E2-55 for age-dependent virulence (206). Viruses with His at E2-55 can overcome the inhibition to apoptosis that is present in AT-3 cells expressing Bcl-2 (210). Neurovirulent strains also induce abundant apoptosis in vivo, as compared with less virulent strains (112).

Cellular Inhibitors of Apoptosis Modulate SV-Induced Cell Death

The molecular mechanisms underlying the effects of maturation of the host on susceptibility to SV or to a number of other viruses are not known. The maturity of the immune system does not appear to be the determining factor for SV (58). A putative SV receptor is more abundant in mouse brains at 16 h than at 96 h after birth, so the efficiency of binding to this vanishing receptor could play a role (208). Another possibility is that anti- and pro-apoptotic genes that are regulated during maturation modulate the outcome of infection (63, 94, 100, 110). Cellular genes that protect neurons from undergoing apoptosis to a variety of stimuli could block SV-induced apoptosis as well. The anti-apoptotic Bcl-2 and Bcl-x$_L$ proteins are normally expressed in neurons of the brain and spinal cord, and they are therefore candidates for this role

(106, 108). Alternatively, downregulation of Bax or a Bax-like protein would be expected to have protective effects.

A potential role for cellular anti-apoptotic genes in the suppression of virus-induced apoptosis was explored by infecting a cell line that stably expressed the Bcl-2 protein. AT-3 cells overexpressing Bcl-2 were protected from cell death following infection with SV, although the control AT-3 cells all died (110). Furthermore, SV replication was suppressed in AT3Bcl-2 cells, which suggests that cellular changes associated with apoptosis may favor SV replication (210). These results suggest the possibility that inhibitors of apoptotic cell death can alter the outcome of virus infection in animals as well as in individual cells.

To test the possibility that other cellular genes modify SV-induced apoptosis, a SV vector with a second subgenomic promoter was generated in which a variety of genes could be inserted into the viral genome. This assures that cellular proteins of interest will be expressed at high levels in the infected cell, permitting their effects on SV-induced apoptosis to be assessed rapidly in a wide variety of cell types and in a mouse model. Recombinant viruses carrying a copy of the *bcl-2* gene or *bcl-x*$_L$ gene under the control of an SV subgenomic promoter are impaired in their ability to kill cells. In contrast, control viruses with a stop codon in the *bcl-2* or *bcl-x* open reading frame, or viruses in which the *bcl-2* or *bcl-x* gene was inserted in reverse orientation, kill cells efficiently. Recombinant viruses expressing the death-promoting *bax* gene kill cells more rapidly than do viruses without this gene (23). These analyses have now been extended to include a wide variety of viral and cellular genes that regulate cell death, including iap-related proteins, ICE family protease inhibitors, and regulators of the cell cycle (45). All of these proteins affect the ability of SV to induce apoptosis (Nava, Friesen, Hardwick, Nargi, Oyler & Griffin, unpublished data). Furthermore, this system permits rapid analysis of a wide variety of wild-type and mutant genes, without the necessity of generating stable cell lines.

SV-Induced Apoptosis Correlates with Neurovirulence

SV causes encephalomyelitis in mice. The ability of SV to induce apoptotic cell death of infected neurons in mouse brains and spinal cords was confirmed by in situ TUNEL assays on mouse tissues, and by the detection of DNA fragmentation in extracts of infected mouse brains (112). To determine if Bcl-2 could protect mice from a lethal SV infection, animals were infected with an SV vector carrying the *bcl-2* gene. Those mice expressing the human Bcl-2 protein in neurons were protected from fatal encephalitis and SV replication was suppressed (106). Induction of apoptosis by SV in neurons of mouse brains and spinal cords correlates with mortality (112).

Strains of SV vary in their ability to cause fatal infection in mice of different ages. Avirulent strains (for example, HRSP) cause fatal encephalitis only in newborn mice (175), whereas the most virulent SV strains (for example, neuroadapted SV) cause fatal encephalitis in 4- to 6-week-old mice (59). Determinants of virulence have been mapped primarily to the E1 and E2 glycoproteins, although there are clearly regions of the nonstructural portion of the genome that also influence virulence (122). Extensive sequence analysis and construction of recombinant viruses have identified several specific determinants of neurovirulence (203, 205, 206). A particularly potent mutation is the change from a glutamine to a histidine at position 55 of the Sindbis virus E2 glycoprotein (206).

Both virulent (E2 His55) and avirulent (E2 Gln55) strains of SV replicate efficiently in newborn mice and cause neuronal apoptosis, resulting in 100% mortality. However, by the time mice reach two weeks of age, the avirulent strain has a reduced ability to replicate and has a low mortality rate compared to the neurovirulent strain (112, 206). Therefore, by definition, host cell factors are responsible for protecting older mice from avirulent viruses.

A similar phenomenon can be observed in cell culture. That is, both the virulent and avirulent strains readily kill freshly explanted cultures of primary dorsal root ganglia neurons. As the neurons mature in culture over the next two to six weeks, they become increasingly resistant to SV-induced cell death when infected with the avirulent strain (E2 Gln55) (110). Just as both virus strains replicate efficiently in newborn mouse brains, these viruses are indistinguishable by many criteria when grown in BHK or AT-3 cells (203, 210). These results suggest that there is an age-dependent factor(s) in mouse brains that is responsible for suppressing virus-induced apoptosis and/or replication of avirulent viruses, which protects mice from a lethal infection. In vitro, E2 His55 viruses are able to overcome Bcl-2–mediated inhibition of virus replication and to induce apoptosis in AT3– or Bcl-2 cells (210).

Thus, it appears that viral determinants allow neurovirulent strains to replicate with increased efficiency in the face of cellular inhibitors of apoptosis, which leads to the induction of apoptosis. However, the mechanism by which a single amino acid mutation in an SV structural protein modulates virus replication in mouse nervous tissues and alters neurovirulence is not known. The accumulating evidence suggests that E2 His55 has little effect on the ability of virus to bind to cells, but that it has a significant effect on an early step in viral replication in neuroblastoma cells (although not in BHK cells) (Dropulic, Tucker, Hardwick & Griffin, unpublished data).

Virus-Induced Apoptosis Versus Viral Persistence

Although SV causes apoptosis in most cultured mammalian cell lines, lines of persistently infected cells have been derived (78, 110, 225), and persistent infection is routinely established in mosquito cells in vivo and in vitro (77, 152, 185). Alphaviruses are arthropod-borne viruses that, in nature, cycle between their invertebrate mosquito and vertebrate avian or mammalian hosts. Persistent infection is typically established in mosquitoes. Studies on *Aedes albopictus* mosquito cells in culture have shown that persistent infection is easily established with a number of alphaviruses (77, 78). Viral replication is typically high early after infection and then is maintained at a moderate level (77, 159). This change is associated with production of a non-interferon antiviral peptide that suppresses SV replication (35, 160). Continuous production of virus is necessary to maintain persistent infection (77, 159). Defective interfering particles may also be produced (77).

In BHK cells, the establishment of a persistent infection was linked to the selection of a temperature-sensitive virus and the formation of defective interfering particles (225). In murine cells, persistence was linked both to mutant autointerfering particles and to in vitro production of interferon (78). Therefore, in these systems, mutations in the infecting viruses that affect replication appear to be primarily responsible for the ability to establish persistence.

More recently, persistent infections have been established in cultured cells expressing inhibitors of apoptosis. AT-3 cells overexpressing Bcl-2 can be persistently infected with a strain of SV that kills AT-3 cells that are not overexpressing Bcl-2 (110). Persistence is associated with restricted replication (210). Likewise, persistent infection can be established in mature but not in immature cultured neurons, presumably because neurons express increasing amounts of cellular inhibitors of apoptosis or decreasing amounts of pro-apoptotic factors as they mature (110). Bcl-2, Bcl- x_L, and Bax are three such modulators (40, 165), but it is likely that others will be defined as changes associated with neuronal maturation are further characterized.

Persistent infection, with continued production of infectious virus, is not usually established in vertebrate hosts, because the immune system limits virus replication. However, as with primary neuronal cultures, persistent infection can be established in the neurons of the brain and spinal cord of mature immunodeficient mice (109). Amino acid sequence changes in the E2 glycoprotein affect the ability of SV to establish persistent infection (111). Antibody controls virus replication in neurons without eliminating the infected cells. This nonlytic mechanism of control of in vivo infection results in persistence of viral RNA in the CNS for the life of the host (107, 108). This is the direct result of SV infecting mature neurons without inducing apoptosis, leading to a persistent productive infection that is controlled by the nonlytic effect of antibody. Thus, antibody functions to preserve a nonrenewable cell, but persistence of viral RNA with the potential for reactivation of viral replication, even if rare, could be the source of progressive late disease (107).

The ability of cellular anti-apoptotic genes to block virus-induced cell death supports the attractive hypothesis that failure to activate the death pathway following virus infection leads to persistent infection (110). Such a mechanism could explain persistence of the RNA of avirulent SV in neurons of the brains of infected mice for extended periods of time (107).

> Visit the *Annual Reviews home page* at
> http://www.annurev.org.

Literature Cited

1. Adamson DC, Dawson TM, Zink MC, Clements JE, Dawson VL. 1996. Neurovirulent simian immunodeficiency virus infection induces neuronal, endothelial, and glial apoptosis. *Mol. Med.* 2:417–28

2. Adamson DC, Wildemann B, Sasaki M, Glass JD, McArthur JC et al. 1996. Immunologic NO synthase: elevation in severe AIDS dementia and induction by HIV-1 gp41. *Science* 274:1917–21

3. Afonso CL, Neilan JG, Kutish GF, Rock DL. 1996. An african swine fever virus *bcl-2* homolog, 5-HL, suppresses apoptotic cell death. *J. Virol.* 70:4858–63

4. Alnemri ES, Livingston DJ, Nicholson DW, Salvesen G, Thornberry NA, et al. 1996. Human ICE/CED-3 protease nomenclature. *Cell* 87:171

5. Deleted in proof

6. Arends MJ, Morris RG, Wyllie AH. 1990. Apoptosis. The role of the endonuclease. *Am. J. Pathol.* 136:593–608

7. Baffy G, Miyashita T, Williamson JR, Reed JC. 1993. Apoptosis induced by withdrawal of interleukin-3 (IL-3) from an IL-3-dependent hematopoietic cell line is associated with repartitioning of intracellular calcium and is blocked by enforced bcl-2 oncoprotein production. *J. Biol. Chem.* 268:6511–19

8. Beg AA, Baltimore D. 1996. An essential role for NF-kappaB in preventing TNF-induced cell death. *Science* 274:782–84

9. Bertin J, Mendrysa SM, LaCount DJ, Gaur S, Krebs JF, et al. 1996. Apoptotic suppression by baculovirus P35 involves cleavage by and inhibition of a virus-induced CED-3/ICE-like protease. *J. Virol.* 70:6251–59

10. Beyaert R, Vanhaesebroeck B, Declercq W, Vanlint J, Vandenabeele P, et al. 1995. Casein kinase-1 phosphorylates the p75 tumor necrosis factor receptor and negatively regulates tumor necrosis factor signaling for apoptosis. *J. Biol. Chem.* 270:23293–99

11. Birnbaum MJ, Clem RJ, Miller LK. 1994. An apoptosis-inhibiting gene from a nuclear polyhedrosis virus encoding a polypeptide with Cys/His sequence motifs. *J. Virol.* 68:2521–28

12. Bissonnette RP, Echeverri F, Mahboubi A, Green DR. 1992. Apoptotic cell death induced by c-myc is inhibited by bcl-2. *Nature* 359:552–54

13. Boise LH, Gonzalez-Garcia M, Postema CE, Ding LY, Lindsten T, et al. 1993. Bcl-x, a bcl-2-related gene that functions as a dominant regulator of apoptotic cell death. *Cell* 74:597–608

14. Boldin MP, Goncharov TM, Goltsev YV, Wallach D. 1996. Involvement of MACH, a novel MORT1/FADD-interacting protease, in Fas/APO-1-and TNF receptor-induced cell death. *Cell* 85:803–15

15. Borden KLB, Boddy MN, Lally J, O'Reilly NJ, Martin S, et al. 1995. The solution structure of the RING finger domain from the acute promyelocytic leukaemia protooncoprotein PML. *EMBO J.* 14:1532–41

16. Boyd JM, Gallo GJ, Elangovan B, Houghton AB, Mastrom S, et al. 1995. Bik, a novel death-inducing protein shares a distinct sequence motif with Bcl-2 family proteins and interacts with viral and cellular survival-promoting proteins. *Oncogene* 11:1921–28

17. Brooks MA, Ali AN, Turner PC, Moyer RW. 1995. A rabbitpox virus serpin gene controls host range by inhibiting apoptosis in restrictive cells. *J. Virol.* 69:7688–98

18. Brun A, Rivas C, Esteban M, Escribano JM, Alonso C. 1996. African swine fever virus gene A179L, a viral homologue of *bcl-2*, protects cells from programmed cell death. *Virology* 225:227–30

19. Bump NJ, Hackett M, Hugunin M, Seshagiri S, Brady K, et al. 1995. Inhibition of

ICE family proteases by baculovirus anti-apoptotic protein p35. *Science* 269:1885–88

20. Caillet-Fauquet P, Perros M, Brandenburger A, Spegelaere P, Rommelaere J. 1990. Programmed killing of human cells by means of an inducible clone of parvoviral genes encoding non-structural proteins. *EMBO J.* 9:2989–95

21. Casano FJ, Rolando AM, Mudgett JS, Molineaux SM. 1996. The structure and complete nucleotide sequence of the murine gene encoding interleukin-1 beta converting enzyme (ICE). *Genomics* 20:474–81

22. Cerretti DP, Kozlosky CJ, Mosley B, Nelson N, Vanness K, et al. 1992. Molecular cloning of the interleukin-1β converting enzyme. *Science* 256:97–100

23. Cheng EH, Levine B, Boise LH, Thompson CB, Hardwick JM. 1996. Bax-independent inhibition of apoptosis by Bcl-xL. *Nature* 379:554–56

24. Chen-Levy Z, Cleary ML. 1990. Membrane topology of the Bcl-2 proto-oncogenic protein demonstrated in vitro. *J. Biol. Chem.* 265:4929–33

25. Chen-Levy Z, Nourse J, Cleary ML. 1989. The bcl-2 candidate proto-oncogene product is a 24-kilodalton integral-membrane protein highly expressed in lymphoid cell lines and lymphomas carrying the t(14;18) translocation. *Mol. Cell Biol.* 9:701–10

26. Chinnaiyan AM, O'Rourke K, Tewari M, Dixit VM. 1995. FADD, a novel death domain-containing protein, interacts with the death domain of Fas and initiates apoptosis. *Cell* 81:505–12

27. Chinnaiyan AM, Orth K, O'Rourke K, Duan HJ, Poirier GG, Dixit VM. 1996. Molecular ordering of the cell death pathway. Bcl-2 and Bcl-xL function upstream of the CED-3-like apoptotic proteases. *J. Biol. Chem.* 271:4573–76

28. Chinnaiyan AM, Tepper CG, Seldin MF, O'Rourke K, Kischfel FC, et al. 1996. FADD/MORT1 is a common mediator of CD95 (Fas/APO-1) and tumor necrosis factor receptor-induced apoptosis. *J. Biol. Chem.* 271:4961–65

29. Chittenden T, Flemington C, Houghton AB, Ebb RG, Gallo GJ, et al. 1995. A conserved domain in Bak, distinct from BH1 and BH2, mediates cell death and protein binding functions. *EMBO J.* 14:5589–96

30. Chittenden T, Harrington EA, O'Connor R, Flemington C, Lutz RJ, et al. 1995. Induction of apoptosis by the Bcl-2 homologue Bak. *Nature* 374:733–36

31. Cleary ML, Smith SD, Sklar J. 1986.

Cloning and structural analysis of cDNAs for bcl-2 and a hybrid bcl-2/immunoglobulin transcript resulting from the t(14;18) translocation. *Cell* 47:19–28

32. Clem RJ, Fechheimer M, Miller LK. 1991. Prevention of apoptosis by a baculovirus gene during infection of insect cells. *Science* 254:1388–90

33. Clem RJ, Miller LK. 1993. Apoptosis reduces both the in vitro replication and the in vivo infectivity of a baculovirus. *J. Virol.* 67:3730–38

34. Clem RJ, Miller LK. 1994. Control of programmed cell death by the baculovirus gene p35 and iap. *Mol. Cell Biol.* 14:5212–22

35. Condreay LD, Brown DT. 1986. Exclusion of superinfecting homologous virus by Sindbis virus-infected *Aedes albopictus* (mosquito) cells. *J. Virol.* 58:81–86

36. Cowan WM, Fawcett JW, O'Leary DD, Stanfield BB. 1984. Regressive events in neurogenesis. *Science* 225:1258–65

37. Crook NE, Clem RJ, Miller LK. 1993. An apoptosis-inhibiting baculovirus gene with a zinc finger-like motif. *J. Virol.* 67:2168–74

38. Darmon AJ, Nicholson DW, Bleackley RC. 1995. Activation of the apoptotic protease CPP32 by cytotoxic T-cell-derived granzyme-B. *Nature* 377:446–48

39. Deckwerth TL, Elliott JL, Knudson CM, Johnson EM, Snider WD, Korsmeyer SJ. 1996. Bax is required for neuronal death after trophic factor deprivation and during development. *Neuron* 17:401–11

40. de Jong D, Prins FA, Mason DY, Reed JC, van Ommen GB, Kluin PM. 1994. Subcellular localization of the bcl-2 protein in malignant and normal lymphoid cells. *Cancer Res.* 54:256–60

41. Deng G, Podack ER. 1993. Suppression of apoptosis in a cytotoxic T-cell line by interleukin 2-mediated gene transcription and deregulated expression of the protooncogene bcl-2. *Proc. Natl. Acad. Sci. USA* 90:2189–93

42. Distelhorst CW, Lam M, McCormick TS. 1996. Bcl-2 inhibits hydrogen peroxide-induced ER Ca^{2+} pool depletion. *Oncogene* 12:2051–55

43. Driscoll M. 1992. Molecular genetics of cell death in the nematode *Caenorhabditis elegans*. *J. Neurobiol.* 23:1327–51

44. Duan H, Orth K, Chinnaiyan AM, Poirier GG, Froehlich CJ, et al. 1996. ICE-LAP6, a novel member of the ICE/Ced-3 gene family, is activated by the cytotoxic T cell protease granzyme B. *J. Biol. Chem.* 271:16720–24

45. Duckett CS, Nava VE, Gedrich RW, Clem RJ, Van Dogen JL et al. 1996. A conserved family of cellular genes related to the baculovirus iap gene and encoding apoptosis inhibitors. *EMBO J.* 15:2685–94

46. Durant S, Homo F, Duval D. 1980. Calcium and A23187-induced cytolysis of mouse thymocytes. *Biochem. Biophys. Res. Commun.* 93:385–91

47. Duvall E, Wyllie AH. 1986. Death and the cell. *Immunol. Today* 7:115–19

48. Enari M, Talanian RV, Wong WW, Nagata S. 1996. Sequential activation of ICE-like and CPP32-like proteases during Fas-mediated apoptosis. *Nature* 380:723–26

49. Esolen LM, Park SW, Hardwick JM, Griffin DE. 1995. Apoptosis as a cause of death in measles-virus–infected cells. *J. Virol.* 69:3955–58

50. Estus S, Zaks WJ, Freeman RS, Gruda M, Bravo R, Johnson EM Jr. 1994. Altered gene expression in neurons during programmed cell death: identification of c-jun as necessary for neuronal apoptosis. *J. Cell Biol.* 127:1717–27

51. Faucheu C, Diu A, Chan AWE, Blanchet AM, Miossec C, et al. 1995. A novel human protease similar to the interleukin-1β converting enzyme induces apoptosis in transfected cells. *EMBO J.* 14:1914–22

52. Freeman RS, Estus S, Johnson EM Jr. 1994. Analysis of cell cycle-related gene expression in postmitotic neurons: selective induction of Cyclin D1 during programmed cell death. *Neuron* 12:343–55

53. Frolov I, Schlesinger S. 1994. Comparison of the effects of Sindbis virus and Sindbis virus replicons on host cell protein synthesis and cytopathogenicity in BHK cells. *J. Virol.* 68:1721–27

54. Gagliardini V, Fernandez P-A, Lee RKK, Drexler HCA, Rotello RJ, et al. 1994. Prevention of vertebrate neuronal death by the crmA gene. *Science* 263:826–28

55. Gajewski TF, Thompson CB. 1996. Apoptosis meets signal transduction: elimination of a BAD influence. *Cell* 87:589–92

56. Garry RF, Bishop JM, Park S, Westbrook K, Lewis G, Waite MRF. 1979. Na$^+$ and K$^+$ concentrations and the regulation of protein synthesis in Sindbis virus-infected chick cells. *Virology* 96:108–20

57. Gibson L, Holmgreen SP, Huang DCS, Bernard O, Copeland NG et al. 1996. Bcl-w, a novel member of the bcl-2 family, promotes cell survival. *Oncogene* 13:665–75

58. Griffin DE. 1976. Role of the immune response in age-dependent resistance of mice to encephalitis due to Sindbis virus. *J. Infect. Dis.* 133:456–64

59. Griffin DE, Johnson RT. 1977. Role of the immune response in recovery from Sindbis virus encephalitis in mice. *J. Immunol.* 118:1070–75

60. Hahn CS, Lustig S, Strauss EG, Strauss JH. 1988. Western equine encephalitis virus is a recombinant virus. *Proc. Natl. Acad. Sci. USA* 85:5997–6001

61. Hammar SP, Mottet NK. 1971. Tetrazolium salt and electron-microscopic studies of cellular degeneration and necrosis in the interdigital areas of the developing chick limb. *J. Cell Sci.* 8:229–51

62. Han J, Sabbatini P, Perez D, Rao L, Modha D, White E. 1996. The E1B 19K protein blocks apoptosis by interacting with and inhibiting the p53-inducible and death-promoting Bax protein. *Genes Dev.* 10:461–77

63. Hanada M, Krajewski S, Tanaka S, Cazalshatem D, Spengler BA, et al. 1993. Regulation of Bcl-2 oncoprotein levels with differentiation of human neuroblastoma cells. *Cancer Res.* 53:4978–86

64. Hawkins CJ, Uren AG, Hacker G, Medcalf RL, Vaux DL. 1996. Inhibition of interleukin 1β-converting enzyme-mediated apoptosis of mammalian cells by baculovirus IAP. *Proc. Natl. Acad. Sci. USA* 93:13786–90

65. Hay BA, Wassarman DA, Rubin GM. 1995. Drosophila homologs of baculovirus inhibitor of apoptosis proteins function to block cell death. *Cell* 83:1253–62

66. Heller RA, Song K, Fan N, Chang DJ. 1992. The p70 tumor necrosis factor receptor mediates cytotoxicity. *Cell* 70:47–56

67. Henderson S, Huen D, Rowe M, Dawson C, Johnson G, Rickinson A. 1993. Epstein-Barr virus-coded BHRF1 protein, a viral homologue of bcl-2, protects human B cells from programmed cell death. *Proc. Natl. Acad. Sci. USA* 90:8479–83

68. Hengartner MO, Ellis RE, Horvitz HR. 1992. *Caenorhabditis elegans* gene ced-9 protects cells from programmed cell death. *Nature* 356:494–99

69. Hengartner MO, Horvitz HR. 1994. *C. elegans* cell survival gene ced-9 encodes a functional homolog of the mammalian proto-oncogene bcl-2. *Cell* 76:665–76

70. Hinshaw VS, Olsen CW, Dybdahl-Sissoko N, Evans D. 1994. Apoptosis: a mechanism of cell killing by influenza A and B viruses. *J. Virol.* 68:3667–73

71. Hiramatsu N, Hayashi N, Katayama K, Mochizuki K, Kawanishi Y, et al. 1994.

Immunohistochemical detection of Fas antigen in liver tissue of patients with chronic hepatitis C. *Hepatology* 19:1354–59

72. Hockenbery D, Nunez G, Milliman C, Schreiber RD, Korsmeyer SJ. 1990. Bcl-2 is an inner mitochondrial membrane protein that blocks programmed cell death. *Nature* 348:334–36

73. Hockenbery DM, Oltvai ZN, Yin XM, Milliman CL, Korsmeyer SJ. 1993. Bcl-2 functions in an antioxidant pathway to prevent apoptosis. *Cell* 75:241–51

74. Hockenbery DM, Zutter M, Hickey W, Nahm M, Korsmeyer SJ. 1991. Bcl-2 protein is topographically restricted in tissues characterized by apoptotic cell death. *Proc. Natl. Acad. Sci. USA* 88:6961–65

75. Hsu H, Shu HB, Pan MG, Goeddel DV. 1996. TRADD-TRAF2 and TRADD-FADD interactions define two distinct TNF receptor 1 signal transduction pathways. *Cell* 84:299–308

76. Hsu H, Xiong J, Goeddel DV. 1995. The TNF receptor 1-associated protein TRADD signals cell death and NF-kB activation. *Cell* 81:495–504

77. Igarashi A, Koo R, Stollar V. 1977. Evolution and properties of *Aedes albopictus* cell cultures persistently infected with Sindbis virus. *Virology* 82:69–83

78. Inglot AD, Albin M, Chudzio T. 1973. Persistent infection of mouse cells with Sindbis virus: role of virulence of strains, auto-interfering particles and interferon. *J. Gen. Virol.* 20:105–10

79. Ishii HH, Gobe GC. 1993. Epstein-Barr virus infection is associated with increased apoptosis in untreated and phorbol ester-treated human Burkitt's lymphoma (AW-Ramos) cells. *Biochem. Biophys. Res. Commun.* 192:1415–23

80. Itoh N, Tsujimoto Y, Nagata S. 1993. Effect of bcl-2 on Fas antigen-mediated cell death. *J. Immunol.* 151:621–27

81. Itoh N, Yonehara S, Ishii A, Yonehara M, Mizushima S, et al. 1991. The polypeptide encoded by the cDNA for human cell surface antigen fas can mediate apoptosis. *Cell* 66:233–43

82. Jackson AC, Moench TR, Griffin DE. 1987. The pathogenesis of spinal cord involvement in the encephalomyelitis of mice caused by neuroadapted Sindbis virus infection. *Lab. Invest.* 56:418–23

83. Jackson AC, Moench TR, Trapp BD, Griffin DE. 1988. Basis of neurovirulence in Sindbis virus encephalomyelitis of mice. *Lab. Invest.* 58:503–9

84. Jackson M, Morrison KE, Al-Chalabi A, Bakker M, Leigh PN. 1996. Analysis of chromosome 5q13 genes in amyotrophic lateral sclerosis: homozygous NAIP deletion in a sporadic case. *Ann. Neurol.* 39:796–800

85. Jacobson MD, Burne JF, King MP, Miyashita T, Reed JC, Raff MC. 1993. Bcl-2 blocks apoptosis in cells lacking mitochondrial DNA. *Nature* 361:365–69

86. Jacobson MD, Raff MC. 1995. Programmed cell death and bcl-2 protection in very low oxygen. *Nature* 374:814–16

87. Jelachich ML, Lipton HL. 1996. Theiler's murine encephalomyelitis virus kills restrictive but not permissive cells by apoptosis. *J. Virol.* 70:6856–61

88. Joe AK, Ferrari G, Jiang HH, Liang XH, Levine B. 1996. Dominant inhibitory ras delays Sindbis virus-induced apoptosis in neuronal cells. *J. Virol.* 70:7744–51

89. Johnson RT, McFarland HF, Levy SE. 1972. Age-dependent resistance to viral encephalitis: Studies of infections due to Sindbis virus in mice. *J. Infect. Dis.* 125:257–62

90. Kamada S, Shimono A, Shinto Y, Tsujimura T, Takahashi T, et al. 1995. Bcl-2 deficiency in mice leads to pleiotropic abnormalities: accelerated lymphoid cell death in thymus and spleen, polycystic kidney, hair hypopigmentation, and distorted small intestine. *Cancer Res.* 55:354–59

91. Katsikis PD, Wunderlich ES, Smith CA, Herzenberg LA. 1995. Fas antigen stimulation induces marked apoptosis of T lymphocytes in human immunodeficiency virus-infected individuals. *J. Exp. Med.* 181:2029–36

92. Kawanishi M. 1993. Epstein-Barr virus induces fragmentation of chromosomal DNA during lytic infection. *J. Virol.* 67:7654–58

93. Kerr JF, Wyllie AH, Currie AR. 1972. Apoptosis: a basic biological phenomenon with wide-ranging implications in tissue kinetics. *Br. J. Can.* 26:239–57

94. Kiefer MC, Brauer MJ, Powers VC, Wu JJ, Umansky SR, et al. 1995. Modulation of apoptosis by the widely distributed Bcl-2 homolgue Bak. *Nature* 374:736–39

95. Knudson CM, Tung KSK, Toutellotte WG, Brown GAJ, Korsmeyer SJ. 1995. Bax-deficient mice with lymphoid hyperplasia and male germ cell death. *Science* 270:96–99

96. Kobayashi N, Hamamoto Y, Yamamoto N, Ishii A, Yonehara M, Yonehara S. 1990. Anti-fas monoclonal antibody is cytocidal to human immunodeficiency virus-infected cells without augmenting

viral replication. *Proc. Natl. Acad. Sci. USA* 87:9620–24

97. Deleted in proof

98. Koyama AH. 1995. Induction of apoptotic DNA fragmentation by the infection of vesicular stomatitis virus. *Virus Res.* 37:285–90

99. Kozopas KM, Yang T, Buchan HL, Zhou P, Craig RW. 1993. MCL1, a gene expressed in programmed myeloid cell differentiation, has sequence similarity to bcl-2. *Proc. Natl. Acad. Sci. USA* 90:3516–20

100. Krajewski S, Krajewsa M, Reed JC. 1996. Immunohistochemical analysis of in vivo patterns of Bak expression, a proapoptotic member of the Bcl-2 protein family. *Cancer Res.* 56:2849–55

101. Krajewski S, Krajewska M, Shabaik A, Wang HG, Irie S, et al. 1994. Immunohistochemical analysis of in vivo patterns of Bcl-X expression. *Cancer Res.* 54:5501–7

102. Krajewski S, Tanaka S, Takayama S, Schibler MJ, Fenton W, Reed JC. 1993. Investigation of the subcellular distribution of the bcl-2 oncoprotein: residence in the nuclear envelope, endoplasmic reticulum, and outer mitochondrial membranes. *Cancer Res.* 53:4701–14

103. Kumar S, Kinoshita M, Noda M, Copeland NG, Jenkins NA. 1994. Induction of apoptosis by the mouse Nedd2 gene, which encodes a protein similar to the product of the *Caenorhabditis elegans* cells death gene *ced-3* and the mammalian IL-1β-converting enzyme. *Genes Dev.* 8:1613–26

104. Lam M, Dubyak G, Chen L, Nunez G, Miesfeld RL, Distelhorst CW. 1994. Evidence that bcl-2 represses apoptosis by regulating endoplasmic reticulum-associated Ca^{2+} fluxes. *Proc. Natl. Acad. Sci. USA* 91:6569–73

105. Laurent-Crawford AG, Krust B, Riviere Y, Desgranges C, Muller S, et al. 1993. Membrane expression of HIV envelope glycoproteins triggers apoptosis in CD4 cells. *AIDS Res. Hum. Retroviruses* 9:761–73

106. Levine B, Goldman JE, Jiang HH, Griffin DE, Hardwick JM. 1996. Bcl-2 protects mice against fatal alphavirus encephalitis. *Proc. Natl. Acad. Sci. USA* 93:4810–15

107. Levine B, Griffin DE. 1992. Persistence of viral RNA in mouse brains after recovery from acute alphavirus encephalitis. *J. Virol.* 66:6429–35

108. Levine B, Hardwick JM, Griffin DE. 1994. Persistence of alphaviruses in ver-

tebrate hosts. *Trends Microbiol.* 2:25–28

109. Levine B, Hardwick JM, Trapp BD, Crawford TO, Bollinger RC, Griffin DE. 1991. Antibody-mediated clearance of alphavirus infection from neurons. *Science* 254:856–60

110. Levine B, Huang Q, Isaacs JT, Reed JC, Griffin DE, Hardwick JM. 1993. Conversion of lytic to persistent alphavirus infection by the bcl-2 cellular oncogene. *Nature* 361:739–42

111. Levine B, Jiang HH, Kleeman L, Yang G. 1996. Effect of E2 envelope glycoprotein cytoplasmic domain mutations on Sindbis virus pathogenesis. *J. Virol.* 70:1255–60

112. Lewis J, Wesselingh SL, Griffin DE, Hardwick JM. 1996. Alphavirus-induced apoptosis in mouse brains correlates with neurovirulence. *J. Virol.* 70:1828–35

113. Lewis M, Tartaglia LA, Lee A, Bennett GL, Rice GC, et al. 1991. Cloning and expression of cDNAs for two distinct murine tumor necrosis factor receptors demonstrate one receptor is species specific. *Proc. Natl. Acad. Sci. USA* 88:2830–34

114. Lin K-I, Lee SH, Narayanan R, Baraban JM, Hardwick JM, Ratan RR. 1995. Thiol agents and Bcl-2 identify an alphavirus-induced apoptotic pathway that requires activation of the transcription factor NF-kappa B. *J. Cell Biol.* 131:1–14

115. Liston P, Roy N, Tamai K, Lefebvre C, Baird S et al. 1996. Suppression of apoptosis in mammalian cells by NAIP and a related family of IAP genes. *Nature* 379:349–53

116. Liu X, Kim CN, Yang J, Jemmerson R, Wang X. 1996. Induction of apoptotic program in cell-free extracts: requirement for dATP and cytochrome c. *Cell* 86:147–57

117. Loetscher H, Pan Y-CE, Lahm H-W, Gentz R, Brockhaus M, et al. 1990. Molecular cloning and expression of the human 55 kd tumor necrosis factor receptor. *Cell* 61:351–59

118. London E. 1992. Diptheria toxin: membrane interaction and membrane translocation. *Biochim. Biophys. Acta* 1113:25–51

119. Lopez S, Yao J-S, Kuhn RJ, Strauss EG, Strauss JH. 1994. Nucleocapsid-glycoprotein interactions required for assembly of alphaviruses. *J. Virol.* 68:1316–23

120. Lowe SW, Ruley HE. 1993. Stabilization of the p53 tumor suppressor is induced by adenovirus 5 E1A and accompanies apoptosis. *Genes Dev.* 7:535–45

121. Lustig S, Halevy M, Ben-Nathan D, Akov

Y. 1992. A novel variant of Sindbis virus is both neurovirulent and neuroinvasive in adult mice. *J. Arch. Virol.* 122:237–48

122. Lustig S, Jackson AC, Hahn CS, Griffin DE, Strauss EG, Strauss JH. 1988. The molecular basis of Sindbis virus neurovirulence in mice. *J. Virol.* 62:2329–36

123. Marikainen P, Kyprianou N, Tucker RW, Isaacs JT. 1991. Programmed death of nonproliferating androgen-independent prostatic cancer cells. *Cancer Res.* 51: 4693–700

124. Martin DP, Schmidt RE, DiStefano PS, Lowry OH, Carter JG, Johnson EM Jr. 1988. Inhibitors of protein synthesis and RNA synthesis prevent neuronal death caused by nerve growth factor deprivation. *J. Cell Biol.* 106:829–44

125. McConkey DJ, Hartzell P, Nicotera P, Orrenius S. 1989. Calcium-activated DNA fragmentation kills immature thymocytes. *FASEB J.* 3:1843–49

126. McConkey DJ, Orrenius S. 1996. The role of calcium in the regulation of apoptosis. *J. Leukocyte Biol.* 59:775–83

127. Middleton G, Nunez G, Davies AM. 1996. Bax promotes neuronal survival and antagonises the survival effects of neurotrophic factors. *Development* 122:695–701

128. Miura M, Friedlander RM, Yuan JY. 1995. Tumor necrosis factor-induced apoptosis is mediated by a CrmA-sensitive cell death pathway. *Proc. Natl. Acad. Sci. USA* 92:8318–22

129. Miura M, Zhu H, Rotello R, Hartwieg EA, Yuan JY. 1993. Induction of apoptosis in fibroblasts by IL-1-beta-converting enzyme, a mammalian homolog of the *C. elegans* cell death gene *ced-3*. *Cell* 75:653–60

130. Moore LL, Bostick DA, Garry RF. 1988. Sindbis virus infection decreases intracellular pH: Alkaline medium inhibits processing of Sindbis virus polyproteins. *Virology* 166:1–8

131. Morey AL, Ferguson DJP, Fleming KA. 1993. Ultrastructural features of fetal erythroid precursors infected with parvovirus B19 in vitro: evidence of cell death by apoptosis. *J. Pathol.* 169:213–20

132. Motoyama N, Wang FP, Roth KA, Sawa H, Nakayama K, et al. 1995. Massive cell death of immature hematopoietic cells and neurons in Bcl-x-deficient mice. *Science* 267:1506–10

133. Muchmore SW, Sattler M, Liang H, Meadows RP, Harlan JE et al. 1996. X-ray and NMR structure of human Bcl-xL, an inhibitor of programmed cell death. *Nature* 381:335–41

134. Muzio M, Chinnaiyan AM, Kischkel FC, O'Rourke K, Shevchenko A, et al. 1996. FLICE, a novel FADD-homologous ICE/CED-3–like protease, is recruited to the CD95 (Fas/APO-1) death-inducing signaling complex. *Cell* 85:817–27

135. Nagata S, Golstein P. 1995. The fas death factor. *Science* 267:1449–56

136. Nakayama K, Nakayama K, Negishi I, Kuida K, Sawa H, Loh DY. 1994. Targeted disruption of Bcl-2 alpha beta in mice: occurrence of gray hair, polycystic kidney disease, and lymphocytopenia. *Proc. Natl. Acad. Sci. USA* 91:3700–4

137. Nakayama K, Nakayama K, Nakayama K, Negishi I, Kuida K, Shinkai Y. 1993. Disappearance of the lymphoid system in Bcl-2 homozygous mutant chimeric mice. *Science* 261:1584–88

138. Neilan JG, Lu Z, Afonso CL, Kutish GF, Sussman MD, Rock DL. 1993. An African swine fever virus gene with similarity to the proto-oncogene bcl-2 and the Epstein-Barr virus gene BHRF1. *J. Virol.* 67:4391–94

139. Nguyen M, Millar DG, Wee Yong V, Korsmeyer SJ, Shore GC. 1996. Targeting of Bcl-2 to the mitochondrial outer membrane by a COOH-terminal signal anchor sequence. *Am. J. Med. Sci.* 268:25265–68

140. Nicholson DW, Ali A, Thornberry NA, Vaillancourt JP, Ding CK, et al. 1995. Identification and inhibition of the ICE/CED-3 protease necessary for mammalian apoptosis. *Nature* 376:37–43

141. Noteborn MHM, Todd D, Verschueren CAJ, Degauw HWFM, Curran WL, et al. 1994. A single chicken anemia virus protein induces apoptosis. *J. Virol.* 68:346–51

142. Nunez G, Clarke MF. 1994. The Bcl-2 family of proteins: regulators of cell death and survival. *Trends Cell Biol.* 4:399–406

143. Nunez G, Hockenbery D, McDonnell TJ, Sorensen CM, Korsmeyer SJ. 1991. Bcl-2 maintains B cell memory. *Nature* 353:71–73

144. Nunez G, London L, Hockenbery D, Alexander M, McKearn JP, Korsmeyer SJ. 1990. Deregulated bcl-2 gene expression selectively prolongs surrvival of growth factor-deprived hemopoietic cell lines. *J. Immunol.* 144:3602–10

145. Oehm A, Behrmann I, Falk W, Pawlita M, Maier G, et al. 1992. Purification and molecular cloning of the APO-1 cell surface antigen, a member of the tumor necrosis factor/nerve growth factor receptor superfamily. *J. Biol. Chem.* 267: 10709–15

146. Oltvai ZN, Milliman CL, Korsmeyer SJ. 1993. Bcl-2 heterodimerizes in vivo with a conserved homolog, Bax, that accelerates programmed cell death. *Cell* 74:609–19

147. Osborne BA, Schwartz LM. 1994. Essential genes that regulate apoptosis. *Trends Cell Biol.* 4:394–98

148. Owen-Schaub LB, Yonehara S, Crump WL, Grimm EA. 1992. DNA fragmentation and cell death is selectively triggered in activated human lymphocytes by Fas antigen engagement. *Cell. Immunol.* 140:197–205

149. Peitsch MC, Georg H, Tschopp M, Tschopp J. 1994. The apoptosis endonucleases: cleaning up after cell death. *Trends Cell Biol.* 4:37–41

150. Peitsch MC, Polzar B, Stephan H, Crompton T, MacDonald HR, et al. 1993. Characterization of the endogenous deoxyribonuclease involved in nuclear DNA degradation during apoptosis (programmed cell death). *EMBO J.* 12:371–77

151. Pekosz A, Phillips J, Pleasure D, Merry D, Gonzalez-Scarano F. 1996. Induction of apoptosis by La Crosse virus infection and role of neuronal differentiation and human *bcl-2* expression in its prevention. *J. Virol.* 70:5329–35

152. Peleg J. 1969. Inapparent persistent virus infection in continuously grown *Aedes aegypti* mosquito cells. *J. Gen. Virol.* 5:463–71

153. Prikhod'ko EA, Miller LK. 1996. Induction of apoptosis by baculovirus transactivator IE1. *J. Virol.* 70:7116–24

154. Qin XQ, Livingston DM, Kaelin WG Jr, Adams PD. 1994. Deregulated transcription factor E2F-1 expression leads to S-phase entry and p-53 mediated apoptosis. *Proc. Natl. Acad. Sci. USA* 91:10918–22

155. Raff MC. 1992. Social controls on cell survival and cell death. *Nature* 356:397–400

156. Rao L, Debbas M, Sabbatini P, Hockenberry D, Korsmeyer S, White E. 1992. The adenovirus E1A proteins induce apoptosis, which is inhibited by the E1B 19-kDa and Bcl-2 proteins. *Proc. Natl. Acad. Sci. USA* 89:7742–46

157. Ray CA, Black RA, Kronheim SR, Greenstreet TA, Sleath PR, et al. 1992. Viral inhibition of inflammation: Cowpox virus encodes an inhibitor of the interleukin-1-beta converting enzyme. *Cell* 69:597–604

158. Rice CM, Bell JR, Hunkapillar MW, Strauss EG, Strauss JH. 1982. Isolation and characterization of the hydrophobic COOH terminal domains of the Sindbis virus glycoproteins. *J. Mol. Biol.* 154:355–78

159. Riedel B, Brown DT. 1977. Role of extracellular virus in the maintenance of the persistent infection induced in *Aedes albopictus* (mosquito) cells by Sindbis virus. *J. Virol.* 23:554–61

160. Riedel B, Brown DT. 1979. Novel antiviral activity found in the media of Sindbis virus persistently infected mosquito (*Aedes albopictus*) cell cultures. *J. Virol.* 29:51–60

161. Rosen A, Rosen-Casciola L, Ahearn J. 1995. Novel packages of viral and self-antigens are generated during apoptosis. *J. Exp. Med.* 181:1557–61

162. Rothe M, Pan MG, Henzel WJ, Ayres TM, Goeddel DV. 1995. The TNFR2-TRAF signaling complex contains two novel proteins related to baculoviral inhibitor of apoptosis proteins. *Cell* 83:1243–52

163. Rothe M, Wong SC, Henzel WJ, Goeddel DV. 1994. A novel family of putative signal transducers associated with the cytoplasmic domain of the 75 kDa tumor necrosis factor receptor. *Cell* 78:681–92

164. Rothe M, Xiong J, Shu HB, Williamson K, Goddard A, Goeddel DV. 1996. I-TRAF is a novel TRAF-interacting protein that regulates TRAF-mediated signal transduction. *Proc. Natl. Acad. Sci. USA* 93:8241–46

165. Rouayrenc JF, Boise LH, Thompson CB, Privat A, Patey G. 1995. Presence of the long and the short forms of Bcl-X in several human and murine tissues. *C. R. Acad. Sci.* 318:537–40

166. Roy N, Mahadevan MS, McLean M, Shutler G, Yaraghi Z, et al. 1995. The gene for neuronal apoptosis inhibitory protein is partially deleted in individuals with spinal muscular atrophy. *Cell* 80:167–78

167. Russell DL, Dalrymple JM, Johnston RE. 1989. Sindbis virus mutations which coordinately affect glycoprotein processing, penetration, and virulence in mice. *J. Virol.* 63:1619–29

168. Savill J, Fadok V, Henson P, Haslett C. 1993. Phagocyte recognition of cells undergoing apoptosis. *Immunol. Today* 14:131–36

169. Sawicki DL, Sawicki SG, Keranen S, Kaariainen L. 1981. Specific Sindbis virus-coded function for minus-strand RNA synthesis. *J. Virol.* 39:348–58

170. Schall TJ, Lewis M, Koller KJ, Lee A, Rice GC, et al. 1990. Molecular cloning and expression of a receptor for human tumor necrosis factor. *Cell* 61:361–70

171. Schnitzler P, Hug M, Handermann M, Janssen W, Koonin EV, et al. 1994.

Identification of genes encoding zinc finger proteins, non-histone chromosomal HMG protein homologue, and a putative GTP phosphohydrolase in the genome of Chilo iridescent virus. *Nucleic Acids Res.* 22:158–66

172. Schoepp RJ, Johnston RE. 1993. Directed mutagenesis of a Sindbis virus pathogenesis site. *Virology* 193:149–59

173. Schulze-Osthoff K. 1994. The Fas/APO-1 receptor and its deadly ligand. *Trends Cell Biol.* 4:421–26

174. Deleted in proof

175. Sherman LA, Griffin DE. 1990. Pathogenesis of encephalitis induced in newborn mice by virulent and avirulent strains of Sindbis virus. *J. Virol.* 64:2041–46

176. Shimizu S, Eguchi Y, Kamiike W, Matsuda H, Tsujimoto Y. 1996. Bcl-2 expression prevents activation of the ICE protease cascade. *Oncogene* 12:2251–57

177. Shimizu S, Eguchi Y, Kamiike W, et al. 1996. Bcl-2 blocks loss of mitochondrial membrane potential while ICE inhibitors act at a different step during inhibition of death induced by respiratory chain inhibitors. *Oncogene* 13:21–29

178. Shimizu S, Eguchi Y, Kosaka H, Kamiike W, Matsuda H, Tsujimoto Y. 1995. Prevention of hypoxia-induced cell death by bcl-2 and bcl-xL. *Nature* 374:811–16

179. Shu HB, Takeuchi M, Goeddel DV. 1996. The tumor necrosis factor receptor 2 signal transducers TRAF2 and c-IAP1 are components of the tumor necrosis factor receptor 1 signaling complex. *Proc. Natl. Acad. Sci. USA* 93:13973–78

180. Simonian PL, Grillot DAM, Merino R, Nunez G. 1996. Bax can antagonize Bcl-x$_L$ during etoposide and cisplatin-induced cell death independently of its heterodimerization with Bcl-x$_L$. *J. Biol. Chem.* 271:22764–72

181. Singh I, Helenius A. 1992. Role of ribosomes in Semliki Forest virus nucleocapsid uncoating. *J. Virol.* 66:7049–58

182. Smith CA, Davis T, Anderson D, Solam L, Beckmann MP, et al. 1990. A receptor of tumor necrosis factor defines an unusual family of cellular and viral proteins. *Science* 248:1019–23

183. Song HY, Rothe M, Goeddel DV. 1996. The tumor necrosis factor-inducible zinc finger protein A20 interacts with TRAF1/TRAF2 and inhibits NF-kappaB activation. *Proc. Natl. Acad. Sci. USA* 93:6721–25

184. Stellar H. 1995. Mechanisms and genes of cellular suicide. *Science* 267:1445–49

185. Stevens TM. 1970. Arbovirus replication in mosquito cell lines (Singh) grown in monolayer or suspension culture. *Proc. Soc. Exp. Biol. Med.* 134:356–61

186. Strasser A, Harris AW, Cory S. 1991. Bcl-2 transgene inhibits T cell death and perturbs thymic self-censorship. *Cell* 67:889–99

187. Strasser A, Whittingham S, Vaux DL, Bath ML, Adams JM, et al. 1991. Enforced *bcl-2* expression in B-lymphoid cells prolongs antibody responses and elicits autoimmune disease. *Proc. Natl. Acad. Sci. USA* 88:8661–65

188. Strauss JH, Strauss EG. 1994. The alphaviruses: gene expression, replication and evolution. *Microbiol. Rev.* 58:491–562

189. Takeuchi M, Rothe M, Goeddel DV. 1996. Anatomy of TRAF2. Distinct domains for nuclear factor-kappaB activation and association with tumor necrosis factor signaling proteins. *J. Biol. Chem.* 271:19935–42

190. Takizawa T, Fukuda R, Miyawaki T, Ohashi K, Nakanishi Y. 1995. Activation of the apoptotic Fas antigen-encoding gene upon influenza virus infection involving spontaneously produced beta-interferon. *Virology* 209:288–96

191. Takizawa T, Matsukawa S, Higuchi Y, Nakamura S, Nakanishi Y, Fukuda R. 1993. Induction of programmed cell death (apoptosis) by influenza virus infection in tissue culture cells. *J. Gen. Virol.* 74: 2347–55

192. Tartaglia LA, Goeddel DV. 1992. Two TNF receptors. *Immunol. Today* 13:151–53

193. Tartaglia LA, Goeddel DV, Reynolds C, Figari IS, Weber RF, et al. 1993. Stimulation of human T-cell proliferation by specific activation of the 75-kDa tumor necrosis factor receptor. *J. Immunol.* 151:4637–41

194. Tartaglia LA, Weber RF, Figari IS, Reynolds C, Palladino MA Jr, Goeddel DV. 1991. The two different receptors for tumor necrosis factor mediate distinct cellular responses. *Proc. Natl. Acad. Sci. USA* 88:9292–96

195. Tewari M, Telford WG, Miller RA, Dixit VM. 1995. CrmA, a poxvirus-encoded serpin, inhibits cytotoxic T-lymphocyte-mediated apoptosis. *J. Biol. Chem.* 270:22705–8

196. Thornberry NA, Bull HG, Calaycay JR, Chapman KT, Howard AD, et al. 1992. A novel heterodimeric cysteine protease is required for interleukin-1β processing in monocytes. *Nature* 356:768–74

197. Tollefson AE, Ryerse JS, Scaria A, Hermiston TW, Wold WSM. 1996. The E3–

11.6-kDa adenovirus death protein (ADP) is required for efficient cell death: characterization of cells infected with adp mutants. *Virology* 220:152–62

198. Tolskaya EA, Romanova LI, Kolesnikova MS, Ivannikova TA, Smirnova EA, et al. 1995. Apoptosis-inducing and apoptosis-preventing function of poliovirus. *J. Virol.* 69:1181–89

199. Tortorella D, Sesardic D, Dawes CS, London E. 1995. Immunochemical analysis shows all three domains of diphtheria toxin penetrate across model membranes. *J. Biol. Chem.* 270:27446–52

200. Trauth BC, Klas C, Peters AMJ, Matzku S, Moller P, et al. 1989. Monoclonal antibody-mediated tumor regression by induction of apoptosis. *Science* 245:301–5

201. Tropea F, Troiano L, Monti D, Lovato E, Malorni W, et al. 1995. Sendai virus and herpes virus type 1 induce apoptosis in human peripheral blood mononuclear cells. *Exp. Cell Res.* 218:63–70

202. Tsujimoto Y, Finger LR, Yunis J, Nowell PC, Croce CM. 1984. Cloning of the chromosome breakpoint of neoplastic B cells with the t(14;18) chromosome translocation. *Science* 226:1097–99

203. Tucker PC, Griffin DE. 1991. The mechanism of altered Sindbis virus neurovirulence associated with a single amino acid change in the E2 glycoprotein. *J. Virol.* 65:1551–57

204. Tucker PC, Griffin DE, Choi S, Bui N, Wesselingh S. 1996. Inhibition of nitric oxide synthesis increases mortality in Sindbis virus encephalitis. *J. Virol.* 70:3972–77

205. Tucker PC, Strauss EG, Kuhn RJ, Strauss JH, Griffin DE. 1992. Viral determinants of age-dependent virulence of Sindbis virus for mice. *J. Virol.* 67:4605–10

206. Tucker PC, Strauss EG, Kuhn RJ, Strauss JH, Griffin DE. 1993. The age-dependent neurovirulence of Sindbis virus for mice is influenced by a single amino acid change at position 55 of the E2 glycoprotein. *J. Virol.* 67:4605–10

207. Tyler KL, Squier MKT, Rodgers SE, Schneider BE, Oberhaus SM, et al. 1995. Differences in the capacity of reovirus strains to induce apoptosis are determined by the viral attachment protein sigma-1. *J. Virol.* 69:6972–79

208. Ubol S, Griffin DE. 1991. Identification of a putative alphavirus receptor on mouse neural cells. *J. Virol.* 65:6913–21

209. Ubol S, Park S, Budihardjo I, Desnoyers S, Montrose MH et al. 1996. Temporal changes in chromatin, intracellular calcium, and poly (ADP-ribose) polymerase during Sindbis virus-induced apotosis of neuroblastoma cells. *J. Virol.* 70:2215–20

210. Ubol S, Tucker PC, Griffin DE, Hardwick JM. 1994. Neurovirulent strains of alphavirus induce apoptosis in bcl-2-expressing cells; role of a single amino acid change in the E2 glycoprotein. *Proc. Natl. Acad. Sci. USA* 91:5202–6

211. Ulug ET, Bose HR Jr. 1985. Effect of tunicamycin on the development of the cytopathic effect in Sindbis virus-infected avian fibroblasts. *Virology* 143:546–57

212. Uren AG, Pakusch M, Hawkins CJ, Puls KL, Vaux DL. 1996. Cloning and expression of apoptosis inhibitory protein homologs that function to inhibit apoptosis and/or bind tumor necrosis factor receptor-associated factors. *Proc. Natl. Acad. Sci. USA* 93:4974–78

213. Van Antwerp DJ, Martin SJ, Kafri T, Green DR, Verma IM. 1996. Suppression of TNF-induced apoptosis by NF-kappaB. *Science* 274:787–89

214. Vandenabeele P, Declercq W, Beyaert R, Fiers W. 1995. Two tumour necrosis factor receptors: structure and function. *Trends Cell Biol.* 5:392–99

215. Vaux DL, Cory S, Adams JM. 1988. Bcl-2 gene promotes haemopoietic cell survival and cooperates with c-myc to immortalize pre-B cells. *Nature* 335:440–42

216. Vaux DL, Hacker G. 1995. Hypothesis: apoptosis caused by cytotoxins represents a defensive response that evolved to combat intracellular pathogens. *Clin. Exp. Pharmacol. Physiol.* 22:861–63

217. Vaux DL, Weissman IL, Kim SK. 1992. Prevention of programmed cell death in *Caenorhabditis elegans* by human bcl-2. *Science* 258:1955–56

218. Veis DJ, Sorenson CM, Shutter JR, Korsmeyer SJ. 1993. Bcl-2-deficient mice demonstrate fulminant lymphoid apoptosis, polycystic kidneys, and hypopigmented hair. *Cell* 75:229–40

219. Wada N, Matsumura M, Ohba Y, Kobayashi N, Takizawa T, Nakanishi Y. 1995. Transcription stimulation of the Fas-encoding gene by nuclear factor for interleukin-6 expression upon influenza virus infection. *J. Biol. Chem.* 270:18007–12

220. Waite MRF, Pfefferkorn ER. 1970. Phospholipid synthesis in Sindbis virus-infected cells. *J. Virol.* 6:637–43

221. Wang CY, Mayo MW, Baldwin AS Jr. 1996. TNF- and cancer therapy-induced apoptosis: potentiation by inhibition of NF-kappaB. *Science* 274:784–87

222. Wang H, Rapp UR, Reed JC. 1996. Bcl-2 targets the protein kinase Raf-1 to mitochondria. *Cell* 87:629–38

223. Wang L, Miura M, Bergeron L, Zhu H, Yuan J. 1994. Ich-1, an *Ice/ced-3*-related gene, encodes both positive and negative regulators of programmed cell death. *Cell* 78:739–50

224. Wang S, Pittman RN. 1993. Altered protein binding to the octamer motif appears to be an early event in programmed neuronal cell death. *Proc. Natl. Acad. Sci. USA* 90:10385–89

225. Weiss B, Rosenthal R, Schlesinger S. 1980. Establishment and maintenance of persistent infection by Sindbis virus in BHK cells. *J. Virol.* 33:463–74

226. Wengler G. 1984. Identification of a transfer of viral core protein to cellular ribosomes during the early stages of alphavirus infection. *Virology* 134:435–42

227. Westendorp MO, Frank R, Ochsenbauer C, Stricker K, Dhein J, et al. 1995. Sensitization of T cells to CD95-mediated apoptosis by HIV-1 Tat and gp120. *Nature* 375:497–500

228. White E. 1993. Regulation of apoptosis by the transforming genes of the DNA tumor virus adenovirus. *Proc. Soc. Exp. Biol. Med.* 204:30–39

229. White E, Sabbatini P, Debbas M, Wold WSM, Kusher DI, Gooding LR. 1992. The 19-kilodalton adenovirus E1B transforming protein inhibits programmed cell death and prevents cytolysis by tumor necrosis factor alpha. *Mol. Cell. Biol.* 12:2570–80

230. White K, Tahaoglu E, Steller H. 1996. Cell killing by the Drosophila gene reaper. *Science* 271:805–7

231. Wu HY, Lozano G. 1994. NF-kappa-B activation of p53: a potential mechanism for suppressing cell growth in response to stress. *J. Biol. Chem.* 269:20067–74

232. Xue D, Shaham S, Horvitz HR. 1996. The *Caenorhabditis elegans* cell-death protein CED-3 is a cysteine protease with substrate specificities similar to those of the human CPP32 protease. *Genes Dev.* 10:1073–83

233. Yamada T, Yamaoka S, Goto T, Nakai M, Tsujimoto Y, Hatanaka M. 1994. The human T-cell leukemia virus type I tax protein induces apoptosis which is blocked by the bcl-2 protein. *J. Virol.* 68:3374–79

234. Yang E, Zha J, Jockel J, Boise LH, Thompson CB, Korsmeyer SJ. 1995. Bad, a heterodimeric partner for Bcl-XL and Bcl-2, displaces Bax and promotes cell death. *Cell* 80:285–91

235. Yin XM, Oltvai ZN, Korsmeyer SJ. 1994. BH1 and BH2 domains of Bcl-2 are required for inhibition of apoptosis and heterodimerization with Bax. *Nature* 369:321–23

236. Yuan JY, Horvitz HR. 1990. The *Caenorhabditis elegans* genes ced-3 and ced-4 act cell autonomously to cause programmed cell death. *Dev. Biol.* 138:33–41

237. Yuan JY, Shaham S, Ledoux S, Ellis HM, Horvitz HR. 1993. The *C. elegans* cell death gene ced-3 encodes a protein similar to mammalian interleukin-1-beta-converting enzyme. *Cell* 75:641–52

238. Zha H, Aime-Sempe C, Sato T, Reed JC. 1996. Proapoptotic proteins Bax heterodimerizes with Bcl-2 and homodimerizes with Bax via a novel domain (BH3) distinct from BH1 and BH2. *J. Biol. Chem.* 271:7440–44

239. Zha J, Harada H, Yang E, Jockel J, Korsmeyer SJ. 1996. Serine phosphorylation of death agonist BAD in response to survival factor results in binding to 14-3-3 not BCL-XL. *Cell* 87:619–28

Annu. Rev. Microbiol. 1997. 51:593–628

CLUES AND CONSEQUENCES OF DNA BENDING IN TRANSCRIPTION

José Pérez-Martín and Víctor de Lorenzo
Centro Nacional de Biotecnología, Consejo Superior de Investigaciones Científicas,
Campus de Cantoblanco, 28049 Madrid, Spain; e-mail: vdlorenzo@samba.cnb.uam.es

KEY WORDS: coregulation, curvature, histone-like proteins, promoters, remote activation

ABSTRACT
This review attempts to substantiate the notion that nonlinear DNA structures allow prokaryotic cells to evolve complex signal integration devices that, to some extent, parallel the transduction cascades employed by higher organisms to control cell growth and differentiation. Regulatory cascades allow the possibility of inserting additional checks, either positive or negative, in every step of the process. In this context, the major consequence of DNA bending in transcription is that promoter geometry becomes a key regulatory element. By using DNA bending, bacteria afford multiple metabolic control levels simply through alteration of promoter architecture, so that positive signals favor an optimal constellation of protein-protein and protein-DNA contacts required for activation. Additional effects of regulated DNA bending in prokaryotic promoters include the amplification and translation of small physiological signals into major transcriptional responses and the control of promoter specificity for cognate regulators.

CONTENTS

INTRODUCTION

The nucleotide sequence of DNA is the ultimate physical support of the information for the buildup of proteins and higher structures. In spite of a growing number of exceptions, the unidirectional flow DNA → RNA → proteins remains one of the basic pillars of modern biology (176). But DNA not only encodes proteins; regulatory sequences without any structural information play a pivotal role in many biological processes as well. Furthermore, beginning in the early 1980s, observations showing an entirely new aspect of DNA started to emerge: the role of nonlinear DNA structures in replication, recombination, and transcription was not directly linked to the proteins encoded or bound by the nucleotide sequence. In this respect, the DNA molecule is unique: it is able to encode in its sequence at least two independent levels of functional information. The first one (what might be called digital information) is for encoding proteins, whether structural or regulatory, and sequence targets for DNA-binding factors. A second level of information is contained in the physical and structural properties of the DNA molecule itself, which—as is reviewed in detail in this article—plays a central role in the coregulation of essential cellular processes. Although such properties are ultimately determined in each case by the nucleotide sequence, they are exploited by the cells in a fashion in which the sequence itself plays no role other than to support or facilitate a certain spatial structure. Along with RNA, this double feature of DNA has no parallel in other biological macromolecules.

DNA bending has now been identified in an entire collection of molecular phenomena, but this review focuses on only a particular and paradigmatic aspect of the subject, namely, its role in transcriptional coregulation of prokaryotic promoters, with some extensions into eukaryotic systems. In this context, we do not address the direct role of DNA curvature and bending in transcription, a controversial issue that has been examined in many different systems (48, 86, 109, 110, 123, 147a, 150, 174, 178, 183, 204) and has been recently reviewed (19, 39, 66, 140). Instead, we concentrate on nonlinear DNA structures as auxiliary regulatory elements that allow prokaryotic cells to process and integrate complex environmental signals. To this end, bendability and intrinsic or protein-induced curvatures become regulatory assets as building blocks for a productive promoter geometry. In this way, bacteria (*a*) afford multiple metabolic control levels by altering promoter architecture, (*b*) amplify and translate minor physiological signals into major transcriptional responses, and (*c*) control promoter specificity.

THE NATURE OF DNA BENDING

Certain nucleotide sequences spontaneously impart a preferred direction of curvature on a DNA molecule, i.e. originate an intrinsic bend (72, 106, 190). Both for natural and artificially designed sequences, any short homopolymeric run of A or T nucleotides longer than 3 bp, when repeated in phase with the helical screw, can introduce a detectable degree of curvature in an isolated DNA molecule. Contiguous AA dinucleotides making minor wedge angles may sum up within the DNA helix and generate a curved structure (72, 190). In addition to AA pairs, certain dinucleotides such as AG, CG, GA, or GC can induce or contribute significantly to DNA curvature (16). Although A or T tracts are believed to be major determinants of DNA curvature, the sequence context also plays a critical role (33). In addition, polyA/polyT tracts are particularly rigid and bending may occur preferentially at the intervening sequences or at the A/T boundaries (180a, 182a). Alternatively (or at the same time), axial deflections may also arise from structural discontinuities at the boundaries between the A tracts assembled in B-DNA and the rest of the nucleotide sequence (23, 73, 90).

Bendability: Consequences for DNA-Protein Interactions

Apart from intrinsic DNA bends, a critical aspect of the structural diversity of DNA is its bendability. DNA does not behave as an isotropic rod: Depending on the sequence, it might bend more easily in one plane than another, indicating it possesses a degree of anisotropic flexibility. In any bending event, both the major and the minor groove of the helix surface inside the curve must narrow to some extent, because of the compression associated with bending, while those on the outside of the curve become correspondingly wider. It is the ability of specific short sequences to assume these conformations that enables the DNA structure to accommodate the deformation associated with bending and, thus, determines the bendability of DNA (34, 49, 87, 188). One important difference between curved DNA and bendable DNA arises from different physical dynamism. Statically curved DNA is deformed even in the absence of external forces, thereby resulting in a very rigid structure. On the other hand, bendable DNA allows a mixture of many different conformational states, the equilibrium of which can be displaced toward one specific form by external forces such as proteins interacting with them (49, 188). This is critical, since interactions between static structures are insufficient to explain DNA-protein recognition events (74). In fact, the specificity of DNA-protein contacts is the result not only of direct recognition of certain bases within the DNA helix by particular amino acids, but also of an array of interactions between protein and DNA surfaces not necessarily linked to a particular nucleotide sequence. This type of interaction has been termed indirect readout (124) or analog recognition (35, 36) as opposed to the direct recognition of individual base pairs. Indirect readout involves protein contacts (e.g. amino acids and their side-groups) with the sugar-phosphate backbone of DNA through ionic bridges, hydrogen bonds, and hydrophobic interactions. Direct recognition requires a mutual DNA-protein reading to form bonds between the amino acid chains and the hydrogen atoms presented by the bases on the major or minor grooves of the double helix. Both types of interaction contribute to the specificity and stability of the complex and require the protein and the DNA to accommodate each other in order to set up adequate contacts. In other words, both the protein and DNA partners seek regions of the other that maximize their interaction, but conformational distortions may also be required in both partners to achieve this optimum fit (181, 182).

Regardless of the type of DNA-binding motif, the number of proteins known to bend their DNA target increases every year. But how is bending actually effected? In some cases it can be explained by the ability of proteins with typical major-groove recognition motifs (for instance, helix-turn-helix structures) to neutralize charges in the DNA backbone (179) or to set up extended DNA-protein contacts (181, 182). Interestingly, some regulators, such as the Cro protein of phage λ, seem to bend DNA efficiently without specific sequence recognition (42, 160). For proteins that interact with the minor groove, a model of the bending mechanism has been proposed that involves the intercalation of a protein side chain in the minor groove that acts like a wedge and eventually causes a deformation of the DNA helix (199). In the case of the integration host factor (IHF) protein, this effect is heavily dependent on indirect readout via the phosphodiester backbone (152a).

The extent to which sequence-dependent bendability of DNA plays a key role in DNA-protein interactions is clearly reflected in the affinity of a number of regulatory proteins for precurved or bending-prone sequences. The pioneering work with the catabolite gene activator protein (CAP) or cyclic AMP receptor protein (CRP) 10 years ago showed the correlation between the bendability of a CAP site within a DNA sequence and the affinity of CRP for the target site (53, 101). Similarly, the *Drosophila* protein Su (HW) requires an intrinsically bent DNA sequence in its target site (173). Also in the prokaryotic world, the bacterial σ^{70}-RNA polymerase induces a strong bend in the promoter sequence upon binding (92, 139), and the correlation between promoter strength and the presence of upstream curved DNA has been extensively shown (see 140 for review). Along the same line, the effects of a DNA bend on the binding of eukaryotic TBP or holo-TFIID to the TATA box were examined (175) with the conclusion that DNA bending is an important component of the eukaryotic promoter.

CLUES OF DNA BENDING

DNA Bending as an Inducer or Inhibitor of Protein-DNA Interactions

Given the importance of DNA conformation for its recognition by some proteins, it is easy to anticipate that additional intrinsic or protein-induced bends may synergize or antagonize the interactions with proteins required to bend their targets for optimal binding. This effect, called structural synergy (70), was first shown in experiments in which an intrinsic bend was placed in different helical phases relative to a CAP-binding site (85). CAP binding to its target sequence increased by two orders of magnitude when the precurved DNA was in the same orientation as the CAP-induced bend. On the other hand, CAP binding was impaired when the bends were located in opposite directions. Moreover, a separate study pointed out that such an effect could even occur at considerable distances (98). A remarkable example of structural synergy involving different regulators is that of the interplay between the *Escherichia coli* repressor PutA and the integration host factor (IHF, see below), which binds to and bends two sequences adjacent to the PutA sites in the intergenic region between the *putA* and the *putP* promoters (Figure 1), thereby facilitating the binding of the repressor to the promoters targets (118). IHF is one major architectural element for nucleoprotein complexes in bacteria (50, 66), and its co-crystal with DNA has recently been resolved (152a). Because IHF participates in a variety of otherwise very different molecular events, its roles and properties are discussed in separate sections of this article.

The effect of DNA bending on the interaction of proteins and DNA has been addressed also in a number of eukaryotic promoters. Insertion of an intrinsically curved DNA upstream of the TATA box was shown to increase the DNA-binding affinity of the TBP protein in the context of a minicircle by a factor of 100 relative to linear DNA (128). Interestingly, this increase was strictly dependent on the specific phase of the curved DNA relative to the TATA box. A similar instance has been described for the eukaryotic histone-like DNA-bending protein HMG1, which facilitates the binding of the human progesterone receptor to its cognate promoter by inducing a structural change in the target DNA (120).

The notion of structural synergy has shed new light on the intimate mechanism of traditional regulatory devices such as down-regulation of prokaryotic promoters by repressors that bind DNA in the proximity of the RNA polymerase. For instance, the CopG protein (formerly named RepA) of the streptococcal

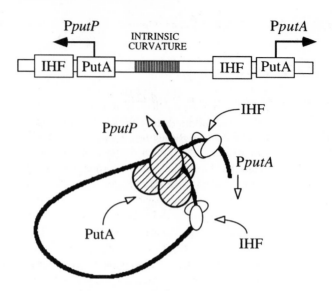

Figure 1   Organization of the PutA/integration host factor (IHF)-mediated repression complex of proline-uptake promoters of *E. coli*. The *PputA/PputP* divergent promoter region includes not only binding sites for the endogenous repressor of the system (PutA) but also an array of intrinsic DNA curvatures and IHF sites. In the absence of proline, a strong repression complex is formed, as sketched in the figure, which involves interactions between PutA molecules bound to distant sites, which are favored and sustained by both intrinsic and IHF-induced DNA bends within the region.

plasmid pLS1 is a small transcriptional repressor that bends its target DNA sharply at the plasmid replication origin (29, 136). A RepA-operator sequence placed artificially approximately 90 bp upstream of a standard *E. coli* promoter enhanced or decreased transcription in a side-of-the-helix–dependent fashion relative to the RNAP-binding site both in vitro and in vitro (137). This suggested that repression follows trapping the DNA region in a conformation not amenable to RNA polymerase binding instead of being the result of a competition between RepA and the enzyme for the same target sites. The notion that factors that bend promoter sequences in a direction unfavorable for RNAP binding can effect transcription repression (93, 205) has been substantiated in other cases. These include the repression exerted by the p4 protein from the *Bacillus subtilis* phage Φ29 on PA2b, a major promoter of early viral genes (156, 157), and the down-regulation caused by IHF on the *ilvPG1* promoter of *E. coli* (125, 196). In these and other (138) cases, there seems to be a transcriptional switch between two divergent or tandem overlapping promoters, an issue that has been reviewed separately (140).

DNA Bending as a Catalyst of Protein-Protein Interplay

Readout of the digital information encoded in DNA is preceded by the assembly of higher-order structures comprising multiple proteins, an event that is particularly complex in eukaryotic promoters (40, 70). Transcriptional factors frequently bind to distinct, sometimes distant, sites in such a way that protein-protein interactions require the distortion of the DNA structure in order to shorten the distance between the factors already bound to DNA. Recruitment of proteins to their specific targets face a dilemma. If the binding sites are too far away from each other, then the probability of contacts between two or more proteins decreases in parallel to the distance. This effect has a thermodynamic basis: For interactions to occur, the free energy resulting from the loss of entropy of the DNA when the sites are brought together through protein-protein contacts should be balanced by the free energy of association between the proteins themselves. However, "the closer the better" is not true, because below a specific DNA distance (called persistence length) the DNA stiffness makes interprotein contacts involving DNA distortions energetically unfavorable (195). Despite these >considerations, the fact is that interactions do occur between proteins bound to DNA sites ranging from few nucleotides to many kilobases. How can this occur?

In prokaryotic promoters, interactions between nonadjacent DNA-bound proteins require systematically the formation of a loop, which is energetically disfavored if the binding sites are separated by 30–140 nucleotides. DNA fragments of that size range display a limited flexibility, and the stiffness of the intervening DNA sequence may prevent the DNA from acquiring the curvature needed to juxtapose nonadjacent binding sites (170, 195). In these cases, one solution consists of altering the conformation of the DNA. The sequence between both protein-binding sites may be already curved in the appropriate direction or may show an increased and directed flexibility (i.e. bendability). Although both curved DNA and bendable DNA may show the same final effect (increasing interactions between proteins bound at both ends of the region), the statically bent DNA originates a very stable conformation, while the bendable DNA is anisotropically flexible and tolerates a more dynamic structure (see above). This concept has been directly tested in an artificial system based on the p4 protein of phage Φ29, a regulator that activates transcription at a certain distance from its cognate promoter (155). In a series of elegant experiments, Serrano et al (167) showed that activation occurred only when the intervening sequence between the binding sites for the RNA polymerase and the p4 protein was engineered to contain either a static DNA bend or a sequence with high bendability, thus facilitating productive protein-protein contacts.

A second solution to the problem of bringing about interactions between proteins bound to distant sites is the participation of extra factors that either bend DNA or increase its flexibility. These proteins may recognize specific sequences in the DNA, as it occurs with the bacterial proteins CAP or IHF or the eukaryotic factor LEF-1, some properties of which are discussed below. The binding of these proteins to DNA provokes a conformational change that has a precise directionality and is thought to line up nonadjacent binding sites, thereby enabling otherwise unlikely protein-protein interactions. A role for the DNA bending induced by these proteins is inferred from various bend swapping experiments, in which the binding site of the DNA-bending protein is exchanged either by the recognition site of unrelated DNA-bending proteins (55, 63, 64, 153) or by curved DNA (63, 112, 141). Other factors that facilitate interactions between proteins bound to distant sites may lack sequence specificity, as is the case for the bacterial histone-like proteins HU or, in eukaryotes, the high-mobility group of histone-like proteins such as HMG1 (see below).

The existence of auxiliary proteins capable of bending the DNA appears to be particularly useful in eukaryotic genomes, where distances between promoters and regulatory sites are frequently measured in kilobases. This is probably one of the many roles of the nucleosomes, the organization of which brings into physical proximity sequences that otherwise would be far apart. This concept has growing experimental support. For example, activation of the promoter of the *hsp26* gene of *Drosophila* requires that two binding sites for the heat-shock transcription factor are brought together by an intervening nucleosome (41, 186). Similarly, addition of reconstituted nucleosomes to a transcription system for the *Xenopus* vitellogenin gene results in a five- to ten-fold stimulation of its activity. This is due to the positioning of the nucleosomes between a distant recognition element for the estrogen receptor and a proximal promoter element that results in the enhancement of interactions between proteins bound to faraway sites (161; see below).

DNA Chaperones: DNA-Bending Proteins Working as Enzymes

One effect of DNA-bending proteins is to stabilize an otherwise loose structure in a particular conformation that is able to sustain the assembly of additional proteins into a higher-order complex. In these cases, the role of DNA-bending proteins resembles the function of certain chaperones, which stabilize polypeptides in a folding step that is critical for subsequent tridimensional assembly of the entire protein (189). Similarly, DNA-bending proteins play an early role in the assembly of nucleoprotein complexes by organizing a longer or shorter DNA segment into a curved configuration, where other proteins may later become docked. At that point, DNA-bending proteins may be displaced from the DNA by those that form the final assembly. The net effect of protein-induced bending is, therefore, that of facilitating formation of the nucleoprotein complex. For this reason, DNA-bending proteins behave operatively in this context as virtual enzymes. To this end, the DNA chaperones should (*a*) bind the nucleotide sequences with little or no specificity, (*b*) induce bending, and finally (*c*) form complexes with the DNA unstable enough to allow their displacement by the components of the final complex (116, 189).

Proteins with DNA chaperone properties exist in both prokaryotic and eukaryotic cells. The archetype of the bacterial proteins of this class is HU, which binds to the minor groove of the DNA without apparent specificity and is able to induce an array of conformational changes in the nucleotide sequence (37, 80). In addition, HU binds to unusual DNA structures, such as cruciforms (18, 147). Although the best-known case of HU protein is that from *E. coli* (37, 114), HU equivalents have been found in virtually every bacterium in which it has been searched for (28, 76), and it has been particularly well studied in the case of *Bacillus stearothermophilus* (184) and *Bacillus subtilis* (6, 7). The eukaryotic counterpart of HU (14) appears to be the chromatin-associated proteins that belong to the high mobility group, namely HMG1 and HMG2 (11). Similarly to HU, these proteins bind and bend DNA with limited (if any) sequence selectivity (144). Both HU and HMG1/HMG2 play crucial roles not only in transcription but also in an entire collection of cellular processes, including DNA replication transposition and recombination, which is consistent with their primary role as DNA chaperones (22, 71).

Despite the functional relationships between HU proteins and HMG1/HMG2 proteins, they are very different structurally. The most salient structural motif of HU proteins is a β-sheet protruding from the rest of the protein in an arm-like configuration (114, 184), whereas the DNA-binding domain of HMG1 consists of three L-shaped α-helices (151, 198). These structural differences hint at a case of divergent evolution for the same function as DNA chaperones. Interestingly, the presence of HMG-type DNA binding domains has been reported in the transcriptional regulator CarD of *Myxococcus xantus* (117), thus suggesting that there is not a barrier for these structural motifs in the eukaryotic and prokaryotic worlds. Along the same line, the yeast mitochondrial histone HM has been suggested as the missing link between bacterial HU and the nuclear HMG1 proteins (108).

According to the view that DNA chaperones act as enzyme-like elements, transient complexes between these proteins and DNA would be finally displaced by the components of the final higher-order structure. This notion has been directly tested in the assembly of the nucleoprotein complex involved in the Hin invertasome, which mediates site-specific gene inversion in *Salmonella* (129), and in which the structural role of the IHF protein could be substituted by either HU or HMG1 proteins. However, while it was possible to detect the presence of IHF in the final complex, DNase I footprinting failed to reveal HU bound to DNA. In fact, only the use of very sensitive agents for detection of short-lived complexes could indicate interactions of HU and the target DNA (99). These results suggest that in contrast to IHF, which acts as an authentic structural element to sustain the complex, the role of DNA chaperones is mostly transient.

The cellular functions of HU/HMG1-like chaperones are not restricted to being catalysts of protein-protein interactions; they seem also to stimulate DNA-protein interactions, at least in some cases. For instance, binding of the CAP protein and the LacI repressor to their target sites can be enhanced 10- to 20-fold by the HU protein in vitro, although HU does not appear to be present in the final DNA-protein complex (49). Similarly, HU assists the binding of IHF to the *E. coli* origin of chromosomal replication, *oriC* (17), in a remarkable case where the effect of HU is not to replace IHF (see next section) but to increase its binding to a target DNA.

Functional Substitutions Between DNA-Bending Proteins

An important property of HU- and HMG1-type DNA chaperones is their ability to replace, at least in part, specific DNA-bending proteins at their target sites. Such an ability was first tested in the assembly of the protein complex termed intasome, which mediates site-specific integration of the λ phage to the bacterial chromosome (64), in which IHF could be replaced by HU. More recently, functional substitutions of IHF by HU have been extended to promoters in which IHF acts as either a negative or positive coregulator by virtue of its structural effect on binding and bending DNA. In one case, both HU and HMG1 could substitute IHF in vitro during the formation of the repression complex of the Mu phage (13). A more intriguing case is the substitution of IHF by HU in the coactivation of σ^{54}-dependent promoters (20, 133, 135). This is one of the few cases in the prokaryotic world of control at a distance, since the activators bind to upstream activating sequences (UAS) placed at considerable distances (> 100 bp) in respect to the RNA polymerase binding site (see below). In a class of σ^{54}-dependent promoters, contacts between the enzyme and the activator bound to the UAS is facilitated by the binding and bending of IHF to a site present within the intervening region (see below). At least in two cases (20, 135), HU and HMG1 could mimic the stimulatory effect of IHF protein during transcription of these promoters in vitro. Interestingly, the substitution of IHF by HU occurs in vitro also in the σ^{54}-dependent promoter *Pu* of *Pseudomonas putida* (1, 26, 141, 131), although the replacement is not fully efficient. Furthermore, in other σ^{54}-dependent promoters lacking an IHF site, HU appears to be the authentic coregulator of the system (20, 133). The corollary to these observations is that the functional substitution of IHF by HU in vivo and in vitro for coactivation of this class of promoters (and perhaps in other systems as well) may not be simply a side effect of their intrinsic properties as DNA chaperones. It could reflect also a functional redundancy of their activities that ensures the basal functioning of the many systems that require the concourse of static or protein-induced bending. In addition, this effect raises interesting possibilities for understanding the evolution of transcriptional control systems (27; see below).

An important issue on the redundancy between specific DNA-bending proteins and DNA chaperones is how nonspecific proteins are able to replace the bends at distinct positions. Studies on the integration of phage λ (166) suggest that although DNA chaperones (such as HU) randomly associate to DNA, only the small fraction of complexes in which the binding sites involved in the final complex are properly aligned will result in a productive nucleoprotein formation. This effect, named structural trapping, is anticipated to be more significant if the DNA site involved is already bent or prone to bend in a particular orientation; the physical properties of the sequence act then as a guiding element for curving the DNA in a certain direction. Along the same line, the mechanism of IHF-mediated stabilization of the nucleoprotein complexes that repress phage Mu (13) reveals the requirement of a such a guiding element, i.e. intrinsically bent DNA, for the preferential action of heterologous DNA chaperones (HU or HMG1) on specific DNA segments.

CONSEQUENCES OF DNA BENDING

Channeling Signals Through Promoter Architecture

An important regulatory consequence of the requirement for proteins acting as architectural elements is that they become regulatory assets through their ability to sustain or inhibit an active promoter configuration. In this way, regulation assisted by protein-induced DNA bending allows the superimposition of different regulatory levels for the control of a single promoter. This is particularly useful for individual prokaryotic promoters, where their relatively simple organization and the few protein targets available (as compared to their eukaryotic counterparts) would not allow responses to simultaneous signals unless coregulation elements such as DNA bending are present in the system.

CAP AS AN ARCHITECTURAL ELEMENT   An archetypical case of the proteins that channel environmental signals into promoters by altering their geometry is that of CAP (cAMP receptor protein or CRP) of E. coli, some aspects of which have been discussed above. Along with IHF, CAP is the best-studied example of the proteins that introduce a bend at the site of interaction with DNA (39, 89). This protein has been the subject of intensive studies, because of its involvement in the phenomenon called catabolite repression, which consists of the lack of expression (i.e. lack of activation) of a number of operons when cells face glucose in the medium. In the absence of this sugar, cAMP levels increase and dimers of the CAP-cAMP complex bind to specific sequences at target promoters, bringing about a sharp bend in the bound DNA (201). In most cases, CAP activates promoters through direct protein-protein contacts with the RNA polymerase (19) and some contribution of the upstream DNA bend caused by the binding of the regulator (see 89, 140 for reviews). However, there are promoters in which catabolite repression is just one of the various signals the system has to integrate and in which altering DNA architecture is the only possibility of entering an additional regulatory level. The diverse ways by which the DNA-bending ability of CAP enters a level of catabolite control in promoters regulated primarily by other signals is truly remarkable. Some of these systems are discussed below.

The activity of the divergent *malEp* and *malKp* promoters of the maltose operon of *E. coli* depends both on MalT, the cognate regulator of the system, and CAP (165). The intervening region between the two promoters (217 bp long) is a nearly continuous stretch of operators for both proteins, in which two series of MalT sites appear separated by various CAP sites (148). Analysis of the mechanism whereby CAP and MalT coactivate both promoters has revealed that CAP induces the repositioning of MalT within the proximal region of *malEp* and *malKp* from nonproductive but high-affinity sites to productive but low-affinity sites, which are displaced by 3 bp from the former (154). In this context, CAP promotes the assembly of the active complex by merely bending the DNA in the intervening region. This facilitates the cooperative binding of MalT and the shifting of the equilibrium toward occupation of the productive MalT sites (Figure 2). This notion has been confirmed by the successful replacement of the CAP binding sites by either intrinsically curved DNA or by IHF-binding sites (153).

Another interesting example of CAP as a coregulator is its role in the repression complex that controls expression the divergent *nagE-nagBACD* operons of *E. coli*, which are involved in the uptake and metabolism of *N*-acetylglucosamine (145). In this system (Figure 3), *nag* genes are induced by growth on *N*-acetylglucosamine or glucosamine but are also subject to catabolite control. In the absence of inducers, the repressor NagC binds to cognate sites overlapping the *nagE* and *nagB* promoters, which are themselves separated by 130 bp, thus resulting in the looping-out of the intervening sequence. A strong CAP-binding site appears positioned between the two NagC operators in such a way that CAP binding and subsequent DNA bending stabilize the loop formed between the two NagC binding sites, thus improving the repression of *nag* genes and placing the system under the control of CRP-cAMP (146).

The repressor called CytR regulates transcription initiation from a number of promoters involved in the metabolism of nucleotides in *E. coli*. Various of these

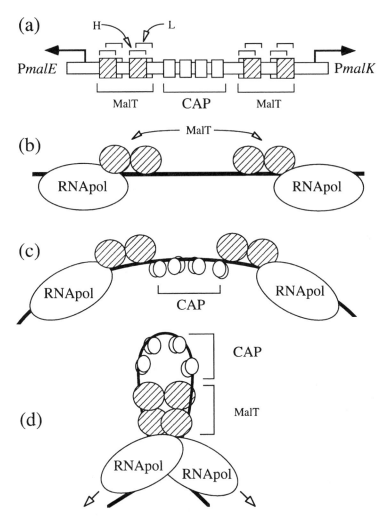

Figure 2 Activation of divergent promoters *malEp* and *malKp* of *E. coli* by MalT and CAP. (*a*) The linear arrangement of all control elements. These include catabolite gene activator protein (CAP) and MalT sites, the later bearing high-affinity (*H*, diagonal lines) or low-affinity (*L*, open) sequences for the regulator. *H* and *L* sites are separated by three nucleotides, so they overlap to an extent. All four schemes summarize the sequence of events leading to the activation of the divergent promoters in the presence of maltose and absence of glucose. To this end, both MalT and CAP must bind the region. The effect of CAP is that of bending the DNA so that MalT becomes repositioned at the productive sites (*L*) that guide its interactions with the RNA polymerase (RNApol).

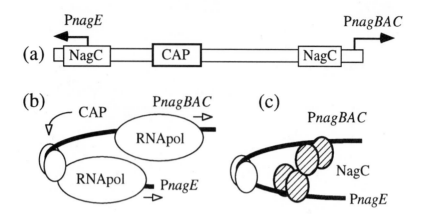

Figure 3 Assembly of a repression/activation complex at the *nag* promoters of *E. coli*. (*a*) The divergent promoter region *PnagE/PnagBAC* includes two sites for the NagC repressor, along with a catabolite gene activator protein (CAP) site more proximal to *PnagE*. (*b*) The CAP protein bound to its site activates *PnagE* by contacting the RNA polymerase (RNApol), whereas it has no effect on *PnagBAC*, which is expressed constitutively. (*c*) The DNA bend caused by CAP binding assists the formation of a repression complex that involves NagC molecules bound to distant sites and prevents the entry of the polymerase.

promoters also bear CAP sites, so the combination of CytR sites with CAP sites originates an entire repertoire of regulatory possibilities (54, 191). One case is that of the *cytR* gene promoter itself. P*cytR* is activated by CAP, the effect of which is counteracted by CytR (130). To this end, CAP binds to a region located around position −64, whereas CytR binds to sequences immediately downstream. Both proteins can bind simultaneously and cooperatively the P*cytR* promoter, a process that implies direct interactions between them. Interestingly, both proteins bend DNA at their target sites, but in different orientations, in such a way that the repression efficiency of CytR lies, at least in part, in its ability to counteract the CAP-induced bend (130). In a further level of complexity, CRP seems also to change the sequence-specificity of CytR for its target site, an effect that could be induced by protein-protein interactions between the two regulators (130a).

The control of the P*araBAD* promoter of the arabinose system of *E. coli* is one of the best-known paradigms of prokaryotic gene expression (162). P*araBAD* is both positively and negatively regulated by the same protein, AraC, which in the absence of arabinose mediates the formation of a repressor loop by binding two half-operator sites separated by 210 bp, namely *araO2* and *araI1* (Figure 4). Alternatively, AraC binds in the presence of arabinose to a complete site (*araI1/araI2*) next to −35 box of P*araBAD* (77). The repression loop

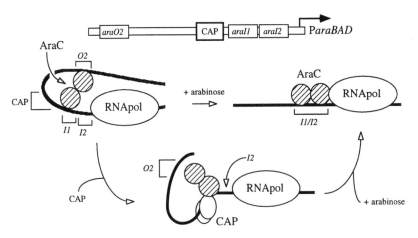

Figure 4   Antirepression of the arabinose promoter of *E. coli* by a catabolite gene activator protein (CAP). The P*araBAD* promoter region contains proximal and distal binding sites for AraC, as well as a CAP-sequence adjacent to the AraC proximal site. In the absence of arabinose, a repression loop is formed between AraC molecules bound to distant half-sites, which prevents occupation of the AraC half-site more proximal to RNA polymerase (RNApol). Arabinose destroys such a loop and allows occupation of the *I2* half-site and the ensuing AraC-RNApol contacts required for transcriptional activation. The absence of glucose further contributes to this effect, since CAP binding bends and misorients the upstream region involved in the repression loop, thereby indirectly favoring occupation of *I2*.

plays an anti-induction role, thus preventing AraC from binding to the activator site *araI1/araI2* (102). Besides the AraC-binding sites, the regulatory region includes a CAP site just upstream from the *araI1* half-operator involved in the loop. The CAP site is centered along the face of the DNA helix opposite *araI1*. It is believed that the role of the bend induced by CAP is not that of helping formation of an activating nucleoprotein complex. On the contrary, CAP inhibits the repression loop formed by AraC by misorienting the intervening sequence, thus stimulating P*araBAD* activity in conditions where glucose is not present (103).

COREGULATION OF σ^{54} PROMOTERS THROUGH DNA BENDING   The positive and negative control of promoter architecture by DNA-bending proteins may become quite sophisticated in some σ^{54}-dependent promoters. These promoters are generally involved in expression of functions for adaptation to harsh metabolic and environmental situations (95), and they are unique in that the sequence recognized by the σ^{54}-RNA polymerase holoenzyme includes GG and GC doublets at positions -24 and -12, respectively, instead of the typical -10 and -35 hexamers of the σ^{70}-promoters. As mentioned above, σ^{54}-promoters are activated at a distance (100–200 bp) by specific regulators bound to upstream enhancer-like sequences that must loop out for productive contact with the polymerase (see 94 for a review). This arrangement of elements makes the intervening DNA sequence between the binding sites for the activator (UAS) and the σ^{54}-RNA polymerase an ideal target for factors that promote or inhibit the correct promoter geometry required for activation.

In most σ^{54}-promoters examined, an IHF site is found at such intervening regions. Since the absence of IHF results in a drastic drop in promoter activity, the role of this protein in these systems appears to be to assist in the formation of a DNA loop between the UAS-bound regulators and the polymerase (1, 21, 26, 60, 62, 81). This notion is supported by the possibility of effecting functional substitutions of the IHF sites by intrinsically curved DNA in the σ^{54}-dependent promoter *Pu* of *Pseudomonas putida* (141) and the *PnifH* promoter of *Klebsiella pneumoniae* (112).

More recently, the intervening region between the UAS and $-12/-24$ of σ^{54}-dependent promoters has been shown to be the target of negative coregulation. One example of this is the down-regulation that the regulator Nac of *Klebsiella aerogenes* exerts on its own transcription. Nac is the activator of a set of genes involved in the metabolism of several amino acids, but it is also the repressor of other genes required for nitrogen assimilation in the absence of amino acids (12). Expectedly, the *nac* gene is transcribed from a σ^{54}-promoter activated by the prokaryotic enhancer-binding protein NtrC, which responds to nitrogen limitation (44). For autoregulation, Nac happens to bind to a specific site located exactly between those for σ^{54}-RNA polymerase and NtrC. Binding of the Nac protein to its target at the *nac* promoter results in the bending of the DNA, but it has no effect on the binding of either σ^{54}-RNA polymerase or NtrC to the region. These results indicate that Nac down-regulates the *nac* promoter through an antiactivation mechanism that prevents the interactions between the activator and the polymerase by misorienting the DNA in a nonproductive geometry (45).

A second example of negative coregulation of σ^{54}-systems through DNA bending is provided by the mechanism at work in the catabolite repression of some *Bacillus subtilis* promoters, which differs substantially from its counterparts in *E. coli*. For instance, expression of the levanase operon of *B. subtilis* is driven by a σ^{54}-promoter that requires the activation of the regulator LevR bound at a distant site (25, 180). In addition, transcription of the levanase genes needs the inactivation of the CcpA repressor, which mediates the catabolite control of the system (107). Since the CcpA protein binds to the intervening region between the LevR binding sites and the $-12/-24$ sequences, it has been suggested that its repressive effect lies in its ability to counteract the activator loop required for transcription of the σ^{54}-dependent promoter (107).

EUKARYOTIC COUNTERPARTS?   As opposed to prokaryotic systems, the complexity of the transcriptional machinery of eukaryotes makes available a large number of potential targets for regulation. However, in these cases, the DNA is employed also as a regulatory asset to channel signals into a particular system. Protein LEF-1, a lymphoid-specific member of the family of factors containing HMG-type DNA-binding domains, binds one specific site in the center of the human TCRa enhancer (55, 57). LEF-1 cannot stimulate transcription by itself but collaborates with other enhancer-binding factors (56). On the basis of the observation that LEF-1 induces a sharp bend in the DNA helix, it has been proposed that, similarly to the prokaryotic counterparts, this protein acts as an architectural element that facilitates the interaction between the proteins bound at the sequences flanking the LEF-1 site on either site (70).

The c-*fos* promoter provides another paradigm for the role of protein-induced DNA bends in eukaryotic transcriptional regulation. One binding site for the Zn-finger protein YY1 is located in this promoter between the TATA box and the upstream site for the CREB protein, a factor that is partly responsible for promoter activation (115). In this system, YY1 bends the DNA and represses promoter function, presumably by preventing the interactions between the activator CREB protein and the components of the basal transcription machinery, in a fashion similar to that described above for the repressors in σ^{54}-dependent promoters (115).

Nucleosomes broaden the repertoire of regulatory devices available to the eukaryotic cells by employing the DNA-bending ability of the histone octamer. As mentioned above, the *Xenopus* vitellogenin B1 promoter (161) contains a nucleosome positioned between an estrogen-responsive enhancer and the binding sites for two transcription factors: the hepatocyte nuclear factor 3 (HNF3) and the nuclear factor 1 (NF1). This nucleosome creates a static loop that brings together the enhancer with the promoter proximal elements, stimulating transcription as a result. An intriguing aspect of this process is the nature of the signal(s) that guide the histones to associate with a particularly strategic position. A recent study (4) has shown that the transcription factor NF1 interacts specifically with the histone H3, thereby directing the positioning of the entire nucleosome in a particular location. These results open the possibility that proteins unable to bend DNA guide the effect of other DNA-bending proteins in order to stimulate transcription as part of an extended regulatory nucleoprotein complex.

DNA Bending and Transcriptional Promiscuity

Typically, enhancers (whether eukaryotic or prokaryotic) control transcription at a distance, in different orientations, and even located downstream from the expressed gene (75, 94). As a consequence, it can be anticipated that the same enhancer sequence may act on various promoters. This is not necessarily a disadvantage, since a degree of nonspecificity is an evolutionary asset for adaptation to novel environments (27). At other times, however, this may become a problem, since nonspecific activation of many promoters would be deleterious (113). Therefore, it seems necessary for cells to suppress, or at least control, what has been termed transcriptional noise (27).

A number of examples can be found in the literature where transcriptional cross-regulation is restrained through structural DNA elements and DNA-bending proteins. The bacterial σ^{54}-dependent promoters provide again an example of the role of protein-induced bending in the suppression of cross-activation in vitro. As mentioned above, it is currently believed that the role of IHF in σ^{54}-promoters is to facilitate contact between the activator protein attached to the UAS and the σ^{54}-polymerase prebound to the promoter. The same loop sustained by IHF that brings the two proteins in close proximity may, at the same time, restrict the flexibility of the region. An IHF-induced bend can inhibit the function of an activator bound to a site that is not correctly placed with respect to the bend (21). Santero et al (159) suggested that IHF-induced bends in σ^{54}-promoters ensure high fidelity and high efficiency of activation by the cognate regulator while disfavoring the action of heterologous activators from solution. This hypothesis has been tested in the σ^{54}-promoter Pu from $P.\ putida$. The Pu promoter is activated at a distance by the XylR protein (132, 134, 149). Similarly to other σ^{54}-systems (see above), the intervening DNA segment between the UAS and those for RNA polymerase binding contains an IHF site that is required for full transcriptional activity (1, 26). When placed in $E.\ coli$ cells lacking IHF, the Pu promoter not only drops its activity, it can also be cross-activated by other members of the family of regulatory proteins that act in concert with σ^{54} (46, 131). Such illegitimate activation does not require the binding of the heterologous regulators to DNA and can be suppressed by bent DNA structures, either static or protein-induced, between the UAS and RNAP recognition sequence (131). The term "restrictor" has been proposed to describe this specificity-enhancing function assigned to protein-induced or static bends in σ^{54}-promoters (27, 46, 131), although the concept may have a more general value.

Indications of the restrictor effect of IHF on other σ^{54} promoters can be traced in the literature, even if the phenomenon was not identified as such at that time. For instance, Claverie-Martín & Magasanik (20a) reported that deletion of the UAS of the IHF-dependent promoter $glnHp2$ of $E.\ coli$ resulted in a promoter variant that could be activated from solution in vitro by NR_I (NtrC), but only in the absence of IHF.

Similarly, IHF inhibited transcriptional activation of $PnifH$ by $Azotobacter$'s NifA on a supercoiled DNA plasmid lacking the UAS but still containing an

IHF binding site (8a). Along the same line, Kustu and coworkers detected that activation of the *PnifH* promoter of *Klebsiella* by the central domain of NifA (i.e. an active form of the protein unable to bind DNA) was inhibited by IHF (12a). On the contrary, IHF stimulated activation of the same promoter by full-size NifA (12a).

We think DNA bending has evolved to control the degree of promiscuity of certain promoters, in particular those of the σ^{54} type. This is based on the observation that some of these promoters have optimal IHF sites (i.e. have a restrictor element), that others have suboptimal binding sequences, and that others do not have IHF sites at all. This suggests that protein-induced bending provides these promoters with an additional level of control for fine-tuning their physiological activity (Figure 5). This concept is consistent with the observations made on the regulation of two σ^{54}-promoters of the TOL plasmid of *P. putida* (149). Both *Pu* and *Ps* are positively regulated by XylR, but *Pu* contains an IHF site (see above) whereas *Ps* does not. Interestingly, the requirement

Activator/UAS

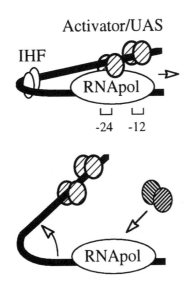

Figure 5   Restrictor effect of integration host factor (IHF) on σ^{54}-dependent promoters. Most (but not all) promoters of this class contain an IHF site placed between the binding sequences for the σ^{54}-containing RNA polymerase (bound to $-12/-24$) and the upstream activating sequences (UAS) for the cognate activator. Besides assisting the assembly of an optimal promoter geometry, IHF seems to sustain a restricted conformation of the entire promoter region that prevents an illegitimate activation of the RNApol by other regulators of the NtrC family. In the absence of IHF, some of these noncognate regulators may gain access to the promoter from solution and cause a degree of basal transcription.

for the products they transcribe are different: In the absence of m-xylene, the restrictor-containing Pu promoter remains totally silent, while the restrictor-less Ps promoter still has a considerable level of basal transcription, owing in part to cross-activation by other regulators of the NtrC family (46, 133). This allows growth on m-xylene through transcription of the upper and the *meta* operons due to the XylR-dependent activity of Pu and Ps. But also, cells can grow on m-toluate alone through the activity of the *meta* pathway caused by the XylR-independent transcription of XylS from Ps, while keeping the upper pathway totally silent. This effect is due also to the presence of a low-constitutive σ^{70} promoter in front of the *xylS* gene (149). In any case, at least in the TOL system, the presence or absence of the DNA bend caused by IHF seems to displace the responsiveness of each promoter, Pu and Ps, either toward specificity or toward promiscuity for the sake of a proper physiological balance.

One potential case of restrictor elements in eukaryotic systems seems to be present in the assembly of the three-dimensional complex of the virus-inducible enhancer of the human β-interferon gene. This enhancer is bound by the transcription factors IRF (interferon regulatory factor), by NF-κB, by ATF-2/c-Jun, and by the DNA-bending protein HMG1(Y). Within this complex, HMG-I(Y) recognizes sequences present in the center of the NF-κB site and flanking the nonadjacent ATF-2/c-Jun binding site through contacts with the minor groove of the DNA (185). Unlike IHF in σ^{54}-systems, HMG1(Y) contacts directly with the other proteins, i.e. NF-κB and ATF2, and appears to induce in them conformational changes that increase their interaction with the DNA and with each other (38). In this case, HMG1(Y) acts as an architectural protein that contributes to formation of a higher-order structure by optimizing DNA geometry and protein conformation. Interestingly, synthetic enhancers lacking HMG1(Y) binding sites were still able to activate transcription but they displayed unusually high levels of basal promoter activity and became less responsive to viral infection than the enhancer with HMG1(Y) sites. Furthermore, the synthetic enhancers were activated also by other non-viral inducers (185), thus suggesting a restrictor effect of HMG1(Y) in the system. It should be noted, however, that in this case such an effect has not been correlated with the DNA-bending ability of HMG1(Y).

DNA Bending in Response-Amplification Mechanisms

A final regulatory aspect of DNA bending is its role in the amplification of minor environmental signals into major transcriptional responses. An interesting example of this type is found in the regulation of the Pe and Pc promoters of phage Mu, the balance between their activities controlling lysis-lysogeny decisions during the developmental cycle of the phage (65). Pe is an early promoter that drives expression of genes required for lytic growth, whereas the Pc

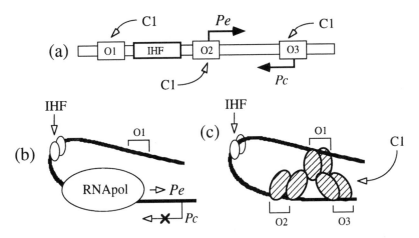

Figure 6   Stimulation of lysis/lysogeny of phage Mu by DNA bending. (*a*) The DNA segment spanning the promoter region *Pe/Pc* contains sites for RNA polymerase (RNApol) binding within the divergent promoters, which overlap two sites (O2 and O3) for the Mu repressor (the C1 protein). The region also contains an IHF site as well as an additional sequence for C1 binding (O1). When C1 concentration is low (*b*), the DNA bending caused by IHF stimulates transcription of lysis genes from the *Pe* promoter, while inhibiting transcription from *Pc*. IHF also assists in relieving the structural repression caused by H-NS on the *Pe* promoter. On the other hand, an increase in C1 levels leads to the formation of a repression complex (*c*) that includes C1 molecules bound to distant sites. This complex is assembled with the structural assistance of the same DNA bend produced by IHF.

promoter transcribes the gene *c1* for the Mu repressor (Figure 6). Normally, this repressor prevents transcription from both *Pe* and *Pc* by binding to cognate sites that overlap in each case the respective promoter sequences, thereby fixing the phage in a lysogenic state. Although the two promoters (along with their cognate operator sequences for the repressor, O1 and O2) are separated within the Mu genome, the repressor must bind both of them simultaneously to shut down transcription from *Pe*. Binding of the repressor to O1 and O2 sites is very cooperative and seems to occur simultaneously rather than sequentially. As is seen in many other cases, bringing two distant elements into close proximity must involve changes in the conformation of the intervening DNA sequence. The region between *Pe* and *Pc* has a minor intrinsic bend, which is further exacerbated upon the binding of IHF to a specific site also present in the sequence. In turn, IHF binding results in a tighter binding of the repressor proteins to O1 and O2 and, therefore, in a decrease of promoter activity. This effect is due to the stabilization by IHF of a loop that engages the occupied O1 and O2 sites, allowing protein-protein interactions between repressor molecules (3, 13, 52).

However, in the absence of Mu repressor, IHF stimulates transcription from *Pe* and inhibits transcription from *Pc*, thus favoring lytic growth (91). Therefore, depending on the status of the repressor, IHF-induced bends may stimulate either lysis by activating *Pe* or lysogeny through stabilization of repressor complexes. At least for Mu, an IHF-induced bend seems to behave as an amplifier of developmental decisions, the ultimate sign of which is determined by other factors, mainly those affecting repressor concentration. Recent observations (193) indicate that under conditions of lytic growth, IHF assists also to relieve the repression of the *Pe* promoter caused by the histone-like protein H-NS. Since H-NS is involved in the maintenance of chromosomal superhelicity (see below), the antagonistic effect of IHF could be one way to bring the system under one more coregulation check.

CHROMOSOME ORGANIZATION AND DNA STRUCTURES

The chromosome of *E. coli* is a single, supercoiled circular molecule of about 4.7 Mb (i.e. about 1 mm long) that is condensed into a nucleoid structure. It has been generally believed (143, 200) that such a nucleoid was organized in large superhelical independent domains, the structure of which is influenced by the action of topoisomerases and the binding of histone-like proteins. There is increasing evidence, however, that the building of a higher-order chromosomal structure is basically a stochastic process (79). Various attempts have been made (111, 129a, 172) to identify in vivo regions of differential supercoiling in the chromosome of *E. coli* by using genetic probes (transposons or minitransposons) bearing *lacZ* fusions to supercoiling-responsive promoters such as those for the genes of the DNA gyrase subunits (*gyrA* and *gyrB*). Every study has consistently reported that the variable expression of promoters placed at different positions in the chromosome is due exclusively to their proximity to the replication origin and not to the effect of differential supercoiling. Although barriers to supercoil diffusion exist (200), the DNA segments engaged in formation of superhelical domains vary from cell to cell or within one cell over time (79). Therefore, a sequence-specific overall architecture of the bacterial chromosome does not seem to exist as such (at least in *E. coli* and *Salmonella*).

DNA Supercoiling and Coregulation of Gene Expression

At a more local level, however, the chromosomal context and the physical structure of the DNA region can influence the architecture of transcription complexes. Supercoiled DNA appears in vitro mostly as a rod of two interwound duplex chains (2, 15). This configuration predicts the appearance of curved apices in which the DNA folds back upon itself. Obviously, such apices are favored to become coincident with intrinsically or protein-induced bent DNA sequences (97). The tendency of bent DNA to nucleate DNA folds in supercoiled DNA suggests an interesting link between DNA bending, supercoiling, and the assembly of transcription complexes, because the assembly of productive geometries may be enhanced or inhibited by supercoiling (152). Since chromosomal superhelicity varies during growth depending on the ATP/ADP ratio (194), the resulting changes in local DNA geometry might be a mechanism of sensing physiological conditions and channeling such general signals into specific promoters. In this respect, it is interesting to note that the movement of interwound supercoils might be affected by sequence-specific and sequence-nonspecific, nucleoid-associated DNA-binding/bending proteins such as HU, IHF, FIS (factor for inversion stimulation), and H-NS (79; see below). Furthermore, there is a change in the abundance of these proteins along the growth curve (9, 10, 30, 32), suggesting that the interplay between supercoiling and DNA-bending proteins provides a major level of gene expression control in connection with the physiological status of the cells. Another level of general metabolic control of promoter activity could be related to the overall effect that internal K^+ glutamate pools could have on the affinity of DNA-binding proteins to chromosomal DNA. It has been reported that the external levels of ammonium determine not only the regulation of nitrogen-related genes in *Salmonella*, they also alter drastically the internal pool of potassium glutamate (201a). It is tempting to speculate that, in turn, this may influence the binding of an entire collection of regulatory DNA-bending and nonbending proteins to their respective targets. In this way, such a major metabolic condition (nitrogen limitation) could end up having an overall effect on virtually every promoter.

Nucleoid-Associated Proteins and DNA Bending

It is remarkable that most bacterial proteins known to bend the DNA are generally associated with the nucleoid, thus suggesting the existence of nucleosome-like structures in the prokaryotic chromosome. The best-studied proteins of this type in *E. coli* (that many authors design generically as histone-like proteins) include not only the above-mentioned HU and IHF factors, but also H-NS and an entire collection of additional small proteins able to bind and/or bend DNA with a variable degree of sequence specificity. These include FIS (47), LRP (leucin-responsive protein; 24), and Dps (DNA-binding protein for starved cells; 5) and their functional equivalents in other genera (76, 168). In this section, we deal exclusively with aspects of some histone-like proteins that are relevant to this article.

One major class of nucleoid-associated proteins includes HU and IHF of *E. coli* (and their functional equivalents in many other genera), as well as the TF1 protein of *Bacillus* phage SP01. The structure and properties of HU and

IHF have been reviewed (50, 114), and some of their coregulation roles are addressed above. The HU protein (158) is a heterodimer of two very similar, but nonidentical, subunits of 9.5 kDa in *E. coli* and *Salmonella typhimurium* but a homodimer of similar size in many other prokaryotic species (28, 163). Interestingly, IHF homodimers of *E. coli* are not synthesized in an active form, whereas HU homodimers are perfectly active, suggesting that bacteria of this genus can tolerate the loss of one of the two HU genes (called *hupA* and *hupB*). In addition, HU displays a significant structural homology with IHF, which makes possible the construction of chimeric active HU-IHF hybrids (61, 67). However, as mentioned above, HU binds DNA, apparently without much sequence specificity, increasing the flexibility of the bound DNA and thus facilitating formation of bent structures (80, 105, 169). It is tempting to speculate on the evolutionary meaning of such similarities in light of the various observations (see above) on the functional substitutions of IHF by HU in vitro and in vivo. It looks as if bacterial cells had evolved a hierarchy of nucleoid-associated factors that cover all needs of DNA bending, with each spanning a different range of sequence specificity, so that in case of need the less specific ones act as functional backups of the more specific ones. An extreme case of specificity would be IHF, which requires the assembly of its two subunits for activity in vitro. In the absence of IHF, HU takes over the basic cellular needs for DNA bending, albeit less efficiently. But even if one of the two HU subunits fails, the other can still work by forming homodimers. The pleiotropism of HU mutations (126) and the presence of HU-like proteins in many bacterial genera (76) pinpoints this type of protein as one key element of overall chromosome functioning. The properties of TF1 are similar to those of IHF and HU, but this factor binds preferentially to SP01 virus DNA, which has 5-hydroxymethyl-uracil instead of thymine (83), thereby showing a strong binding preference for sites at the SPO1 genome (69). TF1 bends sharply its target sequence (164), thus resulting in the preferential packaging of the viral genome into phage heads, although a role in the transcription of SP01 promoters has not been proved.

Recent work made with IHF has revealed two novel aspects of the protein that are important for its role as a global coregulator of transcriptional activity and that could be extended to other nucleoid proteins. One of them is its responsiveness to temperature. Giladi et al (58) have shown that IHF binds to its target sequence at the λP_L promoter more avidly at low temperatures, thus resulting in an increase in transcriptional activity in that particular promoter, where IHF acts directly on the RNA polymerase. If, in general, IHF binding to DNA is dependent on temperature (as with many other proteins), it is likely that such environmental signals can be channeled into every nucleoprotein complex where IHF acts as an auxiliary factor (see above), thus imposing a level of physiological control on the activity of individual promoters. Another interesting finding is the ability of IHF to activate at a distance ($-80/-100$ nucleotides) the *ilvPG* promoter of *E. coli* in the absence of any protein-protein interaction, even when the relative orientation of the IHF-caused bend and the RNA polymerase is reversed (127). That the effect is not observed in a linear template suggests that the combination of negative supercoiling and a severe distant bend causes a conformational change at the downstream promoter site that facilitates transition from a closed to an open complex. IHF can be replaced in the same system by the eukaryotic protein LEF-1 (see above) along with a cognate site, which indicates that besides architectural effects, DNA-bending proteins may coregulate transcription of certain promoters through a structural transmission event separate from their action on DNA geometry.

Another major nucleoid-associated protein is H-NS, an abundant neutral histone-like polypeptide of 15.5 kDa (37, 82, 142, 163). H-NS is not a strong DNA bender, but it does bind and stabilize curved structures without much sequence specificity. H-NS participates actively in the control of chromosomal supercoiling (78, 82) and seems to perform as a general transcriptional silencer (8, 43, 84, 119, 122, 171, 187), acting sometimes over long DNA segments. Since changes in growth conditions (temperature, osmolarity, nutrients) determine DNA topology (30, 82), H-NS protein may bind to local DNA structures characteristic of a certain physiological status, thus activating subsets of genes and silencing others by virtue of its effect on promoter geometry.

FIS is another small (98 amino acids) nucleoid-associated, basic and dimeric protein, which is involved in a whole set of recombination, replication, and transcription reactions. Although it is structurally unrelated to IHF or HU, it shares some of their properties, in particular the ability to bend sharply target DNA, albeit with a relatively low sequence specificity. The role of FIS in transcription is best known for its direct stimulation of ribosomal promoters (51, 68, 100, 151, 203). However, its activity on events unrelated to transcription, such as the stimulation of recombination by DNA invertases (47, 88), indicated also a role as an architectural element in the assembly of nucleoprotein complexes. Similarly to IHF, FIS levels in *E. coli* vary dramatically in response to nutritional conditions, but in an opposite direction: While IHF increases its intracellular concentration at the onset of the stationary phase (9, 32), FIS levels become undetectable during slow growth but quickly reach >30,000 dimers per cell when bacteria are shifted to rapid growth in rich medium (10, 76).

Another ubiquitous DNA-bending factor in *E. coli* is the leucine-responsive regulatory protein (Lrp), which is increasingly recognized as a general regulator of gene expression (24). Similarly to the other nucleoid-associated proteins, Lrp activates the expression of some operons and represses the activity of others (96, 192), and it seems to bind and bend target sequences located at various sites within the corresponding promoters with a loose specificity (197). Lrp is a homodimer (19 kDa each subunit), thus resembling various histone-like proteins. Similarly to IHF, Lrp can regulate transcription through the specific activation of a large collection of individual promoters and by affecting the maintenance of chromosome structure (24, 122).

REP and RIP Sequences

REP sequences, also termed PU (palindromic units), are conserved repetitive extragenic palindromic sequences that account for as much as 1% of the *E. coli* chromosome (104). A REP sequence consists of 30–40 bp arranged as an imperfect inverted repeat separated by a loop of up to 4 bp. They appear in clusters, and over 100 of them have been located on the *E. coli* chromosome (31). Their function is not very clear. They stabilize certain transcripts (177), but they can also play a role in chromosome folding (59). Interestingly, HU stimulates the binding of DNA gyrase to REP sequences (202). Successive REPs within a REP cluster always alternate with respect to their orientation. REP clusters contain additional conserved DNA motifs and were labeled BIMEs (bacterial interspersed mosaic elements). The motifs of the REP clusters allow BIMEs to be divided into classes. One particular motif (called L) is particularly well conserved. Oppenheim et al showed (121) that such an L motif is, in fact, an IHF-binding site. There is, therefore, a subclass of REP clusters (designed RIP sequences) in which two REP elements are separated by a single IHF recognition site. There seem to be about 70 of these RIP elements in the *E. coli* chromosome (121). Although they have not been found in other microorganisms, it is possible that the IHF-binding site can be replaced by another DNA-bending protein or by intrinsic DNA bends. Interestingly, all RIP sequences occur in transcribed but not translated regions of the chromosome (i.e. at the untranslated 3′ ends of mRNAs), suggesting a role in influencing transcript stability. It is possible also that the RIP elements are DNA gyrase-binding sites that help to maintain negative superhelicity in these regions. This would add one more function to the many roles of a simple protein such as IHF, whose only special characteristic is that of bending sharply DNA at distinct sites.

EPILOGUE

A consistent body of evidence supports the concept that static and protein-induced DNA bending has allowed prokaryotic cells to evolve complex signal integration devices for environmental inputs. One raison d'être of regulatory cascades is the possibility of insertion of additional checks, either positive or negative, at every stage of the signal transduction pathway. Since, unlike eukaryotes, the number of channels to enter physiological signals in prokaryotic promoters is limited, bacteria seem to have thoroughly exploited the physical properties of DNA to accomplish additional levels of regulation through changes in promoter architecture. The major consequence of DNA bending in transcriptional regulation is, therefore, that promoter geometry becomes an authentic regulatory asset. In this way, bacteria afford multiple metabolic control levels by just altering promoter architecture, so that positive signals favor an optimal assembly of protein-protein and protein-DNA contacts required for activation, while negative inputs prevent such a productive geometry. Additional effects of regulated DNA bending in prokaryotic promoters include the possibility of amplifying physiological signals and also of increasing or decreasing promoter specificity or promiscuity for cognate regulators. Both aspects appear to play an important role in evolution and adaptation to novel environments.

ACKNOWLEDGMENTS

Work in our laboratory is supported by the ENV4-CT95-0141 (ENVIRON-MENT) contract of the European Union and by grant BIO95-0788 of the Comisión Interministerial de Ciencia y Tecnología. G Bertoni, I Cases, F Rojo, S Kustu, M Carmona, R Giraldo, A Oppenheim, and JC Alonso are gratefully acknowledged for inspiring discussions. We are indebted to Isabel Mendizábal for her help with the manuscript.

> **Visit the *Annual Reviews home page* at**
> **http://www.annurev.org.**

Literature Cited

1. Abril MA, Buck M, Ramos JL. 1991. Activation of the *Pseudomonas* TOL plasmid *upper* pathway operon. Identification of binding sites for the positive regulator XylR and for integration host factor. *J. Biol. Chem.* 266:15832–38

2. Adrian M, ten Heggeler-Bordier B, Wahli W, Stasiak AZ, Stasiak A, Dubochet J. 1990. Direct visualization of super-coiled DNA molecules in solution. *EMBO J.* 9:4551–54

3. Alazard R, Betemier M, Chandler M. 1992. *Escherichia coli* integration host factor stabilizes bacteriophage Mu repressor interactions with operator DNA *in vitro*. *Mol. Microbiol.* 6:1707–14

4. Alevizopoulos A, Dusserre Y, Tsai-Pflugfelderm VDWT, Wahli W, Mermod N. 1995. A proline-rich TGF-b-responsive transcriptional activator interacts with histone H3. *Genes Dev.* 9:3051–66

5. Almiron M, Link D, Furlong D, Kolter

R. 1992. A novel DNA binding protein with regulatory and protective roles in starved *Escherichia coli*. *Genes Dev.* 6:2646–54

6. Alonso JC, Gutierrez C, Rojo F. 1995. The role of the chromatin-associated protein Hbsu in β-mediated DNA recombination is to facilitate the joining of distant recombination sites. *Mol. Microbiol.* 18:471–78

7. Alonso JC, Weise F, Rojo F. 1995. The *Bacillus subtilis* histone-like protein Hbsu is required for DNA resolution and DNA inversion mediated by the β recombinase of plasmid pSM19035. *J. Biol. Chem.* 270:2938–45

8. Atlung T, Sund S, Olesen K, Brondsted L. 1996. The histone-like protein H-NS acts as transcriptional repressor for expression of the anaerobic and growth phase activator AppY of *Escherichia coli*. *J. Bacteriol.* 178:3418–25

8a. Austin S, Buck M, Cannon W, Eydmann

T, Dixon R. 1994. Purification and in vitro activities of the native nitrogen fixation control proteins NifA and NifL. *J. Bacteriol.* 176:3460-65

9. Aviv M, Giladi H, Schreiber G, Oppenheim AB, Glaser G. 1994. Expression of the gene coding for the *Escherichia coli* Integration Host Factor are controlled by growth phase, *rpoS*, ppGpp and by autoregulation. *Mol. Microbiol.* 14:1021-31

10. Ball C, Osuna R, Ferguson K, Johnson R. 1992. Dramatic changes in Fis levels upon nutrient upshift in *Escherichia coli*. *J. Bacteriol.* 174:8043-56

11. Baxevanis AD, Landsman D. 1995. The HMG-1 box protein family, classification and functional relationships. *Nucleic Acid Res.* 23:1604-13

12. Bender RA. 1991. The role of the NAC protein in the nitrogen regulation of *Klebsiella aerogenes*. *Mol. Microbiol.* 5:2575-80

12a. Berger DK, Narberhaus F, Kustu S. 1994. The isolated catalytic domain of NifA, a bacterial enhancer-binding protein, activates transcription in vitro: Activation is inhibited by NifL. *Proc. Natl. Acad. Sci. USA* 91:103-7

13. Betemier M, Rousseau P, Alazard R, Chandler M. 1995. Mutual stabilisation of bacteriophage Mu repressor and histone-like proteins in a nucleoprotein structure. *J. Mol. Biol.* 249:332-41

14. Bianchi ME. 1994. Prokaryotic HU and eukaryotic HMGI: a kinked relationship. *Mol. Microbiol.* 14:1-5

15. Bliska JB, Cozzarelli NR. 1987. Use of site specific recombination as a probe of DNA structure and metabolism *in vivo*. *J. Mol. Biol.* 194:205-18

16. Bolshoy A, McNamara P, Harrington RE, Trifonov EN. 1991. Curved DNA without A-A: experimental estimation of all 16 DNA wedge angles. *Proc. Natl. Acad. Sci. USA* 88:2312-16

17. Bonnefoy E, Rouviere-Yaniv J. 1992. HU, the major histone-like of *E. coli*, modulates the binding of IHF to *oriC*. *EMBO J.* 11:4489-96

18. Bonnefoy E, Takahashi M, Rouviere-Yaniv J. 1994. DNA-binding parameters of the HU protein of *Escherichia coli* to cruciform DNA. *J. Mol. Biol.* 242:116-29

19. Busby S, Ebright R. 1994. Promoter structure, promoter recognition and transcription activation in prokaryotes. *Cell* 79:743-46

20. Carmona M, Magasanik B. 1996. Activation of transcription at σ^{54}-dependent promoters on linear templates requires intrinsic or induced bending of the DNA. *J. Mol. Biol.* 261:348-56

20a. Claverie-Martín F, Magasanik B. 1991. Role of integration host factor in the regulation of the *glnHp2* promoter of *Escherichia coli*. *Proc. Natl. Acad. Sci. USA* 88:1631-35

21. Claverie-Martin F, Magasanik B. 1992. Positive and negative effects of DNA bending on activation of transcription from a distant site. *J. Mol. Biol.* 227:996-1008

22. Crothers DM. 1993. Architectural elements in nucleoprotein complexes. *Curr. Biol.* 3:675-76

23. Crothers DM, Haran TE, Nadeau JG. 1990. Intrinsically bent DNA. *J. Biol. Chem.* 265:7093-96

24. D'Ari R, Lin RT, Newman EB. 1993. The leucine responsive regulatory protein: more than a regulator? *Trends Biochem. Sci.* 18:260-63

25. Debarbouille M, Martin-Verstraete I, Klier A, Rapoport G. 1991. The transcriptional regulator LevR of *Bacillus subtilis* has domains homologous to both σ^{54}- and phosphotransferase system-dependent regulators. *Proc. Natl. Acad. Sci. USA* 88:9092-96

26. de Lorenzo V, Herrero M, Metzke M, Timmis KN. 1991. An upstream XylR and IHF-induced nucleoprotein complex regulates the σ^{54}-dependent *Pu* promoter of TOL plasmid. *EMBO J.* 10:1159-67

27. de Lorenzo V, Pérez-Martín J. 1996. Regulatory noise in prokaryotic promoters: How bacteria learn to respond to novel environmental signals. *Mol. Microbiol.* 19:7-17

28. Delic-Attree I, Toussaint B, Vignais PM. 1995. Cloning and sequence analyses of the genes coding for the integration host factor (IHF) and HU proteins of *Pseudomonas aeruginosa*. *Gene* 154:61-64

29. del Solar G, Pérez-Martín J, Espinosa M. 1990. Plasmid pLS1-encoded RepA protein regulates transcription from *repAB* promoter by binding to a DNA sequence containing a 13-bp symmetric element. *J. Biol. Chem.* 265:12569-75

30. Dersch P, Schmidt K, Bremer E. 1993. Synthesis of the *Escherichia coli* K12 nucleoid-associated DNA-binding protein H-NS is subjected to growth-phase control and autoregulation. *Mol. Microbiol.* 8:875-89

31. Dimri GP, Rudd KE, Morgan MK, Bayat

H, Ames GF. 1992. Physical mapping of repetitive extragenic palindromic sequences in *Escherichia coli* and phylogenetic distribution among *Escherichia coli* strains and other enteric bacteria. *J. Bacteriol.* 174:4583–93

32. Ditto MD, Roberts D, Weisberg RA. 1994. Growth phase variation of Integration Host Factor in *Escherichia coli*. *J. Bacteriol.* 176:3738–48
33. Dlakic M, Harrington RE. 1996. The effects of sequence context on DNA curvature. *Proc. Natl. Acad. Sci. USA* 93:3847–52
34. Drew HR, Calladine, CR. 1987. Sequence-specific positioning of core histones on an 860 base pair DNA. *J. Mol. Biol.* 195:143–73
35. Drew HR, Travers AA. 1985. Structural junctions in DNA: the influence of flanking sequences on nuclease digestion specificities. *Nucleic Acids Res.* 13:4445–67
36. Drew HR, Weeks JR, Travers AA. 1985. Negative supercoiling induces spontaneous unwinding of a bacterial promoter. *EMBO J.* 4:1025–32
37. Drlica K, Rouviere-Yaniv J. 1987. Histone-like proteins of bacteria. *Microbiol. Rev.* 51:301–19
38. Du W, Thanos D, Maniatis T. 1993. Mechanism of transcriptional synergism between distinct virus-inducible enhancer elements. *Cell* 74:887–98
39. Ebright RH. 1993. Transcription activation at Class 1 CAP-dependent promoters. *Mol. Microbiol.* 8:797–802
40. Echols, H. 1986. Multiple DNA-protein interactions governing high-precision DNA transactions. *Science* 233:1050–56
41. Elgin SC. 1988. The formation and function of DNase I hypersensitive sites in the process of gene activation. *J. Biol. Chem.* 263:19259–62
42. Erie DA, Yang G, Schultz HC, Bustamante C. 1994. DNA bending by Cro protein in specific and nonspecific complexes: implications for protein site recognition and specificity. *Science* 266:1562–66
43. Falconi M, Higgins NP, Spurio R, Pon CL, Gualerzi CO. 1993. Expression of the gene encoding the major bacterial nucleoid protein H-NS is subject to transcriptional auto-repression. *Mol. Microbiol.* 10:273–82
44. Feng J, Goss TJ, Bender RA, Ninfa AJ. 1995. Activation of transcription initiation from the *nac* promoter of *Klebsiella aerogenes*. *J. Bacteriol.* 177:5523–34
45. Feng J, Goss TJ, Bender RA, Ninfa AJ. 1995. Repression of the *Klebsiella aerogenes nac* promoter. *J. Bacteriol.* 177:5535–38
46. Fernández S, Pérez-Martín J, de Lorenzo V. 1996. Specificity and promiscuity of catabolic promoters of *Pseudomonas* spp. In *Molecular Biology of Pseudomonas*, ed. T Nakazawa, K Furukawa, D Haas, S Silver, pp. 165–75. Washington: Am. Soc. Microbiol.
47. Finkel SE, Johnson RC. 1992. The Fis protein: it's not just for DNA inversion anymore. *Mol. Microbiol.* 6:3257–65
48. Fisher RF, Long SR. 1993. Interactions of NodD at the *nod* box: NodD binds to two distinct sites on the same face of the helix and induces a bend in the DNA. *J. Mol. Biol.* 233:336–48
49. Flashner Y, Gralla JD. 1988. DNA dynamic flexibility and protein recognition: differential stimulation by bacterial histone-like protein HU. *Cell* 54:713–721
50. Friedman, DI. 1988. Integration Host Factor: a protein for all reasons. *Cell* 55:545–54
51. Gaal T, Rao L, Estrem ST, Yang J, Wartell RM, Gourse RL. 1994. Localization of the intrinsically bent DNA region upstream of the *E. coli rrnB* P1 promoter. *Nucleic Acids Res.* 22:2344–50
52. Gama MJ, Toussaint A, Higgins NP. 1992. Stabilization of bacteriophage Mu repressor-operator complexes by the *Escherichia coli* integration host factor protein. *Mol. Microbiol.* 6:1715–22
53. Gartenberg MR, Crothers DM. 1988. DNA sequence determinants of CAP-induced bending and protein affinity. *Nature* 333:824–29
54. Gerlach P, Søgaard-Andersen L, Pedersen H, Martinussen J, Valentin-Hansen P, et al. 1991. The cyclic AMP (cAMP)-receptor protein functions both as an activator and as a corepressor at the *tsx*-P2 promoter of *Escherichia coli* K12. *J. Bacteriol.* 173:5419–30
55. Giese K, Cox J, Grosschedl R. 1992. The HMG domain of lymphoid enhancer factor 1 bends DNA and facilitates assembly of functional nucleoprotein structures. *Cell* 69:185–95
56. Giese K, Grosschedl R. 1993. LEF-1 contains an activation domain that stimulates transcription only in a specific context of factor-binding sites. *EMBO J.* 12:4667–76
57. Giese K, Kingsley C, Kirshner JR, Grosschedl R. 1995. Assembly and

function of a TCRa enhancer complex is dependent on LEF-1-induced DNA bending and multiple protein-protein interactions. *Genes Dev.* 9:995–1008

58. Giladi G, Goldenberg D, Koby S, Oppenheim AB. 1995. Enhanced activity of the bacteriophage λ PL promoter at low temperature. *Proc. Natl. Acad. Sci. USA* 92:2184–88

59. Gilson E, Saurin W, Perrin D, Bachellier S, Hofnung M. 1991. Palindromic units are part of a new bacterial interspersed mosaic element (BIME). *Nucleic Acids Res.* 19:1375–83

60. Gober JW, Shapiro L. 1990. Integration Host Factor is required for the activation of developmentally regulated genes in *Caulobacter. Genes Dev.* 4:1494–504

61. Goldenberg D, Giladi H, Oppenheim AB. 1994. Genetic and biochemical analysis of IHF/HU hybrid proteins. *Biochimie* 76:941–50

62. Gomada M, Imaidhi H, Miura K, Inouye S, Nakazawa T, Nakazawa A. 1994. Analysis of DNA bend structure of promoter regulatory regions of xylene-metabolizing genes on the *Pseudomonas* TOL plasmid. *J. Biochem.* 116:1096–104

63. Goodman S, Nash HA. 1989. Functional replacement of protein-induced bend in a DNA recombination site. *Nature* 341:251–54

64. Goodman S, Nicholson SC, Nash HA. 1992. Deformation of DNA during site-specific recombination of bacteriophage lambda: replacement of IHF by HU protein or sequence directed bends. *Proc. Natl. Acad. Sci. USA* 89:11910–14

65. Goosen N, van de Putten P. 1987. Regulation of transcription. In *Phage Mu*, ed. N Symonds, A Toussaint, P van de Putte, MM Howe, pp. 41–52. Cold Spring Harbor: Cold Spring Harbor Lab.

66. Goosen N, van de Putte P. 1995. The regulation of transcription initiation by integration host factor. *Mol. Microbiol.* 16:1–7

67. Goshima N, Inagaki Y, Otaki H, Tanaka H, Hayashi N, Imamoto F, Kano Y. 1992. Chimeric HU-IHF proteins that alter DNA binding ability. *Gene* 118:97–102

68. Gosink KK, Ross W, Leirmo S, Osuna R, Finkel SE, et al. 1993. DNA binding and bending are necessary but not sufficient for Fis-dependent activation of *rrnB* P1. *J. Bacteriol.* 175:1580–89

69. Green JR, Geiduscheck EP. 1985. Site-specific binding by a bacteriophage SPO1-encoded type II DNA-binding protein. *EMBO J.* 4:1345–49

70. Grosschedl R. 1995. Higher-order nucleoprotein complexes in transcription: analogies with site-specific recombination. *Curr. Opin. Cell Biol.* 7:362–70

71. Grosschedl R, Giese K, Pagel J. 1994. HMG domain proteins: architectural elements in the assembly of nucleoprotein structures. *Trends Genet.* 10:94–100

72. Hagerman PJ. 1990. Sequence-directed curvature of DNA. *Annu. Rev. Biochem.* 59:755–81

73. Haran TE, Kahn JD, Crothers DM. 1994. Sequence elements responsible for DNA curvature. *J. Mol. Biol.* 244:135–43

74. Harrington RE. 1992. DNA curving and bending in protein-DNA recognition. *Mol. Microbiol.* 6:2549–55

75. Hatzopoulos AK, Schlokat U, Gruss P. 1988. Enhancers and other *cis*-acting sequences. In *Transcription and Splicing*, ed. BD Hames, DM Glover, pp. 43–96. Washington, DC: IRL

76. Hayat MA, Mancarella DA. 1995. Nucleoid proteins. *Micron* 26:461–80

77. Hendrickson W, Schleif R. 1984. Regulation of the *Escherichia coli* L-arabinose operon studied by gel electrophoresis DNA binding assay. *J. Mol. Biol.* 174:611–28

78. Higgins CF, Hinton JCD, Hulton CSJ, Owen-Hughes T, Pavitt GD, et al. 1990. Protein H1: a role for chromatin structure in the regulation of bacterial gene expression and virulence? *Mol. Microbiol.* 4:2007–12

79. Higgins NP, Yang X, Fu Q, Roth JR. 1996. Surveying a supercoil domain by using the γδ resolution system in *Salmonella typhimurium. J. Bacteriol.* 178:2825–35

80. Hodges-Garcia Y, Hagerman PJ, Pettijohn DE. 1989. DNA ring closure mediated by protein HU. *J. Biol. Chem.* 264:14621–23

81. Hoover TR, Santero E, Porter S, Kustu S. 1990. Integration Host factor stimulates interaction of RNA polymerase with NifA, the transcriptional activator for nitrogen fixation operons. *Cell* 63:11–22

82. Hulton CSJ, Seirafi A, Hinton JCD, Sidebotham JM, Wadel L, et al. 1990. Histone-like protein H1 (H-NS), DNA supercoiling and gene expression in bacteria. *Cell* 63:631–42

83. Johnson GG, Geiduscheck EP. 1977. Specificity of the weak binding between the phage SPO1 transcription-inhibitory protein, TF1 and SPO1 DNA. *Biochemistry* 16:1473–85

84. Jordi BJ, van der Zeijst BA, Gaastra W. 1994. Regions of the CFA/I promoter involved in the activation by the transcriptional activator CfaD and repression by the histone-like protein H-NS. *Biochimie* 76:1052–54

85. Kahn JD, Crothers DM. 1992. Protein-induced bending and DNA cyclization. *Proc. Natl. Acad. Sci. USA* 89:6343–47

86. Kahn JD, Crothers DM. 1993. DNA bending in transcription initiation. *Cold Spring Harbor Symp. Quant. Biol.* 58:115–22

87. Kahn JD, Yun E, Crothers DM. 1994. Detection of localized DNA flexibility. *Nature* 368:163–66

88. Kahmann R, Rudt F, Koch C, Mertens G. 1985. G inversion in bacteriophage Mu DNA is stimulated by a site within the invertase gene and a host factor. *Cell* 41:771–80

89. Kolb A, Busby S, Buc H, Garges S, Adhya S. 1993. Transcriptional regulation by cAMP and its receptor protein. *Annu. Rev. Biochem.* 62:749–95

90. Koo HS, Wu HM, Crothers DM. 1986. DNA bending at adenine/thymine tracts. *Nature* 320:501–6

91. Krause HM, Higgins NP. 1986. Positive and negative regulation of the Mu operator by Mu repressor and *Escherichia coli* integration host factor. *J. Biol. Chem.* 261:3744–52

92. Kuhnke GC, Theres C, Fritz KJ, Ehring R. 1989. RNA polymerase and gal repressor bind simultaneously and with DNA bending to the control region of *Escherichia coli* galactose operon. *EMBO J.* 6:1247–55

93. Kur J, Hasan N, Szybalski W. 1989. Repression of transcription from the b2-att region of the coliphage lambda by integration host factor. *Virology* 168:236–44

94. Kustu S, North AK, Weiss DS. 1991. Prokaryotic transcriptional enhancers and enhancer-binding proteins. *Trends Biochem. Sci.* 16:397–402

95. Kustu S, Santero E, Keener J, Popham D, Weiss D. 1989. Expression of σ^{54} (*ntrA*)-dependent genes is probably united by a common mechanism. *Microbiol. Rev.* 53:367–76

96. Landini P, Hajec LI, Nguyen NH, Burguess RR, Volkert R. 1996. The leucine-responsive regulatory protein (Lpr) acts as an specific repressor for σ^S-dependent transcription of the *Escherichia coli aidB* gene. *Mol. Microbiol.* 20:947–55

97. Laundon CH, Griffith JD. 1988. Curved helix segments can uniquely orient the topology of supertwisted DNA. *Cell* 52:545–49

98. Lavigne M, Kolb A, Yeramian E, Buc H. 1994. CRP fixes the rotational orientation of covalently closed DNA molecules. *EMBO J.* 13:4983–90

99. Lavoie BD, Chaconas G. 1993. Site-specific HU binding in the Mu transpososome: conversion of a sequence-independent DNA binding protein into a chemical nuclease. *Genes Dev.* 7:2510–19

100. Lazarus LR, Travers AA. 1993. The *Escherichia coli* Fis protein is not required for the activation of tyrT transcription on entry into exponential growth. *EMBO J.* 12:2483–94

101. Liu-Johnson HN, Gartenberg MR, Crothers DM. 1986. The DNA binding domain and bending angle of *E. coli* CAP protein. *Cell* 47:995–1005

102. Lobell RB, Schleif R. 1990. DNA looping and unlooping by AraC protein. *Science* 250:528–32

103. Lobell RB, Schleif R. 1991. AraC-DNA looping: orientation and distance dependent loop breaking by cyclic AMP receptor protein. *J. Mol. Biol.* 218:45–54

104. Lupski JR, Weinstock GM. 1992. Short interspersed repetitive DNA sequences in prokaryotic genomes. *J. Bacteriol.* 174:4525–29

105. Malik M, Bensaid A, Rouviere-Yaniv J, Drlica K. 1996. Histone-like protein HU and bacterial DNA topology: suppression of an HU deficiency by gyrase mutations. *J. Mol. Biol.* 256:6–76

106. Marini JC, Levene SD, Crothers DM, Englund PT. 1982. Bent helical structure in kinetoplast DNA. *Proc. Natl. Acad. Sci. USA* 79:7664–68

107. Martin-Vestraete I, Stulke J, Klier A, Rapoport G. 1995. Two different mechanisms mediate catabolite repression of the *Bacillus subtilis* levanase operon. *J. Bacteriol.* 177:6919–27

108. Megraw TL, Kao LR, Chae CB. 1994. The mitochondrial histone HM: an evolutionary link between bacterial HU and nuclear HMG1 proteins. *Biochimie* 76:909–16

109. Mencia M, Salas M, Rojo F. 1993. Residues of the *Bacillus subtilis* phage Φ29 transcriptional activator required both to interact with RNA polymerase and to activate transcription. *J. Mol. Biol.* 233:695–704

110. Merkel TJ, Dahl JL, Ebright RH, Kadner RJ. 1995. Transcription activation at

the *Escherichia coli uhpT* promoter by the catabolite gene activator protein. *J. Bacteriol.* 177:1712–18

111. Miller WG, Simons RW. 1993. Chromosomal supercoiling in *Escherichia coli. Mol. Microbiol.* 10:675–84

112. Molina-Lopez J, Govantes F, Santero E. 1994. Geometry of the process of transcriptional activation at the σ^{54}-dependent nifH promoter of *Klebsiella pneumoniae. J. Biol. Chem.* 269:25419–25

113. Muller HP, Schaffner W. 1990. Transcriptional enhancers can act in trans. *Trends Genet.* 6:300–4

114. Nash HA. 1996. The HU and IHF proteins: accessory factors in complex protein-DNA assemblies. In *Regulation of Gene Expression in Escherichia coli*, ed. ECC Lin, A Simon Lyn, pp. 149–79. Austin: RG Landes

115. Natesan S, Gilman MZ. 1993. DNA bending and orientation dependent function of YY1 in the c-fos promoter. *Genes Dev.* 7:2497–509

116. Ner SS, Travers AA, Churchill MEA. 1994. Harnessing the writhe: a role for DNA chaperones. *Trends Biochem. Sci.* 19:185–87

117. Nicolas FJ, Cayela ML, Martinez-Argudo IM, Ruiz-Vazquez RM, Murillo FJ. 1996. High mobility group I(Y)-like DNA binding domains on a bacterial transcription factor. *Proc. Natl. Acad. Sci. USA* 93:6881–85

118. O'Brien K, Deno G, Otrovsky de Spicer P, Gardner JF, Maloy SR. 1992. Integration Host Factor facilitates repression of the put operon in *Salmonella typhimurium. Gene* 118:13–19

119. Olsen A, Arnqvist A, Hammar M, Sukupolvi S, Normark S. 1993. The RpoS sigma factor relieves H-NS-mediated transcriptional expression of csgA, the subunit gene of fibronectin-binding curli in *Escherichia coli. Mol. Microbiol.* 7:523–36

120. Onate SA, Prendergast P, Wagner JP, Nissen M, Reeves R, et al. 1994. The DNA-bending protein HMG-1 enhances progesterone receptor binding to its target DNA sequences. *Mol. Cell Biol.* 14:3376–91

121. Oppenheim AB, Rudd KE, Mendelson I, Teff D. 1993. Integration host factor binds to an unique class of complex repetitive extragenic DNA sequences in *Escherichia coli. Mol. Microbiol.* 10:113–22

122. Oshima T, Ito K, Kabauama H, Nakamura Y. 1995. Regulation of lrp expression by H-NS and Lrp proteins in *Escherichia coli*: dominant negative mutations in lrp. *Mol. Gen. Genet.* 247:521–28

123. Osuna R, Janes BK, Bender RA. 1994. Roles of catabolite activator protein sites centered at −81.5 and −41.5 in the activation of the *Klebsiella aerogenes* histidine utilization operon hutUH. *J. Bacteriol.* 176:5513–24

124. Otwinowski Z, Schevitz RW, Zhang RG, Lawson CL, Joachimiak A, et al. 1988. Crystal structure of trp repressor/operator complex at atomic resolution. *Nature* 355:321–29

125. Pagel JM, Winkelman JW, Adams CW, Hatfield GW. 1992. DNA topology-mediated regulation of transcription initiation from tandem promoters of the ilvGMEDA operon of *Escherichia coli. J. Mol. Biol.* 224:919–35

126. Painbeni E, Mouray E, Gottesman S, Rouviere-Yaniv J. 1993. An imbalance of HU synthesis induces mucoidy in *Escherichia coli. J. Mol. Biol.* 234:1021–37

127. Parekh BS, Hatfield GW. 1996. Transcriptional activation by protein-induced DNA bending: evidence for a DNA structural transmission model. *Proc. Natl. Acad. Sci. USA* 93:1173–77

128. Parvin JA, McCormick RJ, Sharp PA, Fisher DE. 1995. Pre-bending of a promoter sequence enhances affinity for the TATA-binding factor. *Nature* 373:724–27

129. Paull TT, Haykinson MJ, Johnson R. 1993. The nonspecific DNA binding and bending proteins HMG1 and HMG2 promote the assembly of complex nucleoprotein structures. *Genes Dev.* 7:1521–34

129a. Pavitt GD, Higgins CF. 1993. Chromosomal domains of supercoiling in *Salmonella typhimurium. Mol. Microbiol.* 10:685–96

130. Pedersen H, Sogaard-Andersen L, Holst B, Gerlach P, Bremer E, et al. 1992. cAMP-CRP activator complex and the CytR repressor protein bind cooperatively to the cytRP promoter in *Escherichia coli* and CytR antagonizes the cAMP-CRP-induced bend. *J. Mol. Biol.* 227:396–406

130a. Pedersen H, Valentin-Hansen P. 1997. Protein-induced fit: the CRP activator protein changes sequence-specificity DNA recognition by the CytR repressor, a highly flexible LacI member. *EMBO J.* 16:2108-18

131. Pérez-Martín J, de Lorenzo V. 1995.

Integration Host Factor suppresses promiscuous activation of the σ^{54}-dependent promoter Pu of *Pseudomonas putida*. *Proc. Natl. Acad. Sci. USA* 92: 7277–81

132. Pérez-Martín J, de Lorenzo V. 1995. The receiver N-terminal domain of the prokaryotic enhancer-binding protein XylR is an specific intramolecular repressor. *Proc. Natl. Acad. Sci. USA* 92:9392–96

133. Pérez-Martín J, de Lorenzo V. 1995. The σ^{54}-dependent promoter Ps of the TOL plasmid of *Pseudomonas putida* requires HU for transcriptional activation in vivo by XylR. *J. Bacteriol.* 177:3758–63

134. Pérez-Martín J, de Lorenzo V. 1996. ATP binding to the σ^{54}-dependent activator XylR triggers a protein multimerization cycle catalyzed by UAS DNA. *Cell* 86:331–39

135. Pérez-Martín J, de Lorenzo V. 1997. Co-activation in vitro of the σ^{54}-dependent promoter Pu of *Pseudomonas putida* by the prokaryotic histone HU and the mammalian HMG1 protein. *J. Bacteriol.* 179:2757-60

136. Pérez-Martín J, del Solar G, Lurz R, de la Campa A, Dobrinski B, Espinosa M. 1989. Induced bending of plasmid pLS1 DNA by the plasmid encoded protein RepA. *J. Biol. Chem.* 264:21334–39

137. Pérez-Martín J, Espinosa M. 1991. The RepA repressor can act as a transcriptional activator by inducing DNA bends. *EMBO J.* 10:1375–82

138. Pérez-Martín J, Espinosa M. 1993. Protein-induced bending as a transcriptional switch. *Science* 260:805–7

139. Pérez-Martín J, Espinosa M. 1994. Correlation between DNA bending and transcriptional activation at a plasmid promoter. *J. Mol. Biol.* 241:7–17

140. Pérez-Martín J, Rojo F, de Lorenzo V. 1994. Promoters responsive to DNA bending: a common theme in prokaryotic gene expression. *Microbiol. Rev.* 58:268–90

141. Pérez-Martín J, Timmis KN, de Lorenzo V. 1994. Co-regulation by bent DNA: functional substitutions of the IHF site at the σ^{54}-dependent promoter Pu of the upper-TOL operon by intrinsically curved sequences. *J. Biol. Chem.* 269:22657–62

142. Pettijohn DE. 1988. Histone-like proteins and bacterial chromosome structure. *J. Biol. Chem.* 263:12793–96

143. Pettijohn DE, Hecht R. 1973. RNA molecules bound to the folded bacterial genome stabilize DNA folds and segregate domains of supercoiling. *Cold Spring Harbor Symp. Quant. Biol.* 38: 31–41

144. Pil PM, Chow CS, Lippard SJ. 1993. High-mobility-group 1 protein mediates DNA bending as determined by ring closures. *Proc. Natl. Acad. Sci. USA* 90: 9465–69

145. Plumbridge JA. 1989. Sequence of the nagBACD operon in *Escherichia coli* K12 and pattern of transcription with the nag regulon. *Mol. Microbiol.* 3:506–15

146. Plumbridge JA, Kolb A. 1991. CAP and Nag repressor binding to the regulatory regions of the nagEB and manX genes of *Escherichia coli*. *J. Mol. Biol.* 217:661–79

147. Pontiggia A, Negri A, Beltrame M, Bianchi ME. 1993. Protein HU binds specifically to kinked DNA. *Mol. Microbiol.* 7:343–50

147a. Ptashne M, Gann A. 1997. Transcriptional activation by recruitment. *Nature* 386:569–77

148. Raibaud O, Vidal-Ingigliardi D, Richet E. 1989. A complex nucleoprotein structure involved in activation of transcription of two divergent *Escherichia coli* promoters. *J. Mol. Biol.* 205:471–85

149. Ramos JL, Marqués S, Timmis KN. 1997. Transcriptional control of the *Pseudomonas* TOL plasmid catabolic operons is achieved through an interplay of host factors and plasmid-encoded regulators. *Annu. Rev. Microbiol.* 51: In press

150. Rao L, Ross W, Appleman J, Gaal T, Leirmo S, et al. 1994. Factor-independent activation of rrnB P1: an extended promoter with an upstream element that dramatically increases promoter strength. *J. Mol. Biol.* 235:1421–35

151. Read CM, Cary PD, Crane-Robinson C, Driscoll PC, Norman DG. 1993. Solution structure of a DNA-binding domain from HMG1. *Nucleic Acid Res.* 21: 3427–36

152. Revet B, Brahms S, Brahms G. 1995. Binding of the transcription activator NRI (NTRC) to a supercoiled DNA segment imitates association with the natural enhancer: an electron microscopic investigation. *Proc. Natl. Acad. Sci. USA* 92:7535–39

152a. Rice PA, Yang S, Mizuuchi K, Nash HA. 1996. Crystal structure of an IHF-DNA complex: a protein-induced U-turn. *Cell* 87:1295–306

153. Richet E, Sogaard-Andersen L. 1994. CRP induces the repositioning of MalT at the *Escherichia coli* malKp promoter primarily through DNA bending. *EMBO J.* 13:4558–67

154. Richet E, Vidal-Ingigliardi D, Raibaud O. 1991. A new mechanism for coactivation of transcriptional activation: repositioning of an activator triggered by the binding of a second activator. *Cell* 66:1185–95

155. Rojo F, Nuez M, Mencía M, Salas M. 1993. The main early and late promoters of *Bacillus subtilis* phage Φ29 forms unstable open complexes with σ^A-RNA polymerase that are stabilized by DNA supercoiling. *Nucleic Acids Res.* 21:935–40

156. Rojo F, Salas M. 1991. A DNA curvature can substitute phage Φ29 regulatory protein p4 when acting as a transcriptional repressor. *EMBO J.* 10:3429–38

157. Rojo F, Zaballos A, Salas M. 1990. Bend induced by phage Φ29 transcriptional activator in the viral late promoter is required for activation. *J. Mol. Biol.* 211:713–25

158. Rouviere-Yaniv J, Yaniv M. 1979. *E. coli* DNA binding protein HU forms nucleosome-like structure with circular double-stranded DNA. *Cell* 17:265–74

159. Santero E, Hoover TR, North AK, Berger DK, Porter SC, Kustu S. 1992. Role of Integration Host factor in stimulating transcription from the σ^{54}-dependent nifH promoter. *J. Mol. Biol.* 227:602–20

160. Schepartz A. 1995. Nonspecific DNA bending and the specificity of protein-DNA interactions. *Science* 269:989

161. Schild C, Claret FX, Wahli W, Wolffe AP. 1993. A nucleosome-dependent static loop potentiates estrogen-regulated transcription from the *Xenopus* vitellogenin B1 promoter in vitro. *EMBO J.* 12:423–33

162. Schleif R. 1987. The L-arabinose operon. In *Escherichia coli and Salmonella typhimurium: Cellular and Molecular Biology,* ed. FC Neidhart, JL Ingraham, KB Low, B Magasanik, M Schaetcher, HE Umbarger, pp. 1473–81. Washington, DC: Am. Soc. Microbiol.

163. Schmidt MD. 1990. More than just "histone-like" proteins. *Cell* 63:451–53

164. Schneider GJ, Sayre MH, Geiduscheck EP. 1991. DNA-bending properties of TF1. *J. Mol. Biol.* 221:777–94

165. Schwartz M. 1987. The maltose regulon. In *Escherichia coli and Salmonella typhimurium: Cellular and Molecular Biology,* ed. FC Neidhart, JL Ingraham, KB Low, B Magasanik, M Schaetcher, HE Umbarger, pp. 1482–502. Washington, DC: Am. Soc. Microbiol.

166. Segall AM, Goodman SD, Nash HA. 1994. Architectural elements in nucleoprotein complexes: interchangeability of specific and nonspecific DNA-binding proteins. *EMBO J.* 13:4536–48

167. Serrano M, Barthelemy I, Salas M. 1991. Transcription activation at a distance by phage Φ29 protein p4: effect of bent and non-bent intervening DNA sequences. *J. Mol. Biol.* 219:403–14

168. Serrano M, Gutierrez C, Freire R, Bravo A, Salas M, et al. 1994. Phage Φ29 protein p6: a viral histone-like protein. *Biochimie* 76:981–91

169. Shimizu M, Miyake M, Kanke F, Matsumoto U, Shindo H. 1995. Characterization of the binding of HU and IHF, homologous histone-like proteins of *Escherichia coli*, to curved and uncurved DNA. *Biochim. Biophys. Acta* 1264:330–36

170. Shore D, Baldwin RL. 1983. Energetics of DNA twisting. *J. Mol. Biol.* 170:957–81

171. Sledjeski D, Gottesman S. 1995. A small RNA acts as an antisilencer of the H-NS-silenced rcsA gene of *Escherichia coli*. *Proc. Natl. Acad. Sci. USA* 92:2003–7

172. Sousa C, de Lorenzo V, Cebolla A. 1997. Modulation of gene expression through chromosomal positioning. *Microbiology.* In press

173. Spana C, Corces VG. 1990. DNA bending is a determinant of binding specificity for a *Drosophila* zinc finger protein. *Genes Dev.* 4:1505–15

174. Spassky A, Busby S, Buc H. 1984. On the action of the cyclic AMP-cyclic AMP receptor protein complex at the *Escherichia coli* lactose and galactose promoter regions. *EMBO J.* 3:43–50

175. Starr DB, Hoopes BC, Hawley DK. 1995. DNA bending is an important component of site-recognition by the TATA binding protein. *J. Mol. Biol.* 250:434–46

176. Stent G, Calendar R. 1978. *Molecular Genetics: An Introductory Narrative.* San Francisco: Freeman. 2nd ed.

177. Stern MJ, Prossnitz E, Ames GFL. 1988. Role of the intercistronic region in posttranscriptional control of gene expression in the histidine transport operon of *Salmonella typhimurium*: involvement of REP sequences. *Mol. Microbiol.* 2:141–52

178. Strahs D, Brenowitz M. 1994. DNA conformational changes associated with the cooperative binding of cI-repressor of bacteriophage lambda to OR. *J. Mol. Biol.* 244:494–510

179. Strauss JK, Maher LJ. 1994. DNA bending by asymmetric phosphate neutralization. *Science* 266:1829–34

180. Stulke J, Martin-Vestraete I, Charrier V, Klier A, Deutscher J, Rapoport G. 1995. The HPr protein of the phosphotransferase system links induction and catabolite repression of the *Bacillus subtilis* levanase operon. *J. Bacteriol.* 177:6928–36

180a. Suzuki M, Loakes D, Yagi N. 1996. DNA conformation and its changes upon binding transcription factors. *Adv. Biophys.* 32:53-72

181. Suzuki M, Yagi N. 1995. Stereochemical basis of DNA bending by transcription factors. *Nucleic Acids Res.* 23:2083–91

182. Suzuki M, Yagi N. 1996. An in-the-groove view of DNA structures in complexes with proteins. *J. Mol. Biol.* 255:677–87

182a. Suzuki M, Yagi N, Finch JT. 1996. Role of base-backbone and base-base interactions in alternating DNA conformations. *FEBS Lett.* 379:148-52

183. Szalewska-Palasz A, Wegrzyn G. 1994. An additional role of transcriptional activation of *ori* lambda in the regulation of lambda plasmid replication in *Escherichia coli. Biochem. Biophys. Res. Commun.* 205:802–6

184. Tanaka I, Appelt K, Dijk J, White S, Wilson K. 1984. 3Å resolution structure of a protein with histone-like properties in prokaryotes. *Nature* 310:376–81

185. Thanos D, Maniatis T. 1992. The high mobility group protein HMGI(Y) is required for NF-κB-dependent virus induction of the human IFN-b gene. *Cell* 71:777–89

186. Thomas GH, Elgin SC. 1988. Protein/DNA architecture of the DNase I hypersensitive region of the *Drosophila* HSP26 promoter. *EMBO J.* 7:2191–201

187. Tippner D, Afflerbach H, Bradaczek C, Wagner R. 1994. Evidence for a regulatory function of the histone-like *Escherichia coli* protein H-NS in ribosomal RNA synthesis. *Mol. Microbiol.* 11:589–604

188. Travers AA. 1989. DNA conformation and protein binding. *Annu. Rev. Biochem.* 58:427–52

189. Travers AA, Ner SS, Churchill MEA. 1994. DNA chaperones: a solution to

a persistence problem? *Cell* 77:167–69

190. Trifonov EN. 1985. Curved DNA. *CRC Crit. Rev. Biochem.* 19:89–106

191. Valentin-Hansen P, Søgaard-Andersen L, Pedersen H. 1996. A flexible partnership: the CytR anti-activator and the cAMP-CRP activator protein, comrades in transcription control. *Mol. Microbiol.* 20:461–66

192. van der Woude MW, Kaltenbach LS, Low DA. 1995. Leucine-responsive regulatory protein plays dual roles as both an activator and a repressor of the *Escherichia coli pap* fimbrial operon. *Mol. Microbiol.* 17:303–12

193. van Ulsen P, Hillebrand M, Zulianello L, van de Putte P, Goosen N. 1996. Integration host factor alleviates the H-NS-mediated repression of the early promoter of bacteriophage Mu. *Mol. Microbiol.* 21:567–78

194. van Workum M, van Dooren JM, Oldenburg N, Molenaar D, Jensen PR, et al. 1996. DNA supercoiling depends on the phosphorylation potential in *Escherichia coli. Mol. Microbiol.* 20:351–60

195. Wang JC, Giaever GN. 1988. Action at a distance along a DNA. *Science* 250:528–32

196. Wang Q, Albert FG, Fitzgerald DJ, Calvo JM, Anderson JN. 1994. Sequence determinants of DNA bending in the ilvIH promoter and regulatory region of *Escherichia coli. Nucleic Acids Res.* 22:5753–60

197. Wang Q, Calvo J. 1993. Lrp, a major regulatory protein in *Escherichia coli*, bends DNA and can organize the assembly of a higher-order nucleoprotein structure. *EMBO J.* 12:2495–501

198. Weir HM, Draulis PJ, Hill CS, Raine ARC, Laue ED, Thomas JO. 1993. Structure of the HMG box motif in the B-domain of HMG–1. *EMBO J.* 12:1311–19

199. Werner MG, Grononborn AM, Clore GM. 1996. Intercalation, DNA kinking, and the control of transcription. *Science* 271:778–84

200. Worcel A, Burgi E. 1972. On the structure of the folded chromosome of *Escherichia coli. J. Mol. Biol.* 71:127–47

201. Wu HM, Crothers DM. 1984. The locus of sequence-directed and protein-induced DNA bending. *Nature* 308:509–13

201a. Yan D, Ikeda TP, Shauger AE, Kustu S. 1996. Glutamate is required to maintain the steady state potassium pool in *Salmonella typhimurium. Proc. Natl. Acad. Sci. USA* 93:6527–31

202. Yang Y, Ames GFL. 1990. The family of repetitive extragenic palindromic sequences: interaction with DNA gyrase and histone-like protein HU. In *The Bacterial Chromosome,* ed. K Drilica, M Riley, pp. 211–25. Washington, DC: Am. Soc. Microbiol.

203. Zhang X, Bremer H. 1996. Effects of Fis on ribosome synthesis and activity and on rRNA promoter activities in *Escherichia coli. J. Mol. Biol.* 259:27–40

204. Ziegelhoffer EC, Kiley PJ. 1995. In vitro analysis of a constitutively active mutant form of the *Escherichia coli* global transcription factor FNR. *J. Mol. Biol.* 245:351–61

205. Zwieb C, Kim J, Adhya S. 1989. DNA bending by negative regulatory proteins: Gal and Lac repressors. *Genes Dev.* 3:602–11

Annu. Rev. Microbiol. 1997. 51:629–59
Copyright © 1997 by Annual Reviews Inc. All rights reserved

GENETICS OF EUBACTERIAL CAROTENOID BIOSYNTHESIS: A Colorful Tale

Gregory A. Armstrong
Institute for Plant Sciences, Plant Genetics, Swiss Federal Institute of Technology (ETH), Zürich, Switzerland; e-mail: armstrong@wawona.vmsmail.ethz.ch

KEY WORDS: isoprenoids, photooxidative protection, photosynthesis, provitamin A, xanthophylls

ABSTRACT

Carotenoids represent one of the most widely distributed and structurally diverse classes of natural pigments, with important functions in photosynthesis, nutrition, and protection against photooxidative damage. In the eubacterial community, yellow, orange, and red carotenoids are produced by anoxygenic photosynthetic bacteria, cyanobacteria, and certain species of nonphotosynthetic bacteria. Many eukaryotes, including all algae and plants, as well as some fungi, also synthesize these pigments. In noncarotenogenic organisms, such as mammals, birds, amphibians, fish, crustaceans, and insects, dietary carotenoids and their metabolites also serve important biological roles. Within the last decade, major advances have been made in the elucidation of the molecular genetics, the biochemistry, and the regulation of eubacterial carotenoid biosynthesis. These developments have important implications for eukaryotes, and they make increasingly attractive the genetic manipulation of carotenoid content for biotechnological purposes.

CONTENTS

0066-4227/97/1001-0629$08.00

INTRODUCTION

Distribution, Structures, and Biosynthesis of Carotenoids

Carotenoids, a major class of lipophilic isoprenoids that range in color from light yellow to deep red, are perhaps most familiar to us in everyday life as the dominant pigments in many storage roots, fruits, and flowers. Carrots, tomatoes, red peppers, and the petals of daffodils and marigolds are only a few of the more eye-catching examples of carotenoid-containing tissues. Indeed, among eukaryotes, carotenoids are produced by all chlorophyll (Chl)-containing photosynthetic organisms, from algae to higher plants, as well as by red yeasts and many species of fungi (28, 46). However, carotenoid biosynthesis has an ancient evolutionary origin. These pigments are ubiquitously synthesized by both bacteriochlorophyll (Bchl)-containing anoxygenic photosynthetic and Chl-containing oxygenic photosynthetic eubacteria, and they are widely distributed throughout the nonphotosynthetic eubacteria.

About 600 chemically distinct carotenoids have been described to date, a number that has increased 10-fold during the last 50 years (65, 109). The well-known carotenoids discussed here are referred to by their trivial names, for the sake of simplicity. General aspects of the chemical structures, physical properties, proposed biosynthetic schemes, species-specific distribution, cellular localization and functions of carotenoids have been discussed extensively elsewhere (18, 28, 29, 46, 61, 64, 102, 114). The salient features of this review are summarized below, with an emphasis placed on carotenoids in eubacteria.

Most naturally occurring carotenoids are hydrophobic tetraterpenoids that contain a C_{40} methyl-branched hydrocarbon backbone derived from the successive condensations of eight C_5 isoprene units (Figure 1). Also, novel carotenoids with longer C_{45} or C_{50}, or shorter C_{30}, backbones occur sporadically among nonphotosynthetic bacteria. The term carotenoid encompasses both hydrocarbon carotenes and xanthophylls: carotene derivatives that contain one or more oxygen atoms incorporated into hydroxy-, methoxy-, oxo-, epoxy-, carboxy-aldehydic, and glycosidic functional groups. Carotenoids are furthermore described as being acyclic, monocyclic, or bicyclic, depending on the structure of the ends of the hydrocarbon backbone. Although each double bond in the polyene chromophore can theoretically assume either a *cis* or a *trans* conformation, the vast majority of naturally occurring carotenoids exist primarily or exclusively as all-*trans* isomers. Phytoene provides the most notable exception, and typically, although not always, occurs as the 15, 15'-*cis* isomer in eubacteria and eukaryotes.

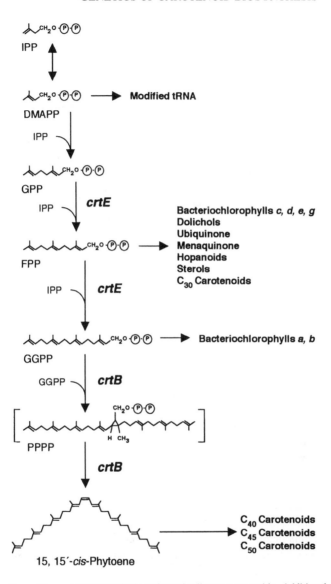

Figure 1 General isoprenoid biosynthesis pathway leading to carotenoids. Additional products of this pathway found in some or all eubacteria are listed. Brackets indicate an unstable interme- diate in the formation of phytoene. Bacterial *crt* genes required for specific biosynthetic conver- sions are given here, and the corresponding enzymes are listed in Table 1. Abbreviations used are as follows: dimethylallyl pyrophosphate (DMAPP); farnesyl pyrophosphate (FPP); geranyl pyrophosphate (GPP); geranylgeranyl pyrophosphate (GGPP); isopentenyl pyrophosphate (IPP); prephytoene pyrophosphate (PPPP).

Colorless C_{40} phytoene, formed by the condensation of two molecules of the C_{20} precursor GGPP, and C_{30} 4, 4′-diapophytoene (also known as dehydrosqalene), formed by the analogous condensation of two molecules of C_{15} FPP (not shown), are the progenitors of all other carotenoids (Figure 1). The reactions leading to their synthesis, in contrast to later steps in colored carotenoid biosynthesis, are carried out by soluble or membrane-associated enzymes. The color of a given carotenoid is determined by the number of conjugated double bonds within the hydrocarbon backbone. Phytoene, for example, contains only three conjugated double bonds and absorbs ultraviolet light, whereas ζ-carotene and more highly unsaturated carotenoids possess seven or more conjugated double bonds and thus absorb visible wavelengths. In anoxygenic photosynthetic bacteria, nonphotosynthetic bacteria, and fungi, three or four consecutive desaturations of phytoene lengthen the polyene chromophore and yield neurosporene or lycopene, respectively (Figure 2). Cyanobacteria, algae, and plants, in contrast, desaturate phytoene four times in two discrete biosynthetic segments, first producing ζ-carotene and subsequently converting it to lycopene. An increasing degree of apparently nonenzymatic isomerization of the carotene products to the all-*trans* conformation usually accompanies the multiple desaturations of 15, 15′-*cis*-phytoene. Lycopene cyclization frequently follows carotene desaturation and precedes xanthophyll formation in the biosynthetic sequence (Figure 2). Carotenoid desaturation and cyclization are the main targets for many known carotenoid biosynthesis inhibitors. During xanthophyll formation, the carotenoid structures are modified such that the end product pigments are often species-specific (Figures 3, 4, and 5).

Localization and Functions of Carotenoids

Carotenoids are relatively hydrophobic molecules, and they are, therefore, typically associated with membranes and/or noncovalently bound to specific proteins. In pigment-protein complexes, carotenoids sometimes display substantial bathochromic shifts. Most carotenoids that are accumulated in anoxygenic photosynthetic bacteria and cyanobacteria are associated with the membrane-bound Bchl- or Chl-binding polypeptides of the photosynthetic apparatus. Among nonphotosynthetic bacteria and, to a lesser extent, among photosynthetic bacteria and cyanobacteria, carotenoids and their glycosides can be found in cytoplasmic and cell wall membranes. There they are thought to influence membrane fluidity.

The paramount function of carotenoids in all photosynthetic organisms, including anoxygenic photosynthetic bacteria and cyanobacteria, is to provide photooxidative protection. Carotenoids protect against the potentially damaging combination of oxygen, light and photosensitizing Bchl or Chl molecules by quenching both the triplet excited states of the photosensitizers and the singlet excited state of oxygen. Among nonphotosynthetic bacteria, carotenoids

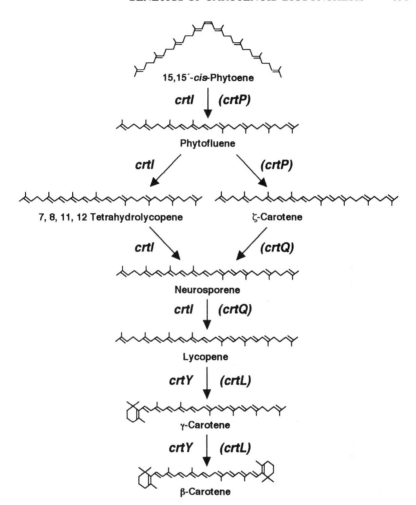

Figure 2   Eubacteria possess one of two discrete genetic systems for the conversion of phytoene to β-carotene. *crt* genes required for specific biosynthetic conversions are given here, either on the left for anoxygenic photosynthetic bacteria and nonphotosynthetic bacteria, or on the right in parentheses for cyanobacteria. The corresponding enzymes are listed in Table 1.

offer similar photooxidative protection against other photosensitizing porphyrin molecules, such as protoporphyrin IX and heme.   In their additional light-harvesting function, carotenoids increase the cross section for the absorption of radiant energy, which is ultimately transferred via Bchl or Chl molecules to the photosynthetic reaction center, the site of primary charge separation.

Besides these functions in eubacteria, certain carotenoids in higher plants and algae participate in the so-called xanthophyll cycle, in order to help dissipate excess radiant energy, and provide the biosynthetic precursors of abscisic acid, an essential plant hormone, and retinal, the photoreceptor pigment for phototaxis in algae. In humans and in a number of animals, carotenoids that contain the half-structure of β-carotene (Figure 2) serve as provitamin A, retinal, and retinoic acid sources for nutrition, vision, and development, respectively. Finally, carotenoids provide natural coloration to birds, reptiles, amphibians, fish, and various invertebrates.

GENETICS AND REGULATION OF CAROTENOID BIOSYNTHESIS

Introduction

During the last few years, a number of thorough reviews of carotenoid biosynthesis in specific groups of eubacteria (5, 49, 51) and eukaryotes (18, 19), as well as more general overviews of the field, have appeared (4, 10, 61, 101). Recent advances in understanding the biochemical and structural properties of carotenoid biosynthesis enzymes are summarized elsewhere (6, 10). This review focuses primarily on the genetics of eubacterial carotenoid biosynthesis, with particular emphasis on new findings and their implications for this rapidly developing field.

The *crt* designation for genes that encode carotenoid biosynthesis enzymes (Tables 1 and 2) has now been adopted by most researchers in the field (4, 26, 49, 124). The parallel *car* nomenclature, originally applied to genetic loci involved in carotenoid synthesis in the fungus *Phycomyces blakesleeanus* (32), and subsequently to loci in the eubacterium *Myxococcus xanthus* (14), has been maintained for carotenoid biosynthesis regulatory genes identified in *M. xanthus* (Table 3).

Anoxygenic Photosynthetic Bacteria

BIOSYNTHESIS GENES   The molecular genetics of carotenoid biosynthesis in anoxygenic photosynthetic bacteria have thus far been analyzed in detail only in the genus *Rhodobacter* (1, 5). Indeed, the isolation of pigment mutants of *Rhodobacter sphaeroides* allowed the identification of the photoprotective role of carotenoids more than 40 years ago. Two closely related Gram-negative species, *Rhodobacter capsulatus* and *R. sphaeroides*, represent genetically well-characterized systems, both with respect to carotenoid biosynthesis and anoxygenic photosynthesis in general (24). These metabolically versatile, purple nonsulfur bacteria normally accumulate a mixture of two specialized acyclic xanthophylls—spheroidene and spheroidenone (Figure 3). Their abundance and distribution, as a function of the amount of dissolved oxygen in the growth medium, strongly influence the color of the bacterial culture (105). Although carotenoids are not absolutely required for viability under strictly anaerobic

Table 1   Carotene biosynthesis genes and enzymes

Gene	Enzymatic function	Species (reference)
Formation of phytoene/4, 4′-diapophytoene from isoprenoid pyrophosphate precursors:		
crtE	GGPP synthase	*E. herbicola* (7, 115), *E. uredovora* (84), *Flavobacterium* sp. strain R1534 (90), *M. xanthus* (26), *R. capsulatus* (8), *R. sphaeroides (68), S. griseus* (107)
crtB	Phytoene synthase	*A. aurantiacum* (85), *E. herbicola* (7, 115), *E. uredovora* (84), *Flavobacterium* sp. strain R1534 (90), *M. xanthus* (26), *R. capsulatus* (8), *R. sphaeroides* (67), *S. griseus* (107), *Synechococcus* sp. strain PCC7942 (33), *Synechocystis* sp. strain PCC6803 (75), *T. thermophilus* (52)
crtM	Diapophytoene synthase	*S. aureus* (119)
Formation of lycopene or neurosporene/4, 4′-diaponeurosporene from phytoene/4, 4′-diapophytoene:		
crtI	Phytoene desaturase (CrtI-type)	*A. aurantiacum* (85), *E. herbicola* (7, 115), *E. uredovora* (84), *Flavobacterium* sp. strain R1534 (90), *M. xanthus* (39), *R. capsulatus* (8, 16), *R. sphaeroides* (67), *S. griseus* (107)
crtP	Phytoene desaturase (CrtP-type)	*Synechococcus* sp. strain PCC7942 (34), *Synechocystis* sp. strain PCC6803 (76)
crtN	Diapophytoene desaturase	*S. aureus* (119)
crtQ[a]	ζ-carotene desaturase	*Anabaena* sp. strain PCC7120 (71)
Formation of β-carotene from lycopene:		
crtY	Lycopene cyclase (CrtY-type)	*A. aurantiacum* (85), *E. herbicola* (56, 115), *E. uredovora* (84), *Flavobacterium* sp. strain R1534 (90), *S. griseus* (107)
crtL	Lycopene cyclase (CrtL-type)	*Synechococcus* sp. strain PCC7942 (38)
ORF6[b]	Lycopene cyclization	*M. xanthus* (26)

[a]The *crtQ* gene does not seem to be present in other species of eubacteria including other cyanobacteria.

[b]Predicted gene product is required for this reaction but is homologous to neither CrtY nor CrtL.

photosynthetic conditions, they are essential during the transition to photosynthesis. These features of the genus *Rhodobacter* have greatly facilitated the isolation of mutants that lack carotenoids, and those with altered pigment compositions (5).

Both *R. capsulatus* and *R. sphaeroides* contain gene superclusters of approximately 50 kilobases that encode most of the functions required for photosynthesis, including all the known Bchl and carotenoid biosynthesis enzymes (1). Isolation of these gene superclusters, by in vivo functional complementation

Table 2   Xanthophyll biosynthesis genes and enzymes/putative carotenoid biosynthesis genes

Gene	Enzymatic function	Species (reference)
Formation of acyclic xanthophylls from neurosporene or lycopene:		
crtC	Hydroxyneurosporene synthase	M. xanthus (26), R. capsulatus (8), R. sphaeroides (68)
crtD	Methoxyneurosporene desaturase	M. xanthus[a] (26), R. capsulatus (8), R. sphaeroides (42, 68)
crtF	Hydroxyneurosporene-O-methyltransferase	R. capsulatus (8), R. sphaeroides (68)
crtA	Spheroidene monooxygenase	R. capsulatus (8, 9, 128), R. sphaeroides (68)
Formation of cyclic xanthophylls and their glycosides from β-carotene:		
crtZ	β-carotene hydroxylase	A. aurantiacum (85), Alcaligenes PC-1 (85), E. herbicola (56), E. uredovora (84), Flavobacterium sp. strain R1534 (90)
crtX	Zeaxanthin glucosylase	E. herbicola (56, 115), E. uredovora (84)
crtW	β–C-4-oxygenase	A. aurantiacum (85), Alcaligenes PC-1 (82)
Postulated biosynthetic function:		
ORF2[a]	Carotenoid desaturase	M. xanthus (26)
crtT [a]	Carotenoid methyltransferase	S. griseus (107)
crtU [a]	Carotenoid desaturase	S. griseus (107)
crtV [a]	Carotenoid methylesterase	S. griseus (107)

[a]Predicted on the basis of protein sequence similarities to other proteins of known function. The crtT, crtU, and crtV gene designations should be considered provisional pending the determination of the carotenoid substrates of the gene products.

of point mutations, led to the generation of transposon- and interposon-tagged pigment-deficient mutants, and to the alignment of physical and genetic maps of chromosomal regions required for carotenoid biosynthesis (36, 43, 44, 74, 94, 108, 113, 124, 129).

Clusters of seven carotenoid biosynthesis genes, *crtA, crtB, crtC, crtD, crtE, crtF* and *crtI*, flanked by Bchl biosynthesis (*bch*) genes, were subsequently molecularly characterized in *R. capsulatus* (8, 9, 16, 128), and shortly thereafter in *R. sphaeroides* (42, 67, 68). The genomic organization of the *crt* genes suggests the existence of a minimum of four operons: *crtA, crtIB, crtDC, crtEF* (8, 68). However, the phenotypes resulting from polar mutations, in addition to transcript mapping experiments, indicate the possible existence of separate promoters for *crtI, crtB, crtD, crtC*, two promoters for *crtE*, and multiple transcripts overlapping several genes (12, 22, 44, 67, 68, 127). Indeed, the operons containing *crtA, crtE,* and *crtF* are embedded within superoperons that allow their cotranscription with *bch* genes, and with the *puf* genes that encode the B870 light-harvesting and reaction-center polypeptides (68, 127, 128). The

Table 3 Carotenoid biosynthesis regulatory genes and gene products

Gene	Gene product function	Mutant phenotype	Species (reference)
carA[a]	DNA-binding repressor	Light-independent carBA expression[b] and carotenoid accumulation, decreased light-inducible crtI expression	M. xanthus (26)
carD[c]	Unknown	Same as carQ with additional pleiotropic developmental defects	M. xanthus (89)
carQ	RNA polymerase σ factor	Loss of light-inducible carBA, crtI and carQRS expression, and carotenoid accumulation	M. xanthus (81)
carR	Anti-σ factor	Light-independent carBA, crtI, and carQRS expression, and carotenoid accumulation	M. xanthus (81)
carS	Transcription factor?	Loss of light-inducible carBA expression, and carotenoid accumulation	M. xanthus (81)
crtS	RNA polymerase σ factor	No carotenoid accumulation	S. setonii (63)
ppsR	DNA-binding repressor	Derepression of crt, bch and puc expression, and carotenoid accumulation under aerobic conditions	R. capsulatus (1), R. sphaeroides (96)
tspO	Outer membrane sensor?	Derepression of crt, bch and puc expression, and carotenoid accumulation under aerobic conditions	R. capsulatus (8), R. sphaeroides (68, 123)

[a]The carA locus may correspond to ORF10 and/or ORF11, both of which encode putative DNA-binding proteins.

[b]The carBA locus encodes two putative regulatory genes (ORF10, ORF11) and all known crt biosynthetic genes, except for the unlinked crtI (formerly carC) gene.

[c]Gene sequence has not been reported.

unlinked *puc* genes that encode the B800-850 peripheral light-harvesting antenna polypeptides are coregulated with *crt* and *bch* genes by separate mechanisms, as described below.

The seven *crt* genes of the genus *Rhodobacter* encode all of the enzymes necessary to produce spheroidene and spheroidenone from isoprenoid pyrophosphate carotenoid precursors (Tables 1 and 2). Most of the genes, *crtE* and *crtB* excepted, were functionally assigned based on the C_{40} carotenoids accumulated in mutants (12, 43, 44, 108). Initial uncertainties in assigning functions to *crtB*

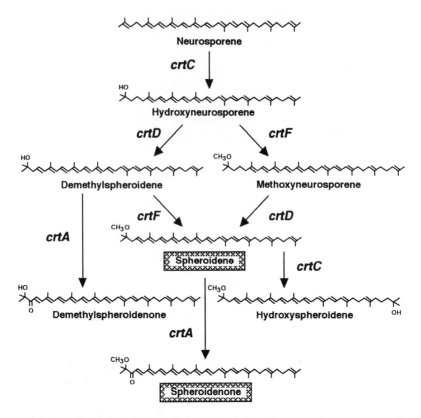

Figure 3 Acyclic xanthophyll biosynthesis pathway in *Rhodobacter* species. *crt* genes required for specific carotenoid biosynthetic reactions are given here and are listed in Table 2. Boxes highlight accumulated end-product carotenoids.

and *crtE* (4, 5) were ultimately resolved by the in vivo and in vitro functional analysis of homologous genes from gram-negative nonphotosynthetic bacteria of the genus *Erwinia* (80, 104, 118), and from the cyanobacterium *Synechococcus* sp. strain PCC7942 (33). *crtB* and *crtI* mutants of *R. capsulatus* have also served as hosts for the in vivo functional analyses of several eukaryotic phytoene synthase and phytoene desaturase genes (15, 17, 20, 21).

Carotenoid biosynthesis in *R. capsulatus* and *R. sphaeroides* begins with the two successive isoprenoid additions of IPP to C_{10} GPP and C_{15} FPP, yielding C_{20} GGPP (Figure 1). *Rhodobacter capsulatus* and *R. sphaeroides crtE* mutants lack colored carotenoids but continue to produce Bchl, and they presumably still contain essential quinones, which suggests that the generally accepted biosynthetic scheme for the formation of GGPP must be more complex than had been supposed (4, 5, 48, 68). One possible explanation for the viability of *crtE* mutants could be the existence of at least two separate pools of GGPP formed by distinct GGPP synthases, including CrtE, a carotenoid-specific enzyme. This model is consistent with the observation that normally noncarotenogenic *Escherichia coli* cells that are transformed with genus *Erwinia crt* genes require an intact *crtE* gene in order to produce more than traces of carotenoids (56, 84).

In those members of the genus *Rhodobacter* examined to date, the first biochemical reaction that is specific to the production of carotenoids is the formation of C_{40} phyotene by the condensation of two molecules of GGPP catalyzed by CrtB, the phytoene synthase (Figure 1). Thereafter, the products of five *crt* genes are thought to be required for a series of reactions that include desaturations mediated by the structurally related CrtI and CrtD proteins, hydration carried out by CrtC, methylation conducted by CrtF, and the addition of an oxo- group catalyzed by CrtA, yielding the colored end products (Figures 2 and 3). Most of these reactions have not yet been demonstrated in vitro. In one relatively well-characterized reaction, CrtI desaturates phytoene in three steps to neurosporene, rather than the usual four steps that yield lycopene in other organisms, both in vivo and in vitro (15, 60, 67, 70, 99). Symmetric ζ-carotene and 7, 8, 11, 12 tetrahydrolycopene, an asymmetric ζ-carotene isomer have both been proposed as possible *Rhodobacter* species desaturation intermediates (43, 46, 99, 105).

REGULATION OF CAROTENOID BIOSYNTHESIS   Carotenoid accumulation in anoxygenic photosynthetic bacteria has long been known to be regulated by two key environmental factors, oxygen and light (5). Carotenoids are usually associated with specific Bchl-containing pigment-protein complexes and thus accumulate preferentially under the low oxygen and low light conditions that favor the formation of the intracytoplasmic photosynthetic membrane. Indeed, there seems to be a strict coupling between carotenoid accumulation and photosynthetic membrane development (16, 23, 36, 129). The expression of most *R. capsulatus* and *R. sphaeroides crt* genes is several-fold higher under anaerobic photosynthetic conditions than it is in semiaerobic or aerobic conditions (9, 44, 67, 98, 123, 127).

Two *R. capsulatus* and *R. sphaeroides* regulatory genes, *ppsR* and *tspO* (Table 3), have recently been demonstrated to encode products that normally repress not only carotenoid and Bchl pigment levels, but also *crt*, *bch* and *puc* gene expression several-fold under aerobic growth conditions (45, 95, 96, 98, 123). *ppsR* (formerly *crtJ* or open reading frame (ORF) 469) and *tspO* (formerly *crtK* or ORF160) were initially thought to be specifically required for carotenoid biosynthesis, based on the results of mutant studies (8, 12, 44, 129). They are, in fact, not directly involved in carotenoid biosynthesis (25, 123). The mechanism by which TspO functions is unknown, although it has been proposed to serve as an environmental sensor. Both this hydrophobic outer membrane protein (123) and its mRNA (9) are more abundant in anaerobic photosynthetic cells than they are in aerobic cells. Interestingly, TspO displays about 35% deduced amino acid sequence identity with mammalian peripheral-type benzodiazepine receptors localized in the mitochondria (5, 13, 123). PpsR is a DNA-binding protein that contains a helix-turn-helix motif conserved in other eubacterial transcriptional repressors (5, 96). The protein recognizes a conserved palindromic sequence found 5' to several operons containing *crt*, *bch*, and *puc* genes (1, 8, 96, 98).

A further mechanism that links the ratio of carotenoid end products to the ratio between the two types of Bchl-binding light-harvesting antenna complexes has recently been identified in *R. sphaeroides* (122). The conversion of spheroidene to spheroidenone catalyzed by CrtA (Figure 3) not only requires molecular oxygen as a source for the oxo-group (105) is also stimulated at high light intensities (122). Intriguingly, at a given oxygen tension and light intensity, the ratio of these two carotenoids correlates with the ratio of the peripheral spheroidene-binding B800-850 light-harvesting complex to the inner-core spheroidenone-binding B875 light-harvesting complex. These data have led to the model that the cellular redox poise either detects or determines the ratio of these carotenoids, and hence couples carotenoid content to the distribution of light-harvesting antenna complexes.

Nonphotosynthetic Bacteria

BIOSYNTHESIS GENES   Carotenoids are produced by many phylogenetically distinct groups of nonphotosynthetic bacteria (46, 61, 114). The genetics of C_{40} carotenoid biosynthesis have been most thoroughly described in the genus *Erwinia*, represented by *E. herbicola* strains Eho10, Eho13, and *E. uredovora*, and in the gram-negative species *Myxococcus xanthus*, in which carotenogenesis is light induced. Also, varying amounts of molecular genetic information about C_{40} carotenoid biosynthesis in *Agrobacterium aurantiacum, Alcaligenes* PC-1, *Thermus thermophilus, Streptomyces griseus, Streptomyces setonii, Flavobacterium* sp. strain R1534, and the C_{30} carotenoid biosynthesis pathway of *Staphylococcus aureus* have been obtained.

The study of carotenoid biosynthesis in phytopathogenic soil bacteria of the genus *Erwinia* has thus far been performed almost exclusively in a genetically amenable noncarotenogenic heterologous host, *E. coli*. The latter unpigmented bacterium is capable of expressing the clustered genus *Erwinia crt* genes, which originally allowed their cloning by screening for yellow pigmented bacterial colonies after transformation with cosmid libraries of genus *Erwinia* DNA (69, 84, 97). In contrast, transfer of the *R. capsulatus crt* gene cluster to *E. coli* did not result in carotenoid accumulation (74).

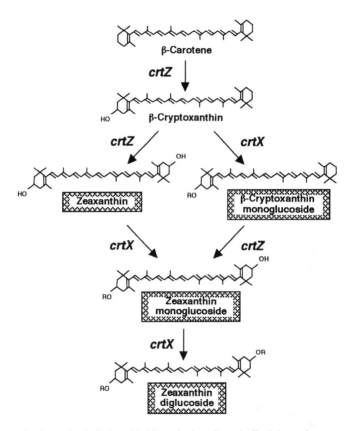

Figure 4 Cyclic xanthophyll glycoside biosynthesis pathway in *Erwinia* species. *crt* genes required for specific carotenoid biosynthetic reactions are given here and are listed in Table 2. R = glucoside. Boxes highlight accumulated end-product carotenoids.

Transformed *E. coli* expressing the *E. herbicola* pigments are resistant to photooxidative killing when exposed to a combination of long wavelength ultraviolet light and lipophilic photosensitizing molecules, which suggests that the carotenoids accumulate in cell membranes (117). Determination of the structures and compositions of carotenoids from *E. herbicola, E. uredovora*, and the corresponding transformed *E. coli* strains revealed the end products to be β-carotene-derived cyclic xanthophylls, cryptoxanthin and zeaxanthin, and glucosides thereof (Figure 4) (57, 84). Thus, *Erwinia* species might be expected to follow a carotenoid biosynthetic pathway that is similar, in part, to that used by cyanobacteria, algae, and plants, in which zeaxanthin is a major pigment (4, 46).

Nucleotide sequence comparisons and mutational analyses of three cloned genus *Erwinia crt* gene clusters have defined six genes, *crtB, crtE, crtI, crtX, crtY,* and *crtZ,* that encode enzymes (Tables 1 and 2) required for the synthesis of zeaxanthin and cryptoxanthin glucosides (Figure 4) (7, 56, 57, 84, 115). The *E. herbicola* strain Eho10 *crt* gene cluster also contains an additional open reading frame, ORF6, of unknown function (56, 57). The organization of the *crt* genes mandates the existence of at least two operons, *crtE*(ORF6)*XYIB* and *crtZ.* The existence of the former has been supported by primer extension and promoter analyses (115).

Sequence comparisons of the *E. herbicola* strain Eho10 and *E. uredovora crt* gene clusters with that of *R. capsulatus* allowed the identification of homologs of *crtB, crtE,* and *crtI* (7, 11, 84). The final assignment of functions to *crtE* and *crtB* was performed by analysis of the products formed upon addition of radiolabeled isoprenoid pyrophosphates to extracts of *E. coli* or *Agrobacterium tumefaciens* transformed with one or both of these genes (80, 104). CrtE, CrtB, and CrtI were, in fact, not only structurally conserved between anoxygenic photosynthetic bacteria and nonphotosynthetic bacteria, but also in eukaryotes (11). CrtE displayed structural homologies with GGPP synthases from the fungus *Neurospora crassa* (31) and red pepper (66); CrtB was identified as the structural homolog of a tomato protein, later demonstrated to represent phytoene synthase, that was preferentially expressed during fruit ripening (7); and CrtI was found to be both structurally and functionally conserved with the phytoene desaturase of *N. crassa* (15).

The functions of the *crtI, crtX, crtY,* and *crtZ* genes were identified or confirmed by analyzing the carotenoids accumulated in *E. coli* strains containing partially deleted or mutated genus *Erwinia crt* gene clusters (56, 84), and by complementing an *R. sphaeroides crtI* mutant in vivo (60). *E. coli* strains that express one or more of the genus *Erwinia crt* genes, referred to here as the *Erwinia-E. coli* system, have also proved invaluable as tools for the isolation and in vivo functional analysis of other eubacterial *crt* genes and eukaryotic carotenoid biosynthesis cDNAs, on the basis of pigment accumulation. Novel carotenoid biosynthetic reactions carried out by genus *Erwinia crt* gene products included cyclization, performed by CrtY, hydroxylation, mediated by CrtZ, and glucosylation, catalyzed by CrtX (Figures 2, 4). CrtE (118), CrtI (41), CrtX (58), CrtY (59, 106, 111) and CrtZ (59) were subsequently overexpressed in *E. coli,* and their functions were confirmed by biochemical analyses.

During the last decade, the fruiting bacterium *Myxococcus xanthus* has become the most important model system for classical and molecular genetic studies of blue light-induced carotenogenesis, a phenomenon found in some nonphotosynthetic bacteria and fungi (51). The molecular details of the regulatory circuits involved in light regulation are discussed in the next section.

Early work with *M. xanthus* demonstrated that carotenoid accumulation was strictly light-dependent and occurred when cells entered the stationary phase, thereby preventing cellular photolysis caused by a 16-fold increase in the accumulation of the photosensitizer protoporphyrin IX in cell membranes. Furthermore, protoporphyrin IX was identified as the blue light photoreceptor for both carotenogenesis and photolysis, thus establishing a direct physiological link between these two processes. The cytoplasmic membrane is a major site of carotenoid accumulation in the genus *Myxococcus*.

Typical C_{40} end-product carotenoids of the genus *Myxococcus* include myxobactone ester and myxobactin ester, both monocyclic carotenoid glucoside fatty acid esters thought to arise from γ-carotene (Figure 2) (51, 112). Surprisingly large amounts of phytoene have also been observed in wildtype strains of some *Myxococcus* species, including *M. xanthus* (77, 112).

In genetic studies of *M. xanthus* carotenogenesis, the analysis of point and transposon Tn5 mutants has revealed two phenotypic classes that are disturbed in colored carotenoid accumulation (51). One class of putative regulatory mutants displays constitutive (i.e. light-independent) synthesis of carotenoids (Car^c), whereas the other group of putative regulatory/biosynthetic mutants does not produce carotenoids under any conditions (Car^-). The loci found to regulate carotenogenesis are referred to below as *car*, except for those that have been molecularly cloned and demonstrated to encode carotenoid biosynthesis enzymes, in which case the designation *crt* has been adopted (26).

Analysis of Car^- mutants initially revealed the existence of two unlinked genetic loci, *carB* and *carC*, representing putative carotenoid biosynthetic functions (14). The positively light-regulated *carB* locus was also linked to *carA*, a regulatory locus associated with the Car^c phenotype. Based on pigment analyses, *carB* and *carC* were proposed to be responsible for the synthesis and desaturation of phytoene, respectively (Figures 1 and 2) (77). The positively light-regulated *carC* gene was subsequently cloned and found to be the *M. xanthus* homolog of *crtI* (Table 1) (39). Transposon mutagenesis and transcriptional analysis of the cloned *carA-carB* region provided evidence for the existence of a light-induced *carBA* operon that coupled transcription of several different carotenoid biosynthesis functions to the *carA* regulatory function (100).

The nucleotide sequence of the entire *carBA* region has recently been determined, allowing the alignment of gene sequences with the genetic map generated by transposon mutagenesis (26). The available data indicate that the *carBA* region encodes 11 genes, including four of known function (*crtB, crtE, crtC*, and *crtD*, Tables 1 and 2), two putative carotenoid biosynthesis genes (ORF2 and ORF6), two putative regulatory genes (ORF10 and ORF11), and three genes of unknown function (ORF7, ORF8, and ORF9), that seem not to be required for carotenoid biosynthesis. These 11 genes are organized in a *crtE*ORF2*crtBD*CORF6-ORF7-ORF8-ORF9-ORF10-ORF11 operon, abbreviated here as the *crt*ORF operon. The *carB* locus, originally associated with a Car⁻ phenotype, corresponds to the *crtE* gene; the identity of *carA* is discussed below. One unusual feature of *M. xanthus* is the physical separation of *crtI* from the other clustered *crt* genes in the *crt*ORF operon (39).

Functions were assigned to some of the genes in the *M. xanthus crt*ORF operon based on mutant phenotypes and structural similarities of the gene products to other known carotenoid biosynthesis enzymes (Tables 1 and 2) (26). The *crtE, crtB, crtC,* and *crtD* genes of this operon, and the unlinked *crtI* gene should be sufficient to direct the synthesis of neurosporene, lycopene and their hydroxy xanthophyll derivatives (Figures 2 and 3). Of the two putative carotenoid biosynthesis genes, ORF6 encodes a novel protein required for lycopene cyclization that lacks sequence similarity to either CrtY- or CrtL-type lycopene cyclases (see below). The product of ORF2 is predicted to encode a Crt-I type carotenoid desaturase that is probably required late in the biosynthetic pathway. Based on structural relationships between carotenoids, the biosynthetic functions provided by the five known *crt* genes, and postulated for ORF6 and ORF2 could be sufficient to produce myxobactin. In order to synthesize the end-product carotenoids found in the genus *Myxococcus*, enzymes that catalyze glucosylation, addition of an oxo- group, and sugar esterification with a fatty acid would have to be encoded by as yet unidentified genes (26, 51, 112).

crt genes from marine bacteria, including a *crt* gene cluster from *A. aurantiacum* and two of the corresponding genes from *Alicaligenes* PC-1, have recently been cloned (82, 85). The major carotenoids accumulated in *A. aurantiacum* cells grown under several different conditions are adonixanthin and astaxanthin (Figure 5) (125). The *A. aurantiacum crt* gene cluster was cloned by screening for altered pigmentation in the *Erwinia-E. coli* system (85). The cloned gene cluster includes five genes, *crtB, crtI, crtW, crtY,* and *crtZ*, potentially organized in a *crtWZYIB* operon. In contrast to other eubacterial *crt* gene clusters, no *crtE* gene was found, although it is possible that a flanking region of the genome might encode this gene. Homologies between the predicted *crtB, crtI, crtY,* and *crtZ* gene products and cognate enzymes of other eubacteria combined with analyses of pigment content in the *Erwinia-E. coli* system allowed assignments of gene function (Tables 1 and 2). The novel *crtW* gene identified in *A. aurantiacum* and *Alcaligenes* PC-1 is required for the addition of an oxo-group to cyclic carotenes and xanthophylls (82, 85). The resulting biosynthetic pathway for the synthesis of astaxanthin from GGPP suggests that CrtZ, β-carotene hydroxylase, and CrtW, β-C-4-oxygenase, can each accept several different substrates (Figure 5). Algal homologs of CrtW (27, 62, 73) and a plant homolog of CrtZ (109a) have recently been identified.

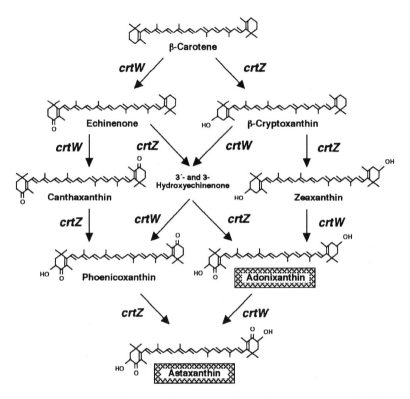

Figure 5 Cyclic xanthophyll biosynthesis pathway in *A. aurantiacum. crt* genes required for specific carotenoid biosynthetic reactions are given here and are listed in Table 2. Structures of intermediates not shown are given elsewhere (85). Boxes highlight accumulated end-product carotenoids.

T. thermophilus, an aerobic gram-negative thermophilic bacterium normally found in hot springs, represents a promising system for future classical and molecular genetic studies of carotenoid biosynthesis. The genus *Thermus* had long been known to produce polar carotenoids of unknown structure localized in the outer membrane (46). Attempts to clone *T. thermophilus crt* genes in *E. coli* by screening for pigment production at mesophilic temperatures were unsuccessful. Therefore, the novel strategy of gene cloning by screening for carotenoid overproduction in the homologous host was adopted, leading to the identification of the *crtB* gene (52). This result suggests that phytoene synthase is a rate-limiting carotenoid biosynthetic enzyme in this organism. In novel carotenoid underproducing and overproducing mutants of *T. thermophilus*, overproduction severely retarded growth at 80°C but not 70°C, while survival rates after ultraviolet irradiation were slightly decreased in underproducers and substantially increased in overproducers (53). The *T. thermophilus crt* genes are clustered but, unusually among eubacteria, seem to be encoded on a megaplasmid rather than on the chromosome (110). The unique end-product carotenoids of *T. thermophilus* have only recently been identified as thermozeaxanthins—novel branched fatty acid esters derived from zeaxanthin glucosides that have been proposed to regulate membrane stability (126). The structures of these pigments suggest that carotenoid biosynthesis in *T. thermophilus* follows a pathway closely related to that of the genus *Erwinia* (Figures 1, 2, and 4).

In the gram-positive genus *Streptomyces*, as in *M. xanthus*, carotenogenesis can be light-induced, constitutive, or completely absent (63). A cryptic *crt* gene cluster has recently been cloned from one Car⁻ strain of *S. griseus* by transformation of another Car⁻ strain in which some of the cryptic genes were expressed (107). Some strains of the genus *Streptomyces* accumulate specialized β-carotene-derived pigments containing aromatic ring structures (Figure 2), such as isorenieratene, β-isorenieratene, and hydroxy xanthophyll derivatives thereof (46, 61, 105). Sequence analysis of the *S. griseus crt* gene cluster exposed the presence of seven genes, *crtB, crtE, crtI, crtU, crtV, crtT*, and *crtY*, potentially organized in *crtEIBV* and *crtYTU* operons. Functional analyses of *S. griseus crtE* in the *Erwinia-E. coli* system, and of *S. griseus crtB* and *crtE* transferred to the colorless mutant of *S. griseus*, confirmed these assignments of gene function (Tables 1 and 2). Novel *S. griseus crt* genes, *crtT, crtU*, and *crtV*, were designated as such because their products displayed homologies to CrtF, CrtI/CrtD, and the protein-glutamate methyltransferase CheB, respectively (107). Although no direct evidence exists, *crtT, crtU*, and *crtV* have been postulated to be involved in the conversion of β-carotene to isorenieratene via β-isorenieratene, which is thought to involve the successive aromatization of both of the aliphatic β-rings accompanied by the migration of a methyl group. Primer extension analysis suggests the presence of a promoter 5′ to *crtE* that regulates expression of the *crtEIBV* operon.

A cluster of five *crt* genes from the Gram-negative *Flavobacterium* R1534 that direct the synthesis of zeaxanthin when expressed in *E. coli* has very recently been identified (90). The five genes are reported to correspond to *crtB, crtE, crtI, crtY*, and *crtZ*, consistent with the known genetic requirements for the synthesis of zeaxanthin in the genus *Erwinia* (Figures 1, 2, and 4), and are organized in at least two operons.

Two *Staphylococcus aureus crt* genes representing part of a larger, as yet uncharacterized, gene cluster involved in C_{30} carotenoid biosynthesis in this gram-positive bacterium have been identified by their expression in *E. coli* and in an unpigmented strain of *Staphylococcus carnosus* (119). Members of the genus *Staphylococcus* accumulate a wide variety of C_{30} carotenoids originating from FPP rather than GGPP (Figure 1). The early reactions of C_{30} carotenoid biosynthesis involving 4, 4'-diapophytoene formation and its desaturation to 4, 4'-diaponeurosporene parallel the reaction sequence well established for the C_{40} pathway (Figure 2). Some of the pigment end-products accumulated in the genus *Staphylococcus* have unusual structures, requiring various oxidation, glucosylation, and acylation reactions (46, 61, 114). The *S. aureus* strain from which the *crt* genes were cloned accumulates 4, 4'-diaponeurosporene during the early stages of growth. This carotenoid is eventually converted to staphyloxanthin, a novel pigment in which the C_{30} carotene backbone is linked via a carboxylic acid to a glucose fatty acid ester (114, 119). Sequence analysis of the two *S. aureus crt* genes, *crtM*, and *crtN*, indicates that their products are the respective C_{30} carotenoid biosynthesis homologs of CrtB and CrtI in the C_{40} pathway (Figure 1) (Table 1), and that they probably form a *crtNM* operon. Confirming these observations, both *E. coli* and the *S. carnosus* mutant carrying the two *S. aureus crt* genes accumulate 4, 4'-diaponeurosporene (119). Furthermore, cell extracts prepared from *E. coli* transformed with either *S. aureus crtM* alone or with both *crtM* and *crtN* are capable of incorporating radiolabeled FPP into 4, 4'-diapophytoene and 4, 4'-diaponeurosporene, respectively.

REGULATION OF CAROTENOID BIOSYNTHESIS   Little information is available about the regulation of carotenoid biosynthesis in the genus *Erwinia*. Pigment accumulation during growth in the presence or absence of glucose was determined for a number of *E. herbicola* strains (97). Carotenoid synthesis was repressed by glucose in both *E. herbicola* strain Eho10 and transformed *E. coli* carrying the *E. herbicola crt* genes and depended on cAMP. These results suggest transcriptional derepression of the *crt* genes in both *E. herbicola* and *E. coli* under conditions of carbon limitation. This type of regulation was not, however, observed in a number of other *E. herbicola* strains including Eho13.

In *M. xanthus*, a number of elegant experiments have led to the most complete genetic description of the regulation of carotenoid biosynthesis thus far available for any organism (47, 51). The genetically unlinked putative regulatory loci *carA* and *carR*, defined by the Car^c phenotypes of the corresponding mutants, were identified during transposon mutagenesis of *M. xanthus* a decade ago (14, 78). Shortly thereafter, a regulatory locus closely linked to *carR* and later termed *carQ* was discovered because of its associated Car$^-$ phenotype (77, 79). Further genetic and molecular characterization of the *carR* region revealed the presence of not two but three translationally coupled genes, *carQRS*, organized in a light-induced operon (50, 81). Mutations in the newly identified regulatory gene *carS* were observed to result in a Car$^-$ phenotype. Recently identified mutants in another unlinked regulatory locus, *carD*, also display a Car$^-$ phenotype in addition to being deficient in developmental processes, such as starvation-induced formation of fruiting bodies and sporulation (89).

A series of genetic studies have defined the interplay between the *carA, carD, carQ, carR,* and *carS* regulatory loci (Table 3), and the light-induced promoters of the *crtI, crt*ORF, and *carQRS* operons (14, 26, 39, 47, 50, 79, 81, 89, 100). Although transcription at all three promoters is induced by light, the mechanisms regulating them are nevertheless distinct. For example, the *crtI* promoter is the most highly regulated, displaying a 400-fold light induction under conditions of carbon starvation or upon cessation of cell growth (39, 51). This regulation explains previous observations that massive carotenoid accumulation in *M. xanthus* occurs only in stationary phase cells and pinpoints phytoene desaturation as a key regulatory step. The *crt*ORF promoter, in contrast, is only 20-fold inducible by light and is insensitive to the carbon supply (14, 51). In addition to these qualitative and quantitative differences in promoter responsiveness to environmental stimuli, the *crtI, crt*ORF, and *carQRS* operons also differ in their direct and indirect endogenous genetic requirements for light induction (47, 51).

Some of the structural features of the deduced *car* gene products have proved helpful in understanding the genetic data regarding light regulation. Whereas CarQ was recently reported to be structurally related to a group of proteins known as extracytoplasmic function σ factors of RNA polymerase, CarR seems to be an inner membrane protein and CarS displays no similarity with other known proteins (47, 81). The identity of CarA has proved more elusive, although the characterization of the *crt*ORF operons from the wildtype and a *carA* mutant has revealed the nature of the nucleotide changes that lead to the loss of CarA function (26). Two closely linked point mutations at the 3' end of the *crt*ORF operon apparently inactivate ORF10 and/or ORF11, both of which encode putative regulatory proteins that contain helix-turn-helix DNA-binding motifs. These findings are consistent with the model that CarA is a transcriptional regulator indirectly involved in upregulation of *crtI* expression in response to light, and directly responsible for repression of *crt*ORF expression in the dark (47).

One recent and comprehensive model incorporating CarA, CarQ, CarR, and CarS as regulatory factors (Table 3) to explain the known features of blue light-induced carotenogenesis in *M. xanthus* can be summarized as follows (47). Basal transcription in the dark of the *carQRS* operon allows small amounts of the inner membrane protein CarR, a putative anti-σ factor, and associated σ factor CarQ to be sequestered at the cytoplasmic membrane. When dark-grown *M. xanthus* cells are exposed to light, membrane-localized protoporphyrin IX generates an unidentified signal leading to the degradation of CarR and the release of CarQ. The latter then interacts with the core RNA polymerase, allowing transcription of the light-induced *crtI* and *carQRS* operons. Newly synthesized CarR is destroyed by the protoporphyrin IX-generated signal, but increasing amounts of CarQ, CarS, and CrtI accumulate. CarS activates transcription of the *crt*ORF operon and thereby allows the accumulation of carotenoid biosynthesis enzymes and CarA. The latter eventually competes with CarS to repress the *crt*ORF operon. Before this happens, however, all of the enzymes needed to produce myxobactone ester and myxobactin ester have been made. Carotenoids then accumulate in the cytoplasmic membrane and quench the protoporphyrin IX-generated signal for the destruction of CarR, thus completing the regulatory cycle. Although the above model does not explicitly discuss the role of CarD, this protein is necessary along with CarQ for the light-induced transcription of the *crtI* and *carQRS* operons, and may help to integrate developmental and environmental signals that activate carotenogenesis (89).

As mentioned previously, carotenogenesis is also light-induced in some strains of the genus *Streptomyces*. In *S. setonii*, the light-inducibility is subject to genetic variation and is lost at a relatively high frequency (63). The isolation of a revertant of a Car⁻ mutant led to the identification of *crtS*, a regulatory gene that restored colored carotenoid accumulation. CrtS seems to represent an alternative σ factor of RNA polymerase that activates carotenoid accumulation by permitting the expression of cryptic *crt* genes in *S. setonii* (Table 3). *S. setonii crtS* also activated carotenoid accumulation in a Car⁻ strain of *S. griseus* bearing cryptic *crt* genes (107). By analogy to light-induced carotenogenesis in *M. xanthus*, one might speculate that CrtS fulfills a role similar to that of the proposed extracytoplasmic function σ factor CarQ (47).

Cyanobacteria

BIOSYNTHESIS GENES   Carotenoids are found universally among the Chl-containing oxygenic photosynthetic cyanobacteria (46, 49). Typical pigments accumulated include β-carotene, the cyclic xanthophylls echinenone, canthaxanthin, zeaxanthin, and the cyclic xanthophyll glycoside myxoxanthophyll, that contains a rhamnose sugar moiety (Figures 4 and 5). The primary functions of cyanobacterial carotenoids are, as in the anoxygenic photosynthetic bacteria, photoprotective quenching of excited state Chl molecules and singlet oxygen, and light-harvesting. Cyanobacterial carotenoids are noncovalently attached to membrane-localized Chl-binding polypeptides of photosystems I and II, to novel lipophilic or soluble carotenoid-binding proteins that lack Chl, or are constituents of the cytoplasmic and outer membranes (49).

Although the genetics of an entire cyanobacterial carotenoid biosynthetic pathway have not yet been established, studies of *Synechococcus* sp. strain PCC7942 (33, 34, 37, 72), *Synechocystis* sp. strain PCC6803 (75, 76), and *Anabaena* sp. strain PCC7120 (71) have provided substantial information about the genes required for the biosynthesis of cyclic carotenes. The former two species have been demonstrated to accumulate the normal spectrum of cyanobacterial carotenoids (35, 72).

In the absence of mutants with altered carotenoid compositions (49), two alternative approaches involving both selection and screening have been taken to isolate cyanobacterial *crt* genes. Four *crt* genes, termed *crtB, crtP, crtQ,* and *crtL* in the current nomenclature, involved in the conversion of GGPP to β-carotene have thus far been identified (Figures 1 and 2). The first strategy was to select for resistance to inhibitors of carotenoid biosynthesis (37, 72, 76). This approach relied on the observations that bleaching (i.e. chlorosis-inducing) herbicides such as norflurazon inhibit the desaturation of phytoene in Chl-containing organisms (102). Genetic selections to identify norfluazon-resistant cyanobacterial colonies led to the cloning and sequencing of herbicide resistance genes from *Synechococcus* sp. strain PCC7942 (34) and *Synechocystis* sp. strain PCC6813 (76). A series of norflurazon-resistant mutants were analyzed with respect to carotenoid content, in vitro phytoene desaturase activity, and nature of the mutation, leading to the conclusion that phytoene desaturation is the rate-limiting step in cyanobacterial carotenoid biosynthesis (35). The most direct proof that the *Synechococcus* sp. strain PCC7942 *crtP* (formerly *pds*) gene encoded phytoene desaturase was obtained by in vivo complementation in the *Erwinia-E. coli* system (70). This and several similar experiments performed with plant homologs of *crtP* (21, 55, 91) demonstrated conclusively that the CrtP-type phytoene desaturase from Chl-containing eubacteria and eukaryotes catalyzes, in contrast to CrtI, only two successive desaturations to yield ζ-carotene (Figure 2). A selection strategy using a lycopene cyclase inhibitor was also employed to clone and characterize the *crtL* (formerly *lcy*) gene encoding the cyanobacterial lycopene cyclase (38).

The second approach used to clone a cyanobacterial *crt* gene was to procede directly to a library-based screen for pigment accumulation (71). To this end, a cDNA library from *Anabaena* sp. strain PCC7120 was transformed into carotene-accumulating *E. coli* strains that carried combinations of *crt* genes from other eubacteria. This resulted in the isolation and sequencing of *crtQ* (formerly *zds*), a gene whose product allowed the in vivo synthesis of lycopene from ζ-carotene and neurosporene, but not from phytoene (Figure 2).

The *crtB* (formerly *psy* or *pys*) genes of *Synechococcus* sp. strain PCC7942 (33) and *Synechocystis* sp. strain PCC6803 (75), were identified because they are adjacent to the respective *crtP* genes of these cyanobacteria (34, 76). Despite this bit of serendipity, a general feature distinguishing cyanobacteria from other eubacteria is the chromosomal dispersal of the *crt* genes identified thus far. Only *crtP* and *crtB* are physically linked, and even they do not seem to form an obligate operon. Function was established for cyanobacterial *crtB* by in vivo functional complementation in the *Erwinia-E. coli* system, and by following the incorporation of radiolabeled GGPP added to cell extracts of the transformed *E. coli* (33, 75).

Although cyanobacterial CrtB is homologous to all other known eubacterial and eukaryotic phytoene synthases (10, 18), CrtP, the cyanobacterial phytoene desaturase, CrtQ, the ζ-carotene desaturase, and CrtL, the cyanobacterial lycopene cyclase, represent structurally and/or functionally unique enzymes not found in other eubacteria (Table 1). CrtP and CrtL do, however, have homologs in algae (92) and higher plants (21, 38, 54, 55, 91, 93). Curiously, *Anabaena* sp. strain PCC7120 CrtQ does not structurally resemble a recently identified higher plant CrtP-type ζ-carotene desaturase (2), but rather CrtI-type carotenoid desaturases from other eubacteria (71). In this regard, it is worth noting that the *crtQ* gene seems to be unique to *Anabaena* sp. strain PCC7120.

Cyanobacterial CrtP and CrtQ have been overexpressed in *E. coli* and isolated in an active state for biochemical studies (3, 40) that have confirmed and expanded previous findings from in vivo complementation with the *Erwinia-E. coli* system (70, 71). Overexpressed and purified *Synechococcus* sp. strain PCC7942 CrtP (40) and *E. uredovora* CrtI (41) have also been compared in terms of their responses to different classes of phytoene desaturation inhibitors (103). Whereas CrtP was found to be strongly inhibited by bleaching herbicides including norflurazon, CrtI activity was highly sensitive to diphenylamine, a classical inhibitor of phytoene desaturation in eubacteria (46).

BIOTECHNOLOGICAL APPLICATIONS

One of the most exciting prospects on the horizon in the field of carotenoid biosynthesis is the development of novel strategies for the production of carotenoids of nutritional, medical, or aesthetic interest (61). As one example, dietary vitamin A deficiency, a severe worldwide nutritional problem in underdeveloped countries and one that is particularly acute among children (116), is currently being approached with the goal of using genetic engineering to achieve β-carotene (i.e. provitamin A) accumulation in staple foods that normally lack carotenoids. The main target of this effort, under the auspices of the Rockefeller Foundation, is the carotenoid-deficient endosperm, or seed storage tissue, of rice, the staple of much of the world's population. Initial efforts have led to the production of transgenic rice plants that accumulate phytoene in their endosperms (30). One future strategy could be to express bacterial *crtI* and *crtY* genes encoding phytoene desaturase and lycopene cyclase, respectively, to allow this phytoene to be converted to β-carotene in the endosperm plastids of rice seeds.

That a bacterial carotenoid biosynthesis gene can function in a higher plant has been demonstrated by the engineering of herbicide resistance based on the known differences between the CrtI- and CrtP-type phytoene desaturases (Table 1) in their sensitivities to norflurazon (103). This inhibitor blocks

CrtP-mediated phytoene desaturation in cyanobacteria and plants (34). Norflurazon-resistant tobacco plants were produced by transformation with a recombinant *E. uredovora crtI* gene (83, 87). The same gene was also used to introduce norflurazon resistance into *Synechococcus* sp. strain PCC7942 (120). Coexpression of the eubacterial CrtI-type and the endogenous eukaryotic CrtP-type phytoene desaturases altered the distribution of the end-product xanthophylls in transgenic tobacco plants (83).

Naturally occurring carotenoids of commercial interest as coloring agents for food, pharmaceuticals, cosmetics, and animal feed include lycopene, β-carotene, lutein, zeaxanthin, canthaxanthin, astaxanthin, and rhodoxanthin (61, 88). Several of these pigments, such as β-carotene, canthaxanthin, and astaxanthin, are currently commercially produced by total chemical synthesis. The recent genetic elucidation of bacterial carotenoid biosynthetic pathways leading to the accumulation of zeaxanthin, canthaxanthin, and astaxanthin (Figures 4 and 5) may offer interesting alternatives for their in vivo production (82, 85, 90). Introduction of genus *Erwinia crt* genes into other eubacteria, such as *A. tumefaciens* and *Zymomonas mobilis* (86), or eukaryotes, such as the yeast *Saccharomyces cerevisiae* (121), has already been used to demonstrate the accumulation of substantial amounts of lycopene and β-carotene in these normally noncarotenogenic organisms.

SUMMARY

In exploring the general principles of eubacterial carotenoid biosynthesis, several recurring themes can be discerned. First, *crt* genes encoding biosynthetic enzymes are usually clustered , with the notable exception of the known cyanobacterial *crt* genes. In particular, potential *crtIB* or *crtNM* operons have been found in all *crt* gene clusters thus far, except for that of *M. xanthus*. Clusters of *crt* genes may provide a selective advantage by allowing the rapid coregulation of multigene operons to activate or shut down several steps in pigment biosynthesis in response to changing environmental conditions.

Second, all bacterial *crt* gene clusters thus far analyzed in their entirety contain a *crtE* gene, with the possible exception of *A. aurantiacum*. This observation, together with mutational analyses, supports the hypothesis that carotenogenic eubacteria contain more than one GGPP synthase, at least one of which supplies FPP and GGPP for essential functions such as quinone production, and another of which, encoded by *crtE*, that is required for carotenoid pigment accumulation (Figure 1).

Third, as evident from eubacterial carotenoid biosynthetic pathways that have been genetically characterized (Figures 2, 3, 4, and 5), and recent biochemical studies with purified enzymes, many of the *crt* genes encode products that display promiscuous substrate requirements. Thus, starting from a relatively small pool of unique biosynthetic functions, it is possible to generate the tremendous structural diversity found among naturally occurring carotenoids by combining different *crt* genes in different organisms.

Finally, as discussed here briefly, comparisons of the deduced amino acid sequences of eubacterial and eukaryotic pigment biosynthesis enzymes suggest a surprising evolutionary dichotomy in some of the reactions of carotenoid biosynthesis (6, 10). Structurally similar soluble GGPP synthases and membrane-associated phytoene synthases (Figure 1) are found among all carotenogenic organisms (Table 1). Unexpectedly, the functionally related membrane-bound carotene biosynthetic enzymes (Figure 2) of Bchl-containing anoxygenic photosynthetic bacteria, nonphotosynthetic bacteria, and fungi on the one hand, and Chl-containing, oxygenic photosynthetic cyanobacteria, algae, and plants on the other, are structurally dissimilar (Table 1). The CrtI-type carotenoid desaturases (CrtI, CrtN, and CrtD), and the CrtY-type lycopene cyclases of organisms that lack Chl, appear almost totally unrelated in their primary sequences to the functionally related CrtP-type phytoene desaturases and the CrtL-type lycopene cyclases of Chl-containing organisms.

The genetics of eubacterial carotenoid biosynthesis promises to remain a fruitful area for basic and applied research for many years to come. Not only will the complexity of biosynthetic pathways and regulatory mechanisms continue to challenge microbiologists, geneticists, biochemists, and biotechnologists, but many of the lessons learned from eubacteria may be directly relevant to a full understanding of eukaryotic carotenoid biosynthesis.

ACKNOWLEDGMENTS

I wish to thank the numerous friends and colleagues who contributed reprints and preprints of their work. Preparation of this review was supported by the Rockefeller Foundation.

> Visit the *Annual Reviews home page* at
> http://www.annurev.org.

Literature Cited

1. Alberti M, Burke DH, Hearst JE. 1995. Structure and sequence of the photosynthesis gene cluster. See Ref. 24, pp. 1083–106
2. Albrecht M, Klein A, Hugueney P, Sandmann G, Kuntz M. 1995. Molecular cloning and functional expression in *E. coli* of a novel plant enzyme mediating ζ-carotene desaturation. *FEBS Lett.* 372:199–202
3. Albrecht M, Linden H, Sandmann G. 1996. Biochemical characterization of purified ζ-carotene desaturase from *Anabaena* PCC 7120 after expression in *Escherichia coli*. *Eur. J. Biochem.* 236:115–20
4. Armstrong GA. 1994. Eubacteria show their true colors: genetics of carotenoid pigment biosynthesis from microbes to plants. *J. Bacteriol.* 176:4795–802

5. Armstrong GA. 1995. Genetic analysis and regulation of carotenoid biosynthesis: structure and function of the *crt* genes and gene products. See Ref. 24, pp. 1135–57
6. Armstrong GA. 1997. Genetics, biochemistry and regulation of carotenoid biosynthesis. In *Comprehensive Natural Products Chemistry, Volume 2: Isoprenoids Including Carotenoids and Steroids*, ed. DE Cane. Oxford: Elsevier Science. In press
7. Armstrong GA, Alberti M, Hearst JE. 1990. Conserved enzymes mediate the early reactions of carotenoid biosynthesis in nonphotosynthetic and photosynthetic prokaryotes. *Proc. Natl. Acad. Sci. USA* 87:9975–79
8. Armstrong GA, Alberti M, Leach F, Hearst JE. 1989. Nucleotide sequence, organization, and nature of the protein products of the carotenoid biosynthesis gene cluster of *Rhodobacter capsulatus*. *Mol. Gen. Genet.* 216:254–68
9. Armstrong GA, Cook DN, Ma D, Alberti M, Burke DH, Hearst JE. 1993. Regulation of carotenoid and bacteriochlorophyll biosynthesis genes and identification of an evolutionarily conserved gene required for bacteriochlorophyll accumulation. *J. Gen. Microbiol.* 139:897–906
10. Armstrong GA, Hearst JE. 1996. Genetics and molecular biology of carotenoid pigment biosynthesis. *FASEB J.* 10:228–37
11. Armstrong GA, Hundle B, Hearst JE. 1993. Evolutionary conservation and structural similarities of carotenoid biosynthesis gene products from photosynthetic and nonphotosynthetic organisms. *Methods Enzymol.* 214:297–311
12. Armstrong GA, Schmidt A, Sandmann G, Hearst JE. 1990. Genetic and biochemical characterization of carotenoid biosynthesis mutants of *Rhodobacter capsulatus*. *J. Biol. Chem.* 265:8329–38
13. Baker ME, Fanestil DD. 1991. Mammalian peripheral-type benzodiazepine receptor is homologous to the CrtK protein of *Rhodobacter capsulatus*, a photosynthetic bacterium. *Cell* 65:721–22
14. Balsalobre JM, Ruiz-Vásquez RM, Murillo FJ. 1987. Light induction of gene expression in *Myxococcus xanthus*. *Proc. Natl. Acad. Sci. USA* 84:2359–62
15. Bartley GE, Schmidhauser TJ, Yanofsky C, Scolnik PA. 1990. Carotenoid desaturases from *Rhodobacter capsulatus* and *Neurospora crassa* are structurally and functionally conserved and contain domains homologous to flavoprotein disulfide oxidoreductases. *J. Biol. Chem.* 265:16020–24
16. Bartley GE, Scolnik PA. 1989. Carotenoid biosynthesis in photosynthetic bacteria. *J. Biol. Chem.* 264: 13109–13
17. Bartley GE, Scolnik PA. 1993. cDNA cloning, expression during development, and genome mapping of *Psy2*, a second tomato gene encoding phytoene synthase. *J. Biol. Chem.* 268:25718–21
18. Bartley GE, Scolnik PA. 1995. Plant carotenoids: pigments for photoprotection, visual attraction, and human health. *Plant Cell* 7:1027–38
19. Bartley GE, Scolnik PA, Giuliano G. 1994. Molecular biology of carotenoid biosynthesis in plants. *Annu. Rev. Plant Physiol. Plant Mol. Biol.* 45:287–301
20. Bartley GE, Viitanen PV, Bacot KO, Scolnik PA. 1992. A tomato gene expressed during fruit ripening encodes an enzyme of the carotenoid biosynthesis pathway. *J. Biol. Chem.* 267:5036–39
21. Bartley GE, Viitanen PV, Pecker I, Chamovitz D, Hirschberg J, Scolnik PA. 1991. Molecular cloning and expression in photosynthetic bacteria of a soybean cDNA coding for phytoene desaturase, an enzyme of the carotenoid biosynthesis pathway. *Proc. Natl. Acad. Sci. USA* 88:6532–36
22. Beatty JT. 1995. Organization of photosynthesis gene transcripts. See Ref. 24, pp. 1209–19
23. Biel AJ, Marrs BL. 1985. Oxygen does not directly regulate carotenoid biosynthesis in *Rhodopseudomonas capsulata*. *J. Bacteriol.* 162:1320–21
24. Blankenship, RE, Madigan MT, Bauer CE, eds. 1995. *Advances in Photosynthesis: Anoxygenic Photosynthetic Bacteria, Volume 2*, Dordrecht: Kluwer. 1331 pp.
25. Bollivar DW, Suzuki JY, Beatty JT, Dobrowolski JM, Bauer CE. 1994. Directed mutational analysis of bacteriochlorophyll *a* biosynthesis in *Rhodobacter capsulatus*. *J. Mol. Biol.* 237:622–40
26. Botella JA, Murillo FJ, Ruiz-Vázquez R. 1995. A cluster of structural and regulatory genes for light-induced carotenogenesis in *Myxococcus xanthus*. *Eur. J. Biochem.* 233:238–48
27. Breitenbach J, Misawa N, Kajiwara S, Sandmann G. 1996. Expression in *Escherichia coli* and properties of the carotene ketolase from *Haematococcus pluvialis*. *FEMS Microbiol. Lett.* 140:241–46

28. Britton G. 1983. Carotenoids. In *The Biochemistry of Natural Pigments*, pp. 23–73. Cambridge: Cambridge Univ. Press

29. Britton G. 1995. UV/visible spectroscopy. In *Carotenoids, Volume 1B: Spectroscopy*, ed. G Britton, S Liaaen-Jensen, H Pfander, pp. 13–62. Basel: Birkhäuser Verlag

30. Burkhardt PK, Beyer P, Wünn J, Klöti A, Armstrong GA, et al. 1997. Transgenic rice (*Oryza sativa* L.) endosperm expressing daffodil (*Narcissus pseudonarcissus*) phytoene synthase accumulates phytoene, a key intermediate of provitamin A biosynthesis. *Plant J.* 11:1071–78

31. Carattoli A, Romano N, Ballario P, Morelli G, Macino G. 1991. The *Neurospora crassa* carotenoid biosynthesis gene (albino 3) reveals highly conserved regions among prenyltransferases. *J. Biol. Chem.* 266:5854–59

32. Cerdá-Olmedo E, Reau P. 1970. Genetic classification of the lethal effects of various agents on heterokaryotic spores of Phycomyces. *Mutat. Res.* 9:369–84

33. Chamovitz D, Misawa N, Sandmann G, Hirschberg J. 1992. Molecular cloning and expression in *Escherichia coli* of a cyanobacterial gene coding for phytoene synthase, a carotenoid biosynthesis enzyme. *FEBS Lett.* 296:305–10

34. Chamovitz D, Pecker I, Hirschberg J. 1991. The molecular basis of resistance to the herbicide norflurazon. *Plant Mol. Biol.* 16:967–74

35. Chamovitz D, Sandmann G, Hirschberg J. 1993. Molecular and biochemical characterization of herbicide-resistant mutants of cyanobacteria reveals that phytoene desaturation is a rate-limiting step in carotenoid biosynthesis. *J. Biol. Chem.* 268:17348–53

36. Coomber SA, Chaudri M, Connor A, Britton G, Hunter CN. 1990. Localized transposon Tn5 mutagenesis of the photosynthetic gene cluster of *Rhodobacter sphaeroides*. *Mol. Microbiol.* 4:977–89

37. Cunningham FX Jr, Pogson B, Sun Z, McDonald KA, DellaPenna D, Gantt E. 1996. Functional analysis of the β and ϵ lycopene cyclase enzymes of *Arabidopsis* reveals a mechanism for control of cyclic carotenoid formation. *Plant Cell* 8:1613–26

38. Cunningham FX Jr, Sun Z, Chamovitz D, Hirschberg J, Gantt E. 1994. Molecular structure and enzymatic function of lycopene cyclase from the cy-anobacterium *Synechococcus* sp. strain PCC7942. *Plant Cell* 6:1107–21

39. Fontes M, Ruiz-Vázquez R, Murillo FJ. 1993. Growth phase dependence of the activation of a bacterial gene for carotenoid synthesis by blue light. *EMBO J.* 12:1265–75

40. Fraser PD, Linden H, Sandmann G. 1993. Purification and reactivation of recombinant *Synechococcus* phytoene desaturase from an overproducing strain of *Escherichia coli*. *Biochem. J.* 291:687–92

41. Fraser PD, Misawa N, Linden H, Yamano S, Kobayashi K, Sandmann G. 1992. Expression in *Escherichia coli*, purification, and reactivation of the recombinant *Erwinia uredovora* phytoene desaturase. *J. Biol. Chem.* 267:19891–95

42. Garí E, Toledo JC, Gibert I, Barbé J. 1992. Nucleotide sequence of the methoxyneurosporene dehydrogenase gene from *Rhodobacter sphaeroides*: comparison with other bacterial carotenoid dehydrogenases. *FEMS Microbiol. Lett.* 93:103–08

43. Giuliano G, Pollock D, Scolnik PA. 1986. The *crtI* gene mediates the conversion of phytoene into colored carotenoids in *Rhodopseudomonas capsulata*. *J. Biol. Chem.* 261:12925–29

44. Giuliano G, Pollock D, Stapp H, Scolnik PA. 1988. A genetic-physical map of the *Rhodobacter capsulatus* carotenoid biosynthesis gene cluster. *Mol. Gen. Genet.* 213:78–83

45. Gomelsky M, Kaplan S. 1995. Genetic evidence that PpsR from *Rhodobacter sphaeroides* 2.4.1 functions as a repressor of *puc* and *bchF* expression. *J. Bacteriol.* 177:1634–37

46. Goodwin TW. 1980. *The Biochemistry of Carotenoids, Volume 1: Plants*, pp. 1–377. London: Chapman & Hall

47. Gorham HC, McGowan SJ, Robson PRH, Hodgson DA. 1996. Light-induced carotenogenesis in *Myxococcus xanthus*: light-dependent membrane sequestration of ECF sigma factor CarQ by anti-sigma factor CarR. *Mol. Microbiol.* 19:171–86

48. Hahn FM, Baker JA, Poulter CD. 1996. Open reading frame 176 in the photosynthesis gene cluster of *Rhodobacter capsulatus* encodes *idi*, a gene for isopentenyl diphosphate isomerase. *J. Bacteriol.* 178:619–24

49. Hirschberg J, Chamovitz D. 1994. Carotenoids in cyanobacteria. In *Advances in Photosynthesis: The Molec-*

ular Biology of Cyanobacteria, Volume 1, ed. D Bryant, pp. 559–79. Dordrecht: Kluwer

50. Hodgson DA. 1993. Light-induced carotenogenesis in *Myxococcus xanthus*: genetic analysis of the *carR* region. *Mol. Microbiol.* 7:471–88

51. Hodgson DA, Murillo FJ. 1993. Genetics of regulation and pathway of synthesis of carotenoids. In *Myxobacteria II*, ed. M Dworkin, D Kaiser, pp.157–81. Washington DC: Am. Soc. Microbiol.

52. Hoshino T, Fujii R, Nakahara T. 1993. Molecular cloning and sequence analysis of the *crtB* gene of *Thermus thermophilus* HB27, an extreme thermophile producing carotenoid pigments. *Appl. Environ. Microbiol.* 59:3150–53

53. Hoshino T, Yoshino Y, Guevarra ED, Ishida S, Hiruta T, et al. 1994. Isolation and partial characterization of carotenoid underproducing and overproducing mutants from an extremely thermophilic *Thermus thermophilus* HB27. *J. Ferm. Bioeng.* 77:131–36

54. Hugueney P, Badillo A, Chen H-C, Klein A, Hirschberg J, et al. 1995. Metabolism of cyclic carotenoids: a model for the alteration of this biosynthetic pathway in *Capsicum annuum* chromoplasts. *Plant J.* 8:417–24

55. Hugueney P, Römer S, Kuntz M, Camara B. 1992. Characterization and molecular cloning of a flavoprotein catalyzing the synthesis of phytofluene and ζ-carotene in *Capsicum* chromoplasts. *Eur. J. Biochem.* 209:399–407

56. Hundle BS, Alberti M, Nievelstein V, Beyer P, Kleinig H, et al. 1994. Functional assignment of *Erwinia herbicola* Eho10 carotenoid genes expressed in *Escherichia coli. Mol. Gen. Genet.* 245:406–16

57. Hundle BS, Beyer P, Kleinig H, Englert G, Hearst JE. 1991. Carotenoids of *Erwinia herbicola* and an *Escherichia coli* HB101 strain carrying the *Erwinia herbicola* carotenoid gene cluster. *Photochem. Photobiol.* 54:89–93

58. Hundle BS, O'Brien DA, Alberti M, Beyer P, Hearst JE. 1992. Functional expression of zeaxanthin glucosyltransferase from *Erwinia herbicola* and a proposed uridine diphosphate binding site. *Proc. Natl. Acad. Sci. USA* 89:9321–25

59. Hundle BS, O'Brien DA, Beyer P, Kleinig H, Hearst JE. 1993. In vitro expression and activity of lycopene cyclase and β-carotene hydroxylase from *Erwinia herbicola. FEBS Lett.* 315:329–34

60. Hunter CN, Hundle BS, Hearst JE, Lang HP, Gardiner AT, et al. 1994. Introduction of new carotenoids into the bacterial photosynthetic apparatus by combining the carotenoid biosynthetic pathways of *Erwinia herbicola* and *Rhodobacter sphaeroides. J. Bacteriol.* 176:3692–97

61. Johnson EA, Schroeder WA. 1995. Microbial carotenoids. *Adv. Biochem. Eng./Biotech.* 53:119–78

62. Kajiwara S, Kakizono T, Saito T, Kondo K, Ohtani T, et al. 1995. Isolation and functional identification of a novel cDNA for astaxanthin biosynthesis from *Haematococcus pluvialis*, and astaxanthin synthesis in *Escherichia coli. Plant Mol. Biol.* 29:343–52

63. Kato F, Hino T, Nakaji A, Tanaka M, Koyama Y. 1995. Carotenoid synthesis in *Streptomyces setonii* ISP5395 is induced by the gene *crtS*, whose product is similar to a sigma factor. *Mol. Gen. Genet.* 247:387–90

64. Koyama Y. 1991. Structures and functions of carotenoids in photosynthetic systems. *J. Photochem. Photobiol. B* 9: 265–80

65. Kull D, Pfander H. 1995. List of new carotenoids. In *Carotenoids, Volume 1A: Isolation and Analysis*, ed. G Britton, S Liaaen-Jensen, H Pfander, pp. 295–317. Basel: Birkhäuser Verlag

66. Kuntz M, Römer S, Suire C, Hugueney P, Weil JH, et al. 1992. Identification of a cDNA for the plastid-located geranylgeranyl pyrophosphate synthase from *Capsicum annuum* : correlative increase in enzyme activity and transcript level during fruit ripening. *Plant J.* 2:25–34

67. Lang HP, Cogdell RJ, Gardiner AT, Hunter CN. 1994. Early steps in carotenoid biosynthesis: sequences and transcriptional analysis of the *crtI* and *crtB* genes of *Rhodobacter sphaeroides* and overexpression and reactivation of *crtI* in *Escherichia coli* and *R. sphaeroides. J. Bacteriol.* 176:3859–69

68. Lang HP, Cogdell RJ, Takaichi S, Hunter CN. 1995. Complete DNA sequence, specific Tn5 insertion map, and gene assignment of the carotenoid biosynthesis pathway of *Rhodobacter sphaeroides. J. Bacteriol.* 177:2064–73

69. Lee L-Y, Liu S-T. 1991. Characterization of the yellow-pigment genes of *Erwinia herbicola. Mol. Microbiol.* 5:217–24

70. Linden H, Misawa N, Chamovitz D, Pecker I, Hirschberg J, Sandmann G. 1991. Functional complementation in *Escherichia coli* of different phytoene

desaturase genes and analysis of accumulated carotenes. *Z. Naturforsch. C. Biosci.* 46:1045–51

71. Linden H, Misawa N, Saito T, Sandmann G. 1994. A novel carotenoid biosynthesis gene coding for ζ-carotene desaturase: functional expression, sequence and phylogenetic origin. *Plant Mol. Biol.* 24:369–79

72. Linden H, Sandmann G, Chamovitz D, Hirschberg J, Böger P. 1990. Generation and biochemical characterization of *Synechococcus* mutants selected against the bleaching herbicide norflurazon. *Pestic. Biochem. Physiol.* 36:46–51

73. Lotan T, Hirschberg J. 1995. Cloning and expression in *Escherichia coli* of the gene encoding β-C-4-oxygenase, that converts β-carotene to the ketocarotenoid canthaxanthin in *Haematococcus pluvialis*: *FEBS Lett.* 364:125–28

74. Marrs B. 1981. Mobilization of the genes for photosynthesis from *Rhodopseudomonas capsulata* by a promiscuous plasmid. *J. Bacteriol.* 146:1003–12

75. Martínez-Férez I, Fernández-González B, Sandmann G, Vioque A. 1994. Cloning and expression in *Escherichia coli* of the gene coding for phytoene synthase from the cyanobacterium *Synechocystis* sp. PCC6803. *Biochim. Biophys. Acta* 1218:145–52

76. Martínez-Férez IM, Vioque A. 1992. Nucleotide sequence of the phytoene desaturase gene from *Synechocystis* sp. PCC6803 and characterization of a new mutation which confers resistance to the herbicide norflurazon. *Plant Mol. Biol.* 18:981–83

77. Martínez-Laborda A, Balsalobre JM, Fontes M, Murillo FJ. 1990. Accumulation of carotenoids in structural and regulatory mutants of the bacterium *Myxococcus xanthus*. *Mol. Gen. Genet.* 223:205–10

78. Martínez-Laborda A, Elías M, Ruiz-Vázquez R, Murillo FJ. 1986. Insertions of Tn5 linked to mutations affecting carotenoid synthesis in *Myxococcus xanthus*. *Mol. Gen. Genet.* 205:107–14

79. Martínez-Laborda A, Murillo FJ. 1989. Genic and allelic interactions in the carotenogenic response of *Myxococcus xanthus* to blue light. *Genetics* 122:481–90

80. Math SK, Hearst JE, Poulter CD. 1992. The *crtE* gene in *Erwinia herbicola* encodes geranylgeranyl diphosphate synthase. *Proc. Natl. Acad. Sci. USA* 89:6761–64

81. McGowan SJ, Gorham HC, Hodgson DA. 1993. Light induced carotenogenesis in *Myxococcus xanthus*: DNA sequence analysis of the *carR* region. *Mol. Microbiol.* 10:713–35

82. Misawa N, Kajiwara S, Kondo K, Yokoyama A, Satomi Y, et al. 1995. Canthaxanthin biosynthesis by the conversion of methylene to keto groups in a hydrocarbon β-carotene by a single gene. *Biochem. Biophys. Res. Commun.* 209:867–76

83. Misawa N, Masamoto K, Hori T, Ohtani T, Böger P, Sandmann G. 1994. Expression of an *Erwinia* phytoene desaturase gene not only confers multiple resistance to herbicides interfering with carotenoid biosynthesis but also alters xanthophyll metabolism in transgenic plants. *Plant J.* 6:481–89

84. Misawa N, Nakagawa M, Kobayashi K, Yamano S, Izawa Y, et al. 1990. Elucidation of the *Erwinia uredovora* carotenoid biosynthetic pathway by functional analysis of gene products expressed in *Escherichia coli*. *J. Bacteriol.* 172:6704–12

85. Misawa N, Satomi Y, Kondo K, Yokoyama A, Kajiwara S, et al. 1995. Structure and functional analysis of a marine bacterial carotenoid biosynthesis gene cluster and astaxanthin biosynthetic pathway proposed at the gene level. *J. Bacteriol.* 177:6575–84

86. Misawa N, Yamano S, Ikenaga H. 1991. Production of β-carotene in *Zymomonas mobilis* and *Agrobacterium tumefaciens* by introduction of the biosynthesis genes from *Erwinia uredovora*. *Appl. Environ. Microbiol.* 57:1847–49

87. Misawa N, Yamano S, Linden H, de Felipe MR, Lucas M, et al. 1993. Functional expression of the *Erwinia uredovora* carotenoid biosynthetic gene *crtI* in transgenic plants showing an increase in β-carotene biosynthesis activity and resistance to the bleaching herbicide norflurazon. *Plant J.* 4:833–40

88. Nelis HJ, De Leenheer AP. 1991. Microbial sources of carotenoid pigments used in foods and feeds. *J. Appl. Bacteriol.* 70:181–91

89. Nicolás FJ, Ruiz-Vázquez R, Murillo FJ. 1994. A genetic link between light response and multicellular development in the bacterium *Myxococcus xanthus*. *Genes Develop.* 8:2375–87

90. Pasamontes L, Hug D, Tessier M, Hohmann H-P, Schierle J, van Loon APGM. 1996. *Isolation and characterization of the carotenoid biosynthesis*

genes of Flavobacterium *R1354.* Presented at 11th Int. Symp. Carotenoids, Leiden, The Netherlands

91. Pecker I, Chamovitz D, Linden H, Sandmann G, Hirschberg J. 1992. A single polypeptide catalyzing the conversion of phytoene to ζ-carotene is transcriptionally regulated during tomato fruit ripening. *Proc. Natl. Acad. Sci USA* 89:4962–66

92. Pecker I, Chamovitz D, Mann V, Sandmann G, Böger P, Hirschberg J. 1993. Molecular characterization of carotenoid biosynthesis in plants: the phytoene desaturase gene of tomato. In *Research in Photosynthesis, Volume III,* ed. N Murata, pp. 11–18. Dordrecht: Kluwer

93. Pecker I, Gabbay R, Cunningham FX Jr, Hirschberg J. 1996. Cloning and characterization of the cDNA for lycopene β-cyclase from tomato reveals decrease in its expression during fruit ripening. *Plant Mol. Biol.* 30:807–19

94. Pemberton JM, Harding CM. 1986. Cloning of carotenoid biosynthesis genes from *Rhodopseudomonas sphaeroides. Curr. Microbiol.* 14:25–9

95. Penfold RJ, Pemberton JM. 1991. A gene from the photosynthetic gene cluster of *Rhodobacter sphaeroides* induces *trans* suppression of bacteriochlorophyll and carotenoid levels in *R. sphaeroides* and *R. capsulatus. Curr. Microbiol.* 23:259–63

96. Penfold RJ, Pemberton JM. 1994. Sequencing, chromosomal inactivation, and functional expression in *Escherichia coli* of *ppsR,* a gene which represses carotenoid and bacteriochlorophyll synthesis in *Rhodobacter sphaeroides. J. Bacteriol.* 176:2869–76

97. Perry KL, Simonitch TA, Harrison-LaVoie KJ, Liu S-T. 1986. Cloning and regulation of *Erwinia herbicola* pigment genes. *J. Bacteriol.* 168:607–12

98. Ponnampalam SN, Buggy JJ, Bauer CE. 1995. Characterization of an aerobic repressor that coordinately regulates bacteriochlorophyll, carotenoid, and light harvesting-II expression in *Rhodobacter capsulatus. J. Bacteriol.* 177:2990–97

99. Raisig A, Bartley G, Scolnik P, Sandmann G. 1996. Purification in an active state and properties of the 3-step phytoene desaturase from *Rhodobacter capsulatus* overexpressed in *Escherichia coli. J. Biochem.* 119:559–64

100. Ruiz-Vázquez R, Fontes M, Murillo FJ. 1993. Clustering and co-ordinated activation of carotenoid genes in *Myxococcus xanthus* by blue light. *Mol. Microbiol.* 10:25–34

101. Sandmann G. 1994. Carotenoid biosynthesis in microorganisms and plants. *Eur. J. Biochem.* 223:7–24

102. Sandmann G, Böger P. 1989. Inhibition of carotenoid biosynthesis by herbicides. In *Target Sites of Herbicide Action,* ed. P Böger, G. Sandmann, pp. 25–44. Boca Raton: CRC Press

103. Sandmann G, Fraser PD. 1993. Differential inhibition of phytoene desaturases from diverse origins and analysis of resistant cyanobacterial mutants. *Z. Naturforsch. C Biosci.* 48:307–11

104. Sandmann G, Misawa N. 1992. New functional assignments of the carotenogenic genes *crtB* and *crtE* with constructs of these genes from *Erwinia* species. *FEMS Microbiol. Lett.* 90:253–58

105. Schmidt K. 1978. Biosynthesis of carotenoids. In *The Photosynthetic Bacteria,* ed. RK Clayton, WR Sistrom, pp. 729–50. New York: Plenum

106. Schnurr G, Misawa N, Sandmann G. 1996. Expression, purification and properties of lycopene cyclase from *Erwinia uredovora. Biochem. J.* 315:869–74

107. Schumann G, Nürnberger H, Sandmann G, Krügel H. 1996. Activation and analysis of cryptic *crt* genes for carotenoid biosynthesis from *Streptomyces griseus. Mol. Gen. Genet.* 252:658–66

108. Scolnik PA, Walker MA, Marrs BL. 1980. Biosynthesis of carotenoids derived from neurosporene in *Rhodopseudomonas capsulata. J. Biol. Chem.* 255:2427–32

109. Straub O. 1987. *Key to carotenoids,* ed. H Pfander, M Gerspacher, M Rychener, R Schwabe, pp. 1–296. Basel: Birkhäuser Verlag

109a. Sun Z, Gantt E, Cunningham FX Jr. 1996. Cloning and functional analysis of the β-carotene hydroxylase of *Arabidopsis thaliana. J. Biol. Chem.* 271:24349–52

110. Tabata K, Ishida S, Nakahara T, Hoshino T. 1994. A carotenogenic gene cluster exists on a large plasmid in *Thermus thermophilus. FEBS Lett.* 341:251–55

111. Takaichi S, Sandmann G, Schnurr G, Satomi Y, Suzuki A, Misawa N. 1996. The carotenoid 7,8-dihydro-ψ end group can be cyclized by the lycopene cyclases from the bacterium *Erwinia uredovora* and the higher plant *Capsicum annuum. Eur. J. Biochem.* 241:291–96

112. Takaichi S, Yazawa H, Yamamoto Y.

1995. Carotenoids of the fruiting gliding myxobacterium, *Myxococcus* sp. MY-18, isolated from lake sediment: accumulation of phytoene and keto-myxocoxanthin glucoside ester. *Biosci. Biotech. Biochem.* 59:464–68

113. Taylor DP, Cohen SN, Clark WG, Marrs BL. 1983. Alignment of the genetic and restriction maps of the photosynthesis region of the *Rhodopseudomonas capsulata* chromosome by a conjugation-mediated marker rescue technique. *J. Bacteriol.* 154:580–90

114. Taylor RF. 1984. Bacterial triterpenoids. *Microbiol. Rev.* 48:181–98

115. To K-Y, Lai E-M, Lee L-Y, Lin T-P, Hung C-H, et al. 1994. Analysis of the gene cluster encoding carotenoid biosynthesis in *Erwinia herbicola* Eho13. *Microbiol.* 140:331–39

116. Toenniessen GH. 1991. Potentially useful genes for rice genetic engineering. In *Rice Biotechnology*, ed. GS Khush, GH Toenniessen, pp. 253–80. Wallingford: CAB Int.

117. Tuveson RW, Larson RA, Kagan J. 1988. Role of cloned carotenoid genes expressed in *Escherichia coli* in protecting against inactivation by near-UV light and specific phototoxic molecules. *J. Bacteriol.* 170:4675–80

118. Wiedemann M, Misawa N, Sandmann G. 1993. Purification and enzymatic characterization of the geranylgeranyl pyrophosphate synthase from *Erwinia uredovora* after expression in *Escherichia coli*. *Arch. Biochem. Biophys.* 306:152–57

119. Wieland B, Feil C, Gloria-Maercker E, Thumm G, Lechner M, et al. 1994. Genetic and biochemical analysis of the biosynthesis of the yellow carotenoid 4, 4'-diaponeurosporene of *Staphylococcus aureus*. *J. Bacteriol.* 176:7719–26

120. Windhövel U, Geiges B, Sandmann G, Böger P. 1994. Expression of *Erwinia uredovora* phytoene desaturase in *Synechococcus* PCC7942 leading to resistance against a bleaching herbicide. *Plant Physiol.* 104:119–25

121. Yamano S, Ishii T, Nakagawa M, Ikenaga H, Misawa N. 1994. Metabolic

engineering for production of β-carotene and lycopene in *Saccharomyces cerevisiae*. *Biosci. Biotech. Biochem.* 58:1112–14

122. Yeliseev AA, Eraso JM, Kaplan S. 1996. Differential carotenoid composition of the B875 and B800–850 photosynthetic antenna complexes in *Rhodobacter sphaeroides* 2.4.1: involvement of spheroidene and spheroidenone in adaptation to changes in light intensity and oxygen availability. *J. Bacteriol.* 178:5877–83

123. Yeliseev AA, Kaplan S. 1995. A sensory transducer homologous to the mammalian peripheral-type benzodiazepine receptor regulates photosynthetic membrane complex formation in *Rhodobacter sphaeroides* 2.4.1. *J. Biol. Chem.* 270:21167–75

124. Yen HC, Marrs B. 1976. Map of genes for carotenoid and bacteriochlorophyll biosynthesis in *Rhodopseudomonas capsulata*. *J. Bacteriol.* 126:619–29

125. Yokoyama A, Miki W. 1995. Composition and presumed biosynthetic pathway of carotenoids in the astaxanthin-producing bacterium *Agrobacterium aurantiacum*. *FEMS Microbiol. Lett.* 128:139–44

126. Yokoyama A, Sandmann G, Hoshino T, Adachi K, Sakai M, Shizuri Y. 1995. Thermozeaxanthins, new carotenoid-glycoside-esters from thermophilic eubacterium *Thermus thermophilus*. *Tetra. Lett.* 36:4901–04

127. Young DA, Bauer CE, Williams JC, Marrs BL. 1989. Genetic evidence for superoperonal organization of genes for photosynthetic pigments and pigment-binding proteins in *Rhodobacter capsulatus*. *Mol. Gen. Genet.* 218:1–12

128. Young DA, Rudzik MB, Marrs BL. 1992. An overlap between operons involved in carotenoid and bacteriochlorophyll biosynthesis in *Rhodobacter capsulatus*. *FEMS Microbiol. Lett.* 95:213–18

129. Zsebo KM, Hearst JE. 1984. Genetic-physical mapping of a photosynthetic gene cluster from *R. capsulata*. *Cell* 37:937–47

SUBJECT INDEX

A

Abcisic acid
  eubacterial carotenoid
    biosynthesis and, 634
ABC transporter protein
  nonfusogenic pathogen
    vacuoles and, 441
Abf1 transcription factor
  DNA replication in yeast and,
    129
Accessory genes
  cell-cell communication in
    Gram-positive bacteria
    and, 544–47
Accessory proteins
  Pseudomonas TOL plasmid
    catabolic operons and,
    360, 366
Acetobacter pasteurianus
  enantiopure epoxides and,
    516–17
Acetogenium kivui
  history of research, 31
Acetyl coenzyme A
  nonfusogenic pathogen
    vacuoles and, 449
N-Acetylneuraminic acid aldolase
  and pyruvate and
    phosphoenolpyruvate,
    295–98
O-Acetylserine
  and sulfur assimilation in
    filamentous fungi and
    yeast, 73, 75, 80
Acidification
  nonfusogenic pathogen
    vacuoles and, 416–17, 425,
    430, 434, 438–39
Acquired immunodeficiency
  syndrome (AIDS)
  intrabodies and, 257, 274–76
  RNA virus fitness and, 169
Actin
  nonfusogenic pathogen
    vacuoles and, 417, 421
Activation
  DNA bending in transcription
    and, 593
Active penetration
  nonfusogenic pathogen
    vacuoles and, 420
Adaptibility
  RNA virus fitness and, 164–67
Adeno-associated virus (AAV)

intrabodies and, 265–66
Adenosine-5′-phosphosulfate
  kinase
  and sulfur assimilation in
    filamentous fungi and
    yeast, 80
Adenosine triphosphate (ATP)
  nonfusogenic pathogen
    vacuoles and, 421, 439,
    444
  Pseudomonas TOL plasmid
    catabolic operons and, 341
Adhesins
  mucosal antigen-antibody
    interactions and, 313–15
  nonfusogenic pathogen
    vacuoles and, 419
Adjuvants
  mucosal antigen-antibody
    interactions and, 312–32
Adonixanthin
  eubacterial carotenoid
    biosynthesis and, 645
Adsorbable organic halogens
  (AOX)
  organohalogen biosynthesis by
    Basidiomycetes and,
    375–78
Aedes albopictus
  alphavirus and, 582
Aerobic chemosynthesis
  history of research, 25
Aerosols
  mucosal antigen-antibody
    interactions and, 331
A factor
  cell-cell communication in
    Gram-positive bacteria
    and, 556
African green monkey kidney cells
  rotaviruses and, 239
Agaricus spp.
  organohalogen biosynthesis by
    Basidiomycetes and,
    379–80, 390
Agrobacterium aurantiacum
  eubacterial carotenoid
    biosynthesis and, 635–36,
    640, 644, 652
Agrobacterium tumefaciens
  eubacterial carotenoid
    biosynthesis and, 642, 652
Agrocybe spp.
  organohalogen biosynthesis by
    Basidiomycetes and, 380

AGR signal
  cell-cell communication in
    Gram-positive bacteria
    and, 545–47
AIDS, see Acquired
  immunodeficiency syndrome
Alanine
  disulfide bonds and, 181, 194
Alcaligenes PC-1
  eubacterial carotenoid
    biosynthesis and, 636, 640,
    644
Alcohol dehydrogenase I
  intrabodies and, 259
Aldolases
  aldol addition reactions,
    287–88, 298, 301
  2-deoxyribose-5-phosphate
    aldolase, 301–3
  DHAP-dependent
    azasugars, 292–94
    chemoenzymatic
      applications, 290
    disaccharides, 290–92
    general background, 288–90
    monosaccharides, 290–92
    synthetic applications,
      290–92
    thiosugars, 292–94
  glycine-dependent, 303
  introduction, 286–87
  ketol and aldol transfer
    reactions
    N-acetylneuraminic acid
      aldolase, 295–98
    3-deoxy-D-*manno*-2-
      octulosonate aldolase,
      298–99
    3-deoxy-D-*manno*-2-
      octulosonate 8-phosphate
      synthetase, 298–99
    2-keto-3-deoxy-
      phosphogluconate
      aldolase, 300
    NeuAc synthase, 295–98
    overview, 294, 300–1
    transaldolases, 305
    transketolases, 304–5
Algae
  eubacterial carotenoid
    biosynthesis and, 629, 633
ali genes
  cell-cell communication in
    Gram-positive bacteria
    and, 537

661

and sulfur assimilation in
  filamentous fungi and
  yeast, 78
Xenorhabdus spp.
  Photorhabdus spp. and
  cell surface properties,
    65–67
  crystalline protein genes,
    56–57
  DNA relatedness studies, 50,
    52
  DNA transfer into, 57
  extracellular enzymes, 56
  fimbriae, 66–67
  flagella, 66–67
  gene characterization, 53–57
  glycocalyx, 66–67
  introduction, 47–50
  low temperature in gene
    induction, 54–55
  lux genes, 55–56
  maltose metabolism, 55
  molecular biology, 53–57
  outer membrane proteins,
    53–54, 66
  pathogenicity, 57–60

phase variation, 60–65
phylogenic studies with 16s
  rRNA analyses, 52–53
taxonomic studies, 50–51
Xerocomus spp.
  organohalogen biosynthesis by
    Basidiomycetes and, 380
Xerula spp.
  organohalogen biosynthesis by
    Basidiomycetes and, 381
xpr genes
  cell-cell communication in
    Gram-positive bacteria
    and, 545
Xylenes
  Pseudomonas TOL plasmid
    catabolic operons and,
    341–42, 345–46
Xyl proteins
  Pseudomonas TOL plasmid
    catabolic operons and,
    341–42, 346–50, 354–62,
    366
XyPs gene
  DNA bending in transcription
    and, 612

Y

Yeasts
  DNA replication and,
    125–35
  filamentous
    sulfur assimilation and,
      73–94
  intrabodies and, 259

Z

Zeaxanthin
  eubacterial carotenoid
    biosynthesis and, 641,
    645
Zinc-finger-like domain
  intrabodies and, 272
Zymomonas mobilis
  aldolases and, 300
  eubacterial carotenoid
    biosynthesis and, 652
Zymosan
  nonfusogenic pathogen
    vacuoles and, 438

CUMULATIVE INDEXES

CONTRIBUTING AUTHORS, VOLUMES 47–51

A

Adams MWW, 47:627–58
Aerne BL, 51:125–49
Allen BL, 48:585–617
Altendorf PK, 50:791–824
Ames GF-L, 47:291–319
Andrew PW, 47:89–115
Andrews NW, 49:175–200
Appleman JA, 50:645–77
Archelas A, 51:491–525
Armstrong G, 51:629–59
Arvin AM, 50:59–100

B

Bartlett MS, 50:645–77
Battista JR, 51:203–24
Baumann L, 49:55–94
Baumann P, 49:55–94
Beachy RN, 47:739–63
Bej AK, 47:139–66
Berberof M, 48:25–52
Berens C, 48:345–69
Bergstrom JD, 49:607–39
Bills GF, 49:607–39
Blair DF, 49:489–522
Blanchard A, 48:687–712
Bobik TA, 50:137–81
Boe L, 47:139–66
Boemare N, 51:47–72
Borst P, 49:427–60
Boulnois GJ, 47:89–115
Bouvier J, 47:821–53
Brock TD, 49:1–28
Bulawa CE, 47:505–34
Burlage RS, 48:291–309
Burleigh BA, 49:175–200
Byrne K, 49:607–39

C

Caldwell DE, 49:711–45
Campbell A, 48:193–222
Cardon LR, 48:619–54
Casey WM, 49:95–116
Chater KF, 47:685–713
Churchward GG, 49:367–97
Citovsky V, 47:167–97
Clark MA, 49:55–94
Condemine G, 50:213–57

Costerton JW, 49:711–45
Cross GAM, 47:385–411

D

de Jong E, 51:375–414
de Lorenzo V, 51:593–628
de Villiers E-M, 48:427–47
Dean DR, 49:335–66
Debono M, 48:471–97
Deckers-Hebestreit G,
  50:791–824
DeLuca NA, 49:675–710
Doige CA, 47:291–319
Domingo E, 51:151–78
Donachie WD, 47:199–230
Donadio S, 47:875–912
Dowds B, 51:47–72
Draths KM, 49:557–79
Dufresne C, 49:607–39
Duncan K, 49:641–73
Dunny GM, 51:527–64
Dybvig K, 50:25–57

E

Eichinger D, 48:499–523
Embley TM, 48:257–89
Englund PT, 49:117–43
Esko JD, 48:139–62
Estes MK, 49:461–87

F

Fauci AS, 50:825–54
Feagin JE, 48:81–104
Felix CR, 47:791–819
Feng P, 48:401–26
Fenical W, 48:559–84
Ferry JG, 49:305–33
Field JA, 51:375–414
Fink DJ, 49:675–710
Fisher K, 49:335–66
Fitchen JH, 47:739–63
Forst S, 51:47–72
Foster JW, 49:145–74
Foster PL, 47:467–504
Francis SE, 51:97–123
Friedrich B, 47:351–83
Frost JW, 49:557–79
Fujii I, 49:201–38

Fuqua C, 50:727–51
Furstoss R, 51:491–525

G

Gaal T, 50:645–77
García-Sastre A, 47:765–90
Gershon AA, 50:59–100
Gillin FD, 50:679–705
Givskov M, 47:139–66
Glorioso JC, 49:675–710
Goldberg DE, 51:97–123
Goldhar J, 49:239–76
Gonzalez-Scarano F,
  47:117–38
Gordee RS, 48:471–97
Gourse RL, 50:645–77
Greenberg EP, 50:727–51
Griffin DE, 51:565–92
Griot C, 47:117–38
Guerinot ML, 48:743–72

H

Hagedorn S, 48:773–800
Hager KM, 48:139–62
Hajduk SL, 48:139–62
Hansen JN, 47:535–64
Hardwick JM, 51:565–92
Harwood CS, 50:553–90
Hengge-Aronis R, 48:53–80
Hernandez-Pando R,
  50:259–84
Hillen W, 48:345–69
Hoch JA, 47:441–65
Holland JJ, 51:151–78
Holloway BW, 47:659–84
Howard RJ, 50:491–512
Hugouvieux-Cotte-Pattat N,
  50:213–57
Hutchinson CR, 49:201–38
Höök M, 48:585–617

J

Jannasch HW, 51:1–45
Janssen DB, 48:163–91
Jensen LB, 47:139–66
Jensen PR, 48:559–84
Jerris RC, 50:707–25
Johnston LH, 51:125–49

689

CHAPTER TITLES, VOLUMES 47–51

698   CHAPTER TITLES

ANNUAL REVIEW OF:	INDIVIDUALS U.S.	Other countries	INSTITUTIONS U.S.	Other countries
ANTHROPOLOGY				
• Vol. 26 (avail. Oct. 1997)	$55	$60	$110	$120
• Vol. 25 (1996)	$49	$54	$49	$54
ASTRONOMY & ASTROPHYSICS				
• Vol. 35 (avail. Sept. 1997)	$70	$75	$140	$150
• Vol. 34 (1996)	$65	$70	$65	$70
BIOCHEMISTRY				
• Vol. 66 (avail. July 1997)	$68	$74	$136	$148
• Vol. 65 (1996)	$59	$65	$59	$65
BIOPHYSICS & BIOMOLECULAR STRUCTURE				
• Vol. 26 (avail. June 1997)	$70	$75	$140	$150
• Vol. 25 (1996)	$67	$72	$67	$72
CELL & DEVELOPMENTAL BIOLOGY				
• Vol. 13, 1997 (avail. Nov. 1997)	$64	$69	$128	$138
• Vol. 12 (1996)	$56	$61	$56	$61
COMPUTER SCIENCE				
• Vols. 3-4 (1988-1989/90) (suspended)	$47	$52	$47	$52
• Vols. 1-2 (1986-1987)	$41	$46	$41	$46
• Vols. 1-4 Price for all four, ordered together.	$100	$115	$100	$115
EARTH & PLANETARY SCIENCES				
• Vol. 25 (avail. May 1997)	$70	$75	$140	$150
• Vol. 24 (1996)	$67	$72	$67	$72
ECOLOGY & SYSTEMATICS				
• Vol. 28 (avail. Nov. 1997)	$60	$65	$120	$130
• Vol. 27 (1996)	$52	$57	$52	$57
ENERGY & THE ENVIRONMENT				
• Vol. 22 (avail. Oct. 1997)	$76	$81	$152	$162
• Vol. 21 (1996)	$76	$81	$76	$81
ENTOMOLOGY				
• Vol. 42 (avail. Jan. 1997)	$60	$65	$120	$130
• Vol. 41 (1996)	$52	$57	$52	$57

ANNUAL REVIEW OF:	INDIVIDUALS U.S.	Other countries	INSTITUTIONS U.S.	Other countries
FLUID MECHANICS				
• Vol. 29 (avail. Jan. 1997)	$60	$65	$120	$130
• Vol. 28 (1996)	$52	$57	$52	$57
GENETICS				
• Vol. 31 (avail. Dec. 1997)	$60	$65	$120	$130
• Vol. 30 (1996)	$52	$57	$52	$57
IMMUNOLOGY				
• Vol. 15 (avail. April 1997)	$64	$69	$128	$138
• Vol. 14 (1996)	$56	$61	$56	$61
MATERIALS SCIENCE				
• Vol. 27 (avail. Aug. 1997)	$80	$85	$160	$170
• Vol. 26 (1996)	$80	$85	$80	$85
MEDICINE				
• Vol. 48 (avail. Feb. 1997)	$60	$65	$120	$130
• Vol. 47 (1996)	$52	$57	$52	$57
MICROBIOLOGY				
• Vol. 51 (avail. Oct. 1997)	$60	$65	$120	$130
• Vol. 50 (1996)	$53	$58	$53	$58
NEUROSCIENCE				
• Vol. 20 (avail. March 1997)	$60	$65	$120	$130
• Vol. 19 (1996)	$52	$57	$52	$57
NUCLEAR & PARTICLE SCIENCE				
• Vol. 47 (avail. Dec. 1997)	$70	$75	$140	$150
• Vol. 46 (1996)	$67	$72	$67	$72
NUTRITION				
• Vol. 17 (avail. July 1997)	$60	$65	$120	$130
• Vol. 16 (1996)	$53	$58	$53	$58
PHARMACOLOGY & TOXICOLOGY				
• Vol. 37 (avail. April 1997)	$60	$65	$120	$130
• Vol. 36 (1996)	$52	$57	$52	$57
PHYSICAL CHEMISTRY				
• Vol. 48 (avail. Oct. 1997)	$64	$69	$128	$138
• Vol. 47 (1996)	$56	$61	$56	$61

ANNUAL REVIEW OF:	INDIVIDUALS U.S.	Other countries	INSTITUTIONS U.S.	Other countries
PHYSIOLOGY				
• Vol. 59 (avail. March 1997)	$62	$67	$124	$134
• Vol. 58 (1996)	$54	$59	$54	$59
PHYTOPATHOLOGY				
• Vol. 35 (avail. Sept. 1997)	$62	$67	$124	$134
• Vol. 34 (1996)	$54	$59	$54	$59
• Vol. 33 (1995) and 10 Year CD-ROM Archive (volumes 24-33)	$49	$54	$49	$54
• 10 Year CD-ROM Archive only	$40	$45	$40	$45
PLANT PHYSIOLOGY & PLANT MOLECULAR BIOLOGY				
• Vol. 48 (avail. June 1997)	$60	$65	$120	$130
• Vol. 47 (1996)	$52	$57	$52	$57
PSYCHOLOGY				
• Vol. 48 (avail. Feb. 1997)	$55	$60	$110	$120
• Vol. 47 (1996)	$48	$53	$48	$53
PUBLIC HEALTH				
• Vol. 18 (avail. May 1997)	$64	$69	$128	$138
• Vol. 17 (1996)	$57	$62	$57	$62
SOCIOLOGY				
• Vol. 23 (avail. Aug. 1997)	$60	$65	$120	$130
• Vol. 22 (1996)	$54	$59	$54	$59

BACK VOLUMES ARE AVAILABLE
Visit www.annurev.org for a list and prices

The Excitement & Fascination Of Science

	INDIVIDUALS U.S.	Other countries	INSTITUTIONS U.S.	Other countries
• Vol. 4, 1995	$50	$55	$50	$55
• Vol. 3 (1990) 2-part set, sold as a set only	$90	$95	$90	$95
• Vol. 2 (1978)	$25	$29	$25	$29
• Vol. 1 (1965)	$25	$29	$25	$29
Intelligence And Affectivity by Jean Piaget (1981)	$8	$9	$8	$9
ANNUAL REVIEWS INDEX on Diskette (updated quarterly) DOS format only. Prices are the same to all locations.	single copy $15	1 yr. (4 eds) $50	single copy $15	1 yr. (4 eds) $50

Annual Reviews

A nonprofit scientific publisher

4139 El Camino Way • P.O. Box 10139
Palo Alto, CA 94303-0139 USA

BB97

STEP 2: ENTER YOUR ORDER

Qty	Annual Review of:	Vol.	Place on Standing Order? SAVE 10% NOW WITH PAYMENT	Price	Total
		#	☐ Yes, save 10% ☐ No	$	$
			☐ Yes, save 10% ☐ No	$	$
		#	☐ Yes, save 10% ☐ No	$	$
			☐ Yes, save 10% ☐ No	$	$
		#	☐ Yes, save 10% ☐ No	$	$
			☐ Yes, save 10% ☐ No	$	$
		#			

30% STUDENT/RECENT GRADUATE DISCOUNT (past 3 years) *Not for standing orders. Include proof of status.*

CALIFORNIA CUSTOMERS: Add applicable California sales tax for your location. | | $ |

CANADIAN CUSTOMERS: Add 7% GST (Registration # 121449029 RT).

STEP 3: CALCULATE YOUR SHIPPING & HANDLING

HANDLING CHARGE (Add $3 per volume, up to $9 max.). Applies to all orders. | | $ |

SHIPPING OPTIONS:
(No UPS to P.O. boxes)

U.S. Mail 4th Class Book Rate (surface). Standard option. FREE. $ N/C

UPS Ground Service ($3/ volume. 48 contiguous U.S. states.) $ _____

Please note expedited shipping preference:
☐ UPS Next Day Air   ☐ UPS Second Day Air   ☐ US Airmail
☐ UPS Worldwide Express   ☐ UPS Worldwide Expedited

Note option at left. We will calculate amount and add to your total

Abstracts and content lists available on the World Wide Web at www.annurev.org. **E-mail orders welcome: service@annurev.org**   **TOTAL $** _____